R. Gupta's®

IIT-JAM

Joint Admission Test for M.Sc.

GEOLOGY

Previous Years' (Solved) Papers
with **Specialised Study Materials**

by
AJHAR HUSSAIN

2026
EDITION

RAMESH PUBLISHING HOUSE, NEW DELHI

Published by
O.P. Gupta *for* Ramesh Publishing House

Admin. Office
12-H, New Daryaganj Road, Opp. Officers' Mess,
New Delhi-110002 ✆ 23275224, 23245124

E-mail: info@rameshpublishinghouse.com
For Online Shopping: www.rameshpublishinghouse.com

Showroom
- Balaji Market, Nai Sarak, Delhi-6 ✆ 23282525 📱 9354373464
- 4457, Nai Sarak, Delhi-6, ✆ 23918938

Book Code: R-1802

ISBN: 978-93-5012-764-3

Price: ₹ 490

Printed at: Deepak Offset, Delhi

CONTENTS

PREVIOUS YEARS' PAPERS

STUDY MATERIAL

Previous Years' Paper

Indian Institute of Technology Joint Admission Test for M.Sc.

IIT-JAM GEOLOGY-2025

SECTION-A

MULTIPLE CHOICE QUESTIONS (MCQ)

Qs. No. 1-10: *Carry one mark each.*

1. The density contrast across which one of the following transitions in the Earth is maximum?

A. Upper crust – lower crust
B. Upper mantle – lower mantle
C. Lower mantle – outer core
D. Outer core – inner core

2. Which one of the following is NOT an ultramafic rock?

A. Wehrlite
B. Olivine websterite
C. Harzburgite
D. Anorthosite

3. The longitude of the person's position at a place 3 hours ahead of UTC (Prime Meridian) is:

A. 15° E B. 30° W
C. 45° E D. 45° W

4. Which one of the following does NOT have a polymorph?

A. Fluorite B. Pyrite
C. Calcite D. Diamond

5. Which one of the following CORRECTLY describes the footwall block relative to the hanging wall of a planar normal fault?

A. Lies above the fault plane and relatively moves down
B. Lies below the fault plane and relatively moves horizontally
C. Lies above the fault plane and relatively moves up
D. Lies below the fault plane and relatively moves up

6. Which one of the following minerals has a characteristic pale yellow ~~

A. Sphalerite
B. Pyrite
C. Hematite
D. Cuprite

7. Which one of the following is a fluorite deposit?

A. Sonapahar, Meghalaya
B. Mangampeta, Andhra Pradesh
C. Belka Pahar, Rajasthan
D. Dongargaon, Maharashtra

1. C **2.** D **3.** C **4.** A **5.** D **6.** A **7.** D

8. According to Dunham's classification, which one of the following limestones is grain-supported and contains mud?

A. Wackestone
B. Packstone
C. Grainstone
D. Mudstone

9. Which one of the following vertebrate fauna is a Proboscidea?

A. *Hipparion*
B. *Mastodon*
C. *Indratherium*
D. *Sivapithecus*

10. Which one of the following is an undifferentiated meteorite?

A. Basaltic achondrite
B. Carbonaceous chondrite
C. Eucrite
D. Pallasite

Qs. No. 11–30: *Carry two marks each.*

11. Which one of the following is the CORRECT sequence of igneous rocks with increasing SiO_2 content?

A. basalt → komatiite → dacite → andesite
B. dacite → basalt → komatiite → andesite
C. basalt → dacite → andesite → komatiite
D. komatiite → basalt → andesite → dacite

12. The schematic diagram given below shows textural relationship among garnet, muscovite, biotite and kyanite in a metapelite. Biotite defines S1 foliation and muscovite defines S2 and S3 foliations. S1, S2 and S3 fabrics were developed during distinct deformation events D1, D2 and D3, respectively. Which one of the following represents the pre-D3 mineral assemblage?

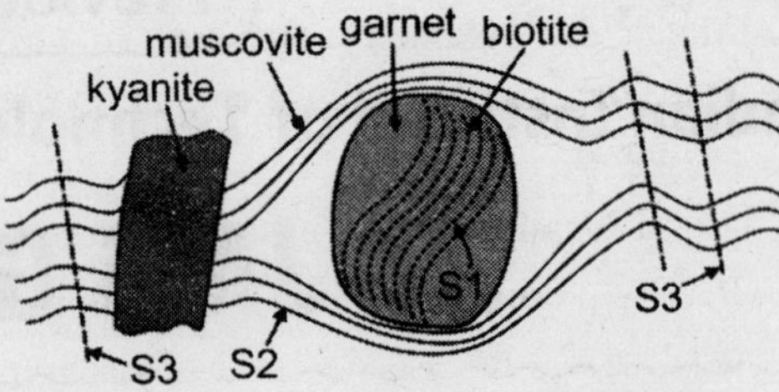

A. kyanite + garnet + biotite
B. garnet + biotite + muscovite
C. garnet + kyanite + muscovite
D. biotite + muscovite + kyanite

13. Which one of the following characteristic mineral assemblages represents eclogite facies metamorphism of a pelitic protolith?

A. garnet + K-feldspar + sillimanite + cordierite
B. talc + kyanite + phengite + garnet
C. chlorite + muscovite + biotite + albite
D. staurolite + chloritoid + biotite + garnet

14. Which one of the following drainage patterns is formed by stream paths that follow circular segments?

A. Annular
B. Trellis
C. Radial
D. Centripetal

15. A mineral displays magenta interference colour (retardation, Δ_m = 550 nm) under crossed polarized light. If an accessory plate adds retardation of 100 nm, the interference colour observed is:

A. 1st order yellow
B. 1st order green
C. 2nd order blue
D. 2nd order red

8. B	9. B	10. B	11. D	12. B	13. B	14. A	15. C

16. Match the symmetry elements in Group I with the corresponding crystal system in Group II.

Group I	**Group II**
P. $3A_2$, 3m, $4A_3$ + i	1. Orthorhombic
Q. i, $1A_4$, m	2. Hexagonal
R. i, $3A_2$, 3m	3. Tetragonal
S. $1A_3$ + m, 3m, $3A_2$	4. Cubic

A. P-3, Q-4, R-1, S-2
B. P-2, Q-3, R-4, S-1
C. P-4, Q-3, R-1, S-2
D. P-4, Q-1, R-2, S-3

17. Match the cycles in Group I with the corresponding processes in Group II.

Group I	**Group II**
P. Hydrological cycle	1. Regolith formation
Q. Biogeochemical cycle	2. Ocean closure
R. Rock cycle	3. Infiltration
S. Wilson cycle	4. Denitrification

A. P-4, Q-2, R-1, S-3
B. P-2, Q-4, R-3, S-1
C. P-3, Q-1, R-4, S-2
D. P-3, Q-4, R-1, S-2

18. Which one of the following is the oldest volcanics?

A. Deccan
B. Panjal
C. Sylhet
D. Rajmahal

19. The following map shows an outcrop of a coal bed that intersects the elevation contours. Which one of the following statements is CORRECT?

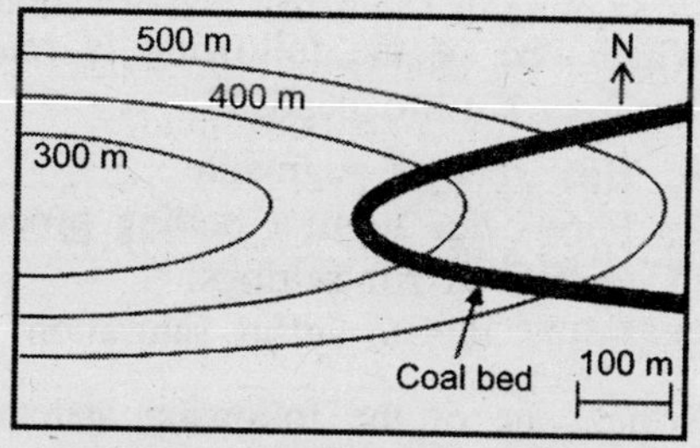

A. Bed intersects a spur and dip of bed is less than gradient of spur axis
B. Bed intersects a valley and dip of bed is greater than valley gradient
C. Bed intersects a spur and dip of bed is equal to gradient of spur axis
D. Bed intersects a valley and dip of bed is less than valley gradient

20. Refer the dam section shown below. Which one of the following conditions represents the most stable state of the dam against rotation about its center of gravity?

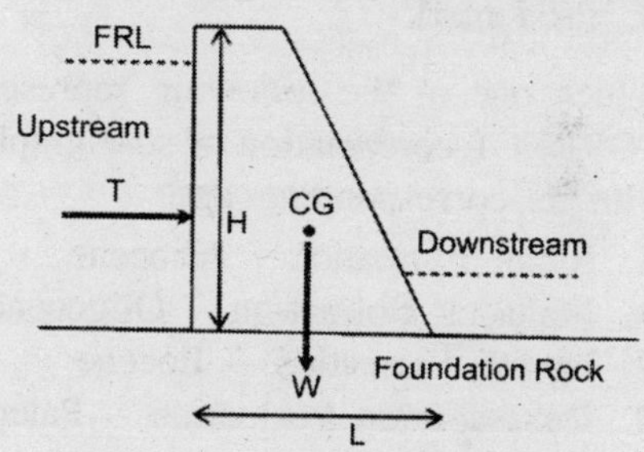

FRL – Full Reservoir Level;
T – horizontal thrust on dam at FRL;
W – weight of dam; CG – center of gravity;
L – base length; H – height of dam

A. T > 3W and L < H
B. Resultant force of T and W passes through the base of the dam
C. H > 3L and T = W
D. Resultant force of T and W passes outside the base of the dam

16. C	**17.** D	**18.** B	**19.** B	**20.** B

21. Which one of the following is related to porphyry Cu-Mo deposit?

A. Host rock is pegmatite
B. Forms due to first boiling process
C. Occurs in rift settings
D. Forms due to sulfur saturation

22. Which one of the following statements is CORRECT for coal?

A. Sapropelic coals are composed of a mixture of macroscopic plant debris
B. Fusain is a microlithotype in coal
C. Clarain is grey in colour and has dull lustre
D. High value of vitrinite reflectance indicates a low rank coal

23. Which one of the following sedimentary structures is formed by rapid release of water?

A. Ripple lamination
B. Gutter cast
C. Convolute lamination
D. Prod mark

24. Which one of the following represents the CORRECT combination of stratigraphic unit with its corresponding age?

A. Barail Formation – Miocene
B. Mahadek Formation – Oligocene
C. Naredi Formation – Eocene
D. Dalmiapuram Formation – Paleocene

25. Match the stratigraphic units in Group I with their corresponding basins in Group II.

Group I	Group II
P. Kajrahat Limestone	1. Cuddapah
Q. Shahabad Limestone	2. Pranhita-Godavari
R. Chanda Limestone	3. Vindhyan
S. Narji Limestone	4. Bhima

A. P-3, Q-2, R-1, S-4
B. P-3, Q-4, R-2, S-1
C. P-2, Q-4, R-3, S-1
D. P-1, Q-3, R-2, S-4

26. Match the morphological features in Group I with their corresponding descriptions given in Group II.

Group I	Group II
P. Hyponome	1. A large plate
Q. Myophore	2. Irregular prismatic crystals
R. Hypostome	3. Long tubular structure
S. Myostracum	4. Shaft with a head

A. P-3, Q-4, R-1, S-2
B. P-3, Q-2, R-4, S-1
C. P-2, Q-4, R-1, S-3
D. P-1, Q-3, R-4, S-2

27. Which one of the following is a Paleozoic flora?

A. *Williamsonia*
B. *Dicroidium*
C. *Gangamopteris*
D. *Nilssonia*

28. Which one of the following lithospheric plates has the least length of convergent boundary?

A. Pacific
B. Indian
C. Antarctic
D. South American

29. Which one of the following elements in the Earth is a chalcophile and shows siderophilic behaviour?

A. Mg
B. Fe
C. Li
D. Pb

30. Which one of the following set of landforms results dominantly due to erosion?

A. Cirque and ventifact
B. Loess and yardangs
C. Barchans and till
D. Moraine and eskers

21. B 22. * 23. C 24. C 25. B 26. A 27. C 28. C 29. B 30. A

SECTION-B

MULTIPLE SELECT QUESTIONS (MSQ)

Qs. No. 31-40: *Carry two marks each.*

31. The shaded region (P) in the given ACF diagram represents the compositional range of mafic rocks that have undergone granulite facies metamorphism. Which of the following equilibrium mineral assemblage(s) is/are identified in these mafic rocks?

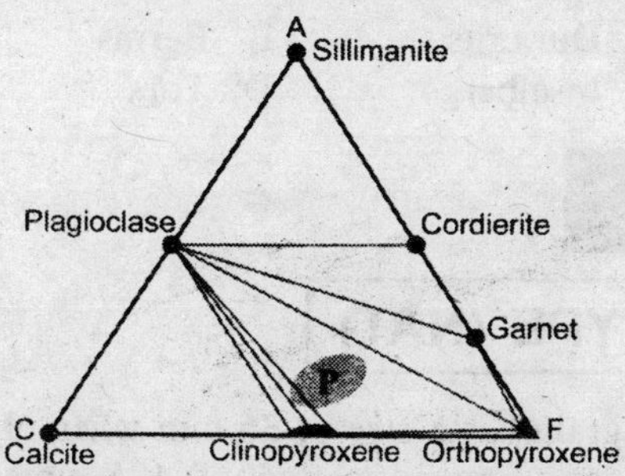

A. plagioclase + clinopyroxene
B. plagioclase + clinopyroxene + orthopyroxene
C. plagioclase + orthopyroxene + garnet
D. plagioclase + garnet + cordierite

32. Which of the following stereographic projections represent(s) an upright plunging fold?

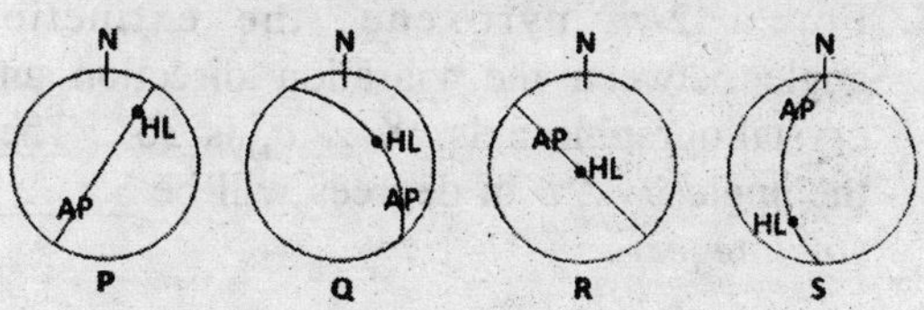

AP: axial plane; HL: hinge line

A. P B. Q
C. R D. S

33. Which of the following property/properties of a mineral indicate(s) uniaxial negative optic sign?

A. Ordinary ray is slow
B. Extraordinary ray is slow
C. $n_\varepsilon < n_\omega$
D. $n_\varepsilon > n_\omega$

34. Which of the following statements related to stratigraphy is/are CORRECT?

A. Grain-size of sediments progressively coarsens upward in a deltaic succession
B. Paraconformity shows a distinct erosional surface
C. Paleogene Period includes Paleocene, Eocene and Oligocene Epochs
D. Sea-level fall causes shifting of river mouth towards deep ocean

35. Which of the following combinations of sedimentary environments and their features is/are CORRECTLY matched?

A. Fluvial – tabular cross-stratification
B. Submarine fan – rain print
C. Beach – planar lamination
D. Supratidal – mud-cracks

36. Which of the following is/are characteristic feature(s) of Brachiopods?

A. Valves equilateral
B. Inequivalved
C. Plane of symmetry between valves
D. Presence of ligament

37. Which of the following characteristics of fossils is/are necessary for biostratigraphic applications?

A. Wide geographic distribution
B. Long vertical range
C. Good preservation of hard shells
D. Facies dependence

31. A, B **32.** A, C **33.** A, C **34.** A, C, D **35.** A, C, D **36.** A, B **37.** A, C

38. Moon is considered to have been formed due to high temperature condensation of post-collisional ejected material. Which of the following statement(s) related to its mantle composition as compared to CI carbonaceous chondrite is/are CORRECT?

A. Depleted in volatile and refractory elements

B. Depleted in volatile and enriched in refractory elements

C. Higher Rb/Sr element ratio

D. Lower K/U element ratio

39. Which of the following statement(s) related to ore-forming processes is/are CORRECT?

A. Lateritization involves eluviation and illuviation processes

B. Stratiform sediment-hosted copper (SSC)-type deposits are formed at mid-oceanic ridges

C. Fluid phase separation is the major process of formation of Mississippi Valley (MV)-type deposits

D. Formation of Superior-type banded iron formation (BIF) is related to the Great Oxidation Event

40. Which of the following landforms is/are weathering dominated?

A. Duricrusts B. Berms

C. Inselbergs D. Tors

SECTION-C

NUMERICAL ANSWER TYPE (NAT)

Qs. No. 41-50: *Carry one mark each.*

41. A metapelite is composed of muscovite + quartz + K-feldspar + kyanite mineral assemblage. If the muscovite and K-feldspar show solid solution between 'Na' and 'K' end-members, the minimum number of components that defines this metapelitic assemblage is ______. *(In integer)*

42. The mineral formula of orthopyroxene calculated on 6 Oxygen atom basis is $(Mg_{1.2}Fe_{0.8})Si_2O_6$. The weight percentage of MgO in the chemical composition of orthopyroxene is ______. *(Round off to one decimal place)*

Use the following data:

Oxide	Molecular weight (g mol^{-1})
SiO_2	60.08
MgO	40.30
FeO	71.84

43. A grain has a size of 36 mm when observed under a magnification of 125 times. The actual size of the grain in μm is ______. *(In integer)*

44. Radiogenic atoms of ^{40}K have a half-life of 1.25×10^9 years. The percentage of ^{40}K atoms left after six half-lives will be ______. *(Round off to two decimal places)*

45. For a 2/*m* pyroxene, the extinction angle between the vibration direction and crystallographic axis, X ∠ *c*, is 36°. Then the angle Y ∠ *b* in degrees will be ______. *(In integer)*

46. A kimberlite pipe has a bulk density of 2.5 g/cc and contains diamond of grade 100 carat/metric ton. If the pipe contains only 1-carat diamond crystals, then the number of diamond crystals per cubic meter in the pipe is ______. *(In integer)*

38. B, D **39.** A, D **40.** A, C, D **41.** 5 **42.** 21.0 to 22.0
43. 288 **44.** 1.50 to 1.60 **45.** 0 **46.** 250

47. Consider the present volume of the Earth's continental crust as 7.5×10^{18} m^3. If continental crust formation started 3.0 Ga ago, then the average rate of continental crustal growth is ______ km^3 yr^{-1}. *(Round off to one decimal place)*

48. The Throw of a fault is 4 m and the Heave is 3 m. The Dip separation of the fault, in meters, is __________. *(In integer)*

49. A Rb–Sr isochron plot of cogenetic volcanic rocks yielded a slope of 0.003. The eruption age of these rocks is ______ Ma. *(Round off to one decimal place)*

(Use $\lambda_{87}Rb = 1.39 \times 10^{-11}$ yr^{-1})

50. The apparent dip amount of a limestone bed in the direction 010° is 30°. If the bed strikes 320°, the value of true dip amount in degrees is ______. *(Round off to nearest integer)*

Qs. No. 51-60: *Carry two marks each.*

51. A magma having density of 2900 kg m^{-3} just reaches the surface through a two-layered crust as shown in the figure below. Assuming isostatic equilibrium, its depth of melting is ______ km. *(Round off to one decimal place)*

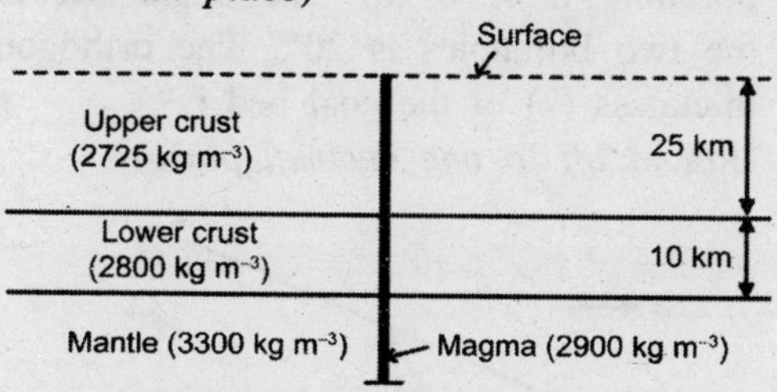

52. The cation exchange equilibrium reaction between end-member components of orthopyroxene and ilmenite is given below:

$MgSiO_3$ + $FeTiO_3$ = $FeSiO_3$ + $MgTiO_3$

(Enstatite) (Fe-Ilmenite) (Ferrosilite) (Mg-Ilmenite)

Considering ideal mixing of Fe and Mg in orthopyroxene and ilmenite solid solutions, the mole fractions are $X_{Mg}^{Orthopyroxene} = 0.60$ and $X_{Mg}^{Ilmenite} = 0.40$. The equilibrium constant of the above reaction is ______. *(Round off to two decimal places)*

53. The isobaric temperature-composition (T–X) phase diagram given below shows the phase relation between components M and N. The equilibrium melting undergone by the rock R to generate the liquid of composition L is ______%. (In integer)

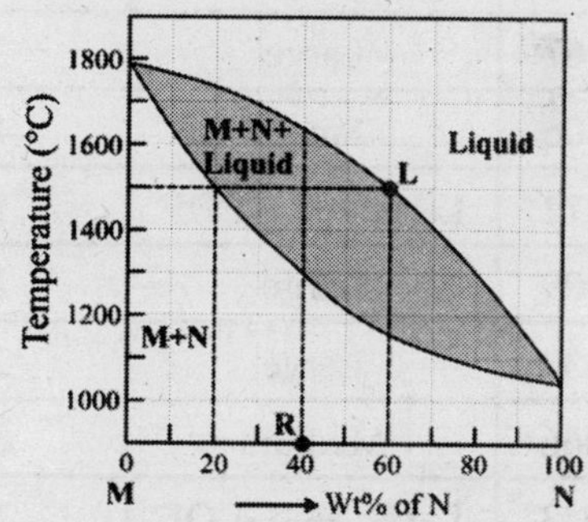

54. A satellite launching vehicle is carrying a lander for Moon mapping. As shown in the figure below, P is the position where the gravitational forces exerted by Earth and Moon on the vehicle balance out. The distance P from the center of the Earth is ______ $\times 10^5$ km. *(Round off to two decimal places)*

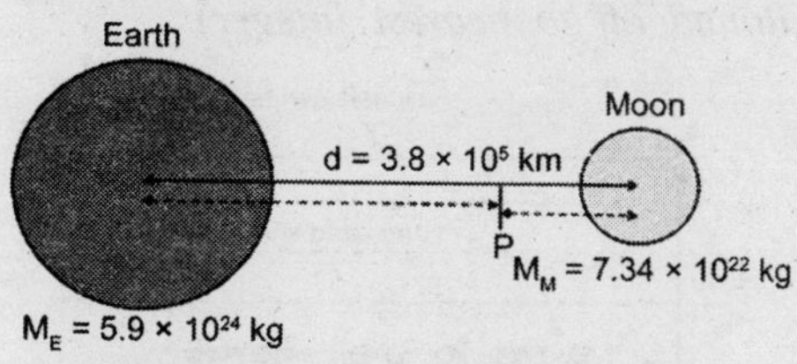

d = distance between the centers of Earth and Moon; M_E = Mass of Earth; M_M = Mass of Moon; $G = 6.67 \times 10^{-11}$ Nm^2 kg^{-2}

47. 2.5 **48.** 5 **49.** 214.0 to 217.0 **50.** 37
51. 48.0 to 49.0 **52.** 0.40 to 0.48 **53.** 50 **54.** 3.37 to 3.45

55. The data tabulated below are for flooding events in the last 400 years. The probability of a large flood accompanied by glacial lake outburst flood (GLOF) in 2025 is ______ $\times 10^{-3}$. (*Round off to one decimal place*)

Year	Flood Size	Magnitude rank
1625	Large	2
1658	Large + GLOF	1
1692	Small	4
1704	Large	2
1767	Large	2
1806	Small	4
1872	Large + GLOF	1
1909	Large	2
1932	Large	2
1966	Medium	3
2023	Large + GLOF	1

56. A well-developed succession of laminated shale is bound by two volcanic ash beds that were precisely dated as shown in the schematic diagram given below. Assuming a constant sedimentation rate, the age of the fossiliferous limestone bed 65 m above the basal volcanic ash bed is ______ Ma. (*Round off to nearest integer*)

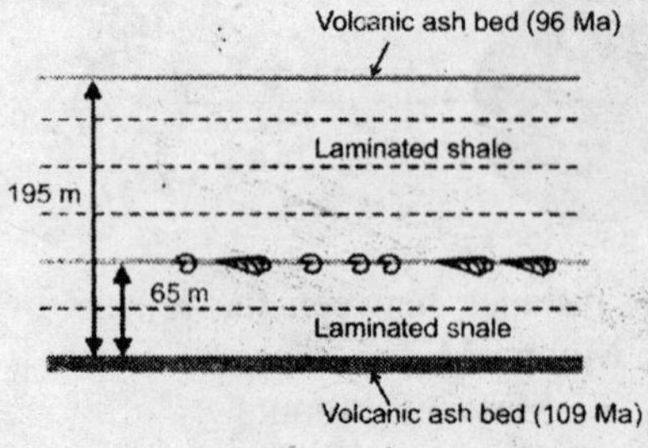

57. Pumping and extraction of groundwater from an unconfined aquifer resulted in a uniform drop of water table by 40 m over an area of 1 km^2. If the porosity of the aquifer is 35% and specific retention is 15%, the volume of water (in m^3) pumped out from the affected area is ______ $\times 10^6$ m^3. (*In integer*)

58. A Cu deposit of 198 tons with average grade of 1.5% Cu contains 60% chalcopyrite, 30% bornite and 10% gangue. The maximum amount of Cu that can be extracted from chalcopyrite is ______ tons. (*Round off to two decimal places*)

59. The orientations of two planar limbs of an overturned antiform are 045°/72°SE and 225°/37°SE. The interlimb angle of the fold in degrees is ______. (*In integer*)

60. Two boreholes A and B, both inclined towards 270°, penetrate a dipping coal bed at the same point and pass through it entirely in the sub-surface as shown in the figure below. The bed dips towards 270°. The thickness of the coal bed, measured along the borehole A is 10 m and along borehole B is 8 m. The angle between the two boreholes is 20°. The orthogonal thickness (x) of the coal bed is ______ m. (*Round off to one decimal place*)

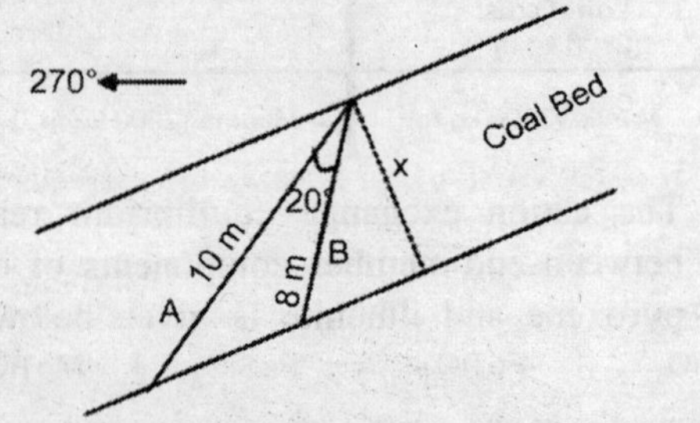

55. 7.0 to 8.0	**56.** 104 to 105	**57.** 8
58. 1.76 to 1.80	**59.** 70	**60.** 7.3 to 7.5

HINTS AND SOLUTIONS

1. (C): The maximum density contrast in Earth's internal structure occurs at the Core-Mantle Boundary (CMB), which separates the solid silicate lower mantle from the liquid metallic outer core. This transition occurs at a depth of about 2,900 km.

- The lower mantle has an average density of about 5.0–5.5 g/cm³ and is primarily composed of silicate minerals like bridgmanite and ferropericlase.
- The outer core is composed mostly of molten iron and nickel, with a density of approximately 9.9–12.2 g/cm³.
- The jump in density across this boundary is therefore about 4.4–6.7 g/cm³, which is the largest among all internal boundaries of the Earth.
- This sharp change is not only due to composition but also due to a phase change from solid to liquid, which also affects seismic wave transmission — P-waves slow down, and S-waves cannot travel through the liquid outer core.

2. (D): Anorthosite is not an ultramafic rock. It is a coarse-grained plutonic igneous rock composed predominantly (90% or more) of plagioclase feldspar, usually labradorite. Its silica content is relatively high, and it contains very little ferromagnesian (mafic) minerals.

Therefore, it does not qualify as ultramafic, which is defined by having more than 90% mafic minerals (olivine, pyroxene) and low silica content.

3. (C): The Earth rotates 360 degrees in 24 hours, so each hour difference in time corresponds to 15° of longitude.

Given:

Time difference = 3 hours ahead of UTC

$\Rightarrow$ Longitude = $3 \times 15° = 45°$

$\Rightarrow$ Since it is ahead of UTC, the longitude lies east of the Prime Meridian.

Therefore, the correct longitude is 45° E.

4. (A): Fluorite (CaF_2) crystallizes in the isometric (cubic) crystal system and does not exhibit polymorphism. Its atomic structure remains consistent under varying temperature and pressure conditions, unlike polymorphic minerals which occur in multiple structural forms of the same composition.

5. (D): Lies below the fault plane and relatively moves up:

In a planar normal fault, the hanging wall moves downward relative to the footwall, due to tensional forces that pull the crust apart.

- The footwall, therefore, lies below the fault plane and, relative to the hanging wall, appears to have moved upward.
- This upward relative movement is not due to actual uplift, but due to the downward movement of the hanging wall in a normal fault setting.

6. (A): Sphalerite (ZnS) is the chief ore of zinc and is known for its characteristic pale yellow to light brown streak, which helps in its identification.

- It usually appears in a variety of colors such as yellow, brown, black, red, or green depending on iron content, but its streak remains typically pale yellow, especially in purer specimens.
- Sphalerite has a resinous to sub-metallic luster and often forms tetrahedral or dodecahedral crystals.

7. (D): Dongargaon in Maharashtra is a significant fluorite (CaF_2) deposit in India.

- Fluorite is used in metallurgical, ceramic, and chemical industries.
- At Dongargaon, fluorite occurs in veins within Precambrian formations, and the deposit has been commercially exploited for its fluorine-rich mineralization.

8. (B): According to Dunham's classification of limestones, Packstone is a grain-supported limestone that contains a considerable amount of micritic (mud) matrix.

- Packstone represents a depositional environment where grains are deposited closely enough to be in contact, but some mud still occupies the space between them.
- It contrasts with grainstone, which is also grain-supported but lacks mud entirely.

9. **(B):** Mastodons belong to the order Proboscidea, which includes elephants and their extinct relatives.

- Mastodons are prehistoric elephant-like mammals that lived from the late Miocene to the Pleistocene epoch.
- They had conical cusps on their molar teeth, unlike the ridged molars of modern elephants, and were primarily forest dwellers.

10. **(B):** Carbonaceous chondrites are classified as undifferentiated meteorites, meaning they have not undergone melting or planetary differentiation.

- They are among the most primitive meteorites, preserving the original solar nebula composition and often contain organic compounds and water-bearing minerals.
- They offer valuable insights into the early solar system.

11. **(D):** Komatiite → Basalt → Andesite → Dacite: This sequence correctly represents igneous rocks arranged in increasing order of silica (SiO_2) content — from ultramafic to felsic compositions.

Let us examine each rock in terms of silica content and magma type:

1. **Komatiite:**
   - ❑ Ultramafic volcanic rock
   - ❑ Extremely low SiO_2 content: <45%
   - ❑ Rich in MgO and FeO, typically formed in high-temperature mantle-derived magmas
   - ❑ Now rare due to their early Earth origin (mostly Archean)
2. **Basalt:**
   - ❑ Mafic volcanic rock
   - ❑ SiO_2 content: ~45–52%
   - ❑ Commonly found at mid-ocean ridges and volcanic arcs
   - ❑ Composed mainly of pyroxene and plagioclase
3. **Andesite:**
   - ❑ Intermediate volcanic rock
   - ❑ SiO_2 content: ~57–63%
   - ❑ Typically forms in subduction zone environments
   - ❑ Contains amphibole and plagioclase
4. **Dacite:**
   - ❑ Felsic-intermediate volcanic rock
   - ❑ SiO_2 content: ~63–68%
   - ❑ Higher quartz and plagioclase content
   - ❑ Represents more evolved, silica-rich magma

Thus, the correct increasing silica sequence is: Komatiite (<45%) → Basalt (45–52%) → Andesite (57–63%) → Dacite (63–68%).

12. **(B):** The question involves identifying the pre-D3 mineral assemblage in a metapelite based on the textural relationships shown in the schematic.

Interpretation of the Diagram:

- S1 foliation is defined by biotite.
- S2 foliation is defined by muscovite.
- S3 foliation is defined by kyanite.
- Garnet appears as a porphyroblast overprinting S1, and possibly predating or coexisting with S2.
- Since S3 is the youngest foliation (D3 event), and kyanite defines it, kyanite is a D3 mineral, and not part of the pre-D3 assemblage.
- The pre-D3 assemblage must include minerals that formed before D3, i.e., before the development of S3.

Now consider each mineral:

- Biotite defines S1 → formed during or before D1. So, present before D3.
- Muscovite defines S2 → formed during D2, i.e., still before D3.
- Garnet overprints S1 → either syn- or post-D1 but pre-D3 (common in regional metamorphism).
- Kyanite defines S3 → formed during D3, not part of the pre-D3 assemblage.

Hence, the minerals that formed before D3 (pre-D3) are:

- Garnet + Biotite + Muscovite.

13. (B): This mineral assemblage is characteristic of eclogite facies metamorphism in a pelitic protolith, which forms under high pressure and moderate to high temperature conditions, typically >1.2 GPa and $>500°C$.

- **Talc:** A hydrous magnesium silicate stable in high-pressure, low-temperature settings, especially in pelitic to mafic systems.
- **Kyanite:** An Al_2SiO_5 polymorph stable at high pressure, typical of eclogite and blueschist facies.
- **Phengite:** A high-silica mica stable in eclogite facies, often present in metapelitic rocks at high pressure.
- **Garnet:** Common in eclogite facies, typically almandine-pyrope solid solution, forms under high pressure.

14. (A): An annular drainage pattern develops when streams follow roughly circular paths, often around structural domes or basins, where alternating resistant and less resistant rocks form concentric bands.

- Streams follow the bedding planes or structural weaknesses, creating ring-like patterns in plan view.

15. (C): Given:

- Original retardation (Δm) = 550 nm
- Accessory plate adds = +100 nm
- Total retardation = 550 + 100 = 650 nm

On a Michel-Levy chart:

- 650 nm corresponds to 2nd order blue interference colour.

16. (C): Match each set of symmetry elements:

P. $3A_2$, 3m, $4A_3$ + i → Cubic system (4)

Cubic crystals exhibit 4 threefold axes, mirror planes, and center of symmetry.

Q. i, $1A_4$, m → Tetragonal system (3)

Tetragonal system has a fourfold rotation axis, mirror plane, and center of symmetry.

R. i, $3A_2$, 3m → Orthorhombic system (1)

Orthorhombic system typically has 3 perpendicular 2-fold axes and 3 mirror planes, matching this set.

S. $1A_3$ + m, 3m, $3A_2$ → Hexagonal system (2)

Characteristic of trigonal subdivision of hexagonal, includes a 3-fold axis and mirror planes.

Hence, correct matching is:

P–4, Q–3, R–1, S–2 → Option C.

17. (D): Matching the cycles with their respective processes:

P. Hydrological cycle → 3. Infiltration

The hydrological cycle (or water cycle) involves processes like evaporation, condensation, precipitation, infiltration, runoff, and groundwater flow.

Infiltration is the process by which water enters the soil, and is a direct component of this cycle.

Q. Biogeochemical cycle → 4. Denitrification

Biogeochemical cycles describe the movement of chemical elements through biological (bio), geological (geo), and chemical processes.

Denitrification is part of the nitrogen cycle, a key biogeochemical cycle, where nitrates are reduced to nitrogen gas by bacteria.

R. Rock cycle → 1. Regolith formation

The rock cycle includes weathering, erosion, transportation, deposition, lithification, metamorphism, and melting.

Regolith formation involves the weathering of rocks to form unconsolidated material, which is part of this cycle.

S. Wilson cycle → 2. Ocean closure

The Wilson cycle explains the opening and closing of ocean basins due to plate tectonics — it includes continental rifting, ocean basin formation, subduction, and continental collision.

Ocean closure through subduction is a major stage of this cycle.

Hence, P–3, Q–4, R–1, S–2 → Option D.

18. (B): The Panjal Traps, located in the northwestern Himalayas (Kashmir region), represent the oldest volcanic formation among the given options. They were formed during the Early Permian period, approximately 280 million years ago (Ma). The Panjal volcanics are associated with continental flood basalt activity and are part of the tectonomagmatic evolution linked with the Gondwana breakup and rifting processes in the Paleo-Tethys Ocean margin.

Here is a comparison of the ages of all options:

- **Panjal Traps (B):**

  Age: ~280 Ma

  Period: Early Permian

  Nature: Mafic volcanic rocks (mainly basalts)
- **Rajmahal Traps (D):**

  Age: ~117–113 Ma

  Period: Early Cretaceous

  Formed during the initial rifting of eastern Gondwana
- **Deccan Traps (A):**

  Age: ~66 Ma

  Period: Late Cretaceous

  Associated with the K–Pg extinction event
- **Sylhet Traps (C):**

  Age: ~66–65 Ma

  Period: Late Cretaceous

  Slightly younger than Deccan, formed in the northeastern part of India

Thus, Panjal is clearly the oldest of the given volcanic formations in terms of geochronology.

19. **(B):** To answer this correctly, we analyze the map provided:

Observations:

- The coal bed outcrop is shown as a thick dark line.
- The coal bed intersects elevation contours at different levels: 300 m, 400 m, 500 m, indicating that the bed is dipping.
- The outcrop of the coal bed is V-shaped and points upstream (toward higher elevation).
- The V-shape points in the opposite direction of the dip, which is a classic feature indicating intersection with a valley, and not a spur.

Interpretation:

- In topographic maps, when a dipping bed intersects a valley, the outcrop forms a "V" pointing upstream (i.e., toward higher elevation).
- This is exactly what is seen in the given map, hence the coal bed intersects a valley.

Now compare the gradient (slope) of the bed and the valley:

- The bed cuts through three contour lines (300 m to 500 m) over a relatively short horizontal distance, indicating a steep dip.
- The valley, by contrast, shows more widely spaced contours, indicating a gentler slope.
- Hence, the dip of the bed is greater than the gradient of the valley.

Therefore, the correct conclusion is: The coal bed intersects a valley and its dip is greater than the valley gradient.

20. **(B):** In the context of dam stability, particularly against rotation about its center of gravity (CG), the goal is to ensure that the resultant of all acting forces (i.e., horizontal thrust T and vertical weight W) passes through the base of the dam and lies within the middle third of the base. This condition prevents overturning and ensures static stability.

Forces acting on the dam:

- W = Vertical weight of the dam (acts through CG, downward)
- T = Horizontal thrust from water pressure acting at full reservoir level (FRL)
- The resultant force (vector sum of W and T) must ideally act within the base width (L) of the dam to avoid tensile stress or overturning.

Why Option B is correct:

- When the resultant force (T + W) passes through the base, it means:
  - The dam is in equilibrium under action of these forces.
  - There is no moment causing rotation about CG, which enhances rotational stability.
  - The structure is most stable against overturning, provided the point lies within the base.

21. (B): Porphyry Cu-Mo deposits are large, low-grade, hydrothermal ore deposits commonly associated with felsic to intermediate calc-alkaline intrusive rocks. They typically form in convergent plate margins and are hydrothermal in origin, with mineralization occurring in and around porphyritic intrusions.

- The "first boiling" process refers to the exsolution of a magmatic fluid phase from a cooling magma, often occurring due to decrease in pressure or temperature, which leads to hydrothermal fluid release.
- This fluid, rich in metals and volatiles, migrates through fractures, precipitating ore minerals such as chalcopyrite ($CuFeS_2$) and molybdenite (MoS_2), leading to the formation of Cu-Mo porphyry systems.

22. (*): None of the given statements is correct

Analysis of each option:

A. Incorrect – Sapropelic coals are algal or amorphous organic matter-dominated (not macroscopic plant debris). Coals with plant debris are humic coals.

B. Incorrect – Fusain is not a microlithotype, it is a maceral type, specifically a fusinite-rich band in coal, resembling charcoal.

C. Incorrect – Clarain is banded, bright and dull, not uniformly grey or dull; it alternates between bright and dull luster due to vitrinite and inertinite.

D. Incorrect – Higher vitrinite reflectance indicates higher rank, i.e., more mature coal (e.g., anthracite). Low-rank coals like lignite have low reflectance.

Hence, correct answer: NONE.

23. (C): Convolute lamination is a sedimentary structure characterized by folded or crumpled laminae found commonly in fine-grained sandstones or siltstones.

It forms due to rapid escape of water during early dewatering or liquefaction, typically right after deposition, often triggered by seismic shaking or rapid sediment loading.

24. (C): The Naredi Formation occurs in western India (Kutch) and is of Eocene age, specifically Lower to Middle Eocene, consisting of shales, marls, and limestones, often fossiliferous.

25. (B): Matching stratigraphic units with their respective basins:

P. Kajrahat Limestone → Vindhyan Basin (3)

Found in the lower Vindhyan sequence, central India.

Q. Shahabad Limestone → Bhima Basin (4)

Located in the Bhima basin of Karnataka, belonging to Proterozoic age.

R. Chanda Limestone → Pranhita-Godavari Basin (2)

Occurs in the Gondwana sequence, part of the Pranhita–Godavari graben.

S. Narji Limestone → Cuddapah Basin (1)

Present in the Kurnool subgroup of the Cuddapah basin, Andhra Pradesh.

Correct matching:

P–3, Q–4, R–2, S–1 → Option B.

26. (A): Matching the morphological features with their descriptions:

P. Hyponome → 3. Long tubular structure

The hyponome is a muscular tubular structure in cephalopods (like squids and nautiloids) used for jet propulsion. Water is expelled through the hyponome for movement.

Q. Myophore → 4. Shaft with a head

A myophore is a bony or shelly projection or process serving as a muscle attachment point, often found in brachiopods or bivalves.

R. Hypostome → 1. A large plate

The hypostome is a rigid, typically calcified plate located on the ventral side of trilobites, over the mouth. It is used in feeding and is diagnostic in trilobite taxonomy.

S. Myostracum → 2. Irregular prismatic crystals

The myostracum is a mineralized layer in mollusc shells, composed of aragonitic prismatic crystals, often appearing irregular and linked with muscle scar formation.

Hence, correct matching:

P–3, Q–4, R–1, S–2 → Option A.

27. **(C):** Gangamopteris is a genus of Glossopterid flora, which is Paleozoic in age, specifically Late Carboniferous to Permian. These plants are typical of Gondwana landmasses, especially well-preserved in the Indian Gondwana coal basins.

28. **(C):** Among the options, the Antarctic Plate has the least length of convergent boundaries. It is largely surrounded by divergent boundaries (like the mid-ocean ridges) and transform faults.

29. **(B):** Iron (Fe) is primarily a siderophile element, meaning it has a strong affinity for metallic iron and preferentially partitions into the core during planetary differentiation.

However, Fe can also behave as chalcophile (affinity for sulfur) under reducing conditions, especially in sulfide-rich environments. This dual behavior is key in ore-forming processes.

30. **(A):** Both cirques and ventifacts are classic landforms produced by erosional processes:

- **Cirque:** Bowl-shaped depression carved by the head of a glacier through plucking and abrasion.
- **Ventifact:** A rock shaped and polished by wind-blown sand, a feature of aeolian erosion.

31. **(A, B):** The shaded region P in the ACF (Al_2O_3–CaO–FeO+MgO) diagram represents the compositional range of mafic rocks that have undergone granulite facies metamorphism. These rocks are expected to exhibit high-temperature, moderate- to high-pressure equilibrium mineral assemblages.

In the given diagram:

- The shaded region P lies in the field bounded by the minerals: Plagioclase, Clinopyroxene, and Orthopyroxene
- It also extends to the binary line between plagioclase and clinopyroxene, suggesting that some compositions within P do not include orthopyroxene, while others do.

This implies that two different equilibrium assemblages are possible within the shaded region:

1. **Plagioclase + Clinopyroxene**
   Found in parts of the field where orthopyroxene is not stable.
   This binary assemblage is characteristic of mafic rocks that have reached granulite facies but lie within the simpler 2-phase field.
2. **Plagioclase + Clinopyroxene + Orthopyroxene**
   Found in the main body of the shaded triangle.
   This ternary equilibrium assemblage is more typical of granulite facies mafic rocks, where both pyroxenes co-exist with plagioclase.

Hence, both assemblages are stable and valid within the granulite facies field depicted by region P.

32. **(A, C):** To determine which stereographic projections represent an upright plunging fold, we analyze the orientation of the axial plane (AP) and the hinge line (HL).

An upright plunging fold is defined by the following criteria:

- The axial plane (AP) is vertical or near-vertical, meaning it appears as a straight north-south line on a stereonet (along the N–S diameter of the projection).
- The hinge line (HL) is plunging, which means it must lie inside the primitive circle, indicating it is inclined and not horizontal.

Let us evaluate each projection individually:

P:

- AP is vertical (along the N–S direction), indicating the fold is upright.
- HL lies within the primitive circle, indicating a plunging hinge line.
- Therefore, this projection represents an upright plunging fold.

Included in answer.

Q:

- AP is inclined, not vertical, suggesting the fold is overturned or inclined, not upright.
- HL is plunging.
- Since the axial plane is not vertical, this is not an upright fold.

Not included

R:

- AP is vertical (along N–S), satisfying the upright condition.
- HL lies inside the primitive circle, indicating a plunging fold axis.
- Thus, this is also an upright plunging fold.

Included in answer.

S:

- AP is vertical (upright).
- HL lies on the primitive circle, indicating a horizontal hinge line.
- This represents an upright horizontal fold, not plunging.

Not included

Correct answer: A, C: P, R

Both diagrams P and R show folds with vertical axial planes and plunging hinge lines, which satisfies the condition of an upright plunging fold.

33. (A, C): To determine the optic sign of a uniaxial mineral, the relationship between the two refractive indices must be evaluated:

- n_ω: Refractive index of the ordinary ray (O-ray)
- n_ε: Refractive index of the extraordinary ray (E-ray)

In a uniaxial negative mineral:

- The extraordinary ray travels faster, meaning it has a lower refractive index (n_ε).
- The ordinary ray travels slower, meaning it has a higher refractive index (n_ω).
- Thus, the condition becomes: $n_\varepsilon < n_\omega$

Let us examine the options:

A. Ordinary ray is slow

→ This means $n_\omega > n_\varepsilon$, so the ordinary ray has a higher refractive index and moves slower.

→ This is the correct physical behavior for a uniaxial negative mineral.

→ Correct.

B. Extraordinary ray is slow

→ If the extraordinary ray is slow, it implies $n_\varepsilon > n_\omega$, which describes a uniaxial positive mineral.

→ Incorrect.

C. $n_\varepsilon < n_\omega$

→ This directly satisfies the definition of a uniaxial negative optic sign.

→ Correct.

D. $n_\varepsilon > n_\omega$

→ This defines a uniaxial positive mineral.

→ Incorrect.

Correct answer: A, C: Ordinary ray is slow; $n_\varepsilon < n_\omega$.

34. (A, C, D): Let's analyze each statement in relation to stratigraphy:

A. Grain-size of sediments progressively coarsens upward in a deltaic succession:

This statement is generally correct. In deltaic successions, the typical vertical sequence shows a fining-downward trend. This is because the bottom of the deltaic sequence starts with coarser sediments (like sands) deposited in distributary channels and the delta front. As you move upward, the sediments become finer (like silts and clays) representing the prodelta and top set beds. Therefore the sediments coarsen upwards.

B. Paraconformity shows a distinct erosional surface:

This statement is incorrect. A paraconformity is a type of unconformity where the layers of rock above and below the unconformity are parallel, and there is no visible erosional surface. The time gap is identified by paleontological or other dating methods.

C. Paleogene Period includes Paleocene, Eocene and Oligocene Epochs:

This statement is correct. The Paleogene Period is a geologic period and is indeed subdivided into the Paleocene, Eocene, and Oligocene epochs.

D. Sea-level fall causes shifting of river mouth towards deep ocean:

This statement is correct. When sea level falls, the base level of rivers is lowered. This causes rivers to incise their channels and extend their mouths further seaward, depositing sediments further into the deep ocean. This process is called basinward shift of facies.

Therefore, the correct statements are A, C and D.

35. **(A, C, D):**

A. Fluvial – tabular cross-stratification

- Fluvial environments often display tabular and trough cross-bedding, created by migrating ripples and dunes in river channels.

B. Submarine fan – rain print

- Rain prints are subaerial features (e.g., in floodplains or tidal flats), not formed in deep marine environments like submarine fans.

C. Beach – planar lamination

- Planar laminations form under swash-backwash processes in beach settings, typically from wave action sorting sands in parallel layers.

D. Supratidal – mud-cracks

- Supratidal flats undergo periodic drying and exposure, leading to formation of mud-cracks due to desiccation.

36. **(A, B):** This question requires evaluating which features correctly describe brachiopods, a phylum of marine invertebrates with hard shells (valves) that are often used as index fossils in Paleozoic strata.

A. Valves equilateral *(Correct)*

- In brachiopods, although the two valves (dorsal and ventral) are inequivalved (i.e., different in size or shape), each individual valve is usually equilateral, meaning it is symmetrical about the midline passing from the beak to the opposite edge.
- The left and right sides of each valve are mirror images of each other.
- This distinguishes brachiopods from bivalves, which often have equivalved but asymmetrical valves.

→ Hence, this statement is correct.

B. Inequivalved *(Correct)*

- Brachiopods have two valves of unequal size and shape – a larger ventral valve and a smaller dorsal valve.
- This property is termed inequivalved, which is one of the defining morphological features of brachiopods.

→ Hence, this statement is correct.

37. **(A, C):** For a fossil to be useful in biostratigraphy, i.e., in correlating and dating rock layers, it must meet certain criteria:

A. Wide geographic distribution

→ Essential for correlation between distant regions. If a fossil is found across different parts of the world, it allows geologists to match strata across those areas.

→ *Correct*

B. Long vertical range

→ *Incorrect.* A long vertical range means the fossil existed over a long period of geological time, which reduces its time resolution and limits its stratigraphic usefulness.

→ Index fossils should have a short vertical range for precise dating.

→ *Incorrect*

C. Good preservation of hard shells

→ *Correct.* Fossils with hard shells or skeletal parts are more likely to be well preserved in the rock record, which is essential for recognition and correlation.

→ *Correct*

D. Facies dependence

→ *Incorrect.* A fossil that is limited to a specific depositional environment or facies-dependent is not ideal for stratigraphic correlation, since it won't occur in all sediment types.

→ Ideal fossils for biostratigraphy are facies-independent and occur in various environments.

→ *Incorrect*

Correct answer: A, C.

38. **(B, D):** The Moon is believed to have formed through a giant impact hypothesis, where Earth collided with a Mars-sized body (Theia), ejecting material that condensed at high temperatures to form the Moon. As a result, the Moon's composition shows specific chemical signatures when compared with CI carbonaceous chondrites, which represent primitive solar system material.

A. Depleted in volatile and refractory elements

→ *Incorrect.* While the Moon is depleted in volatile elements (e.g., Pb, Zn, Na), it is not depleted in refractory elements (like Al, Ca, Ti).

B. Depleted in volatile and enriched in refractory elements

→ *Correct.* High-temperature conditions of moon formation caused loss of volatile elements but preserved/enriched refractory elements, which condense at higher temperatures.

C. Higher Rb/Sr element ratio

→ *Incorrect.* Rb is more volatile than Sr. The Moon, being volatile-depleted, shows lower Rb/Sr ratios than chondrites.

D. Lower K/U element ratio

→ *Correct.* K (a moderately volatile element) is more depleted in the Moon than U (a refractory element), giving the Moon a lower K/U ratio compared to chondritic values.

Correct answer: B, D.

39. (A, D): Lateritization involves eluviation and illuviation processes; Formation of Superior-type banded iron formation (BIF) is related to the Great Oxidation Event.

A. Lateritization involves eluviation and illuviation processes *(Correct)*

- Lateritization is a chemical weathering process occurring in tropical climates, leading to the residual enrichment of iron and aluminum.
- It involves:
  - **Eluviation** – leaching or removal of soluble elements from upper layers (e.g., silica, alkalis)
  - **Illuviation** – accumulation of insoluble elements like Fe and Al oxides in lower zones

→ These vertical geochemical processes lead to the formation of bauxite and laterite ores.

→ Hence, this statement is correct.

D. Formation of Superior-type banded iron formation (BIF) is related to the Great Oxidation Event *(Correct)*

- Superior-type BIFs (2.5 to 1.8 Ga) are closely related to the Great Oxidation Event (GOE), which occurred around 2.4 billion years ago.
- During GOE, increased atmospheric oxygen led to the oxidation and precipitation of dissolved Fe^{2+} as Fe^{3+} oxides (hematite, magnetite), forming BIF layers.

→ Thus, this statement is correct.

Correct answer: A, D.

40. (A, C, D): Let us evaluate which landforms are primarily the result of weathering processes, rather than erosion, deposition, or other geomorphic agents.

A. Duricrusts *(Correct)*

- Duricrusts (e.g., laterite, calcrete, ferricrete) are hard crusts formed by chemical weathering and cementation of soil/saprolite near the surface.
- Formed through precipitation of oxides (Fe, Al, Ca) during weathering under arid to tropical conditions.

→ These are classic weathering-dominated landforms.

B. Berms *(Incorrect)*

- Berms are depositional or erosional beach features — horizontal or gently sloping benches formed by wave action and sediment deposition, not weathering.

→ Hence, not weathering dominated.

C. Inselbergs *(Correct)*

- Inselbergs are isolated steep-sided residual hills formed by prolonged chemical weathering and differential erosion of rock masses in tropical or sub-tropical environments.
- Often composed of resistant crystalline rocks left behind after surrounding material is weathered away.

→ Weathering is a key formative process.

D. Tors *(Correct)*

- Tors are rock outcrops with a piled-up appearance, commonly found in granitic terrains.
- Formed by deep chemical weathering along joints and subsequent removal of the softer material by erosion.

→ The core landform is shaped primarily by weathering, especially spheroidal weathering.

Correct answer: A, C, D.

41. (5): Given mineral assemblage:

- Muscovite
- Quartz
- K-feldspar
- Kyanite

Also mentioned: Muscovite and K-feldspar show Na–K solid solution, so Na and K must both be considered in compositional space.

We are to find the minimum number of chemical components needed to describe this mineral assemblage, considering solid solutions and independent compositional variables.

Let us identify all independent chemical components involved:

1. SiO_2 → Present in quartz, muscovite, and K-feldspar
2. Al_2O_3 → Required for muscovite and kyanite
3. K_2O → Required for K-feldspar, muscovite
4. Na_2O → Enters through solid solution in K-feldspar and muscovite
5. H_2O → Needed for muscovite, which is a hydrous mineral

Therefore, the 5 independent chemical components are:

- SiO_2
- Al_2O_3
- K_2O
- Na_2O
- H_2O

Thus, Answer: 5.

42. (21.4): Given mineral formula:

Orthopyroxene = $(Mg_{1.2}\ Fe_{0.8})Si_2O_6$

Calculate weight percent of MgO, using:

- MgO = 40.30 g/mol
- FeO = 71.84 g/mol
- SiO_2 = 60.08 g/mol

Step-by-step calculation:

MgO content:

$$1.2 \times 40.30 = 48.36 \text{ g}$$

FeO content:

$$0.8 \times 71.84 = 57.47 \text{ g}$$

SiO_2 content:

$$2.0 \times 60.08 = 120.16 \text{ g}$$

Total molecular weight =

48.36 + 57.47 + 120.16

= 225.99 g/mol

Weight % of MgO

$$\left(\frac{48.36}{225.99}\right) \times 100 = 21.4\%$$

Final Answer: 21.4.

43. (288): Given:

- Apparent size under magnification = 36 mm
- Magnification = 125 times
- We are to find actual size in micrometres (μm).

Step 1: Actual size in mm

$$\frac{36}{125} = 0.288 \text{ mm}$$

Step 2: Convert mm to μm:

$$0.288 \text{ mm} = 0.288 \times 1000$$
$$= 288 \text{ μm}$$

→ Answer: 288.

44. (1.56): Given:

- Half-life of $^{40}K = 1.25 \times 10^9$ years
- Number of half-lives = 6
- We are to find the percentage of parent ^{40}K atoms remaining after 6 half-lives

Formula:

$$\text{Fraction remaining} = \left(\frac{1}{2}\right)^n,$$

where n = number of half-lives

$$\left(\frac{1}{2}\right)^6 = \frac{1}{64} = 0.015625$$

$$\text{Percentage remaining} = 0.015625 \times 100$$
$$= 1.56\%.$$

45. (0): Given:

- A 2/m pyroxene, i.e., a monoclinic mineral
- Extinction angle between X and c-axis ($X \angle c$) = 36°
- Required: Angle between Y and b ($Y \angle b$) in degrees

Step-by-step understanding:

In a monoclinic crystal system (2/m) like monoclinic pyroxenes, the symmetry axis (*b*) lies in the optical Y-direction.

Hence, in monoclinic crystals:

- *b*-axis is exactly parallel to vibration direction Y.
- Therefore, the angle between Y and *b* is: $Y\angle b = 0°$

This is a crystallographic property of all monoclinic (2/*m*) minerals — the optical vibration direction Y is parallel to the unique axis (*b*-axis).

Hence, Answer: 0.

46. (250): Given:

- Bulk density of kimberlite = 2.5 g/cc = 2500 kg/m^3
- Diamond grade = 100 carats per metric ton
- Each diamond crystal weighs 1 carat
- Required: Number of 1-carat diamonds per cubic meter

Step-by-step calculation:

Step 1: 1 metric ton = 1000 kg

- Diamond grade = 100 carats per 1000 kg
  $\Rightarrow$ 0.1 carat per kg

Step 2: 1 m^3 of kimberlite = 2500 kg

- So, diamond content
  = 2500 kg × 0.1 carat/kg = 250 carats

Step 3: Each diamond = 1 carat

- So, number of diamonds per m^3
  = 250 carats ÷ 1 carat/diamond
  = 250 diamonds

Hence, Answer: 250.

47. (2.5 in km^3/yr):

Given:

- Present volume of continental crust
  $= 7.5 \times 10^{18}\ m^3$
- Continental crust formation time
  $= 3.0\ Ga = 3.0 \times 10^9$ years

We need to calculate average rate of growth in km^3/yr.

Step 1: Convert m^3 to km^3

$$1\ km^3 = 10^9\ m^3$$

$$\Rightarrow \quad \frac{7.5\times10^{18}\ m^3}{10^9} = 7.5 \times 10^9\ km^3$$

Step 2: Average growth rate

= Total volume / Time

$$\frac{7.5\times10^9\ km^3}{3.0\times10^9\ \text{year}} = 2.5\ km^3/\text{year}$$

Answer: 2.5.

48. (5): Given:

- Throw = 4 m
- Heave = 3 m

We are to calculate dip separation, which is the hypotenuse of the right triangle formed by throw and heave.

Dip separation

$$= \sqrt{(\text{Throw})^2 + (\text{Heave})^2}$$

$$= \sqrt{4^2+3^2} = \sqrt{16+9} = \sqrt{25}$$

= 5 m

Answer: 5.

49. (215.6): (Use slope = 0.003 and decay constant $\lambda_{87}Rb = 1.39 \times 10^{-11}\ yr^{-1}$)

Isochron slope equation for age:

$$t = \frac{1}{\lambda}\ln(\text{Slope}+1)$$

$$t = \frac{1}{1.39\times10^{-11}}\ln(1+0.003)$$

$$= \frac{1}{1.39\times10^{-11}}\ln(1.003)$$

$$\ln(1.003) \approx 0.002996$$

$$\Rightarrow \quad t = \frac{0.002996}{1.39\times10^{-11}}$$

$$\approx 215.6 \times 10^6\ yr$$

$$= 215.6\ Ma$$

Answer: 215.6.

50. (37): Given:

- Apparent dip = 30°
- Apparent dip direction = 010°
- Strike of bed = 320°

We are to find true dip.

Step 1: Calculate angle (θ) between dip direction and apparent dip direction

- Strike = 320°
- So, dip direction = 320° + 90° = 050°
- Angle between apparent dip direction (010°) and true dip direction (050°)

$= |050° - 010°| = 40°$

Step 2: Use formula:

$$\tan(\text{apparent dip}) = \tan(\text{true dip}) \cdot \sin(\theta)$$

$$\tan(30°) = \tan(\text{true dip}) \cdot \sin(40°)$$

$$0.577 = \tan(\text{true dip}) \cdot 0.6428$$

$$\Rightarrow \tan(\text{true dip}) = \frac{0.577}{0.6428} \approx 0.898$$

$$\Rightarrow \text{True dip} = \tan^{-1}(0.898) \approx 37°$$

Answer: 37.

51. **(48.4 km):** Given:

- Density of magma = 2900 kg/m^3
- Upper crust thickness = 25 km, density = 2725 kg/m^3
- Lower crust thickness = 10 km, density = 2800 kg/m^3
- Mantle density = 3300 kg/m^3
- The magma reaches the surface, and we are to find the depth of melting assuming isostatic equilibrium

Let the thickness of mantle traversed by magma = x km

Thus, total vertical column = 25 km (upper crust) + 10 km (lower crust) + x km (mantle)

→ Total depth from surface = 35 + x km

Using weighted average density of the column:

$$\frac{(25 \times 2725) + (10 \times 2800) + (x \times 3300)}{35 + x} = 2900$$

Calculate fixed contributions:

- 25 × 2725 = 68,125
- 10 × 2800 = 28,000

Substitute:

$$\frac{68125 + 28000 + 3300x}{35 + x} = 2900$$

$$\Rightarrow \frac{96125 + 3300x}{35 + x} = 2900$$

Multiply both sides:

$$96125 + 3300x = 2900(35 + x) = 101500 + 2900x$$

Simplify:

$$3300x - 2900x = 101500 - 96125$$

$$\Rightarrow 400x = 5375$$

$$\Rightarrow x = \frac{5375}{400} = 13.4375 \text{ km}$$

Total depth of melting

= 25 km + 10 km + 13.4 km = 48.4 km.

Final Answer: 48.4 km.

52. **(0.44):** We are given the cation exchange reaction:

$$MgSiO_3 + FeTiO_3 \rightleftharpoons FeSiO_3 + MgTiO_3$$

This is a cation exchange between Fe^{2+} and Mg^{2+} in orthopyroxene and ilmenite solid solutions. Assuming ideal mixing, the equilibrium constant K is calculated using mole fractions.

Step 1: Write the equilibrium constant expression

For the above reaction, the equilibrium constant under ideal mixing is:

$$K = \frac{X_{Fe}^{Opx} \cdot X_{Mg}^{Ilm}}{X_{Mg}^{Opx} \cdot X_{Fe}^{Ilm}}$$

Step 2: Use given mole fractions

- $X_{Mg}^{Opx} = 0.60$

$\Rightarrow X_{Fe}^{Opx} = 1 - 0.60 = 0.40$

- $X_{Mg}^{Ilm} = 0.40$

$\Rightarrow X_{Fe}^{Ilm} = 1 - 0.40 = 0.60$

Step 3: Plug into the formula

$$K = \frac{0.40 \times 0.40}{0.60 \times 0.60} = \frac{0.16}{0.36} \approx 0.4444$$

Rounded off to two decimal places:

Final Answer: 0.44.

53. (50): We are given a T–X phase diagram showing the melting relationship between components M and N. The diagram has:

- Temperature on the Y-axis
- Composition in weight % of N on the X-axis
- Two solid phases (M, N) and liquid
- A rock of composition R undergoes partial melting to produce a liquid of composition L

We are to calculate the equilibrium melt fraction (%) produced by rock R to generate liquid L.

Step 1: Use the lever rule for equilibrium melting

In equilibrium partial melting, the lever rule gives the proportion of melt and solid residue based on their relative distances on the horizontal composition (X-axis):

Melt fraction (F)

$$= \frac{\text{Distance from R to solid}}{\text{Total distance between solid and liquid}}$$

From the diagram:

- Composition of R = 40 wt% N
- Composition of liquid L = 60 wt% N
- Composition of solid in equilibrium with melt = 20 wt% N (from the solidus)

Step 2: Apply the lever rule

$$F = \frac{\text{R - Solid}}{\text{Liquid - Solid}}$$

$$= \frac{40-20}{60-20} = \frac{20}{40}$$

$$= 0.50$$

Convert to percentage:

$$F = 0.50 \times 100 = 50\%.$$

Final Answer: 50.

54. (3.40 × 10⁵ km): We are to find the distance P from the center of the Earth, where the gravitational pull by Earth and Moon on a satellite is equal, i.e., where the net gravitational force is zero.

Given:

- Distance between Earth and Moon, $d = 3.8 \times 10^5$ km
- Mass of Earth, $M_E = 5.9 \times 10^{24}$ kg
- Mass of Moon, $M_M = 7.34 \times 10^{22}$ kg
- Gravitational constant, $G = 6.67 \times 10^{-11}$ Nm^{2}kg^{-2}

Let the point P lie at a distance x (in km) from the center of the Earth where gravitational pulls by Earth and Moon balance:

Step 1: Apply gravitational balance condition

At point P,

Gravitational force by Earth = Gravitational force by Moon

$$\frac{GM_E}{x^2} = \frac{GM_M}{(d-x)^2}$$

Cancel G from both sides:

$$\frac{M_E}{x^2} = \frac{M_M}{(d-x)^2}$$

Take square root of both sides:

$$\frac{\sqrt{M_E}}{x} = \frac{\sqrt{M_M}}{d-x}$$

$$\Rightarrow \quad \frac{x}{d-x} = \sqrt{\frac{M_E}{M_M}}$$

Step 2: Plug in values

$$\sqrt{\frac{M_E}{M_M}} = \sqrt{\frac{5.9\times10^{24}}{7.34\times10^{22}}}$$

$$= \sqrt{80.38} \approx 8.96$$

So: $$\frac{x}{d-x} = 8.96$$

$$\Rightarrow \quad x = 8.96(d-x)$$

$$\Rightarrow \quad x = 8.96d - 8.96x$$

$$\Rightarrow \quad x + 8.96x = 8.96d$$

$$\Rightarrow \quad 9.96x = 8.96d$$

$$\Rightarrow \quad x = \frac{8.96}{9.96}d$$

Step 3: Final Calculation

$$x = \frac{8.96}{9.96}\times 3.8\times10^5 \text{ km}$$

$$\Rightarrow \quad x \approx 0.8996 \times 3.8 \times 10^5$$

$$= 3.4185 \times 10^5 \text{ km}$$

Final Answer: 3.42×10^5 km (Rounded to two decimal places).

55. (7.5 × 10^{-3}): We are to calculate the probability of a large flood accompanied by a glacial lake outburst flood (GLOF) occurring in the year 2025, based on historical data from the last 400 years.

Step 1: Total duration of record

From the table, the data starts in 1625 and goes up to 2023, covering:

$$2023 - 1625 = 398 \text{ years}$$

However, as the question states "last 400 years", we use the full 400-year period for probability calculation.

Step 2: Identify events that are Large + GLOF

From the table, the following years are marked as Large + GLOF:

- 1658
- 1872
- 2023

Total number of such events = 3

Step 3: Use the probability formula

Probability of event in any given year

$$= \frac{\text{Number of occurrences}}{\text{Total number of years}}$$

$$= \frac{3}{400}$$

$$= 0.0075$$

$$= 7.5 \times 10^{-3}$$

Final Answer: 7.5×10^{-3}.

56. (104 Ma): We are given:

- The total thickness of laminated shale = 195 m
- Age of basal volcanic ash bed = 109 Ma
- Age of upper volcanic ash bed = 96 Ma
- A fossiliferous limestone bed is present 65 m above the basal ash bed
- Sedimentation rate is assumed constant
- We are to calculate the age of the limestone bed (in Ma)

Step 1: Calculate total time interval

Time between ash beds

= 109 Ma - 96 Ma

= 13 million years

Step 2: Calculate sedimentation rate

$$\text{Sedimentation rate} = \frac{195 \text{ m}}{13 \text{ Ma}}$$

$$= 15 \text{ m/Ma}$$

Step 3: Time taken to deposit 65 m of shale

$$\text{Time to deposit 65 m} = \frac{65 \text{ m}}{15 \text{ m/Ma}}$$

$$= 4.33 \text{ Ma}$$

Step 4: Age of the limestone bed

$$\text{Age} = 109 \text{ Ma} - 4.33 \text{ Ma}$$

$$= 104.67 \text{ Ma}$$

Rounded to nearest integer:

Final Answer: 104 Ma.

57. (8 × 10^6 m^3): We are given:

- Drop in water table = 40 m
- Area affected = 1 km^2 = 1×10^6 m^2
- Porosity (n) = 35%
- Specific retention (Sr) = 15%

Only the drainable porosity contributes to the water that can actually be extracted. That is given by:

Specific yield (Sy)

= Porosity - Specific retention

= 35% - 15%

= 20% = 0.20

Step 1: Calculate volume of saturated zone lost

Volume of drained saturated material

= Area × Drop in water table

= 1×10^6 m^2 × 40 m

= 40×10^6 m^3

Step 2: Multiply by specific yield to get volume of water

Volume of water pumped out

= $40 \times 10^6 \times 0.20$

= 8×10^6 m^3

Final Answer: 8×10^6 m^3.

58. (1.78 tons): We are given:

- Total ore = 198 tons
- Average Cu grade = 1.5%

- Ore composition:

  60% chalcopyrite

  30% bornite

  10% gangue

We are to find the maximum amount of Cu extractable from chalcopyrite.

Step 1: Total copper in the deposit

$$\text{Total Cu} = 198 \times \frac{1.5}{100}$$

$$= 2.97 \text{ tons of Cu}$$

Step 2: Distribution of ore types

- Chalcopyrite proportion = 60% of 198 tons = 198 × 0.60 = 118.8 tons
- The Cu content in chalcopyrite is 34.63% by weight ($CuFeS_2$; molecular weight ≈ 183.5 g/mol, Cu atomic weight = 63.5 g/mol)

$$\text{Max Cu from chalcopyrite} = 118.8 \times \frac{34.63}{100}$$

= 41.15 tons

Step 3: Fraction of total Cu that can be extracted

We know:

- Total Cu = 2.97 tons
- Total chalcopyrite Cu capacity (from 118.8 tons) = 41.15 tons

  But this 41.15 tons of potential Cu is theoretical, and not the actual value needed.

Instead, we proportionally allocate Cu to chalcopyrite:

Cu from chalcopyrite

= 60% of total Cu

= 2.97 × 0.60 = 1.782 tons

Rounded to two decimal places:

Final Answer: 1.78 tons.

59. **(70):** We are given the orientations of two planar limbs of an overturned antiform fold as:

- Limb 1: 045°/72° SE
- Limb 2: 225°/37° SE

We are to calculate the interlimb angle in degrees — the angle between the two limbs of the fold, measured through the hinge zone.

Step 1: Interpret the strike/dip values

Convert each orientation into dip direction and dip angle:

- **Limb 1:** Strike = 045°, Dip = 72°, Dip direction = SE → Azimuth = 135°
- **Limb 2:** Strike = 225°, Dip = 37°, Dip direction = SE → Azimuth = 315°

These dip directions (135° and 315°) are 180° apart, which indicates that the two limbs are dipping in opposite compass directions, though both are labeled "SE" — this is characteristic of an overturned fold, where the axial surface is inclined and both limbs dip in the same general direction on the surface, but geometrically face away from each other.

Step 2: Use the angle between two planes (interlimb angle) formula

We calculate the angle between the poles (normals) to each limb plane.

The formula is:

$\cos\theta = \sin\delta_1 \sin\delta_2 \cos(\alpha_1 - \alpha_2) + \cos\delta_1 \cos\delta_2$

Where:

- $\delta_1 = 72°$, $\alpha_1 = 135°$
- $\delta_2 = 37°$, $\alpha_2 = 315°$

Step 3: Plug in values

- $\alpha_1 - \alpha_2 = 135° - 315° = -180°$

  $\Rightarrow \cos(-180°) = -1$
- sin (72°) = 0.9511
- sin (37°) = 0.6018
- cos (72°) = 0.3090
- cos (37°) = 0.7986

$$\cos\theta = (0.9511 \times 0.6018 \times -1) + (0.3090 \times 0.7986)$$

$$\cos\theta = -0.5725 + 0.2465$$

$$= -0.326$$

$$\theta = \cos^{-1}(-0.326) \approx 109.1°$$

Step 4: Find the interlimb angle

The interlimb angle is the supplementary angle between the two planar limbs:

Interlimb angle = 180° − 109.1° = 70.9°

Rounded to nearest integer:

Final Answer: 70.

60. (7.4 m): We are given:

- Thickness of coal bed along borehole A = 10 m
- Thickness along borehole B = 8 m
- Angle between the two boreholes = 20°
- We are to find the orthogonal (true vertical) thickness of the coal bed, denoted as x

Step 1: Use the Law of Sines in triangle formed by boreholes and orthogonal thickness

This is a triangle problem where:

- Two sides: 10 m (A), 8 m (B)
- Angle between them: 20°
- We drop a perpendicular (x) from the point of intersection to the coal bed (which forms two right triangles)

Let's use the sine area formula for triangle:

$$\text{Area} = \frac{1}{2} ab \sin C$$

Where:

- $a = 10$,

  $b = 8$, $c = 20°$

$$\text{Area} = \frac{1}{2} \times 10 \times 8 \times \sin(20°)$$

$$= 40 \times 0.3420 = 13.68 \text{ m}^2$$

Step 2: Use the formula for area in terms of base and height

$$\text{Area} = \frac{1}{2} \times \text{base} \times \text{height}$$

$$\Rightarrow \quad \text{Area} = \frac{1}{2} \times \text{distance between boreholes at coal bed} \times x$$

To extract orthogonal thickness x, we now find the distance between boreholes at coal bed using the Law of Cosines:

$$c^2 = a^2 + b^2 - 2ab \cos(C)$$

$$\Rightarrow \quad c^2 = 10^2 + 8^2 - 2 \cdot 10 \cdot 8 \cos(20°)$$

$$= 100 + 64 - 160 \cdot 0.9397$$

$$= 164 - 150.35 = 13.65$$

$$\Rightarrow \quad c = \sqrt{13.65} \approx 3.694 \text{ m}$$

Now using:

$$\text{Area} = \frac{1}{2} \times 3.694 \times x = 13.68$$

$$\Rightarrow \quad x = \frac{2 \times 13.68}{3.694}$$

$$\approx \frac{27.36}{3.694} \approx 7.41 \text{ m}$$

Final Answer: 7.4 m.

Previous Years' Paper

Indian Institute of Technology Joint Admission Test for M.Sc.

IIT-JAM GEOLOGY-2024

SECTION-A

Multiple Choice Questions (MCQ)

Qs. No. 1-10: *Carry one mark each.*

1. The plate tectonic setting of Benioff-Wadati zone is:

A. continental rift B. subduction zone
C. passive margin D. mid-oceanic ridge

2. Neutron-rich unstable nuclides undergo:

A. β^- (negatron) decay
B. β^+ (positron) decay
C. α-decay
D. electron capture

3. Which one of the following textures is found in alkali olivine basalt?

A. Rapakivi
B. Graphic
C. Blastoporphyritic
D. Intergranular

4. The mineral assemblage found in a granulite facies metabasalt is:

A. glaucophane + lawsonite + chlorite
B. orthopyroxene + garnet + plagioclase + clinopyroxene + quartz
C. actinolite + albite + chlorite + epidote
D. omphacite + garnet + quartz

5. *Glossopteris* is found in which of the following formations?

A. Raniganj B. Bagra
C. Lameta D. Nimar Sandstone

6. In a sequence of undeformed sedimentary rocks, younger rocks overlie older rocks. This conforms to the principle of:

A. superposition
B. uniformitarianism
C. faunal succession
D. original horizontality

7. Dropstones are found in:

A. Barakar Formation
B. Talchir Formation
C. Raniganj Formation
D. Bijori Formation

8. The sedimentary structure formed by unidirectional current is:

A. trough cross-bedding
B. oscillation ripple
C. concretion
D. hummocky cross-stratification

9. Which of the following is the precursor of petroleum?

A. Sporinite
B. Clarain
C. Kerogen
D. Vitrain

10. Which of the following is an amorphous variety of SiO_2?

A. Quartz B. Citrine
C. Agate D. Opal

Qs. No. 11-30: *Carry two marks each.*

11. The name of an igneous rock having a modal composition of 55% olivine, 40% orthopyroxene and 5% plagioclase, as per the IUGS classification scheme, is:

A. gabbro B. troctolite
C. lherzolite D. harzburgite

12. Which of the following is the correct decreasing order of abundance of elements in our solar system?

A. O > H > Fe > He
B. O > Fe > H > He
C. H > O > Fe > He
D. H > He > O > Fe

13. The suture of a cephalopod having smooth saddles and crenulated lobes is called:

A. orthoceratitic B. goniatitic
C. ceratitic D. ammonitic

14. Which of the following is a body fossil?

A. Coprolite B. Footprint
C. Cast D. Stromatolite

15. Match the morphological features in Group I with the corresponding taxa in Group II.

Group I	Group II
P. Dissepiment	1. Echinodermata
Q. Delthyrium	2. Trilobita
R. Pygidium	3. Brachiopoda
S. Ambulacrum	4. Anthozoa

	P	Q	R	S
A.	4	3	2	1
B.	4	3	1	2
C.	2	3	4	1
D.	3	4	2	1

16. Match the sedimentary features/structures in Group I with the corresponding processes in Group II.

Group I	Group II
P. Stylolite	1. Liquefaction
Q. Pseudonodule	2. Diagenesis
R. Current crescent	3. Organo-sedimentary binding
S. Stromatolite	4. Scouring

	P	Q	R	S
A.	2	1	4	3
B.	2	3	4	1
C.	3	1	4	2
D.	4	1	2	3

17. Match the geomorphic features in Group I with the corresponding environments in Group II.

Group I	Group II
P. Dreikanter	1. Glacial
Q. Cirque	2. Beach
R. Natural levee	3. Eolian
S. Berm	4. Fluvial

	P	Q	R	S
A.	3	1	4	2
B.	2	1	4	3
C.	3	4	1	2
D.	4	2	3	1

18. The correct hierarchy of the given stratigraphic units is:

A. Group > Member > Formation > Bed
B. Eon > Era > Epoch > Period
C. Group > Formation > Member > Bed
D. Eon > Era > Series > Systems

19. Match the minerals in Group I with their highest order of interference colour in Group II (for 0.03 mm mineral thickness).

Group I	Group II
P. Sillimanite	1. First order
Q. Quartz	2. Second order
R. Muscovite	3. Greater than third order
S. Calcite	4. Third order variegated

	P	Q	R	S
A.	2	4	1	3
B.	3	1	2	4
C.	2	1	4	3
D.	4	2	3	1

20. The saturated thickness of an unconfined aquifer is defined by the distance between:

A. the ground surface and the water table
B. the water table and the underlying confining layer
C. the water table and the mean sea level
D. the ground surface and the underlying confining layer

21. Darcy's law quantifies the volume of groundwater flow:

A. per unit surface area of the aquifer
B. per unit time
C. per unit cross-sectional area of the aquifer
D. per unit cross-sectional area of the aquifer per unit time

22. According to Strahler's stream ordering system, what is the highest order of stream in the given diagram?

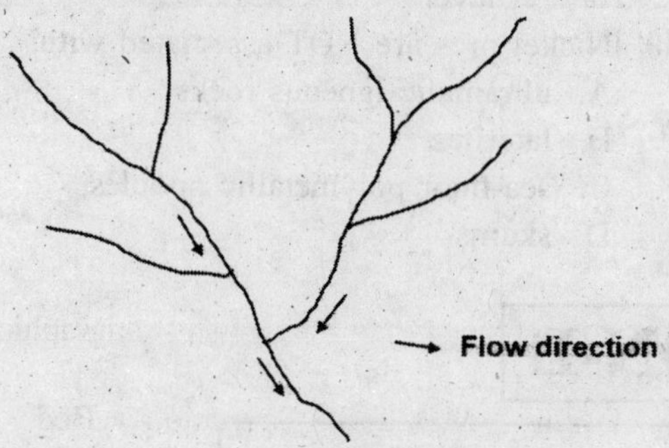

A. 4 B. 3
C. 2 D. 1

23. A tunnel with vertical walls and arched crown is constructed through a set of sedimentary beds. The thickness of individual beds is significantly less than the wall-height of the tunnel. For which of the following conditions, one of the tunnel walls becomes unstable?

A. The tunnel axis is horizontal and the beds are horizontal.
B. The tunnel axis is parallel to the strike of beds and the beds dip 45° – 60°.
C. The tunnel axis is parallel to the strike of beds and the beds are vertical.
D. The tunnel axis is perpendicular to the strike of beds and the beds are vertical.

24. A plunging fold will NOT show a V-shaped outcrop pattern on a planar ground surface if the plunge of the fold axis is:

A. equal to the dip of the ground surface in the same direction
B. steeper than the dip of the ground surface in the same direction
C. equal to the dip of the ground surface in the opposite direction
D. steeper than the dip of the ground surface in the opposite direction

25. The given map shows the outcrop patterns of beds (1 – 4) across a fault plane, F – F', on a flat ground. X and Y refer to the two blocks across F – F'. Which one of the following options is the correct interpretation of the structure depicted on the map?

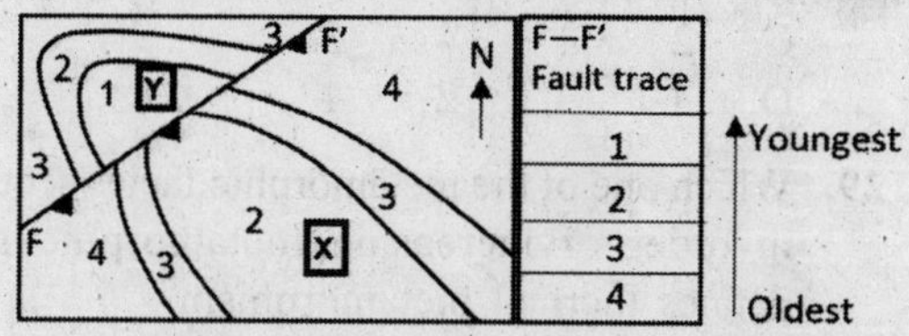

A. A south-easterly plunging synform that was subsequently faulted with block X upthrown
B. A north-westerly plunging antiform that was subsequently faulted with block Y upthrown
C. A south-easterly plunging antiform that was subsequently faulted with block X downthrown
D. A north-westerly plunging synform that was subsequently faulted with block Y downthrown

26. The crystal form 'dome' contains:

A. two parallel faces related by a 2-fold axis of symmetry
B. two non-parallel faces related by a 2-fold axis of symmetry
C. two parallel faces related by a mirror plane
D. two non-parallel faces related by a mirror plane

27. The symbols [100], {100} and (100) in a crystal represent the sequence:

A. form, line and face
B. form, face and line
C. line, face and form
D. line, form and face

28. Match the stratigraphic units in Group I with the corresponding Archean cratons in Group II.

Group I	Group II
P. Bababudan Group	1. Eastern Dharwar
Q. Banded Gneissic Complex-I	2. Western Dharwar
R. Bonai Granite	3. Aravalli
S. Kolar Group	4. Singhbhum

	P	Q	R	S
A.	2	3	4	1
B.	3	2	1	4
C.	2	1	4	3
D.	4	3	2	1

29. Which one of the metamorphic facies sequence in order of increasing metamorphic grade defines thermal metamorphism?

A. Sanidinite → pyroxene hornfels → hornblende hornfels → albite-epidote hornfels

B. Albite-epidote hornfels → hornblende hornfels → sanidinite → pyroxene hornfels

C. Hornblende hornfels → albite-epidote hornfels → pyroxene hornfels → sanidinite

D. Albite-epidote hornfels → hornblende hornfels → pyroxene hornfels → sanidinite

30. Nickel ores are NOT associated with:

A. ultramafic igneous rocks

B. laterites

C. sea-floor polymetallic nodules

D. skarns

SECTION-B

Multiple Select Questions (MSQ)

Qs. No. 31-40: *Carry two marks each.*

31. Which of the following statements on mantle partial melting are correct?

A. Shallow melting produces tholeiitic basalts.

B. Low-degree melting produces alkaline basalts.

C. Presence of CO_2-rich volatiles favors the formation of tholeiitic basalts.

D. Presence of H_2O-rich volatiles favors the formation of alkaline basalts.

32. Which of the following fossil groups are from the Siwalik Group?

A. Proboscidea B. Giraffidae

C. Dinosauria D. Equidae

33. The correct stratigraphic successions arranged from the oldest to the youngest are:

A. Uttatur → Trichinopoly → Ariyalur → Niniyur

B. Chari → Patcham → Umia → Katrol

C. Chinji → Nagri → Dhok Pathan → Tatrot

D. Semri → Rewa → Kaimur → Bhander

34. Which of the following combinations are correctly matched?

A. Photic zone – biogenic carbonate rocks

B. Delta – progradational coarsening-up succession

C. Sabkha – shelf storm deposit

D. Shelf break – submarine fans

35. High drainage density is representative of a terrain with:

A. high relief

B. arid climate

C. impermeable surface layer

D. permeable surface layer

36. Mass-wasting processes are:

A. landslides

B. lahars

C. avalanches

D. sand storms

37. Which ones of the following correspond to the Pyroxene group?

A. $CaMgSi_2O_6$

B. $CaAl_2SiO_6$

C. $Ca_2Si_2O_6$

D. $NaFeSi_2O_6$

38. Which of the following processes are correctly matched with corresponding deformation structures?

A. Pressure solution – rock cleavage

B. Jointing – plumose markings

C. Layer parallel compression – buckle folds

D. Cohesion loss – slickensides

39. For the given Barrovian metamorphic sequence, which of the following statements are correct?

Barrovian sequence

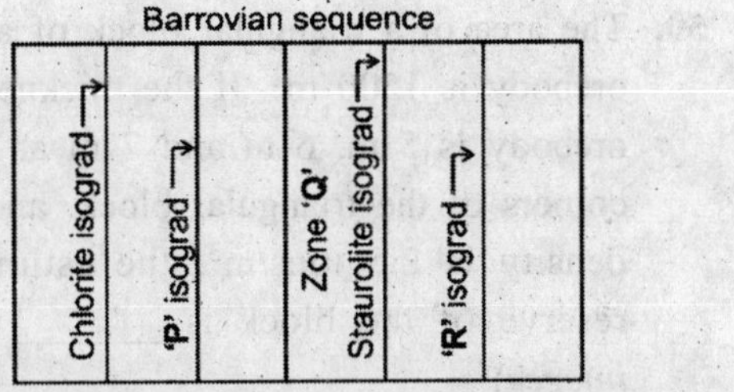

A. Grade of metamorphism increases from left to right.

B. 'P' isograd is the Garnet isograd.

C. Zone 'Q' is the Garnet zone.

D. 'R' isograd is the Kyanite isograd.

40. Which ones of the following are formed by brittle deformation?

A. Cataclasite B. Breccia

C. Mylonite D. Gouge

SECTION-C

Numerical Answer Type (NTA)

Qs. No. 41-50: *Carry one mark each.*

41. The value of φ (*phi*) of a sediment grain having a diameter of 0.125 mm is _______. (*In integer*)

42. The vertical separation of a displaced horizontal stratum along a dip-slip reverse fault is 10 m when measured on a section perpendicular to the fault-strike. If the dip of the fault is 30°, the net slip of the fault will be _______ m. (*In integer*)

43. The dips of the normal and overturned limbs of a horizontal-overturned antiform are 30° and 70°, respectively. The interlimb angle of this fold is _______ degrees. (*In integer*)

44. In a mineral with formula $KAl_3Si_3O_{10}(F_{0.5}OH_x)$, the value of '$x$' is _______. (*Round off to one decimal place*)

45. The atom percent of Fe in pyrrhotite of composition $Fe_{0.77}S$ is _______. (*Round off to two decimal places*)

46. Consider the univariant metamorphic reaction Albite = Jadeite + Quartz. The minimum number of chemical components required to describe the composition of all the phases is _______. (*In integer*)

47. The mean flow velocity of water in an open channel having an average depth of 0.2 m, and with Froude Number 4, is _______ m/s. (*Round off to one decimal place*) (Use $g = 9.8\ ms^{-2}$)

48. An aquifer has a cross-sectional area of 10 m^2 and a hydraulic conductivity of 0.25 cm/s. The volume of water that will flow per second through the aquifer for a hydraulic gradient of 0.04 is _______ cm^3. (*Round off to three decimal places*)

49. The geothermal gradient in the continental crust is 0.02 °C/m. If the surface temperature is 25 °C, the temperature at a depth of 18 km from the surface is _______ °C. (*In integer*)

50. The area of a triangular block of a massive orebody is 1500 m^2. If the thickness of the orebody is 5 m, 6 m and 7 m at the three corners of the triangular block, and the ore density is 2.5 tons/m^3, the estimated ore reserve of the block is ______ tons. (*In integer*)

Qs. No. 51-60: *Carry two marks each.*

51. Clinopyroxene crystallizing from a basaltic magma has Sm concentration of 24 ppm. If the clinopyroxene-melt partition coefficient for Sm is 1.2, the concentration of Sm in the basaltic magma will be ______ ppm. (*In integer*)

52. The lithostatic pressure at a depth of 36.5 km in the continental crust having an average density of 2800 kg/m^3, is ______ GPa. (*Round off to the nearest integer*) (Use g = 9.8 m/s^2)

53. The fraction of ${}^{24}_{11}$Na atoms remaining after a decay interval of 5.0 hours will be ______. (*Round off to three decimal places*) (Use $t_{1/2}$ = 15.0 hours)

54. The thickness of a dipping coal bed measured along a vertical drill hole is 15 m. If the dip of the coal bed is 30°, the orthogonal thickness of the coal bed is ______ m. (*Round off to the nearest integer*)

55. The mole fraction of forsterite in olivine with MgO = 29.17 weight %, FeO = 34.65 weight % and SiO_2 = 36.18 weight % is ______. (*Round off to two decimal places*) (Use molecular weight, in g/mol, of MgO = 40.31, FeO = 71.85 and SiO_2 = 60.00)

56. A partially saturated soil sample has a volume of 1200 cc. The volume of water present in the sample is 300 cc. The mass of solid in the sample is 1908 g and the particle density is 2.65 g/cc. The porosity (n) of the soil sample is ______ %. (*In integer*)

57. A rock element during deformation, experienced a pressure change of 5 × 10^4 N/m^2, due to which its volume changed from 4 cm^3 to 3.9 cm^3. The bulk modulus of the rock is ______ × 10^6 N/m^2. (*In integer*)

58. For an anisotropic crystal of thickness 0.04 mm and refractive indices of 1.636 and 1.486 along the slow and fast directions, respectively, the retardation produced is ______ nm. (*In integer*)

59. An orebody contains pyrite and chalcopyrite in the same molar proportions. The percentage concentration of Cu in the ore will be ______. (*Round off to the nearest integer*) (Use atomic weight, in g/mol, of Cu = 63.55, Fe = 55.85, S = 32.06)

60. In the given isobaric binary temperature-composition (T-X) phase diagram involving solids A and B, the fraction of melt remaining at point Q for a magma having initial composition P will be ______. (*Round off to one decimal place*)

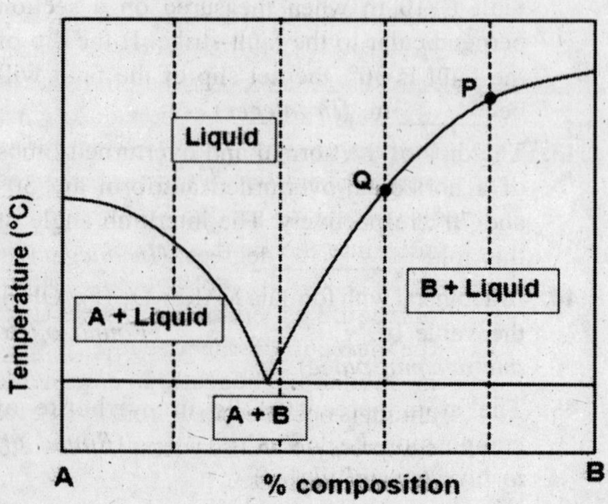

ANSWER KEY

1	2	3	4	5	6	7	8	9	10
B	A	D	B	A	A	B	A	C	D
11	**12**	**13**	**14**	**15**	**16**	**17**	**18**	**19**	**20**
D	D	C	C	A	A	A	C	C	B
21	**22**	**23**	**24**	**25**	**26**	**27**	**28**	**29**	**30**
B	B	B	A	A	D	D	A	D	D
31	**32**	**33**	**34**	**35**	**36**	**37**	**38**	**39**	**40**
A,B	A,B,D	A,C	A,B,D	A,C	A,B,C	A,B,D	A,B,C	A,C,D	A,B,D
41	**42**	**43**	**44**	**45**	**46**	**47**	**48**	**49**	**50**
3	20	40	1.5	43.49 to 43.51	2	5.6	1000	385	22500
51	**52**	**53**	**54**	**55**	**56**	**57**	**58**	**59**	**60**
20	1	0.792 to 0.796	13	0.58 to 0.62	40	2	6000	21	0.5

HINTS AND SOLUTIONS

1. (B): The Benioff-Wadati zone is closely associated with subduction zones. This zone is defined as the area where one tectonic plate is subducted beneath another, resulting in a plane of deep seismic activity that dips into the Earth.

- The Benioff-Wadati zone features a band of earthquakes that occurs at increasing depths beneath a volcanic arc. This pattern of seismicity is typically inclined at an angle ranging from 25° to 75° into the Earth, reflecting the angle of subduction.
- In terms of numerical data supporting this, earthquake depths in these zones can vary widely but generally increase from approximately 40 kilometers near the trench where the subducting plate begins its descent, to several hundred kilometers deep into the Earth's mantle.
- This distinct pattern of deepening earthquakes correlates directly with the descent of the oceanic plate as it is subducted beneath a continental or another oceanic plate, indicative of the compressive and tensional forces at work in these regions.

The identification of a Benioff-Wadati zone as part of a subduction zone rather than any other tectonic setting (like mid-oceanic ridges or continental rifts) is crucial because it specifically involves the interaction of two tectonic plates where one is forced under another, a process not observed in the settings listed in options (A), (C) and (D).

2. (A): Neutron-rich unstable nuclides typically undergo β-decay.

- In β-decay, an excess neutron in the nucleus is transformed into a proton. This transformation involves the emission of an electron (commonly referred to as a beta particle) and an antineutrino.
- The conversion process can be represented by the nuclear reaction: $n \rightarrow p + e^- + \bar{\nu}_e$ where n is a neutron, p is a proton, e^- is the electron, and $\bar{\nu}_e$ is the antineutrino.
- This type of decay helps the nucleus reach a more stable state by reducing the neutron-to-proton ratio. This is particularly important for neutron-rich nuclides, where

the imbalance between neutrons and protons makes the nucleus unstable.

- Numerical example: Consider a neutron-rich isotope such as ^{32}P. This isotope undergoes β-decay to become ^{32}S, changing one of its neutrons to a proton and reducing the neutron excess.

The process of β-decay is most commonly observed in neutron-rich nuclides, distinguishing it from other decay types such as $\beta+$ decay, α-decay, or electron capture, which are characteristic of different nuclear instabilities.

3. (D): Intergranular texture is the correct association with alkali olivine basalt.

- Alkali olivine basalt is characterized by its texture where mineral grains are typically interlocked. This intergranular texture features grains that are roughly equal in size and fit together like a jigsaw puzzle, which is common in volcanic rocks that have cooled relatively quickly.
- The intergranular texture in alkali olivine basalt allows for the close packing of predominantly fine-grained crystals. These crystals are mostly plagioclase and pyroxene, with olivine often present as a significant phase.
- Numerical data relating to textures such as grain size can help characterize the cooling rate of the magma. For instance, the crystals in alkali olivine basalt typically range from less than 1 mm to several millimeters across, indicative of a relatively rapid cooling where crystals do not have much time to grow larger.
- This textural feature differs from other textures like rapakivi, which is characterized by large ovoid crystals surrounded by a rim of another mineral, or graphic texture where minerals intergrow in a way that resembles cuneiform writing.

4. (B): The mineral assemblage of orthopyroxene + garnet + plagioclase + clinopyroxene + quartz is characteristic of a granulite facies metabasalt.

- Granulite facies is a high-grade metamorphic facies characterized by high temperatures and moderate to high pressures. It typically results in the formation of anhydrous minerals due to the dehydration reactions that occur during metamorphism.
- In a metabasalt, the mineral assemblage orthopyroxene, garnet, plagioclase, clinopyroxene, and quartz indicates the conditions of high temperature and sufficient pressure. These minerals are stable under such conditions and signify a deep burial metamorphic environment.
- Orthopyroxene and clinopyroxene in the assemblage reflect the high-temperature conditions, while garnet often indicates a pressure sufficient to stabilize it alongside these pyroxenes.
- Plagioclase and quartz are common in granulite facies rocks and indicate that the original basalt had sufficient silica content to form these minerals during metamorphism.
- Numerical data supportive of these conditions includes the typical temperature range of 700°C to 850°C and pressures of 6 to 10 kbar, which are conducive to the formation of this specific mineral assemblage in granulite facies.

5. (A): Glossopteris is found in the Raniganj Formation. This decision is based on geological and paleontological evidence indicating the presence of Glossopteris fossils in this specific formation:

- The Raniganj Formation is part of the Gondwana Group of sediments in India, which are known for their rich fossil content, particularly the Glossopteris flora. This formation dates back to the Late Permian period, roughly between 255 to 250 million years ago.

- Glossopteris is a significant genus of seed ferns that serves as an index fossil for the Permian geological period in Gondwana continents, indicating cold to temperate climates. Its widespread distribution across Gondwana landmasses is key evidence supporting the theory of continental drift.
- Numerical data related to this topic often include stratigraphic thicknesses and radiometric dating results. For example, the Raniganj Formation typically shows a stratigraphic thickness of several hundred meters, and carbon dating techniques might be used to pinpoint more precise ages within the Permian period.
- The presence of Glossopteris in the Raniganj Formation rather than the other formations listed is due to the specific environmental and depositional conditions of the Raniganj area during the Late Permian, which were suitable for the preservation of such plant fossils.

6. (A): The principle that describes younger rocks overlying older rocks in an undeformed sedimentary sequence is the principle of superposition.

- The principle of superposition states that in any undisturbed sequence of sedimentary rocks, each layer is younger than the one beneath it and older than the one above it. This is a fundamental concept in stratigraphy and is used to determine the relative ages of sedimentary rocks.
- This principle applies specifically to sedimentary rocks and is used to understand the chronological order of geological events by examining the sequence of rock deposition. It does not apply to disturbed sequences where layers may have been overturned or disrupted.
- Numerical data that supports this principle might include measurements of sedimentary layer thicknesses and the dating of material within those layers, like volcanic ash beds, which can provide a timestamp for when layers were formed.
- For example, in a hypothetical sequence of sedimentary rocks containing layers A, B, C, and D, with A being the oldest and D being the youngest, the principle of superposition would tell us that $A < B < C < D$ in terms of their relative ages.

7. (B): Dropstones are found in the Talchir Formation.

- **Geological Context:** The Talchir Formation is part of the Gondwana Supergroup and is well-known for its glacial deposits. This formation primarily comprises sediments that were deposited during the late Carboniferous to early Permian period, a time known for significant glaciation, particularly in the Gondwana regions.
- **Definition and Formation of Dropstones:** Dropstones are isolated stones or boulders that have been dropped into fine-grained sedimentary deposits from floating ice, such as icebergs or glacial meltwater rafts. These stones, which are larger than the surrounding matrix, penetrate the sediment layers below due to their size and weight, often distorting the sedimentary bedding.
- **Evidence and Significance:** The presence of dropstones is indicative of past glacial activity. In the Talchir Formation, these dropstones provide crucial evidence of glaciation in the Gondwana continents. Their study helps paleoclimatologists reconstruct past climates and understand the extent and impact of ice ages on global geography.
- **Numerical Data and Analysis:** The analysis of dropstones involves examining their size, composition, and distribution within the sedimentary layers. For example, dropstones within the Talchir Formation can range in diameter from a few centimeters to several meters, indicating the presence

of substantial glacial activity and iceberg calving events during the deposition of these sediments.

- **Contrast with Other Formations:** While other formations listed, such as the Barakar, Raniganj, and Bijori, also belong to the Gondwana Supergroup, they are characterized by different depositional environments and typically do not show evidence of direct glacial activity in the form of dropstones like the Talchir Formation does.

8. (A): Trough cross-bedding is the sedimentary structure typically formed by unidirectional currents.

- **Formation and Characteristics:** Trough cross-bedding is characterized by sets of cross-strata that are trough-shaped due to the migration of curved ripples or dunes across a depositional surface. The sediments are deposited on the lee side of the ripple or dune as it migrates downstream due to the flow of water or wind in one direction.
- **Geological Indicators:** The cross-beds in trough cross-bedding are inclined layers that intersect the horizontal bedding planes. The direction of the dip of these cross-beds can indicate the flow direction of the current at the time of deposition. The curvature of the troughs in cross-bedding can distinguish them from planar cross-beds, which are typically straight and indicate different flow conditions.
- **Numerical Data and Interpretation:** The angle of cross-bedding (which can vary widely but is often between 10° and 30°) and the dimensions of the troughs (which can be from a few centimeters to several meters in width and depth) provide quantitative data used to interpret past fluid dynamics and sediment transport mechanisms.
- **Application in Geology:** Trough cross-bedding is significant in geological studies as it helps reconstruct past environments, such as river channels, deltas, or shallow marine contexts where unidirectional currents were present. It also aids in determining paleocurrent directions, which are crucial for understanding sedimentary basin evolution.
- **Distinction from Other Options:** Unlike oscillation ripples, which are formed by back-and-forth motion of water typical of wave action, or hummocky cross-stratification, which indicates storm deposits in marine settings, trough cross-bedding specifically indicates the presence of a consistent, unidirectional flow typically found in fluvial and aeolian environments.

9. (C): Kerogen is the precursor of petroleum.

- **Definition and Role:** Kerogen is a complex mixture of organic chemical compounds that forms the principal component of organic matter in sedimentary rocks. It is derived from the diagenetic alteration of biological material like plants, algae, and microscopic marine organisms.
- **Formation Process:** During sedimentation, organic material gets buried and undergoes chemical reactions over millions of years due to increasing heat and pressure. This process, known as diagenesis, converts the organic material into kerogen.
- **Conversion to Petroleum:** As kerogen-rich rocks are subjected to even higher temperatures and pressures during deeper burial (typically in the range of 90°C to 150°C), the kerogen thermally decomposes into liquid and gaseous hydrocarbons — this process is known as catagenesis, leading to the formation of petroleum and natural gas.
- **Numerical Data and Significance:** The quantity and type of kerogen (Type I, II, or III based on the origin and composition

of the organic material) in a rock can determine the potential volume and quality of petroleum that can be generated. Type I and II kerogens are more likely to yield oil, whereas Type III is more prone to produce gas.

- **Distinction from Other Options:** Sporinite, Clarain, and Vitrain are all components or types of coal or coal-related substances, not directly linked to the petroleum generation process as kerogen is. Sporinite is a type of maceral in coal derived from spores and pollen, while Clarain and Vitrain are specific types of coal textures or ranks, indicating different stages of coalification but not directly involved in the petroleum generation.

10. (D): Opal is the amorphous variety of SiO_2.

- **Amorphous Nature:** Unlike crystalline forms of SiO_2 such as quartz and its varieties (including citrine and agate), opal is amorphous. This means it lacks a regular, repeating atomic structure and is instead composed of silica spheres arranged in a disordered network. This structure gives opal its unique play of colour and internal structure.
- **Formation and Composition:** Opal forms from the precipitation of silica from solution, typically in near-surface environments. The process often involves the evaporation of silica-rich water, commonly in sedimentary rock fissures, around hot springs, or as replacements in fossils.
- **Distinctive Features:** Opal is particularly noted for its hydrous composition, containing water within its silica matrix, which can range from 3% to 21% by weight. This water content, along with the microstructure of silica spheres, leads to the diffraction of light and causes the colourful iridescence opal is famous for.
- **Numerical Data and Geological Relevance:** The size of the silica spheres in opal ranges from about 150 to 300 nanometers in diameter, influencing the specific wavelengths of light that are diffracted, and hence the colors seen in precious opal.
- **Contrast with Other Options:** Quartz, citrine, and agate are all crystalline forms of SiO_2 and exhibit a definite crystal structure. Citrine is a variety of quartz colored by impurities, usually iron, and agate is a banded form of finely-grained, microcrystalline quartz.

11. (D): Harzburgite is the correct identification for an igneous rock with a modal composition of 55% olivine, 40% orthopyroxene, and 5% plagioclase, as per the IUGS classification scheme.

- **Composition and Characteristics:** Harzburgite is a type of ultramafic mantle rock primarily composed of olivine and orthopyroxene. It is a member of the peridotite group, which typically has very low silica content and very high magnesium and iron content.
- **Specific Modal Composition:** The specified percentages of 55% olivine and 40% orthopyroxene align well with the typical mineralogical composition of harzburgite. The presence of a small amount (5%) of plagioclase is also consistent with some harzburgite occurrences, though plagioclase is generally more rare in this rock type, suggesting a specific tectonic setting or partial melting process.
- **Formation and Occurrence:** Harzburgite is often formed from the partial melting of the Earth's upper mantle, where the more fusible components, such as basaltic components, are extracted, leaving behind a residue enriched in olivine and orthopyroxene.

- **Geological Significance:** As a residue of partial melting, harzburgite is crucial for understanding mantle dynamics, mantle composition, and the processes of magma genesis and extraction. It is often found in ophiolites, which are sections of the Earth's oceanic crust and the underlying upper mantle that have been uplifted and exposed above sea level.

12. (D): The correct decreasing order of abundance of elements in our solar system is H > He > O > Fe.

- **Elemental Abundance in the Solar System:** The most abundant element in our solar system, and indeed the universe, is hydrogen (H), followed by helium (He). These two elements make up the vast majority of the elemental composition of the universe, particularly evident in the composition of stars and the interstellar medium.
- **Hydrogen and Helium Prevalence:** Hydrogen, as the simplest and most plentiful element, was the first to form after the Big Bang, followed by helium. The abundances of these elements are reflected in their stellar formation and nuclear processes in stars. Hydrogen accounts for about 74% of the elemental mass of the universe, while helium accounts for about 24%.
- **Oxygen and Iron:** Oxygen (O) is the third most abundant element in the universe, and it plays a crucial role in the chemical makeup of planetary bodies and life processes. Iron (Fe), while significantly less abundant than oxygen, is the most abundant of the heavy elements due to its role in stellar processes and its presence in planetary cores.
- **Numerical Data:** Numerically, when considering the abundance by mass in the universe, hydrogen (H) constitutes approximately 73.9%, helium (He) about 24%, oxygen (O) around 1.1%, and iron (Fe) less than 0.1%. This abundance is influenced by stellar nucleosynthesis and the initial mass of elements following the Big Bang.

13. (C): The suture of a cephalopod characterized by smooth saddles and crenulated lobes is called ceratitic.

- **Suture Line:** In cephalopods, the suture line refers to the intersection between the septum (the internal wall dividing the shell into chambers) and the shell wall. The complexity and pattern of this line vary among different cephalopod groups and provide key taxonomic and evolutionary insights.
- **Ceratitic Sutures:** Ceratitic sutures are typical of the order Ceratitida, a group of ammonoid cephalopods. These sutures are intermediate in complexity between goniatitic and ammonitic sutures, featuring smooth, rounded lobes (also known as saddles) and more intricately folded or crenulated lobes.
- **Characteristics and Examples:** The smooth saddles are less complex and provide a contrast to the intricately crenulated lobes, which are more deeply indented and detailed. This combination is distinctive of ceratitic sutures and helps distinguish Ceratitida from other ammonoids.
- **Evolutionary Significance:** The ceratitic suture pattern represents an evolutionary stage between the simpler goniatitic sutures (with simple, rounded lobes) and the highly complex ammonitic sutures (with very intricate, frilled patterns). This evolutionary trait indicates adaptations in shell strength and buoyancy control, crucial for the ecological niches these organisms occupied.

- **Geological and Paleontological Context:** The presence and identification of ceratitic sutures in fossil records help paleontologists date rock strata and reconstruct paleoenvironments, reflecting the distribution and evolution of marine ecosystems through time.

14. **(C):** A cast is a type of body fossil. This type of fossil forms when an organism's body decomposes after being buried in sediment, leaving a mold that is later filled with other minerals or sediments to form a cast. This essentially creates a three-dimensional replica of the shape of the organism.

    Body fossils are direct evidence of ancient organisms, capturing their physical form. In the case of a cast, what we often see is the external morphology preserved in fine detail. This differs from trace fossils such as coprolites (fossilized dung) or footprints, which only provide indirect evidence of the organism's presence by recording its activity, not its actual body.

    Stromatolites, on the other hand, while biogenic and made up of layers of biofilms and precipitated minerals, are not considered body fossils as they do not preserve the direct physical remains of individual organisms. They are more akin to environmental or community structures created by the activity of microbial mats.

    The identification of casts in the fossil record is crucial as it allows paleontologists to understand the external features of extinct species, contributing directly to our knowledge of past life and evolutionary history. This form of fossilization provides insights that are not just limited to the shape and size of organisms but can also include details of their skin texture, which can be invaluable for reconstructive and interpretative purposes in paleobiology.

15. **(A):** Correct matching is:

Morphological Feature	Taxon	Description
Dissepiment	Anthozoa (4)	Thin horizontal partitions found in the skeletons of some corals, supporting the structure.
Delthyrium	Brachiopoda (3)	A triangular opening beneath the beak of a brachiopod shell, where muscles attach.
Pygidium	Trilobita (2)	The posterior segment of the trilobite's body, containing fused segments.
Ambulacrum	Echinodermata (1)	Areas in echinoderms (starfish, sea urchins) bearing tube feet, part of their vascular system.

16. **(A):** Matching sedimentary features with their respective geological processes reveals how different environmental conditions and physical processes shape the sedimentary record.
    - **Stylolite Formation:** Seen in diagenesis, stylolites are jagged, irregular surfaces within rock formed by pressure solution where material has been dissolved away. This occurs under significant compaction and the dissolved material is often precipitated in other areas of the rock, contributing to the overall compaction and lithification process.
    - **Pseudonodule Characteristics:** Formed during events of liquefaction, pseudonodules arise when unconsolidated sediment is suddenly disturbed, often by seismic activity, and then reconsolidated into irregular, nodular forms. This process reflects the dynamic and sometimes abrupt nature of sedimentary deposit changes due to external stresses.

- **Current Crescent:** These are erosional features shaped by the unidirectional flow of water that scours around an object lying on the sediment surface, such as a stone or a clump of vegetation. The sediment downstream of the obstruction is eroded away, leaving a crescent-shaped depression that points in the direction of current flow.
- **Stromatolite Development:** Stromatolites are organo-sedimentary structures produced by the growth of layers of cyanobacteria, a type of photosynthesizing microorganism. These microbes trap and bind sedimentary grains, which over time are cemented together by the calcium carbonate they precipitate. This results in layered, rock-like structures that record both biological activity and sediment accumulation over geological timescales.

17. (A): Correctly matching the geomorphic features with their environments is essential for understanding how physical processes shape different landscapes.

P. Dreikanter: Dreikanters are abrasive features shaped by wind, marking them as eolian (3). They are typically angular rocks that have been faceted by the sand-blasting effects of wind-driven sand. This shaping is most commonly seen in desert environments where wind erosion is significant.

Q. Cirque: Cirques are large, amphitheater-like valleys formed by glacial action (1). They are carved out by the headward erosion due to the plucking and abrasion at the back of a glacier. The distinctive rounded shape of cirques is a classic feature of mountainous regions where glaciers have retreated.

R. Natural levee: Natural levees are formed by fluvial processes (4), specifically by the deposition of sediments during floods. As floodwaters spill over a river's banks, the heaviest materials settle first, building up a ridge adjacent to the river channel, which can help contain subsequent floods.

S. Berm: Berms are features commonly found on beaches (2), created by the wave action depositing sand and other materials during high tide. They mark the highest line of ordinary wave reach and are a typical feature in coastal environments, often seen as slightly elevated platforms of sand parallel to the shoreline.

18. (C): The correct hierarchy of the stratigraphic units listed is as follows: Group > Formation > Member > Bed. This order represents the standard sequence in stratigraphy, which is used to organize and categorize layers of rock by their relative age and composition.

- **Group:** This is the largest unit among those listed. A group consists of several formations, which are linked by their similar geological properties or their geographic proximity.
- **Formation:** A formation is the fundamental unit in stratigraphy. It is a body of rock with a consistent lithology and can be mapped at the earth's surface or traceable in the subsurface.
- **Member:** A member is a smaller unit within a formation and is distinguishable from other parts of the formation by its specific lithologic characteristics or its fossil content.
- **Bed:** The smallest unit, a bed is a layer of rock that has distinct boundaries with layers above and below. Beds can vary in thickness from a few millimeters to several meters but are always smaller than members.

19. (C): Matching the minerals with their highest order of interference color in thin section petrography requires an understanding of how light interacts with the crystalline structure of these minerals.

(P) Sillimanite: Sillimanite typically exhibits second order interference colors (2). This mineral has strong birefringence, which can result in a wide range of colors in polarized light, but the thickness of 0.03 mm generally limits it to second order.

(Q) Quartz: Quartz, with its low birefringence, exhibits first order interference colors (1) for a 0.03 mm thickness. This results in shades typically ranging from grays to whites, which are characteristic of the lower order colors.

(R) Muscovite: Muscovite can show interference colors up to third order variegated (4). Being a mica, it has moderate to high birefringence, and at a typical section thickness, it can display a colorful spectrum that often includes pinks, greens, and yellows.

(S) Calcite: Calcite shows greater than third order interference colors (3). This mineral is highly birefringent, and even at thin section thicknesses, it can display very high order colors, including bright reds, blues, and greens that surpass the third order.

20. (B): The saturated thickness of an unconfined aquifer is correctly defined by the distance between the water table and the underlying confining layer.

- **Unconfined Aquifers:** An unconfined aquifer is one where the water table serves as the upper boundary of the aquifer. Below this water table, the pore spaces in the rock or sediment are completely filled with water, and above it, the pore spaces contain air or are only partially filled with water.
- **Saturated Thickness:** The saturated thickness in an unconfined aquifer specifically refers to the vertical distance between the water table—the top level of the groundwater under natural conditions—and the impermeable layer or confining bed that restricts the downward movement of water. This distance represents the actual thickness of the aquifer where groundwater is fully saturated and capable of being extracted.
- **Practical Implications:** Measuring the saturated thickness is crucial for various hydrological studies and water resource management, including the assessment of groundwater availability and the planning of extraction rates. The saturated thickness can fluctuate with seasonal variations in water table levels or due to prolonged periods of extraction or recharge.

21. (B): Darcy's law quantifies the volume of groundwater flow per unit time.

- **Darcy's Law:** Darcy's law is a fundamental equation in hydrogeology that describes the flow of groundwater through porous media. The law states that the flow rate of groundwater through a medium is proportional to the hydraulic gradient, the cross-sectional area through which the water flows, and the hydraulic conductivity of the material.
- **Clarifying the Components:**
  - **Flow Rate (Q):** In the context of Darcy's Law, Q represents the volume of water flowing through an area per unit of time (e.g., cubic meters per second).
  - **Hydraulic Gradient (*dh*/*dl*):** This is the change in hydraulic head per unit distance, driving the flow of groundwater.
  - **Hydraulic Conductivity (K):** This term represents the ability of the aquifer material to transmit water, typically measured in meters per day.
  - **Cross-sectional Area (A):** This is the area through which the water is flowing.
- **Equation and Application:** The full expression of Darcy's Law can be written as:

$$Q = K \cdot A \cdot \frac{dh}{dl}$$

While this equation indeed involves the hydraulic conductivity, the gradient, and the area, it fundamentally quantifies the flow rate, which is directly measured as a volume per unit time (m^3/s), making option B correct.

- **Practical Usage:** This law is crucial in fields such as groundwater management, environmental engineering, and hydrology, where it is used to estimate the amount of water moving through aquifers and to design systems for water extraction or remediation.

22. (B): According to Strahler's stream ordering system, the highest order of stream in the provided diagram, is 3. Strahler's system is crucial for hydrological studies and environmental management, as it helps in assessing the stream network complexity and potential water flow volume in hydrographic basins.

- **Strahler's Stream Ordering System:** This system assigns a hierarchy of numbers to streams based on the structure of their tributaries. The smallest, unbranched tributaries are classified as order 1. When two streams of the same order join, the resulting stream is the next higher order. However, if two streams of different orders join, the higher order is maintained.
- **Analyzing the Diagram:**
  - The smallest tributaries in the diagram, which have no tributaries themselves, are classified as order 1.
  - When two order 1 streams meet, the resulting stream becomes order 2.
  - The diagram shows at least one instance where two streams of order 2 meet, which then makes the resulting stream order 3.
- **Conclusion:** From the branching pattern visible in the diagram, the highest stream order observed is 3, as there are two streams of order 2 combining into a higher order stream. There are no further mergers that would elevate the stream order to 4 within the visible segments of the diagram, hence the answer is B. 3.

23. (B): The condition where the tunnel wall becomes unstable is when the tunnel axis is parallel to the strike of the beds and the beds dip at an angle of 45° – 60°. The angle of dip (45° – 60°) is significant because it falls within a range that can facilitate sliding, especially when compounded by geological weaknesses like clay layers or faulted zones within the bedding planes. This setup creates a direct pathway for blocks to slide down, making the wall along the dip downward direction particularly vulnerable.

- **Tunnel Stability:** The stability of a tunnel constructed through sedimentary beds largely depends on the orientation of the tunnel axis relative to the geological structures, specifically the bedding planes. The alignment affects how forces are transmitted through the rock mass and, consequently, the potential failure modes.
- **Analysis of Each Condition:**

  (A) Horizontal Tunnel and Horizontal Beds: This scenario typically provides a stable configuration as the loading is uniformly distributed across the beds, which are horizontal and continuous along the tunnel's length.

  (B) Tunnel Axis Parallel to Strike, Beds Dip 45° – 60°: In this case, the dip of the beds can lead to instability. The steep dip angle results in a scenario where gravitational forces can promote sliding along the dip direction. This is particularly concerning on the wall where the dip direction is downward, as blocks of material can detach or slide along the bedding planes, leading to wall collapse.

  (C) Tunnel Axis Parallel to Strike, Beds are Vertical: Although this can pose challenges for wall stability, it is generally more stable than steeply dipping beds because the potential for sliding along the bedding planes is less critical compared to scenario B.

(D) Tunnel Axis Perpendicular to Strike, Beds are Vertical: This configuration is typically stable since the bedding planes intersect the tunnel walls in a way that does not promote sliding. The vertical beds act like pillars supporting the tunnel.

24. (A): A plunging fold will not show a V-shaped outcrop pattern on a planar ground surface if the plunge of the fold axis is equal to the dip of the ground surface in the same direction. A plunging fold has an axis that tilts into the earth rather than lying horizontal. This tilt, or plunge, gives rise to characteristic V-shaped outcrop patterns on the surface, as the fold structure intersects the ground at an angle.

- **Interpretation of Each Condition:**

(A) Equal Plunge and Dip in Same Direction: When the plunge of the fold axis is equal to the dip of the ground surface, and both are in the same direction, the fold will appear to continue straight across the surface rather than forming a V-shaped pattern. This is because the fold's plunge is aligned with the slope of the ground, causing the outcrop pattern to align directly with the surface contour, effectively masking the typical V-shaped pattern.

(B) Plunge Steeper Than Dip in Same Direction: In this scenario, the plunging fold will still show a V-shape because the fold axis descends into the ground at a steeper angle than the ground surface dips. This creates a visible intersection that results in a V-shaped outcrop pattern.

(C) Equal Plunge and Dip in Opposite Direction: Here, the V-shaped pattern might still be observable, but it could be less pronounced or distorted due to the opposing directions of the fold plunge and ground dip.

(D) Plunge Steeper Than Dip in Opposite Direction: The fold will exhibit a very pronounced **V-shaped** pattern because the steep **plunge in** the opposite direction to the **ground** dip will accentuate the **angular** intersection between the fold and the ground surface.

- **Geological Implications:** The relationship between the direction and steepness of a fold's plunge relative to the ground dip significantly influences how geological structures are displayed at the surface.

The analysis shows that Option A **provides** the specific condition under which the **typical** V-shaped outcrop pattern of a plunging **fold** would not occur, highlighting the **importance** of considering both the plunge and **dip** directions in geological studies.

25. (A): The correct interpretation of the structure depicted on the map is a south-easterly plunging synform that was subsequently faulted with block X upthrown.

- **The Structural Setting:**
  - **Synform Indication:** The beds are arched with the youngest layer (1) at the center, flanked by progressively older layers (2, 3, 4) towards both ends. In the map, this arrangement **combined** with the dip directions **suggests a** synform structure where layers **dip** towards a central axis, implying a **fold** that is convex downward.
  - **Plunge Direction:** The synform plunges towards the southeast, as indicated by the orientation of the beds which tilt and point towards this direction. The plunging nature suggests that the axis of the synform extends into the ground in that direction.
- **Fault Dynamics:**
  - **Block Movement:** The fault between blocks X and Y shows that block X is higher relative to Y. This can be

inferred from the positioning of the same numbered beds on either side of the fault, with block X having its beds visibly above those of block Y when moving across the fault line.

- **Type of Fault:** This displacement suggests a reverse fault, where block X has been thrust up relative to block Y, typical in compressional stress environments.

- **Geological Context:**
  - **Stratigraphic Sequence:** The sequence from oldest to youngest is correctly preserved on both sides of the fault, which supports the interpretation of the fault movement without a reversal of stratigraphy which would indicate a more complex history.
  - **Numerical Data:** The precise angle of plunge and dip might not be specified in the diagram but can be estimated from the steepness of the lines depicting the beds. Typically, plunges subtle enough to form clear synclinal structures suggest angles less than 45 degrees.

26. (D): The crystal form 'dome' contains two non-parallel faces related by a mirror plane. In crystallography, a dome is a form where two faces are inclined relative to each other, not parallel, and are related by a mirror plane. This means the faces are symmetric reflections of each other across a plane, but they do not extend in the same direction or angle, which distinguishes them from forms related by a 2-fold axis of symmetry.

- **Mirror Plane and Crystal Faces:**
  - **Mirror Plane:** A mirror plane in crystallography is an imaginary plane that divides a crystal into two symmetrical halves. Any feature on one side of the plane has a mirror image on the other side.
  - **Non-Parallel Faces:** In the context of a dome, the faces are oriented at different angles such that they converge but are not parallel. This angular relationship combined with their mirror symmetry creates a characteristic dome-shaped appearance.
- **Crystal Form Terminology:**
  - **Parallel vs. Non-Parallel Faces:** Parallel faces would extend in the same direction and maintain a constant angle between them, which is not the case in domes. Non-parallel faces that converge or diverge emphasize the dome's unique structural aspect.
  - **2-Fold Axis vs. Mirror Plane:** A 2-fold axis of symmetry implies that the structure can be rotated 180 degrees around an axis to yield an indistinguishable configuration, which differs from a mirror plane that reflects faces directly across from each other.

27. (D): The symbols [100], {100} and (100) in crystallography represent a line, a form, and a face, respectively.

- **[100]:** This symbol indicates a line in crystallography, specifically a crystallographic direction along the *x*-axis in Cartesian coordinate system. Directions in crystals are described by vectors within the unit cell and [100] specifically points along the primary axis without deviation.
- **{100}:** This symbol is used to denote a form, which is a set of crystal faces related by symmetry. In the case of {100}, it refers to all faces that have the same crystallographic indices irrespective of their orientation sign. This includes all normal variations like the front, back, left, right, top and bottom of a cubic crystal.
- **(100):** This notation represents a face of a crystal, particularly one that is perpendicular to the x-axis in a standard coordinate system. It describes a specific plane in the crystal, distinguishing it from other planes that might have similar orientations but different positional contexts within the crystal lattice.

28. (A): The correct matching of the stratigraphic units in Group I with the corresponding Archean cratons in Group II is detailed below:

(P) Bababudan Group: Matches with Western Dharwar (2). The Bababudan Group, well-known for its iron ore deposits, is geographically and geologically part of the Western Dharwar Craton. This region is characterized by extensive greenstone belts and associated sedimentary sequences.

(Q) Banded Gneissic Complex-I: Corresponds to Aravalli (3). The Banded Gneissic Complex represents some of the oldest rock formations and is predominant in regions like the Aravalli range, showcasing early Archean features.

(R) Bonai Granite: Aligns with Singhbhum (4). Bonai Granite is associated with the Singhbhum Craton, which is known for its significant granitic and metamorphic activities dating back to the Archean era.

(S) Kolar Group: Matches with Eastern Dharwar (1). The Kolar Group, famous for its gold-bearing greenstone belts, is located within the Eastern Dharwar Craton, a region rich in ancient volcanic and sedimentary sequences.

29. (D): The correct sequence of metamorphic facies in order of increasing metamorphic grade for thermal metamorphism is:

Albite-epidote hornfels → Hornblende hornfels → Pyroxene hornfels → Sanidinite

- **Albite-epidote hornfels:** Represents a lower-grade metamorphism where temperature conditions are relatively moderate. This facies is characterized by minerals such as albite and epidote, which are stable at lower temperatures.
- **Hornblende hornfels:** As temperatures increase, hornblende becomes stable, marking a medium-grade metamorphic condition. This facies indicates higher temperatures but not enough to destabilize silicate structures completely.
- **Pyroxene hornfels:** At higher temperatures, pyroxene minerals become stable, indicating a higher grade of metamorphism before the peak conditions of thermal metamorphism are reached. This facies is indicative of conditions where temperatures are sufficient to form pyroxene but not yet at the maximum thermal metamorphic grade.
- **Sanidinite:** This is the highest-grade facies in the context of thermal metamorphism, characterized by the stability of minerals such as sanidine and other high-temperature phases. Sanidinite facies typically forms under conditions near magmatic temperatures, representing the peak of thermal metamorphic grades.

30. (D): Nickel ores are not typically associated with skarns as skarns are generally not a source for nickel ores compared to the other types listed, which are more directly related to nickel accumulation either through magmatic differentiation, direct precipitation, or lateritic weathering processes.

(A) Ultramafic Igneous Rocks: Nickel ores, especially those rich in minerals like pentlandite, are commonly associated with ultramafic igneous rocks. These rocks, due to their high magnesium and iron content, are a primary source for nickel-bearing minerals.

(B) Laterites: Lateritic nickel deposits form in tropical climates where intense weathering of ultramafic rocks occurs. The weathering process leads to the concentration of nickel along with other elements like iron and aluminum in the soil.

(C) Sea-Floor Polymetallic Nodules: Nickel is also found in sea-floor polymetallic nodules, which are rounded lumps of minerals precipitated from sea water on the ocean floor. These nodules can contain manganese, copper, cobalt, and nickel.

(D) **Skarns:** Skarns are formed by the interaction of magmatic fluids with surrounding carbonate rocks, leading to the formation of calcium-rich silicates. While skarns are rich in metals like iron, copper, and tungsten, they are not typically associated with nickel deposits.

31. **(A & B):** Both statements (A) and (B) are correct regarding mantle partial melting.

(A) **Shallow melting produces tholeiitic basalts:** This statement is correct. Tholeiitic basalts typically form from partial melting of the mantle at relatively shallow depths, often near mid-ocean ridges or in rift zones. The melting process at these shallower depths and lower pressures allows for the formation of tholeiitic magma, which is characterized by its low content of incompatible elements and relatively high iron and magnesium content.

(B) **Low-degree melting produces alkaline basalts:** This statement is also correct. Alkaline basalts generally result from low degrees of partial melting of the mantle. This type of melting tends to occur at greater depths or in geologic settings with higher pressures, which allows for the retention of more volatile components such as H_2O and CO_2 in the melt. These conditions promote the enrichment of alkaline and incompatible elements (elements that preferentially remain in the melt rather than crystallizing in minerals), leading to the distinctive composition of alkaline basalts.

(C) **Presence of CO_2-rich volatiles favors the formation of tholeiitic basalts:** This statement is incorrect. The presence of CO_2-rich volatiles generally leads to the formation of more silica-undersaturated and potentially more alkaline magmas, rather than tholeiitic basalts, which are more silica-saturated.

(D) **Presence of H_2O-rich volatiles favors the formation of alkaline basalts:** This statement is incorrect. The presence of H_2O-rich volatiles can influence melting temperatures and extents but does not necessarily lead to the production of alkaline basalts specifically. H_2O generally lowers the melting temperature of the mantle, which can promote melting but the magma type produced (alkaline versus tholeiitic) depends on other factors such as mantle source composition and the degree of partial melting.

32. **(A, B, & D):** The correct fossil groups from the Siwalik Group include Proboscidea, Giraffidae, and Equidae.

(A) **Proboscidea:** This group includes mammals like elephants and their ancestors. Fossils of various proboscideans, such as gomphotheres and early elephants, have been discovered in the Siwalik Group sediments. These fossils help trace the evolutionary history of elephants and their relatives across the Indian subcontinent.

(B) **Giraffidae:** The Giraffidae family, which includes giraffes and their relatives, is also represented in the Siwalik Group. Fossils of Sivatherium, an extinct genus of Giraffidae that resembled a massive, robust giraffe, are well-documented within these deposits.

(D) **Equidae:** This family includes modern horses and their extinct relatives. Fossils of various members of the Equidae, such as Hipparion, which is a three-toed horse, have been frequently found in the Siwalik Group, indicating a diverse assemblage of grassland-adapted herbivores during the time these sediments were deposited.

(C) **Dinosauria:** This option is incorrect for the Siwalik Group, which is of Neogene age (spanning from about 23 to 26 million years ago). By this time, dinosaurs had been extinct for over 60 million years (since the end of the Cretaceous Period, about 66 million years ago).

33. **(A & C):** Both options (A) and (C) correctly list stratigraphic successions from the oldest to the youngest for their respective regions.

(A) Cretaceous Formations in South India

- **Uttatur Group:** This is among the older Cretaceous formations in South India, containing rich deposits of marine fossils.
- **Trichinopoly Group:** Follows the Uttatur Group, representing a sequence that contains younger sedimentary layers, typically characterized by different fossil assemblages.
- **Ariyalur Group:** This group overlies the Trichinopoly Group, noted for its more extensive and varied sediment types and fossil content.
- **Niniyur Formation:** Represents the youngest in this sequence, typically consisting of sediments deposited in a shallower marine environment than its predecessors.

(C) Siwalik Group in the Sub-Himalayan Region

- **Chinji Formation:** Known for its fluviatile sediments, marking the beginning of significant sedimentation in the Siwalik Group.
- **Nagri Formation:** Overlies the Chinji Formation, with more varied sedimentology reflecting changing environmental conditions.
- **Dhok Pathan Formation:** Successor to the Nagri Formation, noted for a greater diversity of both fluvial environments and fossil mammalian species.
- **Tatrot Formation:** The youngest formation in this sequence, characterized by even more recent fluviatile deposits and a continuation of diverse faunal records.

34. **(A, B & D):** The combinations (A), (B) and (D) are correctly matched, each representing a geological or ecological setting accurately paired with its characteristic feature or depositional environment.

(A) Photic Zone – Biogenic Carbonate Rocks

- **Photic Zone:** This is the upper layer of water bodies exposed to sunlight, enabling photosynthesis.
- **Biogenic Carbonate Rocks:** These rocks are typically formed in the photic zone where organisms capable of photosynthesis, such as corals and algae, are abundant. These organisms contribute to carbonate sedimentation through their skeletal structures, leading to the formation of biogenic carbonates like limestone.

(B) Delta – Progradational Coarsening-Up Succession

- **Delta:** A deltaic environment is formed where a river meets a standing body of water, depositing sediments.
- **Progradational Coarsening-Up Succession:** This type of sedimentary succession is characteristic of deltaic environments where finer sediments deposited further from shore progressively give way to coarser sediments nearer to the river's mouth as the delta progrades outward into the body of water.

(D) Shelf Break – Submarine Fans

- **Shelf Break:** This is the point where a continental shelf steepens significantly and descends into the deeper waters of a continental slope.
- **Submarine Fans:** These are large, fan-shaped sediment deposits found below the shelf break, formed as sediments are transported from the continental shelf down the slope, often through turbidity currents. These features are typical of deep-water depositional environments.

Incorrect Match:

(C) Sabkha – Shelf Storm Deposit: This match is incorrect because a sabkha is a coastal flat area characterized by evaporitic conditions, not associated with shelf storm deposits. Storm deposits are more typically related to deeper, offshore environments where storm-driven processes redistribute sediments along a continental shelf.

35. (A & C): High drainage density is representative of a terrain with high relief and an impermeable surface layer.

(A) High Relief: Terrain with high relief typically features steep slopes. These conditions promote rapid surface runoff as water is less likely to infiltrate into the soil, leading to a higher concentration of drainage channels to accommodate the flow. This creates a network of closely spaced streams and rivers, increasing the drainage density.

(C) Impermeable Surface Layer: An impermeable surface layer, such as clay or bedrock, restricts water infiltration into the soil, causing more water to remain on the surface and flow downhill. This leads to the formation of more numerous and closely spaced drainage networks, thus increasing the drainage density.

Incorrect Options:

(B) Arid Climate: In arid climates, there is typically less rainfall, which generally leads to lower drainage density because there is less surface water to create and maintain extensive drainage networks. Thus, arid conditions are typically associated with lower drainage density.

(D) Permeable Surface Layer: Terrain with a permeable surface layer, such as sandy or gravelly soils, allows water to infiltrate easily, reducing the volume of water available for surface runoff. This tends to lower the drainage density as fewer channels are needed to carry surface water.

36. (A, B & C): Mass-wasting, also known as slope movement or mass movement, encompasses several types of ground movements related to gravity's pull on earth materials like rock, soil, artificial fill, or a combination of these. These processes typically occur on slopes and are driven by the force of gravity, often exacerbated by factors such as water saturation, river undercutting, seismic activity, and human activities.

- **Landslides:** Landslides are characterized by the movement of a mass of rock, debris, or earth down a slope. They can be triggered by rains, earthquakes, volcanic activity, or man-made disturbances such as construction. For instance, the Oso landslide in Washington state in 2014 involved approximately 8 million cubic meters of material, emphasizing the scale and impact of such events.
- **Lahars:** Lahars are volcanic mudflows or debris flows composed of a slurry of pyroclastic material, rocky debris, and water. They originate on the slopes of a volcano, typically during an eruption when rapidly melting snow or ice mixes with volcanic material. A notable historical example is the 1985 lahar from Nevado del Ruiz in Colombia, which tragically killed about 23,000 people and had a flow volume of roughly 40 million cubic meters.
- **Avalanches:** These are rapid flows of snow down a sloping surface, which can incorporate rock, soil, and vegetation, exemplifying mass-wasting in colder climates. Avalanches are particularly common in mountainous areas and can be initiated by factors such as snowpack, slope angle, weather, and human activity. Statistical data shows that avalanches kill more than 150 people worldwide each year.

Thus, each of these processes fits the criteria of mass-wasting, driven by gravity and involving the downward and outward movement of earth materials on slopes. Sand storms, however, involve the movement of sand by wind, not gravity, and thus do not constitute mass-wasting.

38. (A, B & C): Deformation structures in geology are specific features formed in rocks due to various geological processes. Each listed process has a characteristic effect on rock formations, which can be observed and studied.

(A) Pressure solution – rock cleavage: Pressure solution is a diagenetic process where mineral grains partially dissolve under high stress, usually in the presence of a fluid phase. The dissolved material may later precipitate in areas of lower stress. This process often contributes to the development of rock cleavage, which is a type of foliation typically found in fine-grained rocks like shales and slates, where minerals are aligned due to differential stress.

(B) Jointing – plumose markings: Jointing refers to the natural formation of fractures or separations in rock without displacement of the sides. Plumose markings are a type of surface feature found on joints, characterized by feather-like patterns that indicate the direction of crack propagation. These markings help in understanding the sequence of joint formation and the stresses that caused them.

(C) Layer parallel compression – buckle folds: Buckle folds occur when layers of rock are compressed parallel to their layering, causing the material to buckle and fold. This type of folding is common in sedimentary sequences and thin-skinned tectonic settings where horizontal compressive stresses or layer-parallel shortening forces the layers to warp into folds.

39. (A, C & D): In the context of Barrovian metamorphism, which is a type of regional metamorphism characterized by progressive increases in temperature and pressure, several key isograds and zones are typically recognized based on the appearance of specific index minerals.

(A) Grade of metamorphism increases from left to right: This is a fundamental concept in metamorphic geology, especially in Barrovian sequences. As you move from left (lower-grade zones such as the Chlorite isograd) to right (higher-grade zones such as the Kyanite isograd), both temperature and pressure conditions typically increase, leading to the formation of different minerals indicative of these higher pressures and temperatures.

(C) Zone 'Q' is the Garnet zone: In Barrovian metamorphism, different minerals appear as metamorphic grade increases. Garnet typically first appears at the boundary defined as the 'Garnet isograd'. This zone is marked by the stable presence of garnet in the rock, indicating a specific range of temperature and pressure suitable for garnet formation. Zone 'Q' corresponds to this isograd in the given sequence, denoting an intermediate metamorphic grade where garnet becomes a stable mineral phase.

(D) 'R' isograd is the Kyanite isograd: Kyanite appears at higher metamorphic grades and is typically associated with high pressure and moderate to high temperature conditions. The 'R' isograd, marking the onset of the Kyanite zone, indicates that the rocks have undergone sufficient metamorphic changes to stabilize kyanite. This isograd is a critical marker in Barrovian sequences, delineating an advanced stage of metamorphism.

40. (A, B & D): Brittle deformation refers to the process where rocks break or fracture without significant plastic deformation. Rocks subjected to brittle deformation typically exhibit characteristics such as cracking, fracturing, or complete loss of cohesion, which leads to the formation of certain types of geological structures.

(A) Cataclasite: Cataclasites are rocks formed by mechanical crushing and grinding of rocks within fault zones. This process occurs at relatively low temperatures and pressures where rocks behave in a brittle manner. The resulting rock is typically composed of angular fragments of the original rock embedded in a finer-grained matrix of crushed material. Cataclasites are a direct result of the intense, localized stress that causes the rock to fracture and shatter.

(B) Breccia: Breccia is a rock type composed of larger fragments of pre-existing rocks that are angular in shape, cemented together by a finer-grained matrix. This formation is typically associated with brittle conditions such as faulting, volcanic eruptions, or collapse structures where rapid deformation results in the fracturing of rock. The angularity of the fragments indicates that they have not been transported far from their source, which is characteristic of brittle deformation.

(D) Gouge: Gouge is composed of very fine-grained, powdered rock material formed by the grinding action along a fault. It is essentially the extreme end-product of cataclasis in fault zones where the rock has been ground into fine particles. This type of deformation structure is indicative of very high strain localized along fault surfaces and is associated with significant brittleness in the rock structure.

Mylonite (C), on the other hand, is typically formed under conditions that involve both high pressure and sufficient temperature to allow for ductile deformation of the rock, which includes stretching and recrystallization rather than just fracturing and brecciation. Thus, it does not fall into the category of structures formed by brittle deformation.

41. (3): To determine the φ (phi) value of a sediment grain with a diameter of 0.125 mm, the phi scale, a logarithmic scale based on base 2, is used. The formula to calculate phi is:

$$\varphi = -\log_2\left(\frac{D}{D_0}\right)$$

where D is the grain diameter and D_0 is a reference diameter, commonly set at 1 mm. Applying the values:

$$\varphi = -\log_2\left(\frac{0.125}{1}\right)$$

Calculating the log base 2 of 0.125:

$$0.125 = 2^{-3}$$

Thus,

$$\log_2 (0.125) = -3$$

and therefore,

$$\varphi = -(-3) = 3$$

This calculation confirms that the φ value for a sediment grain with a diameter of 0.125 mm is indeed 3. This phi value indicates how fine the sediment is on a scale where higher values denote finer particles.

42. (20): To determine the net slip of a fault with a given vertical separation and fault dip, one uses the relationship between vertical separation (V), net slip (S), and the angle of dip (θ). The formula to find net slip is:

$$S = \frac{V}{\sin(\theta)}$$

Given:

- V = 10 m (vertical separation)
- θ = 30°

Plugging the values into the formula:

$$S = \frac{10 \text{ m}}{\sin(30°)}$$

Since sin(30°) = 0.5,

$$S = \frac{10 \text{ m}}{0.5} = 20 \text{ m}$$

Thus, the net slip on the fault is 20 m, demonstrating how the vertical separation scales with the sine of the dip angle.

43. (40): To calculate the interlimb angle of an overturned antiform, where the limbs have different dips and orientations, we consider the absolute difference between the two dip angles. For a normal and overturned limb with dips of 30° and 70° respectively, the interlimb angle can be calculated directly as the acute angle between these two limbs.

Given:

- Dips of normal limb = 30°
- Dips of overturned limb = 70°

The interlimb angle is obtained by subtracting the smaller dip from the larger dip:

Interlimb Angle = 70° – 30° = 40°

This calculation correctly indicates the acute angle between the two limbs of the fold, which is 40 degrees. This represents the angular separation visible or measurable from a cross-sectional view perpendicular to the axis of the fold, fitting the typical geological interpretations of such structures.

44. (1.5): To determine the value of 'x' in the formula $KAl_3Si_3O_{10}(F_{0.5}OH_x)$, we need to ensure that the overall chemical formula of the mineral is electrically neutral.

First, let's consider the charges of each component in the formula:

- K^+ contributes +1 charge.
- Al^{3+} contributes +3 charge each, and there are 3 Al atoms, so +9 charge total.
- Si^{4+} contributes +4 charge each, and there are 3 Si atoms, so +12 charge total.
- O^{2-} contributes –2 charge each, and there are 10 O atoms, so –20 charge total.
- F^- contributes –1 charge, and there is 0.5 F, so –0.5 charge total.
- OH^- contributes –1 charge per group.

The sum of the charges contributed by the non-hydroxyl components is calculated as follows:

+1 (from K^+) + 9 (from 3 × Al^{3+}) + 12 (from 3 × Si^{4+}) – 20 (from 10 × O^{2-}) – 0.5 (from 0.5 × F^-) = +1.5

The overall charge of the mineral must be zero, so the negative charges from the hydroxide ions must balance the remaining positive charge. Since the remaining positive charge is +1.5, we need 1.5 negative charges from the hydroxide ions:

$$1.5 = x \times 1 \text{ (as each OH is } -1)$$

Therefore, the value of 'x' must be:

$$x = 1.5$$

Thus, the value of 'x' in the mineral formula $KAl_3Si_3O_{10}(F_{0.5}OH_x)$ is 1.5, rounded to one decimal place.

45. (43.49 to 43.51): To find the atom percent of Fe in pyrrhotite of composition $Fe_{0.77}S$, one must calculate the ratio of iron atoms to the total number of atoms in the formula, which includes both iron and sulfur atoms.

Given the composition:

- $Fe_{0.77}$
- S

The total number of atoms per formula unit is the sum of the moles of iron and sulfur:

- Moles of Fe = 0.77
- Moles of S = 1

Total moles in the formula:

Total moles = 0.77 + 1 = 1.77

The atom percent of Fe is then calculated by the ratio of moles of Fe to total moles, multiplied by 100 to get a percentage:

$$\text{Atom percent Fe} = \left(\frac{0.77}{1.77}\right) \times 100$$

Performing the calculation:

$$\text{Atom percent Fe} = \left(\frac{0.77}{1.77}\right) \times 100 \approx 43.50$$

This value falls within the given range of 43.49 to 43.51, accurately reflecting the atomic percentage of iron in the mineral pyrrhotite, $Fe_{0.77}S$. This computation shows the proportion of iron atoms relative to the total atoms (iron + sulfur) in the mineral.

46. (2): To determine the minimum number of chemical components required to describe the composition of all phases in the univariant metamorphic reaction Albite = Jadeite + Quartz, we need to consider the chemical compositions of each phase and the concept of a component in petrology, which refers to the minimum number of independent chemical constituents necessary to describe all phases of a system.

Albite ($NaAlSi_3O_8$) consists of Na, Al, Si, and O.

Jadeite ($NaAlSi_2O_6$) contains Na, Al, Si, and O.

Quartz (SiO_2) contains Si and O.

To analyze this, we observe that:

- Oxygen (O) is common to all minerals.
- Silicon (Si) is also shared across all minerals.
- Sodium (Na) and Aluminum (Al) vary between Albite and Jadeite but are essential for both, and they often substitute for each other in silicate minerals, thus can be grouped into a single system component.

Considering the role of components:

1. **Silica (SiO_2)** can act as one component. This is justified because SiO_2 alone can account for Quartz and contributes to the silica content in both Albite and Jadeite.
2. **Sodium alumino-silicate ($NaAlSiO_4$)** can be another component, which can effectively combine the rest of the compositions. This component can cover the sodium and aluminium requirements in both Albite and Jadeite, and in conjunction with the SiO_2 component, the whole system's composition is accounted for.

Hence, two components, SiO_2 and $NaAlSiO_4$, are sufficient to describe all phases in this metamorphic reaction. This is consistent with the principles of component calculation in petrology, where the objective is to minimize the number of components needed to describe all constituents and variations in a system comprehensively. Thus, the minimum number of components required is 2.

47. (5.6): To calculate the mean flow velocity of water in an open channel using the Froude Number (Fr) and the average depth of the water, the following formula can be used:

$$Fr = \frac{V}{\sqrt{g \cdot d}}$$

where:

- V is the flow velocity,
- g is the acceleration due to gravity (9.8 m/s^2),
- d is the depth of the channel.

Given:

- Froude Number (Fr) = 4,
- Average depth (d) = 0.2 m.

Rearranging the formula to solve for V:

$$V = Fr \cdot \sqrt{g \cdot d}$$

Plugging in the given values:

$$V = 4 \cdot \sqrt{9.8 \cdot 0.2}$$

Calculating the square root:

$$\sqrt{9.8 \cdot 0.2} = \sqrt{1.96} \approx 1.4$$

Now, calculate V:

$$V = 4 \cdot 1.4 = 5.6 \text{ m/s}$$

This calculation shows that the mean flow velocity in the channel is 5.6 m/s, which is derived by utilizing the known Froude Number and the depth of the channel. This velocity represents the rate at which water flows through the channel, characterized by the relation of inertial to gravitational forces as described by the Froude Number.

48. (1000): To calculate the volume of water that flows through an aquifer per second, one can use Darcy's Law, which relates the discharge rate through a porous medium to the hydraulic conductivity, the cross-sectional area through which the water flows, and the hydraulic gradient.

The formula for Darcy's Law is given by:

$$Q = K \cdot A \cdot i$$

Where:

- Q is the discharge (volume per unit time, in cm^3/s),
- K is the hydraulic conductivity (in cm/s),
- A is the cross-sectional area (in m^2),
- i is the hydraulic gradient (dimensionless).

Given:

- Hydraulic conductivity K = 0.25 cm/s,
- Cross-sectional area A = 10 m^2,
- Hydraulic gradient i = 0.04.

First, convert the area from square meters to square centimeters (since 1 m^2 = 10,000 cm^2):

$$A = 10\ m^2 = 10 \times 10{,}000\ cm^2 = 100{,}000\ cm^2$$

Now plug these values into Darcy's Law:

$$Q = 0.25\ cm/s \times 100{,}000\ cm^2 \times 0.04$$

Calculate the discharge Q:

$$Q = 0.25 \times 100{,}000 \times 0.04 = 1000\ cm^3/s$$

This calculation results in a discharge rate of 1000 cubic centimeters per second, indicating that 1000 cm^3 of water flows through the aquifer every second under the given conditions. This discharge rate is derived from applying the hydraulic properties of the aquifer and the gradient driving the flow.

49. (385): To calculate the temperature at a certain depth using the geothermal gradient, one needs to multiply the depth by the geothermal gradient and add the result to the surface temperature.

Given:

- Geothermal gradient = 0.02 °C/m
- Surface temperature = 25 °C
- Depth = 18 km

First, convert the depth from kilometers to meters (since 1 km = 1000 m):

$$\text{Depth in meters} = 18\ km \times 1000\ m/km = 18{,}000\ m$$

Now, calculate the increase in temperature at this depth due to the geothermal gradient:

Increase in temperature = 0.02 °C/m × 18,000 m

Increase in temperature = 360°C

Finally, add this increase to the surface temperature to find the temperature at the depth:

Temperature at depth = 25 °C + 360 °C = 385°C.

This calculation shows that the temperature at a depth of 18 km, given the specified geothermal gradient, is 385 °C. This result is consistent with the understanding that the Earth's interior warms up by approximately 0.02 °C for every meter of increased depth, starting from an initial surface temperature of 25 °C.

50. (22500): To estimate the ore reserve of a triangular block of an orebody, calculate the average volume by taking the average of the thicknesses at the three corners and multiplying by the area. The volume is then multiplied by the ore density to estimate the total mass of the ore.

Given:

- Area of the triangular block = 1500 m^2
- Thicknesses at the three corners = 5 m, 6 m, 7 m
- Ore density = 2.5 tons/m^3

First, calculate the average thickness:

$$\text{Average thickness} = \frac{5\,m + 6\,m + 7\,m}{3} = \frac{18\,m}{3} = 6\,m$$

Then, calculate the volume of the orebody:

$$\text{Volume} = \text{Area} \times \text{Average thickness} = 1500\,m^2 \times 6\,m = 9000\,m^3$$

Now, calculate the total mass of the ore:

$$\text{Ore reserve} = \text{Volume} \times \text{Density} = 9000\ m^3 \times 2.5\ tons/m^3 = 22500\ tons$$

Thus, the estimated ore reserve of the triangular block is 22500 tons, which is calculated by averaging the variations in thickness across the triangular block to estimate a uniform volume, then applying the density to find the total mass. This approach ensures a realistic and practical estimation of the ore reserves within the geometric constraints provided.

51. (20): To determine the concentration of Sm (Samarium) in the basaltic magma from the concentration in clinopyroxene and the partition coefficient, we use the relationship provided by the partition coefficient formula:

$$\text{Partition Coefficient (D)} = \frac{\text{Concentration in the crystal } (C_{crystal})}{\text{Concentration in the melt } (C_{melt})}$$

Given:

- Concentration in clinopyroxene $C_{crystal}$ = 24 ppm
- Partition coefficient D = 1.2

Rearranging the formula to find the concentration in the melt:

$$C_{melt} = \frac{C_{crystal}}{D}$$

Plugging in the given values:

$$C_{melt} = \frac{24 \text{ ppm}}{1.2} = 20 \text{ ppm}$$

Therefore, the concentration of Sm in the basaltic magma is calculated to be 20 ppm. This value indicates that the basaltic magma has a lower concentration of Sm compared to the clinopyroxene that crystallized from it, as reflected by the partition coefficient greater than 1, showing enrichment in the crystalline phase.

52. (1): To calculate the lithostatic pressure at a depth in the Earth's crust, use the equation:

$$P = \rho \cdot g \cdot h$$

where:

- ρ is the density of the rock (2800 kg/m³),
- g is the acceleration due to gravity (9.8 m/s²),
- h is the depth (36.5 km, converted to meters).

Converting the depth from kilometers to meters:

$$h = 36.5 \text{ km} = 36{,}500 \text{ m}$$

Calculating the pressure in Pascals:

$$P = 2800 \text{ kg/m}^3 \times 9.8 \text{ m/s}^2 \times 36{,}500 \text{ m}$$

$$P = 1{,}001{,}660{,}000 \text{ Pa}$$

Convert Pascals to Gigapascals (GPa), noting that 1 GPa = 10^9 Pa:

$$P = \frac{1{,}001{,}660{,}000 \text{ Pa}}{10^9 \text{ Pa/GPa}} \approx 1.002 \text{ GPa}$$

Rounding to the nearest integer:

$$P \approx 1 \text{ GPa}$$

Thus, the lithostatic pressure at a depth of 36.5 km in the continental crust with an average density of 2800 kg/m³ is approximately 1 GPa, when rounded to the nearest integer. This estimate indicates the substantial pressure exerted by the overlying rock mass at such a depth.

53. (0.792 to 0.796): To determine the fraction of ^{24}Na atoms remaining after a specified decay interval, we use the formula for radioactive decay, which is based on the half-life ($t_{1/2}$) of the isotope. The formula to calculate the remaining fraction of a radioactive substance after time t is:

$$N(t) = N_0 \cdot \left(\frac{1}{2}\right)^{\frac{t}{t_{1/2}}}$$

where:

- $N(t)$ is the number of undecayed atoms remaining after time t,
- N_0 is the initial number of atoms,
- $t_{1/2}$ is the half-life of the substance,
- t is the time elapsed.

Given:

- $t = 5.0$ hours,
- $t_{1/2} = 15.0$ hours.

Plug these values into the decay formula to find the fraction of ^{24}Na remaining:

$$\text{Fraction remaining} = \left(\frac{1}{2}\right)^{\frac{5.0}{15.0}} = \left(\frac{1}{2}\right)^{\frac{1}{3}}$$

Calculate $\left(\frac{1}{2}\right)^{\frac{1}{3}}$:

$$\left(\frac{1}{2}\right)^{\frac{1}{3}} \approx 0.793$$

Rounding to three decimal places, the fraction of ^{24}Na atoms remaining after 5 hours is approximately 0.794. This value, falling within the given range of 0.792 to 0.796, shows that approximately 79.4% of the ^{24}Na atoms are still undecayed after this time period, a direct consequence of the exponential nature of radioactive decay based on the provided half-life.

54. (13): To calculate the orthogonal (true) thickness of a dipping coal bed from the thickness measured along a vertical drill hole and the dip angle, use the trigonometric relationship:

True Thickness = Measured Thickness × cos(Dip Angle)

where:

- The Measured Thickness is the thickness measured along the drill hole (15 m).
- The Dip Angle is the angle at which the bed dips from the horizontal (30°).

Applying the formula:

$$\text{True Thickness} = 15\,\text{m} \times \cos(30°)$$

The cosine of 30° is approximately 0.866. Therefore:

$$\text{True Thickness} = 15\,\text{m} \times 0.866 \approx 12.99\,\text{m}$$

Rounding 12.99 m to the nearest integer gives:

$$\text{True Thickness} = 13\,\text{m}$$

This calculation shows that the orthogonal thickness of the coal bed, which accounts for the dipping angle, is 13 meters. This method ensures that the thickness is adjusted for the angle of dip, providing a true vertical thickness that is useful for accurate geological and resource estimation.

55. (0.58 to 0.62): To calculate the mole fraction of forsterite (Mg_2SiO_4) in olivine, which also contains fayalite (Fe_2SiO_4), we first need to convert the weight percentages of MgO, FeO, and SiO_2 to moles using their respective molecular weights, and then calculate the moles of forsterite and fayalite.

1. Convert weight percentages to moles:

Given weights and molecular weights:

- **MgO:** 29.17% weight, molecular weight = 40.31 g/mol
- **FeO:** 34.65% weight, molecular weight = 71.85 g/mol
- SiO_2: 36.18% weight, molecular weight = 60.00 g/mol

Convert to moles:

$$\text{Moles of MgO} = \frac{29.17}{40.31}$$

$$\text{Moles of FeO} = \frac{34.65}{71.85}$$

$$\text{Moles of } SiO_2 = \frac{36.18}{60.00}$$

2. Calculate moles for Mg and Fe (ignoring SiO_2 for this step):

$$\text{Moles of MgO} \approx \frac{29.17}{40.31} \approx 0.723 \text{ moles of Mg}$$

$$\text{Moles of FeO} \approx \frac{34.65}{71.85} \approx 0.482 \text{ moles of Fe}$$

3. Calculate the mole fraction of forsterite and fayalite:

Forsterite and fayalite are ideally Mg_2SiO_4 and Fe_2SiO_4 respectively. Therefore, each mole of MgO or FeO contributes to one mole of Mg or Fe in these minerals, respectively. The stoichiometry requires twice as much Mg or Fe as SiO_2, so the limiting reagent is SiO_2. Using the moles of SiO_2 (which is also the moles of SiO_4 groups available):

$$\text{Moles of } SiO_2 \approx \frac{36.18}{60.00}$$

$$\approx 0.603 \text{ moles of } SiO_4$$

Calculate the total possible moles of Mg_2SiO_4 and Fe_2SiO_4 if all available Mg and Fe were used:

$$\text{Moles of } Mg_2SiO_4 = \frac{0.723}{2}$$

$$= 0.3615 \text{ moles}$$

$$\text{Moles of } Fe_2SiO_4 = \frac{0.482}{2}$$

$$= 0.241 \text{ moles}$$

Add these to get the total moles of olivine:

$$\text{Total moles of olivine} = 0.3615 + 0.241$$

$$= 0.6025 \text{ moles}$$

Finally, the mole fraction of forsterite:

$$\text{Mole fraction of forsterite} = \frac{0.3615}{0.6025} \approx 0.60$$

Thus, the mole fraction of forsterite in the olivine is approximately 0.60, which falls within the given range of 0.58 to 0.62. This calculation assumes that all Mg and Fe contribute to forming olivine and that SiO_2 is fully utilized to bind with Mg and Fe to form the silicate structure of olivine.

56. (40): To calculate the porosity (n) of a soil sample, the formula used is:

$$n = \left(\frac{\text{Volume of Voids}}{\text{Total Volume}}\right) \times 100\%$$

1. Calculate the volume of the solids:

Given:

- Particle density (ρ) = 2.65 g/cc
- Mass of solids (m) = 1908 g

The volume of solids (V_{solids}) can be calculated using the density formula:

$$V_{solids} = \frac{m}{p} = \frac{1908 \text{ g}}{2.65 \text{ g/cc}} \approx 720 \text{ cc}$$

2. Calculate the volume of voids:

Given:

- Total volume (V_{total}) = 1200 cc

The volume of voids (V_{voids}) is the total volume minus the volume of solids:

$$V_{voids} = V_{total} - V_{solids}$$

$$= 1200 \text{ cc} - 720 \text{ cc}$$

$$= 480 \text{ cc}$$

3. Calculate the porosity:

$$n = \left(\frac{V_{voids}}{V_{total}}\right) \times 100\%$$

$$= \left(\frac{480 \text{ cc}}{1200 \text{ cc}}\right) \times 100\% \approx 40\%$$

Thus, the porosity of the soil sample is 40%. This value is a measure of the fraction of the total volume of the soil that is occupied by voids (air and water spaces), and it has been calculated based on the given physical properties of the soil sample.

57. (2): To determine the bulk modulus (K) of a rock during deformation, we use the formula for bulk modulus:

$$K = -\frac{\Delta P}{\frac{\Delta V}{V_0}}$$

where:

- ΔP is the change in pressure,
- ΔV is the change in volume,
- V_0 is the initial volume.

Given:

- $\Delta P = 5 \times 10^4$ N/m² (increase in pressure),
- $V_0 = 4$ cm³,
- $V = 3.9$ cm³.

First, convert the volume from cm³ to m³ for consistency in units since 1 cm³ = 1×10^{-6} m³:

$$V_0 = 4\,\text{cm}^3 = 4 \times 10^{-6}\,\text{m}^3$$
$$V = 3.9\,\text{cm}^3 = 3.9 \times 10^{-6}\,\text{m}^3$$

Now, calculate ΔV:

$$\Delta V = V - V_0$$
$$= 3.9 \times 10^{-6}\,\text{m}^3 - 4 \times 10^{-6}\,\text{m}^3$$
$$= -0.1 \times 10^{-6}\,\text{m}^3$$

Substitute these values into the bulk modulus formula:

$$K = -\frac{5\times10^4\,\text{N/m}^2}{\frac{-0.1\times10^{-6}\,\text{m}^3}{4\times10^{-6}\,\text{m}^3}}$$

$$= -\frac{5\times10^4\,\text{N/m}^2}{-0.025}$$

$$K = 2 \times 10^6\,\text{N/m}^2$$

Thus, the bulk modulus of the rock is 2×10^6 N/m², or 2 expressed in terms of 10^6 N/m². This reflects how the volume of the rock responds to changes in pressure, indicating a relatively incompressible material typical of many geological materials.

58. (6000): To determine the retardation produced by an anisotropic crystal, we use the difference in refractive indices between the slow and fast directions, multiplied by the thickness of the crystal. Retardation is typically expressed in nanometers and calculated using the formula:

$$\text{Retardation} = (n_{slow} - n_{fast}) \times d \times 10^6$$

where:

- n_{slow} is the refractive index along the slow direction,
- n_{fast} is the refractive index along the fast direction,
- d is the thickness of the crystal in millimeters,
- 10^6 converts the result from meters to nanometers (since 1 mm = 10^{-3} meters)

Given:

- $n_{slow} = 1.636$,
- $n_{fast} = 1.486$,
- $d = 0.04$ mm.

Calculate the retardation:

$$\text{Retardation} = (1.636 - 1.486) \times 0.04 \times 10^6$$
$$\text{Retardation} = 0.15 \times 0.04 \times 10^6$$
$$\text{Retardation} = 0.006 \times 10^6 = 6000\text{ nm}$$

Thus, the retardation produced by the crystal, given the differences in refractive indices and the thickness of the crystal, is 6000 nm. This measurement indicates the phase difference created by the crystal due to the differing speeds at which light travels through the crystal in the two perpendicular directions defined by the refractive indices.

59. (21): To determine the percentage concentration of copper (Cu) in an orebody containing pyrite (FeS_2) and chalcopyrite ($CuFeS_2$) in the same molar proportions, we calculate the total mass of each component and their contribution to the total ore mass.

First, calculate the molar mass of each mineral:

- **Pyrite (FeS_2):**
  - Iron (Fe): 55.85 g/mol
  - Sulfur (S): 32.06 g/mol
  - Molar mass of FeS_2 = 55.85 + 2 × 32.06 = 119.97 g/mol
- **Chalcopyrite ($CuFeS_2$):**
  - Copper (Cu): 63.55 g/mol
  - Iron (Fe): 55.85 g/mol
  - Sulfur (S): 32.06 g/mol
  - Molar mass of $CuFeS_2$ = 63.55 + 55.85 + 2 × 32.06 = 183.52 g/mol

Given the ore contains pyrite and chalcopyrite in the same molar proportions, consider 1 mole of each for simplicity:

- **Total moles of Cu in 1 mole of chalcopyrite:**
  - 1 mole of Cu contributes 63.55 g of Cu
- **Total mass of the ore mixture:**
  - 1 mole of FeS_2 = 119.97 g
  - 1 mole of $CuFeS_2$ = 183.52 g
  - Total = 119.97 g + 183.52 g = 303.49 g

Calculate the percentage concentration of Cu:

$$\text{Percentage of Cu} = \left(\frac{63.55 \text{ g of Cu}}{303.49 \text{ g total}}\right) \times 100$$

$$\approx 20.94\%$$

Rounding to the nearest integer gives 21%. This represents the percentage of copper by weight in the ore, assuming equal molar contributions from pyrite (which contains no copper) and chalcopyrite (which does contain copper).

60. (0.5): To determine the fraction of melt remaining at point Q in an isobaric binary temperature-composition phase diagram, we utilize the lever rule. The lever rule is used in phase diagrams to calculate the proportion of each phase in a two-phase region based on the relative distances on the diagram.

Given:

- Point P represents the initial composition of the magma.
- Point Q represents the temperature and composition at which both solid B and liquid are present (in the B + Liquid field).

Steps to Use the Lever Rule:

1. **Identify tie line:** Locate the tie line at point Q that connects the solid phase and the liquid phase in equilibrium. The tie line extends horizontally at the temperature of point Q.
2. **Measure distances:** The fraction of each phase is proportional to the distance from the point of interest (point Q) to the phase boundary on the opposite side of the tie line.
   - L = Distance from point Q to the solid B boundary.
   - S = Distance from point Q to the liquid boundary on the tie line.
3. **Calculate the fraction of liquid:** The fraction of liquid (melt) remaining is given by the formula:

$$\text{Fraction of liquid} = \frac{L}{L+S}$$

If the distances L and S are equal, which is suggested but not explicitly measured due to the constraints of text description, then the calculation simplifies to:

$$\text{Fraction of liquid} = \frac{1}{2} = 0.5$$

Hence, assuming L and S are approximately equal based on the typical use of the lever rule for such diagrams (as it appears that Q might be centrally located on the tie line between solid and liquid phases), the fraction of melt remaining at point Q would be 0.5, or 50%. This estimate assumes symmetrical partitioning of phases, which is common in many phase diagram assessments without more precise graphical data available.

Previous Papers (Solved)

Indian Institute of Technology Joint Admission Test for M.Sc.

IIT-JAM GEOLOGY-2023

SECTION-A

Multiple Choice Questions (MCQ)

Qs. No. 1-10: *Carry one mark each.*

1. Hollandite is an ore mineral of which one of the following elements?

A. Fe B. Mn
C. Pt D. Cr

2. The transition from spinel to perovskite structure occurs between ________.

A. lower mantle and outer core
B. outer core and inner core
C. upper mantle and lower mantle
D. lower crust and upper mantle

3. Which one of the following textures shows cuneiform-shape intergrowth between alkali feldspar and quartz?

A. Spherulite texture
B. Graphic texture
C. Porphyritic texture
D. Spinifex texture

4. A pelitic rock consisting of cordierite + garnet + K-feldspar + sillimanite belongs to which one of the following metamorphic facies?

A. Granulite B. Eclogite
C. Greenschist D. Blueschist

5. Which one of the following dams resists external forces by its own weight?

A. Earthen dam B. Gravity dam
C. Storage dam D. Detention dam

6. Which one of the following minerals is NOT a framework silicate?

A. Feldspar B. Zeolite
C. Chlorite D. Quartz

7. Crustal thickness is maximum at the ______.

A. ocean-ocean convergent plate boundary
B. ocean-continent convergent plate boundary
C. continent-continent convergent plate boundary
D. continent-continent divergent plate boundary

8. Which one of the following causes sediment movement parallel to shoreline in the coastal area?

A. Longshore current
B. Rip current
C. Backwash
D. Edge wave

9. Which one of the following dinosaur fossils is a theropod?

A. *Kotasaurus* B. *Titanosaurus*
C. *Rajasaurus* D. *Barapasaurus*

10. Spiti Shale was deposited during the ______ time.

A. Palaeozoic B. Mesozoic
C. Cenozoic D. Proterozoic

Qs. No. 11–30: *Carry two marks each.*

11. Which one of the following is a gently sloping (< 10°) volcanic landform resulting from eruption of basaltic lava?

A. Shield volcano
B. Composite volcano
C. Lava dome
D. Caldera

12. On the magnetic polarity time scale, the present day epoch/chron is called _______.

A. Matuyama B. Gilbert
C. Gauss D. Brunhes

13. Which one of the following options is the CORRECT sequence of seismic waves in order of arrival time recorded on a seismogram after an earthquake?

A. P-waves, S-waves, Rayleigh waves, Love waves
B. P-waves, Rayleigh waves, S-waves, Love waves
C. S-waves, P-waves, Love waves, Rayleigh waves
D. P-waves, S-waves, Love waves, Rayleigh waves

14. Match the geomorphic agents in Column-I with their corresponding landforms in Column-II.

Column-I	Column-II
P. Wind	1. Backswamp
Q. Groundwater	2. Yardang
R. Glacier	3. Doline
S. River	4. Drumlin

	P	Q	R	S
A.	2	3	4	1
B.	3	1	2	4
C.	2	1	4	3
D.	4	3	2	1

15. Which one of the following processes is NOT a mechanism for bedload sediment transport in a river channel?

A. Cavitation B. Sliding
C. Rolling D. Saltation

16. Which one of the following relationships between the topographic contour value (t) and the stratum contour value (x) of a bed must be TRUE for an outcrop of the bed to occur on the topographic surface?

A. $t = x$ B. $t = 2x$
C. $t = 3x$ D. $t = 4x$

17. As per Ramsay's classification of folds, the maximum thickening of fold hinge and the maximum thinning of the fold limbs are observed in ___________.

A. Class 1A B. Class 1B
C. Class 2 D. Class 3

18. The number of hinge(s) in a monocline is _______.

A. 0 B. 1
C. 2 D. 3

19. Which one of the following Gondwana flora is a seed?

A. *Dadoxylon* B. *Cordaicarpus*
C. *Taeniopteris* D. *Palaeovittaria*

20. Which one of the following gastropod genera displays sinistral coiling?

A. *Physa* B. *Cypraea*
C. *Murex* D. *Conus*

21. Which one of the following was emplaced in the Neoproterozoic time?

A. Singhbhum Granite
B. Dongargarh Granite
C. Closepet Granite
D. Erinpura Granite

22. Match the lithostratigraphic groups in Column-I with their corresponding formations in Column-II.

Column-I	Column-II
P. Papaghni Group	1. Kajrahat Formation
Q. Uttatur Group	2. Nagri Formation
R. Siwalik Group	3. Vempalle Formation
S. Semri Group	4. Karai Formation

	P	Q	R	S
A.	4	3	1	2
B.	3	4	2	1
C.	3	1	2	4
D.	2	4	3	1

23. Which one of the following symmetry elements is an INCORRECT representation of rotoinversion operation?

A. $1A_3$ + inversion centre = $\bar{3}$
B. $1A_2 = \bar{4}$
C. Mirror plane = $\bar{2}$
D. $1A_3/m = \bar{6}$

24. A plutonic igneous rock is composed of 50% orthopyroxene, 45% olivine and 5%

clinopyroxene. What is the appropriate name of the rock according to the IUGS classification?

A. Norite B. Wehrlite
C. Troctolite D. Harzburgite

25. Which one of the following is NOT a sediment-gravity flow?

A. Hypopycnal flow B. Cohesive debris flow
C. Turbidity flow D. Mud flow

26. Which one among the following mineral pairs crystallise early during the cooling of a basaltic melt?

A. Forsterite and albite
B. Biotite and anorthite
C. Enstatite and bytownite
D. Forsterite and quartz

27. Match the ore deposits in Column-I with their corresponding ores in Column-II.

Column-I	Column-II
P. Malanjkhand	1. Uranium ore
Q. Tummalapalle	2. Gold ore
R. Bhukia	3. Tin ore
S. Tosham	4. Copper ore

	P	Q	R	S
A.	4	3	2	1
B.	3	1	4	2
C.	4	1	2	3
D.	2	4	1	3

28. Which one of the following statements is CORRECT?

A. Banded Iron Formations are of chemogenic origin
B. Porphyry-type deposits are formed purely by sedimentary processes
C. Quartz-Pebble Conglomerate hosted gold deposits are formed by supergene enrichment
D. Fullerene is formed by residual concentration process

29. Which one of the following statements about the hydrological cycle is CORRECT?

A. Groundwater represents the largest share of fresh water on Earth
B. 'Precipitation rate greater than infiltration rate' is a necessary condition to generate surface runoff
C. All precipitation falling on the land finally ends up as groundwater
D. Groundwater flows in curved and concave-upward path

30. Which one of the following mineral deposits is NOT related to the mining for energy production?

A. Narwapahar B. Rampura-Agucha
C. Jaduguda D. Turamdih

SECTION-B

Multiple Select Questions (MSQ)

Qs. No. 31-40: *Carry two marks each.*

31. At which of the following locations do lignite deposits occur in India?

A. Raniganj B. Singrauli
C. Barmer D. Neyveli

32. Which of the following types of dunes form(s) primarily by uni-directional wind?

A. Linear dunes
B. Parabolic dunes
C. Barchan dunes
D. Star dunes

33. The attitude of a fault plane was measured to be 350°, 75°E. The rake of the slickenline on the fault plane was found to be 90°. Which of the faults listed below satisfy(ies) these observations?

A. Dip-slip fault
B. Normal fault
C. Reverse fault
D. Strike-slip fault

34. What type(s) of fossil remains is/are studied in ichnology?

A. Fishes and amphibians

B. Spores and pollens
C. Tracks and trails
D. Burrows and bioturbation

35. Which of the following combinations of Basin and Formation is/are CORRECT?
A. Cauvery Basin – Niniyur Formation
B. Assam Basin – Tipam Formation
C. Bengal Basin – Jalangi Formation
D. Kutch Basin – Dhok Pathan Formation

36. Which of the following optical properties CORRECTLY identify(ies) the apatite {0001} section?
A. Isotropic under crossed nicols
B. Second-order interference colour
C. Centered uniaxial interference figure
D. High birefringence

37. The AFM diagram given below shows stability of minerals in the garnet zone. If P, Q, R and S represent the compositions of different pelitic rocks, which of the following is/are characterised by the equilibrium assemblage of muscovite + garnet + biotite + quartz?

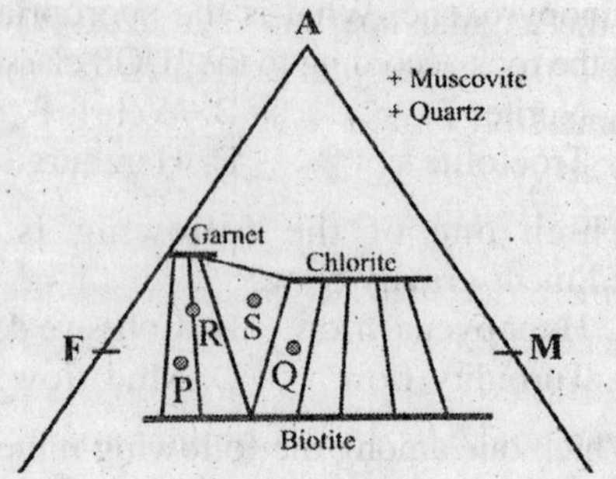

A. P. B. Q
C. R D. S

38. Which of the following sedimentary structures is/are tool marks?
A. Bounce marks B. Wrinkle marks
C. Prod marks D. Skip marks

39. Which of the following is/are NOT copper-bearing mineral(s)?
A. Bornite B. Chalcocite
C. Braunite D. Chrysocolla

40. Which of the following is/are used to estimate the strength of a rock mass?
A. API gravity B. Resistivity
C. Kriging D. RQD

SECTION-C

Numerical Answer Type (NAT)

Qs. No. 41-50: *Carry one mark each.*

41. The amplitude recorded at a station for a magnitude 5 earthquake is x. If another earthquake recorded at the same station has an amplitude of $15x$, then the magnitude of this earthquake is _______. *(Round off to two decimal places)*

42. If the intercepts of crystallographic axes are 0.5a : 1b : 0.75c on a crystallographic plane $\{h\ k\ l\}$, the value of 'l' is _________. *(In integer)*

43. An ocean wave with a wavelength of 200 m approaches the coast. If water depth at the observation point is 75 m, the wave velocity is _______ m/s. *(Round off to two decimal places)* (Use g = 10 m/s^2)

44. A bed with an attitude 045°, 20°SE is rotated 60° clockwise (looking down) about a vertical axis. The strike value (in the azimuthal convention following right hand rule) of the rotated bed is _______ degrees. *(In integer)*

45. A one-meter deep and sheet-like waterflow on a sandy beach developed antidunes. The minimum velocity of the waterflow was _________ m/s. *(Round off to two decimal places)* (Use g = 10 m/s^2)

46. If the angular aperture of a 20X objective is 46°, the numerical aperture of the water immersion objective is ________. *(Round off to two decimal places)*
(Use RI of water = 1.33)

47. A metamorphic rock is composed of grossular garnet ($Ca_3Al_2Si_3O_{12}$), kyanite (Al_2SiO_5), anorthite ($CaAl_2Si_2O_8$) and quartz (SiO_2). If these minerals show an univariant reaction relationship, the number of components in this assemblage is ________. *(In integer)*

48. If the dip separation vector on a normal fault plane has an attitude 60° → 040° and a magnitude of 6 m, the heave on the fault is ________ m. *(In integer)*

49. A hillslope with an angle of 40° consists of soil having an internal friction angle of 30°. The factor of safety of the hillslope is ______. *(Round off to two decimal places)*

50. The water table over an area of 1 km^2 was lowered by 4 m. If the porosity of rock is 30% and the specific retention is 10%, the change in the groundwater storage is _____ × 10^3 m^3. *(In integer)*

Qs. No. 51-60: *Carry two marks each.*

51. The $\frac{^{143}Nd}{^{144}Nd}$ and $\frac{^{147}Sm}{^{144}Nd}$ ratios of a rock are 0.516 and 0.389, respectively. The rock evolved as a closed system. As per the exact parent-daughter relationship equation, the $\frac{^{143}Nd}{^{144}Nd}$ ratio of the rock 4.6 × 10^9 years ago was ______. *(Round off to three decimal places)*

(Use decay constant for ^{147}Sm = 6.54 × 10^{-12} y^{-1})

52. A longitudinal profile of a river is shown in the figure below. If the average discharge of the river at reach AB is 200 m^3/s and increases to 300 m^3/s at reach CD, then the stream power from the reach AB to CD will change by a factor of ________. *(Round off to two decimal places)*

(Use g = 10 m/s², ρ_{water} = 1000 kg/m³)

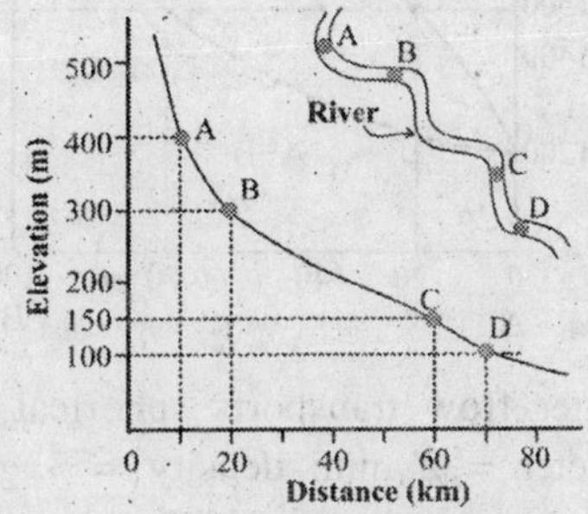

53. An underground vertical dyke is intercepted by an inclined borehole as shown in the figure below. The length of the dyke core intercepted by the borehole is 4 m. If the true thickness of the dyke is 2 m, the inclination of the borehole from the vertical is ________ degrees. *(In integer)*

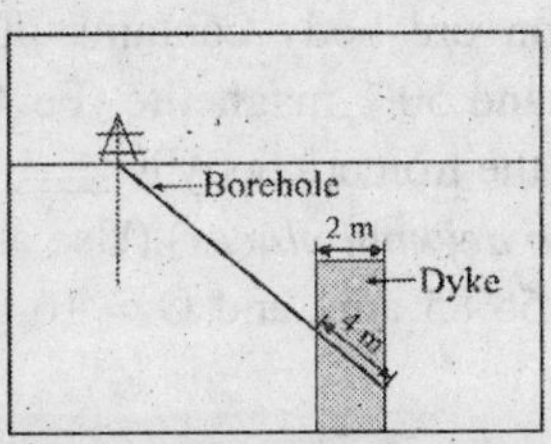

54. A cylindrical copper ore body has a vertical thickness of 45 m and a diameter of 14 m with a density of 2.9 g/cm³. The reserve of the copper ore body is ______ tons. *(In integer)*

55. The density of a FCC unit cell is 6.5 g/cm³. If the mass of a single atom is 60 g/mol, the diagonal length of the face {100} is _____ Å. *(Round off to two decimal places)* (Use N_A = 6.022 × 10^{23})

56. The following figure shows an isobaric temperature-composition (T-X) phase diagram for the binary system A-B. If *'P'* is the initial composition of liquid, the amount of liquid that remains in the system when the liquid cools from 1800°C (point L_1) to 1500°C (point L_2) is ________ %. *(In integer)*

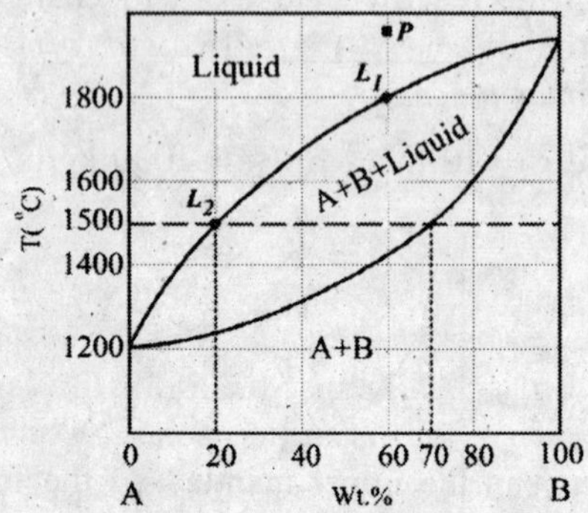

57. A water flow transports spherical particles (diameter = 2 mm; density = 3 g/cm^3) in suspension mode. If additional particles of density 2 g/cm^3 are added into the flow, then the diameter of the particles that can be transported without a change in terminal fall velocity, using Stokes law, is ________ mm. (*Round off to two decimal places*) (Use density of water = 1 g/cm^3)

58. If an iron ore body contains 50% hematite (Fe_2O_3) and 50% magnetite (Fe_3O_4), then the grade of the iron ore body is ______%. (*Round off to two decimal places*) (Use atomic weight of Fe = 55.85 amu and O = 16 amu).

59. The schematic stereographic projection below shows dip amount and dip direction of three sets of joints (J1, J2 and J3) on a hillslope. If the internal friction angle of the hillslope material is 30°, the strike of the potential failure joint plane (in azimuthal convention following right hand rule) is ______ degrees. (*In integer*)

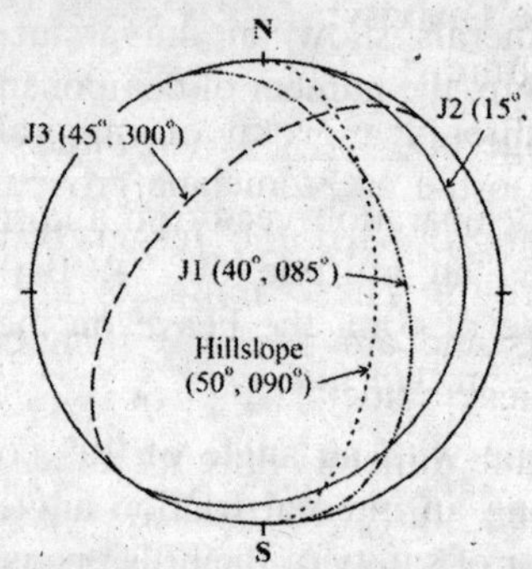

60. The hydraulic conductivity of a 100 cm long cylindrical core is estimated as 1.2 cm/min when hydraulic head difference is 20 cm in an experimental setup. If the effective porosity of the core is 20%, then, assuming steady state Darcy flow, the average interstitial velocity of groundwater through the core is _____ m/day. (*Round off to two decimal places*)

ANSWER KEY

1	**2**	**3**	**4**	**5**	**6**	**7**	**8**	**9**	**10**
B	C	B	A	B	C	C	A	C	B
11	**12**	**13**	**14**	**15**	**16**	**17**	**18**	**19**	**20**
A	D	D	A	A	A	D	C	B	A
21	**22**	**23**	**24**	**25**	**26**	**27**	**28**	**29**	**30**
D	B	B	D	A	C	C	A	D	B
31	**32**	**33**	**34**	**35**	**36**	**37**	**38**	**39**	**40**
C,D	B,C	A,B,C	C,D	A,B,C	A,C	A,C	A,C,D	C	D
41	**42**	**43**	**44**	**45**	**46**	**47**	**48**	**49**	**50**
6.15 to 6.19	4	17.50 to 18.00	105	3.10 to 3.20	0.50 to 0.54	3	3	0.65 to 0.75	800
51	**52**	**53**	**54**	**55**	**56**	**57**	**58**	**59**	**60**
0.502 to 0.506	0.75	30 or 150	20070 to 20100	5.55 to 5.59	20	2.80 to 2.84	70.50 to 71.50	355	17.20 to 17.40

HINTS AND SOLUTIONS

1. Hollandite: Mn ore mineral. **Formula:** $Ba(Mn_6^{4+}Mn_2^{3+})O_{16}$ Formerly defined as $Ba(Mn^{4+}, Mn^{2+})_8O_{16}$.

- **Colour:** Silvery-grey to greyish black, black; **Lustre:** Metallic; **Hardness:** 4 – 6; **Specific Gravity:** 4.95; **Crystal System:** Monoclinic
- Hollandite is a group of minerals that belong to the cryptomelane group, which is a subgroup of the larger birnessite supergroup. These minerals are oxide minerals and are primarily composed of manganese oxide.
- Hollandite minerals have a unique crystal structure that consists of tunnel-like channels. These channels can host various cations and water molecules. The specific composition of hollandite minerals may vary depending on the presence of different cations in the tunnels.
- One of the essential characteristics of hollandite minerals is their ability to incorporate various cations, such as barium (Ba), calcium (Ca), potassium (K), and others, into their crystal structure. This characteristic allows them to serve as host minerals for certain elements and ions.
- The name "hollandite" is derived from the type locality, which is the Ojuela Mine in Mapimi, Durango, Mexico. Hollandite minerals are also found in other locations around the world, and they are significant in various geological settings.
- Hollandite minerals are of interest to researchers due to their unique crystal structure and their potential role in hosting certain elements. They are also studied for their geochemical and environmental significance. Additionally, hollandite minerals have been found in association with some manganese ore deposits, and their presence can be significant for the economic extraction of manganese.

2. Between Upper mantle and lower mantle the transition from spinel to perovskite structure found.

- The transition from the spinel structure to the perovskite structure is observed in the Earth's mantle, specifically in the region known as the "transition zone," which lies between the upper mantle and the lower mantle.
- The Earth's mantle is primarily composed of various minerals, and at different depths within the mantle, there are significant phase changes due to the variations in pressure and temperature. The transition zone is located at depths between approximately 410 kilometers (about 250 miles) and 660 kilometers (about 410 miles) below the Earth's surface.
- At the upper part of the transition zone (around 410 km depth), there is a mineral phase transition from a mineral called spinel to a mineral called perovskite. Spinel is a mineral with a cubic crystal structure, and the most common mineral of this type in the mantle is $MgAl_2O_4$ (magnesium aluminum oxide). As the pressure and temperature increase with depth, the spinel structure transforms into a new mineral called perovskite.
- Perovskite has a different crystal structure, which is characterized by a cubic arrangement of oxygen ions with larger cations, typically magnesium, iron, or calcium, located in the center of the oxygen octahedra. The general chemical formula of perovskite is ABX_3, where A and B are cations, and X is an anion, usually oxygen.
- This phase transition from spinel to perovskite is a critical boundary in the Earth's mantle and plays a significant role in the behavior and dynamics of the planet's interior. The transition zone acts

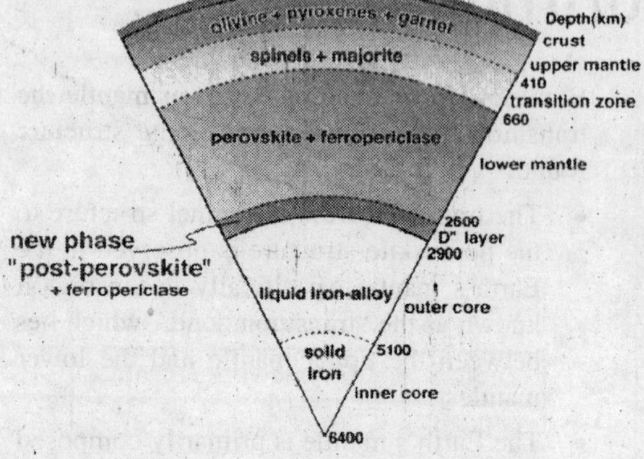

as a barrier for the movement of material between the upper and lower mantle, affecting the flow of heat and material within the Earth. The mineral phase changes in this region can have implications for seismic wave propagation, mantle convection, and the overall geodynamic processes occurring deep within the Earth.

3. ***Graphic texture*** – The texture shows cuneiform-shape intergrowth between alkali feldspar and quartz. This texture is most commonly observed in pegmatites.

4. A pelitic rock consisting of cordierite + garnet + K-feldspar + sillimanite belongs to Granulite facies. Types of metamorphic facies:

Facies - Protolith—Mineral Assemblage Table

Facies	Pelitic	Calcareous	Mafic
Zeolite 100-200°C	interlayered smectite/ chlorite calcite	calcite	Laumonite, thompsonite, calcite, interlayered smectite/ chlorite
Prehnite-Pumpellyite 150-300°C	Prehnite, pumpellyite, calcite, chlorite, albite	calcite	Prehnite, pumpellyite, calcite, chlorite, albite
Greenschist 300-450°C	muscovite, chlorite, quartz, albite, biotite, garnet	calcite, dolomite, quartz, epidote, tremolite	albite, chlorite, quartz, epidote, actinolite, sphene
Epidote Amphibolite 450-550°C	muscovite, biotite, garnet, albite, quartz	calcite, quartz, tremolite, epidote, diopside	albite, epidote, hornblende, quartz
Amphibolite 500-700°C	garnet, biotite, muscovite, quartz, plagioclase, staurolite, kyanite or sillimanite	calcite, diopside quartz, wollastonite	hornblende, plagioclase, garnet, quartz, sphene, biotite
Granulite 700-900°C	garnet, Kspar, sillimanite or kyanite, quartz, plagioclase, hypersthene	calcite, quartz, plagioclase, diopside, hypersthene	plagioclase, augite, hypersthene, hornblende, garnet, olivine
Blueschist 150-350°C P > 5-8 Kb	Jadeite, albite, quartz, lawsonite, aragonite, paragonite	aragonite, white mica	Glaucophane, albite, lawsonite, sphene, ± garnet
Eclogite 350-750°C P > 8-10 Kb	coesite, Kspar, sillimanite, plagioclase	aragonite, quartz, plagioclase, diopside, hypersthene	omphacite (px), pyrope garnet

5. **Dam and its direction:**

Dam: A dam is a hydraulic structure built across a river or a stream to create a reservoir for the available water. A dam is also referred to as a barrier which is built across a river or a stream to restrict and control the flow of water.

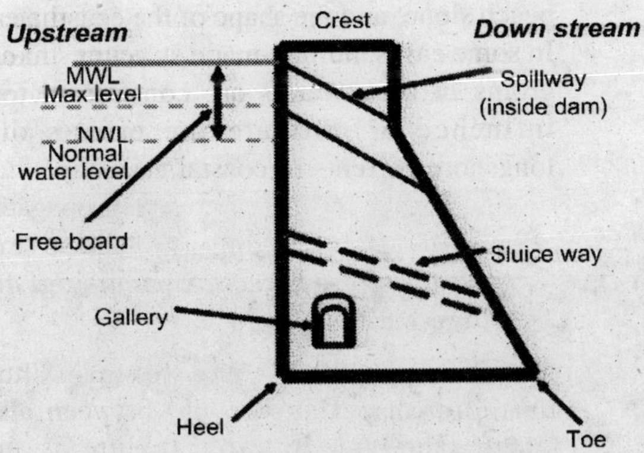

Earthen dam: They are made with clay, silt, and alluvial materials that are available in the river valleys and adjoining areas of the proposed site of construction.

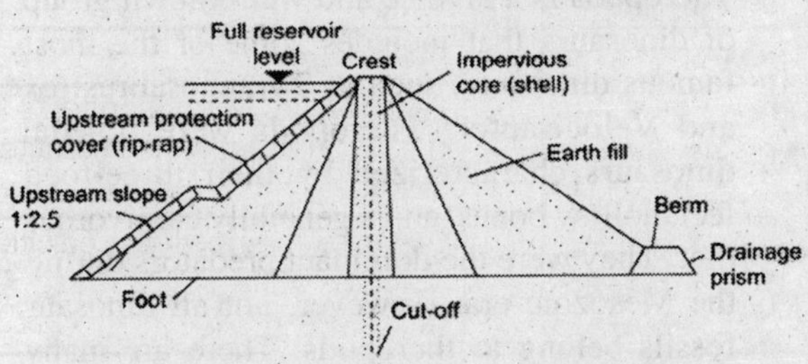

An earth dam with impervious core and gentle slope

Gravity dam: It is also known as a concrete dam. In this type of dam, the axis is straight or slightly curved. These dams resist the thrust of water by their weight; gravity dams are massive concrete structures. The foundation rock of the dam should be very strong so that the load of the dam may be borne by the rock. **Example:** Bhakra Dam.

Storage dam: Storage dams are constructed to store water during the monsoon season when the water is available in excess amount and the flow of water in the river is considerably good. The water stored in the dam is utilized for various activities like human utilization, agriculture, hydropower generation, etc. If not specified, in general terms, dam means storage dam only.

Detention dam: Detention dams are constructed for flood control. The detention dam slows down the flow of flood water in the river on its downstream. It also stores some of the flood water to slow down the flow and then releases the water gradually at a controlled rate for its required use. Detention dams protect the downstream area from floods.

6. Minerals with silicate structure:

Minerals	Types of silicates	Si:O
Feldspar	Framework silicate	1:2
Zeolite	Framework silicate	1:2
Chlorite	Phyllosilicate	2:5
Quartz	Framework silicate	1:2

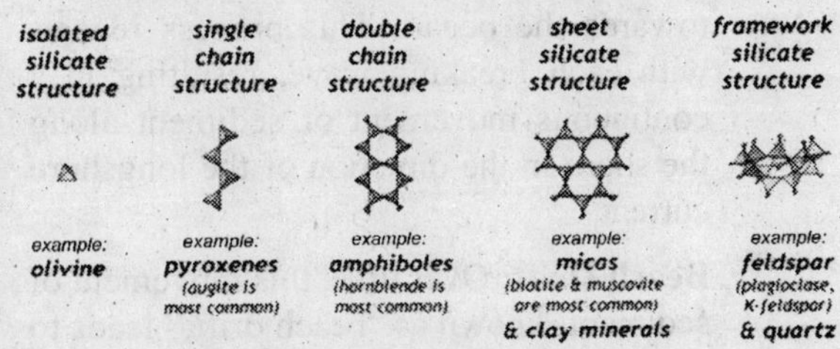

8. Longshore current causes sediment movement parallel to shoreline in the coastal area.

- Longshore currents cause sediment movement parallel to the shoreline in coastal areas due to the interaction of waves with the seafloor and the shoreline.
- A longshore current is a current of water that flows parallel to the shoreline, driven by the breaking and approaching waves. It occurs when waves approach the shore at an angle rather than directly perpendicular to the shoreline. The process can be explained as follows:

  Wave Approach: Ocean waves are generated by winds, and as they travel towards the coastline, they often arrive at an angle due to the prevailing wind direction. These obliquely approaching waves strike the coastline at an angle.

  Swash and Backwash: When a wave breaks onto the shoreline, it rushes up the beach in a forward motion. This is called the "swash." After the swash, the water returns back towards the ocean as "backwash." The swash and backwash together create a net movement of water parallel to the shore.

Zigzag Motion: As the swash and backwash move up and down the beach, they form a zigzag motion along the coastline. This zigzag motion effectively carries sediment along the shore in the direction of the longshore current.

Sediment Transport: The swash carries sediment particles up the beach, and the backwash returns the sediment back towards the ocean. This process repeats with each breaking wave, resulting in a continuous movement of sediment along the shore in the direction of the longshore current.

Beach Drift: Over time, this movement of sediment, known as "beach drift," leads to the gradual transport of sand and other sediment along the shoreline, parallel to the coast.

- The longshore current and beach drift play a crucial role in shaping coastal landforms, such as beaches, spits, and barrier islands. They are also essential for the transport of sediment along the coast, which contributes to the natural processes of erosion and deposition in coastal environments.
- It is worth noting that the intensity and direction of longshore currents can vary based on factors such as wave energy, beach slope, and the shape of the coastline. In some cases, human-made structures like groins or breakwaters are constructed to influence or mitigate the effects of longshore currents in coastal areas.

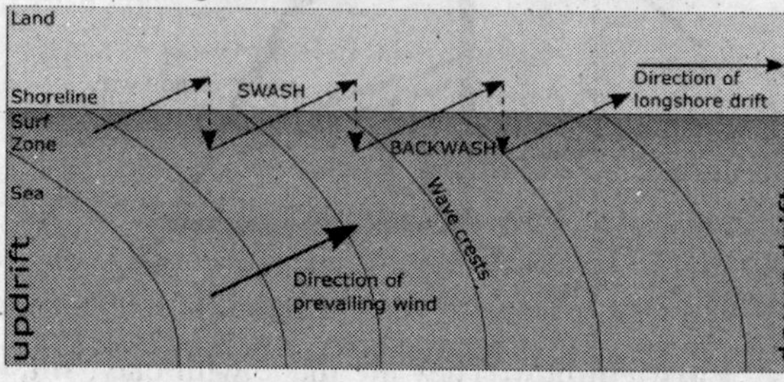

9. **Dinosaur fossils:** Dinosaur fossils can belong to various groups, and theropods are one of the many types of dinosaurs. The group Theropoda is a diverse and well-known group of dinosaurs that includes some of the most famous dinosaurs, such as Tyrannosaurus rex and Velociraptor. Theropods were bipedal dinosaurs characterized by their three-toed feet, hollow bones, and a generally carnivorous diet. They were the dominant predators during the Mesozoic era. However, not all dinosaur fossils belong to theropods. There are many other groups of dinosaurs, such as sauropods (long-necked herbivores) and ornithischians (another group of herbivores with bird-like hips). Paleontologists classify dinosaur fossils based on their anatomical features, bone structures, and other characteristics to determine which group they belong to. When excavating and studying dinosaur fossils, paleontologists often rely on a combination of these features to identify the dinosaur's group and species accurately.

Dinosaur fossils	Characteristics
Kotasaurus	Kotasaurus (meaning "Kota Formation lizard" is a genus of sauropod dinosaur from the Early Jurassic period (Sinemurian–Pliensbachian). The only known species is *Kotasaurus yamanpalliensis*. It was discovered in the Kota Formation of Telangana, India and shared its habitat with the related Barapasaurus. So far the remains of at least 12 individuals are known. The greater part of the skeleton is known, but the skull is missing, with the exception of two teeth. Like all sauropods, it was a large, quadrupedal herbivore with long neck and tail.

Titanosaurus	Titanosaurus (or titanosaurians; members of the group Titanosauria) were a diverse group of sauropod dinosaurs, including genera from all seven continents. The titanosaurs were the last surviving group of long-necked sauropods, with taxa still thriving at the time of the extinction event at the end of the Cretaceous. This group includes some of the largest land animals known to have ever existed, such as Patagotitan—estimated at 37 m (121 ft) long with a weight of 69 tonnes (76 tons) and the comparably-sized Argentinosaurus and Puertasaurus from the same region.
Rajasaurus	Rajasaurus (meaning "King lizard") is a genus of carnivorous abelisaurid theropod dinosaur from the Late Cretaceous of India, containing one species: *Rajasaurus narmadensis.* The bones were excavated from the Lameta Formation in the Gujarat state of Western India, probably inhabiting what is now the Narmada River Valley. It was formally described by palaeontologist Jeffrey A. Wilson and colleagues in 2003 based on a partial skeleton comprising the braincase, spine, hip bone, legs, and tail–a first for an Indian theropod. The dinosaur likely measured 6.6 metres (22 ft), and had a single horn on the forehead which was probably used for display and head-butting. Like other abelisaurids, Rajasaurus was probably an ambush predator.
Barapasaurus	Barapasaurus is a genus of basal sauropod dinosaur from Jurassic rocks of India. The only species is *B. tagorei.* Barapasaurus comes from the lower part of the Kota Formation, which is of Early to Middle Jurassic age. It is therefore one of the earliest known sauropods. Barapasaurus is known from approximately 300 bones from at least six individuals, so that the skeleton is almost completely known except for the anterior cervical vertebrae and the skull. This makes Barapasaurus one of the most completely known sauropods from the early Jurassic.

10. Spiti Shale was deposited during the Mesozoic time.

Era	Time (million years)	Age	Group/Formation	Lithology
Cenozoic	1.8	Quaternary	Fluvial / Fluvio-glacial / Alluvium	Sand, silt, pebbles, cobbles, boulders
Mesozoic	65	Cretaceous	Giumal-Chikkim Formation	Sandstone, siltstone, shale, conglomerate
Jurassic		**Spiti Formation**		Shale, sandstone and siltstone
Triassic-Jurassic	243	Lilang Group / Kalhel Formation		Limestones, shales, dolomites and sandstones.
Permo-Triassic-Jurassic		Tandi Group		Limestone
Palaeozoic	270	Devonian to Permian		Kuling group; Quartzites, sandstones and black shales
Lower Cambrian to Silurian	570	Kunamla-Thango-Takche Formation		Shale, schist, sandstone, dolomite
		Rohtang Granitoid		Granitoids

Proterozoic	670	Neoproterozoic to Terminal Proterozoic	Batal formation, shales, phyllites and quartzites
Neoproterozoic	900	Manjir Formation	Sandstone, limestone, shale
Neoproterozoic	1600	Bhalai Formation (Salkhala Group)	Slates, schists, slates, phyllites, quartzites
Mesoproterozoic	2500	Vaikrita Group	Slates, phyllites, quartzites, schists

Lithostratigraphy		Metres	Age
Giumal Formation		~400	Albian –Berriasian
Spiti Shale Fm	upper member (Lochambel Beds)	<300	Berriasian –Early Tithonian
	middle member (Chidamu Beds)		Early Tithonian –Kimmeridgian
	lower member (B. gerardi Beds)		Late Oxfordian –Middle Callovian
Ferruginous Oolite Fm		0-20	Early Callovian –Early Bathonian
Tagling & Para FM		~600	Middle Jurassic –Late Triassic

11. Types of Lava flow:

Shield volcano: A gently sloping (< 10°) volcanic landform resulting from eruption of basaltic lava.

Lava dome: A lava dome is a type of volcanic feature that forms when lava of relatively high viscosity (thick and sticky) erupts from a volcano. Unlike the more fluid lava flows that can cover large distances, lava domes are composed of slow-moving lava that piles up near the vent of the volcano. This viscous lava tends to build up and solidify around the vent, creating a dome-shaped structure. The lava that forms a lava dome is usually rich in silica, which increases its viscosity and makes it less likely to flow quickly. As a result, the lava accumulates around the vent, layer by layer, over time. Lava domes can vary in size and shape, ranging from small mounds to large and steep-sided structures. Lava domes can be found at the summits or on the flanks of some strato-volcanoes and composite volcanoes. They are often associated with explosive volcanic activity and can be hazardous due to their potential to collapse or produce pyroclastic flows when they become unstable. In some cases, the pressure of magma beneath the dome can cause explosive eruptions, leading to dangerous volcanic events.

Caldera: A caldera is a large, basin-like depression that forms on the Earth's surface after a massive volcanic eruption or the collapse of a volcanic structure. It is typically much larger than the original volcanic crater and can range from a few kilometers to tens of kilometers in diameter. Calderas are some of the most dramatic and awe-inspiring geological features on our planet.

Types of Basalt	Characteristics	Example
Flood or Plateau Basalt	Very liquid lava; flows very wide spread, emitted from factures	Columbia river plateau
Shield Volcano	Liquid lava emmited from a central vent, large, sometimes has a collapse caldera	Hawaiian volcanoes
Cinder Cone	Explosive liquid lava, small emmited from central vent, if continiued long enough, may build up a shield volcano	Mount Zion, Mount Tabor
Composite or Strato-volcano	More viscous lavas, much explosive (pyroclastic) debris; large, emitted from central vent	Mount Baker

12. On the magnetic polarity time scale, the present day epoch/chron is called Brunhes.

- The Brunhes, also known as the Brunhes-Matuyama reversal, refers to a specific geomagnetic reversal event in Earth's history. Geomagnetic reversals are periods when the Earth's magnetic field flips, with the North and South magnetic poles exchanging their positions. These events are recorded in the Earth's rocks as magnetic anomalies, where the magnetic minerals align in the opposite direction to the current magnetic field.
- The Brunhes-Matuyama reversal occurred approximately 780,000 years ago during the Pleistocene epoch. It marks the transition from the Brunhes Chron, a period of normal magnetic polarity (similar to today's magnetic field orientation), to the Matuyama Chron, characterized by reversed magnetic polarity.

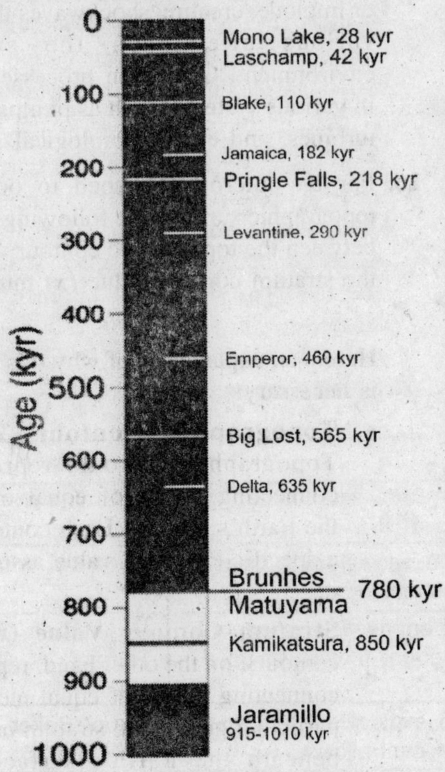

- Chron is short for "chronological unit," and it is a specific segment of time during which the Earth's magnetic field remained relatively constant with a particular polarity. The Brunhes-Matuyama reversal was a significant event in Earth's history, and it is often studied in paleomagnetism to better understand the behaviour and changes in the planet's magnetic field over time. These geomagnetic reversals are crucial for geological dating and correlation of rocks, as they provide important time markers in the geological record.

13. P-waves, S-waves, Love waves, Rayleigh waves are correct sequence of seismic waves in order of arrival time recorded on a seismogram after an earthquake. The sequence of seismic waves is as follows:

1. **P-waves (Primary waves):** P-waves are the first seismic waves to arrive at a seismograph station after an earthquake. They are compressional waves, meaning that they cause particles to move in the same direction as the wave is traveling. P-waves can travel through both solid and liquid layers of the Earth and are the fastest seismic waves, moving at speeds of about 1 to 14 kilometers per second (km/s) in the Earth's crust.
2. **S-waves (Secondary waves):** S-waves arrive after P-waves but before surface waves. S-waves are shear waves, meaning they cause particles to move perpendicular to the direction of the wave propagation. Unlike P-waves, S-waves cannot travel through liquids, and they are slower than P-waves, typically moving at speeds of about 1 to 8 km/s in the Earth's crust.
3. **Love waves:** Love waves are a type of surface wave and arrive after P-waves and S-waves. They are guided horizontally

along the Earth's surface and cause horizontal shaking. Love waves are slower than P-waves and S-waves, moving at speeds of about 1 to 5 km/s.

4. **Rayleigh waves:** Rayleigh waves are also a type of surface wave and are the last to arrive on the seismogram after an earthquake. They cause both vertical and horizontal motion, with a rolling or undulating motion. Rayleigh waves are slower than P-waves, S-waves, and Love waves, typically moving at speeds of about 1 to 4 km/s.

It is important to note that the exact arrival times of these waves can vary depending on the distance of the seismograph station from the earthquake's epicenter and the properties of the Earth's subsurface through which the waves travel. Seismologists use the differences in arrival times of these waves at multiple seismograph stations to triangulate the earthquake's location and understand its source and characteristics.

Types of waves with direction of propagation:

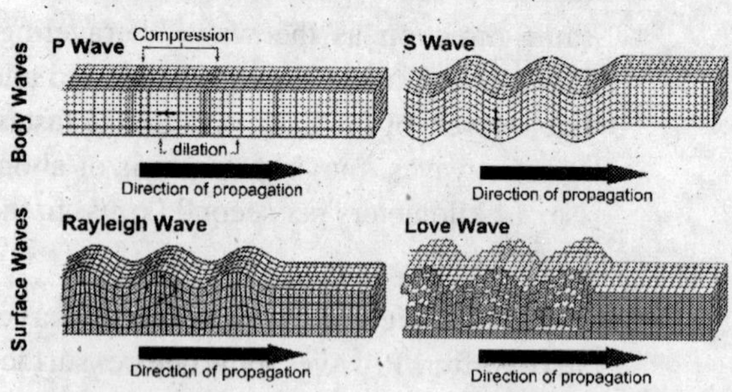

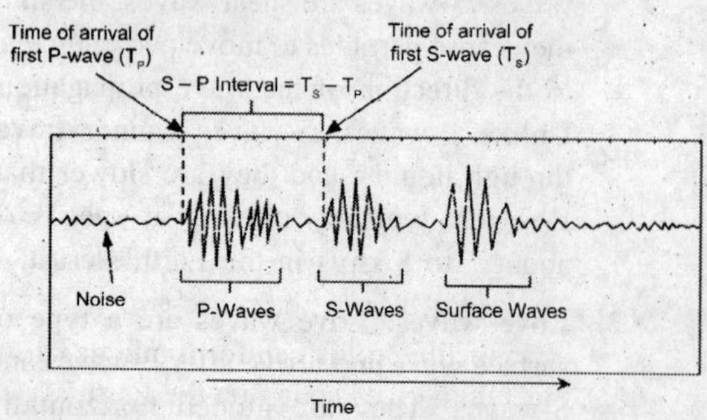

14. Geomorphic agents in Column-I with their corresponding landforms in Column-II.

Column-I	Column-II
Wind	Sand Dunes, Yardang, Loess Deposits, Desert Pavement, Ventifacts, Deflation Hollows, Blowouts, Sand Sheets and Sand Ripples, Desert Varnish.
Groundwater	Stalactites and Stalagmites, sinkhole, Doline
Glacier	Drumlin, esker
River	Backswamp

15. **Cavitation processes** is NOT a mechanism for bedload sediment transport in a river channel. Cavitation is a phenomenon that occurs in fluid dynamics when the pressure of a liquid drops below its vapor pressure, causing the formation of vapor bubbles within the liquid. These vapor bubbles can then collapse or implode, creating shockwaves that can have various effects on the surrounding environment. Cavitation processes can occur in various systems, such as pumps, propellers, turbines, and even in biological systems.

16. For an outcrop of a bed to occur on the topographic surface, the following relationship between the topographic contour value (t) and the stratum contour value (x) must be true:

$$t > x$$

Here's an explanation of why this relationship is necessary:

- **Topographic Contour Value (t):** Topographic contours represent lines connecting points of equal elevation on the Earth's surface. Each contour line has a specific elevation value associated with it.
- **Stratum Contour Value (x):** Stratum contours, on the other hand, represent lines connecting points of equal elevation of a particular geological stratum or rock layer beneath the Earth's surface. Stratum

contours are used to illustrate the elevation of a specific geological layer in the subsurface.

- Now, for an outcrop of a bed to occur on the topographic surface, the bed must be exposed or visible at the Earth's surface. This means that the elevation of the bed (x) must be higher than the elevation of the topographic surface (t).
- If $t \leq x$, it means that the topographic contour at a particular point is at an equal or higher elevation than the elevation of the bed (stratum). In this case, the bed would not be exposed on the surface; it would be below the ground and covered by other layers.
- On the other hand, if $t > x$, it means that the topographic contour is at a lower elevation than the bed's elevation. This scenario allows the bed to be exposed on the surface, creating an outcrop where the geological layer is visible and accessible.
- In summary, for an outcrop of a bed to occur on the topographic surface, the topographic contour value (t) must be greater than the stratum contour value (x), ensuring that the bed is exposed and visible on the Earth's surface.

17. As per Ramsay's classification of folds, the maximum thickening of fold hinge and the maximum thinning of the fold limbs are observed in Class 3.

- **Ramsay's Classification of fold:** John Ramsay proposed a classification scheme for folds that is used to describe folds in profile based upon curvature of the inner and outer lines of a fold, and the behaviour of dip isogons.

  Dip isogon is a line that connects points of equal inclination or dip on the outer and inner bounding surfaces of a folded layer
- Class 1 - Dip isogons converge downward towards axial surface, signifying that the curvature of the outer arc is less than that of the inner arc
- Class 1A - Limbs thicker than hinges
- Class 1B - Layer thickness constant; parallel fold
- Class 1C - Limbs thinner than hinges
- Class 2 - Dip isogons are parallel, signifying that the curvature of the outer arc exactly matches the curvature of the inner arc; similar fold
- Class 3 - Dip isogons diverge downward towards axial surface, signifying that the curvature of the outer arc is greater than that of the inner arc

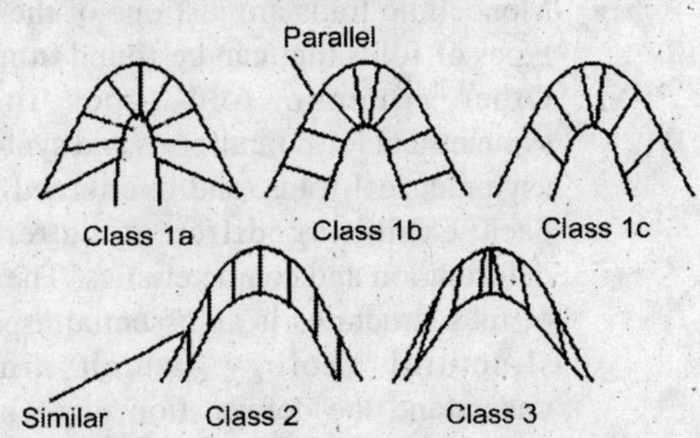

18. The number of hinge(s) in a monocline is 2. Monoclines are the simplest types of folds. Monoclines occur when horizontal strata are bent upward so that the two limbs of the fold are still horizontal.

- A monoclinic fold is a geological structure that refers to a type of rock layer deformation in which the layers of rocks are folded, creating a non-horizontal, tilted fold with one axis of fold symmetry. Monoclinic folds are one of the several types of fold structures that can be observed in geological settings.
- In a monoclinic fold, the fold axis is tilted, but not vertical, meaning that the two limbs of the fold are not parallel to each other. The term "monoclinic" comes from crystallography, where it describes a crystal system with one axis that is not perpendicular to the other two.
- Monoclinic folds can form due to various tectonic forces acting on the Earth's crust,

including compression and lateral stress. These forces cause the rocks to bend and deform, resulting in folds that have a tilted axis of fold symmetry.

- Geologists study fold structures like monoclinic folds to understand the tectonic history and deformation processes that have shaped a particular region. The study of folds provides valuable information about the forces and pressures that have acted on the Earth's crust over geological time scales.
- Monoclinic folds are just one of the many types of folds that can be found in nature. Other common fold types include symmetrical folds (antiforms and synforms), asymmetrical folds, and overturned folds, each exhibiting different patterns of deformation and characteristics. The study of fold structures is an essential aspect of structural geology, which aims to understand the deformation of rocks and how geological structures have evolved over time.

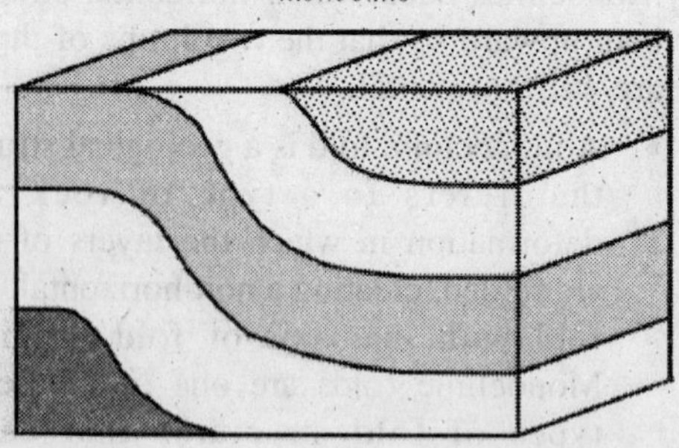

19. The Gondwana flora seed comprised various seed-bearing plant groups that were present on the supercontinent of Gondwana during the Paleozoic and Mesozoic eras. Some of the significant seed plant groups found in the Gondwana flora include:

1. **Glossopterids:** Glossopterids were one of the dominant seed plant groups during the Permian and Triassic periods. These gymnosperms were characterized by their tongue-shaped leaves and were widespread across Gondwana.
2. **Bennettitales:** Bennettitales were an extinct group of seed plants that had a considerable presence during the Mesozoic era. They resembled cycads in appearance and were an important part of the Gondwana seed flora.
3. **Cycads:** Cycads, a group of gymnosperms, were abundant during the Mesozoic era and were diverse components of the Gondwana seed flora.
4. **Conifers:** Conifers, which include pine, fir, spruce, and other related trees, were also part of the Gondwana seed flora. They played a significant role in the terrestrial ecosystems of that time.
5. **Ginkgo:** Ginkgo biloba, often referred to as a "living fossil," traces its evolutionary history back to the Mesozoic era and was part of the Gondwana seed flora.

These seed-bearing plant groups were crucial in shaping the ancient ecosystems of Gondwana and contributed to the rich diversity of plant life during that era. Many of these ancient seed plant groups are now extinct, but their fossil remains have provided valuable insights into the past vegetation and the evolution of modern plant lineages. Some of their living descendants, such as cycads and Ginkgo biloba, have survived to the present day, providing a glimpse into the ancient past of the Gondwana seed flora.

20. Physa Gastropod genera displays sinistral coiling. Some of the gastropod genera that display sinistral coiling include:

- **Lymnaea:** Lymnaea is a genus of freshwater snails in the family Lymnaeidae. Some species within this genus exhibit sinistral coiling.
- **Pomatias:** Pomatias is a genus of land snails in the family Pomatiidae. Certain species in this genus have sinistral shells.

- **Oxyloma:** Oxyloma is a genus of small air-breathing land snails in the family Succineidae. Some species in this genus show sinistral coiling.
- **Amphidromus:** Amphidromus is a diverse genus of tropical land snails in the family Camaenidae. Some species within this genus are sinistral.
- **Opisthostoma:** Opisthostoma is a genus of small land snails belonging to the family Diplommatinidae. Several species in this genus exhibit sinistral coiling.
- **Thysanophora:** Thysanophora is a genus of terrestrial gastropods in the family Spiraxidae. Some species within this genus have sinistral shells.
- It's important to note that sinistral coiling is relatively rare compared to the more common dextral coiling found in the majority of gastropod species. The occurrence of sinistral coiling is often due to genetic mutations affecting the direction of shell coiling in certain lineages.

21. Erinpura Granite: Erinpura Granites form the basement for the Neoproterozoic Malani Igneous Suite in NW India. Based on this study, the Erinpura granites can be divided into Erinpura-East (gneissic fabric), a belt parallel to the southern sector of the South Delhi Fold Belt, and Erinpura-West (magmatic fabrics).

23. In crystallography and symmetry operations, the rotoinversion operation is a combination of rotation and inversion. It involves rotating a crystal or an object by a certain angle and then inverting it through a center of symmetry. The symbol for a rotoinversion operation is represented as n/m, where "n" denotes the rotation angle in degrees, and "m" indicates the order of the operation.

Symmetry elements of the crystal system: 32 crystal classes, their symmetry, Hermann-Mauguin symbol, and class name.

Crystal System	Crystal Class	Symmetry	Name of Class
Triclinic	1	None	Pedial
	$\bar{1}$	I	Pinacoidal
	2	$1A_2$	Sphenoidal
Monoclinic	M	1m	Domatic
	2/m	i, $1A_2$, 1m	Prismatic
	222	$3A_2$	Rhombic-disphenoidal
Orthorhombic	mm^2 (2mm)	$1A_2$, 2m	Rhombic-pyramidal
	$2/m^2/m^2/m$	i, $3A_2$, 3m	Rhombic-dipyramidal
	4	$1A_4$	Tetragonal- Pyramidal
	$\bar{4}$	$\bar{A}_4$	Tetragonal-disphenoidal
	4/m	i, $1A_4$, 1m	Tetragonal-dipyramidal
Tetragonal	422	$1A_4$, $4A_2$	Tetragonal-trapezohedral
	4mm	$1A_4$, 4m	Ditetragonal-pyramidal
	$\bar{4}2m$	$1\bar{A}_4$, $2A_2$, 2m	Tetragonal-scalenohedral
	$4/m^2/m^2/m$	i, $1A_4$, $4A_2$, 5m	Ditetragonal-dipyramidal
	3	$1A_3$	Trigonal-pyramidal
	$\bar{3}$	$1\bar{A}_3$	Rhombohedral
	32	$1A_3$, $3A_2$	Trigonal-trapezohedral
	3m	$1A_3$, 3m	Ditrigonal-pyramidal

	$\bar{3}2/m$	$1\bar{A}_3$, $3A_2$, 3m	Hexagonal-scalenohedral
Hexagonal	6	$1A_6$	Hexagonal-pyramidal
	$\bar{6}$	$1\bar{A}_6$	Trigonal-dipyramidal
	6/m	i, $1A_6$, 1m	Hexagonal-dipyramidal
	622	$1A_6$, $6A_2$	Hexagonal-trapezohedral
	6mm	$1A_6$, 6m	Dihexagonal-pyramidal
	$\bar{6}m^2$	$1\bar{A}_6$, $3A_2$, 3m	Ditrigonal-dipyramidal
	$6/m^2/m^2/m$	i, $1A_6$, $6A_2$, 7m	Dihexagonal-dipyramidal
	23	$3A_2$, $4A_3$	Tetaroidal
	$2/m\bar{3}$	$3A_2$, 3m, $4\bar{A}_3$	Diploidal
Isometric	432	$3A_4$, $4A_3$, $6A_2$	Gyroidal
	$\bar{4}3m$	$3\bar{A}_4$, $4A_3$, 6m	Hextetrahedral
	$\bar{4}/m\bar{3}2/m$	$3A_4$, $4\bar{A}_3$ $6A_2$, 9m	Hexoctahedral

24. Harzburgite: A plutonic igneous rock is composed of 50% orthopyroxene, 45% olivine and 5% clinopyroxene. Harzburgite is a type of ultramafic rock that is primarily composed of two main minerals: olivine and orthopyroxene. It is part of the peridotite group and is considered one of the key rock types found in the Earth's upper mantle. Harzburgite is a significant constituent of the Earth's mantle and is also found in certain tectonic settings such as ophiolites.

Here's the general composition of harzburgite:

1. **Olivine:** Olivine is the dominant mineral in harzburgite. It is a green mineral composed of magnesium, iron, and silica. The chemical formula of olivine is $(Mg, Fe)2SiO_4$.

2. **Orthopyroxene:** Orthopyroxene is the second most abundant mineral in harzburgite. It belongs to the pyroxene group and has the general chemical formula $(Mg, Fe)2Si_2O_6$.

These two minerals together constitute more than 90% of the rock's composition. Harzburgite is an essential part of the Earth's mantle, which lies beneath the crust, and it is thought to represent the residual mantle after the extraction of basaltic magma during partial melting processes. It's worth noting that harzburgite can contain minor amounts of other minerals, such as spinel, chromite, and amphibole, depending on specific geological conditions and the history of the rock's formation. However, olivine and orthopyroxene remain the dominant minerals in this type of rock.

25. Classification of sediment flow:

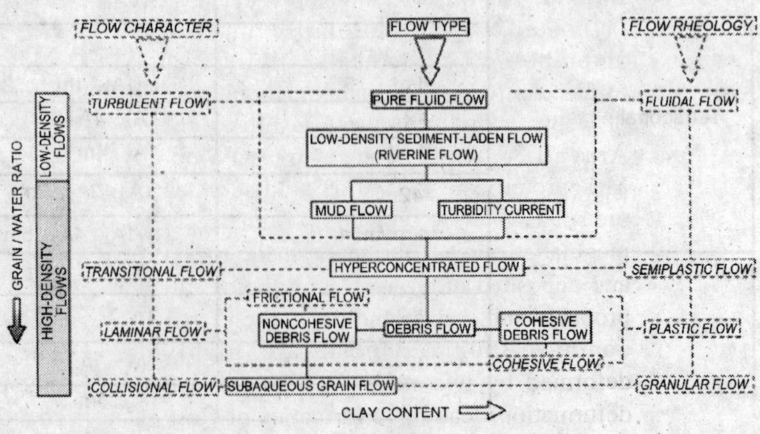

27. • **Malanjkhand:** The Malanjkhand copper deposit is one of the largest copper deposits in India. It is located in the Balaghat district of Madhya Pradesh state in central India. The deposit is significant in terms of both its size and grade of copper ore, making it an important resource for the country.

The Malanjkhand deposit is classified as a sediment-hosted stratiform copper deposit. It is believed to have formed during the Proterozoic Eon (around 1.6 billion years ago) as a result of hydrothermal processes that enriched the surrounding rocks with copper.

Tummalapalle: The Tummalapalle uranium deposit is a significant uranium ore deposit located in the Kadapa district of Andhra Pradesh, India. It is one of the largest uranium reserves in the country and is of strategic importance for India's nuclear energy program. The Tummalapalle deposit was discovered in 1971 during the exploration for copper. Subsequent investigations revealed the presence of significant uranium mineralization. The uranium ore deposit at Tummalapalle is classified as a sedimentary type. It is a sandstone-hosted deposit, where uranium minerals are concentrated in the pore spaces of sandstone rocks.

- **Bhukia:** The Bhukia gold-sulphide deposit is hosted within dolomitic marble, amphibolite, calc-gneiss, albite rich tuffaceous rock, mica schist and quartzite of Delwara Group (Lower Aravalli) of the Aravalli Supergroup of Paleo-Mesoproterozoic age. The gold-sulphide ore solutions were brought up from deeper magmatic sources (ultramafic-mafic rocks) and deposited along with the lithology in probably sub aerial conditions. Strata that contain Bhukia deposit are highly deformed by two major and progressive deformations leading to formation of cross folds and conjugate limb shear zones followed by extensional regime. The ores were metamorphosed along with the host rocks in consequence to ductile deformations. The Bhukia deposit is, therefore, a metamorphosed gold-sulphide deposit.

- **Tosham:** The Tosham granite/Rhyolite is associated with tin-Tungsten and copper mineralization. Tin-Tungsten and copper mineralization at the Tosham hill resulted from hydrothermal fluids of high to low salinity that were active over a temperature range of 150° to 450°C. Within this period, records of three stages of fluid activity were preserved in fluid inclusions. The first stage is characterized by high temperature (300° to 450°C) and high salinity (>37 wt% NaCl) fluids. In the second stage fluid temperature ranges 260° to 380°C and salinity ranges 12 to 14 wt% NaCl; these hydrothermal fluid is related to a major precipitation of ore minerals during this stage. Further, hydrothermal fluids cooled down at the last stage of fluid evolution in temperature range of 150° to 190°C and salinity range of 10.5 to 16 wt% NaCl. The associated sulphides were deposited in the last event.

28. The following statements:

(A) Banded Iron Formations are of chemogenic origin: Banded Iron Formations (BIFs) are sedimentary rock sequences that consist of alternating layers of iron-rich minerals, predominantly hematite or magnetite, and chert (silica-rich sedimentary rock). These formations are unique and important as they represent one of the major sources of iron ore in the world. The origin of BIFs has been a subject of debate among geologists for many years.

The term "chemogenic" refers to the precipitation or deposition of minerals from chemical processes. In the context of BIFs, the chemogenic origin suggests that the iron and silica minerals forming the bands were precipitated from chemical solutions rather than being directly derived from igneous or volcanic processes.

It is important to note that the formation of BIFs likely involved a combination of different processes, and the specific mechanism may vary depending on the geological setting and age of the BIF deposit. The study of BIFs provides valuable insights into the geological history of the Earth, ancient ocean chemistry, and the evolution of life on our planet.

(B) Porphyry-type deposits are formed purely by magmatic processes: Porphyry-type deposits are a specific class of mineral deposits that are typically associated with large-scale, low-to-medium-grade hydrothermal ore systems. These deposits are economically significant sources of copper, gold, and molybdenum, and they are commonly found in association with certain types of igneous rocks known as porphyries. The origin of porphyry-type deposits is believed to involve complex geological processes, and the most widely accepted model for their formation is known as the "Porphyry Copper Model."

Porphyry-type deposits are often associated with large mineralized systems that can span several kilometers in diameter. The geological processes involved in their formation can take millions of years, and their discovery and exploitation are complex endeavors that require careful exploration and evaluation by geologists and mining experts.

(C) Quartz-Pebble Conglomerate hosted gold deposits are formed by Hydrothermal process.

(D) Fullerene is formed by metamorphic process

30. Narwapahar: Narwapahar is a significant uranium mining project located in the East Singhbhum district of Jharkhand, India. It is operated by the Uranium Corporation of India Limited (UCIL), which is a public-sector enterprise under the Department of Atomic Energy, Government of India. Narwapahar is an underground uranium mining operation. Uranium ore is extracted from the underground mine through shafts and tunnels.

Rampura-Agucha: Rampura-Agucha is one of the largest and most productive zinc-lead mines in the world. It is located in the Bhilwara district of Rajasthan, India. The mine is operated by Hindustan Zinc Limited (HZL), a subsidiary of Vedanta Resources Limited. Rampura-Agucha is known for its massive and high-grade zinc-lead deposits. The ore bodies at the mine contain significant amounts of zinc and lead minerals, primarily sphalerite (zinc) and galena (lead).

Jaduguda: Jaduguda is a village located in the Singhbhum district of Jharkhand, India. It is well-known for being the location of the Jaduguda Uranium Mine, which is the first uranium mine in India. The mine is operated by the Uranium Corporation of India Limited (UCIL), a public-sector enterprise under the Department of Atomic Energy, Government of India.The Jaduguda Uranium Mine is an underground mining operation that extracts uranium ore from the Earth's crust. Uranium is a vital element used as fuel in nuclear power plants, and it plays a crucial role in India's nuclear energy program.

Turamdih: Turamdih is another significant location in the Singhbhum district of Jharkhand, India, known for its association with uranium mining. Like Jaduguda, Turamdih is also a prominent site for uranium mining and processing, operated by the Uranium Corporation of India Limited (UCIL), a public-sector enterprise under the Department of Atomic Energy, Government of India.

31. Raniganj: Raniganj coalfield is one of the oldest and most significant coal mining regions in India. It is located in the Bardhaman district of the state of West Bengal. Raniganj coalfield has a long history of coal mining that dates back to the late 18th century during British colonial rule.

Raniganj coalfield is part of the Gondwana coalfields and is situated within the Damodar Valley Basin. The basin is known for its vast reserves of high-quality coal, which plays a significant role in India's energy sector.

The coal extracted from the Raniganj coalfield mainly belongs to the bituminous coal category. Bituminous coal is a high-energy coal with relatively high carbon content, making it suitable for various industrial and energy applications.

Singrauli: Singrauli is a region located in the Sidhi and Singrauli districts of the state of Madhya Pradesh and Sonbhadra district of the state of Uttar Pradesh, India. It is known for its vast reserves of coal and is often referred to as the "Energy Capital of India" due to the extensive presence of coal-based power plants and mining operations in the area. This coalfield is part of the Gondwana coalfields, which are significant sources of coal in the country.

Barmer: Barmer district in the state of Rajasthan, India, is not a significant lignite coal mining region. The area is known for its oil and gas resources rather than coal reserves. The Mangala oil field, one of the largest onshore oil fields in India, is located in Barmer district.

Neyveli: Lignite deposits of coal. Neyveli Lignite Corporation (NLC) is a public-sector lignite mining and power generation company in India. It operates lignite mines in the Neyveli region of Tamil Nadu, which is located in the southern part of India. Neyveli is well-known for its abundant lignite deposits and is one of the largest lignite mining regions in the country.

32. Types of Dunes: Dunes are fascinating landforms created by the wind's action, and they come in various shapes and sizes. The type of dune that forms depends on factors such as wind strength, sand availability, and local topography. Here are some common types of dunes:

1. **Barchan Dunes:** Barchan dunes are crescent-shaped dunes with their tips pointing downwind. They form in areas with limited sand supply and moderate wind. The horns of the crescent point in the direction of the prevailing wind.
2. **Transverse Dunes:** Transverse dunes are long, straight or slightly curved dunes that form perpendicular to the prevailing wind direction. They typically occur in areas with a moderate sand supply and consistent wind direction.
3. **Seif Dunes:** Seif dunes are large, linear dunes with long, straight ridges. They can stretch for several kilometers and are commonly found in areas with abundant sand and steady winds.
4. **Star Dunes:** Star dunes are complex dunes with multiple arms radiating from a central point. They form in areas with variable wind directions and abundant sand supply. Star dunes can be quite tall and have a pyramidal shape.
5. **Parabolic Dunes:** Parabolic dunes have a U-shaped or horseshoe-shaped appearance, with the tips pointing upwind. They typically form in areas with vegetation or some form of surface stabilization, which partially covers the sand and allows it to accumulate in the open space.
6. **Dome Dunes:** Dome dunes are rounded, isolated mounds of sand that form in areas with limited sand supply and variable wind directions. They are relatively small compared to other dune types.
7. **Longitudinal Dunes:** Longitudinal dunes are elongated ridges that run parallel to the prevailing wind direction. They form in areas with moderate sand supply and consistent wind patterns.

8. **Draa or Linear Dunes:** Draa dunes are long, narrow dunes that form in extensive sand seas or deserts. They are linear and can extend for great distances.

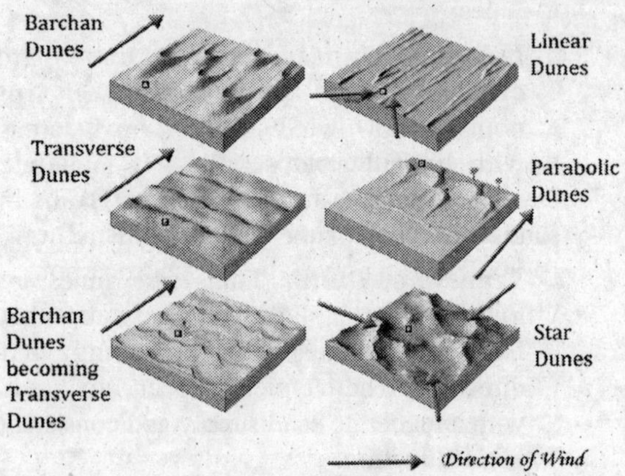

These are some of the common types of dunes found in deserts and sandy environments around the world. The shape and orientation of dunes are continually changing due to wind patterns, sand supply, and other geological factors, creating dynamic and ever-changing landscapes.

33. The altitude of a fault plane = 350°, 75°E This measurement tells us the strike and dip of the fault plane. The strike is the compass direction of the line formed by the intersection of the fault plane and a horizontal plane (measured clockwise from North). The dip is the angle of the fault plane from the horizontal plane.

- The rake of the slickenline on the fault plane = 90°: The slickenline is the line representing the direction of movement along the fault plane. The rake is the angle between the slickenline and the strike of the fault plane.
- If the rake of the slickenline on the fault plane is maximum (true dip) direction; the fault show dip slip fault. And normal and reverse fault also found due to movement with dip direction of the fault plane.
- Strike slip fault never found due to movement of the maximum angle direction.

34. What is ichnology: The branch of paleontology concerned with the study of fossilized tracks, trails, burrows, borings, or other trace fossils as evidence of the occurrence or behaviour of the organisms that produced them. Ichnology is the scientific study of traces and tracks left by living organisms in the geological record. These traces are known as "ichnofossils," and they provide valuable information about the behaviour, activities, and interactions of ancient organisms.

Ichnology involves the study and interpretation of various types of trace fossils, including:

Tracks: Footprints, trails, or burrows left by animals as they moved across or through sediment or other surfaces.

Burrows: Cavities or tunnels created by organisms as they burrowed into sediment or other substrates.

Feeding Traces: Marks or borings left behind by organisms while feeding on other organisms (e.g., predation traces, drill holes).

Resting Marks: Impressions or marks left by organisms while resting or settling on a surface.

- Ichnologists examine the size, shape, preservation, and distribution of these trace fossils to infer the behaviour and lifestyle of the organisms that made them. For example, ichnology can provide insights into how ancient animals moved, their feeding habits, how they interacted with their environment, and even their evolutionary relationships.
- Ichnology is an interdisciplinary field that incorporates aspects of paleontology, sedimentology, biology, and ecology. It helps paleontologists reconstruct ancient ecosystems and understand the ecological dynamics of past environments. Ichnological studies are particularly important when body fossils (the remains of the organisms themselves) are rare or

poorly preserved, as trace fossils can provide an alternative means of understanding ancient life.

- Overall, ichnology is a valuable tool in the study of the history of life on Earth and provides a unique perspective on the behavior and activities of organisms that lived long before the appearance of complex life forms like dinosaurs and mammals.

35. The following combinations of Basin and Formation:

Cauvery Basin – Niniyur Formation = Cretaceous of Trichinopoly Tamil Nadu: The Cauvery Basin is a significant geological and hydrological region located in the southern part of India, primarily in the state of Tamil Nadu. The basin is named after the Cauvery River, which is one of the major rivers in peninsular India and plays a crucial role in the water supply and irrigation in the region.

The Cauvery Basin covers a vast area and is spread across several states, including Karnataka, Tamil Nadu, Kerala, and Puducherry. However, the basin's largest portion lies in the state of Tamil Nadu.

The Cauvery Basin is home to diverse flora and fauna. The river and its tributaries support various aquatic species, and the surrounding forests and wetlands provide habitats for a range of wildlife.

Assam Basin – Tipam Formation = Tertiary of Assam: The Assam Basin is a significant sedimentary basin located in the northeastern part of India. It is one of the prolific hydrocarbon-bearing basins in the country. Within the Assam Basin, the Tipam Formation is a well-known geological unit that belongs to the Tertiary period.

The Assam Basin is an elongated sedimentary basin situated in the northeastern states of Assam, Arunachal Pradesh, Nagaland, and Meghalaya. It is bounded by the Himalayas in the north and the Patkai and Naga Hills in the east. The basin covers an area of about 72,500 square kilometers. The Assam Basin and its surrounding areas are rich in biodiversity, featuring diverse flora and fauna. The region's forests and wetlands are home to various species of plants, animals, and birds.

- **Tipam Formation:** The Tipam Formation is a prominent stratigraphic unit within the Tertiary sedimentary rocks of the Assam Basin. It is primarily composed of sandstone, shale, and coal deposits. The formation is known for its hydrocarbon potential and is a major target for oil and gas exploration in the region.

Bengal Basin – Jalangi Formation: The Jalangi Formation is a significant stratigraphic unit within the sedimentary rocks of the Bengal Basin. It is part of the Quaternary period, which extends from about 2.6 million years ago to the present. The Quaternary is the most recent geological period and is characterized by significant climatic fluctuations.

- The Jalangi Formation is primarily composed of fluvial and alluvial sediments. It includes a variety of deposits such as sand, silt, clay, and gravel, which were laid down by river systems that were active during the Quaternary period.
- The Bengal Basin is a large sedimentary basin located in the northeastern part of India and southwestern Bangladesh. It is one of the most extensive and geologically significant basins in the Indian subcontinent. Within the Bengal Basin, the Jalangi Formation is a notable geological unit.
- The Bengal Basin is a Ganges-Brahmaputra deltaic basin formed by the sediment deposition from the Ganges, Brahmaputra, and other rivers that drain the eastern Himalayas. It covers a vast area of about 569,000 square kilometers, with approximately two-thirds of it lying in Bangladesh and the rest in the Indian states of West Bengal, Tripura, and Assam.

Kutch Basin – Patcham – chari- katrol – umia = Jurassic of Kutch Basin: The Kutch Basin is a sedimentary basin located in the western part of India, primarily in the state of Gujarat. It is a region with diverse geological formations and has significant hydrocarbon potential. Within the Kutch Basin, the Patcham Formation, Chari Formation, Katrol Formation, and Umia Formation are notable geological units that belong to the Jurassic period.

The Kutch Basin is an elongated and relatively narrow sedimentary basin situated in the western part of India. It is surrounded by the Aravalli Range to the northeast and the Gulf of Kutch to the south. The basin covers an area of approximately 30,000 square kilometers.

The geological history of the Kutch Basin is complex and spans millions of years. It is a rift basin formed during the Mesozoic era due to the rifting of the Indian subcontinent from the eastern margin of the African plate.

- **Patcham Formation:** The Patcham Formation is a geologic unit within the Kutch Basin that belongs to the Jurassic period. It is known for its sedimentary rocks, including sandstone, shale, and limestone.
- **Chari Formation:** The Chari Formation is another important Jurassic unit within the Kutch Basin. It comprises sedimentary rocks, including sandstone, shale, and conglomerate.
- **Katrol Formation:** The Katrol Formation is also a part of the Jurassic rocks in the Kutch Basin. It consists of sedimentary deposits like sandstone, shale, and siltstone.
- **Umia Formation:**The Umia Formation is one more significant unit from the Jurassic period within the Kutch Basin. Like the other formations, it comprises various sedimentary rocks.
- The Kutch Basin is known for its hydrocarbon potential, and several oil and gas fields have been discovered and exploited in the region. The Jurassic formations, including the Patcham, Chari, Katrol, and Umia, are some of the targets for hydrocarbon exploration in the basin.

Dhok Pathan Formation = Siwalik group: The Dhok Pathan Formation is a geologic unit that belongs to the Siwalik Group, which is a prominent stratigraphic sequence in the Siwalik Hills of northern India. The Siwalik Group is known for its extensive sedimentary rocks that were deposited during the Miocene and Pliocene epochs of the Neogene period. These rocks are composed of material eroded and transported from the rising Himalayan mountain range.

- **Dhok Pathan Formation:** The Dhok Pathan Formation is one of the distinct units within the Siwalik Group. It represents a specific interval of sedimentation during the Miocene and/or Pliocene epochs. The rocks in this formation are primarily composed of sandstones, siltstones, mudstones, and conglomerates.
- The sedimentary rocks of the Dhok Pathan Formation, like other units within the Siwalik Group, provide valuable insights into the ancient environments of the region during the Miocene and Pliocene epochs. The deposition occurred in a fluvial (river) environment, and the rocks contain a wealth of information about the ancient flora and fauna that inhabited the area.

37. The AFM diagram given below shows stability of minerals in the garnet zone. If P, Q, R and S represent the compositions of different pelitic rocks.

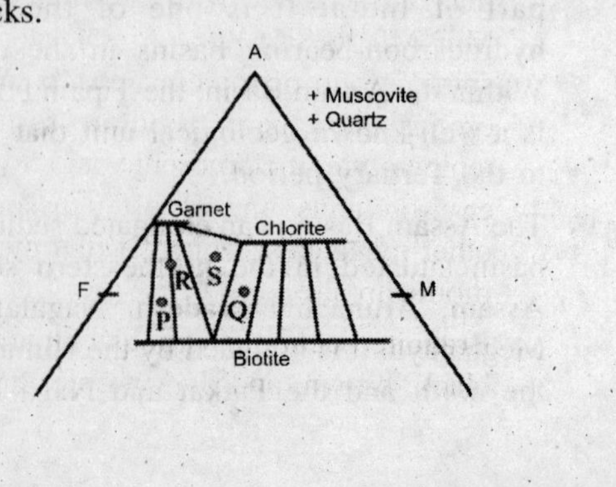

The following P, Q, R, S zone represent the different equilibrium assemblage:

- **At zone P:** Muscovite + garnet + quartz + biotite
- **At zone Q:** Biotite + chlorite + muscovite + quartz
- **At zone R:** Muscovite + garnet + quartz + biotite
- **At zone S:** Garnet + biotite + chlorite + muscovite + quartz

38. The following sedimentary structures is/are:

- **Bounce marks:** Impact of tool at low angle
- **Wrinkle marks:** This may be due to activity of microbial mats (Cyanobacteria).
- **Prod marks:** This may be due to impact of tool due to high angle.
- **Skip marks:** Multiple bounce of tool.

39. Bornite: Bornite is a mineral that is also known as peacock ore due to its vibrant iridescent colours. It is a sulfide mineral with the chemical composition Cu_5FeS_4, which means it is a copper iron sulfide. Bornite is a valuable ore of copper and is mined for its copper content.

Chalcocite: Chalcocite is a copper sulfide mineral with the chemical composition Cu_2S. It is one of the most important secondary copper minerals and serves as a significant source of copper ore. Chalcocite is often found in association with other copper minerals like bornite and chalcopyrite.

Braunite: Braunite is a mineral composed of manganese oxide with the chemical formula $Mn^{2+}Mn^{3+}6(SiO_4)O_{12}$. It is a member of the spinel group of minerals and is an important ore of manganese. Braunite forms in high-temperature metamorphic and hydrothermal environments and is often associated with other manganese minerals like pyrolusite and rhodochrosite.

Braunite occurs in various colours, including black, brown, and dark purple. It often has a sub-metallic to vitreous luster, giving it a shiny appearance.

Braunite is an important source of manganese, an essential metal used in the production of steel, batteries, and various other industrial applications. Manganese is a critical component in the steel-making process, as it improves the strength, toughness, and wear resistance of steel.

- Braunite contains both divalent (Mn^{2+}) and trivalent (Mn^{3+}) manganese cations, which contribute to its complex crystal structure.

Chrysocolla: Chrysocolla is a mineral that is a hydrous copper silicate with a variable chemical composition. It is often found in association with other copper minerals and forms as a secondary mineral in the oxidation zones of copper ore deposits. Chrysocolla is known for its beautiful blue and green colours, and it is prized by collectors and used in jewellry and ornamental objects.

- Chrysocolla is typically blue to green and can have various shades of these colours. It often occurs as botryoidal (grape-like) or stalactitic masses and may have a vitreous to waxy luster.

40. API gravity: A measure of density of petroleum. API gravity, often referred to as just API, is a measure used to determine the density of liquid petroleum products, especially crude oil and refined petroleum products. The API gravity scale was developed by the American Petroleum Institute (API) and is widely used in the oil and gas industry to classify and compare different types of crude oil. API gravity is expressed in degrees. The API gravity value provides crucial information for the oil and gas industry. It helps in assessing the quality and characteristics of different crude oils, determining their market value, and understanding their behaviour during transportation, refining, and processing. Lighter oils with higher API gravity are generally more desirable because they yield a higher percentage of valuable products during refining and have lower environmental impacts during extraction and processing.

Resistivity: Related to electrical property. Resistivity is a fundamental electrical property that measures the resistance of a material to the flow of electric current. It is denoted by the symbol "ρ" (rho) and is expressed in ohm-meters (Ω m). Resistivity plays a crucial role in understanding and analyzing the behaviour of materials in electrical circuits, as well as in various applications in science and engineering.

Kriging: Geostatistical technique. Kriging, also known as Gaussian process regression, is a geostatistical technique used for spatial interpolation and estimation. It is widely applied in various fields, including geology, mining, environmental science, agriculture, and remote sensing. Kriging provides a method to estimate unknown values at unmeasured locations based on the values observed at nearby measured locations.

Rock Quality Designation (RQD): Rock Quality Designation is a quantitative measure used in geotechnical engineering and geology to assess the quality of rock masses. It provides an indication of the degree of fracturing and jointing in a rock mass, which is essential for various engineering projects involving rock, such as tunneling, mining, slope stability analysis, and foundation design.

- RQD is based on the concept of core recovery and is typically determined from the analysis of drill core samples obtained during geological investigations. The RQD value is expressed as a percentage and represents the proportion of intact and sound rock recovered from the total length of the core.

The calculation of RQD involves the following steps:

- **Measurement of Core Length:** The total length of the drill core sample is measured in meters.
- **Measurement of Intact Rock Length:** The length of intact and sound rock pieces greater than a specified minimum size (typically 10 cm or 4 inches) is measured. These pieces are considered to be representative of the unfractured rock.
- **Calculation of RQD:** RQD is calculated as the sum of the intact rock lengths divided by the total core length and then multiplied by 100 to express it as a percentage. Mathematically, RQD is calculated as:
- RQD = (Sum of Intact Rock Lengths/Total Core Length) × 100
- The RQD value can range from 0% to 100%. A high RQD value indicates a rock mass with good quality and limited fracturing, whereas a low RQD value suggests a rock mass with significant fractures and jointing.
- RQD is a valuable parameter for characterizing the rock mass properties and is used in various geotechnical analyses and engineering designs. It helps geologists and engineers to assess the rock mass stability, design suitable support systems for tunnels and excavations, and understand the behaviour of the rock during construction activities. In addition to RQD, other rock mass classification systems, such as the Rock Mass Rating (RMR) and the Geological Strength Index (GSI), are often used in conjunction with RQD to provide a comprehensive evaluation of the rock mass characteristics.

41. Given, Earthquake A1 = x

Earthquake A2 = $15x$

Magnitude M1 = 5

$$M2 - M1 = \text{Log}_{10}\ (A2/A1)$$

$$M2 = M1 + \text{Log}_{10}\ (A2/A1)$$

$$= 5 + \log_{10}\ (15x/x)$$

$$= 5 + \log_{10} 15$$

$$= 6.176$$

$$= 6.18$$

42. Given the intercepts of crystallographic axes $= 0.5a : 1b : 0.75c$

On crystallographic plane = $\{h\ k\ l\}$

Given, $ia = 0.5a$, $ib = 1b$, $ic = 0.75c$

Crystallographic plane $(h\ k\ l) = a/ia : b/ib : c/ic$

$= a/0.5a : b/1b : c/0.75c$

$= 1/0.5 : 1 : 1/0.75$

$= 3/0.5 : 3 : 3/0.75$

$= 6 : 3 : 4$

The value of 'l' is = 4.

43. Given to find the wave velocity, we can use the deep water wave equation, which relates the wave velocity (V) to the wavelength (λ) and the acceleration due to gravity (g) in deep water conditions (where the water depth is much greater than the wavelength):

$$V = \sqrt{\left(\frac{g * \lambda}{(2\pi)}\right)}$$

where:

$g =$ acceleration due to gravity

$= 10\ m/s^2$ (given)

$\lambda =$ wavelength of the wave

$= 200$ m (given)

Now, let's plug in the values and calculate the wave velocity:

$$V = \sqrt{\left(\frac{10 * 200}{(2\pi)}\right)}$$

$$V = \sqrt{\left(\frac{2000}{(2\pi)}\right)}$$

$$V = \sqrt{\left(\frac{2000}{6.28318530718}\right)}$$

$$V \approx \sqrt{318.31}$$

$$V \approx 17.84\ m/s$$

44. To find the strike value of the rotated bed, we need to determine the new strike direction after a 60° clockwise rotation.

- The original bed has an attitude of 045°, 20°SE. This means the strike direction of the bed is 045° and it is dipping at an angle of 20° towards the southeast.
- When the bed is rotated 60° clockwise (looking down) about a vertical axis, we need to find the new strike direction after the rotation.
- To find the new strike direction, we need to add the rotation angle (60°) to the original strike direction (045°) and then adjust it based on the right-hand rule convention.

New Strike = Original Strike + Rotation Angle

New Strike = 045° + 60°

New Strike = 105°

However, we need to adjust the value to be within the range of 0° to 360° (due to strike being a circular measurement). So, we subtract 360° from the calculated value:

New Strike = 105° – 360°

New Strike = –255°

Now, we need to convert the negative value to a positive value within the range of 0° to 360°:

New Strike = 360° + (–255°)

New Strike = 105°

45. To find the minimum velocity of the waterflow that developed antidunes, we can use the relationship between the flow velocity (V), the flow depth (h), and the acceleration due to gravity (g) for antidunes.

For antidunes, the minimum flow velocity occurs when the flow depth is equal to half the wavelength of the antidunes. (Round off to two decimal places) (Use $g = 10\ m/s^2$)

The flow velocity (V) can be calculated using the following formula:

$$V = \sqrt{(g * h)}$$

where:

$g =$ acceleration due to gravity

$= 10\ m/s^2$ (given)

h = flow depth = 1 meter (given)

Now, let's plug in the values and calculate the minimum flow velocity:

$$V = \sqrt{(10*1)}$$

$$V = \sqrt{10}$$

$$V \approx 3.16 \text{ m/s}$$

46. Given, the angular aperture of a 20X objective AA = 46° (Use RI of water = 1.33)

To find the numerical aperture (NA) of the water immersion objective, we can use the formula for calculating numerical aperture:

Numerical Aperture (NA) = n * sin(AA/2)

where:

n = refractive index of the medium between the objective and the specimen

θ = angular aperture of the objective

For a water immersion objective, the refractive index (n) is the refractive index of water, which is given as 1.33.

Angular aperture (θ) of the 20X objective = 46°

Now, let's calculate the numerical aperture (NA):

$$NA = 1.33 * \sin(46°/2)$$

$$= 1.33 * \sin 23^0$$

$$= 0.52$$

47. To determine the number of components in this metamorphic rock assemblage, we need to apply the phase rule, which relates the number of components, phases, and degrees of freedom in a system at equilibrium.

The phase rule is given by:

$$F = C - P + 2$$

where:

F = degrees of freedom

C = number of components

P = number of phases

In this case, we have four minerals: grossular garnet, kyanite, anorthite, and quartz. Each mineral represents a phase in the assemblage. So, P = 4.

Now, we need to identify the number of components (C). A component is an independent chemical species that can vary independently. To determine the components, we need to look at the chemical formulas of the minerals and see if there are any constraints or relationships among the elements.

For the given minerals:

1. Grossular garnet ($Ca_3Al_2Si_3O_{12}$) contains Ca, Al, and Si.
2. Kyanite (Al_2SiO_5) contains Al and Si.
3. Anorthite ($CaAl_2Si_2O_8$) contains Ca, Al, and Si.
4. Quartz (SiO_2) contains Si.

If we analyze the chemical formulas, we find that Ca, Al, and Si are the only chemical species present, and they can vary independently among the minerals. So, the number of components (C) is 3.

48. The dip separation vector represents the horizontal displacement or separation across the fault plane due to the faulting. It is the horizontal component of the relative movement along the fault plane. The magnitude of the dip separation vector is the actual displacement across the fault.

Given:

Dip separation vector attitude = 60° → 040°

Magnitude of the dip separation vector = 6 m

Let's break down the attitude of the dip separation vector:

The dip direction is 040°, which means the dip separation vector points in the direction of 040° from the horizontal plane (measured clockwise from the north direction).

The dip angle is 60°, which represents the inclination of the dip separation vector from the horizontal plane.

Now, we can use trigonometry to find the horizontal component (heave) of the dip separation vector. The horizontal component is given by:

Heave = Magnitude × cos(dip angle)

Heave = 6 m × cos(60°)

Heave ≈ 6 m × 0.5

Heave ≈ 3 m

49. To calculate the factor of safety of the hillslope, we'll use the concept of the "resisting force" and the "driving force." The factor of safety (FoS) is defined as the ratio of the resisting force to the driving force. In the case of a hillslope, the resisting force is the shear strength of the soil, while the driving force is the force acting downhill due to the slope angle.

Given:

Angle of the hillslope (slope angle) = 40°

Internal friction angle of the soil = 30°

First, we need to find the resisting force (R) and the driving force (D).

Resisting Force (R): The resisting force is the component of the weight of the soil that acts perpendicular to the slope. It can be calculated as follows:

R = Weight of the soil × sin(φ)

where φ is the internal friction angle of the soil.

Driving Force (D): The driving force is the component of the weight of the soil that acts parallel to the slope. It can be calculated as follows:

D = Weight of the soil × cos(θ)

where θ is the slope angle.

Now, the factor of safety (FoS) can be expressed as:

FoS = R/D

Let's calculate it step by step:

Calculate the resisting force (R):

R = Weight of the soil × sin(30°)

Calculate the driving force (D):

D = Weight of the soil × cos(40°)

Calculate the factor of safety (FoS):

FoS = R/D

Since we have no specific information about the weight of the soil, we can assume a constant value for simplicity (*e.g.*, 1 unit). The weight units will cancel out when calculating the factor of safety.

R = 1 × sin(30°) ≈ 0.5

D = 1 × cos(40°) ≈ 0.766

FoS ≈ 0.5 / 0.766 ≈ 0.6528.

50. To calculate the change in groundwater storage, we need to consider the volume of water that was originally stored in the rock and the volume that remains after the water table was lowered.

Given:

Area over which the water table was lowered:

$1\ km^2 = 1{,}000{,}000\ m^2$

Lowering of the water table: 4 m

Porosity of the rock: 30% (expressed as a decimal, 30% = 0.30)

Specific retention: 10% (expressed as a decimal, 10% = 0.10)

Specific yield = 0.30 – 0.10 = 0.20

Volume of water drained = Area × Depth × Porosity (Specific yield)

$= 1{,}000{,}000\ m^2 \times 4\ m \times 0.20$

$= 10^6 \times 4 \times 0.2$

$= 0.8 \times 10^6$

$= 800 \times 10^3$.

51. Given, 143Nd/144Nd = 0.516

147Sm/144Nd = 0.389

As per the exact parent-daughter relationship equation, the 143Nd/144Nd of the rock

$= 4.6 \times 10^9$ years

(Use decay constant for 147Sm λ

$= 6.54 \times 10^{-12}\ y^{-1}$)

To calculate the Nd isotopic ratio (143Nd/144Nd) of the rock $t = 4.6 \times 10^9$ years ago,

$$\lambda t = 6.54 \times 10^{-12} \times 4.6 \times 10^{9}$$
$$= 0.030084$$

$$(143Nd/144Nd) = (143Nd/144Nd)_0 + (147Sm / 144Nd)(e^{\lambda t} - 1)$$

$$(143Nd/144Nd)_0 = (143Nd/144Nd) - (147Sm / 144Nd)(e^{\lambda t} - 1)$$
$$= 0.516 - 0.389\ (e^{0.30084} - 1)$$
$$= 0.516 - (0.389 \times 0.0305 - 1)$$
$$= 0.516 - 0.0118804$$
$$= 0.5041196$$
$$= 0.504$$

53. Given, An underground vertical dyke is intercepted by an inclined borehole as shown in the figure below.

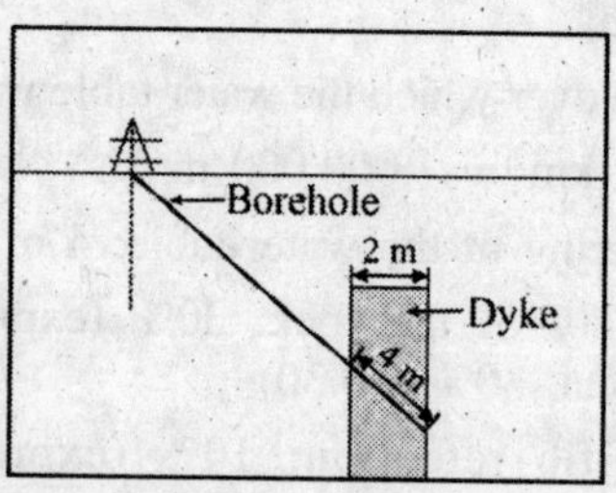

The length of the dyke core intercepted by the borehole = 4 m.

If the true thickness of the dyke = 2 m,

The inclination of the borehole from the vertical:

In triangle ABC, sinθ = True thickness of dyke/ The length of the dyke core intercepted by the borehole

$$= \frac{2}{4}$$

$$\theta = \text{Sin}^{-1}\left(\frac{2}{4}\right)$$

$$= 30^0 \text{ (In integer)..}$$

54. To find the reserve of the copper ore body, we need to calculate its volume and then convert it to tons. Here's the step-by-step solution:

1. Calculate the volume of the cylindrical copper ore body:

The volume of a cylinder is given by the formula $V = \pi * r^2 * h$, where r is the radius and h is the height (vertical thickness in this case).

Given diameter = 14 m,

$$\text{so the radius } (r) = \frac{\text{Diameter}}{2}$$

$$= \frac{14 \text{ m}}{2} = 7 \text{ m.}$$

Vertical thickness (h) = 45 m.

$$V = \pi * (7 \text{ m})^2 * 45 \text{ m}$$
$$V \approx 3.14 * 49 \text{ m}^2 * 45 \text{ m}$$
$$V \approx 6923.7 \text{ m}^3$$

2. Convert the volume to cubic centimeters (cm^3) because the density is given in g/cm^3:

$$1 \text{ m}^3 = 1{,}000{,}000 \text{ cm}^3$$

So, $6923.7 \text{ m}^3 = 6923.7 * 1{,}000{,}000 \text{ cm}^3$

$$= 6{,}923700000 \text{ cm}^3$$

3. Calculate the mass of the copper ore body:

Mass (M) = Volume * Density

$$M = 6{,}923700000 \text{ cm}^3 * 2.9 \text{ g/cm}^3$$
$$M = 20078730{,}000 \text{ g}$$

4. Convert the mass to tons:

$$1 \text{ ton} = 1{,}000{,}000 \text{ g}$$

$$\text{Tons} = \frac{20078730{,}000 \text{ g}}{1{,}000{,}000 \text{ g/ton}}$$

$$= 20{,}078.73 \text{ tons Ans.}$$

The reserve of the copper ore body is approximately 20,078.73 tons.

55. Given, the density of a FCC unit cell

$$= 6.5 \text{ g/cm}^3.$$

Avogadro's number = (Round off to two decimal places) (Use NA = 6.022×10^{23})

The mass of a single atom = 60 g/mol/ (6.022×10^{23})

No of atom in FCC = 4

Total mss per FCC, m = $\frac{(1\times 60)}{(6.022\times 10^{23})}$

Density of FCC, = $\frac{m}{v}$ = 6.5 g/cm^3

$$V = \frac{4\times 60}{6.022\times 10^{23}\times 6.5}$$

$= 61.313 \times 10^{-24}$ cm^3

A3 = 61.313 A^{0}3

A = 3.943 A^0

Diagonal length (d) = $a \times \sqrt{2}$

= 3.943 A$^0x\sqrt{2}$

= 5.58.

56. Composition at L_2 = 70 – 20 = 50

Percentage of liquid = $\left[\frac{(70-60)}{(70-20)}\right]\times 100$

$= \left(\frac{10}{50}\right)\times 100 = 20.$

57. To find the diameter of the particles that can be transported without a change in terminal fall velocity, we can use Stokes' law. Stokes' law relates the terminal fall velocity of a particle in a fluid to its diameter, density, and the density of the fluid.

The formula for terminal fall velocity (V) in Stokes' law is given by:

$$V = (2/9) * (g * (d^2) * (\rho p - \rho f))/\eta$$

Where:

V = Terminal fall velocity of the particle

g = Acceleration due to gravity (approximately 9.81 m/s^2)

d = Diameter of the particle

ρp = Density of the particle

ρf = Density of the fluid (water in this case)

η = Dynamic viscosity of the fluid (assumed constant)

Let's calculate the terminal fall velocity for both types of particles and equate them to find the diameter of the added particles.

For the original particles (density = 3 g/cm^3):

ρp = 3 g/cm^3

ρf = 1 g/cm^3

d = 2 mm = 0.2 cm

For the added particles (density = 2 g/cm^3):

ρp = 2 g/cm^3

ρf = 1 g/cm^3

d = ? (to be determined)

We can set up the equation as follows:

$(2/9) * (g * (d^2) * (\rho p - \rho f)) / \eta$ (for original particles)

$= (2/9) * (g * (d^2) * (\rho p - \rho f)) / \eta$ (for added particles)

Since the terminal fall velocity remains the same for both cases, we can equate the two expressions:

$(2/9) * (g * (0.2^2) * (3 - 1)) / \eta$

$= (2/9) * (g * (d^2) * (2 - 1)) / \eta$

Simplifying the equation:

$0.04 * 2 / \eta = d^2$

$d^2 = 0.08 / \eta$

Now, we need to find the value of η (dynamic viscosity of water) and substitute it into the equation to solve for d.

The dynamic viscosity of water at room temperature (approximately 20°C) is around 0.01 g/(cm·s) or 0.01 g/(cm·ms).

Let's use η = 1 g/(cm·s) (rounded value):

$$d^2 = \frac{0.08}{1} = 0.08$$

$$d = \sqrt{0.08} = 0.283473 \text{ cm}$$

To convert centimeters (cm) to millimeters (mm), you simply multiply the value in centimeters by 10, as there are 10 millimeters in one centimeter.

0.283473 cm * 10 mm/cm = 2.83473 mm.

58. Given, To calculate the grade of the iron ore body, we need to determine the percentage of iron (Fe) in the ore, as the grade is typically expressed as the percentage of the valuable metal in the ore.

The molecular formulas for hematite and magnetite are as follows:

Hematite (Fe_2O_3): 2 Fe atoms + 3 O atoms

Magnetite (Fe_3O_4): 3 Fe atoms + 4 O atoms

Given that the iron ore body contains 50% hematite and 50% magnetite, we can calculate the percentage of iron in the ore as follows:

1. Calculate the molar masses of Fe and O:

Molar mass of Fe (Fe): 55.85 amu

Molar mass of O (O): 16 amu

2. Calculate the total molar mass of the ore:

Total molar mass = (0.5 * molar mass of hematite) + (0.5 * molar mass of magnetite)

Molar mass of hematite (Fe_2O_3):

= [(2 * molar mass of Fe) /(2 * molar mass of Fe) + (3 * molar mass of O)] × 100

= [(2 * 55.85 amu)/ 2 * 55.85 amu + (3 * 16 amu)] × 100

= 70.25 amu

Molar mass of magnetite (Fe_3O_4):

= [(3 * molar mass of Fe) /3 * molar mass of Fe + (4 * molar mass of O)] × 100

= [(3 * 55.85 amu)/ (3 * 55.85 amu) + (4 * 16 amu)] × 100

= 71.05 amu

Total molar mass of the ore:

= (0.5 * 70.25 amu) + (0.5 * 71.05 amu)

= 35.125 amu + 35.525 amu

= 70.65 amu

When the ore body has 50 percenage of both:

Grade = 0.5 × 71.05 + 0.5 × 70.65

= 70.85.

59. Given, the schematic stereographic projection below shows dip amount and dip direction of three sets of joints (J1, J2 and J3) on a hillslope.

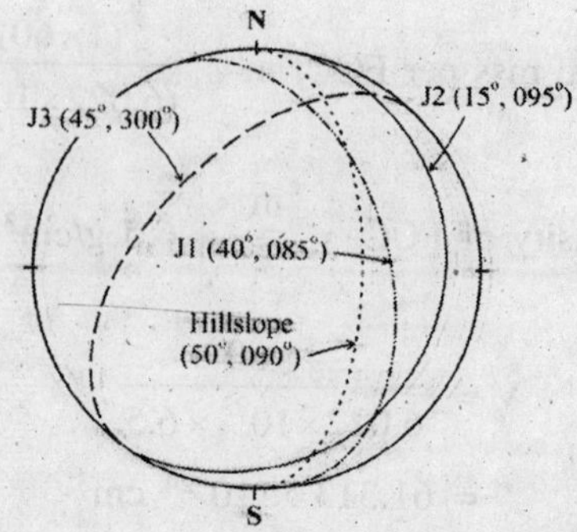

- If the internal friction angle of the hillslope material = 30°
- The strike of the potential failure joint plane (in azimuthal convention following right hand rule) =
- The hillslope Dotted line in circle) = (50, 090 degree)
- The value of J1 = (40, 085 degree)
- The value of J2 = (15, 095 degree)
- The value of J3 = (45, 300 degree)
- In shaded region J1 have dip less than slope and more than internal friction angle, the strike of J1 = 355 degree
- Dip of the joint is less than the dip of slope it favors failure of joint plane, when dip direction of joint and slope are same.
- J2 is not a potential failure plane because it has dip less than internal friction angle.
- J3 has a opposite dip direction, not potential failure joint plane.

60. Given, the hydraulic conductivity of a L = 100 cm long cylindrical core is estimated k = 1.2 cm/min

$$K = 1.2 \times 60 \times \frac{24}{100} \text{ m/day}$$
$$= 17.28 \text{ m/day}$$

Hydraulic head difference Δh = 20 cm

The effective porosity of the core = 20%

$$\text{Darcy velocity} = \frac{k\Delta h}{L} = 17.28 \times \frac{20}{100}$$
$$= 3.456 \text{ m/day}$$

$$\text{Interstitial velocity} = \frac{3.456}{0.2} = 17.28 \text{ m/day.}$$

Previous Papers (Solved)

Indian Institute of Technology Joint Admission Test for M.Sc.

IIT-JAM GEOLOGY-2022

SECTION-A

Multiple Choice Questions (MCQ)

Qs. No. 1-10: *Carry one mark each.*

1. Which one of the following is a geochronologic unit?

A. System B. Period
C. Member D. Formation

2. Which one of the following must have thickness less than 1 cm?

A. Lamina B. Bed
C. Stratum D. Layer

3. Which one of the following organisms became extinct during the Cretaceous-Tertiary mass extinction event?

A. Trilobite B. Ammonite
C. Brachiopod D. Echinoderm

4. Match the geomorphic features in Group I with the related processes in Group II.

Group-I	Group-II
P. Cirque	1. Fluvial
Q. Ventifact	2. Glacial
R. Point bar	3. Volcanic
S. Tephra	4. Aeolian

	P	Q	R	S
A.	2	4	1	3
B.	2	3	1	4
C.	4	2	3	1
D.	1	2	3	4

5. Which one of the given textural features results from exsolution?

A. Ophitic
B. Perthitic
C. Graphic
D. Glomeroporphyritic

6. In the holosymmetric class of the Cubic System, how many more faces does the {110} form have compared to the {111} form?

A. 2 B. 4
C. 6 D. 8

7. Which one of the following seismic waves involves compression and rarefaction (but not rotation) of the material that it passes through?

A. P-waves B. S-waves
C. Rayleigh waves D. Love waves

8. Realgar and orpiment are both minerals of arsenic (As) and have the same chemical composition. Which one of the following properties can be used to distinguish between the two minerals in hand specimen?

A. Lustre B. Hardness
C. Colour D. Fracture

9. Buckle folds result from:

A. Layer parallel shortening
B. Layer perpendicular slip
C. Layer parallel shearing
D. Layer perpendicular shortening

10. Sandstone beds above a magmatic body are domal in shape, while the beds below are horizontal. The magmatic body is a:

A. Batholith B. Laccolith
C. Lopolith D. Sill

Qs. No. 11–30: *Carry two marks each.*

11. Match the morphological features in Group I with the corresponding fossil groups in Group II.

Group-I	Group-II
P. Pedicle Foramen	1. Trilobita
Q. Pallial Sinus	2. Cephalopoda
R. Pygidium	3. Pelecypoda
S. Siphuncle	4. Brachiopoda

	P	Q	R	S
A.	4	3	1	2
B.	4	1	2	3
C.	3	4	1	2
D.	2	1	4	3

12. The Triassic-Jurassic boundary lies within which one of the following stratigraphic units?

A. Panchet Formation
B. Dharmaram Formation
C. Pachmarhi Formation
D. Denwa Formation

13. Which one is the correct order of stability of the minerals (arranged from the most stable to the least stable) during chemical weathering?

A. Muscovite > Amphibole > Quartz > Olivine
B. Quartz > Amphibole > Olivine > Muscovite
C. Quartz > Muscovite > Amphibole > Olivine
D. Muscovite > Olivine > Quartz > Amphibole

14. Match the following sedimentary rocks in Group I with their compositions in Group II.

Group-I	Group-II
P. Packstone	1. <15% matrix, > 25% rock fragments
Q. Grainstone	2. >15% matrix, >25% feldspar
R. Lithic arenite	3. Grain supported, contains no mud
S. Arkosic wacke	4. Grain supported, contains mud

	P	Q	R	S
A.	4	3	1	2
B.	3	4	2	1
C.	3	1	4	2
D.	2	4	1	3

15. Match the parameters in Group I with their corresponding dimensions in Group II.

Group-I	Group-II
P. Shear modulus	1. $M^0L^0T^0$
Q. Hydraulic conductivity	2. $M^1L^{-3}T^0$
R. Volumetric strain	3. $M^0L^1T^{-1}$
S. Dry density	4. $M^1L^{-1}T^{-2}$

	P	Q	R	S
A.	4	3	1	2
B.	3	1	2	4
C.	2	3	4	1
D.	3	4	1	2

16. Match the countries in Group I with the plate tectonic features in Group II that cause seismic activity in them.

Group-I	Group-II
P. Iceland	1. Subduction Zone
Q. Indonesia	2. Transform Fault
R. Nepal	3. Mid-Oceanic Ridge
S. New Zealand	4. Continental Collision

	P	Q	R	S
A.	3	1	4	2
B.	3	1	2	4
C.	1	3	4	2
D.	2	1	4	3

17. Which one of the magnitude scales given below DOES NOT saturate while estimating size of earthquakes?

A. Local magnitude scale (M_L)
B. Body wave magnitude scale (M_b)
C. Surface wave magnitude scale (M_S)
D. Moment magnitude scale (M_W)

18. What is the minimum number of forms that an actual crystal must contain in Class 1 (Pedial) of the Triclinic System?

A. 1   B. 2
C. 3   D. 4

19. The apparent dip of a plane is measured to be 45° towards NE. The true dip of the plane is:

A. 55° towards SSW
B. 40° towards NNE
C. 48° towards ENE
D. 40° towards E

20. A horizontal upright fold will have a:
A. vertical fold axis and horizontal axial plane
B. horizontal fold axis and vertical axial plane
C. horizontal fold axis and axial plane with any dip
D. plunging fold axis on a vertical axial plane

21. The displacement of the bed shown in the figure below is caused by a single movement along fault XYZ. Of the options given below, which fault-type can explain the observed displacement?

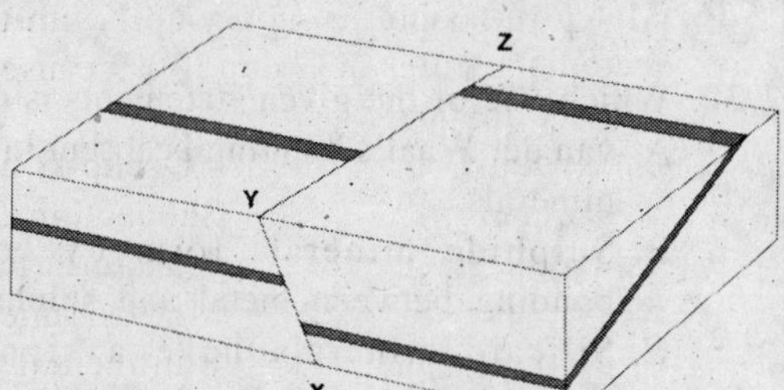

A. Strike-slip B. Reverse
C. Normal D. Trace-slip

22. In the AFM diagram below, W, X, Y and Z represent four pelitic assemblages. Which combinations of assemblages contain the same mineral phases in equilibrium in the staurolite zone?

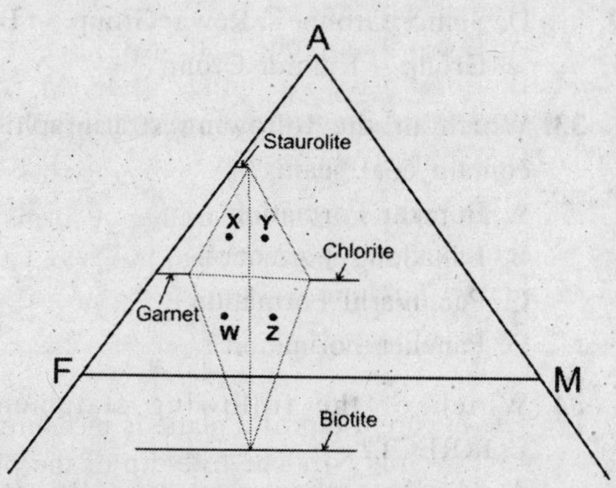

A. (X-Y) & (W-Z) B. (X-W) & (Y-Z)
C. (X-Z) & (Y-W) D. (W-X-Y) & (Z)

23. Match the rock types in Group I with their characteristic mineral assemblages in Group II.

Group-I	Group-II
P. Diorite	1. Plagioclase – orthopyroxene ± clinopyroxene
Q. Tonalite	2. Olivine – orthopyroxene – clinopyroxene
R. Norite	3. Plagioclase – hornblende ± quartz
S. Lherzolite	4. Quartz – plagioclase ± K-feldspar

	P	Q	R	S
A.	4	3	2	1
B.	2	1	3	4
C.	3	4	1	2
D.	1	3	4	2

24. Match the mineral deposit types in Group I with the water types in Group II considered dominantly responsible for their origin.

Group-I	Group-II
P. Porphyry copper deposits	1. Meteoric water
Q. Mississippi Valley Type deposits	2. Groundwater
R. Roll-front uranium deposits	3. Magmatic water
S. Epithermal gold deposits	4. Connate water

	P	Q	R	S
A.	4	3	2	1
B.	3	4	1	2
C.	3	4	2	1
D.	4	1	2	3

25. Match the minerals in Group I with their optical properties in Group II.

Group-I	Group-II
P. Sodalite	1. Mottled extinction
Q. Tourmaline	2. Isotropic
R. Calcite	3. Pleochroic from blue to brown
S. Muscovite	4. Twinkling effect

	P	Q	R	S
A.	4	3	2	1
B.	2	3	4	1
C.	3	1	2	4
D.	1	3	4	2

26. The contact between the Talchir Formation and the underlying Precambrian basement is:
A. An angular unconformity
B. A disconformity
C. A paraconformity
D. A nonconformity

27. Increased diversity of siphonate bivalves occurred in response to:
A. The Cambrian explosion in the Paleozoic
B. Increased temperature in the Cenozoic
C. Increased predation pressure in the Mesozoic
D. Increased oxygen level in the Proterozoic

28. An index fossil should have
A. Large geographic range and small temporal range
B. Small geographic range and large temporal range
C. Small geographic range and small temporal range
D. Large geographic range and large temporal range

29. Match the formations in Group I with corresponding stratigraphic periods in Group II.

Group-I	Group-II
P. *Syringothyris* Limestone	1. Permian
Q. Karai Shale	2. Jurassic
R. Chari	3. Carboniferous
S. Barren Measures	4. Cretaceous

	P	Q	R	S
A.	1	2	3	4
B.	2	4	1	3
C.	3	4	2	1
D.	4	1	2	3

30. Which one of the given statements is correct?
A. van der Waal's bonding is absent in silicate minerals
B. Sulphide minerals form by covalent bonding between metal and sulphur
C. Silicate minerals have a significant component of metallic bonding
D. Metal-sulphide formation does not involve splitting of *d*-orbitals

SECTION-B

Multiple Select Questions (MSQ)

Qs. No. 31–40: *Carry two marks each.*

31. Which of the following structures form in marine environment?
A. Lateral accretionary surfaces
B. Hummocky cross stratification
C. Herringbone cross stratification
D. Barchanoids

32. Identify the correct stratigraphic successions ordered from oldest to youngest.
A. Papaghni Group – Kurnool Group – Nallamalai Group – Chitravati Group
B. Semri Group – Kaimur Group – Rewa Group – Bhander Group
C. Papaghni Group – Chitravati Group – Nallamalai Group – Kurnool Group
D. Semri Group – Rewa Group – Bhander Group – Kaimur Group

33. Which of the following stratigraphic units contain coal seams?
A. Barakar Formation
B. Lakadong Formation
C. Pachmarhi Formation
D. Panchet Formation

34. Which of the following statements are CORRECT?
A. *Mytilus* represents byssally attached bivalves
B. *Nautilus* is the only living cephalopod genus with a coiled external shell

C. The cidaroids are the only echinoid group still living

D. Trilobites did not moult

35. Which of the following genera are stem fossils?

A. *Dadoxylon* B. *Dicroidium*

C. *Vertebraria* D. *Ptilophyllum*

36. Which of the following statements are correct?

A. Abutments are the sides of the valley supporting the dam structure

B. Spillways can control the release of water from the reservoir

C. The toe of the dam is the upstream edge of the base of the dam structure

D. Galleries serve as passages through the dam

37. The acceleration due to gravity on the Earth's surface depends on:

A. Latitude

B. Longitude

C. Elevation

D. Topography of the surrounding terrain

38. A metamorphosed basaltic assemblage can include the minerals

A. Garnet-omphacite

B. Hornblende-plagioclase

C. Garnet-staurolite

D. Glaucophane-lawsonite

39. Which of the following pairs represent correct plutonic - volcanic equivalents?

A. Granodiorite - dacite

B. Norite - basalt

C. Dunite - komatiite

D. Nepheline syenite - Phonolite

40. Based on the given stereographic projection, the fold can be classified as

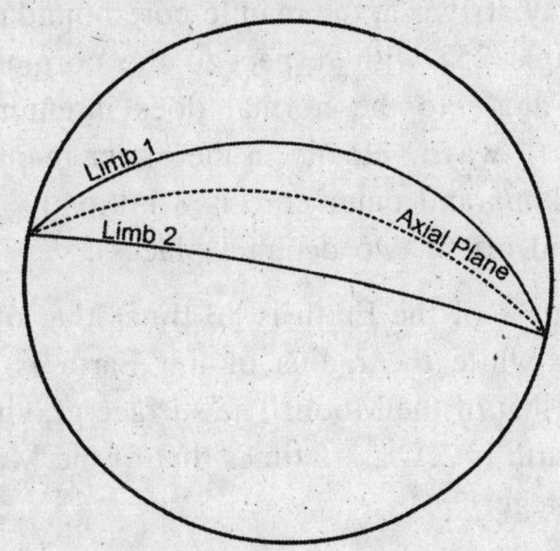

A. Reclined fold B. Vertical fold

C. Overturned fold D. Non-plunging fold

SECTION-C

Numerical Answer Type (NAT)

Qs. No. 41-50: *Carry one mark each.*

41. The Ripple Symmetry Index (RSI) for the given hypothetical asymmetric ripple is .... (Round off to one decimal place)

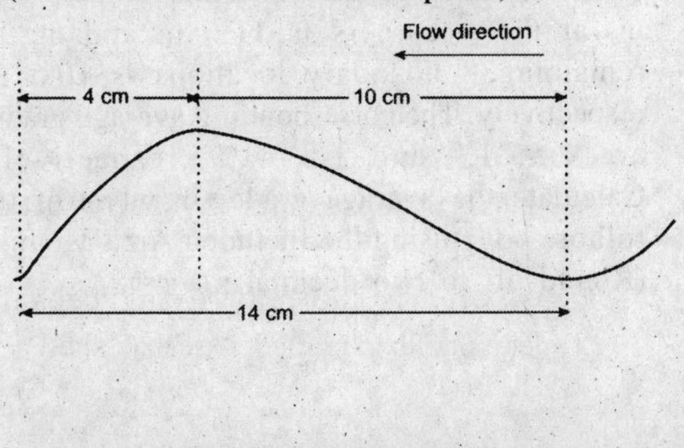

42. Within a fourth order drainage basin, the total lengths of the 1st, 2nd, 3rd and 4th order streams are 10.5 km, 7.5 km, 5.5 km and 1.5 km, respectively. If the drainage density of the basin is 0.5 km^{-1}, the basin area is ...... km^2. (In integer)

43. A soil has a void ratio of 0.5. The total porosity of the soil is .... (Round off to two decimal places)

44. The average unit weight of the uppermost part of the crust is 25000 N/m^3. The vertical stress at a depth of 1 km would be .... MPa. (In integer)

45. The radius of the Earth's circular orbit round the Sun is 149×10^6 km. The Earth takes 365 days to orbit the Sun. The tangential velocity of the Earth is .... km/hour. ($\pi = 3.14$) (Round off to one decimal place)

46. A borehole inclined at 60° to the horizontal pierces a vertical basaltic dyke of uniform thickness. If the length of the basaltic drill core along the core axis is 12 m, the thickness of the dyke is .... m. (In integer).

47. A P-ray arrives at the mantle-core boundary at an angle 25° with respect to the normal. At what angle to the normal does it enter the core? (P-wave velocity in the lower mantle is 13.7 km/s and outer core is 8.1 km/s) (Round off to two decimal places).

48. The mass of the Earth is 80 times that of the Moon while the radius of the Earth is four times that of the Moon. The surface gravity of the Earth is .... times that of the Moon? (In integer)

49. A hypothetical rock contains the assemblage kyanite, sillimanite and quartz. The variance (degree of freedom) of the assemblage is ..... . (In integer)

50. The cut-off grade of copper is 0.45 wt%. A mine has 1 million tonne of waste with a grade of 0.25 wt%. The mine also has stock of high grade ore with a grade of 1.8 wt%. How much of this high grade ore (in million tonne) must be blended with the waste to sell the blended ore at a grade of 0.5 wt%? (Round off to three decimal places)

Qs. No. 51-60: *Carry two marks each.*

51. The maximum and minimum principal stresses in a zone of active normal faulting are 28 MPa and 8 MPa, respectively. The fault plane strikes N30°E and dips 60° towards SE. Considering Anderson's theory of faulting, the normal stress on the fault plane is .... MPa. (In integer)

52. A granite block starts sliding on a slope (inclination of 30° with the horizontal) under the effect of gravity only, along the true direction of inclination of the slope and hits the ground in 4 seconds. Considering zero friction and zero cohesion during sliding, the vertical height of the point (with respect to the ground) from where the block was dislodged is .... m. ($g = 10\ m/s^2$) (In integer)

53. A cylindrical soil sample is encased in an open-ended inclined tube with a diameter of 100 mm. There is a constant supply of water from the upper end of the sample and the outflow from the other end is collected in a beaker. The average amount of water collected is 1000 mm^3 every 10 sec. The average outflow velocity is .... mm/sec. ($\pi = 3.14$) (Round off to three decimal places)

54. Using Airy's hypothesis, calculate the thickness of the root beneath a 4 km high mountain in isostatic equilibrium with a 40 km thick continental crust of density 2800 kg/m^3 and a mantle of density 3300 kg/m^3. Express your answer in km. (Round off to one decimal place)

55. Given atomic weights of Cu, Fe and S as 63.55, 55.85 and 32.10, respectively, find out the weight of copper (in gram) metal in an ore (no associated gangue) of 1 kg weight constituting of bornite, chalcopyrite and chalcocite present in weight fractions of 0.4, 0.4 and 0.2, respectively. (Round off to one decimal place)

56. An ore body defined by a 300 m × 300 m area is shown in the figure in which the drill hole locations on equally spaced square grid are marked (numbers 1-16). The average thickness of the ore body at the 4 interior points is 10.8 m, at the 4 corners is 11.0 m and at the remaining 8 boundary locations is 10.5 m, respectively. The corresponding average grades are 1.5, 1.9 and 1.8 wt%, respectively. Calculate the average grade (in wt%) of the full ore body using the Included Area Method. (Round off to two decimal places).

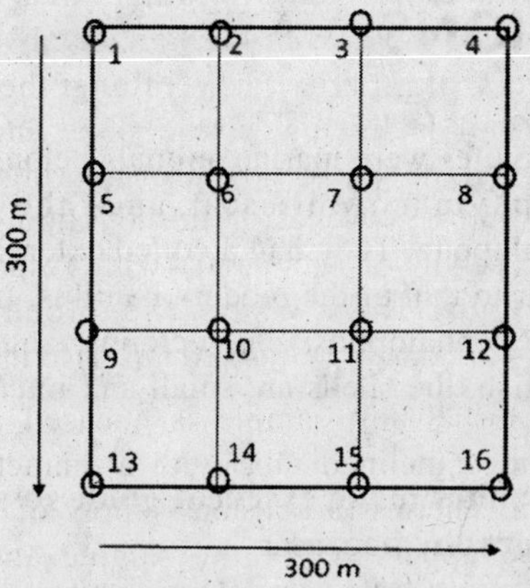

57. The $^{87}Sr/^{86}Sr$ ratio of a 1000 Ma granite was measured as 0.8001. If its $^{87}Rb/^{86}Sr$ ratio is 2.499, what was the Sr isotopic ratio of the source at the time of derivation of the granite? (decay constant of $^{87}Rb = 1.39 \times 10^{-11}\ yr^{-1}$) (Round off to three places of decimals)

58. The coefficients of permeability of two aquifers—1 and 2, are 60 m/day and 40 m/day, respectively. Their saturated thicknesses are 30 m and 15 m, respectively. Assuming steady state Darcian flow, the transmissivity of aquifer 1 is .... times that of aquifer 2. (In integer)

59. Assume that ^{218}Po, with a half-life of 138 days, is in secular equilibrium with ^{238}U whose half-life is 4.5×10^9 y. How many grams of ^{218}Po will be present for each gram of ^{238}U in the mineral? Express your answer in logarithm (to the base 10). (Round off to two decimal places)

60. The figure below is an isobaric binary temperature-composition (T-X) plot. What amount (in %) of the equilibrium melting of rock R will generate a melt of composition L? (Round off to one decimal place)

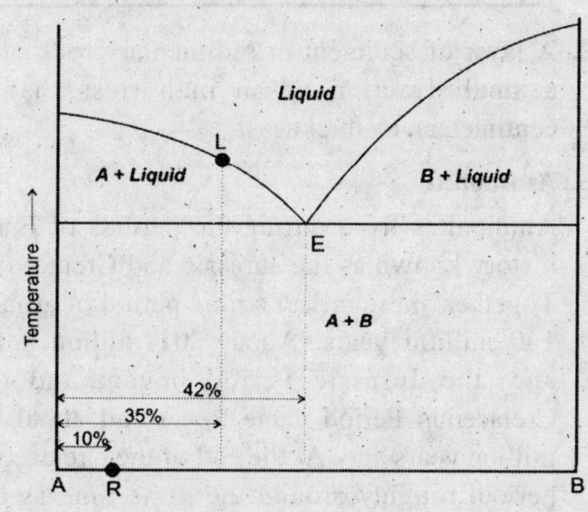

ANSWER KEY

1	2	3	4	5	6	7	8	9	10
B	A	B	A	B	B	A	C	A	B
11	**12**	**13**	**14**	**15**	**16**	**17**	**18**	**19**	**20**
A	B	C	A	A	A	D	D	C	B
21	**22**	**23**	**24**	**25**	**26**	**27**	**28**	**29**	**30**
*	B	C	C	B	*	C	A	C	B
31	**32**	**33**	**34**	**35**	**36**	**37**	**38**	**39**	**40**
A,B,C	B,C	A,B	A,B	A,C	A,B,D	A,C,D	A,B,D	A,D	C,D
41	**42**	**43**	**44**	**45**	**46**	**47**	**48**	**49**	**50**
2.5 to 2.5	50	0.33 to 0.34	25	106817.0 to 106818.0	6	14.35 to 14.50	5	1	0.191 to 0.193
51	**52**	**53**	**54**	**55**	**56**	**57**	**58**	**59**	**60**
13	20	0.012 to 0.013	22.4 to 22.4	550.0 to 553.0	1.67 to 1.68	0.765 to 0.766	3	−10.08 to −10.07	28.0 to 29.0

HINTS AND SOLUTIONS

1.

Chronostratigraphic	Geochronologic
Eonothem	Eon
Erathem	Era
System	Period
Series	Epoch
Subseries	Subepoch
Stage	Age
Substage	Subage

2. A layer of sediment or sedimentary rock only a small fraction of an inch (less than a centimeter) in thickness.

3. **Ammonite**

Ammonites lived during the periods of Earth history known as the Jurassic and Cretaceous. Together, these reflect a time period of around 140 million years. About 201 million years ago, the Jurassic Period began, and the Cretaceous Period came to an end about 66 million years ago. At the end of the Cretaceous period, roughly around the same time as the dinosaurs became extinct, the ammonites also went extinct. But because they are frequently discovered as fossils, produced after the animal's remnants or traces were buried by sediments that eventually crystallised into rock, we know a lot about them.

The subclass Ammonoidea, a group that is often referred to as ammonites, first appeared about 450 million years ago. Ammonoidea includes a more exclusive group called Ammonitida, also known as the true ammonites. These animals are known from the Jurassic Period, from about 200 million years ago. Most ammonites died out at the same time as the non-avian dinosaurs, at the end of the Cretaceous Period, 66 million years ago. A few species continued into the Palaeogene in the Western Interior Seaway before dying out.

Ammonites were marine animals belonging to the phylum Mollusca and the class Cephalopoda. They had a coiled external shell similar to that of the modern nautilus. In other living cephalopods, e.g. octopus, squid and cuttlefish, the shells are small and internal, or absent.

Ammonites make excellent guide fossils for stratigraphy because:

- they evolved rapidly so that each ammonite species has a relatively short life span
- they are found in many types of marine sedimentary rocks
- they are relatively common and reasonably easy to identify
- they have a worldwide geographical distribution
- the rapidity of ammonite evolution is the single most important reason for their superiority over other fossils for the purposes of correlation. Such correlation can be on a worldwide scale.
- Ammonites can be used to distinguish intervals of geological time of less than 200000 years' duration. In terms of Earth history, this is very precise.

Trilobite

The group Trilobita existed from early in the Cambrian Period (520 million years ago) until the end of the Permian Period (250 million years ago). The name Trilobita is derived from the three (tri-) lobed structure of the exoskeleton, which has a raised central lobe (or axis) and a pair of side lobes, called pleurae. The trilobite body is also divided lengthwise into three regions or tagmata: a head or cephalon, a middle region (thorax) composed of several to many articulated segments, and a tail plate called a pygidium, which consists of fused segments.

- They appeared abruptly in the early part of the Cambrian Period and came to dominate the Cambrian and early Ordovician seas. A prolonged decline then set in before they finally became extinct at the end of the Permian Period, about 250 million years ago.
- Like many invertebrate animals living today, including crustaceans, spiders and insects, trilobites were arthropods, belonging to the phylum Arthropoda. Geologists know that they were marine animals because of the rocks in which they are found and the other types of fossils associated with them.

The exoskeleton is divided from anterior (front) to posterior (back) into:

- Cephalon (head-shield)
- Thorax
- Pygidium (tail-shield)
- It is also divided from side-to-side into a central axis, with two side regions (lobes) called pleurae. The word 'trilobite' refers to this side-to-side partition, rather than the cephalon, thorax and pygidium.

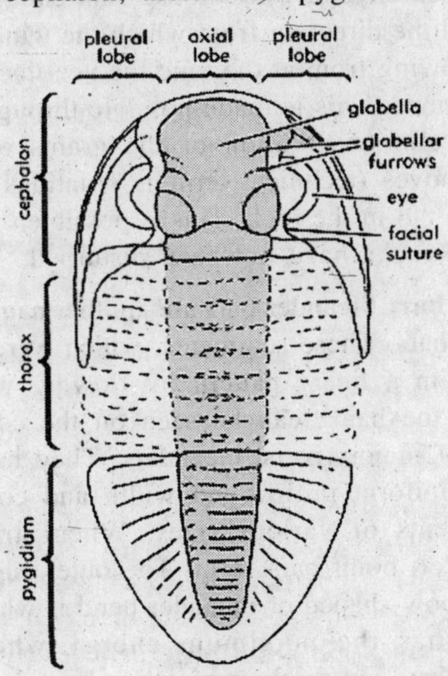

Trilobite

Brachiopod

Brachiopods have a very long history of life on Earth; at least 550 million years. They first appear as fossils in rocks of earliest Cambrian age and their descendants survive, albeit relatively rarely, in today's oceans and seas. They were particularly abundant during Palaeozoic times (248-545 million years ago) and are often the most common fossils in rocks of that age. Brachiopods vary considerably in size. The aptly named Carboniferous genus Gigantoproductus may grow up to 30 cm across. In contrast, the Jurassic rhynchonellid genus Nannirhynchia is minute, generally only two or three millimetres across - the size of a large pin head.

- Brachiopods are marine animals belonging to their own phylum of the animal kingdom, Brachiopoda. Although relatively rare, modern brachiopods occupy a variety of seabed habitats ranging from the tropics to the cold waters of the Arctic and, especially, the Antarctic.
- All but a few brachiopods fall into two basic types, the rhynchonelliform (or articulate) brachiopods and the lingulate (or inarticulate) brachiopods. Rhynchonelliform brachiopods have shells made calcium carbonate and interlocking pegs (teeth) and sockets that form a hinge between the valves. The teeth are in one valve (the pedicle or ventral valve) and the sockets are in the other (the brachial or dorsal valve). Many species were anchored to the sea floor by a muscular stalk (pedicle) which emerged through a hole in the pedicle valve (pedicle opening).

Echinoderm

Since the Late Ordovician, approximately 450 million years ago, echinoids have been in the oceans. This is roughly 220 million years before the emergence of dinosaurs. These animals' remnants and traces were buried in sediment, which subsequently solidified into rock and preserved them as fossils. The well-known sea urchins that live in many shallow coastal waters throughout the world are the living embodiments of echinoids. The fact that some modern sea urchins closely resemble some fossil echinoids enables us to appreciate how they must have lived. Echinoids are marine creatures that are members of the class Echinoidea and phylum Echinodermata. In live echinoids, little knobby bumps called tubercles that are coated in a hard shell called a test are attached with spines. Normally, the pieces are the test and the spines.

4. **Cirque** is a glacial large-scale erosional features common to many mountainous regions. Classic cirques take the form of armchair-shaped hollows, with a steep headwall (which often culminates in a sharp ridge, or arête) and a gently-sloping or overdeepened valley floor.

Ventifact: Simply defined, a ventifact is a rock or a stone that has one or more highly polished and flattened sides that are a direct consequence of sand or ice crystals driven by the wind. These features, which can be eye-catching at times, are mostly found in arid places where there is little to no vegetation to break the wind. Due to the low vegetation, the winds in these places are stronger and more frequent. Another characteristic of environments where ventifacts form is a steady supply of sand but not an overwhelming amount.

- Mostly, ventifacts are formed from hard rocks with fine grains such as quartz, chert and obsidian. As the wind blows towards a rock while carrying its load of sand grains, the rock outcropping acts as a windbreaker. The side of the rock that acts as the windbreaker depends on the direction from which the wind is blowing from. This side becomes abraded over time by the sand particles, which leads to the formation of a ventifact. As one surface is continually exposed to the wind, the rock structure might become unstable over time thus falling over and exposing another side to the wind. The direction of the winds might also change over time leading to a ventifact with multiple facets.
- Depending on the size of the rock, ventifacts have been known to resemble anything from a disc to jaw-dropping shapes such as that of a mushroom. The formation of such a shape is simple. If the rock protrusion is taller than the height where the sand or ice particles are carried, then most of the particles will work on the base of the rock. Very few particles will reach the top, if at all. Over time, a mushroom like shape is formed.
- Depending on the number of facets formed, ventifacts can have different names to show this number. For instance, a ventifact shaped as a pyramid with three facets is referred to as a dreikanter (one of the sides will be round and buried). Some of the stones just have one facet, for example desert pavements.
- It is also possible for experts, or pretty much anyone who is observant enough, to tell the direction from which the wind was blowing from at the time the ventifact was formed. This is made possible through the careful observation of the grains or the grooves (technical term is striations) that are cut in the rock. This is possible only if the rocks have not been disturbed.

Point bar: Meander bars are another name for point bars. These sediments, which were laid down in a linear pattern by flowing waters along the bank, can be seen on the convex side of major rivers' meanders. They have a very uniform profile and width and contain sediments of various sizes. When streams meander, point bars form. The outer edge of an oxbow-shaped or meander bend is where a river has the maximum energy when it meanders, and this is typically where erosion occurs (the cut bank). On the other side of the stream, at the point, the energy cannot carry as much sediment and shifts to depositional flow. The point bar gets worse as the stream erodes the cut bank.

- Point bars are also known as meander bars. They are found on the convex side of meanders of large rivers and are sediments deposited in a linear fashion by flowing waters along the bank. They are almost uniform in profile and in width and contain mixed sizes of sediments. Point bars are formed in meandering streams. When a river meanders the highest energy is near the outer edge of the oxbow shape or meander bend and is generally where erosion takes place (the cut bank). Stream energy on the opposite side (the point) is much lower and cannot carry as much sediment and becomes depositional. As the stream erodes the cut bank the point bar aggrades accordingly.
- If there more than one ridge, narrow and elongated depressions am found in between the point bars. Rivers build a series of them depending upon the water flow and supply of sediment. As the rivers build the point bars on the convex side, the bank on the concave side will erode actively.

Tephra: The term tephra defines all pieces of all fragments of rock ejected into the air by an erupting volcano. Most tephra falls back onto the slopes of the volcano, enlarging it. But, billions of smaller and lighter pieces less than 2 mm diameter (less than one tenth of an inch), termed ash, are carried by winds for thousands of miles. Smaller ejecta such as *lapilli* (2-64 mm) and *ash* (<2 mm) which are convected upward by the heat of the eruption will fall out farther from the volcano.

5. Perthitic texture formed by Exsolution. Perthite is used to describe an intergrowth of two feldspars: a host grain of potassium-rich alkali feldspar (near K-feldspar, $KAlSi_3O_8$, in composition) includes exsolved lamellae or irregular intergrowths of sodic alkali feldspar (near albite, $NaAlSi_3O_8$, in composition).

6. In the holosymmetric class of the Cubic System, four (4) more faces does the {110} form have compared to the {111} form.

Crystal Class in Cubic System: Following are the forms that commonly develop in the crystals belonging to Isometric System:

- **Cube:** A form bounded by six similar square faces, each of which is parallel to two of three crystallographic axes and meets the third axis. Symbol (100).
- **Octahedron:** A form bounded by eight similar faces, each of the shape of an equilateral triangle, each meeting the three crystallographic axes at equal distances. Symbol (111).
- **Dodecahedron:** It is a form with twelve similar faces each of which is parallel to one of the three crystallographic axes and meets the other two at equal distances, Symbol (101).
- **Trisoctahedron (hhl):** A form of twenty four (24) faces; each face meeting two axes at unit length and to the third at greater than unity. Faces occur in eight groups of three each.
- **Trapezohedron (hll):** A form of twenty four (24) faces each face meeting one axes at unit length and to the other two at greater than unity. Each face is a trapezium.
- **Tetra-Hexahedron (hol):** Twenty four (24) faces; each face is parallel to one axis and meets other two at unequal distance which are simple multiple of each other; faces occur in six groups of four each.
- **Hexaocatahedron (hkl):** Forty eight (48) faces; each face meets the three axis at unequal distances.

Beside the normal class, following three classes are also represented among the minerals:

1. Pyritohedral Class (Pyrite Type)
2. Tetrahedral Class (Tetrahedrite Type)
3. Plagiohedral Class (Cuprite Type)

Examples of Isometric Minerals:

- Galena
- Pyrite
- Sphalerite
- Spinel
- Magnetite
- Sodalite
- Garnet
- Diamond
- Flourite
- Leucite.

7. P-Waves (P stands for primary or pressure or push-pull). These waves are also called longitudinal waves or compressional waves due to particle compression during their transport. These waves involve compression and rarefaction of the material as the wave passes through is but not rotation. P-wave is transmitted by particle movement back and forth along the direction of propagation of the wave. The most correct description of P-waves is it is a dilational or irrotational waves. A P-wave is a sound wave travelling through rock. In a P-wave, the rock particles are alternately squished together and pulled apart (called compressions and dilatations), so P-waves are also called compressional waves. These waves can travel through solids, liquids, and gases. P-waves can travel through the liquid outer core.

S-Waves: An S-wave is a different beast. In an S-wave, the rock particles slide past one another, undergoing shear—so an S-wave is also called a shear wave. You can make shear waves by, for example, tying a rope to a tree and shaking the free end of the rope up and down or side-to-side. The waves themselves will travel forward, toward the tree. But the rope particles will stay in one place, sliding back and forth past each other. Shear waves cannot travel in liquids or gases—so, for example, S waves don't travel through the ocean or through the outer core.

Surface waves are called surface waves because they are trapped near the Earth's surface, rather than travelling through the "body" of the earth like P and S waves. There are two major kinds of surface waves: Love waves, which are shear waves trapped near the surface, and Rayleigh waves, which have rock particle motions that are very similar to the motions of water particles in ocean waves.

8. Colour because Realgar is red to orange and orpiment is yellow, golden, yellow-brown.

Identification (realgar)

Colour	Red to yellow-orange; in polished section, pale gray, with abundant yellow to red internal reflections
Crystal habit	Prismatic striated crystals; more commonly massive, coarse to fine granular, or as incrustations
Twinning	Contact twins on {100}
Cleavage	Good on {010}; less so on {101}, {100}, {120}, and {110}
Tenacity	Sectile, slightly brittle
Mohs scale hardness	1.5-2
Luster	Resinous to greasy
Streak	Red-orange to red
Diaphaneity	Transparent
Specific gravity	3.56
Optical properties	Biaxial (-)
Refractive index	$n_\alpha = 2.538$ $n_\beta = 2.684$ $n_\gamma = 2.704$
Birefringence	$\delta = 0.166$
Pleochroism	Nearly colourless to pale golden yellow
2V angle	40°
Dispersion	r > v, very strong
Other characteristics	Toxic and carcinogenic. Disintegrates on long exposure to light to a powder composed of pararealgar or arsenolite and orpiment.

Orpiment: Identification properties

Identification (Orpiment)

Colour	Lemon-yellow to golden or brownish yellow
Crystal habit	Commonly in foliated columnar or fibrous aggregates; may be reniform or botryoidal; also granular or powdery; rarely as prismatic crystals
Twinning	On {100}
Cleavage	Perfect on {010}, imperfect on {100};
Tenacity	Sectile
Mohs scale hardness	1.5-2
Luster	Resinous, pearly on cleavage surface
Streak	Pale lemon-yellow
Diaphaneity	Transparent
Specific gravity	3.49
Optical properties	Biaxial (–)
Refractive index	n_α = 2.400 n_β = 2.810 n_γ = 3.020
Birefringence	δ = 0.620
Pleochroism	In reflected light, strong, white to pale gray with reddish tint; in transmitted light, Y = yellow, Z = greenish yellow
2V angle	Measured: 30° to 76°, Calculated: 62°
Dispersion	r > v, strong

9. Layer parallel shortening because Buckling occurs when the deforming force is applied parallel to rock layers. This is usually caused by horizontal compressional tectonic forces and results in layer-parallel shortening of rocks and thickening (relief) of the rock body perpendicular to stress direction.

- Single layer situation where a more competent layer is surrounded by an incompetent material. The former is free to form buckle folds that have a parallel character, while the former will show another shape.
- Biot (1957)—correspondence principle between elastic and viscoelastic materials (Newtonian viscosity), looks at stress-strain relationships in buckled layer.
- Ramberg (1959)—approaches problem as one of fluid dynamics - particular function which satisfies both the biharmonic equation of a stream function and boundary conditions around a buckled layer.

10. Sandstone beds above a magmatic body are domal in shape, while the beds below are horizontal. The magmatic body is a Laccolith.

- Some upward-moving magma reaches the surface, resulting in volcanic eruptions, but most cools within the crust. The resulting body of rock is known as a pluton. Plutons can have various different shapes and relationships to the surrounding country rock.
- Large irregular-shaped plutons are called either stocks or batholiths. The distinction between the two is made on the basis of the area that is exposed at the surface: if the body has an exposed surface area greater than 100 km^2, then it's a batholith; smaller than 100 km^2 and it's a stock. Batholiths are typically formed only when a number of stocks coalesce beneath the surface to create one large body. One of the largest batholiths in the world is the Coast Range Plutonic Complex, which extends all the way from the Vancouver region to southeastern Alaska.
- Tabular (sheet-like) plutons are distinguished on the basis of whether or not they are concordant with (parallel to) existing layering (e.g., sedimentary bedding or metamorphic foliation) in the country rock. A sill is concordant with existing layering, and a dyke is discordant. If the country rock has no bedding or foliation, then any tabular body within it is a dyke. Note that the sill-versus-dyke designation is not determined simply by the orientation of the feature. A dyke can be horizontal and a sill can be vertical (if the bedding is vertical).
- A laccolith is a sill-like body that has expanded upward by deforming the overlying rock.

- Finally, a pipe is a cylindrical body (with a circular, elliptical, or even irregular cross-section) that served as a conduit for the movement of magma from one location to another. Most known pipes fed volcanoes, although pipes can also connect plutons. It is also possible for a dyke to feed a volcano.

11. **Pedicle Foramen:** A rounded, slit-like, or closed opening in the pedicle valve of a brachiopod (Brachiopoda) for the passage of the pedicle to the exterior.

Pallial Sinus: The pallial sinus is an indentation or inward bending in the pallial line on the interior of a bivalve mollusk shell's valves that corresponds to the position of the siphons in those types of clams which have siphons (i.e. siphonate). The position of the pallial sinus is often clearly visible as a shiny line on the inside of the bivalve shell. The pallial sinus is an etched "bent line" present on the inside of some bivalve shells, at their posterior end. The sinus is a continuation of the pallial line, a scar that indicates the limit of the bivalve body that is attached internally to its shell. The pallial sinus corresponds to a "pouch" in the animal body that houses the siphons when they are retracted. The siphons are fleshy tubes that can be expanded for the intake and outtake of water for respiration and feeding. Usually, bivalves that burrow deep in the sand or mud have long siphons; the longer the siphons, the deeper the pallial sinus.

Pygidium: Part of the trilobite. The pygidium is a tail piece composed of a variable number of fused segments at the posterior end of a trilobite. While the cephalon of a trilobite is typically wider than the thorax, and one of the single largest major features of a trilobite body, the pygidium can range from extremely small to larger than the cephalon.

Siphuncle: Found in Pelecypoda. The siphuncle is a strand of tissue passing longitudinally through the shell of a cephalopod mollusk. Only cephalopods with chambered shells have siphuncles, such as the extinct ammonites and belemnites, and the living nautiluses, cuttlefish, and Spirula. In the case of the cuttlefish, the siphuncle is indistinct and connects all the small chambers of that animal's highly modified shell; in the other cephalopods it is thread-like and passes through small openings in the septa (walls) dividing the camerae (chambers). Some older studies have used the term siphon for the siphuncle, though this naming convention is uncommon in modern studies to prevent confusion with a mollusc organ of the same name.

12. **Dharmaram Formation**

The Upper Dharmaram Formation is an Early Jurassic geologic formation found in Andhra Pradesh, India. It is of Sinemurian age (Triassic-Jurassic), and dinosaur remains are among the fossils that have been recovered from the formation.

13. Quartz > Muscovite > Amphibole > Olivine

- Minerals in rocks react with their new environment to produce new minerals that are stable under conditions near the surface. Minerals that are stable under P, T, H_2O, and O_2 conditions near the surface are, in order of most stable to least stable:
- Iron oxides, Aluminum oxides - such as hematite Fe_2O_3, and gibbsite $Al(OH)_3$.

Bowen's reaction series:

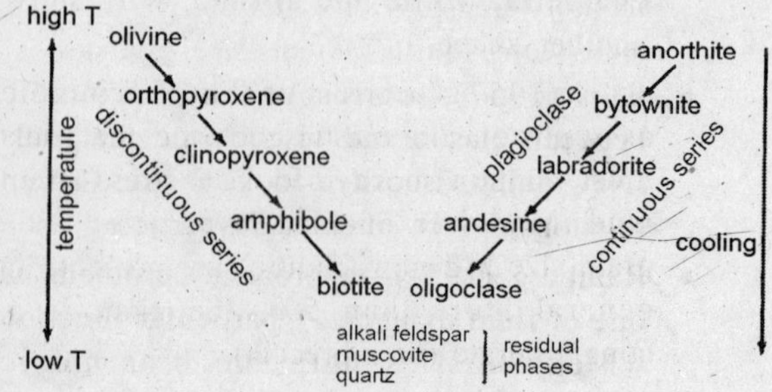

14. **Packstone** is defined as a grain-supported carbonate rock that contains 1% or more mud-grade fraction.

A grainstone is a carbonate rock in which grains larger than 0.25 mm comprise more than 10 volume% of the rock and have a matrix of sparry calcite and having no mud.

Lithic arenite is sedimentary rock composed of grains 0.06-2 mm [0.0024-0.08 inch] in diameter containing over 25 percent rock fragments and less than 15% matrix.

Arkosic wacke comprising more than 5% feldspar, and more feldspar than rock fragments, with more than 15% matrix.

Classification of Sandstone

Sandstones are a type of clastic (terrigenous) sedimentary rocks deriving, as the name suggest, from the consolidation of sandy sediments.

- Sand consists of clastic particles produced by the erosion of pre-existing rocks with grain size comprised between 0.0625 and 2 mm.
- The upper boundary of 2 mm arbitrarily separates sand (and sandstones) from the coarser gravel and their lithified counterparts: conglomerate and breccia.
- The lower boundary set at 0.0625 mm may apparently sound arbitrary but represents the limit of the human eye to discern grains. Indeed; our eye can only recognize grains larger than 0.0625 mm in diameter.
- Smaller particles, known as mud (silt + clay), are not visible by the human eye and the rocks consisting of them, like mudstones, appear homogeneous to us. To qualify itself as sandstone, a clastic sedimentary rock must contain more than 25% sand over mud. Moreover, if enough particles with grain size > 2 mm (ruditic) are present, in general more than 5%, the rock is a conglomerate (or a breccia).
- There are more than 50 classification schemes for sandstones but the diagram by Dott (1964) is the most widely used.
- This classification can be used on sand and sandstones. It is based on
  - ❑ the percentage of matrix, defined as the grain size < 30 μm and
  - ❑ the proportion of quartz, feldspar, and lithic fragments in the > 30 μm framework grains.
- If the percentage of matrix > 75%, the rock is a mudstone. Sandstones are divided in arenites (< 15% matrix) and wackes or graywackes (> 15% matrix).
- Arenites and wackes can be further classified based on their composition in terms of quartz (Q), feldspar (F), and rock or lithic fragments (L).
- If quartz > 95% (F + L < 5%), they can be classified as quartzarenite and quartzwacke respectively.
- Arkosicarenite and arkosic wacke occur when feldspars dominate, whereas if lithic fragments are more abundant, sandstones are classified as lithic arenite and lithic wacke.
- Feldspathic can be used instead of arkosic. Further subdivisions of arenites are subarkose (quartz between 75 and 95%, feldspar > lithic fragments) and sublitarenite (quartz between 75 and 95%, lithic fragments > feldspar).

15. **Shear modulus:** The shear modulus is defined as the ratio of shear stress to shear strain. It is also known as the modulus of rigidity and may be denoted by G or less commonly by S or μ. The SI unit of shear modulus is the Pascal (Pa), but values are usually expressed in gigapascals (GPa). In English units, shear modulus is given in terms of pounds per square inch (PSI) or kilo (thousands) pounds per square in (ksi).

- A large shear modulus value indicates a solid is highly rigid. In other words, a large force is required to produce deformation.
- A small shear modulus value indicates a solid is soft or flexible. Little force is needed to deform it.
- One definition of a fluid is a substance with a shear modulus of zero. Any force deforms its surface.
- The shear modulus is determined by measuring the deformation of a solid from applying a force parallel to one surface of a solid, while an opposing force acts on its opposite surface and holds the solid in place. Think of shear as pushing against one side of a block, with friction as the opposing force. Another example would be attempting to cut wire or hair with dull scissors.

The equation for the shear modulus is:

$G = \tau_{xy}/\gamma_{xy} = F/A \ / \ \Delta x/l = Fl \ / \ A\Delta x$

Where:

- G is the shear modulus or modulus of rigidity
- τ_{xy} is the shear stress
- γ_{xy} is the shear strain
- A is the area over which the force acts
- Δx is the transverse displacement
- l is the initial length

Shear strain is $\Delta x/l = \tan\theta$ or sometimes $= \theta$, where θ is the angle formed by the deformation produced by the applied force.

Hydraulic conductivity: "Hydraulic Conductivity" (K), in hydrogeology and hydrology, represents the capacity of a porous medium (such as soil) to transmit water, as per Darcy's Law. To illustrate, the rate at which fluid can move through a permeable medium (such as soil), depends on the properties of that soil (such as intrinsic permeability) and the fluid (such as viscosity and specific weight).

Hydraulic Conductivity via Darcy's Law

The principles and theory of Darcy's Law are the same in all fields. However, in the field of geology and hydrogeology, the following formulas apply to represent Darcy's Law. Solving for "K" in each of these formulas yields the value for "Hydraulic Conductivity."

$Q = K \bullet A\ [(h_1 - h_2) \div L]$

The following variables and coefficients apply to this formula: "K" = Hydraulic Conductivity; "Q" = Discharge Rate; "A" = Area of Cross-Section that Water Flows; "h" = Hydraulic Head ($h = p/\rho g + z$); and "L" = Length/Distance of Water Table Elevation Change.

$q = Q \div A = -K\ [(dh) \div (dl)]$

In this formula, the following variables apply: "K" = Hydraulic Conductivity; "Q" = Discharge Rate; "q" = Specific Discharge [L/T] ($q = Q/A$); "dh/dl" = Hydraulic Gradient; and "K" = Hydraulic Conductivity.

Volumetric strain: When an elastic body is subjected to external deforming forces such that there is a change in volume of the body, then the ratio of change in volume to original volume is called volumetric strain.

Mathematically it can be represented as,

$$eV = \Delta V/V$$

Volumetric Strain of Rectangular Bar or Prismatic Member

The initial volume of the rectangular bar,

$$V = L \times B \times D$$

Then,

Change in volume of the rectangular bar,

$\Delta V = (\Delta l \times B \times D) + (\Delta b \times L \times D) + (\Delta d \times L \times B)$

So,

The volumetric strain of rectangular bar (eV),

$\Delta V/V = (\Delta l/L) + (\Delta b/B) + (\Delta d/D)$

So it can be written as,

$eV = \Sigma xx + \Sigma yy + \Sigma zz$

Where,

Σxx = Strain in x-x direction

Σyy = Strain in y-y direction

Σzz = Strain in z-z direction

As we know,

$\Sigma xx = (\sigma x/E) + (\sigma y/E) - \mu(\sigma z/E)$

$\Sigma yy = (\sigma y/E) - \mu(\sigma x/E) - \mu(\sigma z/E)$

$\Sigma zz = (\sigma z/E) - \mu(\sigma x/E) - \mu(\sigma y/E)$

Now putting these value,

$eV = \{(\sigma x/E) + (\sigma y/E) - \mu(\sigma z/E)\} + \{(\sigma y/E) - \mu(\sigma x/E) - \mu(\sigma z/E)\} + \{(\sigma z/E) - \mu(\sigma x/E) - \mu(\sigma y/E)\}$

After calculating these values the final result will be obtained.

$eV = \{(\sigma x + \sigma y + \sigma z)/E\} \times (1 - 2\mu)$

If loading is triaxial with equal and alike stresses then,

$\sigma x = \sigma y = \sigma z = ?$

So,

$eV = \{(3\sigma)/E\} \times (1 - 2\mu)$

As we know,

$E = 3K \times (1 - 2\mu)$

So,

$eV = \sigma/K$

Dry density: Moisture content of soil is defined as the ratio of mass of water to the mass of solids present in the soil sample. It is represented by "w".

$$w = \frac{\text{Mass of water}}{\text{Mass of solids}} = \frac{M_w}{M_s}$$

Dry density of soil is defined as the ratio of Mass of solids to the total volume of the soil. It is represented by "ρ_d"

$$\rho_d = \frac{\text{Mass of solids}}{\text{Total volume of soil}} = \frac{M_s}{V}$$

To obtain relationship between the moisture content and dry density, multiply the numerator and denominator of expression of dry density with "M" which is mass of soil sample.

$$\rho_d = \frac{M_s}{V} \cdot \frac{M}{M} = \frac{M}{V} \cdot \frac{M_s}{M}$$

Mass of the soil (M) is nothing but the addition of Mass of solids (M_s) and Mass of water (M_w).

$$\rho_d = \frac{\frac{M}{V}}{\frac{M_s + M_w}{M_s}}$$

Ratio of mass to the volume is bulk density of soil which is denoted as "ρ". Therefore,

$$\rho_d = \frac{\rho}{1 + w}$$

Hence to obtain dry density, we need bulk density and moisture content of soil sample.

16. Subduction Zone: Example is Indonesia. Where two tectonic plates converge, if one or both of the plates is oceanic lithosphere, a subduction zone will form. An oceanic plate will sink back into the mantle. Remember, oceanic plates are formed from mantle material at midocean ridges. Young oceanic lithosphere is hot and buoyant (low density) when it forms at a midocean ridge. But as it spreads away from the ridge and cools and contracts (becomse denser) it is able to sink into the hotter underlying mantle. There is a deep ocean trench where the oceanic plate bends downward.

- **Volcanic Arcs:** The basaltic ocean crust contains hydrous minerals like amphiboles, some of which formed by hydrothermal alteration as seawater seeped through hot, fractured, young ocean crust at the mid-ocean ridge. As the ocean crust sinks deeper into the mantle the pressure increases (the temperature of the ocean crust rocks increases more slowly because rocks are poor conductors of heat). At depths of around 100 km beneath the surface, the pressure is great enough for the hydrous minerals to undergo metamorphism. The resulting minerals are denser and they don't contain the bonded water. This metamorphic dewatering process liberates water from the descending crust. The water gradually seeps upward into the overlying wedge of hot mantle. The addition of water to the already hot mantle rocks lowers their melting temperature resulting in partial melting of ultramafic mantle rocks to yield mafic magma. Melting aided by the addition of water or other fluid is called flux melting. It is somewhat more complicated than this, but metamorphic dewatering of suducting crust and flux melting of the mantle wedge appears to account for most of the magma at subduction zones.

- Magma formed above a subducting plate slowly rise into the overriding crust and finally to the surface forming a volcanic arc, a chain of active volcanoes which parallels the deep ocean trench. Beneath the active volcanic arc lie intrusive igneous rocks formed from magma that didn't make it all the way to the surface before crystallizing. The volcanic arcs may be volcanic island arcs (e.g., Aleutians, Mariannas), where one oceanic plate subducts beneath another oceanic plate, or continental volcanic arcs (e.g., Andes, Cascades), where oceanic plates subduct under a continental plate. The most abundant igneous rock formed at volcanic arcs is andesite (or intrusive diorite), though volcanic arc rocks may range in composition from basalt to rhyolite (mafic to felsic).
- **Benioff Zones:** Earthquakes in and around deep ocean trenches are principally produced by motions on thrust faults, indicating compression (converging plates). A plane of earthquake focci descend from the area around the trench underneath the overriding plate. The farther from the trench, the deeper the earthquakes are. These earthquakes of the Benioff Zone (or Wadati-Benioff Zone) occur near the upper surface of the descending plate (or slab). They occur down to depths of around 670 km at some subduction zones.
- **Note:** The volcanic arc lies where the Benioff Zone earthquakes are around 100 km beneath the surface but Benioff zone earthquakes continue past this, landward down to 60 km; therefore the slab has not melted away!

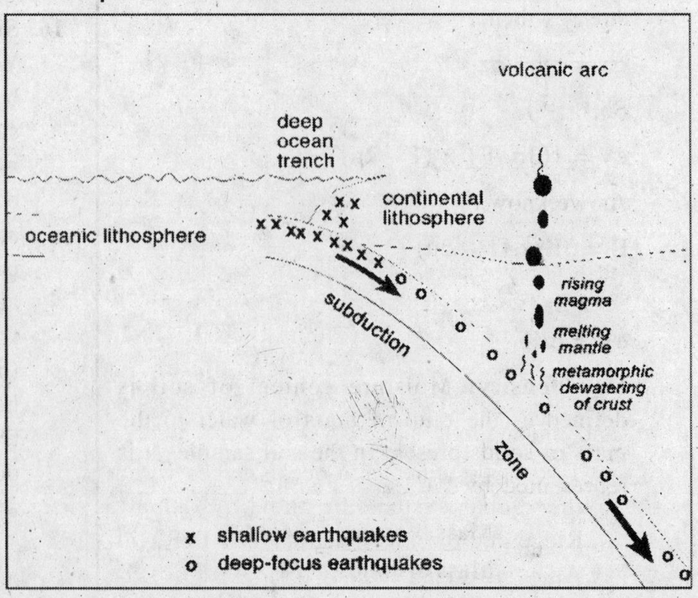

Transform Fault: A transform fault is a type of strike-slip fault wherein the relative horizontal slip is accommodating the movement between two ocean ridges or other tectonic boundaries. They are connected on both ends to other faults. A transform fault or transform boundary, also known as conservative plate boundary since these faults neither create nor destroy lithosphere, is a type of fault whose relative motion is predominantly horizontal. Example is New Zealand.

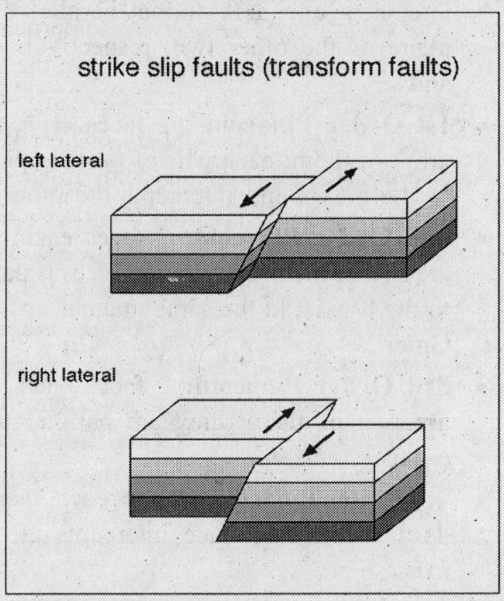

Mid-Oceanic Ridge: Iceland, Mid-ocean ridges occur along divergent plate boundaries, where new ocean floor is created as the Earth's tectonic plates spread apart. As the plates separate, molten rock rises to the seafloor, producing enormous volcanic eruptions of basalt. The speed of spreading affects the shape of a ridge - slower spreading rates result in steep, irregular topography while faster spreading rates produce much wider profiles and more gentle slopes.

- Two well-studied mid-ocean ridges within the global system are the Mid-Atlantic Ridge and the East Pacific Rise. The Mid-Atlantic Ridge runs down the center of the Atlantic Ocean, slowly spreading at a rate of 2 to 5 centimeters (0.8 to 2 inches) per year and forming a rift valley that is about the depth and width of the Grand Canyon. In contrast, the East Pacific Rise is spreading quickly, at rates of 6 to 16 centimeters (3 to 6 inches) per year. Due to the fast spreading rates, there is no rift valley in the Pacific, just a smooth volcanic summit with a crack along the crest that is much smaller than the Atlantic rift valley.
- The massive mid-ocean ridge system is a continuous range of underwater volcanoes that wraps around the globe like seams on a baseball, stretching nearly 65,000 kilometers (40,390 miles). The majority of the system is underwater, with an average water depth to the top of the ridge of 2,500 meters (8,200 feet).
- Dynamic geological processes are continuously reshaping the Earth's surface. From the creation of new igneous rock comprised of volcanic lava, to the formation of new sedimentary rock via the compaction of rock fragments, geological processes are constantly transforming Earth's features. While many of these processes can be observed, other processes occur beneath the Earth's surface.

Continental Collision: Nepal, Continental collision occurs as two continental tectonic plates run into each other. One of the colliding continental plates will typically dive beneath the other. Unlike a collision involving an oceanic plate, though, the other continental plate tends to break and is forced upward. This upward force occurs during a continental collision because these plates tend to be low in density. Anything low in density will have a tendency to lift upward during a collision.

- The collision creating the mountains incorporates a finite volume of rock that is not augmented following the collision.

- The orogenic rock mass is subject to isostatic uplift during denudation; in general, sedimentary rock types are exposed first, followed by crystalline varieties.
- The collision that initiates such orogenesis ultimately adds rock to the adjacent craton, and in thickening the adjacent crust often initiates nearby cratonic tilting and/or uplift.
- Because such mountains develop between continents and are thus elevated in the midst of a consequent megacontinent (Pangaea in the case of the Appalachians), they are far from oceanic evaporation sources and therefore often undergo initial denudation under arid geomorphic systems in the manner of the present mountains of central Asia.
- As the climatic setting of such mountains is largely established tectonically, it may endure in the same climate for scores of millions of years and, as noted in 1901 by the American geomorphologist Douglas W. Johnson, a desert mountain range tends to bury itself in its own waste.
- Re-exposure of such mountains to nearby precipitation sources by plate adjustments may result in dramatic climate changes from arid to humid, so that perennial fluvial erosion is widely initiated on a relict arid, alluvial cover mass with resulting transverse drainage by superimposition. In illustration, one can compare the Appalachian Mountains of North America and the Zagros Mountains of Iran, as described by the American geomorphologist Theodore M. Oberlander in 1965.

17. Moment magnitude scale (M_W) is a measure of an earthquake's magnitude ("size" or strength) based on its seismic moment.

18. Order in the Triclinic System:

In designating forms in the Triclinic System, most crystallographers use 1st through 4th order labels. 1st, 2nd, and 3rd order labels are applied to pinacoids that intercept two of the axes and are parallel to the third axis. 1st order denotes forms parallel to the a-axis. 2nd order is parallel to the b-axis. 3rd order is parallel to the c-axis. And 4th order intercepts all three axes, not being parallel to any of them.

Triclinic Pinacoid: 2 faces, parallel to each other

- **Front, Side, Top Pinacoids:** 2 faces, there are three pinacoids; each pair of faces intersects one axis and is parallel to the plane of the other two, respectively a, b, and c.
- **1st Order Pinacoids:** 2 faces each, there are 2 of them; each pair of faces is parallel to the a-axis and intercepts the other two.
- **2nd Order Pinacoid:** 2 faces each, there are two of them, each pair of faces is parallel to the b-axis; in the same manner as for 1st Order.
- **3rd Order Pinacoid:** 2 faces each, there are two of them; they are parallel to the c-axis.
- **4th Order Pinacoid:** 2 faces each, there are 4 of them; each face intercepts all three axes.

Pedion: 1 face - There is not a similar face on the opposite side of the crystal.

The symbols for the pinacoids and the pedions are the same. The pedion is considered to be derived as half of a pinacoid. In place of each possible pinacoid there are two possible pedions.

A pedion is an open, one faced form. Pedions are the only forms that occur in the Pedial class (1). Since a pedion is not related to any other face by symmetry, each form symbol refers to a single face.

The table below shows the 32 crystal classes, their symmetry, Hermann-Mauguin symbol, and class name.

Crystal System	Crystal Class	Symmetry	Name of Class
Triclinic	1	none	Pedial
	$\bar{1}$	i	Pinacoidal
Monoclinic	2	$1A_2$	Sphenoidal
	m	1m	Domatic
	2/m	i, $1A_2$, 1m	Prismatic
Orthorhombic	222	$3A_2$	Rhombic-disphenoidal
	mm^2 (2mm)	$1A_2$, 2m	Rhombic-pyramidal
	$2/m^2/m^2/m$	i, $3A_2$, 3m	Rhombic-dipyramidal
Tetragonal	4	$1A_4$	Tetragonal- Pyramidal
	$\bar{4}$	$\bar{A}_4$	Tetragonal-disphenoidal
	4/m	i, $1A_4$, 1m	Tetragonal-dipyramidal
	422	$1A_4$, $4A_2$	Tetragonal-trapezohedral
	4mm	$1A_4$, 4m	Ditetragonal-pyramidal
	$\bar{4}2m$	$1\bar{A}_4$, $2A_2$, 2m	Tetragonal-scalenohedral
	$4/m^2/m^2/m$	i, $1A_4$, $4A_2$, 5m	Ditetragonal-dipyramidal
Hexagonal	3	$1A_3$	Trigonal-pyramidal
	$\bar{3}$	$1\bar{A}_3$	Rhombohedral
	32	$1A_3$, $3A_2$	Trigonal-trapezohedral
	3m	$1A_3$, 3m	Ditrigonal-pyramidal
	$\bar{3}2/m$	$1\bar{A}_3$, $3A_2$, 3m	Hexagonal-scalenohedral
	6	$1A_6$	Hexagonal-pyramidal
	$\bar{6}$	$1\bar{A}_6$	Trigonal-dipyramidal
	6/m	i, $1A_6$, 1m	Hexagonal-dipyramidal
	622	$1A_6$, $6A_2$	Hexagonal-trapezohedral
	6mm	$1A_6$, 6m	Dihexagonal-pyramidal
	$\bar{6}m^2$	$1\bar{A}_6$, $3A_2$, 3m	Ditrigonal-dipyramidal
	$6/m^2/m^2/m$	i, $1A_6$, $6A_2$, 7m	Dihexagonal-dipyramidal
Isometric	23	$3A_2$, $4A_3$	Tetaroidal
	$2/m\bar{3}$	$3A_2$, 3m, $4\bar{A}_3$	Diploidal
	432	$3A_4$, $4A_3$, $6A_2$	Gyroidal
	$\bar{4}3m$	$3\bar{A}_4$, $4A_3$, 6m	Hextetrahedral
	$\bar{4}/m\bar{3}2/m$	3A4, $4\bar{A}_3$ $6A_2$, 9m	Hexoctahedral

19. The apparent dip of a plane is measured to be 45° towards NE. The true dip of the plane is 48° towards ENE.

True dip: When a vertical cross-section is perpendicular to the strike of the beds, the inclination seen in the cross section is called the true dip.

Apparent dip: Apparent dip is the inclination of geologic beds as seen from any vertical cross section not perpendicular to the strike of the geologic beds.

Calculating Apparent Dip: To calculate apparent dip, use the following formulas:

$$\tan(d_a) = \tan(d_t) \times \sin(S_{xs} - S_b)$$

or

$$\tan(d_a) = \tan(d_t) \times \cos(S_{xs} - T_{dt})$$

Where:

- d_a is the apparent dip,
- d_t is the true-dip of the beds,
- S_b is the strike of the beds,
- T_{dt} is the true dip direction, and
- S_{xs} is the bearing of your cross-section.

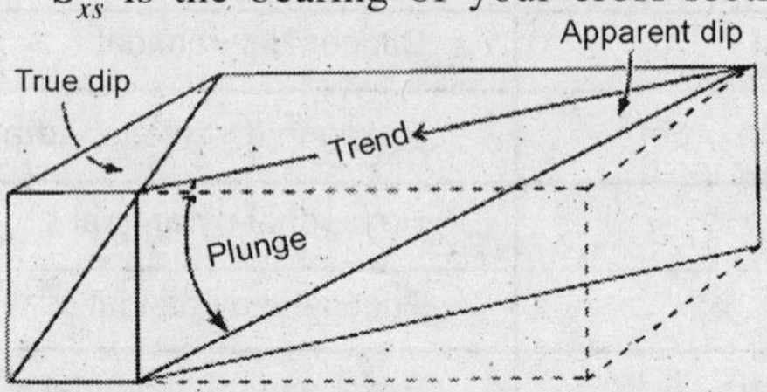

21. Given, The displacement of the bed shown in the figure below is caused by a single movement along fault XYZ.

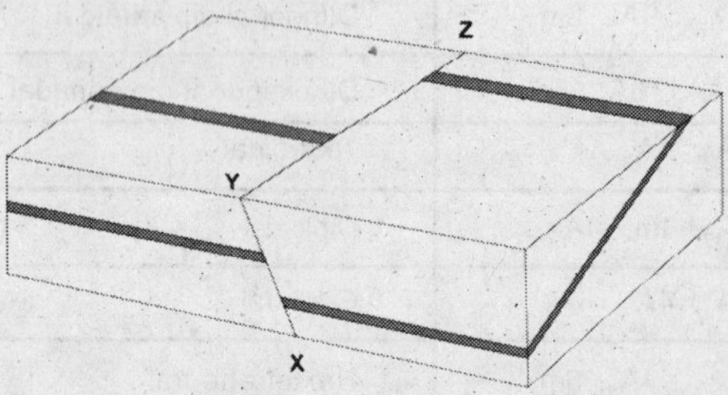

Reverse fault because black band on the right side are come closer it means right block go upward relative to left block.

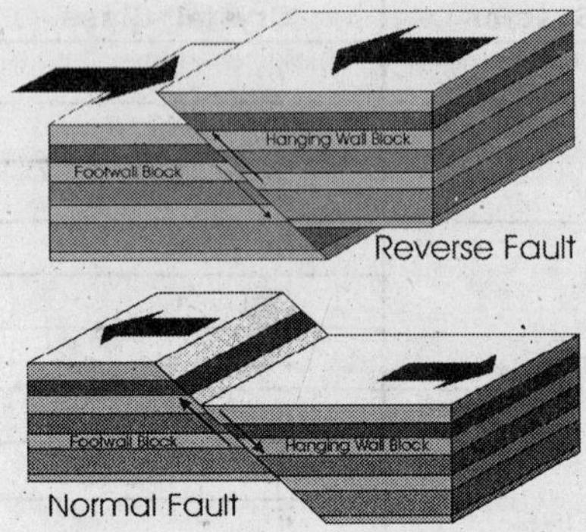

22. Given, In the AFM diagram below, W, X, Y and Z represent four pelitic assemblages.

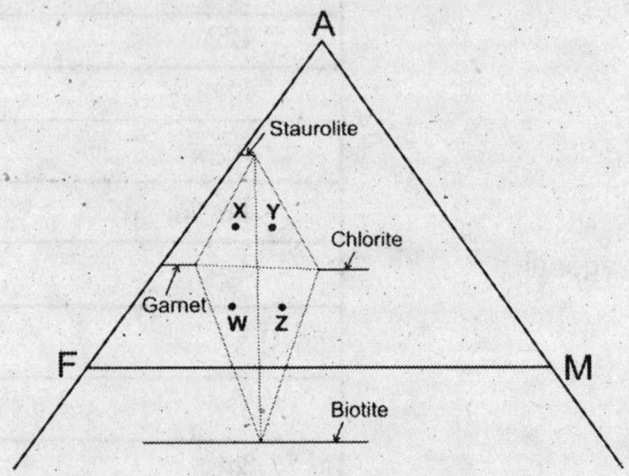

(X-W) & (Y-Z) combinations of assemblages contain the same mineral phases in equilibrium in the staurolite zone.

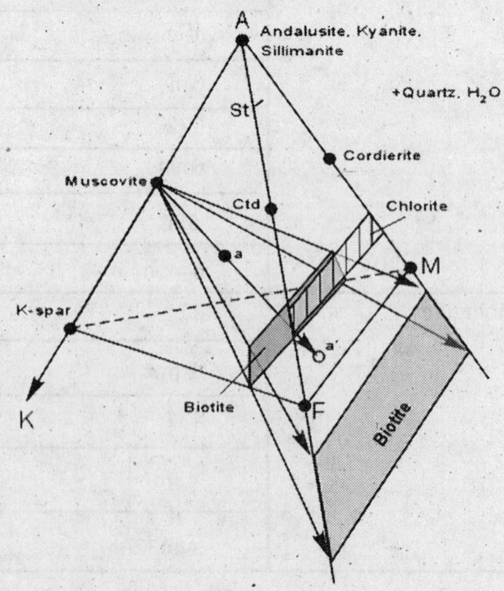

23. Diorite is an intrusive igneous rock composed principally of the silicate minerals plagioclase feldspar (typically andesine), biotite, hornblende, and sometimes pyroxene. The chemical composition of diorite is intermediate, between that of mafic gabbro and felsic granite.

Tonalite is an intermediate igneous rock with essential quartz (>20%) and plagioclase feldspar (oligoclase or andesine), mafic minerals (biotite, pyroxene, hornblende) and minor alkali feldspar (<10% vol).

Norite: A coarse-grained, basic igneous rock consisting of essential plagioclase feldspar, orthopyroxene (hypersthene or bronzite), and clinopyroxene (augite), with accessory ilmenite.

Lherzolite consists mainly of olivine, orthopyroxene, clinopyroxene, phlogopite, minor serpentine, carbonate and accessory chrom spinel, pentlandite, apatite, magnetite.

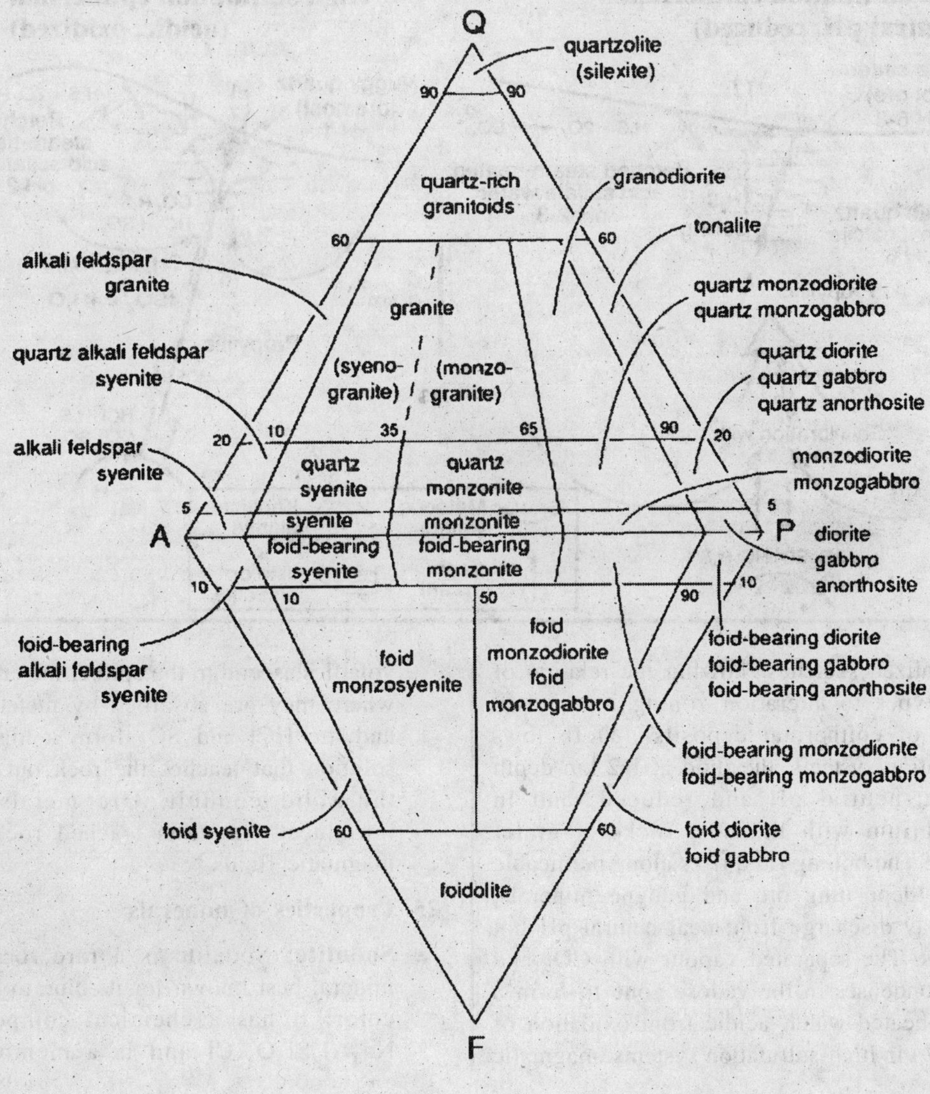

24. Porphyry copper deposits are formed by aqueous fluids exsolved by differentiated, mantle-derived magmas variably mixed with crustal melts.

- Mississippi Valley Type deposits are formed in connate water.
- Uranium occurs in two valence states, U^{4+} and U^{6+}. Weathering of rocks converts uranium into the +6 state, in which state it forms the uranyl ion $(UO_2)^{2+}$. Uranyl compounds tend to be soluble in groundwater, whereas U^{4+} compounds are not. So long as the groundwater remains oxidizing, uranyl ions are stable and uranium can be transported by groundwater; however, when uranyl ions encounter a reducing agent such as organic matter, U^{4+} uranium is precipitated as uraninite and coffinite.

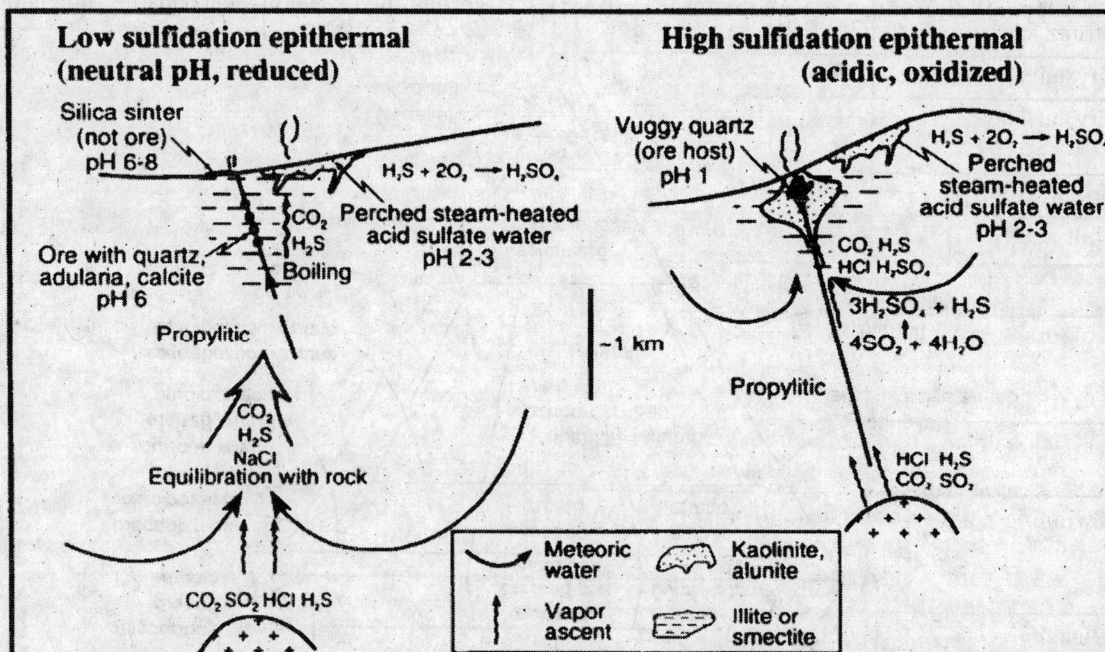

Generalized sketches showing the relation of fluid types to alteration zoning in the two styles of epithermal deposits. (*a*) In low-sulfidation systems, the fluid at 1-2 km depth is near-neutral pH and reduced, and in equilibrium with the host rocks at grater depths. The boiling fluid rises along permeable zones, depositing ore and gangue minerals, and may discharge from near-neutral pH hot springs. The separated vapour with CO_2 and H_2S condenses in the vadose zone to form a steam-heated water, acidic from oxidation of H_2S. (*b*) In high-sulfidation systems, magmatic volatiles ascend to the epithermal environment where they are absorbed by meteroic water, and the HCl and SO_2 form a highly acidic solution that leaches the rock outword from the fluid conduit. Ore metals may be introduced into this leached rock by later magmatic fluids.

25. Properties of minerals

Sodalite: Sodalite is a rare rock-forming mineral best known for its blue to blue-violet color. It has a chemical composition of $Na_4Al_3Si_3O_{12}Cl$ and is a member of the feldspathoid mineral group. High-quality sodalite is used as a gemstone, a sculptural material, and an architectural stone.

General	
Category	Tectosilicates without zeolitic H_2O
Formula (repeating unit)	$Na_8(Al_6Si_6O_{24})Cl_2$
IMA symbol	Sdl
Strunz classification	9.FB.10
Crystal system	Cubic
Crystal class	Hextetrahedral (43m) H-M symbol: (43m)
Space group	P43n
Unit cell	a = 8.876(6) Å; Z = 1
Identification	
Colour	Rich royal blue, green, yellow, violet, white veining common
Crystal habit	Massive; rarely as dodecahedra
Twinning	Common on {111} forming pseudohexagonal prisms
Cleavage	Poor on {110}
Fracture	Conchoidal to uneven
Tenacity	Brittle
Mohs scale hardness	5.5-6
Luster	Dull vitreous to greasy
Streak	White
Diaphaneity	Transparent to translucent
Specific gravity	2.27-2.33
Optical properties	Isotropic
Refractive index	n = 1.483 - 1.487
Ultraviolet fluorescence	Bright red-orange cathodoluminescence and fluorescence under LW and SW UV, with yellowish phosphorescence; may be photochromic in magentas
Fusibility	Easily to a colourless glass; sodium yellow flame
Solubility	Soluble in hydrochloric acid and nitric acid

Tourmaline: Tourmaline is the name of a large group of boron silicate minerals. These minerals share a common crystal structure and similar physical properties - but vary tremendously in chemical composition. The wide range of compositions and colour zoning within crystals causes tourmaline to occur in more colours and colour combinations than any other mineral group.

General	
Category	Cyclosilicate
Formula (repeating unit)	$(Ca, K, Na)(Al, Fe, Li, Mg, Mn)_3(Al, Cr, Fe, V)_6(BO_3)_3(Si, Al, B)_6O_{18}(OH, F)_4$
IMA symbol	Tur
Crystal system	Trigonal
Crystal class	Ditrigonal pyramidal (3m) H-M symbol: (3m)
Space group	*R3m* (no. 160)
Identification	
Colour	Most commonly black, but can range from colourless to brown, red,

	orange, yellow, green, blue, violet, pink, or hues in between; can be bi-coloured, or even tri-coloured; rarely can be neon green or electric blue
Crystal habit	Parallel and elongated. Acicular prisms, sometimes radiating. Massive. Scattered grains (in granite).
Cleavage	Indistinct
Fracture	Uneven, small conchoidal, brittle
Tenacity	Brittle
Mohs scale hardness	7.0-7.5
Luster	Vitreous, sometimes resinous
Streak	White
Diaphaneity	Translucent to opaque
Specific gravity	3.06+0.20-0.06
Density	2.82-3.32
Polish luster	Vitreous
Optical properties	Double refractive, uniaxial negative
Refractive index	n_ω = 1.635-1.675 n_ε = 1.610-1.650
Birefringence	–0.018 to –0.040; typically about –0.020 but in dark stones it may reach –0.040
Pleochroism	• Typically moderate to strong • **Red:** definite; dark red, light red • **Green:** strong; dark green, yellow-green • **Brown:** definite; dark brown, light brown • **Blue:** strong; dark blue, light blue
Dispersion	0.017
Ultraviolet fluorescence	Pink stones; inert to very weak red to violet in long and short wave
Absorption spectra	Strong narrow band at 498 nm, and almost complete absorption of red down to 640 nm in blue and green stones; red and pink stones show lines at 458 and 451 nm as well as a broad band in the green spectrum.

Calcite: Calcite is a rock-forming mineral with a chemical formula of $CaCO_3$. It is extremely common and found throughout the world in sedimentary, metamorphic, and igneous rocks. Some geologists consider it to be a "ubiquitous mineral"—one that is found everywhere. The properties of calcite make it one of the most widely used minerals. It is used as a construction material, abrasive, agricultural soil treatment, construction aggregate, pigment, pharmaceutical and more. It has more uses than almost any other mineral.

General	
Category	Carbonate minerals
Formula (repeating unit)	$CaCO_3$
Strunz classification	5.AB.05
Crystal system	Trigonal
Crystal class	Hexagonal scalenohedral (3m) H-M symbol: (3 2/m)

Space group	R3c
Unit cell	a = 4.9896(2) Å, c = 17.0610(11) Å; Z = 6
Identification	
Colour	Typically colourless or white - may have shades of various colours
Crystal habit	Crystalline, granular, stalactitic, concretionary, massive, rhombohedral
Twinning	Common by four twin laws
Cleavage	Perfect on {1011} three directions with angle of 74° 55'
Fracture	Conchoidal
Tenacity	Brittle
Mohs scalehardness	3 (defining mineral)
Luster	Vitreous to pearly on cleavage surfaces
Streak	White
Diaphaneity	Transparent to translucent
Specific gravity	2.71
Optical properties	Uniaxial (–)
Refractive index	n_ω = 1.640-1.660 n_ε = 1.486
Birefringence	δ = 0.154-0.174
Fusibility	Infusible (decrepitates energetically)
Solubility	Soluble in dilute acids
Other characteristics	May fluoresce red, blue, yellow, and other colours under either SW and LW UV; phosphorescent.

Muscovite: Muscovite is the most common mineral of the mica family. It is an important rock-forming mineral present in igneous, metamorphic, and sedimentary rocks. Like other micas it readily cleaves into thin transparent sheets. Muscovite sheets have a pearly to vitreous luster on their surface. If they are held up to the light, they are transparent and nearly colourless, but most have a slight brown, yellow, green, or rose-colour tint.

General	
Category	Phyllosilicate
Formula (repeating unit)	$KAl_2(AlSi_3O_{10})(F,OH)_2$
IMA symbol	Ms
Strunz classification	9.EC.15
Dana classification	71.02.02a.01
Crystal system	Monoclinic
Crystal class	Prismatic (2/m) (same H-M symbol)
Space group	C2/c
Unit cell	a = 5.199 Å, b = 9.027 Å, c = 20.106 Å, β = 95.78°; Z = 4
Identification	
Colour	White, grey, silvery
Crystal habit	Massive to platy
Twinning	Common on the [310], less common on the {001}
Cleavage	Perfect on the {001}
Fracture	Micaceous
Tenacity	Elastic
Mohs scale hardness	2-2.5 parallel to {001} 4 right angle to {001}
Luster	Vitreous, silky, pearly
Streak	White
Diaphaneity	Transparent to translucent

Specific gravity	2.76-3
Optical properties	Biaxial (-)
Refractive index	n_α = 1.552-1.576 n_β = 1.582-1.615 n_γ = 1.587-1.618
Birefringence	δ = 0.035 - 0.042
Pleochroism	Weak when coloured
Dispersion	r > v weak
Ultraviolet fluorescence	None

27. Increased diversity of siphonate bivalves occurred in response to increased predation pressure in the Mesozoic.

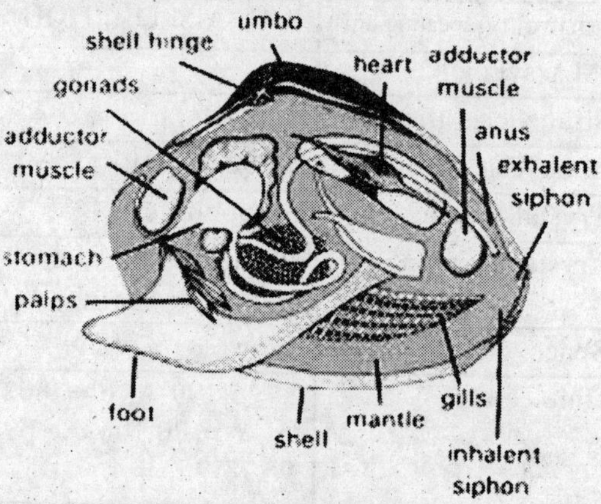

28. An index fossil should have large geographic range and small temporal range. A useful index fossil must be distinctive or easily recognizable, abundant, and have a wide geographic distribution and a short range through time.

The fossil remains of an organism that lived in a particular geologic age, used to identify or date the rock or rock layer in which it is found. The best type of index fossils are usually those of swimming or floating organisms that evolved quickly (and therefore did not cover a long span of geologic history) and were able to spread over large areas. Ammonites and graptolites are good index fossils.

29. Syringothyris Limestone: The geological age is Carboniferous.

Karai Shale: Cretaceous of Trichinopolly, Cauvery basin Tamil Nadu, India. This shale composed of phosphatic nodules.

Chari: Jurassic of Cutch, Gujarat.

Barren Measures: Permian.

30. Sulphide minerals form by covalent bonding between metal and sulphur.

In virtually all of the sulphides, bonding is covalent, but some have metallic properties. The covalent property of sulphur allows sulphur-sulphur bonds and the incorporation of S2 pairs in some sulphides such as pyrite

31. The following structures form in marine environment:

- Lateral accretionary surfaces
- Hummocky cross stratification
- Herringbone cross stratification

Lateral accretionary surfaces formed at Fault zone.

- Hummocky cross-stratification is a type of sedimentary structure found in sandstones. It is a form of cross-bedding usually formed by the action of large storms, such as hurricanes. It takes the form of a series of "smile"-like shapes, crosscutting each other. It is only formed at a depth of water below fair-weather wave base and above storm-weather wave base. They are not related to "hummocks" except in shape.
- During the conventional formation process of cross-stratification, sand grains saltate up the upstream side of the dune, collecting at the peak until the angle of repose is reached. At this point, the crest of granular material has grown too large and will be overcome by the force of the depositing fluid, falling down the downstream side of the dune. Repeated avalanches will eventually form the sedimentary structure known as cross-stratification, with the structure dipping in the direction of the paleocurrent.

32. The correct stratigraphic successions ordered from oldest to youngest.

- Semri Group – Kaimur Group – Rewa Group – Bhander Group
- Papaghni Group – Chitravati Group – Nallamalai Group – Kurnool Group

Classification of Vindhyan:

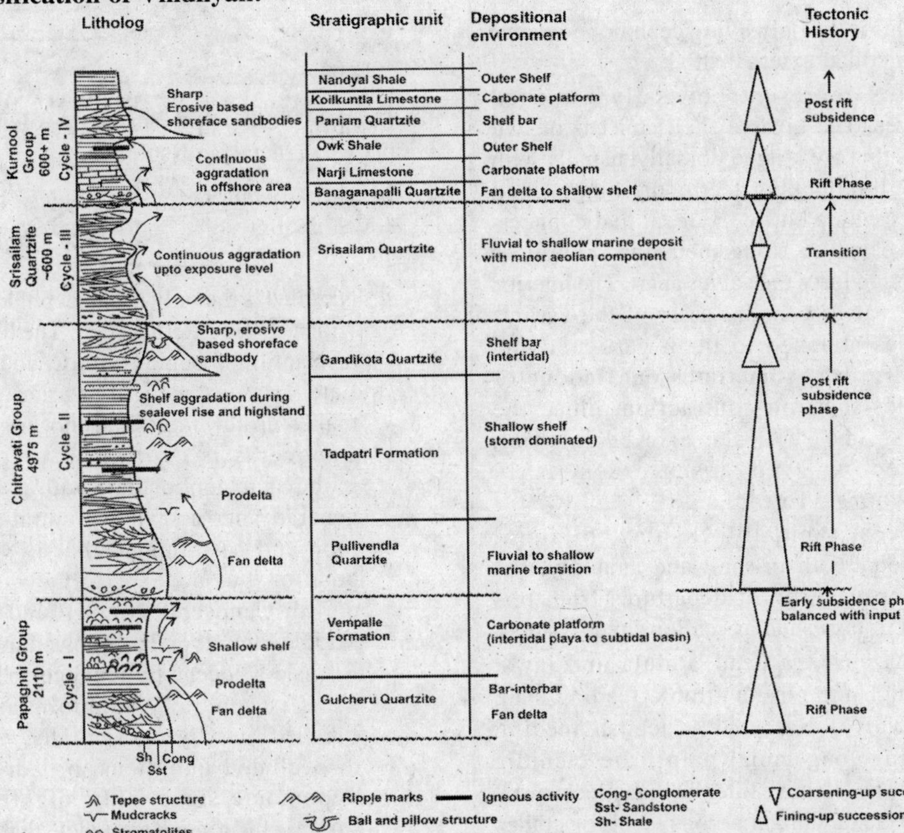

		Rajasthan	Son Valley	Age Constraints
Vindhyan Supergroup	Upper	Bhander Group (~1200m)	Bhander Group (~1200 m)	Bhander Ls: 750 Ma (Sr Isotopic Curve) Chuaria- Tawuwia fossils: 700-1100 Ma
		Rewa Group (285m)	Rewa Group (296m)	Rewa Group: 700-1100 Ma Chuaria-Tawuwia Fossils
		Kaimur Group (180m)	Kaimur Group (425m)	← Majhgawan Kimberlite: 1073 ± 13.7 Ma
	Lower (Semri Series)	Khorhip Group (475m)	Rhotas Group (454m)	Rhotas Ls: 1599 ±48 Ma Pb-Pb Isochron; 1601±130 Ma Pb-Pb Isochron
		Lasrawan Group (272m)	Kheinjua Group (1050m)	Rampur Shale: 1599 ± 8 Ma U-Pb (Zircon) Porcellanite Fm: 1628 ± 8 Ma U-Pb (Zircon) Kajraht Ls: 1721±90 Ma Pb-Pb Isochron
		Sand Group (210m)	Deonar Group (1200m)	
		Satola Group (835m)	Mizapur Group (990m)	

33. Barakar Formation and Lakadong Formation contain coal seams.

34. The following statements are correct:

- Mytilus represents byssally attached bivalves.
- Nautilus is the only living cephalopod genus with a coiled external shell.
- Mytilus represents byssally attached bivalves. The bivalve shell consists of two valves that are hinged dorsally, usually with shelly interlocking teeth (the hinge), and always with a horny ligament that connects the two valves along their dorsal surfaces and acts to force the valves apart. The interior of the valves contains scars of the various muscles attached to it, in particular the (usually two, sometimes one) adductor muscles that, on contraction, close the valves. The shell can also be internal, reduced or even absent such as in "shipworms" (Teredo).
- Bivalves typically display bilateral symmetry both in shell and anatomy, but there are significant departures from this theme in taxa such as scallops and oysters.
- Bivalves lack a head, radula and jaws. Although the plesiomorphic feeding state for bivalves is probably deposit feeding utilizing long labial palps, the ctenidia provide an effective filter feeding mechanism in most taxa with numerous levels or grades of organization. In most bivalves, the mantle pallial cavity contains a pair of very large gills that are used to capture food particles suspended in the inhalant water current.

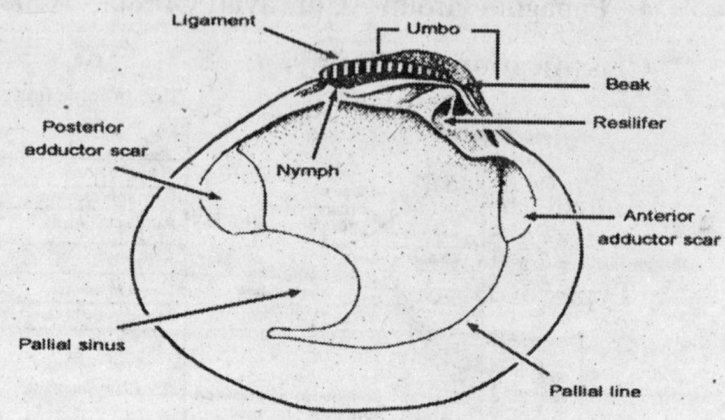

- Nautilus is the only living cephalopod genus with a coiled external shell. The fossil record of Nautilidae begins with Cenoceras in the Late Triassic, a highly varied genus that makes up the Jurassic Cenoceras complex. Cenoceras is evolute to involute, and globular to lentincular; with a suture that generally has a shallow ventral and lateral lobe and a siphuncle that is variable in position but never extremely ventral or dorsal. Cenoceras is not found above the Middle Jurassic and is followed by the Upper Jurassic-Miocene Eutrephoceras. Eutrephoceras is generally subglobular, broadly rounded laterally and ventrally, with a small to occluded umbilicus, broadly rounded hyponomic sinus, only slightly sinuous sutures, and a small siphuncle that is variable in position.

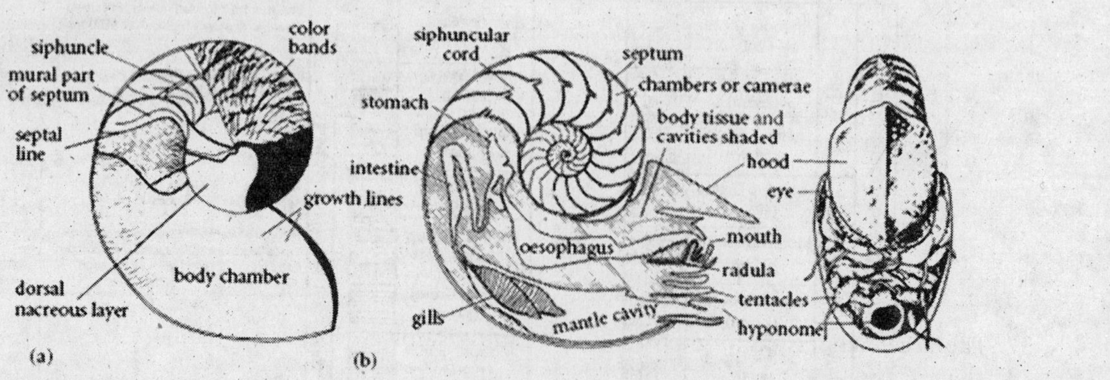

35. The following genera are stem fossils: 1. Rhynia 2. Lepidodendron 3. Lepidostrobus 4. Lepidocarpon 5. Calamites 6. Calamostachys 7. Palaeostachya 8. Sphenophyllum 9. Lyginopteris 10. Kaloxtlon 11. Lagenostoma 12. Crossotheca 13. Heterangium 14. Mudullosa 15. Glossopteris 16. Vertebraria 17. Gangamopteris 18. Caytonia 19. Nipanioxylon 20. Pentoxylon 21. Sahnia 22. Carnoconite 23. Dadoxylon 24. Vertebraria

36. The following statements are correct:
- Abutments are the sides of the valley supporting the dam structure
- Spillways can control the release of water from the reservoir
- Galleries serve as passages through the dam

Types of Dam:

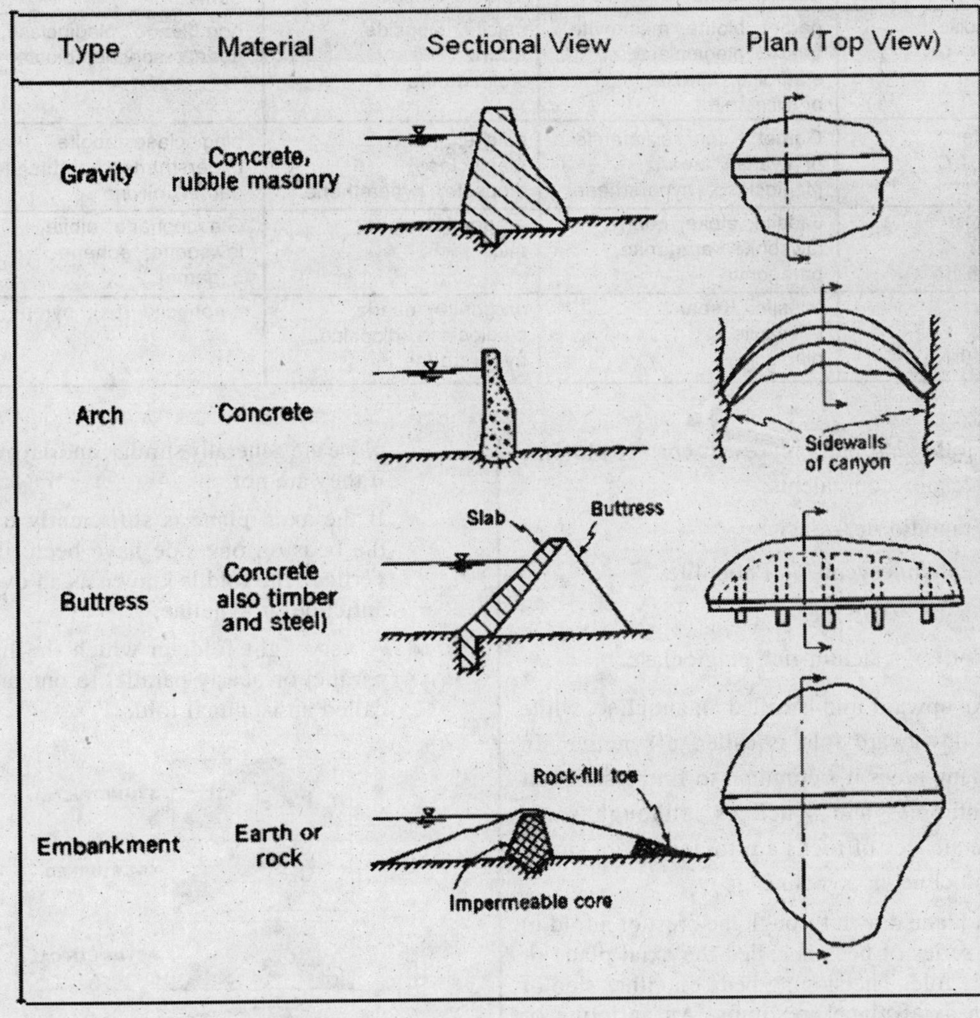

37. The acceleration due to gravity on the Earth's surface depends on: latitude, elevation, topography of the surrounding terrain.

38. Garnet-omphacite, hornblende-plagioclase, glaucophane-lawsonite.

Facies-Protolith-Mineral Assemblage Table:

Facies	Pelitic	Calcareous	Mafic
Zeolite 100-200°C	interlayered smectite/ chlorite calcite	calcite	Laumonite, thompsonite, calcite, interlayered smectite/chlorite
Prehnite-Pumpellyite 150-300°C	Prehnite, pumpellyite, calcite, chlorite, albite	calcite	Prehnite, pumpellyite, calcite, chlorite, albite
Greenschist 300-450° C	muscovite, chlorite, quartz, albite, biotite, garnet	calcite, dolomite, quartz, epidote, tremolite	albite, chlorite, quartz, epidote, actimolite, sphene
Epidote Amphibolite 450-550° C	muscovite, biotite, garnet, albite, quartz	calcite, quartz, tremolite, epidote diopside	albite, epidote, hornblende, quartz
Amphibolite 500-700° C	garnet, biotite, muscovite, quartz, plagioclase, staurolite, kyanite or sillimanite	calcite, diopside quartz, wollastonite	hornblende, plagioclase, garnet, quartz, sphene, biotite
Granulite 700-900° C	Garnet, Kspar, sillimanite or kyanite, quartz, plagioclase, hypersthene	calcite, quartz, plagioclase, diopside, hypersthene	plagioclase, augite, hypersthene, hornblende, garnet, olivine
Blueschist 150-350° C P > 598 Kb	Jadeite, albite, quartz, lawsonite, aragonite, paragonite	aragonite, white mica	Glaucophane, albite, lawsonite, sphene, ± garnet
Eclogite 350-750° C P > 8-10 Kb	coesite, Kspar, sillimanite, plagioclase	aragonite, quartz, plagioclase, diopside, hypersthene	omphacite (px), pyrope garnet

39. The following pairs represent correct plutonic – volcanic equivalents:

- Granodiorite – dacite
- Nephelinesyenite – Phonolite
- Dunite – Olivine
- Norite – calcium-rich plagioclase

40.
- An upward fold is called an anticline, while a downward fold is called a syncline. In many areas it's common to find a series of anticlines and synclines, although some sequences of rocks are folded into a single anticline or syncline.
- A plane drawn through the crest of a fold in a series of beds is called the axial plane of the fold. The sloping beds on either side of an axial plane are limbs. An anticline or syncline is described as symmetrical if the angles between each of limb and the axial plane are generally similar, and asymmetrical if they are not.
- If the axial plane is sufficiently tilted that the beds on one side have been tilted past vertical, the fold is known as an overturned anticline or syncline.
- A very tight fold, in which the limbs are parallel or nearly parallel to one another is called an isoclinal fold.

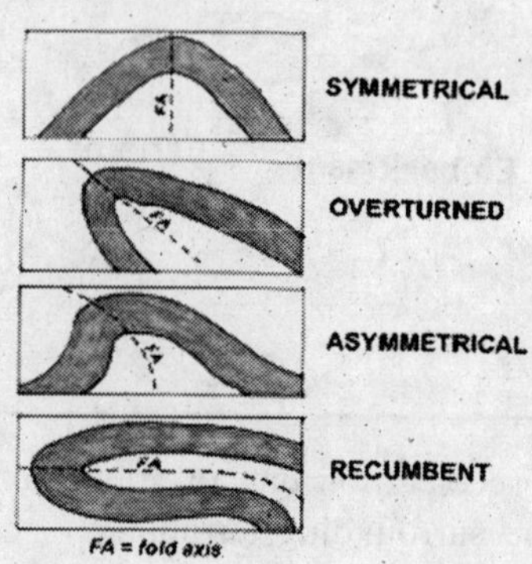

- Isoclinal folds that have been overturned to the extent that their limbs are nearly horizontal are called recumbent folds.
- Anticlines = folds pointing upwards.
- Synclines = folds pointing downwards.
- Upright (Open) Fold: vertical/near vertical axial surface and limbs have similar dips (though in opposite directions).
- Inclined Fold: axial surface is neither vertical nor horizontal.
- Overturned Fold: beds on one limb are upside down.
- Recumbent Fold: axial surface is horizontal.
- Isoclinal Fold: limbs are parallel.
- Non-Plunging Fold: has a horizontal/near horizontal hinge line.
- Anticline: older beds are in the core (center) of the fold and the beds dip away from the core.
- Syncline: younger beds are in the core of the fold and the beds dip into the core.
- Plunging Fold: has an inclined hinge line.
- Plunging Anticline: beds point in the direction of the plunge; older beds are in the core (center) of the fold and the beds dip away from the core.
- Plunging Syncline: beds point in the opposite direction of the plunge; younger beds are in the core of the fold and the beds dip into the core.
- Domes – a circular anticline (oldest beds in the core, beds dip away from the core).
- Basins – a circular syncline (youngest beds in the core, beds dip into the core).

41. The Ripple Symmetry Index (RSI) for the given hypothetical asymmetric ripple is RSI = 10/4 = 2.5.

42. Drainage density = Av. Length of Channels/ Area of drainage basin =

(10.5 + 7.5 + 5.5 + 1.5)/0.5 = 50.

43. Porosity = Void ratio/Void ratio + 1

= 0.5/0.5 + 1 = 0.33

44. The vertical stress at a depth of 1 km =

25000 × 1000/1000 × 1000 = 25MPa

45. The Tangential Velocity

= Angular Velocity × Radius

= 2 × π × r/t

= 2 × 3.14 × 149 × 1000 × 1000/365 × 24

= 106817.935

46. The thickness of the dyke = cos 60

= thickness/12

Thickness = cos 60 × 12 = $\frac{1}{2} \times 12$ = 6m

48. The surface gravity of the Earth

= Fg = GMm/r^2 and g = GM/r^2

(GM/r^2) for earth/(GM/r^2) for moon = (80/16)/(1/1) = 5

49. The variance (degree of freedom) of the assemblage = $Al_2SiO_5 - Al_2O_3$ and SiO_3

So, Component = 2 and Phase = 3

F = C – P + 2 = 2 – 3 + 2 = 1

50. Amount of High grade ore = *x*

(0.25 × 1/100) + (1.8 × *x*/100)/1 + *x* = 0.5/100

X = 0.192 Mt.

51. Normal stress = $((n_1 + n_3)/2) + ((n_1 - n_3)/2)$. cos20 = ((28 + 8)/2) + ((28 – 8)/2).cos20 = 13

52. v = u + at

V = 10 × 0.5 × 4

= 20 m/s

$1/2mv^2$ = mgh

½ × 20 × 10^2 = 10 × h

Hv = 20 m.

53. The average outflow Velocity = Flow Rate/ Area = 100/π × r^2 = 100/7850 = 0.013

54. Let the thickness of the root = x

Then, according to Airy Hypothesis rule,

$(40 \times 2800) + (r + 3300) = (44 + x) \times 2800$

$\therefore x = 22.4$ m.

55. Bornite – Cu_5FeS_4 = 502 at weight

Chalcopyrite = $CuFS_2$ = 1836

Chalcocite = Cu_2S = 159.2

Weight friction of Bornite = 0.4

Mass of bornite = 0.4 × 1000 gm = 400 gm

Similarly, mass of chalcopyrite = 0.4 × 1000 gm = 400 gm

Mass of chalcocite = 0.2 × 1000 gm = 200 gm

Per cent of copper in 400 gm of bornite = 253-187

Per cent of copper in 400 gm of chalcopyrite = 138-45

Percent of copper in 200 gm of chalcocite = 159-67

Total percent of copper in ore of 1 kg weight = 253 – 187 + 138 – 45 + 159 – 67 = 551.31 gm.

56.

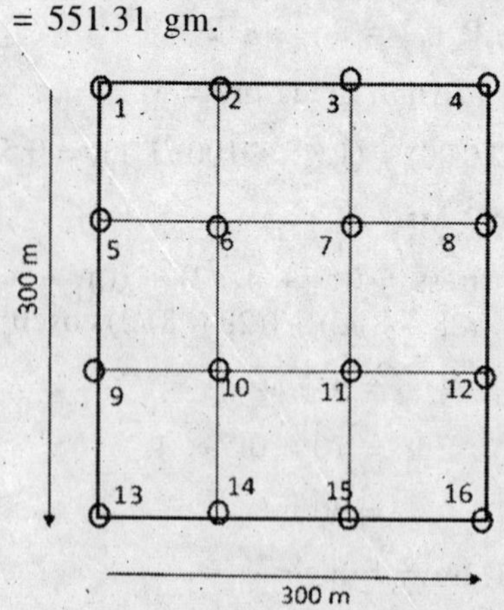

Thickness at point 1 = 11 m

Thickness at point 2 = 10.5 m

Total volume of blocks with t as 10.8m = 108000 m^3

(4 total) total volume of blocks with t as 10.7 m = 428000 m^3

(total 4) total volume of blocks with t as 10.63 m = 426000 m^3

Total volume of all blocks = 962000 m^3

So, total weight of copper,

For 10.8 m thick blocks, (1.5/100) × 108000 = 1620 kg

For 10.7 m thick blocks, (1.75/100) × 428000 = 7490 kg

For 10.65 m thick blocks (1.65/100) × 426000 = 7029 kg

So total copper = 16139 kg

= 16139 kg/962000 m^3

= 0.0167 kg m^3

= 1.67 wt%.

57. $$\left(\frac{^{87}Sr}{^{86}Sr}\right)_t = \left(\frac{^{87}Sr}{^{86}Sr}\right)_0 + \left(\frac{^{87}Rb}{^{86}Sr}\right)_t (e^{\lambda t} - 1)$$

The Sr isotopic ratio of the source at the time of derivation of the granite

$= 0.8001 = x + 2.499 \times (e^{1.39} \times 10^{-9} - 1)$

X = 0.765

58. Assuming steady state Darcian flow,

T = kb

Given, T1 = 60 × 30 = 1800

T2 = 40 × 15 = 600

T1/T2 = 1800/600 = 3 times

60. The required amount (in %)

= 10/35 × 100

= 28.57%.

Previous Papers (Solved)

Indian Institute of Technology Joint Admission Test for M.Sc.

IIT-JAM GEOLOGY-2021

SECTION-A

Multiple Choice Questions (MCQ)

Qs. No. 1-10: *Carry one mark each.*

1. Among the following rocks, the one with highest metamorphic grade is:
A. Chlorite schist B. Glaucophane schist
C. Phyllite D. Gneiss

2. The Earth's radius is maximum at which one of the following latitudes?
A. 0° B. 40° N
C. 60° S D. 90°

3. The closest value to the percentage of the Earth's surface covered by the oceans is:
A. 30% B. 50%
C. 70% D. 90%

4. Which is the shallowest among the marine environments listed below?
A. Neritic B. Littoral
C. Abyssal D. Bathyal

5. Among the following, the tungsten-bearing mineral is:
A. Bornite B. Cassiterite
C. Scheelite D. Greenockite

6. The host rock of Pb-Zn deposit at Zawar is:
A. Quartzite B. Phyllite
C. Dolomite D. Gneiss

7. Bituminous coal deposits in India occur in which one of the following formations?
A. Barren Measures Formation
B. Barakar Formation
C. Naredi Formation
D. Cuddalore Formation

8. Identify the plant fossil from the following list.
A. *Glossopteris* B. *Fenestella*
C. *Productus* D. *Cidaris*

9. The igneous body with dome or mushroom-like shape is known as a:
A. Lopolith B. Ring dike
C. Sill D. Laccolith

10. Which one of the following stratigraphic units belongs to the Cretaceous?
A. Lameta Formation
B. Talchir Boulder Bed
C. *Fenestella* Shale
D. Kasauli Formation

Qs. No. 11–30: *Carry two marks each.*

11. Select the youngest volcanic event out of the following.
A. Rajmahal volcanics
B. Dalma volcanics
C. Panjal volcanics
D. Deccan volcanics

12. Which among the following is the only possible plunge for a lineation located on the foliation plane striking 20° N and dipping 40° southeasterly?
A. 20° SE B. 20° NW
C. 45° SE D. 40° NW

13. Which one of the following tectonic plates has the maximum average velocity?
A. Eurasian B. Pacific
C. African D. North American

14. A limestone contains lime mud and around 25% allochems, which are separated from each other. The name of the rock as per Dunham's classification is:

A. Mudstone B. Wackestone
C. Packstone D. Grainstone

15. Find the **CORRECT** statement out of the following.

A. Convolute laminae form by desiccation.
B. Load cast is an erosional structure.
C. Prod mark is found at the bottom of a bed.
D. Wave ripple occurs at the top of a turbidite deposit.

16. Which one of the following crystal forms **DOES NOT** belong to the di-tetragonal pyramidal class?

A. c-Pedion
B. Prism of 1st order
C. Di-tetragonal prism
D. Tetragonal dipyramid

17. If a coarse-grained igneous rock is composed of >90% of plagioclase and <10% of olivine and pyroxene, then the name of this rock according to the IUGS classification is:

A. Anorthosite B. Olivine gabbro
C. Tonalite D. Olivine websterite

18. Which one of the following represents the compositional change in plagioclase during the crystallization of mafic magma?

A. Na/Ca ratio decreases; Al/Si ratio increases
B. Both Na/Ca and Al/Si ratios increase
C. Na/Ca ratio increases; Al/Si ratio decreases
D. Both Na/Ca and Al/Si ratios decrease

19. Choose the **CORRECT** sequence of older to younger formations in the stratigraphy of the Cuddapah Supergroup

A. Pulivendla-Gulcheru-Vempalle-Tadpatri
B. Gulcheru-Vempalle-Pulivendla-Tadpatri
C. Gulcheru-Pulivendla-Tadpatri-Vempalle
D. Vempalle-Gulcheru-Tadpatri-Pulivendla

20. Match the economic deposits **(Group-I)** with their places of occurrence **(Group-II)**.

Group-I	**Group-II**
P. Iron ore	1. Bhatan
Q. Base metal	2. Sukinda
R. Chromite	3. Rampur-Agucha
S. Uranium	4. Bellary

	P	Q	R	S
A.	4	1	2	3
B.	2	4	3	1
C.	2	3	1	4
D.	4	3	2	1

21. Select the answer that is a **CORRECT** match for the three types of unconformities. (Grey bands = sediments, [+] = igneous rock).

P. Disconformity 1. 

Q. Non-conformity 2. 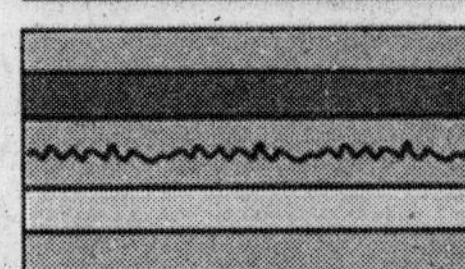

R. Angular Unconformity 3. 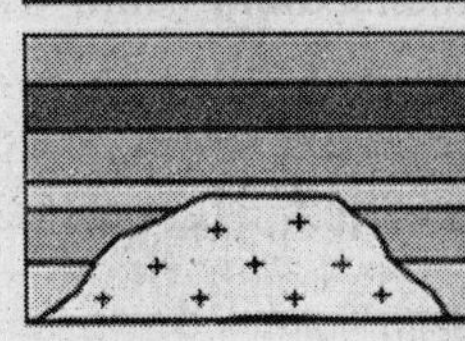

	P	Q	R
A.	2	3	1
B.	2	1	3
C.	1	3	2
D.	3	2	1

22. Which one of the following statements is **FALSE**?

A. Ammonites have fluted septa
B. Brachiopods have a pedicle
C. Echinoids have genal spines
D. Trilobites have a pygidium

23. In the context of phylogeny of horses, the **CORRECT** chronological order from old to young is:

A. *Hyracotherium, Mesohippus, Merychippus, Equus*

B. *Hyracotherium, Merychippus, Mesohippus, Equus*

C. *Equus, Merychippus, Mesohippus, Hyracotherium*

D. *Merychippus, Equus, Mesohippus, Hyracotherium*

24. Choose the **CORRECT** match between items in **Group-I** with the items in **Group-II**.

Group-I	Group-II
P. Polarity zone	1. Biostratigraphy
Q. Formation	2. Chronostratigraphy
R. Biozone	3. Magnetostratigraphy
S. Epoch	4. Lithostratigraphy

	P	Q	R	S
A.	1	2	3	4
B.	2	3	1	4
C.	3	4	1	2
D.	3	4	2	1

25. The following diagram represents the prograde sequence of metamorphic zones that develop during Buchan-type of metamorphism (<4 kbar) of typical pelites. Identify the zones labelled P, Q and R.

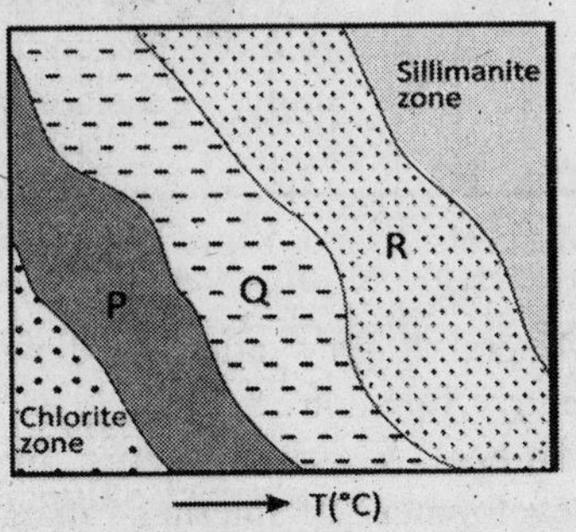

A. P - Biotite zone; Q - Garnet zone; R - Kyanite zone

B. P - Garnet zone; Q - Kyanite zone; R - Staurolite zone

C. P - Biotite zone; Q - Cordierite zone; R - Andalusite zone

D. P - Andalusite zone; Q - Biotite zone; R - Cordierite zone

26. Match the minerals in **Group-I** with the corresponding composition in **Group-II**.

Group-I	Group-II
P. Leucite	1. $(K, Na)AlSi_3O_8$
Q. Andradite	2. $NaAlSi_2O_6$
R. Sanidine	3. $KAlSi_2O_6$
S. Jadeite	4. $Ca_3Fe_2Si_3O_{12}$

	P	Q	R	S
A.	3	1	4	2
B.	2	3	1	4
C.	1	4	2	3
D.	3	4	1	2

27. Choose the **CORRECT** pair of crystal systems that represents the optic orientation shown in the figure. (X, Y, Z are the principal optical vibration directions and a, b, c are the crystallographic axes).

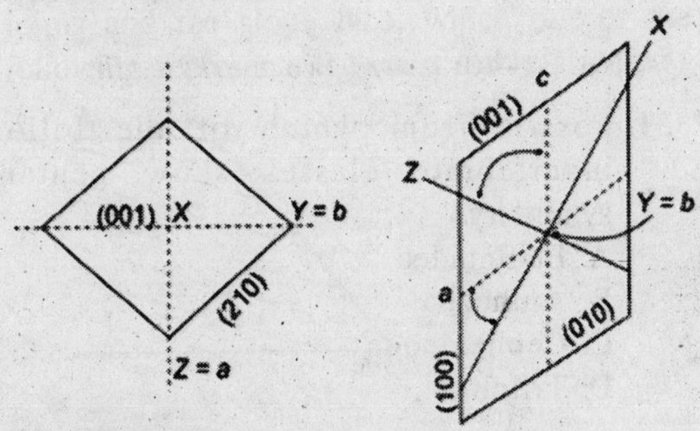

A. Orthorhombic - Triclinic

B. Orthorhombic - Monoclinic

C. Cubic - Hexagonal

D. Tetragonal - Monoclinic

28. Match the environment representing physical geological processes in **Group-I** with the corresponding geomorphic landform/feature in **Group-II**.

Group-I	Group-II
P. Aeolian	1. Drumlin
Q. Glacial	2. Tombolo
R. Fluvial	3. Yardang
S. Coastal	4. Natural levee

	P	Q	R	S
A.	2	1	4	3
B.	2	4	1	3
C.	3	1	4	2
D.	3	4	1	2

29. Match the items in **Group-I** with the corresponding items in **Group-II**.

Group-I	Group-II
P. Chalcocite	1. Supergene enrichment
Q. Bauxite	2. Mechanical accumulation
R. Monazite	3. Magmatic crystallization
S. Chromite	4. Residual processes

	P	Q	R	S
A.	1	4	2	3
B.	2	3	4	1
C.	2	4	1	3
D.	1	3	2	4

30. Which one of the following statements is **FALSE**?

A. Perched water table exists within the zone of aeration

B. Juvenile water is derived from sediment diagenesis

C. Zone of aeration lies above the zone of saturation

D. Both aquiclude and aquifuge are impermeable

SECTION-B

Multiple Select Questions (MSQ)

Qs. No. 31–40: *Carry two marks each.*

31. Fossils from which of the following invertebrate classes show pentameral symmetry?

A. Echinoidea

B. Anthozoa

C. Cephalopoda

D. Trilobita

32. The attitude of the two limbs of a fold was measured as striking 4° N, dipping 85° easterly and striking 30° N, dipping 60° easterly. Which of the following is/are **TRUE** for describing the geometry of the fold?

A. Synform

B. Antiform

C. Overturned

D. Plunging

33. Which of the following statement(s) is/are **CORRECT** regarding ophitic texture?

A. Plagioclase laths are completely enclosed by large pyroxene crystals.

B. Intergrowth occurs between quartz and alkali-feldspar.

C. It is a variety of poikilitic texture.

D. It is a texture observed in peridotite.

34. On the basis of the following schematic diagram, choose the **CORRECT** statement(s).

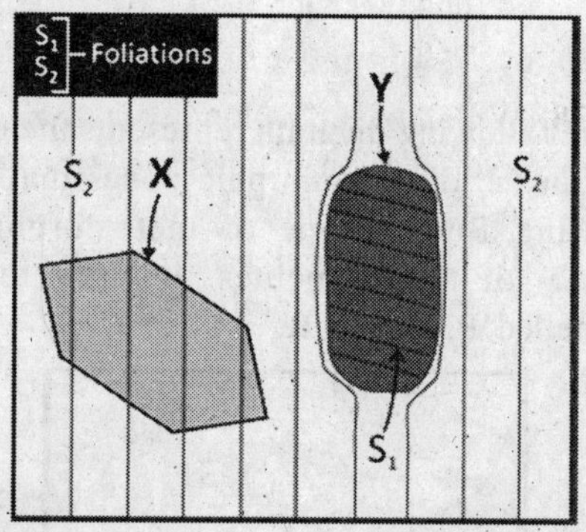

A. Minerals **X** and **Y** predate matrix foliation (S_2).

B. Mineral **X** grew after the development of matrix foliation (S_2).

C. Mineral **Y** grew during or after the development of internal foliation (S_1).

D. Minerals **X** and **Y** postdate matrix foliation (S_2).

35. A marine organic-rich, black shale with tiny pyrite crystals shows complete absence of body or trace fossils. Which of the following statement(s) is/are **TRUE**?

A. The sediments were deposited in low-energy conditions.

B. The deposition took place in dysoxic to anoxic conditions.

C. The rate of sedimentation was high.

D. The environment was stressful for survival of living organisms.

36. Which mineral(s) among the following represent(s) AB_2O_4 composition?

A. Spinel B. Magnetite
C. Chromite D. Ilmenite

37. Which among the following statement(s) is/are **TRUE**?

A. Attrition is more dominant in aeolian than in glacial environment.
B. Centrifugal force drives the sediment-laden water outward when the river channel meanders.
C. U-shaped valley is a common fluvial geomorphic feature.
D. The downstream water velocity in a river channel increases upward from the channel bed towards the water surface.

38. Which of the following statement(s) regarding hydrocarbon occurrence is/are **CORRECT**?

A. Gandhar field is in Cambay basin.
B. Oil and gas occur in Mesozoic reservoir rocks in Bombay High field.
C. Digboi field is in Assam basin.
D. Hydrocarbon occurs in limestone reservoir in Ankleshwar field.

39. Following are the statements regarding types of sandstone as per Pettijohn's classification. Which is/are the **CORRECT** statement(s) out of the following?

A. Arkose contains more than 25% feldspar.
B. Greywacke contains more than 90% matrix.
C. Litharenite contains more than 25% lithic fragment.
D. Quartz arenite contains more than 95% quartz.

40. Choose the **CORRECT** statement(s) out of the following.

A. Shoreline shifts landward during transgression.
B. Shoreline shifts seaward during transgression.
C. Delta deposits preserve the record of transgression.
D. Incised river valley forms because of transgression.

SECTION-C

Numerical Answer Type (NAT)

Qs. No. 41-50: *Carry one mark each.*

41. The given section with uniform lithology and sedimentation rate records two ash layers dated at 77 Ma and 76 Ma, respectively. An index fossil species present in the lower part of the section becomes extinct at a horizon 7 m above the base. The estimated age of the extinction event is ________ Ma. (Answer in integer).

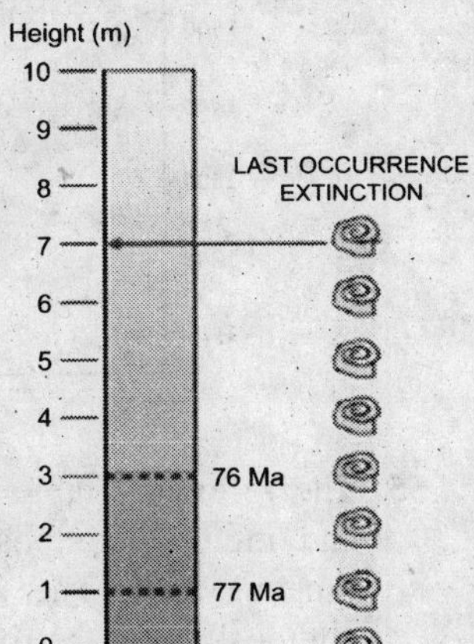

42. A hollow discoid (cylindrical) microfossil has an outer diameter of 20 μm, height 10 μm and wall thickness 1 μm. The internal volume that can be occupied by the organism is ______ μm^3. (use $\pi = 3.14$) (Round off to one decimal place).

43. In the following isobaric temperature-composition diagram, the number of common

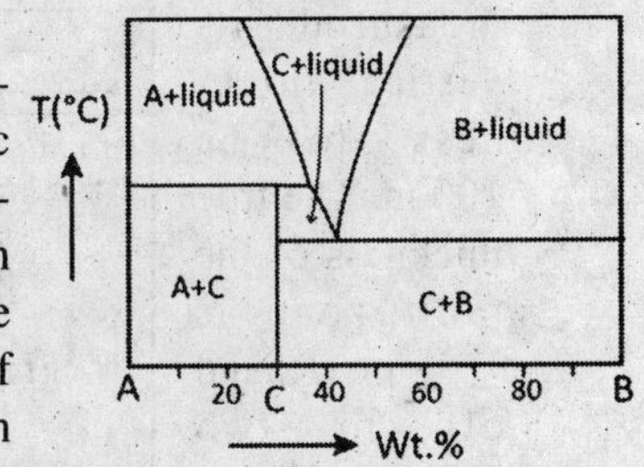

phases in all the invariant points is ______. (Answer in integer).

44. A muscovite has the following composition in which iron is ferrous. The amount of 'Al' in the tetrahedral site is ________ (per formula unit). (Round off to two decimal places).

Muscovite composition:
$KAl_{2.50}Fe_{0.25}Si_{3.25}O_{10}(OH)_2$

45. The density of a 200 g gabbro sample, cut in the form of a cube, is 3125 kg/m^3. The length of the sample is _____ mm. (Answer in integer).

46. A drill run of 3 m was carried out in a coalfield site, where rock core samples were recovered only for a cumulative length of 255 cm. The core loss in percentage is equal to ______. (Answer in integer).

47. During concretionary growth of a spherical grain of radius 2Å, the rate of change of surface area with respect to change in radius of the grain is ______ × 10^{-8} cm (use π = 3.14) (Round off to two decimal places).

48. The weight loss during the conversion of 1 mole of gypsum to anhydrite is ____ % (atomic weights of Ca = 40.0, S = 32.0, O = 16.0, H = 1.0). (Round off to two decimal places).

49. A bed with an attitude 020°, 30° NW is rotated 55° counter-clockwise (looking northerly) along its strike line. The dip of the bed after rotation will be ______° NW. (Answer in integer).

50. The width of the outcrop of a fault zone on a flat surface is 100 m as shown in the figure. A vertical borehole through the fault zone measured its vertical thickness to be 100 m. The true thickness of the fault zone is ____ m. (Round off to two decimal places).

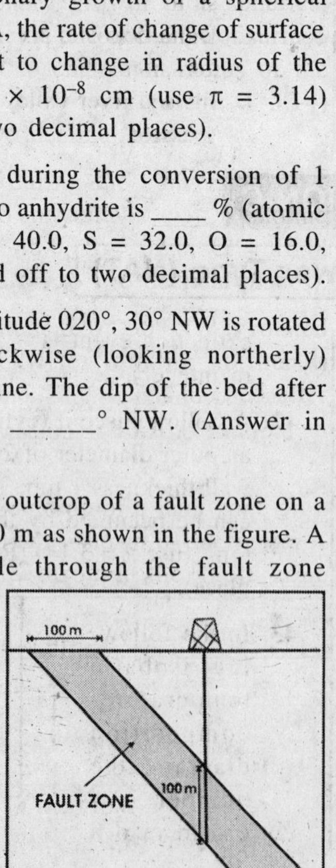

Qs. No. 51-60: *Carry two marks each.*

51. In an oblique slip fault with an attitude 000, 30° E, the net slip vector has a length of 20 m and a rake of 30° S on the fault plane. The displacement of a horizontal bed along the fault trace in a plane perpendicular to the strike of the fault is _______ m. (Answer in integer).

52. If the activity of a radioactive mineral falls from 800 counts/s to 500 counts/s in 80 minutes, half-life of the mineral is _______ minutes. (Round off to two decimal places).

53. In a laboratory experiment, water discharge through a porous rock sample in 2 hours was 10 cm^3. The cylindrical rock sample is 10 cm long and has a diameter of 50 mm. If the discharge occurred at a constant head of 300 cm, the coefficient of permeability of the rock sample is _____ × 10^{-6} cm/s. (Round off to two decimal places).

54. The following diagram represents a binary phase diagram for the system A–B at atmospheric pressure. If 'X' is the initial composition of melt, then the amount of melt that converts to solid when the magma cools from 1400°C to 1250°C is ________ %. (Round off to two decimal places).

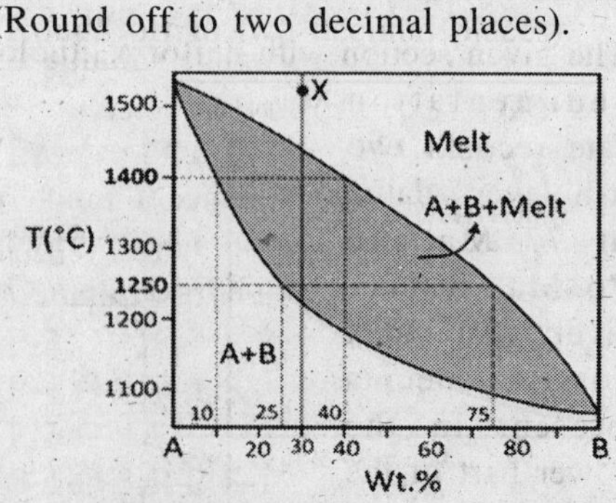

55. The following table shows modal abundance and mineral composition data of a plutonic igneous rock. The amount of SiO_2 in bulk composition of the rock is _______ %. (Round off to two decimal places).

Mineral	Mode (%)	SiO_2 (Wt. %)	CaO (Wt. %)	MgO (Wt. %)
Olivine	45	34	-	66
Clinopyroxene	35	55	25	20
Orthopyroxene	20	58	-	42

56. Refer to the schematic sketch given (not to scale). Assume average saturated density of oceanic crustal rocks = 3200 kg/m^3, density of ocean water = 1000 kg/m^3, and acceleration due to gravity = 10 m/s^2. The overburden pressure at a point (P) located 2 km below seabed and 4 km below the ocean surface is _________ MPa. (Answer in integer).

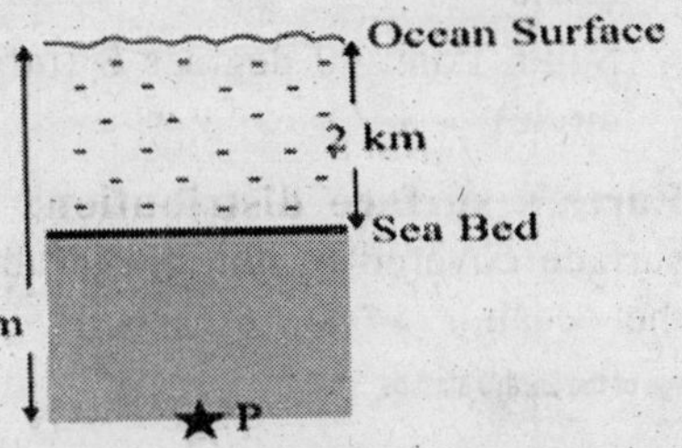

57. If the indices of refraction of a uniaxial section are $\varepsilon = 1.653$ and $\omega = 1.544$, and the retardation between the two rays is 550 nm, then the thickness of the section is _______ μm. (Round off to two decimal places).

58. A crystal has lattice parameters of a : 4.26 Å, b : 10 Å and c : 3.44 Å, respectively. A plane intercepts on the a, b and c axes at 2.13 Å, 10 Å and 1.72 Å, respectively. The Miller Indices for the plane, written as an integer, is _____ (Answer in integer).

59. In the uvavorite garnet ($Ca_3^{+2}Cr_2^{+3}Si_3^{+4}O_{12}$), Ca is in cubic coordination, Cr is in octahedral coordination and Si is in tetrahedral coordination. The electrostatic bond strength of the Ca^{2+} central ion is _______. (Round off to two decimal places).

60. In a structurally controlled fluvial setting, an asymmetric flight of river terraces T1, T2, T3 shown in the figure was sampled at location L1. The age of the sample at L1 was 30 ka (kiloyears). Assuming that the terraces were formed entirely due to deformation related uplift, the average uplift rate in the past 30 ka in the region was ____ mm/yr. (Answer in integer).

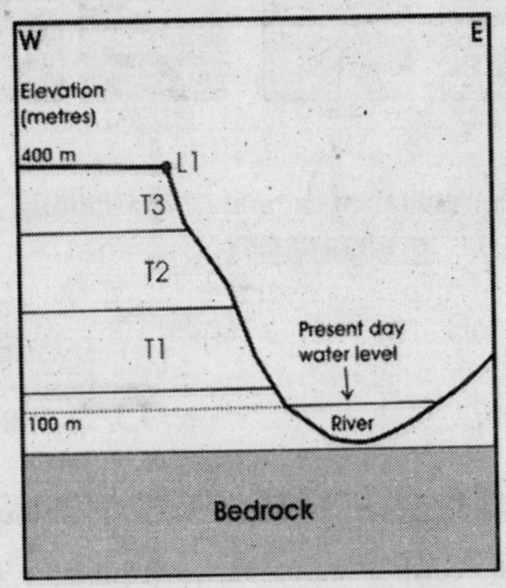

ANSWER KEY

1	2	3	4	5	6	7	8	9	10
D	A	C	B	C	C	B	A	D	A
11	**12**	**13**	**14**	**15**	**16**	**17**	**18**	**19**	**20**
D	A	B	B	C	D	A	C	B	D
21	**22**	**23**	**24**	**25**	**26**	**27**	**28**	**29**	**30**
A	C	A	C	C	D	B	C	A	B
31	**32**	**33**	**34**	**35**	**36**	**37**	**38**	**39**	**40**
A	A, B, C, D	A, C	B, C	A, B, D	A, B, C	A, B, D	A, C	A, C, D	A

41	**42**	**43**	**44**	**45**	**46**	**47**	**48**	**49**	**50**
74	2543.0 to 2546.0	2	0.75	40	15	50.20 to 50.30	20.90 to 20.95	85	70.00 to 71.50

51	**52**	**53**	**54**	**55**	**56**	**57**	**58**	**59**	**60**
10	117.40 to 118.00	2.30 to 2.40	56.40 to 56.80	45.00 to 47.00	84	5.00 to 5.08	212	0.25	10

HINTS AND SOLUTIONS

1. Metamorphic Grade: It is refers to the relative temperature and pressure to metamorphic change (Low grade to high grade) conditions to metamorphic rocks form.

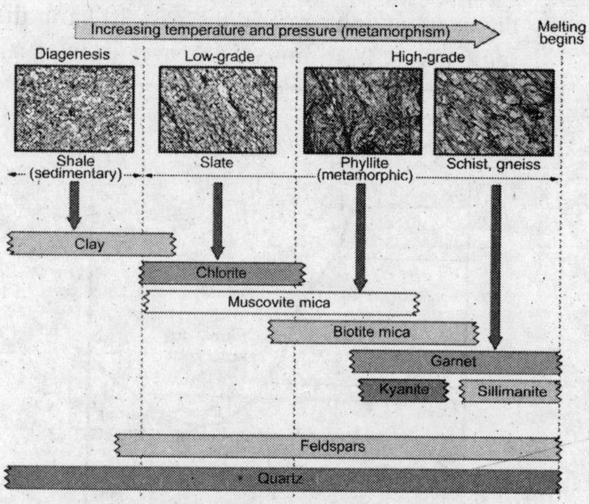

2. The earth's radius varies from 0 to 90 degree latitudes. The maximum and minimum radius of the earth's radius is 0 and 90 degree respectively. **Latitude** is actually an angular measurement north or south of the equator. So 30 degrees north means a point that is 30 degrees north of the equator. Latitude is also expressed in degrees, minutes, and seconds; 360 degrees in a circle, 60 minutes (') in a degree, and 60 seconds (") in a minute.

- Equator, 0 degrees
- Tropic of Cancer, 23.5 degrees N
- Tropic of Capricorn, 23.5 degrees S
- Arctic Circle, 66.5 degrees N
- Antarctic Circle, 66.5 degrees S
- North Pole, 90 degrees N (infinitely small circle)
- South Pole, 90 degrees S (infinitely small circle)

3. Earth's surface distribution: The earth's surface covered by the ocean about 70% by the ocean.

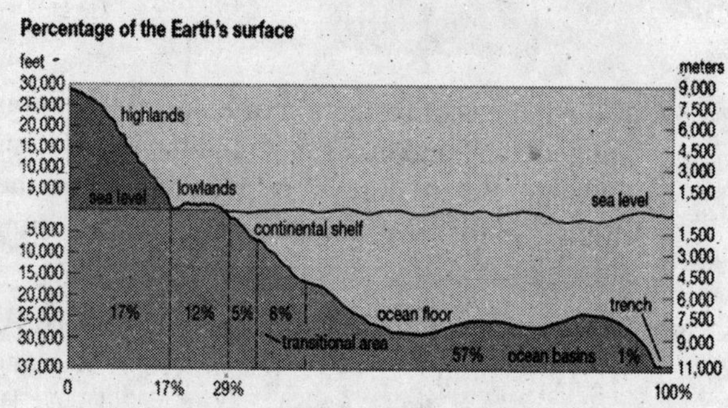

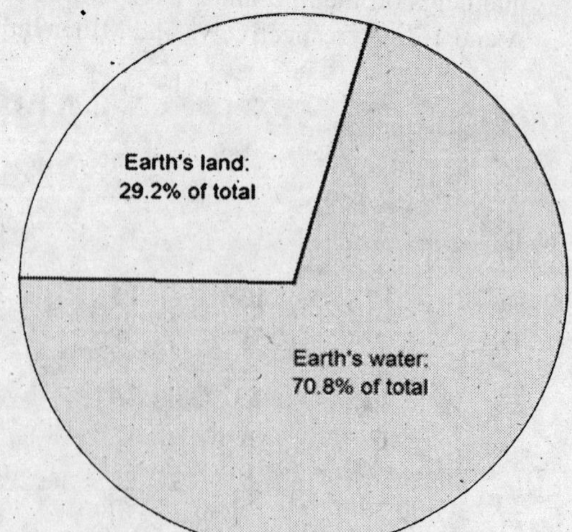

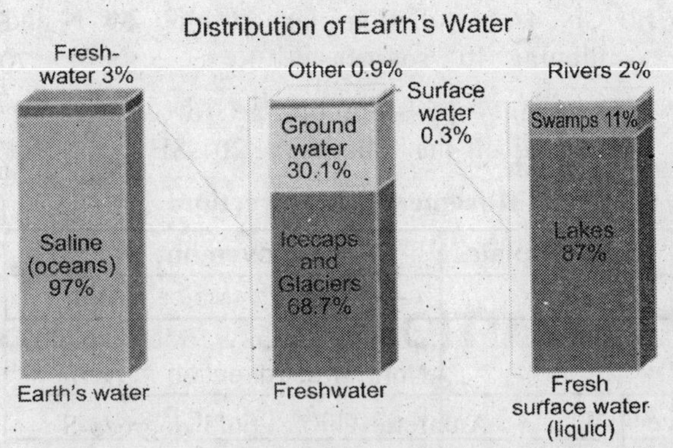

4. Classification of ocean with depth:

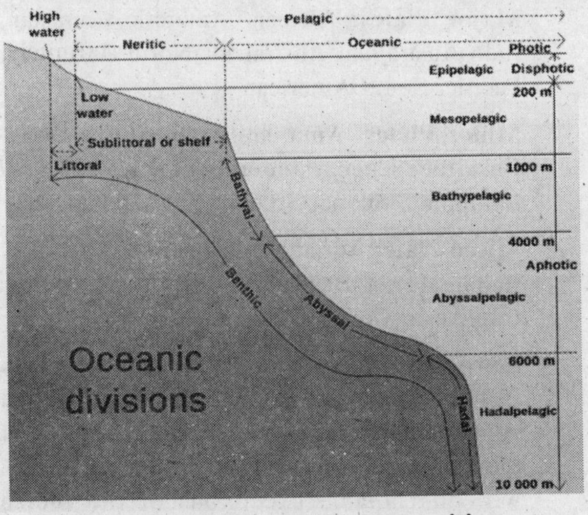

5. Minerals with chemical composition:

- **Bornite:** It is also known as peacock ore, is a sulphide mineral with chemical composition Cu_5FeS_4 that crystallizes in the orthorhombic system (pseudo-cubic).
- **Cassiterite:** SnO_2 (Tin oxide) tetragonal crystal.
- **Scheelite:** $CaWO_4$ (Calcium tungstate mineral) crystallize in tetragonal crystal.
- **Greenockite:** CdS (Cadmium bearing metal sulphide) crystallize in hexagonal crystal.

6. Pb-Zn deposit of Zawar: The **Zawar Belt** comprises a group of carbonate hosted deposits spread over a strike length of up to 20 km, within dolomites and phyllites of the Lower to Middle Proterozoic Aravalli Supergroup (2400 to 1700 Ma). The individual deposits, which include Zawar, Mochia and Balaria, are hosted by dolomites as disseminations, stringers and veinlets and together contain >40 Mt @ 4.5% Zn, 1.9% Pb (Res., 1992). The three main **Zawar** lead-zinc deposits are **Mochia, Balaria** and **Zawarmala** are located around 24° 20' N, 73° 42' E, some 25 km due south of Udaipur in Rajasthan. As such they are approximately 100 km to the south-west of the *Rajpura-Dariba* Zn-Pb mine and 215 km south-west of the *Rampura-Agucha* Zn-Pb open pit. The mines are operated by Hindustan Zinc Limited, a division of the Vedanta Group. Hindustan Zinc was originally formed in 1966 by the nationalisation of the lead-zinc mines of the Zawar district. These mines were in financial difficulties at that stage, and nationalisation was seen as the only way to keep them operating (Jain and Rakesh, 1984). Resource, Indian Bureau of Mines - 250 Mt @ 0.75 to 1.5% Pb, 1.5% Zn, (Krishnaswamy and Sinha, 1988).

7. Coal deposits in India:

- **Bareen Measures Formation:** Non coaliferous lithounit of Jharia basin Jharkhand.
- **Barakar Formation:** Bituminous coal formation of lower Gondwana.
- **Naredi Formation:** The **Naredi Formation** of Kutch, Gujarat, India, is early Eocene in age. The shallow marine shelf sediment deposition of lignite deposit.
- **Cuddalore Formation:** Lignite deposit of Miocene – Pliocene of Tamil Nadu India.

8. Fossils with remark:

Fossils	Remarks
Glossopteris	Lower Gondwana plant fossils
Fenestella	Fenestella, genus of extinct bryozoans, small colonial animals, especially characteristic of the Early Carboniferous Period.
Productus	Brachiopoda invertebrate fossils
Cidaris	Phylum of Echinodermata

9. Igneous body with name:

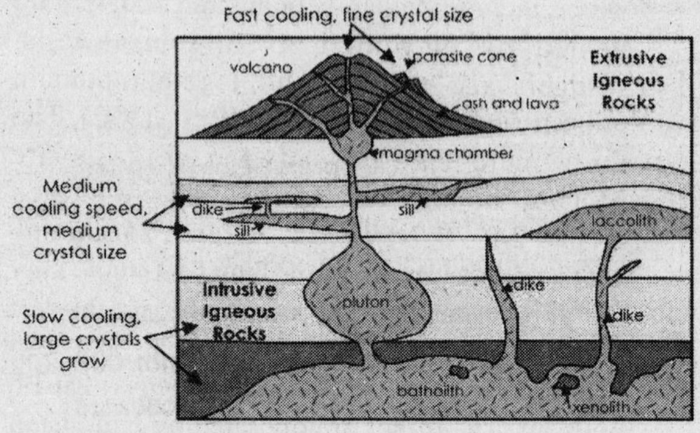

Igneous forms	Explanations
Concordant bodes	A igneous body which parallel to the rocks strats.
Disconcordant bodies	A igneous body which cross cut relation to the country rocks.
Lopolith	A igneous body which is lenticular in shape (Convex downward).
Lacolith	A mashroom shape body which is convex upward.

10. Stratigraphic Units:

Stratigraphic Units	Geological Age
Lameta Formation	Cretaceous of Infratrappean Beds, is a sedimentary rock formation found in Madhya Pradesh, Gujarat, and Maharashtra, India.
Talchir boulder bed	Lower Gondwana of Upper Carboniferous in Jharkhand and Orissa.
Fenestella Shale	Carboniferous age
Kasauli Formation	Early Miocene age of Himalayan

11. Volcanic Event:

Stratigraphic Units	Geological Age
Rajmahal Volcanics	Cretaceous
Dalma Volcanics	Precambrian
Panjal Volcanics	Permian
Deccan Volcanics	Tertiary

12. Given, the foliation plane striking 20° N and dipping 40° southeasterly

The only possible plunge for a lineation located on the foliation = 20° SE

13. Plate Movement with direction:

Tectonic plate	Movement
Eurasian	7–14 mm/yr (Major plate)
Pacific	566–102 mm/yr (Major plate) North–west direction movement
African	2.15 cm/yr (Major Plate)
North America	15–25 mm/yr

Major Plates: African, Antarctic, Eurasian, Indo-Australian, Australian, North American, Pacific, South American

Minor Plates: Amurian, Arabian, Caribbean, Caroline, Cocos, Indian, Nazca, Okhotsk, Philippine, Scotia, Somali, Sunda, Yangtze.

Micro Plate: Adriatic, Aegean Sea, Anatolian, Balmoral Reef, Banda Sea, Bird's Head, Burma, Capricorn, Coiba, Conway Reef etc.

Eurasian Plate: The Eurasian Plate is a tectonic plate which includes most of the continent of Eurasia (a landmass consisting of the traditional continents of Europe and Asia), with the notable exceptions of the Indian subcontinent, the Arabian subcontinent, and the area east of the Chersky Range in East Siberia. It also includes oceanic crust extending westward to the Mid-Atlantic Ridge and northward to the Gakkel Ridge.

Pacific Plate: The Pacific Plate is an oceanic tectonic plate that lies beneath the Pacific plate. At 103 million km^2 (40 million sq mi), it is the largest tectonic plate. The Pacific Plate contains an interior hot spot forming the Hawaiian Islands.

African Plate: The African Plate is a major tectonic plate straddling the Equator as well as the prime meridian. It includes much of the continent of Africa, as well as oceanic crust which lies between the continent and various surrounding ocean ridges.

North American Plate: The North American Plate is a tectonic plate covering most of North America, Cuba, the Bahamas, extreme northeastern Asia, and parts of Iceland and the Azores. With an area of 76 million km^2, it is the Earth's second largest tectonic plate, behind the Pacific Plate (which borders the plate to the west).

14. Dunham's classification of Limestone:

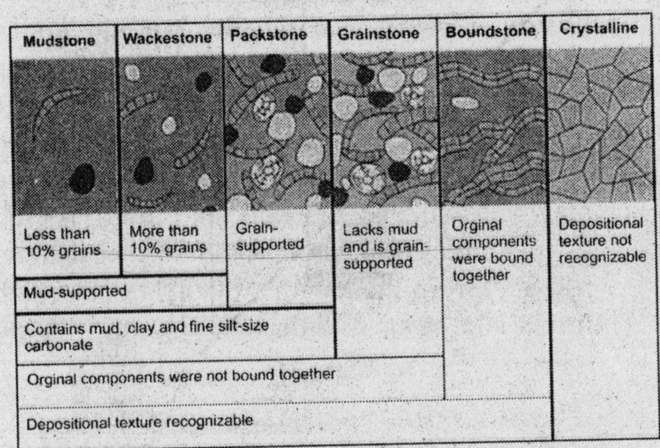

15. Convolute Laminae: Mud cracks formed by desiccation. Convolute laminae due to complete or partial erosion of the newly deposited **laminae**.

Load Cast: Load casts are bulges, lumps, and lobes that can form on the bedding planes that separate the layers of sedimentary rocks. The lumps "hang down" from the upper layer into the lower layer, and typically form with fairly equal spacing. These features form during soft-sediment deformation shortly after sediment burial, before the sediments lithify. They can be created when a denser layer of sediment is deposited on top of a less-dense sediment. This arrangement is gravitationally unstable, which encourages formation of a Rayleigh-Taylor instability if the sediment becomes liquefied. Once the sediments can flow, the instability creates the "hanging" lobes and knobs of the load casts as plumes of the denser sediment descend into the less-dense layer.

Prod mark: Asymmetrical impact mark is found at the bottom of a bed.

Wave ripple: The ripples induced by wave action are called "wave ripples"; their characteristics being different from those of the ripples generated by steady flows. The most striking difference between wave ripple fields and current ripple fields is the regularity of the former. Indeed, regular long-crested wave ripple fields are often observed on tidal beaches from which the sea has withdrawn at low water. This occurs at the bottom of a deposit.

16. Crystal Forms with properties:

Crystal System	Crystal Class	Name of class	Symmetry Content
Triclinic	1	Pedial	None
	1	Pinacoidal	I
Monoclinic	2	Sphenoidal	1A2
	M	Domatic	1 m
	2/m	Prismatic	i, 1 A2, 1 m
Orthorho-mbic	222	Rhombic-disphenoidal	3A2
	mm^2	Rhombic-pyramidal	1A2, 2 m
	2/m^2/m^2/m	Rhombic-dipyramidal	i, 3A2, 3 m
Tetragonal	4	Tetragonal-pyramidal	1A4
	4	Tetragonal-disphenoidal	1A2
	4/m	Tetragonal-dipyramidal	i, 1A4, 1 m
	422	Teragonal-trapezohedral	1A4, 4A2
	4 mm	Ditetragonal-pyramidal	1A4, 4 m
	42 m	Tetragonal-scalenohedral	3A2, 2 m
	4/m^2/m^2/m	Ditetragonal-dipyramidal	i, 1A4, 4A2, 5 m
Hexagonal	3	Trigonal-pyramidal	1A3
Rhombo-hedral Division	3	Rhombo-hedral	i, 1A3

	32	Trigonal-trapezohedral	1A3, 3A2
	3 m	Ditrigonal-pyramidal	1A3, 3 m
	32/m	Trigonal (Hex.)-scalenohedral	i, 1A3, 3A2, 3 m
Hexagonal division	6	Hexagonal-pyramidal	1A6
	6	Trigonal-dipyramidal	1A3, 1 m
	6/m	Hexagonal-dipyramidal	i, 1A6, 1 m
	622	Hexagonal-trapezohedral	1A6, 6A2
	6 mm	Dihexagonal-pyramidal	1A6, 6 m
	62 m	Ditrigonal-dipyramidal	1A3, 3A2, 4 m
	6/m²/m²/m	Dihexagonal-dipyramidal	i, 1A6, 6A2, 7 m
Isometric	23	Tetartoidal	3A2, 4A3
	2/m³	Diploidal	i, 3A2, 3 m, 4A3
	432	Gyroidal	3A4, 4A3, 6A2
	43 m	Hextetra-hedral	4A3, 3A2, 6 m
	4/m 32/m	Hexocta-hedral	i, 3A4, 4A3, 6A2, 9 m

17. IUGS Classification of Igneous rocks:

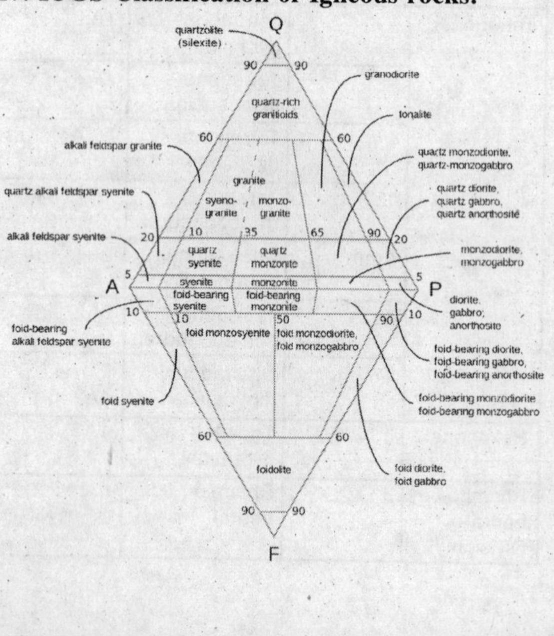

18. Crystallization of mafic magma: The sequence in which minerals crystallize from a magma is known as the **Bowen reaction series**. The common silicate minerals, olivine normally crystallizes first, at between 1200° and 1300°C. As the temperature drops, and assuming that some silica remains in the magma, the olivine crystals react (combine) with some of the silica in the magma to form pyroxene. As long as there is silica remaining and the rate of cooling is slow, this process continues down the discontinuous branch: olivine to pyroxene, pyroxene to amphibole, and (under the right conditions) amphibole to biotite. At about the point where pyroxene begins to crystallize, plagioclase feldspar also begins to crystallize. At that temperature, the plagioclase is calcium-rich (anorthite). As the temperature drops, and providing that there is sodium left in the magma, the plagioclase that forms is a more sodium-rich variety.

Bowen's Reaction Series

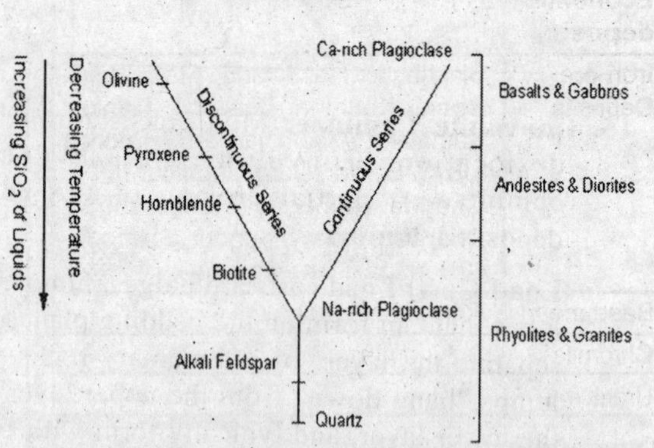

As a *mafic* magma starts to cool, some of the silica combines with iron and magnesium to make olivine. As it cools further, much of the remaining silica goes into calcium-rich plagioclase, and any silica left may be used to convert some of the olivine to pyroxene. Soon after that, all of the magma is used up and no further changes takes place. The minerals present will be olivine, pyroxene, and calcium-rich plagioclase. If the magma cools slowly underground, the product will be **gabbro**; if it cools quickly at the surface, the product will be **basalt**. The compositional change in plagioclase during the crystallization of mafic magma, Na/Ca ratio increases and Al/Si ratio decreases.

19. Classification of Cuddapah Supergroup:

	Group	Formations
Cuddapah Supergroup	**Kurnool Group** (500 + m)	Nandyal Shale Koilkuntala Limestone Paniam Quartzite Owk Shale Narji Limestone Banganapalle Quartzite
	~~ unconformity	
	Chitravati Group (4975 m)	Gandikota Quartzite Tadpatri Formation Pulivendla Quartzite
	~~ unconformity	
	Papaghni Group (2110 m)	Vempalle Formation Gulcheru Quartzite
	~~ unconformity	
	Peninsular Gneiss (Dharwar batholith)	

20. Major Economic deposits in India:

Economic deposits	Remarks
Iron ore Deposits	Singhbhum (Jharkhand), Mayurbhanj, Bonai, Keonjhar (Orissa), Tumkur Karnataka, Assam Jantia Hill, Poonch (J&K), Baster, Durg, Jabalpur (MP), Chanda Ratnagiri (Maharashtra), Dharwar, Bellary, Sundur, Shimoga, Chikmagalur (Karnataka)
Base metal	Rampura-Agucha
Chromite	Sukinda Chromite Deposit
Uranium	Bhatan

21. Type of Unconformity:

Characteristics	Unconformity
Contact effect below and above and plane of unconformity	Disconformity
Pebbles of non-sedimentary rocks over the plane of unconformity	Nonconformity
Paraconformable bedding planes	Paraconformity
Angular discordance at the plane of unconformity	Angular unconformity

22. Morphology of Invertebrates:

A. Morphology of Ammonoites:

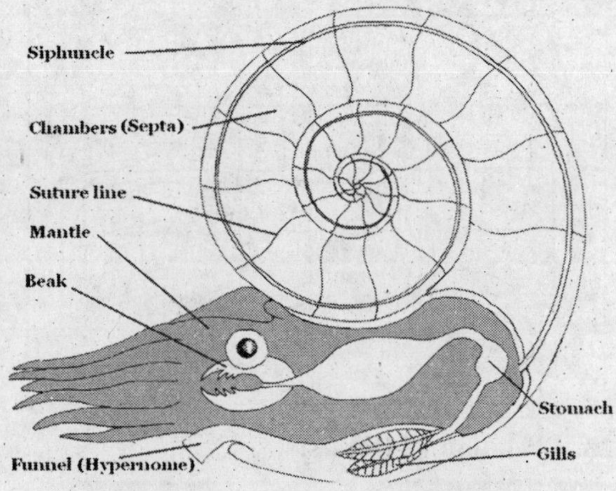

B. Morphology of Brachiopods:

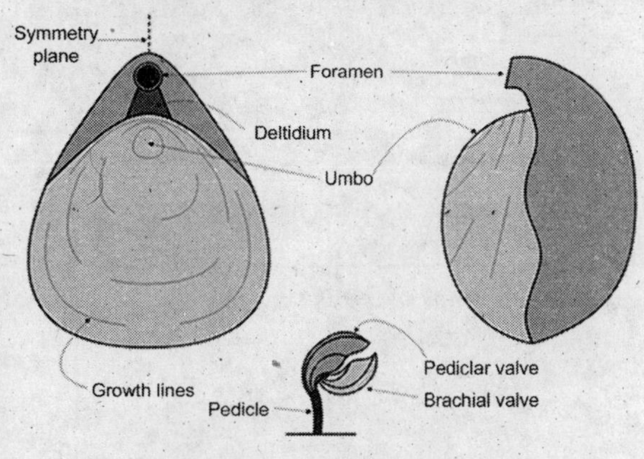

C. Morphology of Echinoids:

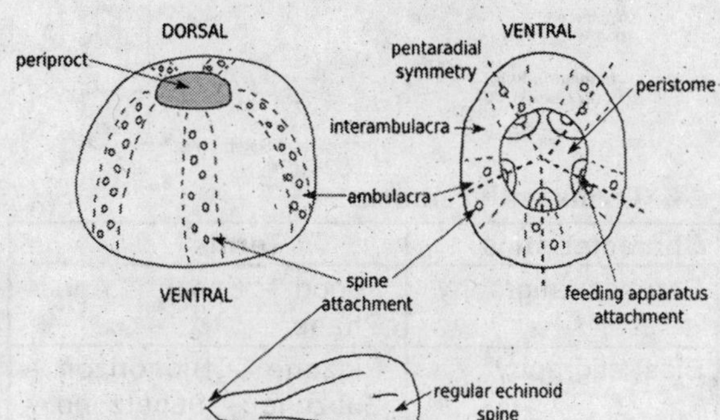

D. Morphology of Trilobites:

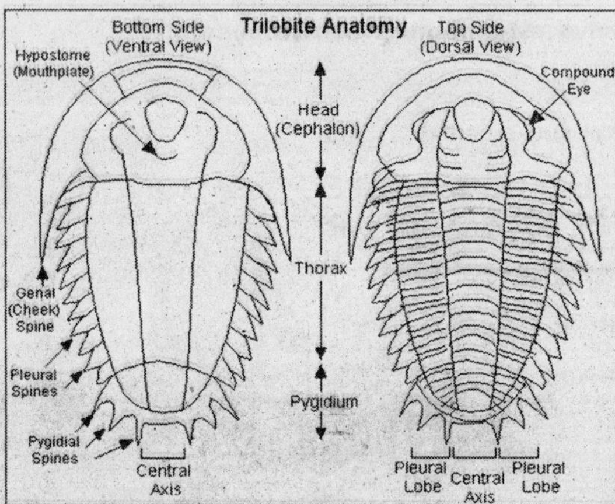

23. Evolution of Horses:

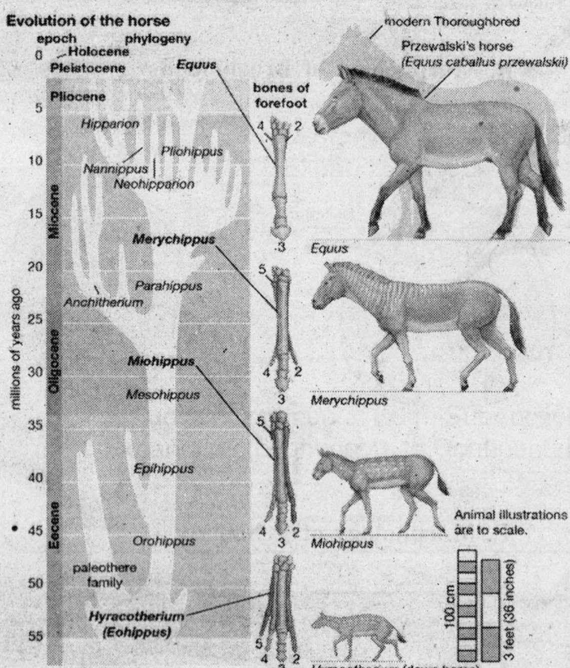

24. Terms with remarks:

Characteristics	Terms
Chronostratigraphy	Period – Epoch – Age – Phase
Biostratigraphy	Biozone – Biohorizon – Subzone – Superzone – Zonul
Lithostratigraphy	Group – Formation – member – bed – Flow
Magnetostratigraphy	Magnetic Polarity of the rocks with respect of earth magnetic zone

25. Buchan Type of Metamorphism:

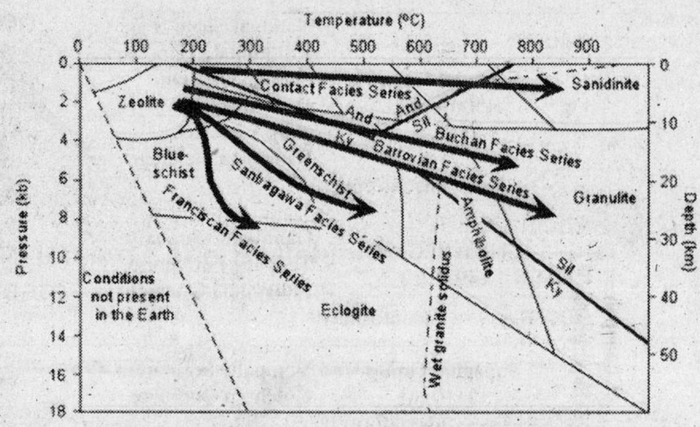

26. Minerals with composition:

Minerals	Compositions
Leucite (Feldspathoids)	$KAlSi_2O_6$
Andradite (Garnet)	$Ca_3Fe_2Si_3O_{12}$
Sanidine (Feldspar)	$(K,Na) AlSi_3O_8$
Jadeite (Pyroxene)	$NaAlSi_2O_6$

27. Optic orientation of Crystal System:

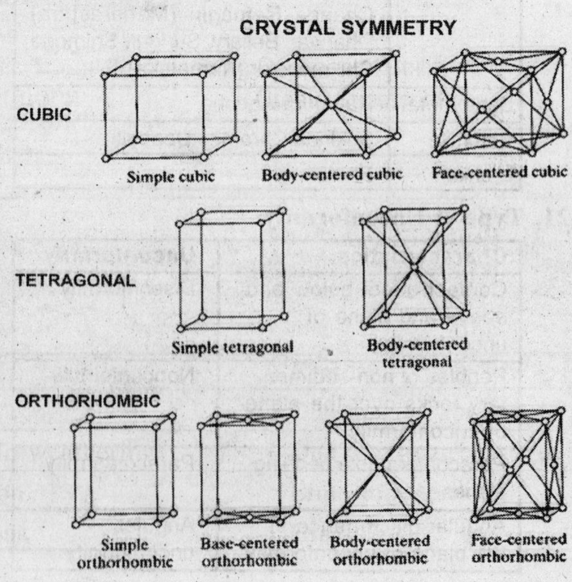

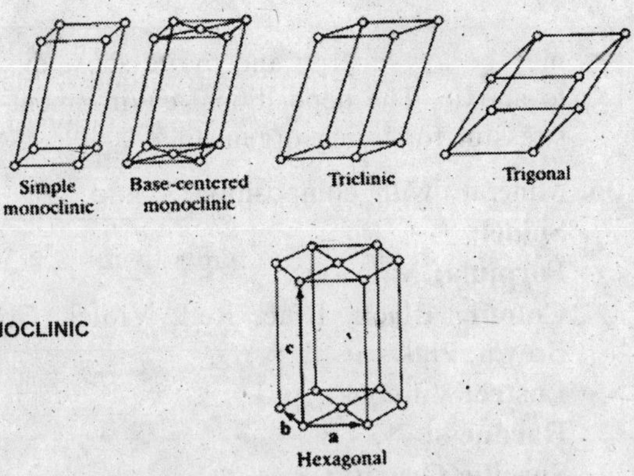

MONOCLINIC

28. Landforms with Process:

Processes	Landforms
Aeolian	Seif, Barchans, Parabolic dunes, Yardang, Ventifact
Glacial	Horn, Hanging valley, Fijord, Morain, Esker, Kames
Fluvial	Water falls, Cuesta, Alluvial fan, Bajada, Delta, Meander
Coastal	Bars, Cliff, Tombolo, Beach

29. Types of deposits with minerals example:

Minerals	Process of formation
Bauxite	Sedimentary deposit
Monazite	Placer deposit
Chromite	Magmatic deposit
Graphite	Regional/contact metamorphism
Chalcocite, Sphelerite, Enargite, Chalcopyrite, Bornite, Galena, Gold, Silver	Hydrothermal deposits

30. Types of aquifer:

Confined/Unconfined Aquifers

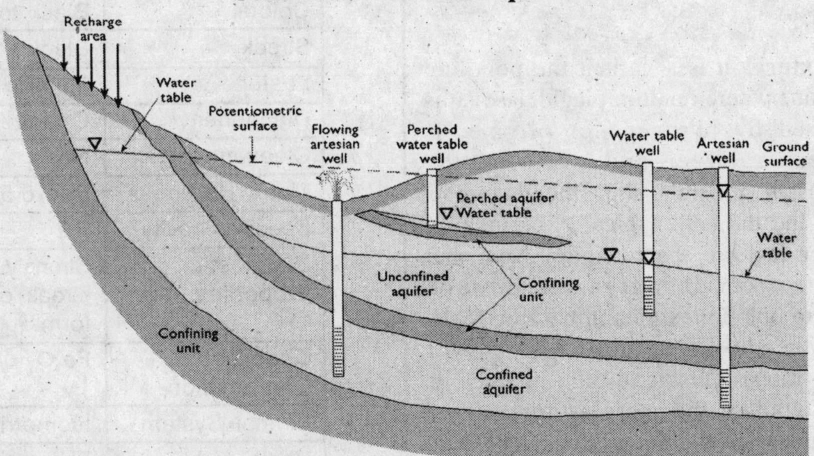

31. Pentameral Symmetry of Invertebrate: In invertebrate only Echinoidea show pentameral symmetry. In the animal kingdom, there is only one known phylum that exhibits pentaradial symmetry, and that's the Phylum Echinodermata. Echinoderms include sea stars (or starfish), sea urchins, sea lilies, sand dollars, and sea cucumbers. They're a weird and wonderful group that are members of the superphylum Deuterostomata — a group we also belong to in which the anus forms first during embryological development (as opposed to the mouth).

Some animals, however, have a special kind of radial symmetry that is based on the number five. **Pentaradial symmetry** ('penta-' means five) follows a pattern where the parts of the animal branch out into five distinct compartments, or arms. Pentaradial symmetry is also known as **pentamerism**, another term referring to the number five.

32. Geometry of folds: The given attitude of the first limb of the fold = Striking 4 degree N, Dipping 85 degree easterly.

Attitude of the second limb of the fold = Striking 30 degree N, Dipping 60 degree easterly.

The given attitude shows the limbs of the fold, same as a strike direction as well as dip direction. So these geometry may be Synform, Antiform, Overturned or plunging folds.

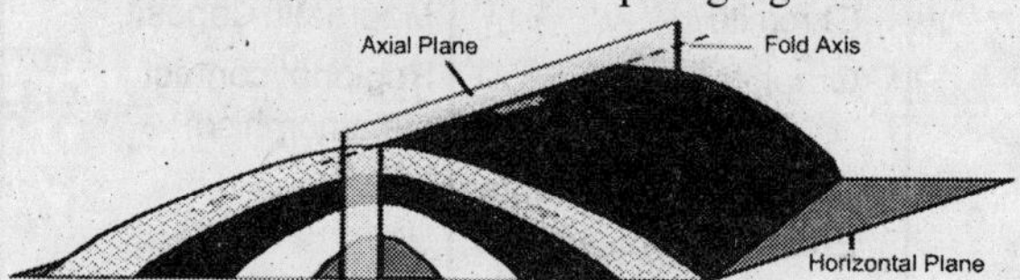

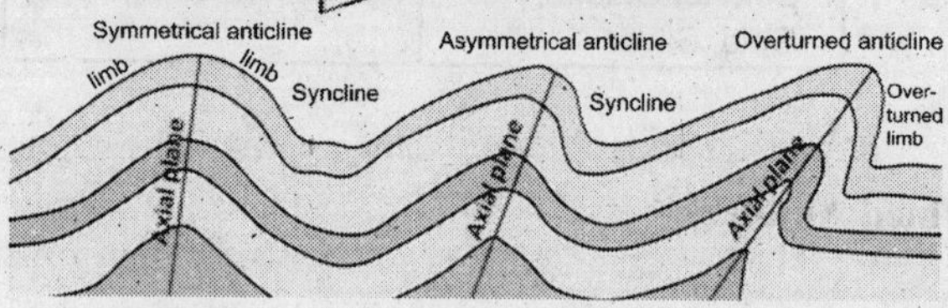

33. Ophitic texture: It is a variant the poikilitic texture, is one where random plagioclase laths are enclosed by pyroxene or olivine. If plagioclase is larger and encloses the ferromagnesian minerals, then the texture is subophitic and the laths typically impinge on one another to form sharp angles. Note that the change from intergranular through subophitic to ophitic textures in basaltic rocks results from slower cooling and slower nucleation rates. This textural sequence is typically found at the margins toward the center in diabasic or doleritic rocks (basaltic dikes) or from the chilled surface to depth of basaltic flows. If the cooling rate is very fast, the interstitial material between plagioclase laths may be quenched to glass to form an intersertal texture.

34. From the given diagram: The first develop matrix foliation (S_2) then mineral X grew. Then mineral Y grew during or after the development of internal foliation (S_1).

35. The given statement, a marine organic rich, black shale with tiny pyrite shows complete absence of body or trace fossils. This condition may be low–energy and pyrite show dysoxic to anoxic. The depositional environment was stressful for living organism.

36. Minerals with composition:

Spinel:

Formula: $MgAl_2O_4$

Colour: Black, Blue, Red, Violet, Green, Brown, Pink

Lustre: Vitreous

Hardness: 7½ - 8

Specific Gravity: 3.6 - 4.1

Crystal System: Isometric

Member of: Spinel Subgroup > Oxyspinel Group > Spinel Supergroup

Physical Properties of Magnetite	
Chemical Classification	Oxide
Colour	Black to silvery gray
Streak	Black
Luster	Metallic to submetallic
Diaphaneity	Opaque
Cleavage	None
Mohs Hardness	5 to 6.5
Specific Gravity	5.2
Diagnostic Properties	Strongly magnetic, colour, streak, octahedral crystal form.
Chemical Composition	Fe_3O_4
Crystal System	Isometric
Uses	The most important ore of iron. Heavy media separation. Studies of Earth's magnetic field.

Physical Properties of Chromite	
Chemical Classification	Oxide
Colour	Dark gray to black, rarely brownish black
Streak	Dark brown
Luster	Metallic to submetallic
Diaphaneity	Opaque

Cleavage	None
Mohs Hardness	5.5 to 6
Specific Gravity	4.0 to 5.1 (variable)
Diagnostic Properties	Luster, streak
Chemical Composition	$FeCr_2O_4$ with magnesium substituting for iron in significant amounts
Crystal System	Isometric
Uses	An ore of chromium

Physical Properties of Ilmenite	
Chemical Classification	Oxide
Colour	Black
Streak	Black
Luster	Metallic, submetallic
Diaphaneity	Opaque
Cleavage	None
Mohs Hardness	5.5 to 6
Specific Gravity	4.7 to 4.8
Diagnostic Properties	Streak; sometimes weakly magnetic.
Chemical Composition	Iron titanium oxide - $FeTiO_3$. Sometimes has significant amounts of magnesium and manganese in solid solution with the iron to yield a composition of $(Fe, Mg, Mn)TiO_3$
Crystal System	Hexagonal
Uses	The primary ore of titanium. A minor source of iron. Used to make titanium dioxide.

37. Terms with explanation:

Attrition: Process of erosion which particles undergo mutual rubbing, grinding. This type of process more active in Aeolian (desert region) environment other than cold (glacial) environment.

Channel Meanders: A **meander** is one of a series of regular sinuous curves, bends, loops, turns, or windings in the channel of a river, stream, or other watercourse. It is produced by a stream or river swinging from side to side as it flows across its floodplain or shifts its channel within a valley. When a fluid is introduced to an initially straight channel which then bends, the sidewalls induce a pressure gradient that causes the fluid to alter course and follow the bend. From here, two opposing processes occur: (1) irrotational flow and (2) secondary flow. For a river to meander, secondary flow must dominate.

U-shaped valley: U-shaped valleys form through glacial erosion. Glaciation develops in established V-shaped river valleys where the ice erodes the surrounding rocks to create a "U" shaped valley with a flat bottom and steep sides. Glacier movement is driven by gravity. So this is not common geomorphic feature.

38. Gandhar Field: This basin for oil in Cambay Basin in India.

Bombay High: Limestone reservoir rock of Miocene age.

Digboi Oil field: It is located in Tinsukia District of Assam. Digboi is known as the Oil City of Assam where the first oil well in Asia was drilled. The first refinery was started here as early as 1901.

Ankleshwar Oil field: This is situated in Cambay Basin, India, which is one of the main onshore Cenozoic **oil** basins of India. The entire area supposed to have subsided during late Paleocene, resulting in deposition of monotonous Cambay shale, which is composed of grey shales in thin alternations with silty and carbonaceous shales, considered as the main source rock.

39. Classification of Sandstone by Pettijohn: Arkose contains more than 25% feldspar, Greywacke contain up to 15% matrix, Litharenite contains more than 25% lithic fragment and quartz arenite contains more than 90% quartz.

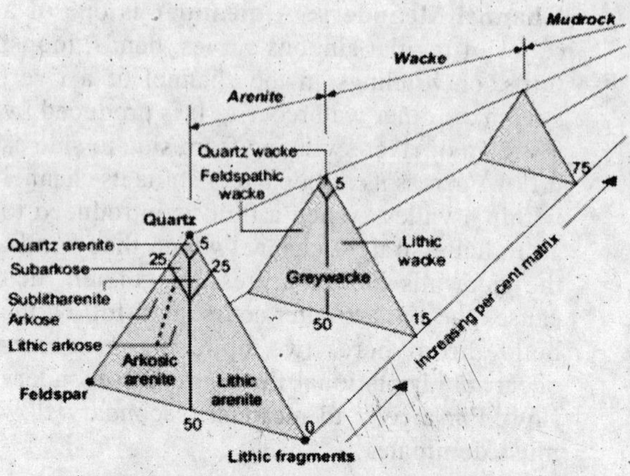

40. Transgression: A marine **transgression** is a **geologic** event during which sea level rises relative to the land and the shoreline moves toward higher ground, which results in flooding. **Transgressions** can be caused by the land sinking or by the ocean basins filling with water or decreasing in capacity.

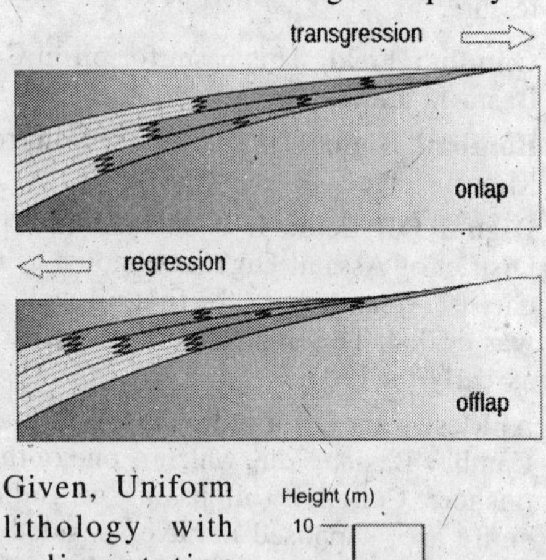

41. Given, Uniform lithology with sedimentation rate record first = 77 Ma Second = 76 Ma

An index fossil species in the lower part of the section became extinct at a = 7 m horizon

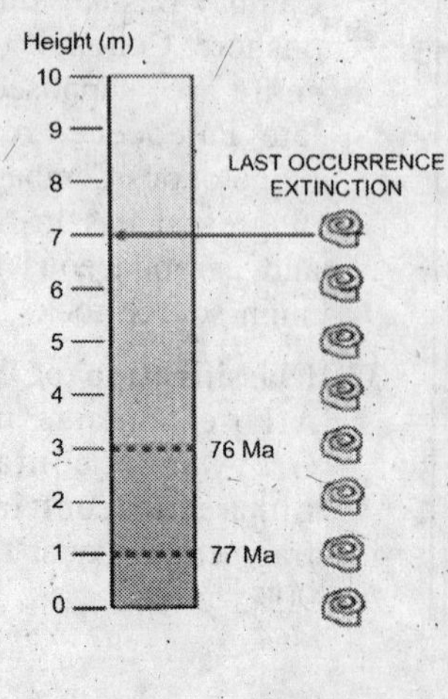

At 1 m age of fossil = 77 Ma

At 3 m age of fossil = 76 Ma

At 5 m age of fossil is = 75 Ma

At 7 m age of fossil is = 74 Ma.

42. Given, A hallow discoid (cylindrical) microfossil outer diameter = 20 μm

Height of the discoid is = 10 μm

Wall thickness of the discoid is = 1.0 μm

The value of pie(π) = 3.14

Larger radius r_1 of the discoid = 10 μm

Smaller radius r_2 of the discoid = 10 – 1.0 = 9.0 μm

Then volume of discoid = pie(π) × r_1 × r_1 × height

= 3.14 × 9.0 × 9.0 × 10.0

= 2543.4 μm^3.

43. Given isobaric temperature-composition diagram,

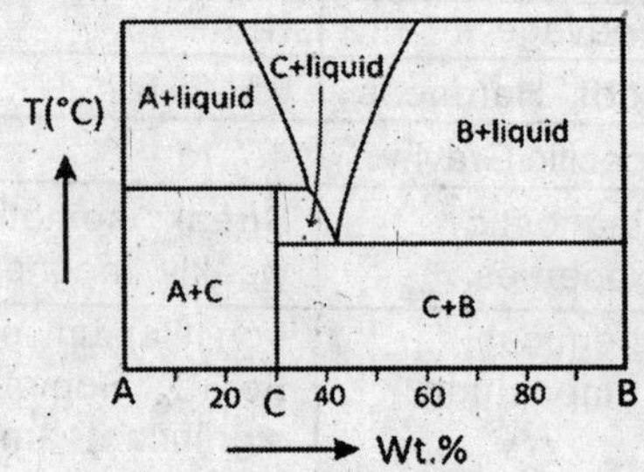

The number of common phase in given diagram = 2

Degree of freedom at invariant point F = 0

Then the two point occur at degree of freedom = 0

Now the number of common phases in all the invariant points is = 2.

44. Given, The muscovite composition

$= KAl_{2.50}Fe_{0.25}Si_{3.25}O_{10}(OH)_2$

General formula of muscovite

$= KAl_2(AlSiO_3)O_{10}(OH)_2$

The amount of Al in the tetrahedral site

= X + 3.25 = 4

X = 4 – 3.25 = 0.75.

45. Given, The density of the gabbro sample (mass) = 200 g = 0.2 kg

Density of the form of a cube (D) = 3125 kg/m^3

Density (D) V = m/D = 0.2/3125 = 6.4×10^{-5}

or V = $4 \times 4 \times 4 \times 10^{-6}$ m^3

$\therefore$ $l = \sqrt[3]{4 \times 4 \times 4 \times 10^{-6}}$

$\therefore$ $l = 4 \times 10^{-2}$ = 0.04 m = 40 mm.

46. Given, A drill run in coalfield = 3 m = 300 cm

Core sample recovered for cumulative length = 255 cm

The core lose in % = (300 – 225/300) × 100 = 15%.

47. Given, Concretionary growth of a spherical grain of radius r = 2A

Surface area of the sphere = $4\pi r^2$

Rate of change of surface area

= $d(A)/dr = [d/dr\ (4\pi r^2)]$

= $4\pi \times 2r$

= $4\pi(2 \times 2)$ = 50.24.

48. Give, Atomic weights of Ca = 40.0

S = 32.0

O = 16.0

H = 1.0

$CaSaO_4.2H_2O - CaSO_4$

40 + 32 + 16 × 4. 2(1 × 2 + 16)

1 mole of gypsum to anhydrite =

Total weight loss%

= (36/40 + 32 + 64 + 36) × 100

= 20.93%.

49. Given, Attitude of the bed = 20°, 30° NW

Rotation of the bed in counter-clockwise (northerly) along strike line = 55°

The dip of the bed after rotation = 55 + 30 = 85°.

50. Given, diagram the width of the fault zone on a flat surface = 100 m

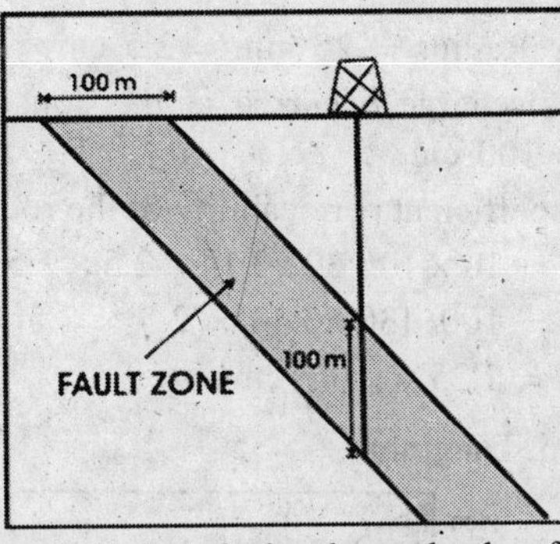

A vertical borehole through the fault zone measured its vertical thickness = 100 m

tan θ = 100/100 = 1

θ = 45°

The true thickness of the fault zone

= sin 45 × 100

= 0.707 × 100 = 70.71 m.

51. Given, An oblique slip fault with attitude = 000, 30° E

The net slip vector = 20 m

A rake of the fault plane = 30° S

The displacement of a horizontal bed along the fault trace in a plane perpendicular to the strike of the fault

sin 30° = Dip slip/20

Dip slip = sin 30° × 20 = 10 m.

52. Given, The activity of a radioactive mineral falls from = 800 counts/s to 500 counts/s

Time = 80 minutes = 4800 Sec

Half-life of the mineral N = Noe$^{\lambda - t}$

$500 = 800 \times e^{-\lambda(4800)}$

$\lambda = 9.75 \times 10^{-5}$

$t1/2 = 0.693/\lambda$

$= 0.693/9.75 \times 10^{-5}$

= 7077.44 sec = 117.45 min.

53. Given, In a laboratory experiment, water discharge through porous rock sample

= 2 hours = 7200 Sec

Volume = 10 cm^3

The cylindrical rock sample length l = 10 cm

The diameter of the cylindrical = 50 mm

Radius = 25 mm = 2.5 cm

The discharge occurred at the constant head (h) = 300 cm

The coefficient permeability of the rock sample

k = 10 × 10/ [300(3.14 × 2.5 × 2.5 × 7200)]

= 100/ [300(3.14 × 2.5 × 2.5 × 7200)

= 2.35 × 10^{-6} cm/s

54. Given, Diagram

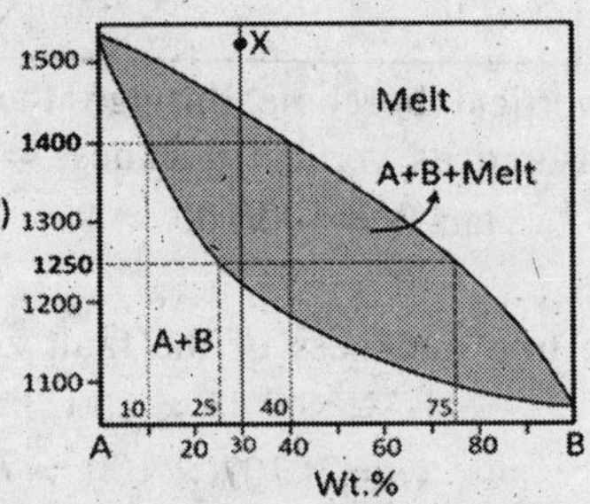

By lever rule: melt to solid at 1400°C

= (20/30) × 100 = 66.66%

Melt to solid at point 1250°C

= (5/50) × 100 = 10%

Difference = 66.66% – 10% = 56.66%.

55. Given, Silica in Olivine = 0.34

Silica in Clinopyroxene = 0.55

Silica in Orthopyroxene = 0.58

Bulk composition of silica

= (0.34 × 45 + 0.55 × 35 + 0.58 × 20/100) × 100

= 46.15%.

56. Given, Average saturated density of oceanic crustal rocks = 3200 kg/m^3

Density of ocean water = 1000 kg/m^3

Mean density = 3200 +1000/2 = 4200/2

= 2100 kg/m^3

Acceleration due to gravity = 10 m/s^2

The overburden at point P = 2100 × 4000 × 10

= 84MPa.

57. Given, The indices of refraction of a uniaxial section ε = 1.65

ω = 1.544

Retardation between the two rays = 550 nm

Retardation = thickness (ε – ω)

Thickness = 550/(1.653 – 1.544)

= 50.458 μm.

58. Given, Lattice parameters of a : 4.26 Å, b : 10 Å and c : 3.44 Å, respectively.

A plane intercepts on the a, b and c axes at 2.13 Å, 10 Å and 1.72 Å, respectively.

The Miller Indices for the plane = Crystal lattice/ intercept

= 4.26/2.13 10/10 3.44/1.72 = (212).

59. Given, In the uvavorite garnet ($Ca_3^{+2}Cr_2^{+3}Si_3^{+4}O_{12}$)

Ca is in cubic coordination

Cr is in octahedral coordination and Si is in tetrahedral coordination.

The electrostatic bond strength of the Ca^{2+} central ion is = Charge/C.N

CN of Ca^{2+} is in = +2/8 = 0.25.

60. Given, In a structurally controlled fluvial setting, an asymmetric flight of river terraces T1, T2, T3 shown in the figure was sampled at location L1.

The age of the sample at L1 was 30 ka (kiloyears).

Assuming that the terraces were formed entirely due to deformation related uplift.

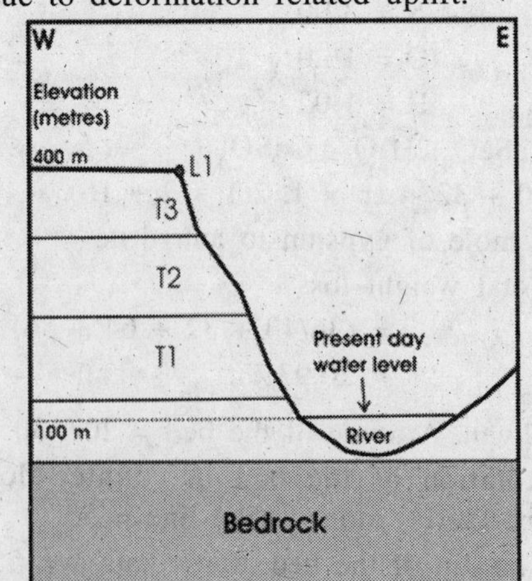

The rate of upliftment

= 400 – 100 × 10^3/30 × 10^3 mm/yr

= 10 mm/yr.

Previous Papers (Solved)

Indian Institute of Technology Joint Admission Test for M.Sc.

IIT-JAM GEOLOGY-2020

SECTION-A

Multiple Choice Questions (MCQ)

Qs. Nos. 1-10 : *Carry one mark each.*

1. The Mesozoic Era approximately ranges from:

A. 1000-540 Ma B. 250-65 Ma
C. 540-250 Ma D. 65 Ma-Present

2. A trench is found at a:

A. divergent plate boundary
B. convergent plate boundary
C. transform boundary
D. passive margin

3. The strike (in degree notation) of a bed dipping 30° towards N45W is:

A. 045-225 B. 025-205
C. 020-200 D. 030-210

4. Randomly oriented ore-bearing veinlets in a mass of rock are known as:

A. en echelon veins B. sheeted veins
C. lode D. stockworks

5. The diagnostic amphibole in a blueschist facies metabasalt is:

A. hornblende B. anthophyllite
C. glaucophane D. actinolite

6. Atoll is a geomorphic feature formed by:

A. glacial erosion
B. wind flow abrasion
C. fluvial deposition
D. coral reef accumulation

7. Approximately 71% of the planetary mass in the solar system is concentrated in:

A. Uranus B. Mercury
C. Saturn D. Jupiter

8. Which of the following oil field is NOT located in the western part of India?

A. Bombay High B. Ankleshwar
C. Gandhar D. Moran

9. The most abundant element in the Earth's continental crust is:

A. silicon B. aluminium
C. oxygen D. iron

10. Mammalian fossils are commonly found in:

A. Haimanta Group B. Jabalpur Group
C. Siwalik Group D. Uttatur Group

Qs. Nos. 11-30 : *Carry two marks each.*

11. Hardness of groundwater is determined by:

A. Mohs' scale of hardness
B. Concentration of calcium and magnesium
C. Bernoulli equation
D. Darcy's law

12. Which one of the following defines a hexagonal dipyramid?

A. A vertical six fold axis of symmetry
B. A horizontal mirror plane
C. Six vertical mirror planes at an angle of 30° with each other
D. A mirror plane that is perpendicular to the vertical six fold axis of symmetry

13. Match the twinning in **Group I** with the corresponding mineral in **Group II.**

Group I	*Group II*
P. Cross-hatched	1. Plagioclase
Q. Carlsbad	2. Microcline
R. Polysynthetic	3. Sanidine
S. Brazil	4. Quartz

	P	Q	R	S
A.	4	2	1	3
B.	2	3	1	4
C.	1	2	3	4
D.	2	3	4	1

14. Evidence of Late Paleozoic glaciation is recorded in:

A. Panchet Formation B. Talchir Formation
C. Motur Formation D. Barakar Formation

15. Match the landform in **Group I** with the corresponding geological process in **Group II.**

Group I	*Group II*
P. Cinder Cone	1. Eolian
Q. Ox-bow lake	2. Glacial
R. Draas	3. Volcanic
S. Drumlin	4. Fluvial

	P	Q	R	S
A.	3	4	1	2
B.	4	2	1	3
C.	2	4	3	1
D.	3	2	4	1

16. Identify the fold in the given figure, where T_0 and T_α represent the axial plane thickness at the hinge and limb, respectively.

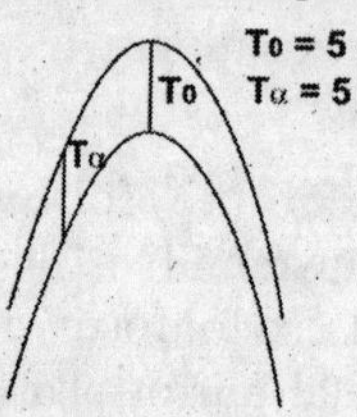

A. Parallel fold
B. Similar fold
C. Supratenuous fold
D. Flattened parallel fold

17. Identify the type of fault in the given figure.

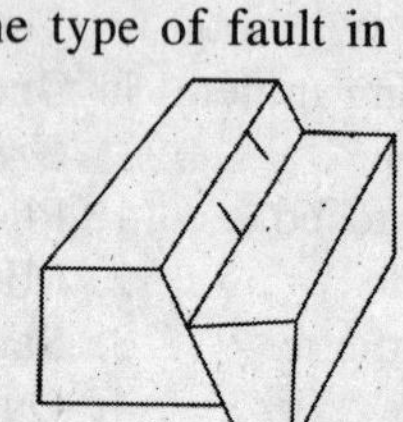

A. Normal fault B. Strike-slip fault
C. Reverse fault D. Thrust fault

18. In clastic sediments, the correct order of decreasing grain size is:

A. Boulder > pebble > silt > sand
B. Granule > pebble > clay > silt
C. Cobble > granule > silt > clay
D. Granule > pebble > sand > silt

19. Match the primary sedimentary structure in **Group I** with the corresponding process of formation in **Group II.**

Group I	*Group II*
P. Asymmetric ripples	1. Water escape
Q. Dish and pillar	2. Bed load transportation of sediments
R. Flute cast	3. Deposition from alternate traction and suspension load
S. Wavy bedding	4. Scouring by turbulent eddy

	P	Q	R	S
A.	1	3	2	4
B.	2	1	4	3
C.	2	4	1	3
D.	3	1	4	2

20. The correct pair of metal and the ore mineral is:

A. Nickel - *Sphalerite*
B. Tin - *Cassiterite*
C. Zinc - *Pyrolusite*
D. Lead - *Bornite*

21. Match the mineral/metal deposit in **Group I** with the corresponding Indian occurrence in **Group II.**

Group I	*Group II*
P. Diamond	1. Malanjkhand
Q. Iron	2. Wajrakarur
R. Fluorite	3. Dalli Rajhara
S. Copper	4. Ambadongar

	P	Q	R	S
A.	4	3	2	1
B.	3	4	1	2
C.	2	1	4	3
D.	2	3	4	1

22. The mineral sequence arranged in order of increasing degree of sharing of $(SiO_4)^{4-}$ tetrahedra is:

A. Olivine → Tremolite → Enstatite → Quartz
B. Olivine → Enstatite → Tremolite → Quartz
C. Quartz → Tremolite → Enstatite → Olivine
D. Olivine → Quartz → Enstatite → Tremolite

23. The metamorphic facies series that explains oceanic subduction zone metamorphism is:

A. Prehnite-Pumpellyite → Blueschist → Eclogite
B. Greenschist → Epidote amphibolite → Amphibolite → Granulite
C. Albite-epidote hornfels → Hornblende hornfels → Pyroxene hornfels → Sanidinite
D. Greenschist → Amphibolite → Granulite → Eclogite

24. The correct combination of textural feature of magmatic rocks with corresponding petrological process is:

A. Exsolution lamellae – *Sub-solidus cooling*
B. Reaction rim – *Eutectic crystallization*
C. Graphic intergrowth – *Quenching of ultramafic lava*
D. Spinifex texture – *Peritectic crystallization*

25. A volcanic rock consisting of alkali feldspar (70%), sodic plagioclase (10%) and nepheline (20%) is named as:

A. Phonolite B. Tephrite
C. Trachyte D. Andesite

26. Match the name of granitoid in **Group I** with the corresponding Craton in **Group II.**

Group I	*Group II*
P. Closepet	1. Singhbhum
Q. Berach	2. Bastar
R. Dongargarh	3. Dharwar
S. Mayurbhanj	4. Aravalli

	P	Q	R	S
A.	4	2	1	3
B.	3	4	2	1
C.	3	4	1	2
D.	4	2	3	1

27. Match the seismic discontinuity in **Group I** with their occurrence in Earth's interior in **Group II.**

Group I	*Group II*
P. Conrad	1. Between lower mantle and outer core
Q. Mohorovićić	2. Between crust and upper mantle
R. Gutenberg	3. Between inner and outer core
S. Lehmann	4. Between lower and upper crust

	P	Q	R	S
A.	4	2	1	3
B.	4	2	3	1
C.	3	2	4	1
D.	2	4	1	3

28. Choose the correct statement:

A. Porosity of weathered granite is less than a crystalline granite.
B. Coarse sands have high porosity and high permeability.
C. Clays have high porosity and high permeability.
D. Ground water table does not fluctuate with water recharge.

29. In the given profile section, **P, Q, R, S** and **T** are sedimentary rocks. Identify the type of unconformity in the sedimentary sequence.

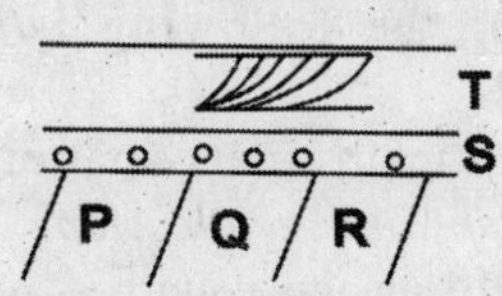

A. Angular unconformity
B. Disconformity
C. Nonconformity
D. Paraconformity

30. Which of the following pairs is **NOT** correctly matched?

A. *Productus* - Brachiopoda
B. *Redlichia* - Arthropoda
C. *Belemnites* - Cephalopoda
D. *Gryphea* - Gastropoda

SECTION-B

Multiple Select Questions (MSQ)

Qs. Nos. 31-40: *Carry two marks each.*

31. Which of the following statements in relation to the solar system is/are correct?

A. The most abundant elements are H and He.

B. The abundance of elements with atomic numbers 1-50 show an overall decreasing trend.

C. The abundance of heavier elements (atomic number >50) are mostly higher than that of lighter elements (atomic number <50).

D. Elements having odd atomic numbers are more abundant than their immediate neighbours.

32. Find out the correct statement(s).

A. Authigenic minerals form during diagenesis of sandstone.

B. Heavy minerals in sandstone are good indicators of provenance.

C. An arkose is mineralogically more mature than a quartz arenite.

D. Matrix in sandstone may form by post-depositional infiltration and/or authigenic filling.

33. Which of the following is/are correctly matched?

A. Dendritic drainage pattern – *uniform substrate and gentle slope*

B. Trellis drainage pattern – *parallel valleys and ridges*

C. Radial drainage pattern – *uniform flat topography*

D. Rectangular drainage pattern – *joints and faults*

34. A northerly dipping fault (F-F) has displaced beds **P, Q** and **R**. The thickness of the beds across the fault is same. Identify the fault type(s).

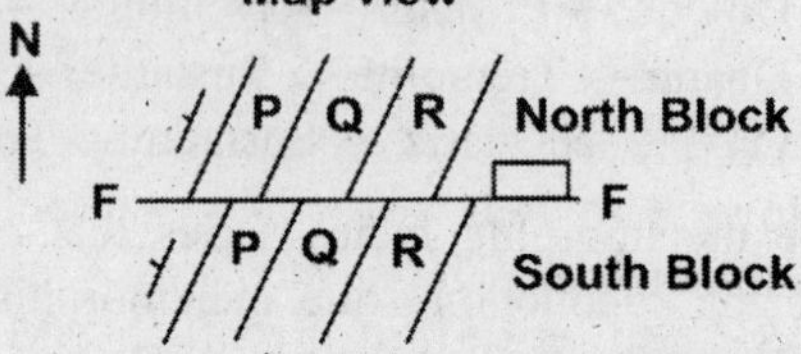

A. Dextral fault
B. Reverse fault
C. Normal fault
D. Sinistral fault

35. A bedding plane, pictorially represented at X, will be plotted in stereonet as:

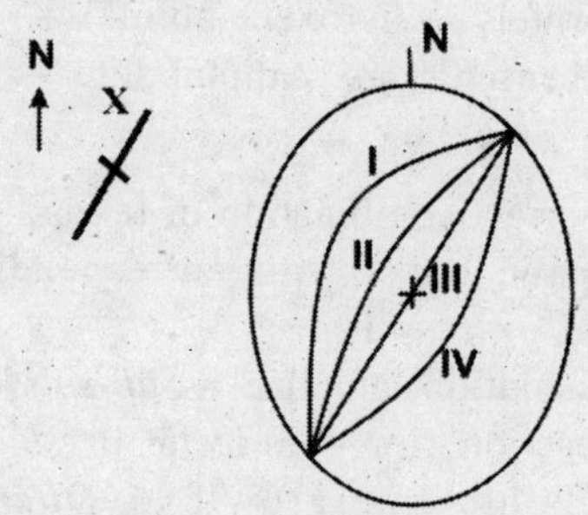

A. I
B. II
C. III
D. IV

36. Choose the correct statement(s):

A. Stratiform chromite deposits form by weathering of ultramafic rocks.

B. Skarn deposits mostly form in calcareous host rocks in contact with felsic intrusives.

C. Nickeliferous laterite is a type of placer deposit.

D. Gold-bearing vein deposits can form from hydrothermal solutions.

37. Under crossed polars, a mineral remains dark at all stages of rotation of the microscope stage. Which of the following statement(s) is/are **NOT** correct?

A. The mineral is isotropic.

B. The mineral is biaxial with an optic orientation corresponding to the principal section of the indicatrix.

C. The mineral is biaxial with an optic orientation containing the Y-vibration direction only.

D. The mineral is uniaxial with an orientation perpendicular to the optic axis.

38. A schematic deformation (D)-metamorphism (M)-time map of a metamorphic belt is shown below. The belt recorded two phases of deformation, D_1-D_2 and attending schistosities, S_1-S_2 and metamorphism, M_1-M_2. The metamorphic minerals stable during these events are shown by solid lines.

On the basis of this information, which of the following statement(s) is/are correct?

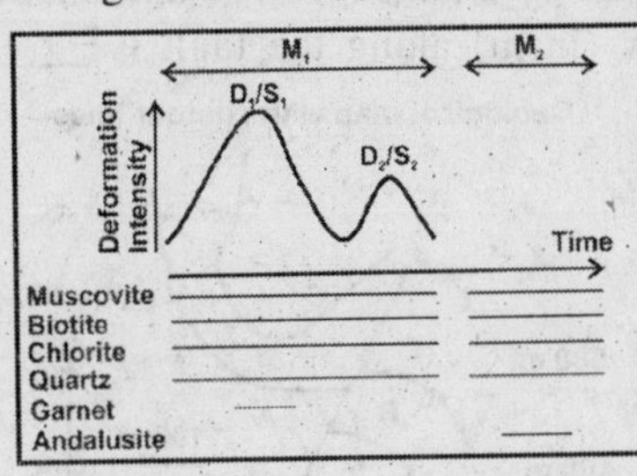

A. M_1 and M_2 events refer to regional and contact metamorphism, respectively.

B. The S_1 schistosity developed as a crenulation cleavage.

C. Garnet grew syntectonic with D_1.

D. Andalusite grew during the waning phase of D_2.

39. Choose the correct statement(s):

A. Gravity dam is a rigid structure, straight or slightly curved in places.

B. Arch dams are commonly built on very strong foundations.

C. Buttress dam chiefly consists of reinforced concrete slab that slopes upstream.

D. Earth dams have a concrete wall curved with convex face pointed towards the upstream.

40. Choose the correct pair(s):

A. Proterozoic – *Origin of flowering plants*

B. Paleozoic – *Origin of amphibians*

C. Mesozoic – *Acme of trilobites*

D. Cenozoic – *Dominance of mammals*

SECTION-C

Numerical Answer Type (NAT)

Qs Nos. 41-50: *Carry one mark each.*

41. The wt.% (correct to two decimal places) of Cu in chalcopyrite ($CuFeS_2$) (atomic weight of Cu = 63.55, Fe = 55.85, S = 32.07) is ____.

42. The general formula of an amphibole mineral is $A_{0-1}B_2C_5T_8O_{22}(OH)_2$, where A, B, C and T are cationic sites with different co-ordination numbers as stated below:

A = 12, B = 6 – 8, C = 6, T = 4.

The amount of octahedral Al in an amphibole of composition

$Na_{0.6}Ca_2Mg_{3.8}Al_{3.0}Si_{6.2}O_{22}(OH)_2$ is ____.

43. In the block diagram, the net slip (= 100 m) is resolved into strike slip (s) and dip slip (d) components. The value (in m, correct to two decimal places) of "s" is ____

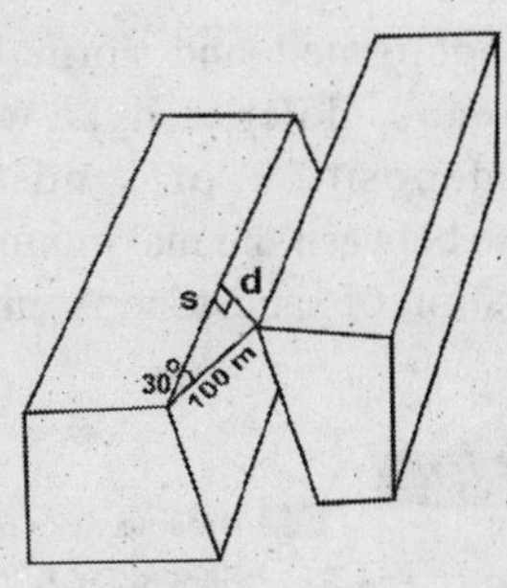

44. The true thickness (*t*, in *m*) of bed B in the given diagram is ____.

Profile view

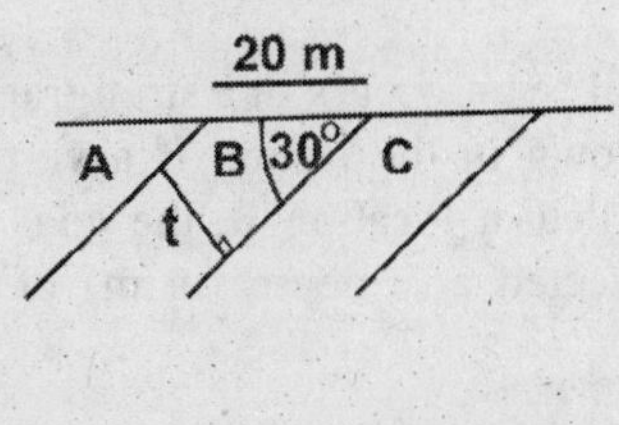

45. If (326) is the Miller Index of a crystal face, then the value of **x** in the corresponding Weiss Parameter of the same face, **xa**: **yb**: **zc** is ____.

46. The value of h in the Miller-Bravais Index $(\bar{4}.1h0)$ is ____.

47. In an ocean basin, a 4 Ma old oceanic crust lies 40 km away from the ridge axis. The average velocity (in cm/yr) of the oceanic lithosphere is ____

48. An aquifer has a cross sectional area of 1000 m^2 and a hydraulic gradient of 0.01. If water is flowing from the aquifer at a rate of 10 m^3/sec, the hydraulic conductivity (in m/sec) of the aquifer is ____.

49. According to the mineralogical phase rule, the number of minerals that can coexist at equilibrium in a 8 component chemical system with 2 degrees of freedom is ____.

50. The grain density (of solids only) and bulk density (solids + voids) of a sandstone sample are 2.7 gm/cm^3 and 2.3 gm/cm^3, respectively. The total porosity (in %, correct to two decimal places) of the sample is ____.

Qs Nos. 51-60: *Carry two marks each.*

51. In an undeformed and normal stratigraphic succession, a dolerite dyke was emplaced before deposition of sandstone B. The difference between the maximum ages (in Myr) of deposition of sandstone A and sandstone B is ____

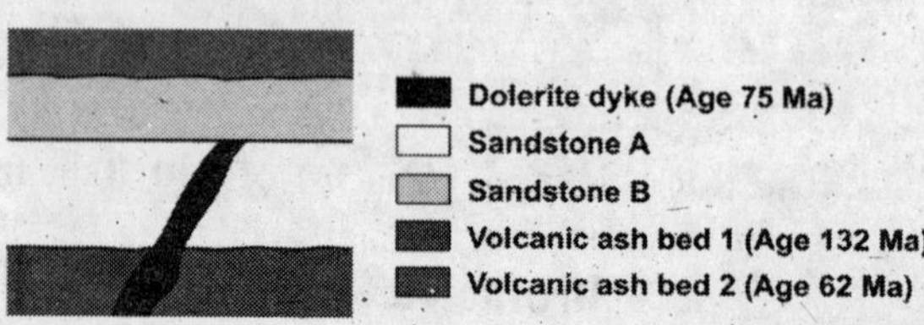

52. A coal seam occurs in a stratigraphic sequence as shown in the figure. If a vertical borehole is drilled at location B, the coal seam will be intersected at a depth (in m) of ____

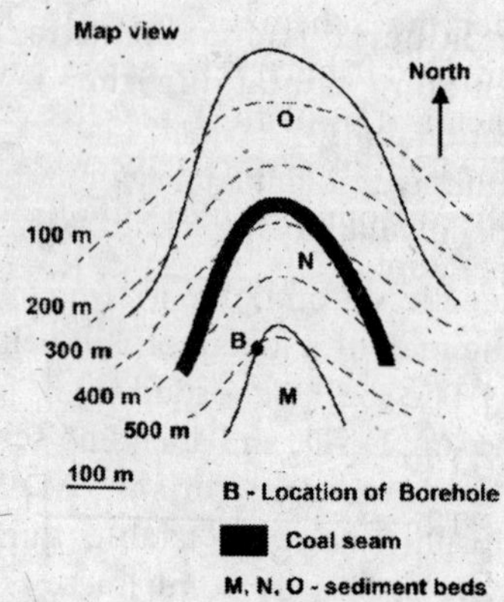

53. A set of sedimentary rocks **A, B** and **C** are affected by a fault F-F. The amount of vertical throw (in m) along the fault is ______.

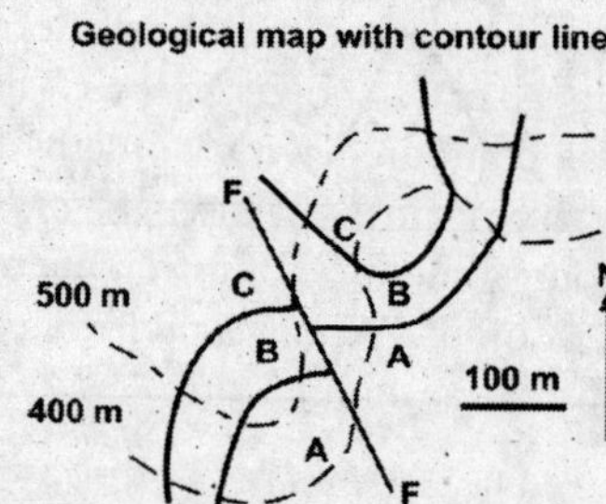

54. The retardation of a uniaxial negative mineral of thickness 0.03 mm is 5160 nm in its principal section of indicatrix. If the refractive index corresponding to the E-ray is 1.486, the value of the refractive index (correct to three decimal places) of the O-ray is ______ .

55. A spherical ore body (diameter = 40 m) has 7% metal content and density of 3300 kg/m^3. The reserve (in tonne) of the ore body is ____.

56. From the data shown in the table, the weighted mean size (in micrometer, correct to two decimal places) of the sediment population is ______.

Grain size (micrometer)	Dry sediment weight (in gram)
4	50
20	75
40	125
60	50

57. Consider the schematic isobaric T-X phase diagram in the binary forsterite (Fo)-fayalite (Fa) chemical system. If there is equilibrium crystallization of melt (L), the wt.% of olivine crystallized from a melt of composition "*a*" at a temperature T2 is ______.

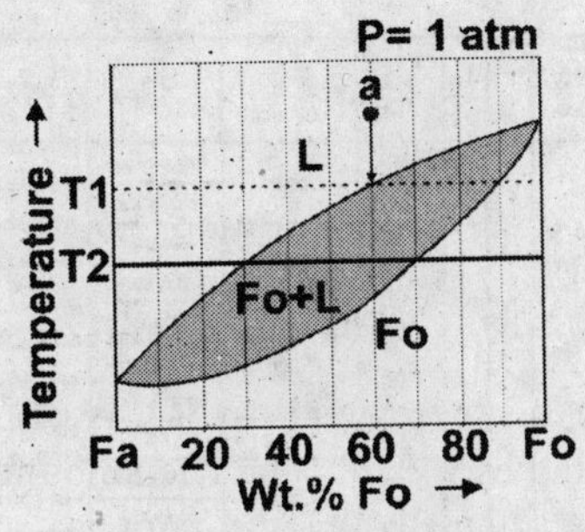

58. Consider a schematic isobaric ternary phase diagram A-B-C, shown below. The diagram, which is contoured with isopleths of liquidus temperatures (in °C), reveals crystallization behaviour of melt (L) of different compositions during cooling. When a melt of composition "*a*" lies at a temperature of 1800 °C, the variance (or degree of freedom) of the magmatic system is ______.

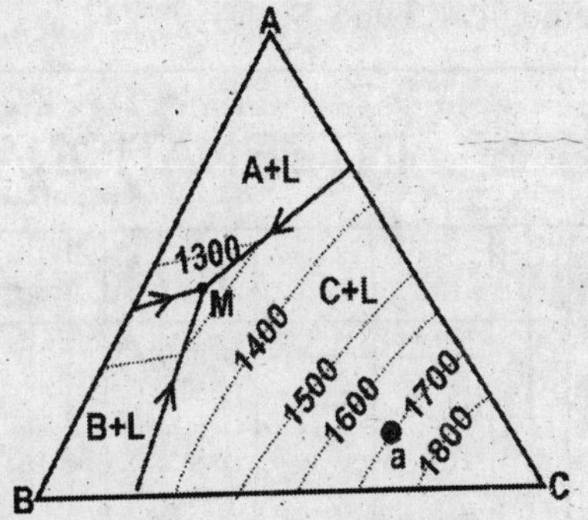

59. An eclogite consists of garnet (60%) and omphacite (40%), where the mineral abundances are in mole %. X_{Mg} [$=Mg/(Mg + Fe^{2+})$] of garnet and omphacite is 0.50 and 0.75, respectively. The X_{Mg} of eclogite is ______ .

60. A harzburgite contains pure forsterite and pure enstatite in a molecular ratio of 60 : 40. The mole % of MgO in the rock is ______ .

ANSWER KEY

1	2	3	4	5	6	7	8	9	10
B	B	A	D	C	D	D	D	C	C
11	**12**	**13**	**14**	**15**	**16**	**17**	**18**	**19**	**20**
B	D	B	B	A	B	A	C	B	B
21	**22**	**23**	**24**	**25**	**26**	**27**	**28**	**29**	**30**
D	B	A	A	A	B	A	B	A	D
31	**32**	**33**	**34**	**35**	**36**	**37**	**38**	**39**	**40**
A, B	A, B, D	A, B, D	B, D	C	B, D	B	A, C	A, C	B, D
41	**42**	**43**	**44**	**45**	**46**	**47**	**48**	**49**	**50**
34.60 to 34.64	1.2	86.00 to 87.00	10	2	3	1	1	8	14.00 to 15.00
51	**52**	**53**	**54**	**55**	**56**	**57**	**58**	**59**	**60**
57	100	100	1.658	110528 to 110629	32.00 to 33.00	75	3	0.60	60

HINTS AND SOLUTIONS

1. Geological Time Scale:

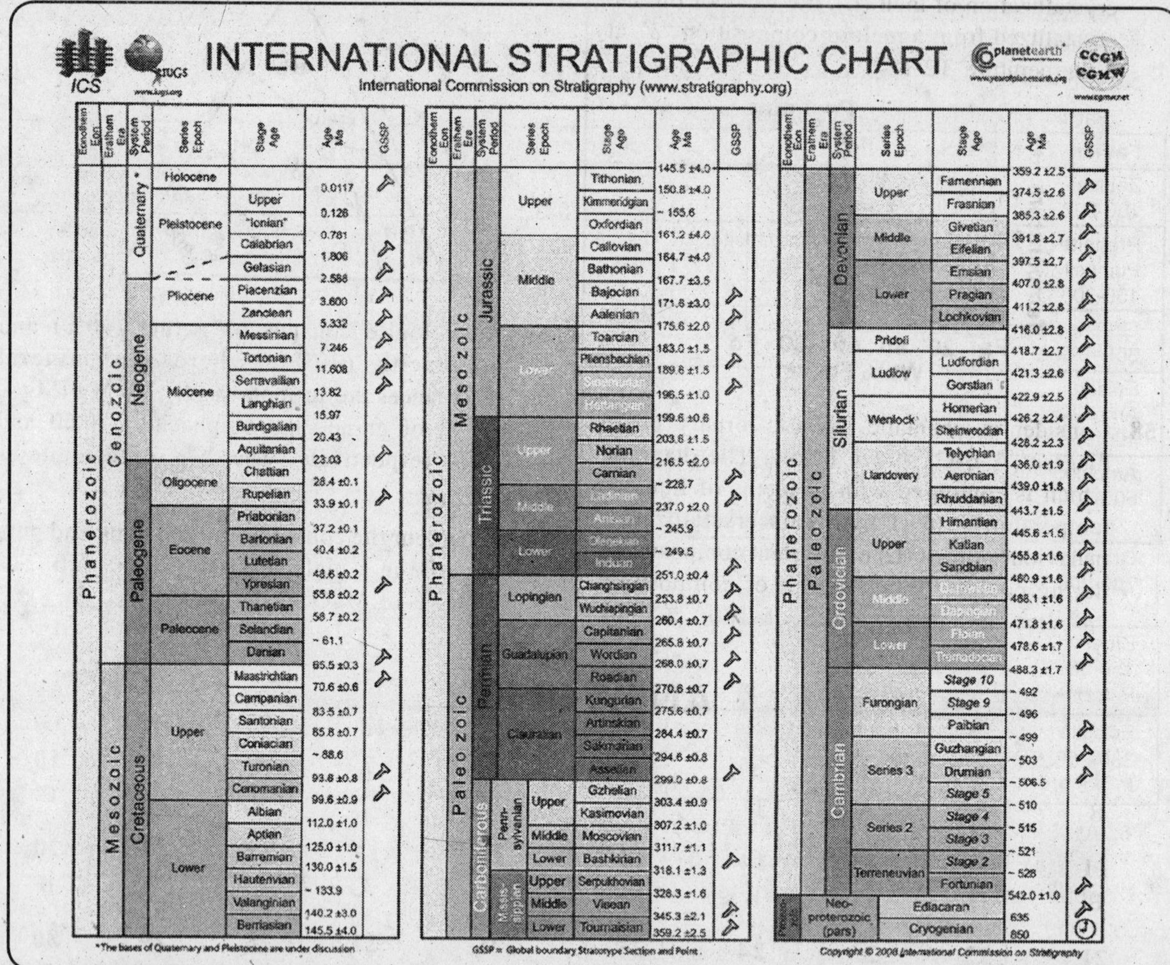

INTERNATIONAL STRATIGRAPHIC CHART

International Commission on Stratigraphy (www.stratigraphy.org)

Eonothem Eon	Erathem Era	System Period	Series Epoch	Stage Age	Age Ma
Phanerozoic	Cenozoic	Quaternary*	Holocene		0.0117
			Pleistocene	Upper	0.126
				"Ionian"	0.781
				Calabrian	1.806
				Gelasian	2.588
		Neogene	Pliocene	Piacenzian	3.600
				Zanclean	5.332
			Miocene	Messinian	7.246
				Tortonian	11.608
				Serravallian	13.82
				Langhian	15.97
				Burdigalian	20.43
				Aquitanian	23.03
		Paleogene	Oligocene	Chattian	28.4 ±0.1
				Rupelian	33.9 ±0.1
			Eocene	Priabonian	37.2 ±0.1
				Bartonian	40.4 ±0.2
				Lutetian	48.6 ±0.2
				Ypresian	55.8 ±0.2
			Paleocene	Thanetian	58.7 ±0.2
				Selandian	~ 61.1
				Danian	65.5 ±0.3
	Mesozoic	Cretaceous	Upper	Maastrichtian	70.6 ±0.6
				Campanian	83.5 ±0.7
				Santonian	85.8 ±0.7
				Coniacian	~ 88.6
				Turonian	93.6 ±0.8
				Cenomanian	99.6 ±0.9
			Lower	Albian	112.0 ±1.0
				Aptian	125.0 ±1.0
				Barremian	130.0 ±1.5
				Hauterivian	~ 133.9
				Valanginian	140.2 ±3.0
				Berriasian	145.5 ±4.0
		Jurassic	Upper	Tithonian	150.8 ±4.0
				Kimmeridgian	~ 155.6
				Oxfordian	161.2 ±4.0
			Middle	Callovian	164.7 ±4.0
				Bathonian	167.7 ±3.5
				Bajocian	171.6 ±3.0
				Aalenian	175.6 ±2.0
			Lower	Toarcian	183.0 ±1.5
				Pliensbachian	189.6 ±1.5
				Sinemurian	196.5 ±1.0
				Hettangian	199.6 ±0.6
		Triassic	Upper	Rhaetian	203.6 ±1.5
				Norian	216.5 ±2.0
				Carnian	~ 228.7
			Middle	Ladinian	237.0 ±2.0
				Anisian	~ 245.9
			Lower	Olenekian	~ 249.5
				Induan	251.0 ±0.4
	Paleozoic	Permian	Lopingian	Changhsingian	253.8 ±0.7
				Wuchiapingian	260.4 ±0.7
			Guadalupian	Capitanian	265.8 ±0.7
				Wordian	268.0 ±0.7
				Roadian	270.6 ±0.7
			Cisuralian	Kungurian	275.6 ±0.7
				Artinskian	284.4 ±0.7
				Sakmarian	294.6 ±0.8
				Asselian	299.0 ±0.8
		Carboniferous (Pennsylvanian)	Upper	Gzhelian	303.4 ±0.9
				Kasimovian	307.2 ±1.0
			Middle	Moscovian	311.7 ±1.1
			Lower	Bashkirian	318.1 ±1.3
		Carboniferous (Mississippian)	Upper	Serpukhovian	328.3 ±1.6
			Middle	Visean	345.3 ±2.1
			Lower	Tournaisian	359.2 ±2.5
		Devonian	Upper	Famennian	374.5 ±2.6
				Frasnian	385.3 ±2.6
			Middle	Givetian	391.8 ±2.7
				Eifelian	397.5 ±2.7
			Lower	Emsian	407.0 ±2.8
				Pragian	411.2 ±2.8
				Lochkovian	416.0 ±2.8
		Silurian	Pridoli		418.7 ±2.7
			Ludlow	Ludfordian	421.3 ±2.6
				Gorstian	422.9 ±2.5
			Wenlock	Homerian	426.2 ±2.4
				Sheinwoodian	428.2 ±2.3
			Llandovery	Telychian	436.0 ±1.9
				Aeronian	439.0 ±1.8
				Rhuddanian	443.7 ±1.5
		Ordovician	Upper	Hirnantian	445.6 ±1.5
				Katian	455.8 ±1.6
				Sandbian	460.9 ±1.6
			Middle	Darriwilian	468.1 ±1.6
				Dapingian	471.8 ±1.6
			Lower	Floian	478.6 ±1.7
				Tremadocian	488.3 ±1.7
		Cambrian	Furongian	Stage 10	~ 492
				Stage 9	~ 496
				Paibian	~ 499
			Series 3	Guzhangian	~ 503
				Drumian	~ 506.5
				Stage 5	~ 510
			Series 2	Stage 4	~ 515
				Stage 3	~ 521
			Terreneuvian	Stage 2	~ 528
				Fortunian	542.0 ±1.0
Proterozoic	Neo-proterozoic (pars)			Ediacaran	635
				Cryogenian	850

*The bases of Quaternary and Pleistocene are under discussion

GSSP = Global boundary Stratotype Section and Point.

2. Formation of trench: Trench form at convergent boundary.

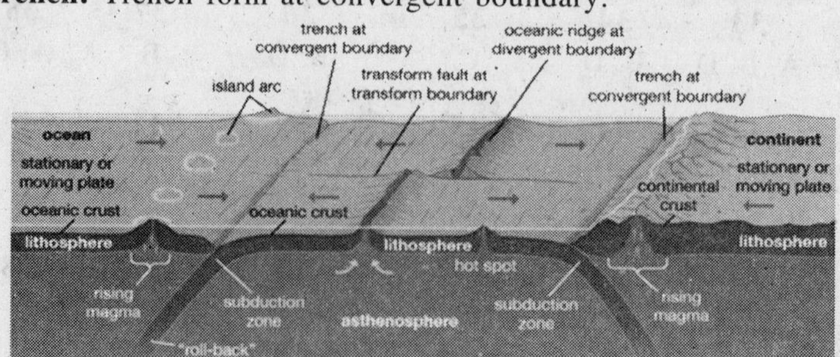

3. The strike (in degree notation) of a bed dipping 30° towards N45W is 045-225.

4. Randomly oriented ore-bearing veinlets in a mass of rock are known as **stockworks.**

en echelon veins: The term 'en echelon' refers to closely-spaced, parallel or subparallel, overlapping or step-like minor structural features in rock (faults, tension fractures), which lie *oblique* to the overall structural trend.

5. The diagnostic metamorphic facies and its minerals:

Facies–Protolith–Mineral Assemblage Table:

Facies	Pelitic	Calcareous	Mafic
Zeolite 100-200°C	Interlayered smectite/ chlorite calcite	Calcite	Laumonite, thompsonite, calcite, interlayered smectite/chlorite
Prehnite-Pumpellyite 150-300°C	Prehnite, pumpellyite, calcite, chlorite, albite	Calcite	Prehnite, pumpellyite, calcite, chlorite, albite
Greenschist 300-450°C	Muscovite, chlorite, quartz, albite, biotite, garnet	Calcite, dolomite, quartz, epidote, tremolite	Albite, chlorite, quartz, epidote, actinolite, sphene
Epidote Amphibolite 450-550°C	Muscovite, biotite, garnet, albite, quartz	Calcite, quartz, tremolite, epidote, diopside	Albite, epidote, hornblende, quartz
Amphibolite 500-700°C	Garnet, biotite, muscovite, quartz, plagioclase, staurolite, kyanite or sillimanite	Calcite, diopside, quartz, wollastonite	Hornblende, plagioclase, garnet, quartz, sphene, biotite
Granulite 700-900°C	Garnet, Kspar, sillimanite or kyanite, quartz, plagioclase, hypersthene	Calcite, quartz, plagioclase, diopside, hypersthene	Plagioclase, augite, hypersthene, hornblende, garnet, olivine
Blueschist 150-350°C P > 5.8 Kb	Jadeite, albite, quartz, lawsonite, aragonite, paragonite	Aragonite, white mica	Glaucophane, albite, lawsonite, spherene, ± garnet
Eclogite 350-750°C P > 8.10 Kb	Coesite, Kspar, sillimanite, plagioclase	Aragonite, quartz, plagioclase, diopside, hypersthene	Omphacite (px), pyrope garnet

6. Atoll is a geomorphic feature formed by Coral reef accumulation. An atoll is a ring-shaped coral reef, island, or series of islets. An atoll surrounds a body of water called a lagoon. Sometimes, atolls and lagoons protect a central island. Channels between islets connect a lagoon to the open ocean or sea. Atolls develop with underwater volcanoes, called seamounts. First, the volcano erupts, piling up lava on the seafloor. As the volcano continues to erupt, the seamount's elevation grows higher, eventually breaking the surface of the water. The top of the volcano becomes an oceanic island. In the next stage, tiny sea animals called corals begin to build a reef around the island. The type of corals that build reefs are called hermatypic corals, or hard corals. Hermatypic corals create a hard exoskeleton of limestone (calcium carbonate). Billions of these limestone exoskeletons are the reef.

This coral reef, called a fringing reef, surrounds the island just below the ocean surface. The thin, shallow strip of water between the fringing reef and the island is the lagoon.

7. Approximately 71% of the planetary mass in the solar system is concentrated in Jupiter.

Planet	Mass (Earth Masses)	Radius (Earth Radii)	Density gm/cm^3
Mercury	0.055	0.38	5.5
Venus	0.815	0.95	5.2
Earth	1.000	1.00	5.5
Mars	0.107	0.53	3.9
Jupiter	318	10.8	1.4

Planet	Mass (Earth Masses)	Radius (Earth Radii)	Density gm/cm^3
Saturn	95	9.0	0.7
Uranus	14.5	3.93	1.3
Neptune	17.2	3.87	1.6
Pluto	0.002	0.178	2.1

8. The following oil field and its location:

Bombay High Oil Field: Bombay High is an offshore oil field 176 km off the west coast of Mumbai, India, in about 75 m of water. The oil operations are run by India's Oil and Natural Gas Corporation (ONGC). The source rock is limestone of Miocene age.

Ankleshwar Oil Field: Ankleshwar, (Ankleshwar) is a city and a municipality in the Bharuch district of the state of Gujarat, India.

Gandhar Oil Field: Gandhar field is one of the major fields Cambay basin, India. The field was discovered in 1983 and put on production in 1984. Oil and gas have been produced mainly from different sands in Hazad member of Ankleshwar formation stacked one over the other with intervening shale units.

Moran Oil Field: Moran Town is in Dibrugarh district in the Indian state of Assam. Moran is an important industrial town in India. It is a major oil field area.

9. The most abundant element in the Earth's continental crust is Oxygen.

Elements in different zone	Percentage decreasing order
Crust	O > Si > Al > Fe > Ca > Na > K > Mg
Lithosphere	O > Si > Al > Fe > Ca > Na > K > Mg
Earth	Fe > Ni > O
Universe	H > He > Fe > O
Atmosphere	N_2 > O > Ar > CO_2

10. Mammalian fossils are commonly found in Siwalik Group. **Shivalik Fossil Park,** also known as the **Suketi Fossil Park,** is a notified National Geo-heritage Monument fossil park in the Sirmaur district in the Indian state of Himachal Pradesh. It has a collection of prehistoric vertebrate fossils and skeletons recovered from the upper and middle Siwaliks geological formations of sandstones and clay at Suketi. The park has a display of the fossil finds and an open-air exhibition of six life-sized fiberglass models of extinct mammals in a recreation of the Sivalik Hills environment of the Plio-Pleistocene era (circa 2.5 million years).

11. Hardness of groundwater is determined by Concentration of calcium and magnesium.

Hardness = [Ca* combining weight of $CaCO_3$/ Combining weight of Ca + Mg* combining weight of $CaCO_3$/Combing weight of Mg].

12. **Hexagonal dipyramid:** A mirror plane that is perpendicular to the vertical six fold axis of symmetry.

Hexagonal System

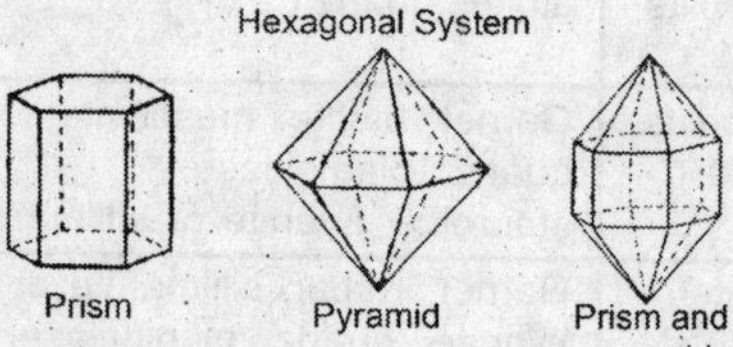

13. Match the twinning with the corresponding mineral:

- Cross-hatched – Microcline
- Carlsbad – Sanidine
- Polysynthetic – Plagioclase
- Brazil – Quartz

14. Evidence of Late Paleozoic glaciation is recorded in Talchir Formation. The Talchir Series is the lowest member and has a tillite or boulder bed, at its own base, which is of great stratigraphical significance. It is 15 to 60 m thick and is considered to be of fluvioglacial origin indicating extension of glacial climate to those areas of Peninsula at the time of its deposition. The boulder bed is followed upwards by greenish, fragmentary type shales and greenish coloured sandstones in that sequence.

15. The landform in Group I with the corresponding geological process in Group II.

- Cinder Cone – Volcanic
- Ox-bow lake – Fluvial
- Draas – Eolian
- Drumlin – Glacial

16. Fold classification:

Classification of layer types based on dip isogons

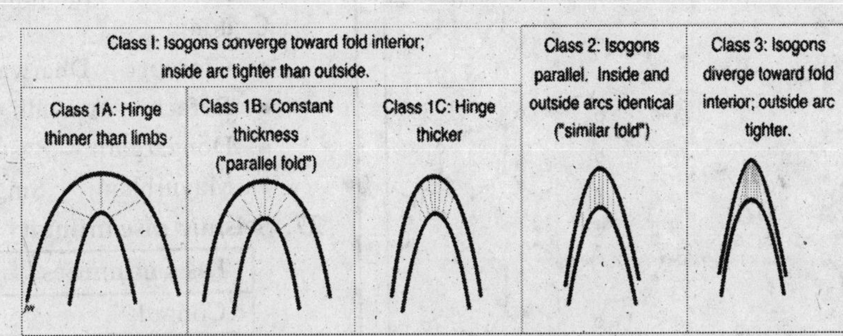

17. Type of Faults:

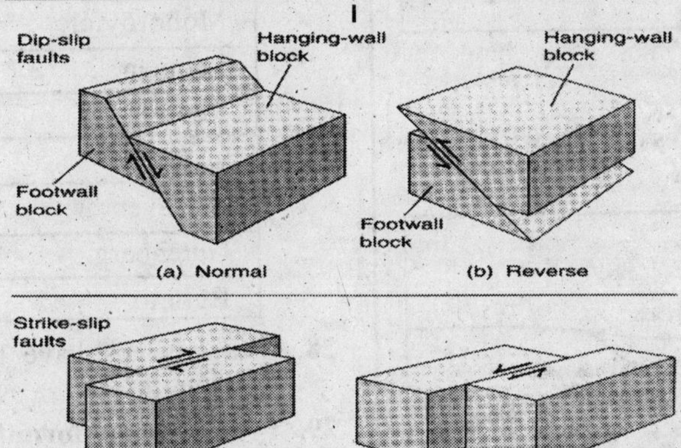

18. Classification of rocks based on grain size:

Millimeters (mm)		Micrometers (μm)	Phi (ϕ)	Wentworth size class	
	4096		-12.0	Boulder	Gravel
	256		-8.0	Cobble	Gravel
	64		-6.0	Pebble	Gravel
	4		-2.0	Granule	Gravel
	2.00		-1.0	Very coarse sand	Sand
	1.00		0.0	Coarse sand	Sand
1/2	0.50	500	1.0	Medium sand	Sand
1/4	0.25	250	2.0	Fine sand	Sand
1/8	0.125	125	3.0	Very fine sand	Sand
1/16	0.0625	63	4.0	Coarse silt	Silt
1/32	0.031	31	5.0	Medium silt	Silt
1/64	0.0156	15.6	6.0	Fine silt	Silt
1/128	0.0078	7.8	7.0	Very fine silt	Silt
1/256	0.0039	3.9	8.0	Clay	Mud
	0.00006	0.06	14.0		

20. The correct pair of metal and the ore mineral is:

- **Nickel:** Pentlandite and pyrrhotite (iron nickel sulfides, the two most important nickel ores), garnierite (hydrous nickel silicate), millerite (nickel sulfide), and niccolite (nickel arsenide).
- **Tin:** *Cassiteritestannite,* cylindrite.
- **Zinc:** Sphalerite, which is a form of zinc sulfide, smithsonite (zinc carbonate), hemimorphite (zinc silicate), wurtzite (another zinc sulfide), and sometimes hydrozincite (basic zinc carbonate).
- **Lead:** Galena

21. The mineral metal deposit with Indian occurrence:

- Diamond – Wajrakarur
- Iron – Dalli Rajhara
- Fluorite – Ambadongar
- Copper – Malanjkhand

22. The mineral sequence arranged in order of increasing degree of sharing of $(SiO_4)^{4-}$

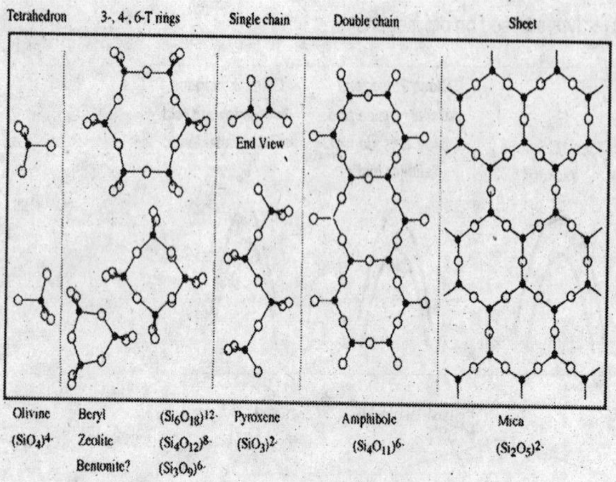

23. The metamorphic facies series with plate setting:

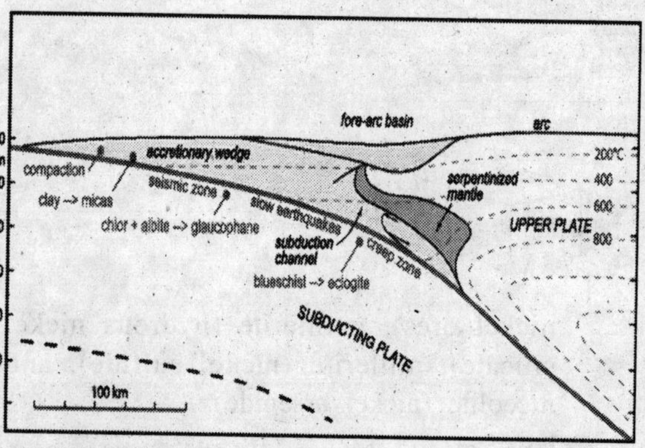

25. Classification of rocks:

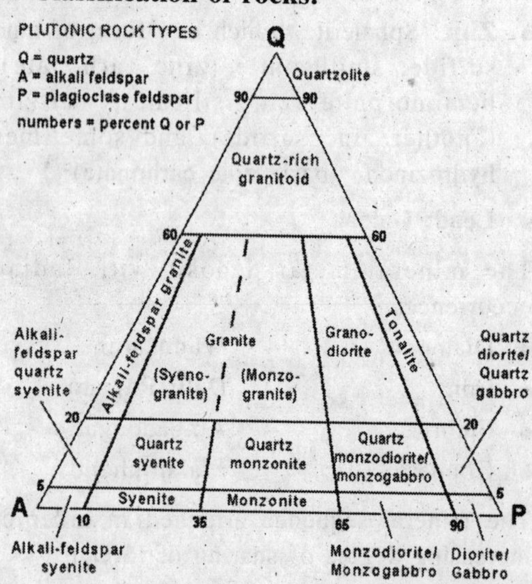

26. The name of granitoid with the corresponding Craton.

- Closepet – Dharwar – Karnataka
- Berach – Aravalli – Rajasthan
- Dongargarh – Bastar – MP
- Mayurbhanj – Singhbhum – Jharkhand

27. Seismic discontinuity in Earth's interior:

Discontinuities	Depths Km
Conrad	10
Mohorovicic	35
Lehman	220
Repetti	750
LVZ	50 – 200
D" layer	2870
Gutenberg	2900
Bullen	5140

28. Coarse sands have high porosity and high permeability.

29. Type of unconformity:

Characteristics	Unconformity
Contact effect below and above and plane of unconformity	Disconformity
Pebbles of non-sedimentary rocks over the plane of unconformity	Nonconformity
Paraconformable bedding planes	Paraconformity
Angular discordance at the plane of unconformity	Angular unconformity

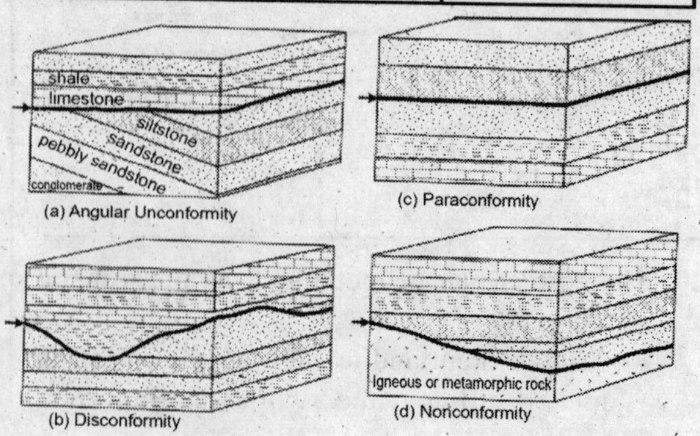

(a) Angular Unconformity (c) Paraconformity (b) Disconformity (d) Nonconformity

30. The following pairs is correctly matched:

- *Productus* – Brachiopoda
- *Redilichia* – Arthropoda
- *Belemnites* – Cephalopoda
- *Gryphea* – Bivalve

31. The following statements in relation to the solar system is/are correct:

- The most abundant elements are H and He.
- The abundance of elements with atomic number 1-50 show an overall decreasing trend.

32. The correct statement(s):

- Authigenic minerals from during diagenesis of sandstone.
- Heavy minerals in sandstone are good indicators of provenance.
- An arkose is mineralogically less mature than a quartz arenite.
- Matrix in sandstone may form by post-depositional infiltration and or authigenic filling.

33. Drainage pattern with example:

- **Dendritic drainage pattern:** uniform substrate and gentle slope
- **Trellis drainage pattern:** parallel valleys and ridges
- **Radial drainage pattern:** streams radiate outwards from a central high point. Volcanoes usually display excellent radial drainage. Other geological features on which radial drainage commonly develops are domes and laccoliths.
- **Rectangular drainage pattern:** joints and faults

34. Faults with movements:

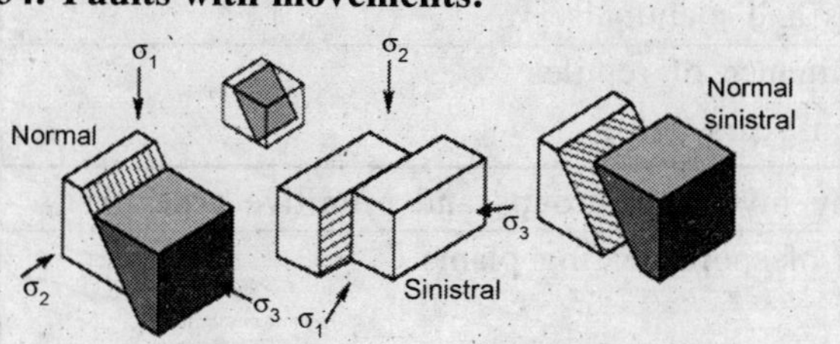

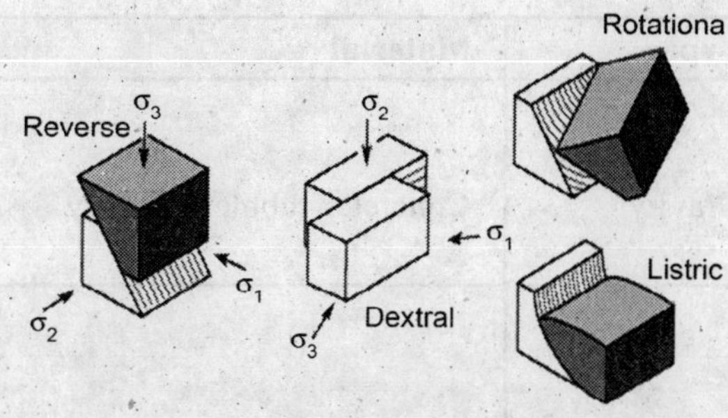

36. Choose the correct statements(s):

- Skarn deposits mostly form in calcareous host rocks in contact with felsic intrusives.
- Gold-bearing vein deposits can form from hydrothermal solutions.

38. A schematic deformation (D)-metamorphism (M)-time map of a metamorphic belt is shown below. The belt recorded two phases of deformation.

D_1-D_2 and attending schistosities, S_1-S_2 and metamorphism, M_1-M_2. The metamorphic minerals stable during these events are shown by solid lines.

On the basis of this information, which of the following statement(s) is/are correct?

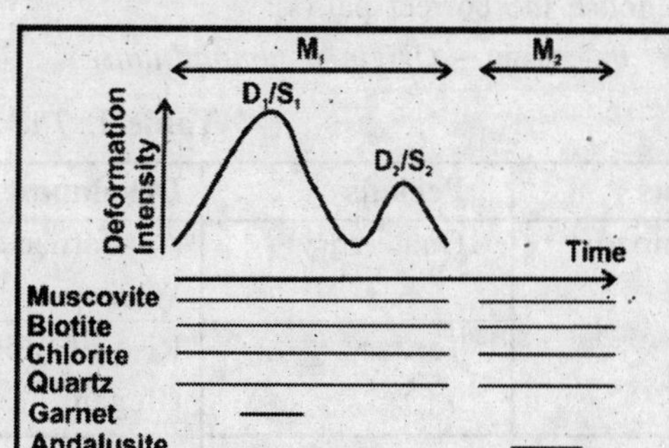

- M_1 and M_2 events refer to regional and contact metamorphism, respectively.
- Garnet grew syntectonic with D_1.

39. Choose the correct statement(s):

- Gravity dam is a rigid structures, straight or slightly curved in places.
- Buttress dam chiefly consists of reinforced concrete slab that slopes upstream.

Type	Material	Sectional View	Plan (Top View)
Gravity	Concrete, rubble masonry		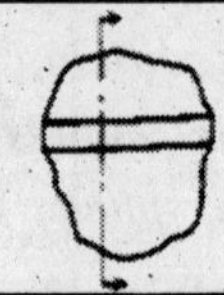
Arch	Concrete	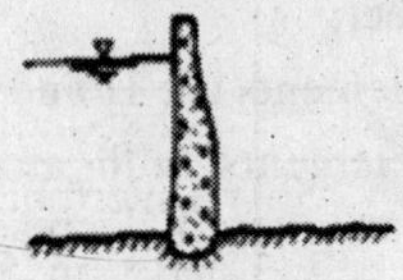	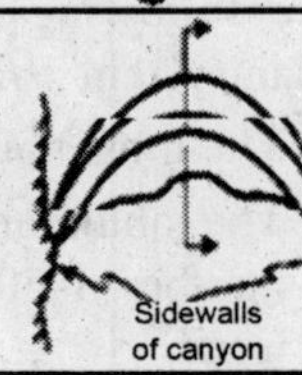
Buttress	Concrete also timber and steel	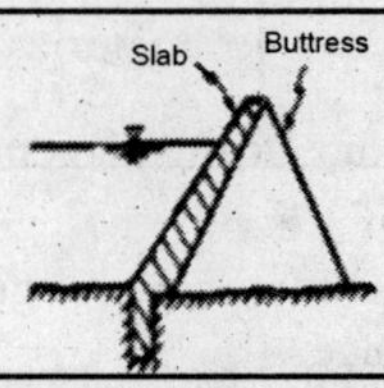	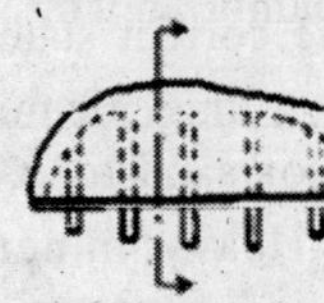
Embankment	Earth or rock	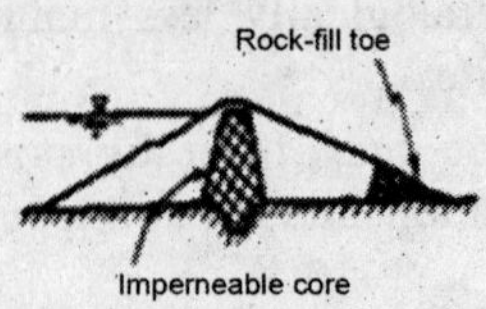	

40. Choose the correct pair(s):

- Paleozoic - *Origin of amphibians*
- Cenozoic - *Dominance of mammals*

Table I. *The Geologic Time Scale*

Eras	Periods	Development of Life
Cenozoic	Quaternary	Appearnace of man Widespread continental glaciation
	Tertiary	Rise of modern mammals and birds (Began 70 million years ago.)
Mesozoic	Cretaceous	Extinction of dinosaurs and many other reptile groups
	Jurassic	Domination by giant reptiles Beginning of birds and mammals
	Triassic	Beginning of dominance of reptiles (Began 200 million years ago).
Paleozoic	Permian	Extinction of many invertable groups and primitive type plants
	Pennsylvanian	Many coal forests of spore bearing plants First reptiles

Eras	Periods	Development of Life
	Mississippian	Abundant spore bearing plants Many insects
	Devonian	First amphibians Numerous marine fishes
	Silurian	First land plants First land animals (scorpians)
	Ordovician	First indication of vertebrates (fishes)
	Cambrian	Invertebrates numerous and varied, trilobites dominant (Began 500 million years ago)
Precambrian Time		Bacteria, algae, and a few primitive invertebrates (Began 4.5 billion years ago)

41. Given, Atomic weight of Cu = 63.55, Fe = 55.85, S = 32.07

General formula of chalcopyrite = $CuFeS_2$

Wt % of Cu in $CuFeS_2$ = (Atomic weight of Cu × mole / At. Wt of Cu × mole + At wt of Fe × mole + At wt of S × mole) × 100

= (63.55 × 1 / 63.55 × 1 + 55.85 × 1 + 32.07 × 2) × 100

= 63.55/ 183.54 × 100 = 34.62%.

42. Si^{+4} and Al^{+3} form in tetrahedral site

But Si^{+4} is given = 6.2

Then, $Al^{+3} = 8 - 6.2 = 1.8$

Amount of Al^{+3} in octahedral site

$= 3 - 1.8 = 1.2.$

43. Apply Cos θ to find '*s*' where, θ = 30 degree

$$\text{Cos } \theta = \frac{B}{H} = \frac{s}{100}$$

$$\text{Cos } 30° = \frac{s}{100} \Rightarrow \sqrt{\frac{3}{2}} = \frac{s}{100}$$

$$s = 50\sqrt{3} \Rightarrow s = 86.60.$$

44. True thickness (*t*) of bed B at on angle of 30°

Then, by applying

$$\sin \theta = \frac{P}{h} = \frac{t}{20}$$

$$\sin 30° = \frac{t}{20} \Rightarrow \frac{1}{2} = \frac{t}{20} \Rightarrow t = 10 \text{ m}.$$

45. Given, (326) is the Miller Index of a crystal face,

Weiss Parameter of the same face, *xa* : *yb* : *zc*

= (326) is the Miller index

= 1/3, 1/2, 1/6

= Multiply by 6

= 6/3, 6/2, 6/6

= [2 3 1]

Then, 2*a* : *yb* : *zc*

$x = 2.$

46. Given, The value of *h* in the Miller-Bravais Index $(\bar{4}.1h0)$

Rule: $h + k + i = 0$ for hexagonal (hkil)

$\bar{1}$ = one bar indicate it is negative as (–1)

Then, $-4 + 1 + h = 0$

$-3 + h = 0$

$h = 3.$

47. Given, In an ocean basin, a 4 Ma old oceanic crust lies 40 km away from the ridge axis. The average velocity (in cm/year) of the oceanic lithosphere is:

Distance = 40 km

= 40 × 1000

= 4×10^6 cm

Age = 4 Ma = 4×10^6 year

Av. Velocity = [distance / displacement] / Age

= 4×10^6 cm/4*x* 10^6 year = 1 cm/year.

48. Given, an aquifer has a cross sectional area of 1000 m^2 and a hydraulic gradient of 0.01. If water is flowing from the aquifer at a rate of 10 m^3/sec, the hydraulic conductivity (in m/sec) of the aquifer is:

(A) Area = 1000 cm^2

(h) hydraulic gradient (h/l) = 0.01

(Q) Rate of aquifer = 10 m^3/sec

Acc. To Darcy law,

$$Q = [kh/l] \times A$$
$$10 = [k \times 0.01/1] \times 1000$$
$$k = 1.$$

49. According to phase Rule

$$F = C + 2 - Q$$

Given, F = 2, C = 8

$$2 = 8 + 2 - Q$$
$$Q = 8.$$

50. Given, Grain density = 2.7 gm/cm^3

Bulk density = 2.3 gm/cm^3

Porosity = [grain density – bulk density/partial density] × 100%

= (2.7 – 2.3) / 2.7

= [0.4 / 2.7] × 100% = 14.81%.

51. The max age difference of deposition of sandstone A and Sandstone B is equal to difference of Bed 1 and Dolerite dyke

= 132 – 75 = 57 Myr.

52. Given, vertical borehole is drilled at location B, at which contour cut at 500 m.

The contour cut the coalseam firstly at 400 m from location B.

Then, the coal seam will be intersected at the depth (in m) = 500 – 400 = 100 m.

53. Given, The amount of vertical throw (in m) along the fault is 100 m.

54. Given t = 0.03 mm,

R = 5160 nm = 0.00516 mm

E-ray = 1.486

According to formula, R = t × birefrienges

= t × (ω – t) for uniaxial negative.

$$0.00516 = 0.03 \times (\omega - t)$$
$$(0.00516/0.03) + 1.486 = \omega$$
$$\omega = 1.65.$$

55. Given, Diameter = 40 m

Radius = 20 m

Density = mass (reserve) / volume

Mass (reserve) = volume × density

= $4/3\pi r^3$ × density (for spherical body)

= 4/3 × 22/7 × 20 × 20 × 20 × 3300

= 110628571.428 kg = 110628 Tonnes.

56. The weight mean size = $\Sigma fixi/\Sigma fi$

= (4 × 50 + 20 × 75 + 40 × 125 + 60 × 50)/ (50 + 75 + 125 + 50) (mm)

= 9700/300 = 32.33.

57. Acc. to lever rule:

Mole % of $x = x/(x + y) \times 100$

Wt% of olivine, we take forsterite because forsterite are dominance crystal in the system

Then, 30/40 × 100 = 75%.

58. At the point 'a' all composition are melt then, C = 3, Q = 1

Acc to phase rule,

F = C + 1 – Q (At equilibrium)

F = 3 + 1 – 1 = 3.

59. Eclogite consist

= (60/100) × 0.5+ (40/100) × 0.75

X mg = 0.60.

60. Forsterite (Fo) = Mg_2SiO_4

= 2MgO + SiO_2

Enstatite = Mg SiO_3

= MgO + SiO_2

Mole% of MgO in Forsterite

= 2/3 × 60 = 40%

Mole% of MgO in Enstatite

= 1/2 × 40 = 20%

Mole% of MgO in the rock

= 40 + 20 = 60%.

Previous Papers (Solved)

Indian Institute of Technology Joint Admission Test for M.Sc.

IIT-JAM GEOLOGY-2019

SECTION-A

Multiple Choice Question (MCQ)

Qs. Nos. 1-10 : *Carry one mark each.*

1. Which of the following is associated with a divergent plate boundary?

A. Ridge B. Trench

C. Island arc D. Accretionary prism

2. Shear waves do not travel through the

A. Upper continental crust

B. Upper mantle

C. Lower mantle

D. Outer core

3. Fossils of burrows and footprints are known as

A. Pseudofossils B. Coprolites

C. Body fossils D. Trace fossils

4. Horst and graben structures are typically formed by

A. Normal faulting

B. Strike-slip faulting

C. Reverse faulting

D. Thrust faulting

5. *Dicroidium* is known from the

A. Pachmarhi Formation

B. Raniganj Formation

C. Panchet Formation

D. Denwa Formation

6. Permian is a/an

A. Eon B. Era

C. Epoch D. Period

7. Polymorphic minerals have

A. Different crystal forms and identical composition

B. Different crystal forms and different compositions

C. Identical crystal form and different compositions

D. Identical crystal form and identical composition

8. Turbidites commonly form in

A. Fluvial environment

B. Deep marine environment

C. Tidal flat environment

D. Beach environment

9. Which of the following is the fundamental constituent of humic coal?

A. Mineral matter

B. Maceral

C. Lithotype

D. Kerogen

10. Which of the following mineral assemblages characterizes blueschist facies metamorphism of a mafic rock?

A. Glaucophane + lawsonite

B. Hornblende + plagioclase ± epidote ± garnet

C. Omphacite + garnet

D. Phengite + chlorite + garnet

Qs. Nos. 11-30 : *Carry two marks each.*

11. Match the geomorphic features in Group-I with corresponding environments in Group-II.

Group-I	Group-II
P. Eskers	1. Beach
Q. Natural levee	2. Volcanic
R. Chenier ridge	3. Glacial
S. Plug dome	4. Fluvial

A. P-3 Q-4 R-1 S-2

B. P-3 Q-1 R-4 S-2

C. P-2 Q-1 R-4 S-3
D. P-2 Q-4 R-1 S-3

12. Match the bivalves in Group-I with their modes of life in Group-II.

Group-I	*Group-II*
P. Gryphaea	1. Swimming
Q. Mya	2. Unattached, free-lying
R. Mytilus	3. Burrowing
S. Pecten	4. Sessile byssate

A. P-4 Q-1 R-2 S-3
B. P-4 Q-3 R-2 S-1
C. P-2 Q-3 R-4 S-1
D. P-2 Q-1 R-4 S-3

13. Which of the following changes occurred during the evolution of Equidae?
A. Number of lateral digits or toes increases
B. Decrease in hypsodonty
C. Lengthening of skull in front of the orbit
D. Limb ratios remained constant

14. Which of the following rocks occurs typically in a ductile shear zone?
A. Gouge
B. Breccia
C. Pseudotachylite
D. Mylonite

15. Figures I, II and III are the outcrop patterns of three inclined beds P, Q and R on a flat ground. P is the oldest and R is the youngest amongst these beds. Identify the correct option that explains repetition of beds in the figures.

I II III

A. I-anticline; II-syncline; III-fault
B. I-syncline; II-anticline; III-fault
C. I-fault; II-anticline; III-syncline

16. The figure shows stereographic projections of the axial plane and the hinge line of a fold. Which one of the following folds is represented in the figure?

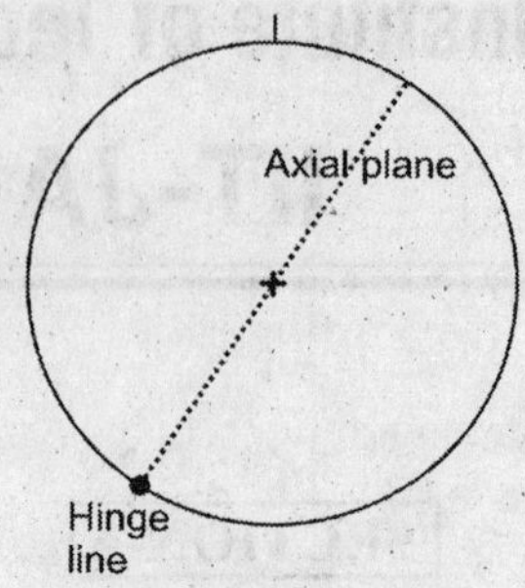

A. Upright plunging
B. Upright non-plunging
C. Reclined
D. Recumbent

17. Figures X and Y show profile sections of folds traced by two cross-bedded sandstone beds. Which one of the following is the correct interpretation?

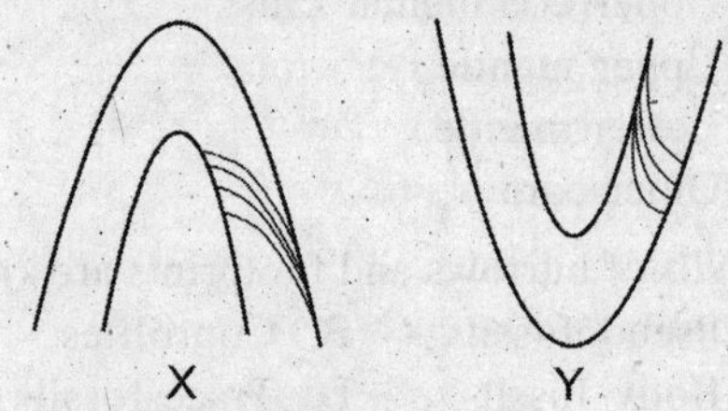

A. X- Antiformal anticline; Y- Antiformal syncline
B. X- Antiformal anticline; Y- Synformal syncline
C. X- Overturned antiform; Y- Overturned synform
D. X-Antiformal syncline; Y- Synformal anticline

18. On a geological map of a flat area, the plunge direction of an antiformal anticline is
A. Towards the fold closure
B. Away from the fold closure
C. Towards the dip direction of the left limb
D. Towards the dip direction of the right limb

19. Match the Precambrian lithounits in Group-I with corresponding cratons in Group-II.

Group-I	*Group-II*
P. Kolhan Group	1. Aravalli
Q. Debari Group	2. Bastar

R. Bababudan Group	3. Singhbhum
S. Sukma Group	4. Dharwar

A. P-2 Q-4 R-1 S-3
B. P-3 Q-4 R-1 S-2
C. P-3 Q-1 R-4 S-2
D. P-2 Q-1 R-4 S-3

20. The stereographic projection given below represents a

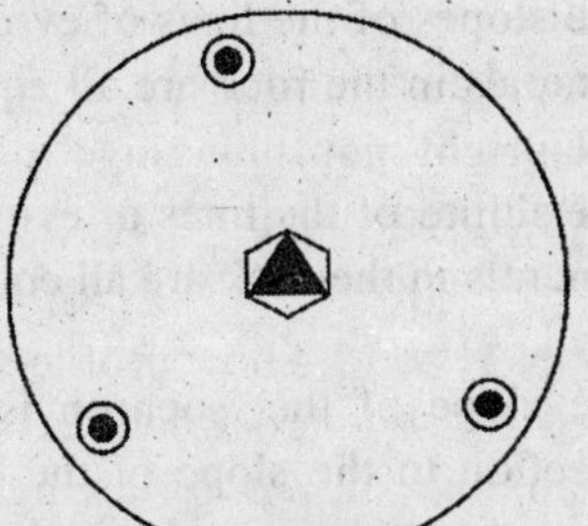

A. trigonal rhombohedron
B. hexagonal scalenohedron
C. trigonal dipyramid
D. hexagonal dipyramid

21. Which of the following minerals exhibits the optic interference figure given below?

A. Quartz
B. Diopside
C. Garnet
D. Orthoclase

22. Match the minerals in Group-I with corresponding characteristic properties in Group-II.

Group-I	*Group-II*
P. Labradorite	1. High specific gravity
Q. Actinolite	2. Play of colours
R. Fluorite	3. Acicular habit
S. Barite	4. Four sets of cleavage

A. P-4 Q-1 R-2 S-3
B. P-4 Q-1 R-3 S-2
C. P-2 Q-3 R-4 S-1
D. P-2 Q-3 R-1 S-4

23. Match the metamorphic rocks in Group-I with corresponding parent rocks in Group-II.

Group-I	*Group-II*
P. Amphibolite	1. Marl
Q. Khondalite	2. Shale
R. Calc-gneiss	3. Sandstone
S. Quartzite	4. Basalt

A. P-4 Q-3 R-1 S-2
B. P-4 Q-2 R-1 S-3
C. P-2 Q-4 R-1 S-3
D. P-4 Q-3 R-2 S-1

24. Match the fuels in Group-I with corresponding areas of occurrence in Group-II.

Group-I	*Group-II*
P. Uranium	1. Vastan, Gujarat
Q. Lignite	2. Singrauli, Madhya Pradesh
R. Bituminous Coal	3. Digboi, Assam
S. Petroleum	4. Jaduguda, Jharkhand

A. P-4 Q-1 R-3 S-2
B. P-4 Q-1 R-2 S-3
C. P-3 Q-4 R-2 S-1
D. P-2 Q-4 R-1 S-3

25. Match the features in Group-I with their characteristic rocks/rock suite in Group-II.

Group-I	*Group-II*
P. Continental arc	1. Olivine tholeiite
Q. Mid-oceanic ridge	2. Alkaline rock
R. Collision zones	3. Andesite
S. Continental rift	4. Ophiolite

A. P-2 Q-1 R-3 S-4
B. P-1 Q-3 R-4 S-2
C. P-3 Q-1 R-4 S-2
D. P-3 Q-2 R-1 S-4

26. Which of the following statements are correct?

I. In a planar tabular cross-bedding, the upper and lower bounding surfaces are planar and parallel
II. In a planar tabular cross-bedding, the upper and lower bounding surfaces are planar but not parallel

III. In a trough cross-bedding, both the upper and the lower bounding surfaces are curved
IV. In a trough cross-bedding, the upper bounding surface is planar and the lower bounding surface is curved

A. II and III B. II, III and IV
C. I, II and III D. I, III and IV

27. Match the sedimentary structures in Group-I with the corresponding processes in Group-II.

Group-I	*Group-II*
P. Stromatolite	1. Desiccation
Q. Flute cast	2. Liquefaction
R. Mud crack	3. Biostratification
S. Dish and pillar	4. Erosion

A. P-3 Q-4 R-1 S-2
B. P-3 Q-4 R-2 S-1
C. P-4 Q-3 R-1 S-2
D. P-2 Q-3 R-4 S-1

28. An ore mineral has the following physical properties:
(1) metallic lustre, (2) steel-grey colour, (3) two sets of octahedral cleavage, (4) high specific gravity, and (5) makes mark on paper. Identify the ore mineral.

A. sphalerite B. magnetite
C. galena D. bornite

29. Which of the following are the characteristics of a typical porphyry copper deposit?
I. Associated with granitic rocks
II. Low-grade and high tonnage of ore
III. Presence of alteration zones
IV. Occurrence in convergent tectonic setting

A. II only
B. I and IV only
C. I, II, III and IV
D. I, III and IV only

30. Match the metal deposits in Group-I with the corresponding processes in Group-II.

Group-I	*Group-II*
P Aluminum	1. Mechanical concentration
Q Iron	2. Contact metasomatism
R. Tungsten	3. Residual concentration
S. Thorium	4. Chemical sedimentation

A. P-1 Q-3 R-2 S-4 B. P-3 Q-4 R-2 S-1
C. P-3 Q-4 R-1 S-2 D. P-2 Q-3 R-4 S-1

SECTION-B

Multiple Choice Question (MCQ)

Qs. Nos. 31-40 : *Carry two marks each.*

31. Which of the following statement(s) is/are correct for a plot of (87Sr/86Sr) versus (87Rb/86Sr) of a rock which has evolved in a closed system?

A. The slopes of the lines of evolution for the minerals in the rock are all equal and positive
B. The slopes of the lines of evolution for the minerals in the rock are all equal and negative
C. The slope of the isochron is identical in direction to the slope of the lines of evolution for the minerals in the rock
D. The slope of the isochron is opposite in direction to the slope of the lines of evolution for the minerals in the rock

32. Which of the following statements is/are correct with reference to subsurface water?

A. The perched water table occurs below the water table
B. Vadose water occurs in the zone of aeration
C. For significant groundwater underflow, the medium must be highly permeable
D. Intersection of the water table with the land surface can result in the formation of a spring

33. Which of the following geomorphic features indicate(s) the presence of a fault?

A. Triangular facets
B. Sudden topographic jump
C. Ox-bow lake
D. Point bar

34. Which of the following gastropods has/have siphonal canal?

A. Cerithium B. Cypraea
C. Fusus D. Murex

35. Which of the following is/are mode(s) of preservation of fossils?

A. Small organisms get trapped on viscous resins secreted by trees
B. Leaves preserved as thin films of carbon
C. Negative impressions of hard parts of organisms
D. Mineral growths that branch like ferns

36. Select the pair(s) that is/are correctly matched.
A. Lathi Formation — Jurassic wood fossils
B. Pinjor Formation — Cretaceous ammonoid fossils
C. Panchet Formation — Triassic reptile fossils
D. Kota Formation — Permian glossopterid fossils

37. Choose the reaction(s) indicating metamorphism of pelites at T > 600 °C and P < 8 kbar.
A. albite + actinolite → hornblende + quartz
B. muscovite + quartz → sillimanite + K-feldspar + H_2O
C. muscovite + biotite + quartz + H_2O → sillimanite + melt
D. albite → jadeite + quartz

38. Select the correct statement(s).
A. Laccoliths are convex upward whereas lopoliths are convex downward bodies
B. Laccoliths can form from any type of magma whereas lopoliths form only from basaltic magma
C. Laccoliths show layered structures whereas lopoliths lack any layering
D. Laccoliths are smaller in dimensions than lopoliths

39. Select the correct statement(s).
A. Phenocrysts are found in igneous rocks whereas porphyroblasts are found in metamorphic rocks
B. Phenocrysts are indicators of crystallization history whereas porphyroblasts indicate metamorphic history
C. Porphyroblasts are relict grains, relatively larger than matrix minerals in deformed igneous rocks
D. Poikiloblastic texture is found in igneous rocks whereas poikilitic texture is found in metamorphic rocks

40. Based on the given schematic ternary phase diagram, choose the correct statement(s).

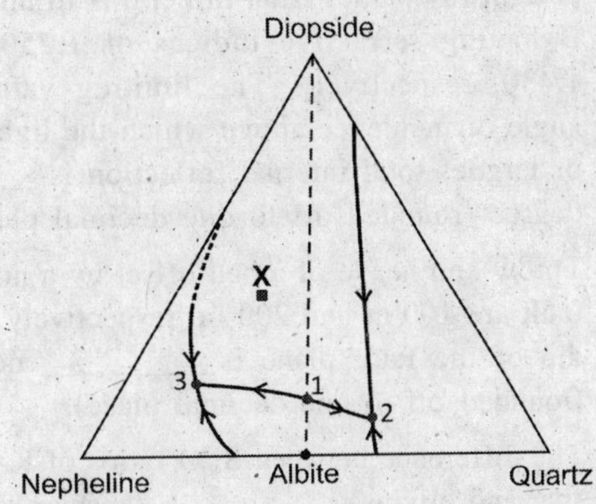

A. Point 1 is the thermal minimum in the system.
B. The diopside-albite join acts as a thermal barrier in the system.
C. Point 2 and 3 are ternary eutectics.
D. Melt of composition X will give rise to a rock containing nepheline + albite.

SECTION-C

Numerical Answer Type (NAT)

Qs. Nos. 41-50 : *Carry one mark each.*

41. The intensity of an earthquake of magnitude 8 on the Richter scale is greater than the intensity of an earthquake of magnitude 5 on the same scale by ___________ times.

42. Assume: (i) geothermal gradient = 25 °C/km in the crust, (ii) density of the crustal rocks = 3000 kg/m^3, and (iii) acceleration due to gravity = 10 m/s^2. Based on these values, the lithostatic pressure at a point where temperature is 400 °C will be __________ MPa.

43. The radii of A+2 and Bions are 1.12Å and 1.31Å, respectively. The coordination number of A+2 in mineral AB2 is____________.

44. In a sedimentary rock, the diameters of two grains A and B are 1ϕ and 3ϕ, respectively. The difference in diameters between A and B is ______ mm (rounded off to two decimal places).

45. A light ray passes from mineral A to mineral B having refractive indices of 1.750 and 1.430, respectively. The limiting value of angle of incidence above which the light ray undergoes total internal reflection is ______ degree (rounded off to one decimal place).

46. Throw and heave of a bed offset by a normal fault are 100 m and 200 m, respectively. The dip of the fault plane is ________ degree (rounded off to one decimal place).

47. The difference between Si/O ratios of K-feldspar and olivine is ________ (answer in two decimal places).

48. If a crystal contains 5 faces and 8 edges, the number of vertices in the crystal is ______.

49. At a temperature of 298.15 Kelvin, the free energy change of a reaction (ΔG^0) is 19.737 kCal/mole. If the universal gas constant (R) = 1.98717 Calorie/degree/mole, the log10 of the equilibrium constant K is ________ (rounded off to two decimal places).

50. A normal fault displaces a sandstone bed such that the dip-slip and the strike-slip components are 3 m and 4 m, respectively. The net-slip of the displacement is ______ m.

Qs. Nos. 51-60 : *Carry two marks each.*

51. In the given diagram, a 6 km high plateau is supported by a crustal root of thickness *r*. The system is in isostatic equilibrium as per Airy's hypothesis of isostasy. Densities of the crust and the mantle are 2800 kg/m^3 and 3100 kg/m^3, respectively. Assuming the acceleration due to gravity to be same throughout the region, the thickness of the root (r) is ______ km

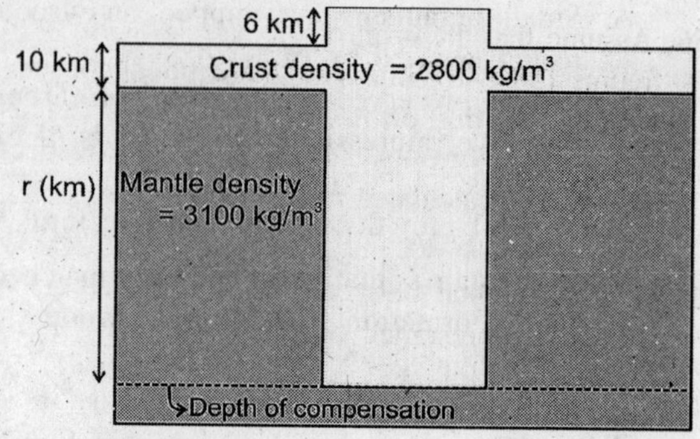

52. A 50 kg granite boulder gets dislodged from a cliff of height 20 m and undergoes an absolute vertical free fall. If the acceleration due to gravity is 10 m/s^2, the boulder will hit the ground with a velocity of _________ m/s.

53. Mass and volume of a fully dried soil sample are 500 g and 250 cm^3, respectively. The average density of the particles in the soil sample is 2.5 g/cm^3. The void ratio of the soil sample is __________%.

54. The geological map shows the contact between sandstone and limestone. The two dotted curves are the contours of 400 m and 500 m, respectively. The difference between the dip angles of the contact surface along the AB and AC directions is _________ degree (rounded off to two decimal places).

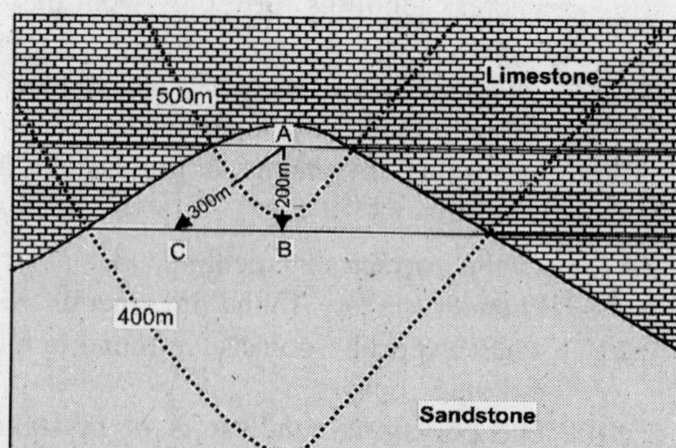

55. A tabular ore body of 9 km^2 area and an average thickness of 9 m has a density of 3000 kg/m^3. The tonnage (in million tonnes) of the ore body is __________.

56. Assume that the orbit of the earth is a circle of radius 150×10^6 km. The gravitational constant and the earth's orbital velocity are given as 6.7×10^{-11} Nm^2/kg^2 and 30×10^3 m/s, respectively. The calculated mass of the sun is ______ $\times 10^{30}$ kg (rounded off to two decimal places).

57. The difference between XAn and XAb in a feldspar represented by X in the given triangular diagram is ____________ (answer in one decimal place).

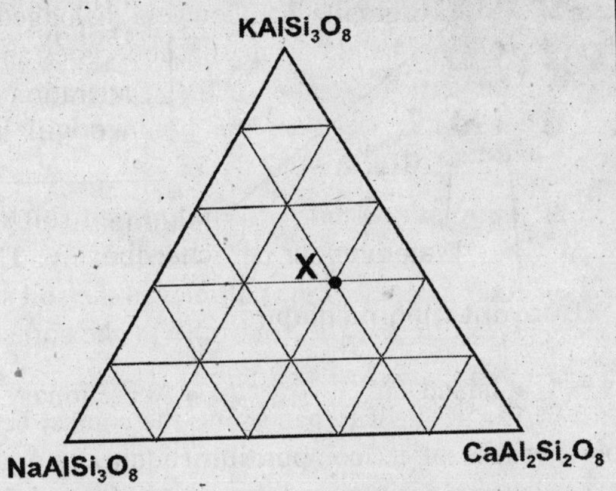

58. Two vertical wells penetrating a confined aquifer are 200 m apart. The water surface elevations in these wells are 35 m and 40 m above a common reference datum. The discharge per unit area through the aquifer is 0.05 m/day. Using Darcy's law, the coefficient of permeability is ____________ m/day.

59. The given P-T diagram shows the relative stability ranges of andalusite, sillimanite and kyanite. The difference in degrees of freedom at points X and Y is ____________.

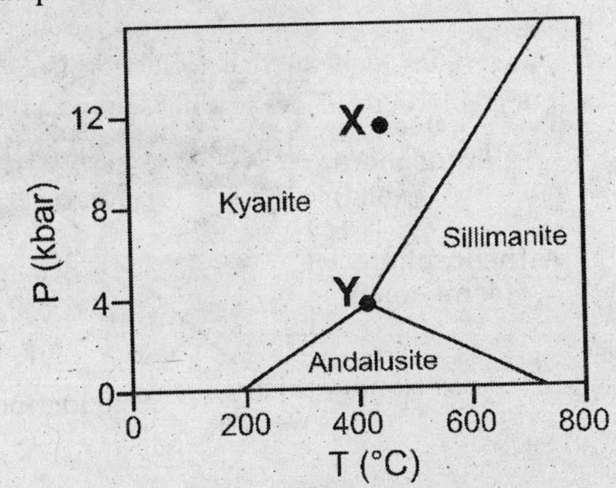

60. The core and rim compositions of garnet are $(Fe_{0.75}Ca_{0.90}Mn_{1.35})\ Al_2Si_3O_{12}$ and $(Fe_{0.90}Ca_{1.35}Mn_{0.75})\ Al_2Si_3O_{12}$, respectively. The difference in mole fractions of spessertine between the core and the rim is ____________ (answer in one decimal place).

ANSWER KEY

1	2	3	4	5	6	7	8	9	10
A	D	D	A	C	D	A	B	B	A
11	**12**	**13**	**14**	**15**	**16**	**17**	**18**	**19**	**20**
A	C	C	D	D	B	D	A	C	C
21	**22**	**23**	**24**	**25**	**26**	**27**	**28**	**29**	**30**
A	C	B	B	C	D	A	C	C	B
31	**32**	**33**	**34**	**35**	**36**	**37**	**38**	**39**	**40**
B,D	B,C,D	A,B	A,B,C,D	A,B,C	A,C	B,C	A,B,D	A,B	B,C,D
41	**42**	**43**	**44**	**45**	**46**	**47**	**48**	**49**	**50**
1000	480	8	0.37-0.39	54.5-55.0	26.0-27.0	0.25/0.12-0.13	5	–14.40 to –14.50	5
51	**52**	**53**	**54**	**55**	**56**	**57**	**58**	**59**	**60**
56	20	25	8.00-8.20	245	2.00-2.02	0.2	2	2	0.2

HINTS AND SOLUTIONS

1. Features of the following associated with a divergent plate boundary?

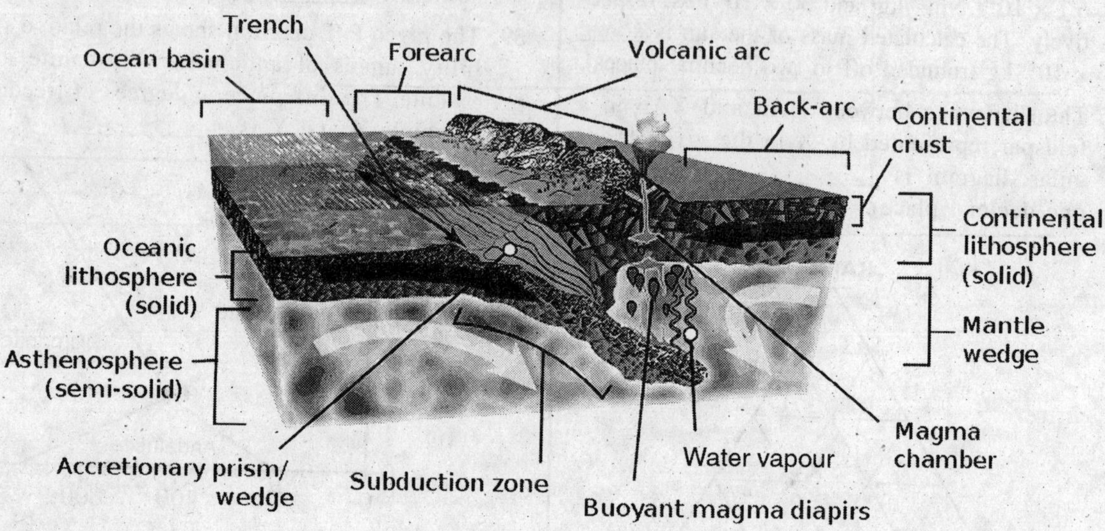

- Ridge
- Trench
- Island arc
- Accretionary prism

2. Shear waves do not travel through the outer core because of its composition (liquid).

Types of Waves:

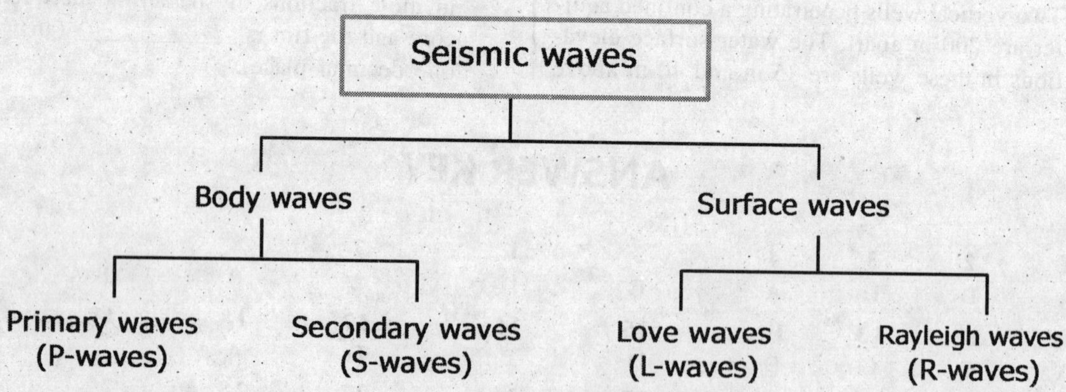

Seismic Wave	Abbreviation	Description	Ground Motion
Primary wave	P	• Type of body wave • First to arrive (fastest) • Ground squeezes and stretches in direction of wave travel. • Travels through solids, liquids and gases.	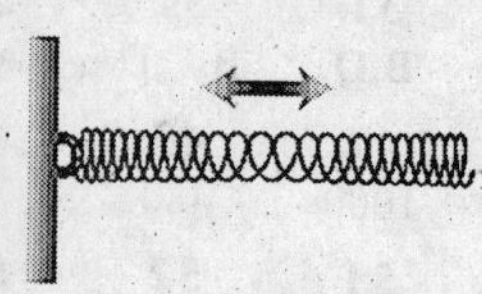

Seismic Wave	Abbreviation	Description	Ground Motion
Secondary wave	S	• Type of body wave • Second to arrive (slower) • Ground motion is perpendicular to direction of wave travel. • Travels through solids but not liquids.	
Surface wave	L	• Travels along Earth's surface • Last to arrive (slowest) • Ground motion is a rolling action, like ripples on a pond.	

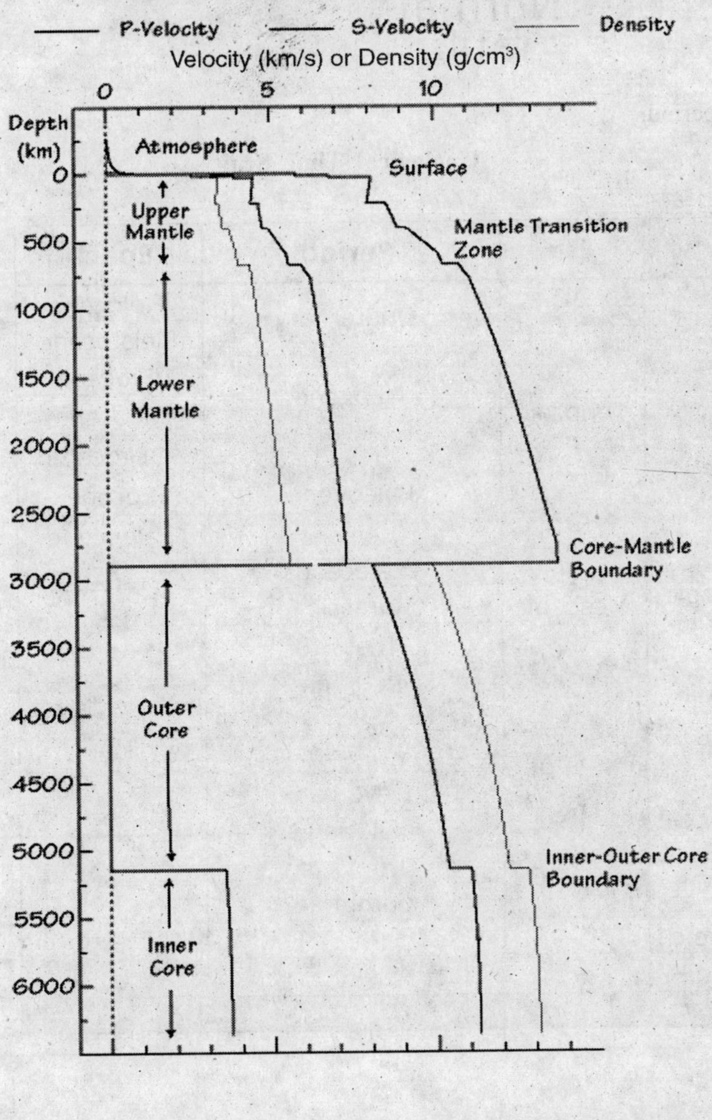

4. Horst and graben structures are typically formed by normal faulting.

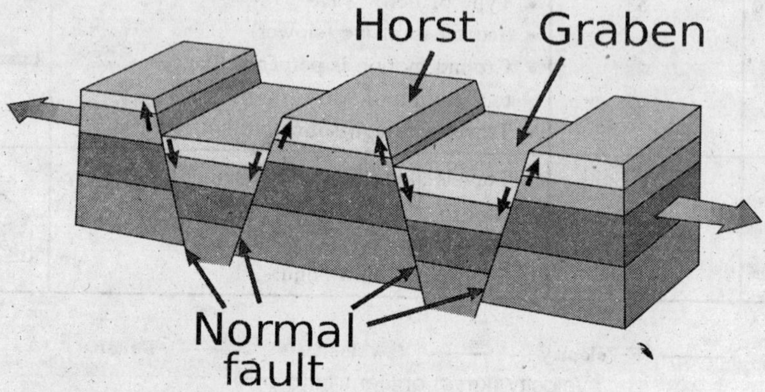

6. Permian is a/an period.

Geological Time Scale

Eon	Era	Period		Epoch	
Phanerozoic	Cenozoic	Quaternary		Holocene	← Today
				Pleistocene	← 11.8 Ka
		Neogene		Pliocene	
				Miocene	
		Paleogene		Oligocene	
				Eocene	
				Paleocene	
	Mesozoic	Cretaceous		~	← 66 Ma
		Jurassic		~	
		Triassic		~	
	Paleozoic	Permian		~	← 252 Ma
		Carboniferous	Pennsylvanian	~	
			Mississippian	~	
		Devonian		~	
		Silurian		~	
		Ordovician		~	
		Cambrian		~	
Proterozoic	~	~		~	← 541 Ma
Archean	~	~		~	← 2.5 Ga
Hadean	~	~		~	← 4.0 Ga
					← 4.54 Ga

Younger ↕ Older

10. Metamorphic facies:

FACIES-PROTOLITH—MINERAL ASSEMBLAGE TABLE:

Facies	Pelitic	Calcareous	Mafic
Zeolite **100-200° C**	**interlayered smectite/chlorite calcite**	**calcite**	**laumonite, thompsonite, calcite, interlayered smectite/chlorite**
Prehnite-Pumpellyite **150-300° C**	**prehnite, pumpellyite, calcite, chlorite, albite**	**calcite**	**prehnite, pumpellyite, calcite, chlorite, albite**
Greenschist **300-450° C**	**muscovite, chlorite, quartz, albite, biotite, garnet**	**calcite, dolomite, quartz, epidote, tremolite**	**albite, chlorite, quartz, epidote, actinolite, sphene**
Epidote Amphibolite **450-550° C**	**muscovite, biotite, garnet, albite, quartz**	**calcite, quartz, tremolite, epidote, diopside**	**albite, epidote, hornblende, quartz**
Amphibolite **500-700° C**	**garnet, biotite, muscovite, quartz, plagioclase, staurolite, kyanite or sillimanite**	**calcite, diopside quartz, wollastonite**	**hornblende, plagioclase, garnet, quartz, sphene, biotite**
Granulite **700-900° C**	**garnet, Kspar, sillimanite or kyanite, quartz, plagioclase, hypersthene**	**calcite, quartz, plagioclase, diopside, hypersthene**	**plagioclase, augite, hypersthene, hornblende, garnet, olivine**
Blueschist **150-350° C** **P > 5-8 Kb**	**Jadeite, albite, quartz, lawsonite, aragonite, paragonite**	**aragonite, white mica**	**glaucophane, albite, lawsonite, sphene, ± garnet**
Eclogite **350-750° C** **P > 8-10 Kb**	**coesite, Kspar, sillimanite, plagioclase**	**aragonite, quartz, plagioclase, diopside, hypersthene**	**omphacite (px), pyrope garnet**

32. The following statements is/are correct with reference to subsurface water.

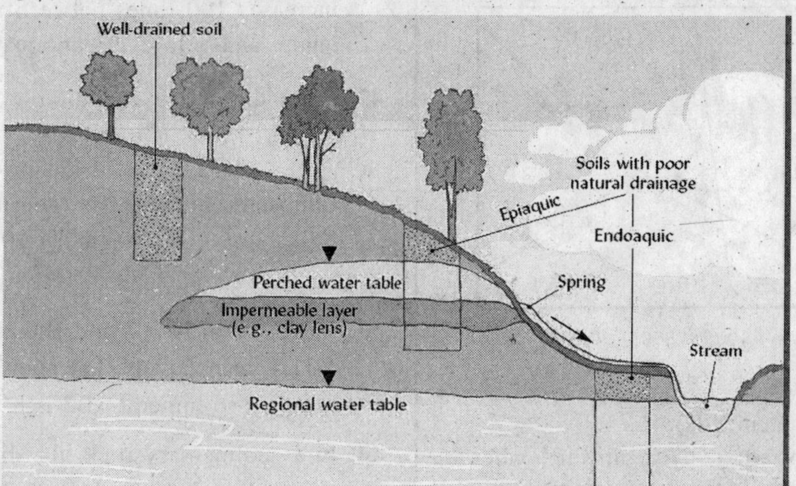

34. **Morphology of Gastropods**

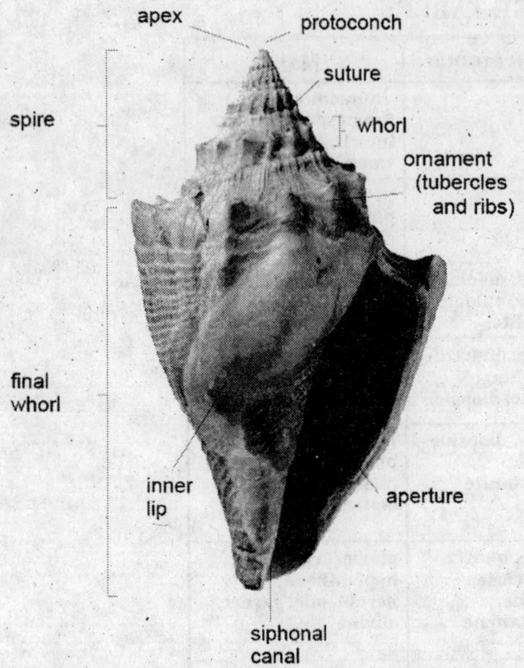

37. The reaction(s) indicating metamorphism of pelites at T > 600 °C and P < 8 kbar.

(B) muscovite + quartz → sillimanite + K-feldspar + H_2O

(C) muscovite + biotite + quartz + H_2O → sillimanite + melt

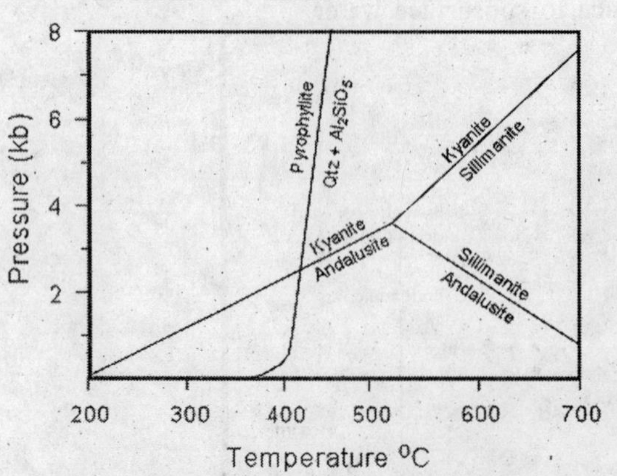

38. The correct statement(s).

A. Laccoliths are convex upward whereas lopoliths are convex downward bodies

B. Laccoliths can form from any type of magma whereas lopoliths form only from basaltic magma

D. Laccoliths are smaller in dimensions than lopoliths

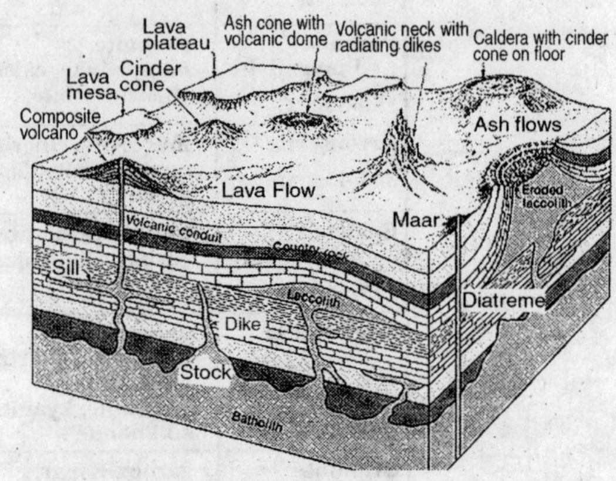

41. The Richter scale is a base-10 logarithmic scale, meaning that each order of magnitude is 10 times more intensive than the last one. In other words, a two is 10 times more intense than a one and a three is 100 times greater. In the case of the Richter scale, the increase is in wave amplitude. That is, the wave amplitude in a level 6 earthquake is 10 times greater than in a level 5 earthquake, and the amplitude increases 100 times between a level 7 earthquake and a level 9 earthquake.

42. From the above condition:

$$h = 400/25$$
$$= 16 \text{ km}$$

Lithostatic pressure P = density × g × height

$$= 3000 \times 10 \times 16$$
$$= 480 \text{ MPa}$$

43. The radii of A + 2 and Bions are 1.12Å and 1.31Å, respectively. The coordination number of A + 2 in mineral AB2 is 8.

44. In a sedimentary rock, the diameters of two grains A and B are 1ϕ and 3ϕ, respectively.

The difference in diameters between A and B is 0.375 mm (rounded off to two decimal places). From the given condition, as we know the relation between:

Millimeters (mm)		Micrometers (µm)	Phi (φ)	Wentworth size class		Rock type
	4096		-12.0			
				Boulder	Gravel	Conglomerate/ Breccia
	256		-8.0			
				Cobble		
	64		-6.0			
				Pebble		
	4		-2.0			
				Granule		
	2.00		-1.0			
				Very coarse sand	Sand	Sandstone
	1.00		0.0			
				Coarse sand		
1/2	0.50	500	1.0			
				Medium sand		
1/4	0.25	250	2.0			
				Fine sand		
1/8	0.125	125	3.0			
				Very fine sand		
1/16	0.0625	63	4.0			
				Coarse silt	Silt	Siltstone
1/32	0.031	31	5.0			
				Medium silt		
1/64	0.0156	15.6	6.0			
				Fine silt		
1/128	0.0078	7.8	7.0			
				Very fine silt		
1/256	0.0039	3.9	8.0			
				Clay	Mud	Claystone
	0.00006	0.06	14.0			

1 = 0.50 mm

3φ = 0.125 mm

The difference = 0.375 mm

45. Given, $\sin i_c = n_1/n_2$

$= 1.43/1.750$

$i_c = \sin^{-1} (0.81)$

$= 54.5$

46. Given,

Heave = 200 m

Throw = 100 m

Dip = 90 – 63.43

(from the tanθ calculation in degree)

= 27 degree.

47. K-feldspar belong to tectosilicate group of minerals. In this group the Si : O = 1 : 2 Olivine belong to nesosilicate group of minerals. In this group the Si : O = 1 : 4.

The difference = 1/2 – 1/4

= 1/4

= 0.25

50. Given, the dip slip = 3 m

Strike slip = 4 m

Net slip = $\sqrt{3^2 + 4^2}$

= 5m

51.

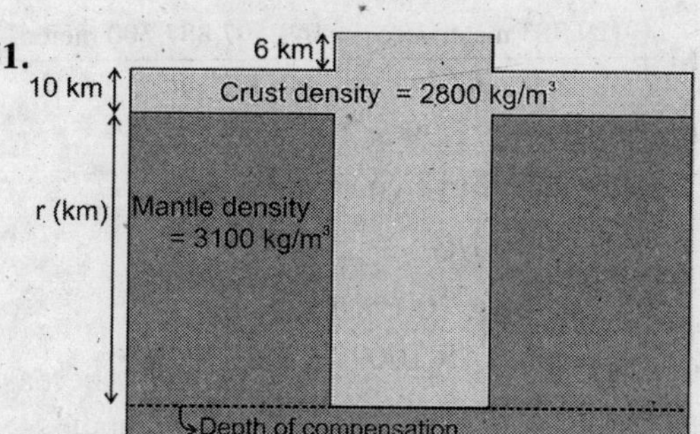

Given,

$H = 6$ km

Mantle density = 3100 kg/cm^3

Crust density = 2800 kg/m^3

$r = h \times$ density of crust/density difference between crust and mantle

$= 6 \times 2800/3100 - 2800$

$= 56$ km

53. Given, mass of the fully dried soil sample

$= 500$ g

Volume of the dried sample = 250 cm^3

Void ratio e = volume of the voids/volume of the solids = 25%

55. Given,

$V = 9 \times 0.009 \times 3000$

$= 243{,}000{,}000{,}000$ kg

$= 243{,}000{,}000$ ton

$= 243$ Mt

56. Since $F_{(gravity)} = F_{(centripetal)}$

it follows that

G = Gravitational Constant: 6.674×10^{-11} m^3/kg/sec^2
M = Mass of Sun
m = Mass of Earth
d = Distance between Masses (Same as Radius of Orbit)
r = Radius of Orbit
v = Velocity

$$\frac{GM \times m}{d^2} = \frac{m \times v^2}{r}$$

it follows that

$$M = \frac{v^2 \times r}{G}$$

therefore

$$M = \frac{(29{,}783 \text{ meters/sec})^2 \times 149{,}597{,}887{,}500 \text{ meters}}{6.674 \times 10^{-11} \text{m}^3/\text{kg}/\text{sec}^2}$$

therefore M = 1.99×10^{30} kg

Just the above conditions, the

$M = v^2 \times r/G$

$= (30 \times 10^3)^2 \times 150 \times 10^6 \times 1000 / 6.7 \times 10^{-11}$

$= 135 \times 10^{29} / 6.7$

$= 2.01 \times 10^{30}$ kg.

57.

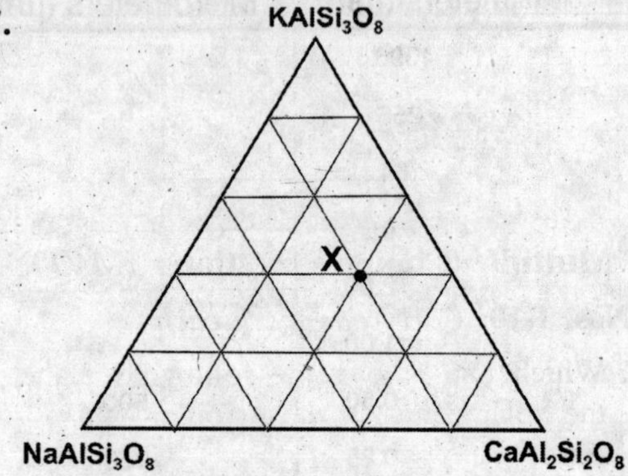

From the above diagram,

The difference between XAn and XAb

$= 40 - 20$

$= 20\%$

$= 0.2$

58. From the above conditions,

$Q = KA\ dh/dl$

$0.05 = K \times 1 \times 5/200$

$K = 2$

60.

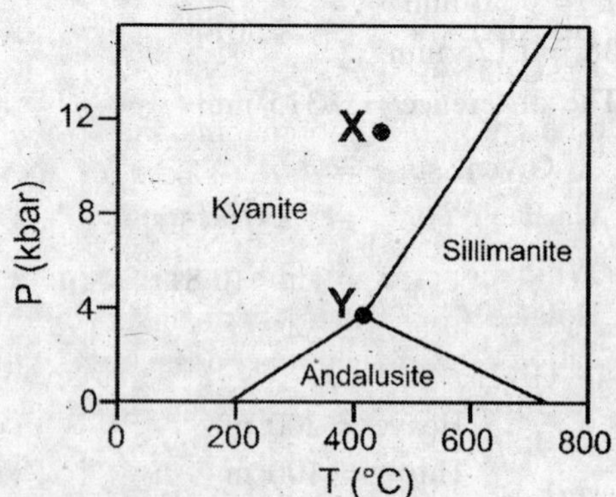

The given conditions,

The degree of freedom at point X = 3

The degree of freedom at point Y = 1

The difference = 3 − 1 = 2

Previous Papers (Solved)

Indian Institute of Technology Joint Admission Test for M.Sc.

IIT-JAM GEOLOGY (GG)-2018

SECTION-A

Multiple Choice Questions (MCQ)

Q. Nos. 1-10: *Carry one mark each.*

1. Which one among the following planets in the Solar system is most similar in size to the Earth?

A. Mercury B. Venus
C. Neptune D. Uranus

2. In which one of the following tectonic settings are the highest mountain chains and thickness crust found?

A. Island arc
B. Continental arc
C. Continental collision
D. Transcurrent

3. The second-most abundant oxide in the Earth's crust is

A. Al_2O_3 B. SiO_2
C. CaO D. Na_2O

4. The type of dentition found in *Trigonia* is

A. Schizodont B. Taxodont
C. Pachydont D. Isodont

5. Which one of the following minerals has isolated $(SiO_4)^{4-}$ tetrahedra linked by divalent cations in octahedral coordination?

A. Muscovite B. Quartz
C. Beryl D. Olivine

6. Which one of the following is NOT found in an extensional setting?

A. Normal faults
B. Horsts
C. Rifts
D. Thrust faults

7. The texture characterized by exsolved lamellae of albite in K-feldspar is known as

A. Myrmekite B. Graphic
C. Perthite D. Antiperthite

8. Fissility is best shown by

A. Sandstone B. Siltstone
C. Shale D. Limestone

9. Petroleum is NOT commercially produced from

A. Krishna-Godawari Basin
B. Cauvery-Palar Basin
C. Cambay Basin
D. Vindhyan Basin

10. Among the following, the mineral showing acicular habit is

A. Kyanite B. Tourmaline
C. Biotite D. Sillimanite

Q. Nos. 11-30: *Carry two marks each.*

11. Isostasy involves continental mountain belts.

A. Compensation in
B. Creation of
C. Destruction of
D. Thrusting

12. Identify the pair from the following list that is NOT correctly matched

A. Caldera - stratovolcano
B. Pillow basalt - subaerial eruption
C. Ropy lava - pahoehoe flow
D. Amygdales - filled vesicles

13. Wilson orogenic cycle in continents is initiated by

A. Collision B. Rifting
C. Drifting D. Subduction

14. Match the processes in Group-I with corresponding geomorphic features in Group-II.

Group I	*Group II*
P. Dissolution	1. Mushroom rocks
Q. Abrasion	2. Exfoliation domes
R. Deposition	3. Sinkholes
S. Onion skin weathering	4. Moraines

A. P-1 Q-3 R-2 S-4
B. P-3 Q-1 R-2 S-4
C. P-3 Q-1 R-4 S-2
D. P-2 Q-1 R-4 S-3

15. The orientations of the fold axis and axial plane in the given figure indicate

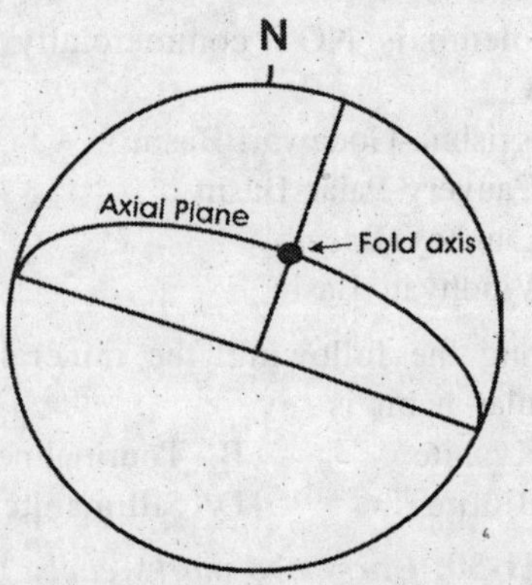

A. Reclined fold B. Vertical fold
C. Recumbent fold D. Horizontal fold

16. Identify the correct morphological features corresponding to numbers I - IV in the echinoid illustrated below:

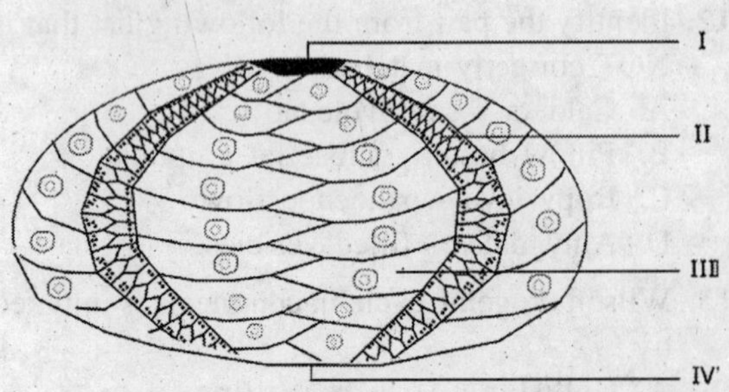

A. I-Periproct II-Ambulacra III-Interambulacra IV-Peristome
B. I-Periproct II-Interambulacra III-Ambulacra IV-Peristome
C. I-Peristome II-Interambulacra III-Ambulacra IV-Periproct
D. I-Peristome II-Ambulacra III-Interambulacra IV-Periproct

17. The correct order of marine benthic habitats with increasing water depths is
A. Abyssal, bathyal, neritic
B. Neritic, abyssal, bathyal
C. Neritic, bathyal, abyssal
D. Bathyal, abyssal, neritic

18. Which one of the following invertebrates has the most primitive visual system?
A. Ammonites B. Brachiopods
C. Gastropods D. Trilobites

19. The correct chronological sequence (older to younger) of the Precambrian stratigraphic units listed below is
A. Sargur Group, Chitradurga Group, Alwar Group, Kaimur Group
B. Chitradurga Group, Sargur Group, Kaimur Group, Alwar Group
C. Sargur Group, Alwar Group, Chitradurga Group, Kaimur Group
D. Sargur Group, Chitradurga Group, Kaimur Group, Alwar Group

20. Match the Formations in Group-I with corresponding characteristic fossils in Group-II.

Group I	*Group II*
P. Barakar Formation	1. Stegodon
Q. Uttatur Formation	2. Sauropoda
R. DhokPathan Formation	3. Belemnites
S. Lameta Formation	4. Glossopteris

A. P-3 Q-1 R-4 S-2
B. P-4 Q-3 R-1 S-2
C. P-3 Q-4 R-1 S-2
D. P-4 Q-3 R-2 S-1

21. Which one of the following sedimentary structures is NOT used for determining top and bottom of beds?
A. Mud cracks
B. Load and flame structures

C. Sharp-crested wave ripples
D. Plane lamination

22. Identify the rocks P and Q in the diagram as per the IUGS classification.

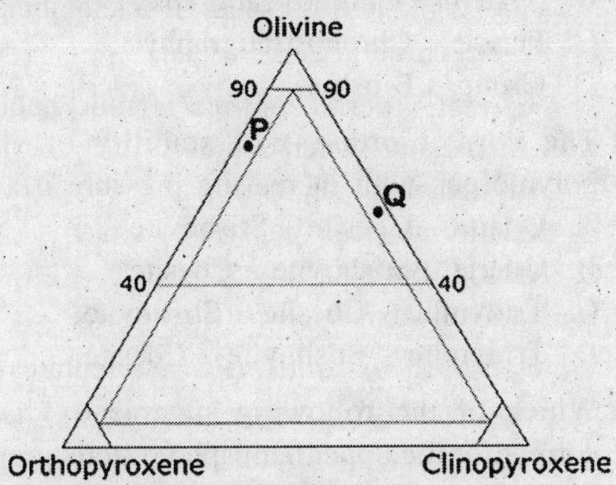

A. P-Websterite, Q-Wehrlite
B. P-Dunite, Q-Websterite
C. P-Websterite, Q-Dunite
D. P-Harzburgite, Q-Wehrlite

23. Which one of the following is produced by a closed-system metamorphic reaction between muscovite and quartz?
A. Orthoclase + sillimanite
B. Orthoclase + biotite
C. Plagioclase + biotite
D. Plagioclase + sillimanite

24. The assemblage staurolite + garnet + biotite + muscovite + quartz in pelites is stable in
A. Green schist facies
B. Amphibolite facies
C. Granulite facies
D. Pyroxene hornfels facies

25. Conglomerates are commonly deposited in
A. Aeolian dunes
B. Tidal flats
C. Alluvial fans
D. River flood plains

26. Match the mineral deposits in Group I with corresponding Indian occurrences in Group II.

Group I	*Group II*
P. Iron	1. Mangampet, Andhra Pradesh
Q. Uranium	2. Balaghat, Madhya Pradesh
R. Manganese	3. NarwaPahar, Jharkhand
S. Baryte	4. Hospet, Karnataka

A. P-1 Q-3 R-4 S-2
B. P-4 Q-1 R-3 S-2
C. P-3 Q-4 R-2 S-1
D. P-4 Q-3 R-2 S-1

27. Which one of the following processes is responsible for the formation of syngenetic Ni-Cu sulphide ore in gabbro-noritic rocks?
A. Hydrothermal replacement
B. Volcanic exhalation
C. Liquid immiscibility
D. Contact metamorphism

28. Dashed lines in the figures given below represent joints. Considering only the orientations of the joints and the slope face, which one of the following represents the most stable slope?

A.
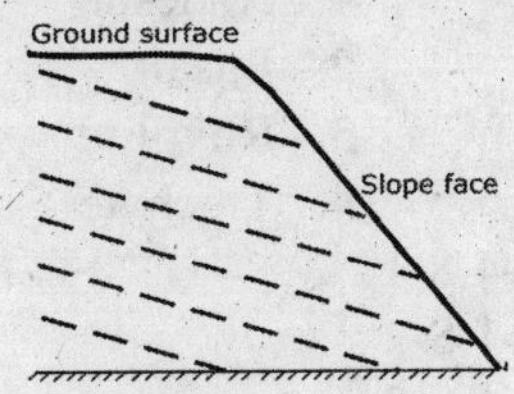

B.
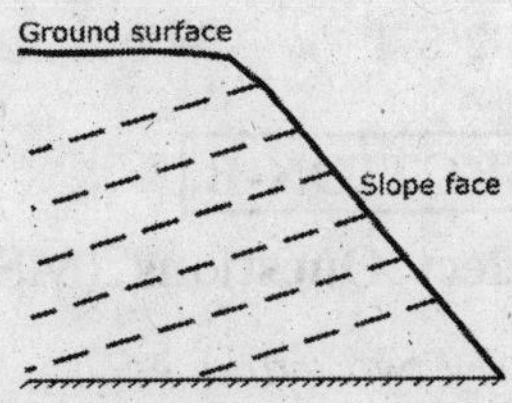

C.
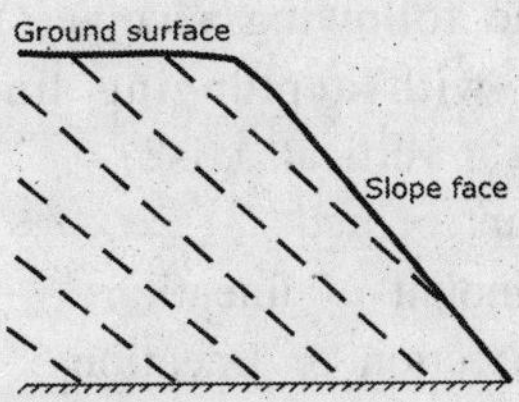

D.
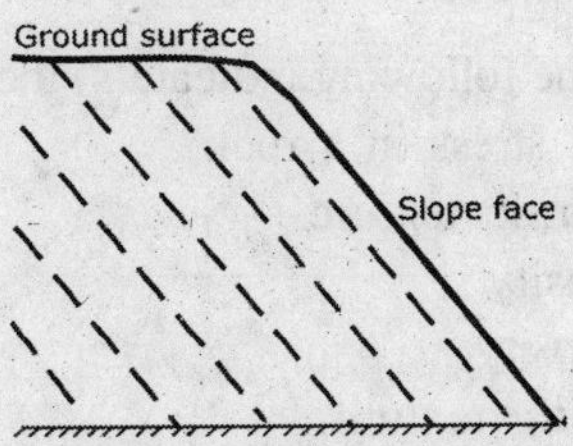

29. Match the morphological features/life processes in Group I with corresponding organisms in Group II.

Group I	*Group II*
P. Water vascular system	1. *Cephalopoda*
Q. Moulting	2. *Echinodermata*
R. Jet propulsion locomotion	3. *Brachiopoda*
S. Lophophore	4. *Trilobita*

A. P-2 Q-3 R-1 S-4
B. P-3 Q-4 R-2 S-1
C. P-2 Q-4 R-1 S-3
D. P-4 Q-3 R-2 S-1

30. Match the plutonic rocks in Group-I with corresponding volcanic equivalents in Group-II.

Group I	*Group II*
P. Granite	1. Andesite
Q. Syenite	2. Basalt
R. Diorite	3. Rhyolite
S. Gabbro	4. Trachyte

A. P-2 Q-3 R-4 S-1
B. P-3 Q-4 R-1 S-2
C. P-4 Q-3 R-1 S-2
D. P-3 Q-4 R-2 S-1

SECTION-B

Multiple Select Questions (MSQ)

Q. Nos. 31-40: *Carry two marks each.*

31. Which of the following change(s) when a dipping bed with a plunging lineation is rotated about a vertical axis?
A. Dip amount of bed
B. Plunge amount of lineation
C. Plunge direction of lineation
D. Strike of bed

32. Which of the following indicate(s) the presence of directed stress in a rock?
A. Porphyritic texture
B. Schistosity
C. Gneissosity
D. Mylonitic texture

33. The correct combination(s) of ranks and corresponding categories of stratigraphic units is/are
A. Formation - Lithostratigraphy
B. System - Chronostratigraphy
C. Period - Chronostratigraphy
D. Group - Biostratigraphy

34. The correct order(s) of stability of silica polymorphs with increasing pressure is/are
A. Quartz - Coesite - Stishovite
B. Quartz - Stishovite - Coesite
C. Tridymite - Coesite - Stishovite
D. Tridymite - Stishovite - Coesite

35. Which of the following statement(s) is/are correct for the upper hemisphere stereographic projection of a crystal given below?

A. Angle between the axes, $\alpha = \beta = \gamma = 90^\circ$
B. Crystal contains 1 tetrad
C. Crystal contains 4 diads
D. Crystal contains 5 mirror planes

36. Which of the following statement(s) is/are correct?
A. An isotropic mineral remains dark through 360° rotation of stage under crossed polars
B. Pleochroism is the change of colour of a mineral during rotation under crossed polars
C. Minerals of the Triclinic system are optically uniaxial
D. Melatope in an interference figure marks the emergence of an optic axis

37. Hermatypic corals are typically found in
A. The photic zone
B. Warm and clear water
C. Cool deep water
D. Reefs

38. Choose the characteristic mineral(s) formed in the supergene enriched zone of a sulphide deposit.
A. Psilomelane B. Covellite
C. Cassiterite D. Chalcocite

39. Which of the following is/are true for crystallization of plagioclase phenocrysts from a basic magma forming a layered intrusion?
A. Cumulus texture at the base
B. Anorthite-rich early plagioclase at the base
C. Albite-rich late plagioclase at the top
D. Quench texture at the base

40. Major mass extinction events occurred in the
A. End Silurian
B. End Carboniferous
C. End Permian
D. Early Devonian

SECTION-C

Numerical Answer Type (NAT)

Q. Nos. 41-50: *Carry one mark each.*

41. When plotted on a map of 1:50000 scale, a 2 km long dyke exposed on a horizontal surface has a length of _______ cm (answer in one decimal place).

42. The valency of iron in hematite is _______.

43. A crustal rock is at a lithostatic pressure of 3 kbar and a temperature of 275°C. If the lithostatic pressure increases at a uniform rate of 0.3 kbar/km, and the surface temperature is 25°C, the geothermal gradient (in °C/km) is _______ (answer in one decimal place).

44. The absolute difference in the Moh's hardness values of the two silicates among the minerals listed below is _______.

Apatite, Corundum, Gypsum, Talc, Topaz

45. Attitudes of beds in sequences A (younger) and B (older), separated by an unconformity UU, are given in the following sectional view. If UU was horizontal when sequence A was deposited, the dip amount of beds in sequence B at that time was _______ (answer in one decimal place).

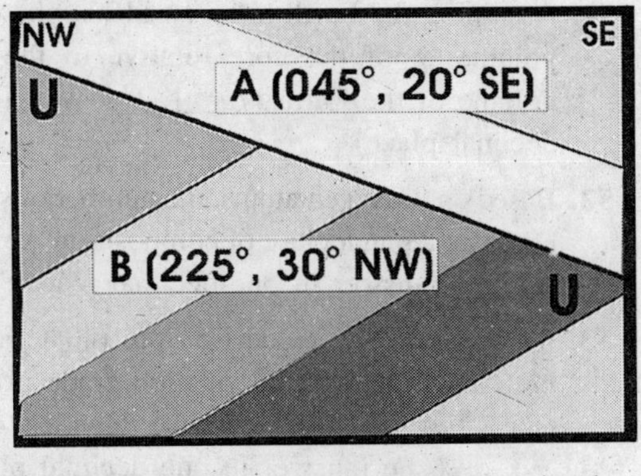

46. The number of alpha (?) particles emitted to produce a daughter isotope of 206Pb from a parent isotope of 238U by radioactive decay is ________.

47. The dip slip on a fault 000°, 30°E is 10 m. Assuming slip equals separation here, the throw on the fault is _____ m (answer in one decimal place).

48. A continuous 10 m thick sequence of shale was deposited in 10,000 years at uniform rate of sedimentation. The number of samples that must be collected at equal stratigraphic intervals to sample the succession every 500 years is ______.

49. Attitudes of the two limbs of a non-plunging kink fold shown below are 045°, 20°SE and 045°, 40°NW. The dip amount (in degrees) of the axial plane of the kink fold is _____ (answer in one decimal place).

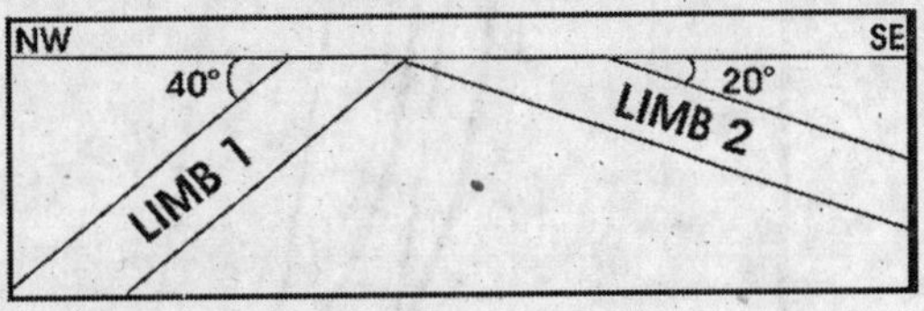

50. In the garnet formula $(Fe_{2.5}Mg_{0.3}Ca_x)Al_2Si_3O_{12}$, x represents the number of atoms of Ca. The mole % of grossular in the garnet is _____ (answer in one decimal place).

Q. Nos. 51-60: *Carry two marks each.*

51. Assuming the Earth to be an ideal sphere, the volume % of the core relative to the total volume of the Earth is _____ (answer in one decimal place).

52. Based on 8 oxygen atoms, the number of silicon atoms in a plagioclase of composition $Ab_{20}An_{80}$ is ____ (answer in one decimal place).

53. 600 tons of low grade iron ore (40% Fe) are blended with 400 tons of high grade iron ore (65% Fe). The grade of the blended ore is _____ % Fe (answer in one decimal place).

54. The mass of a fully dried rock sample of volume 100 cm^3 is 300 g. The mass of the sample, when fully saturated with water of density 1.00 g/cm^3, is 325 g. Assuming no volume change, the computed porosity of the rock is _____ % (answer in one decimal place).

55. When a dunite comprising pure forsterite undergoes melting, the weight % of MgO in the melt is _______________ (answer in one decimal place; given molecular weights of SiO_2 = 60.08; MgO = 40.30).

56. A block of rock with a mass of 72 kg slides on a surface inclined at an angle of 30º as shown in the figure. Assuming no cohesion and friction, the force 'F' is ____________ Newton (answer in one decimal place; acceleration due to gravity = 9.8 m/s^2).

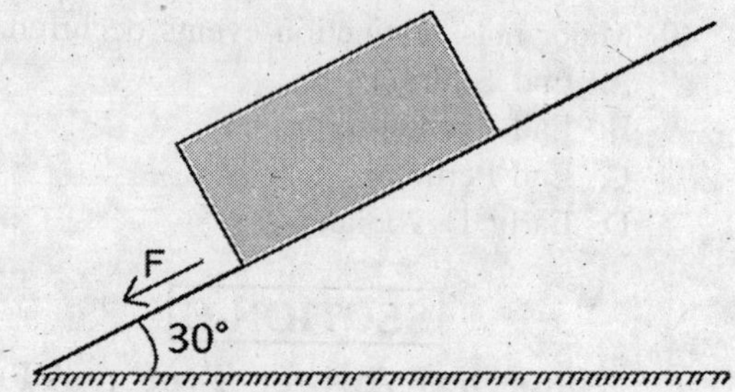

57. The true thickness of Bed A in the map given below is ____ m (answer in one decimal place).

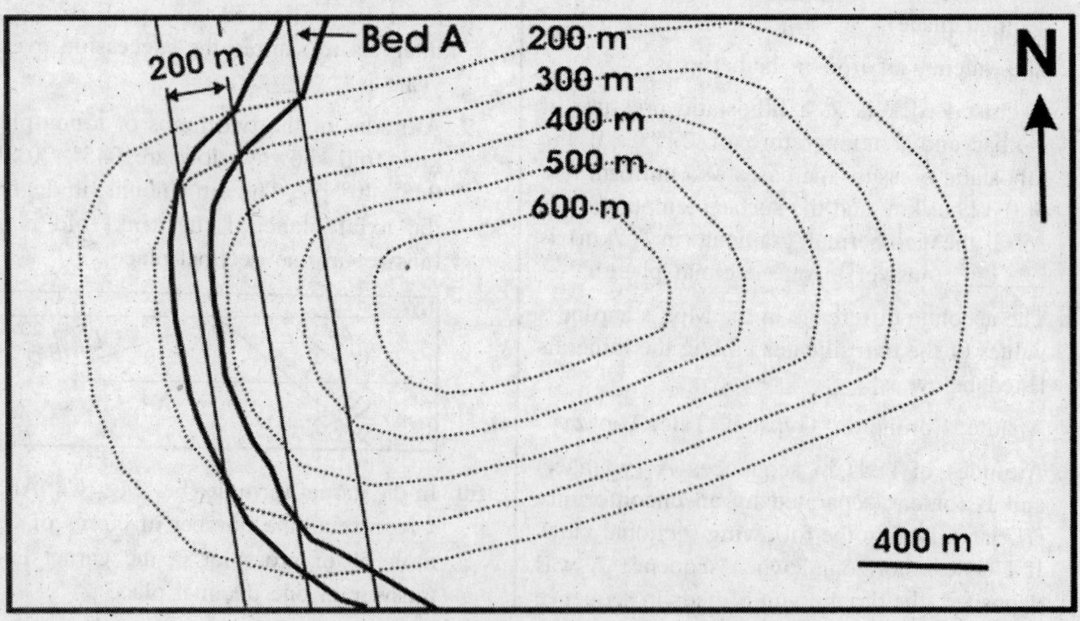

58. A melt containing 900 moles of anorthite and 100 moles of diopside undergoes crystallization. The number of moles of anorthite that crystallizes as the melt composition moves from P to Q is ____.

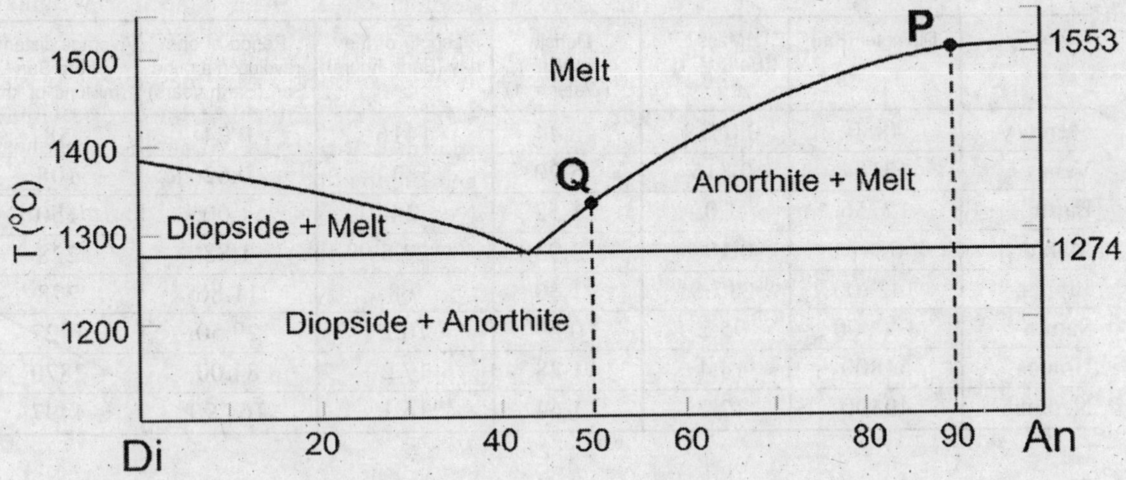

59. A confined sandstone aquifer with a uniform cross-sectional area of 7 m^2 and a hydraulic conductivity of 2 m/s, transmits water across a hydraulic gradient of 3.2. Assuming steady state Darcian flow, the volumetric flow rate through the aquifer is _____ m^3/s (answer in one decimal place).

60. A diamondiferous lamproite is ultrapotassic and has a molar K_2O/Na_2O ratio of 11. If the Na_2O content of the rock is 0.62 wt%, the K_2O content is ________ wt% (answer in one decimal place; molecular weight of Na_2O = 61.98, and K_2O = 94.20).

ANSWER KEY

1	2	3	4	5	6	7	8	9	10
B	C	A	A	D	D	C	C	D	D
11	**12**	**13**	**14**	**15**	**16**	**17**	**18**	**19**	**20**
A	B	B	C	A	A	C	D	A	B
21	**22**	**23**	**24**	**25**	**26**	**27**	**28**	**29**	**30**
D	D	A	B	C	D	C	B	C	B
31	**32**	**33**	**34**	**35**	**36**	**37**	**38**	**39**	**40**
C,D	B,C,D	A,B	A,C	A,B,C,D	A,D	A,B,D	B,D	A,B,C	C
41	**42**	**43**	**44**	**45**	**46**	**47**	**48**	**49**	**50**
4.0	3	25.0	7	50.0	8	5.0	20	70.0 to 90.0	6.6 to 6.8
51	**52**	**53**	**54**	**55**	**56**	**57**	**58**	**59**	**60**
14.0 to 18.0	2.2	50.0	25.0	57.0 to 57.5	352.8	89.0 to 90.0	800	44.8	10.3 to 10.5

HINTS AND SOLUTIONS

1. Size of the Planets of Solar System:

Planets	Diameter (Km)	Mass (Earth = 1)	Density (gm/cm^3) (Water = 1)	Length of the day (Earth hours)	Period of one revolution around Sun (Earth years)	Average distance from Sun (millions of km)
Mercury	4880	0.0558	5.44	1416	0.24	58
Venus	12104	0.815	5.20	5832	0.62	108
Earth	12756	1.0	5.52	24	1.00	150
Mars	6787	0.108	3.93	24.6	1.88	228
Jupiter	142800	317.8	1.30	9.8	11.86	778
Saturn	120000	95.2	0.69	10.2	29.50	1427
Uranus	51800	14.4	1.28	17.2	84.00	2870
Neptune	49500	17.2	1.64	16.1	164.90	4497

2. Plate setting and its features:

Plate setting	*Features develop*
Convergent plate movement	Island Arc, Beni-off zone
Divergent plate movement	Mid Oceanic Ridge, Sea mount
Transform plate movement	Strike - slip faults

3. Composition of the Earth's Lithosphere:

Major oxides	*Per cent (%)*
SiO_2	59.07
Al_2O_3	15.22
CaO	5.10
FeO	3.71
Na_2O	3.71
MgO	3.45
K_2O	3.11
Fe_2O_3	3.10
H_2O	1.30
TiO_2	1.03
Minor oxides	1.20
Total	**100**

4. Dentition pattern and its example:

Dentition pattern	*Example*
Taxodont	Glycimeris
Dysodont	Mytilus
Anomalodont	Allorisma
Isodont/Pachydont	Pecten
Schizodont	Trigonia
Teleodont	Venus
Desmodont	Ostrea
Pantadont	Allodesma
Edentulous	Acharax

5. Minerals and its linked cation:

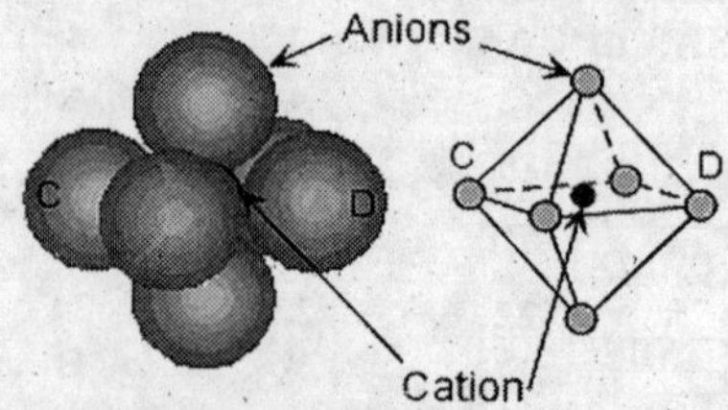

Minerals	*Remarks*	*Si:O*
Muscovite	Phyllosilicate	4:10
Quartz	Tectosilicate	1:2
Beryl	Cyclosilicate	1:3
Olivine	Neso-silicate	1:4

6. Features and its setting:

Features	*Remarks*
Normal faults	Horst and Graben
Horsts	Normal fault
Thrust faults	Nappes
Strike-slip faults	Wrench, tear fault

7. Textures and its characteristics:

Textures	*Remarks*
Myrmekite	Intergrowth between quartz and feldspar
Graphic	Intergrowth between quartz and orthoclase feldspar
Perthite	Intergrowth between orthoclase and albite

8. Rocks and its textures:

Rocks	*Textures/structures*
Sandstones	They consist detrital particles of the sand size.
Siltstone	Silt size particles
Shale	Fissility property of the rocks
Limestone	Limestone is a sedimentary rock composed primarily of calcium carbonate ($CaCO_3$) in the form of the mineral calcite.

9. Vindhyan basin: This basin presently not in commercially for the purpose of the petroleum, but in future it may be a huge amount of the hydrocarbon production.

10. Minerals and its Habit:

Minerals	*Habits*
Kyanite	Thin bedded
Tourmaline	Columnar
Biotite	Tabular
Sillimanite	Acicular

11. Isostasy: The isostasy is defined as the compensation of the various parts of the Earth. The concept based on the mountains are compensated by masses of smaller density under the mountains while oceans are compensated by masses of higher density under the water. It is totally view like Archimedes' principle.

There are several models based on different assumptions regarding the compensation mentioned under the first question above. The two most commonly mentioned models are:

- Airy's model
- Pratt model

12. Types of Volcanoes:

Types of volcanoes	*Products*	*Example*
Strato/ Composite volcano	More viscous lava with alternate band of proclastic materials	Mount baker
Flood/Plateau volcano	Lava flow with very wide spread	Columbia river plateau
Shield volcano	Liquid lava from the central vent	Hawaiian volcanoes
Cinder	Explosive liquid lava with small vent	Mount Tabor
Caldera	Very large composite volcanoes	Crater Lake
Volcanic dome	Very viscous lava	Mono Crater

13. Wilson Orogenic Cycle: The Wilson cycle is a model where a continent rifts, forms an ocean basin in-between, and then begins a process of convergence that leads to the collision of the two plates and closure of the ocean.

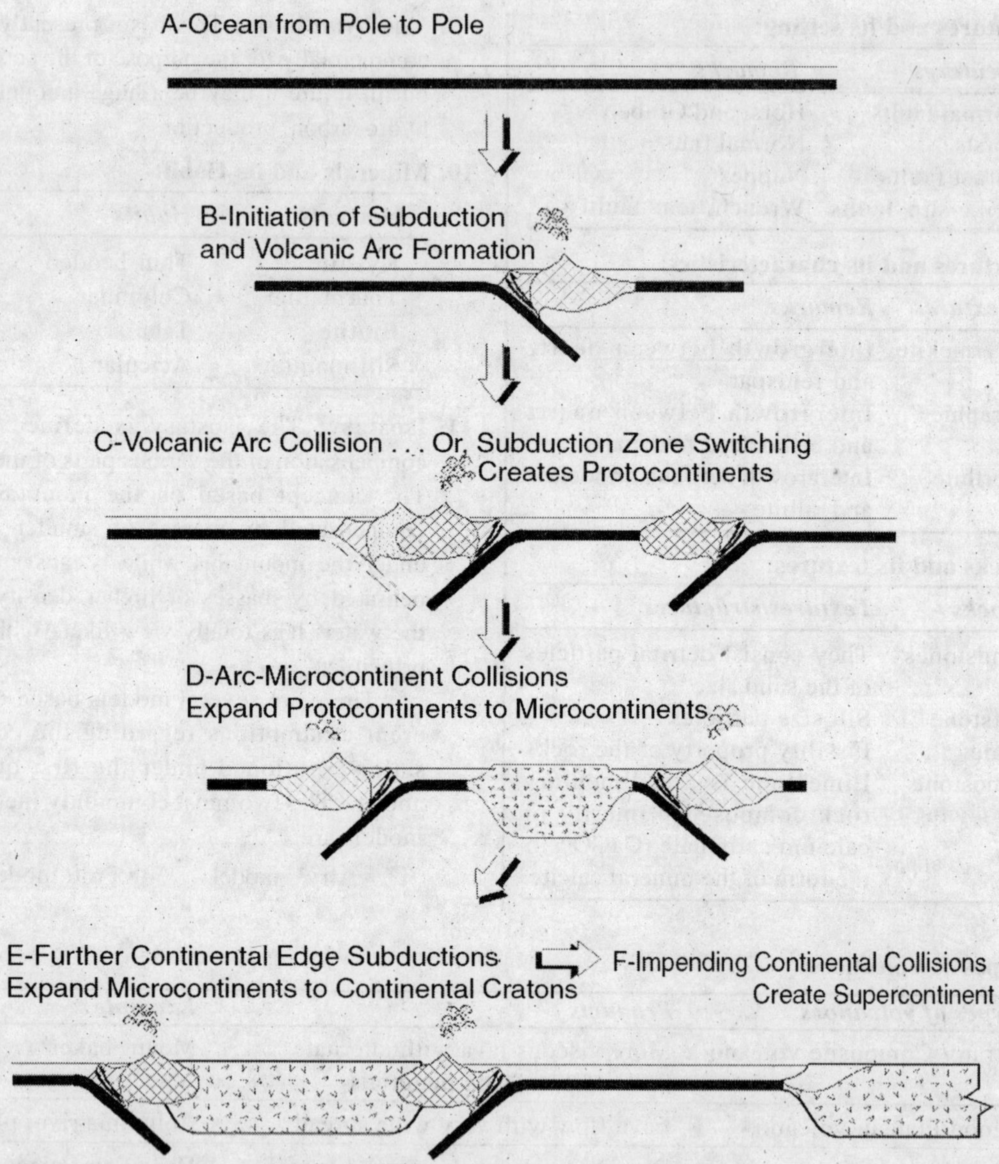

14. Geomorphic features and its characters:

Features	*Processes*	*Remarks*
Mushroom rocks	Abrasion	Due to wind abrasion
Exfoliation domes	Onion skin weathering	Due to climate change
Sinkholes	Dissolution	Limestone terrains
Morains	Depositions	Glacial

15. Fold types and its axial plane:

Types of folds	*Axial planes*
Reclined fold	Plunge axis directly to dip of the axial plane
Recumbent fold	In this fold the axial plane is Horizontal

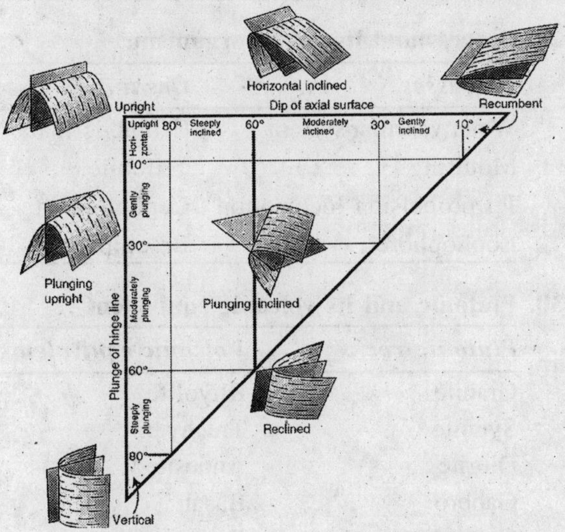

16. Morphological features of Echinoid:

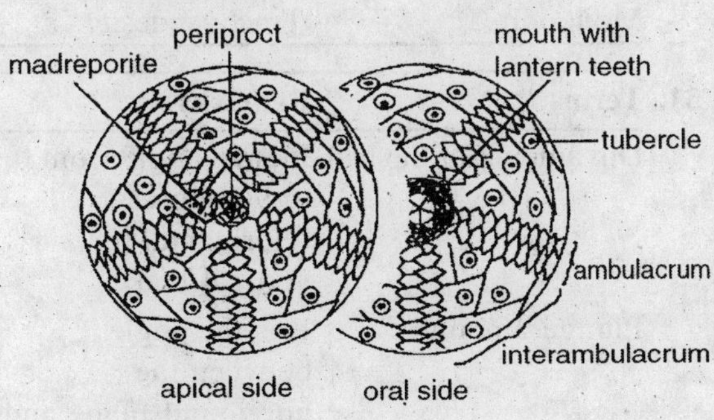

17. Depth of the ocean:

Zones	*Depth (m)*
Photic zone	200 m
Shelf	200 – 100 m
Bathyal	1000 – 2000 m
Abyssal plain	2000 – 6000 m
Hadal	> 6000 m

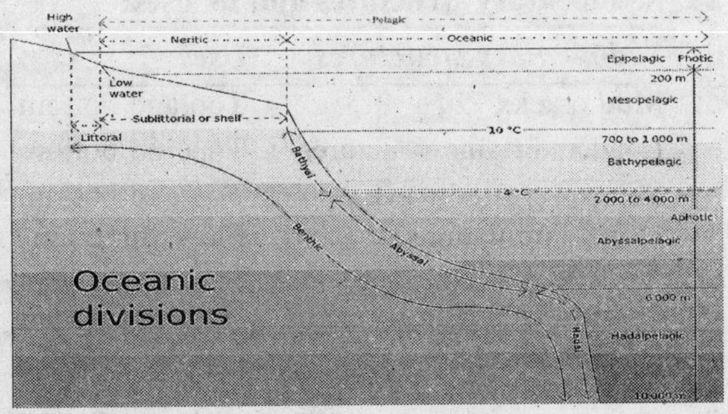

18. Invertebrates:

Invertebrates	*Remarks*
Ammonites	Index fossils from lower Devonian to upper cretaceous
Brachiopods	Exclusively marine with Precambrian to recent
Gastropods	Terrestrial, freshwater and marine from Cambrian to present day
Trilobites	Extinct form from Cambrian to Permian

19. Stratigraphic units and its geological ages:

Stratigraphic units	*Geological ages*
Sargur Group	Schistose rocks of Karnataka about 3,300 – 3,000 million years
Chitradurga Group	Dharwar super group
Alwar Group	Delhi super group
Kaimur Group	Vindhyan super group

20. Fossils and its formations:

Formations	*Fossils*	*Remarks*
Barakar formations	Glossopteris	Gondwana plant fossils
Uttatur Formations	Belemnites	Cephalopod
Dhok-Pathan formation	Stegodon	Elephantidae
Lameta Formation	Sauropoda	Dinosaur

21. Sedimentary structures and its uses:

Sedimentary structures	*Uses*
Mud cracks	Top and bottom
Load and flame structures	Top and bottom
Sharp – crested wave ripples	Top and bottom
Plane laminations	Flow direction

22. Olivine – Orthopyroxene – Clinopyroxene classifications:

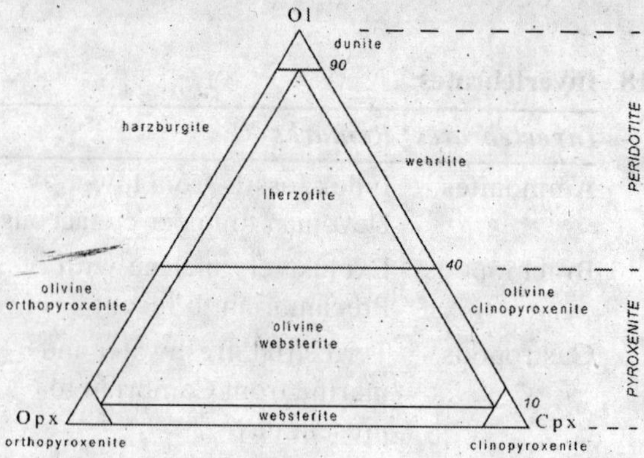

24. Metamorphic facies and its assemblage :

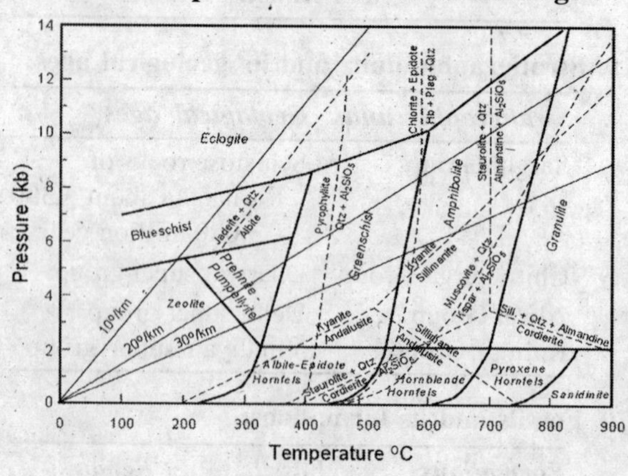

26. Minerals deposits and its Indian location:

Mineral deposits	*Occurrence*
Iron	Hospet, Karnataka
Uranium	Narwa Pahar, Jharkhand
Manganese	Balaghat, Madhya Pradesh
Barite	Mangampet, Andhra Pradesh

27. Processes and its minerals deposits:

Processes	*Remarks*
Hydrothermal replacement	Au – Ag deposits
Volcanic exhalation	Massive sulphide minerals
Liquid immiscibility	Syngenetic Ni–Cu sulphide
Contact metamorphism	Graphite

29. Process and its life of organism:

Processes	*Organisms*
Water vascular system	Echinodermata
Moulting	Trilobite
Jet propulsion locomotion	Cephalopoda
Lophophore	Brachipoda

30. Plutonic and its volcanic equivalent :

Plutonic rocks	*Volcanic equivalents*
Granite	Rhyolite
Syenite	Trachyte
Diorite	Andesite
Gabbro	Basalt
Granodiorite	Dacite
Nepheline-syenite	Phonolite
Monzonite	Trachyandesite

31. Terms	**Remarks**
Dip amount of the bed	It is an angle from 0 to 90 degree (Horizontal to vertical bed)
Plunge amount	It is the angle between the horizontal plane and the axis or line of maximum elongation of a feature.
Plunge direction	It is the direction
Strike of bed	It is the direction of the bed, inclined bed intersects horizontal plain

32. Texture/structures and stress direction:

Textures/structures	*Stress directed*
Porphyritic textures	Fractional crystallizations
Schistosity	Direct stress with respect to the parallel of the grains
Gneissosity	Direct stress with respect to the parallel of the grains
Mylonitic texture	Direct stress with respect to the parallel of the grains

33. Terms	**Remarks**
Interval zone-biozone	Biostratigraphy
Group-formation-member-bed	Lithostratigraphy
Erathem-system-series-stage-zone	Charonostratigraphy

34. P – T diagram of the Quartz:

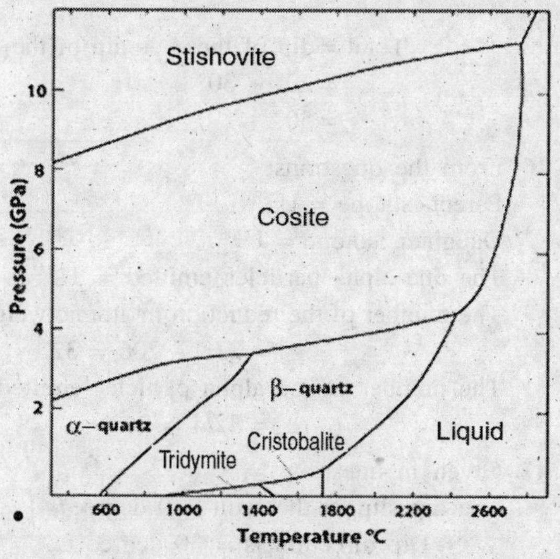

36. Isotropic Minerals: Isotropic minerals are easy to recognize in thin section because they are the only group to go extinct under crossed polarized light (analyzer in). In plane polarized light (analyzer out), these minerals can be identified by using properties such as cleavage, relief, and colour.

- All isotropic minerals will appear dark, and stay dark on rotation of the stage.
- Anisotropic minerals will allow some light to pass, and thus will be generally light, unless in specific orientations.
- Isotropic minerals do no affect the polarization direction of the light which has passed through lower polarizer;
- Light which passes through the mineral is absorbed by the upper polar.

38. Minerals and its depositions environments:

Processes	*Minerals*
Magmatic concentrations process	Diamond, chromite, magnetite
Pegmatitic process	Thorium
Contact metasomatic process	Copper, iron, lead, zinc, cassiterite
Hydrothermal process	Iron, zinc, nickel, gold, silver, lead
Sublimation process	Sulphur
Residual concentration process	Bauxite, kyanite, hematite
Mechanical concentration	Gold, platinum, magnetite
Oxidation and supergene enrichment process	Galena, sphelerite, chalcopyrite
Metamorphic process	Graphite, garnet, kyanite, asbestos

39. Crystallization of the magma: The texture of an igneous rock (fine-grained vs coarse-grained) is dependent on the rate of cooling of the melt: slow cooling allows large crystals to form, fast cooling yields small crystals. Magmas and their resultant plutonic rock bodies cool and crystallize slowly and are characterized by coarse-grained texture, in which the mineral crystals are visible to the unaided eye. On the other hand, lavas cool quickly at the earth's surface and are characterized by fine-grained texture, in which the crystals are too small to be seen by the unaided eye.

Very quickly cooled lavas, typically those quenched in water, will have a glassy texture. They cool too quickly to form crystals. Glasses do not have an orderly arrangement of atoms and there are therefore no minerals, in the strict sense, in them. Volcanic glass is called obsidian.

40. Major mass extinction in geological records:

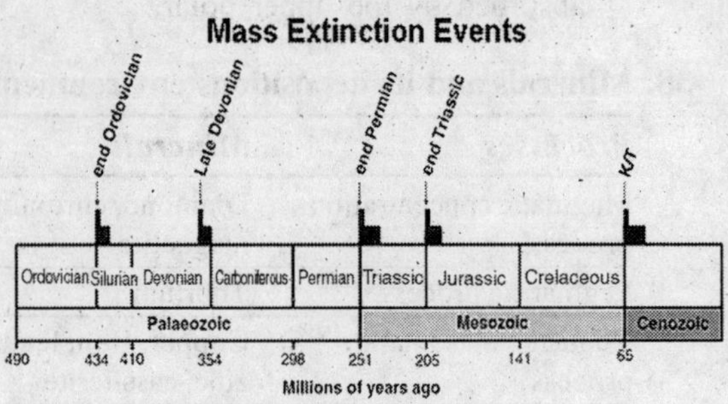

41. From the given question:

Map Scale = 1:50000

1 cm = 50000 cm

= 500 m

= 0.5 km

The dyke exposed on the horizontal is 2 km

The length is = 2/0.5 = 4.0 cm.

42. From the question : The chemical formula of the Hematite is = Fe_2O_3

The valency of the iron is = 3.

43. Given:

Lithostatic pressure is = 3 kbar

Temperature = 275°C

Rate of the pressure increase = 0.3 kbar/km

The depth is = 0.3 × 10 = 3 kb (given)

At temperature 25° C (given)

250 °C = 10 km

1 km = 250/10

= 25° C/km.

44. The Moh's scale of the Hardness:

Hardness	*Minerals*	*Mineral group*
1	Talc	Silicate
2	Gypsum	Sulphate
3	Calcite	Carbonate
4	Fluorite	Halides
5	Apatite	Phosphate
6	Orthoclase	Silicate
7	Quartz	Silicate
8.	Topaz	Silicate
9.	Corundum	Oxide
10.	Diamond	Native element

45. Given, the attitude of the bed A = (45°, 20° SE)

The attitude of the bed B = (225°, 30° NW)

The separation of the UU by unconformity sequence A and B, then the strike of the both beds is same *i.e.*,

When sequence A was deposited, the dip amount of the beds in sequence at that time was:

Total = dip of the A + dip of the B

= 20° + 30°

= 50°.

46. From the questions:

Parent isotope = U^{238}

Daughter isotope = U^{206}

The one alpha particles emitted = $_2He^4$

The number of the reduction in atomic weight

= 238 − 206 = 32

The number of the alpha particles emitted

= 32/4 = 8.

47. Given in question:

The dip slip of the fault = 0 degree

Dip direction is = 30 degree E

Slip = 10 m

The throw of the fault =

Sin 30° = Throw /10

½ = throw /10

Throw = 5.0 m.

48. From the questions:

10 m thick sequence deposited = 10,000 years

The number of the sample must be collected in equal interval = 10,000/ 500

= 20.

49. From the figure:

Limb = non-plunging (Horizontal)

First limb = 045°, 20° SE

Second limb = 045°, 40° NW

When draw the limb with axial plane of the kink fold = 90° – 20° = 70°.

50. Given, $(Fe_{2.5}Mg_{0.3}Ca_x)\ Al_2Si_3O_{12}$

Where x = number of atoms of Ca.

The chemical composition of the grossular

$= Ca_3Al_2(SiO_4)_3$

$= (Fe_{2.5}Mg_{0.3}Ca_x)$

$x = 0.2$

Mole % = 0.2/3

= 6.6.

51. From the question: The earth is an ideal sphere, then

The volume of the sphere = $4/3\pi r^3$

The radius of the earth = 6371 km

The radius of the core = 1200 km

Volume % = $4/3\pi r^3$ (Core) /$4/3\pi r^3$ (Earth)

= $4/3\pi(1200)^3/\ 4/3\pi(6371)^3$

= 18%

53. From the question,

600 tons low grade iron ore of 40% Fe

= 600 × 40/100

= 240 tones

400 tones high grade iron ore of 65 % Fe

= 400 × 65/100

= 260 tones

Total = 500 tones

The grade of the blended ore of Fe%

= 500/1000 × 100

= 50%.

54. Given,

The mass of the fully dried rock sample of the volume = 100 cm³ = 300 g (given)

Mass of the fully saturated = 325 g

Density = 1.00/cm³

Porosity (%)

= Ww-Wd/d

= weight in water – weight in dry/density

= 325 – 300/1 = 25%.

56. From the figure:

The block of rock with mass = 72 kg

Inclined angle = 30 degree

Gravity = 9.8 m/s²

Cohesion and friction = 0 (Assuming)

Force F = Mg Sinθ (From figure then you change in horizontal and vertical components with respect to the surface.

= 72 × 9.8 × Sin 30°

= 72 × 9.8 × ½

= 705.6/2

= 352.8.

58. From the diagram,

Melt containing = 900 moles Anorthite

= 100 mole diopside

From the Lever Rule:

= (900 – 500 /50) × 100

= 800.

59. From the question,

Cross sectional area (A) = 7 m²

Hydraulic conductivity (K) = 2 m/s

Hydraulic gradient (dh/dl) = 3.2

From Darcy's law in steady state:

The volumetric flow rate of the aquifer

Q = K.A. dh/dl

= 2 × 7 × 3.2

= 44.8 m³/sec

Previous Papers (Solved)

Indian Institute of Technology Joint Admission Test for M.Sc.

IIT-JAM GEOLOGY (GG)-2017

SECTION-A

Multiple Choice Question (MCQ)

Qs. Nos. 1-10 : *Carry one mark each.*

1. Which one of the following minerals exhibits luminescence when exposed to ultraviolet light?
A. Cassiterite B. Wolframite
C. Tantalite D. Scheelite

2. In which one of the following mass extinction periods trilobites became extinct?
A. Devonian B. Permian
C. Triassic D. Cretaceous

3. En-echelon sigmoidal 'gash' veins indicate
A. Ductile shear zone
B. Brittle-ductile shear zone
C. Brittle shear zone
D. Saddle reef structure

4. Which one of the following primary sedimentary structures is NOT used for paleocurrent analysis?
A. Current cresent
B. Flute marks
C. Symmetrical wave ripples
D. Imbrication of pebbles

5. The age of the Pathcam Formation is
A. Permian B. Triassic
C. Jurassic D. Cretaceous

6. Rivers that receive water from groundwater seepage are termed as
A. Effluent rivers B. Consequent rivers
C. Influent rivers D. Braided rivers

7. Conservative plate boundary is represented by
A. Normal fault B. Growth fault
C. Transform fault D. Reverse fault

8. Which one of the following prismatic crystal forms belongs to the hexagonal crystal system?
A. $\{11\bar{2}0\}$ B. $\{h0\bar{h}l\}$
C. $\{0001\}$ D. $\{hk0\}$

9. The characteristic rock of contact metamorphism is
A. Hornfels B. Blueschist
C. Eclogite D. Granulite

10. The volcanic equivalent of nepheline syenite is
A. Rhyolite B. Basanite
C. Phonolite D. Andesite

Qs. Nos. 11-30 : *Carry two marks each.*

11. Identify the correct match between mineral/ ore and its physical property.
A. Hematite Yellow streak
B. Barite High specific gravity
C. Psilomelane Comb stricture
D. Azurite Distinctive green colour

12. Match the mineral deposits in Group-I with their India occurrence in Group-II.

Group-I	*Group-II*
P. Chromite	1. Jhamarkotra, Rajasthan
Q. Magnesite	2. Gudur, Andhra Pradesh
R. Mica	3. Byrapur, Karnataka
S. Phosphorite	4. Chalk Hills, Tamil Nadu

A. P-1, Q-4, R-3, S-2 B. P-3, Q-2, R-4, S-1
C. P-3, Q-4, R-2, S-1 D. P-2, Q-1, R-3, S-4

13. A helically coiled ammonite *Turrilites* is differentiated from externally resembling Gastropoda *Turritella* by
A. Apical angle
B. Number of whorls
C. Direction of coiling
D. Chambered shell

14. The facial suture of trilobites running through the genal angle is known as

A. Proparian B. Marginal

C. Gonatoparian D. Opisthoparian

15. Which one of the following statements is correct for Class 1B (Parallel) folds?

A. Orthogonal thickness at hinge > that at limb

B. Axial planar thickness at hinge = that at limb

C. Dip isogons are parallel

D. Dip isogons are convergent

16. In the given map, the X - Y surface has the same orientation as in the Palaeozoic sequence. X - Y represents :

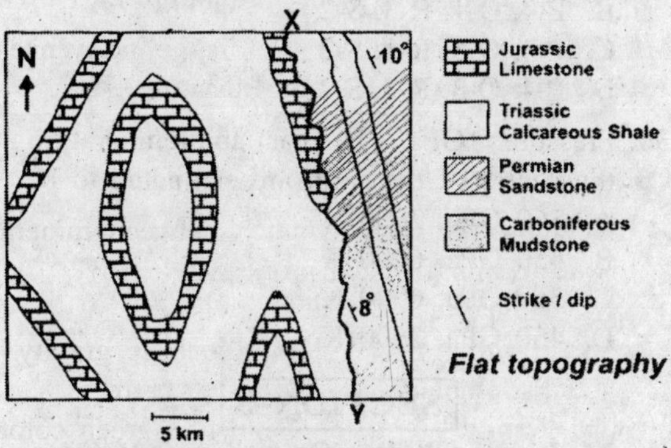

A. Angular unconformity

B. Non - conformity

C. Normal thrust

D. Thrust

17. Match the sedimentary features in Group-I with the corresponding sedimentary environments of their formation in Group-II:

Group-I	*Group-II*
P. Point Bar	1. Tidal
Q. Barchans	2. Meandering Fluvial Channel
R. Dropstone	3. Aeolian
S. Herring - bon cross stratification	4. Glacial

A. P-3, Q-4, R-1, S-2

B. P-2, Q-3, R-4, S-1

C. P-2, Q-4, R-3, S-1

D. P-2, Q-3, R-1, S-4

18. Which one of the following lithostratigraphic units is of Phanerozoic Eon?

A. Sargur Group

B. Semri Group

C. Uttatur Group

D. Papaghani Group

19. Match the geological processes (Group-I) with their example in Indian stratigraphy (Group-II).

Group-I	*Group-II*
P Permo-Carboniferous glaciation	1. Ariyalur Group
Q Cretaceous marine transgression	2. Siwalik Group
R Neogene fluvial sedimentation	3. Talchir Formation
S Cretaceous inter-trappean sedimentation	4. Lameta Formation

A. P-3, Q-1, R-2, S-4

B. P-2, Q-3, R-4, S-1

C. P-3, Q-2, R-1, S-4

D. P-4, Q-1, R-2, S-3

20. Check dams are constructed in association with main dam in the

A. Upstream of the main dam to check the situation of the reservoir

B. Downstream of the main dam to check the situation of the reservoir

C. Upstream of the main dam to check the seepage from the reservoir

D. Downstream of the main dam to check the seepage from the reservoir

21. A geological formation neither containing nor transmitting water is termed as

A. Aquiclude

B. Aquitard

C. Aquifer

D. Aquifuge

22. Which one of the following sequence of silicate structures indicates an increasing degree of sharing of corners of $(SiO_4)^{4-}$ tetrahedra?

A. Nesosilicate - Single chain inosilicate - Phyllosilicate - Tectosilicate

B. Tectosilicate - Phyllosilicate - Single chain inosilicate - Nesosilicate
C. Nesosilicate - Phyllosilicate - Single chain inosilicate - Tectosilicate
D. Single chain inosilicate - Nesosilicate - Phyllosilicate - Tectosilicate

23. Match the igneous bodies in Group-I with their ages in Group-II.

	Group-I		*Group-II*
P	Singhbhum granite	1	Neoproterozoic
Q	Malani rhyolite	2	Cretaceous
R	Deccan volcanics	3	Permian
S	Panjal Traps	4	Archaean

A. P-3, Q-2, R-4, S-1
B. P-4, Q-1, R-2, S-3
C. P-4, Q-3, R-2, S-1
D. P-3, Q-4, R-2, S-1

24. The tubefeet in echinoids emerge through
A. Interambulacral plates
B. Ambulacral plates
C. Bourrelets
D. Plastron

25. A mineral with a point group symmetry 2/m
A. Has two optic axes
B. Show inclined extinction in (100) section
C. Show straight extinction in (010) section
D. Is uniaxial

26. The progressive metamorphic isograde sequence that explains Barrovian metamorphism in pelite is
A. Chlorite - staurolite - biotite - kyanite - sillimanite
B. Chlorite - andalusite - cordierite - sillimanite
C. Chlorite - biotite - garnet - staurolite - kyanite - sillimanite
D. Sillimanite - kyanite - staurolite - garnet - biotite - chlorite

27. In metabasic rocks, plagioclase is not stable in
A. Granulite facies
B. Epidote amphibolite facies
C. Amphibolite facies
D. Eclogite facies

28. A sandstone has < 5 % matrix. The recalculated model compositions of feldspar, quartz and rock fragments are 45%, 35%, 20%, respectively. The sandstone is classified as
A. Feldspathic wacke B. Quartz wacke
C. Lithic arkose D. Subfeldsarenite

29. Match the earth layers (Group-I) with corresponding approximate thickness (Group-II).

Group-I	*Group-II*
P. Lithosphere	1. 2900 km
Q. Mantle	2. 2250 km
R. Outer Core	3. 1200 km
S. Inner Core	4. 100 km

A. P-4, Q-1, R-2, S-3
B. P-4, Q-1, R-3, S-2
C. P-4, Q-3, R-1, S-2
D. P-4, Q-3, R-1, S-2

30. Pressure (1GPa = 10 kbar) and temperature at the centre of the earth are estimated to be
A. 360 GPa, 2600 K
B. 450 GPa, 6000 K
C. 360 GPa, 6000 K
D. 450 GPa, 2600 K

SECTION-B

Multiple Choice Question (MCQ)

Qs. Nos. 31-40 : *Carry two marks each.*

31. Choose the landform(s) resulting from glacial erosion.
A. Fjords B. Moraines
C. Drumlins D. Cirques

32. Choose the correct combination(s) of type of dentition of Bivalvia and the corresponding representative genus.
A. Taxodont *Nucula*
B. Spondylus *Spondylus*
C. Pachydont *Hippurites*
D Desmodont *Mya*

33. Shown below is an isobaric binary temperature - composition phase diagram in the system X - Y with complete miscibility between X and Y.

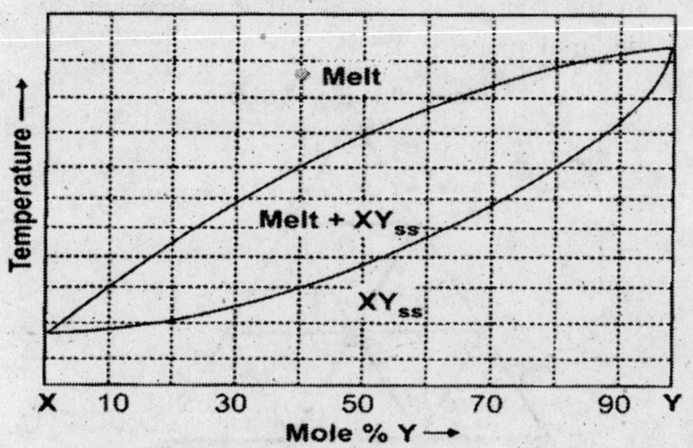

Which of the following statements is/are correct for crystallization of a starting melt of composition X_{60} Y_{40} (the dot in the diagram)?

A. The first formed crystal has a composition of X_{20} Y_{80}

B. The final melt composition during equilibrium crystallization is X_{90} Y_{10}

C. In case of fractional crystallization, the final melt is enriched in X than X_{90} Y_{10}

D. For fractional crystallization, the final crystal composition is X_{60} Y_{40}

34. Choose the correct combination(s) of textural features of magmatic rocks with corresponding petrological processes from the following.

A. Ophitic texture in dolerite — *Peritectic crystallization*

B. Perthite in granite — *Slow subsolidus cooling*

C. Spinifex texture in komatiite — *Eruption of ultramafic lava*

D. Orthopyroxene rim around olivine in peridotite — *Eutectic crystallization*

35. Which of the following statements is/are NOT correct?

A. (110) lies in zone [001]

B. (021) lies in zone [100]

C. (101) lies in zone [010]

D. (111) lies in zone $[1\bar{1}1]$

36. In an outcrop we find that the bedding planes are vertical and cleavage surfaces are horizontal. Which of the following folds types is/are inferred from this observation?

A. Upright fold B. Recumbent fold

C. Vertical fold D. Neutral fold

37. Which of the following stratigraphic units(s) is/are coal/lignite bearing?

A. Barakar formation

B. Barail Group

C. Cuddalore Formation

D. Ariyalur formation

38. The Toposheet No(s). immediately adjacent to toposheet No. 55J/8 is/are

A. 55K/2 B. 55J/12

C. 55J/6 D. 55K/5

39. Which of the following is/are NOT true for texturally immature sandstone?

A. Clay content is high

B. Little or no clay present

C. Grains are well sorted

D. Grains are rounded

40. P and S waves originate at earthquake focus and travel through the earth. Which of the following statements for these waves is/are correct?

A. S-wave shadow zone is 145 degree wide

B. P-wave shadow zone are 49 degree wide

C. P-wave velocity abruptly increases downward at mantle - core boundary

D. P-wave velocity abruptly drops downward at mantle-core boundary

SECTION-C

Numerical Answer Type (NAT)

Qs. Nos. 41-50 : *Carry one marks each.*

41. An object is spotted at S60^0 E front bearing from the observer. If the position is interchanged, the front bearing value in degree from North (measured clockwise) is------------.

42. The mode% of forsterite component in olivine with chemical formula $Mg_{1.8}Fe_{0.2}SiO_4$ is ------.

43. The Weiss symbol of a crystal face is 4a: 2b: c. The value of *h* in the corresponding Miller Index (*hkl*) is ------------.

44. In a minerals with chemical formula AT_4O_8, the ionic radii of A and O are 1.12 Å and 1.40 Å, respecviely. The co-ordination number of cation A is ------------.

45. Aluminium (Al) can occur in both tetrahedral and octahedral co-ordination in silicates. The amount of octahedral Al in a pyroxene crystal composition $Mg_{1.4}Fe_{0.4}Al_{0.4}Si_{1.8}O_6$ is ------ (give answer in one decimal place).

46. The birefringence of a mineral of thickness 30 μm and retardation 0.27 μm is -------. (give answer in three decimal places).

47. Two limbs of a vertical chevron folds strike S70° E and N55° E. The value of the interlimb angle of the fold is ------(degree).

48. The schematic map given below shows intersecting strike lines of the same lithological contact. In the map, AB and CD are 5 cm and 3.5 cm, respectively. The scale of the map is 1 cm = 100 m. the plunge of the fold axis is ------- degrees (give answer in one decimal place).

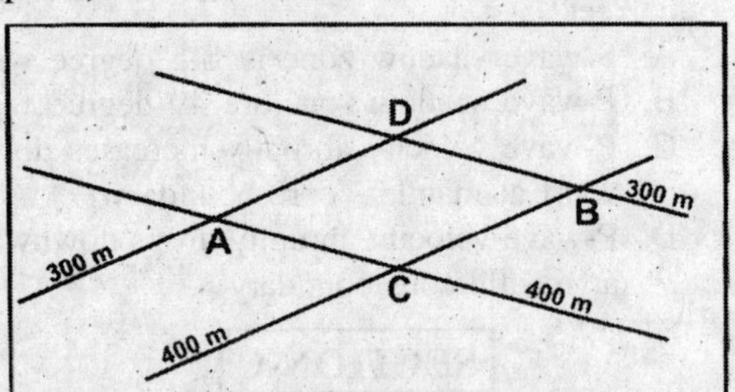

49. The core - rim composition of a normally zoned plagioclase crystal are as follows:

Core: $Ca_{0.6}\ Na_x\ Al_{1.6}\ Si_{2.4}\ O_8$

Rim: $Ca_{0.4}\ Na_y\ Al_{1.4}\ Si_{2.6}\ O_8$

The amount of increase of Na atom from core to rim per formula unit of plagioclase is ---- (give answer in one decimal place).

50. Considering garnet chemical formula in 12 oxygen basis, the number of Mg cations in a garnet of chemical composition P (as shown in the figure) is -------- (give answer in two decimal places).

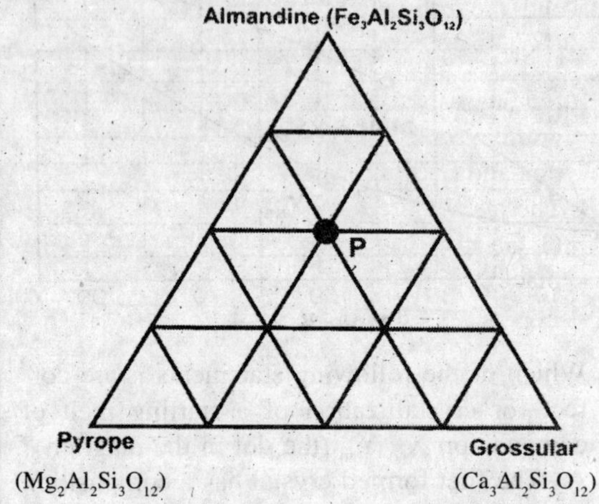

Qs. Nos. 51-60 : *Carry two marks each.*

51. A fault surface in an outcrop has slickenside lineation whose pitch is 30 degree. The horizontal slip on the fault is 1.25 m, as determined from displaced vein. The net slip on the fault is ------- meter (give answer in two decimal places).

52. In an outcrop, we find a Belemnite fossil broken into five rectangular pieces (boudins) of equal size. Long dimension of each boudin is 1.35 cm. Gap between adjacent boudins in all cases is 0.25 cm. Note that the long dimensions of boudins are perfectly aligned. The % elongation is ----- (give answer in one decimal place).

53. A horizontal cylindrical ore body (diameter = 20 m, length = 200 m) has 5% metal content and density of 3500 kg/m³. The reserve of the ore body is ------- million ton(s) (give answer in two decimal places).

54. A drainage basin of fourth order covers an area of 40 sq. km. With in the basin, total length of 1st order drainage is 12.5 km, 2nd order drainage is 8.8 km, 3rd order drainage is 4.7 km and 4th order drainage is 4.0 km. The drainage density of the basin is -------- km^{-1} (give answer in two decimal places).

55. Age of granite rocks can be determined using Rb - Sr whole rock radioactive dating method and the following age equation,

$(^{87}Sr/^{86}Sr) = (^{87}Sr/^{86}Sr)_I + (^{87}Rb/^{86}Sr)(e^{\lambda t} - 1)$

For a suite of representative co-magmatic granitic rocks, the Rb - Sr whole rock isochron plot and relevant data are shown in the diagram. The age of granite is calculated at ---------Ga (1 Ga = 10^9 yrs, give answer in one decimal place).

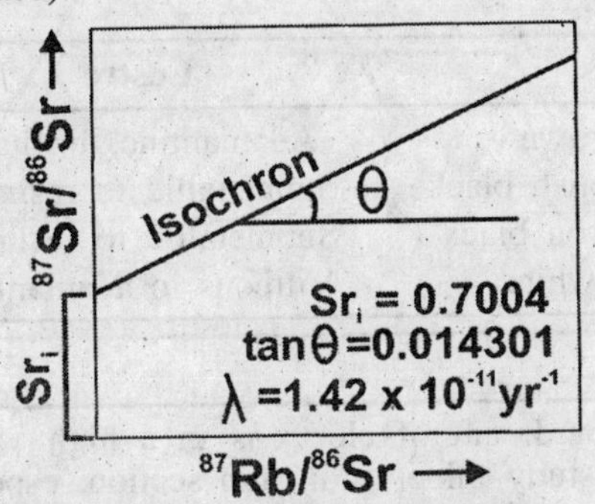

56. Consider a granulite facies metamorphic rock with peak metamorphic condition at 9 kbar, 850°C. Assume a single layer crust of ρ = 3000 kg/m^3 and g = 10 m/sec^2 during metamorphism. The depth of burial during peak metamorphism is ------km. 1 Pascal = 1 kg/m/sec^2 and 1 bar = 10 Pascals.

57. Consider four minerals P, Q, R and S in a three components chemical system (A – B – C) as shown in the figure. For a crossing tie-line relationship, the variance (degree of freedom) of the equilibrium mineral assemblage at X is ---------.

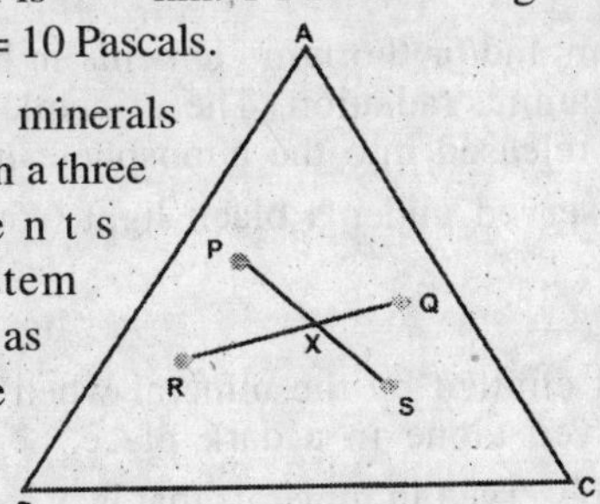

58. The refractive indices of four minerals (P, Q, R, S) are as follows:
P (α = 1.712, β = 1.721, γ = 1.727)
Q (ε = 1.553, ω = 1.544),
R (α = 1.664, β = 1.672, γ = 1.694) and
S (ω = 1.658, ε = 1.486)
The value of maximum birefringence among all the minerals is ---------- (give answer in three decimals places).

59. In a sedimentary succession shown in the figure, the last occurrence of the fossils species Q (dated 50 Ma) and the first occurrence of the fossils species R (dated 30 Ma) are recorded at Y and X, respectively. The estimated rate of sedimentation is ------------- m/million yrs (assume constant rate of sedimentation).

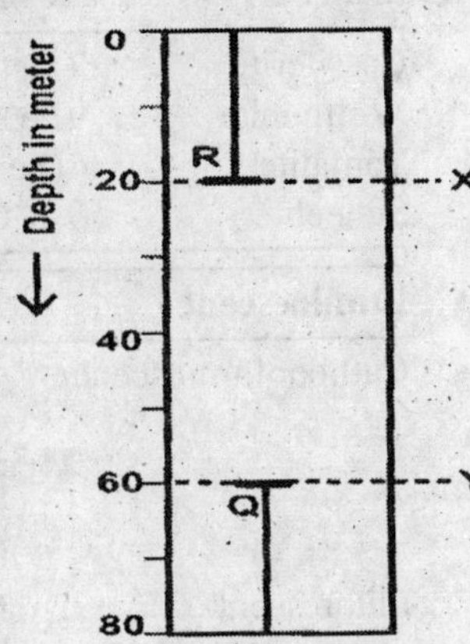

60. The top surface of a coal seam is exposed at 150 m contour level on a hill top at location A. The same surface of the seam is also exposed on a river bed at location B at the 50 m contour level. The aerial distance A-B is 1 km. The amount of dip of the coal seam along A-B is --------- (degree). Give answer in one decimal place.

ANSWER KEY

1	2	3	4	5	6	7	8	9	10
D	B	B	C	C	A	C	A	A	C
11	**12**	**13**	**14**	**15**	**16**	**17**	**18**	**19**	**20**
B	C	D	C	D	D	B	C	A	A

21	22	23	24	25	26	27	28	29	30
D	A	B	B	A	C	D	C	A	C
31	**32**	**33**	**34**	**35**	**36**	**37**	**38**	**39**	**40**
A,D	A,B,C,D	A,B,C	B,C	D	B,D	A,B,C	B,D	B,C,D	A,D
41	**42**	**43**	**44**	**45**	**46**	**47**	**48**	**49**	**50**
300	90	1	8	0.2	0.009	55	15.8-16.0	0.2	0.75
51	**52**	**53**	**54**	**55**	**56**	**57**	**58**	**59**	**60**
1.42-1.46	14.7-14.9	0.21-0.23	0.75	0.9-1.1	30	1	0.172	2	5.0-6.0

HINTS AND SOLUTIONS

1. Minerals	Composition	Colours	Lustre
Cassiterite	Tinstone (SnO_2)	Black or brown	Adamantine, brilliant
Wolframite	(Fe,Mn) (WO_4)	Greyish to brownish black	Submetallic to resinous
Tantalite	(Fe,Mn) T_2O_6	Dark black to iron black	Submetallic to resinous
Scheelite	$CaWO_4$	Colourless, white	Vitreous to adamantine

Luminescence	Description
Cathodolumiescence	Colour generated under the bombardment of electrons in a high vacuum environment. Used in geology to study minerals in thin section, especially dolomite. Generally not observed or reported in routine mineral analysis. This is the term used to describe the colour generated in television picture tubes (cathode ray tubes).
Fluorescent	Defined as the absorption and re-emission of electromagnetic radiation at a lower wavelength. A very useful property that has may striking examples in the mineral kingdom where ultra-violet radiation is re-emitted in the visible spectrum.
Highly Radioactive	Minerals containing uranium and/or thorium elements in high concentrations that emit alpha, beta and gamma radiation. The principle hazard to mineral collectors is the radon gas released into the atmosphere in a closed cabinet.
Non-fluorescent	No visible fluorescence observed under a black light (source of ultra-violet radiation).
None	No luminescence observed.
Phosphorescent	After- glow of visible light emitted by the mineral when it is stimulated by ultra-violet light and observed alone in a dark place.
Piezoelectric	Mechanico-electrical effect observed in minerals that lack a center of symmetry element in the crystallographic parameters.
Radioactive	Minerals containing uranium and/or thorium elements in moderate to low concentrations that emit alpha, beta and gamma radiation.
Sometimes Fluorescent	Minerals that are not normally fluorescent but because of impurities can be fluorescent.
Triboluminescent	Emission of light due to mechanical stress on the crystal lattice. Commonly observed in Wint-O-Green Lifesavers when crushed in the mouth.
Weakly Fluorescent	Minerals that exhibit very weak fluorescence under the ultra-violet light.

2. Evolution of Trilobites:

Kingdom	Animalia
Phylum	Arthropoda
Subphylum	Trilobitomorpha
Class	Trilobite
Order	Agnostida
Geological age	Cambrian to Permian
Remarks	Index fossils

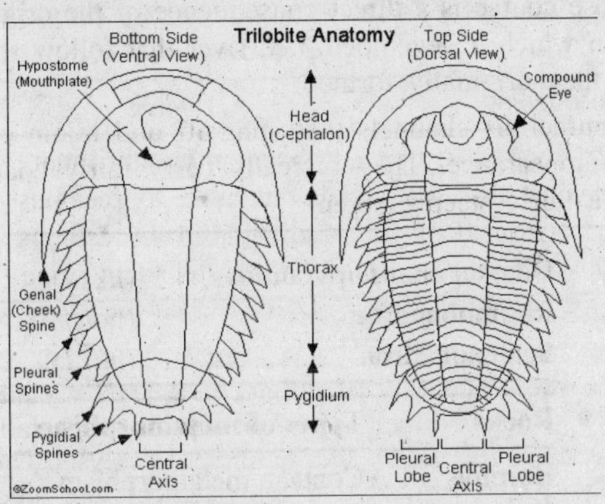

3. Types of Shear Zone:

Types of shear zone	Remarks
Brittle shear zone	Fracture develop by brittle deformation.
Ductile shear zone	Structures developed by shearing under ductile flow
Semi brittle shear zone	A process by pressure solution and cataclastic flow.
Brittle - ductile shear zone	These are formed by brittle and ductile deformation.

En- echelon: This refers to closely-spaced, parallel or subparallel minor structural features in rock that are oblique to the overall structural trend.

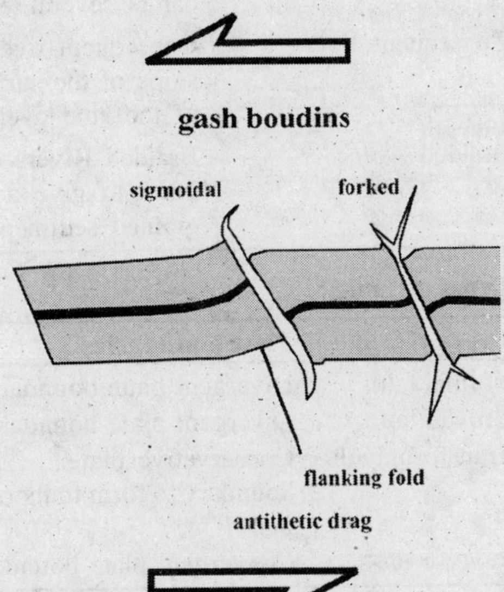

4. Sedimentary Structures:

Sedimentary structures	Applications
Current crescent	Paleocurrent analysis
Flute marks	Paleocurrent analysis
Symmetrical wave ripples	Top and bottom
Imbrication of pebbles	Paleocurrent analysis

5. Classification of Jurassic of Kutch:

Formations	Thickness	Remarks
Bhuj formation	....	*Ptillophyllum flora*
Umia formation	500 m	*Umia ammonite beds*
Katrol formation	700 m	*Belemnites*
Chari formation	400 m	*Dhosa oolite*
Patcham formation	300 m	*Macrocephalites*

6. Terms	Character
Effluent river	Effluent rivers are streams which get their water from the groundwater. The surface of the stream directly relates to the surface of the groundwater (called the water table), and the stream will rise and fall as the water table rises and falls.
Influent river	Influent streams do not usually get deeper and wider downstream. In fact, most actually lose water as they flow towards their ultimate destiny. There can be several reasons for the loss of water. They are mostly in arid climate.
Consequent river	Consequent river are those course is a direct consequence of the original slope of the surface upon which it developed, *i.e.*, river that follow slope of the land over which they originally formed.
Braided river	Braided Rivers exhibit numerous channels that split off and rejoin each other to give a braided appearance. They typically carry fairly coarse-grained sediment down a fairly steep gradient.

7. Types of faults:

Types of faults	Plate boundaries
Normal fault	Divergent plate boundary
Growth fault	Divergent plate boundary
Transform fault	Conservative plate boundary to form transform fault
Reverse fault	Convergent plate boundary

8. Crystal forms of hexagonal system

Crystal form name	Crystal form
Basal Pinacoid	(0001)
Prism of first order	$(10\bar{1}0)$
Prism of second order	$(11\bar{2}0)$
Dihexagonal prism	$(hk\bar{i}o)$
Dipyramid of first order	$(ho\bar{h}l)$
Dipyramidal of second order	$(hh2\bar{h}l)$
Dihexagonal Dipyramid	$(hk\bar{i}l)$
Rhombohedron	$(ho\bar{h}l)$
Scalenohedron	$(hk\bar{i}l)$

9. Rocks	Types of metamorphism
Hornfels	Contact metamorphism
Cataclastic	Cataclastic metamorphism
Granulose	Thermal metamorphism
Schistose	Dynamothermal metamorphism

10. Plutonic equivalent	Volcanic equivalent
Granite	Rhyolite
Adamellite	Rhyodacite
Syenite	Trachyte
Diorite	Andesite
Gabbro	Basalt
Nepheline syenite	Phonolite
Granodiorite	Dacite

11. Minerals and its physical properties:

Properties	Hematite	Barite	Psilomelane	Azurite
Colour	Metallic grey	Colourless, grey	Black with grey	Azure blue
Streak	Bright to dark red	White	Brownish black	Light blue
Specific gravity	5.26	4.3 - 5	3.7 - 4.7	3.7
Hardness	5.5-6.5	3 - 3.5	5.0 - 6.0	3.5 -4
Cleavage	Absent	Perfect	None	Perfect
Remark	Iron ore	Barium sulphate	Manganese oxide	Carbonate minerals

12. Minerals deposits	Location	Remarks
Chromite	Byrapur, Karnataka	Schistose rocks
Magnesite	Chalk Hills, Tamil Nadu	Vein association with peridotite, serpentinite
Mica	Gudur, Andhra Pradesh	Mica schist
Phosphorite	Jhamarkotra, Rajasthan	Dolomitic rocks

13. ***Ammonite Turrilites:*** *Turrilites* is a genus of helically coiled *ammonoid cephalopods* from the lower part of the Upper Cretaceous generally included in the Ancyloceratina. Previously (Arkell, 1957) it was included in the *ammonoid* suborder, *Lytoceratina.*The shell of *Turrilites* is tightly wound in a high trochospiral, with an acute angle at the apex. Ribs are weak to strong and may have 3 or 4 rows of equal numbers of tubercles.

Gastropoda Turritella: *Turritella* is a genus of medium-sized sea snails with an operculum, marine gastropod mollusks in the family Turritellidae. They have tightly coiled shells, whose overall shape is basically that of an elongated cone. The name *Turritella* comes from the Latin word turritus meaning "turreted" or "towered" and the diminutive suffix—*ella.*

14. Terms	Character
Proparian	Proparian forward of the genal angle as in most Phacopina.
Gonatoparian	At the genal angle as in most Calymenina.
Opisthoparian	It is along the hind cephalic margin adaxial to the genal angleas in most Ptychopariina

proparian gonatoparian opisthoparian

15. Classification of folds based on dip:

Classification of Layer Types based on dip isogons

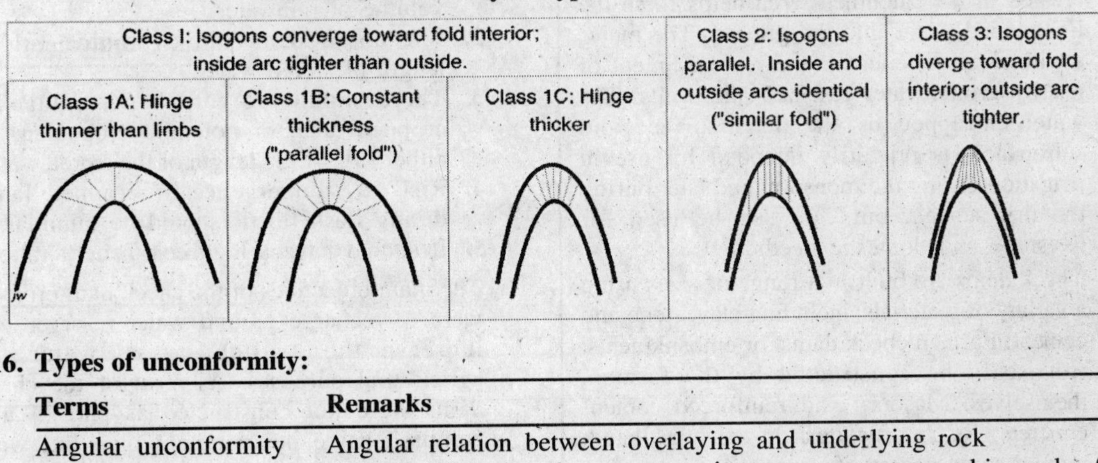

16. Types of unconformity:

Terms	Remarks
Angular unconformity	Angular relation between overlaying and underlying rock
Non-conformity	An unconformity that separate igneous or metamorphic rocks from overlaying sedimentary rocks.
Disconformity	An erosion surface between two package of sediment, but the lower was not tilted.

17.

Sedimentary features	Environments
Point Bar	Meandering fluvial channel
Barchans	Aeolian
Dropstone	Glacial
Herring - bone cross stratification	Tidal

18. Lithostratigraphic Units:

Lithostratigraphic units	Classification	Remarks
Sargur Group	Sargur Schist - PGC - Dharwar super group	More than 3200 Ma
Semri Group	Semri - Kaimur - Rewa - Bhander	Upper purana
Uttatur Group	Uttatur - Trichinopoly - Ariyalur - Niniyur	Cretaceous
Papaghani Group	Papaghani - Cheyair - Nallamalai - Kistna	Proterozoic

19.

Geological processes	Indian stratigraphy	Remarks
Permo – Carboniferous glaciation	Talchir Formation	Gondwana supergroup
Cretaceous marine transgression	Ariyalur Group	Cretaceous of Trichinopolly
Neogene fluvial sedimentation	Siwalik Group	Siwalik of Neogene
Cretaceous inter-trappean sedimentation	Lameta Formation	Cretaceous of Narmada Valley

20. Check dams: "Check dams" are small barriers built across the direction of water flow on shallow rivers and streams for the purpose of water harvesting. The small dams retain excess water flow during monsoon rains in a small catchment area behind the structure. Pressure created in the catchment area helps force the impounded water into the ground. The major environmental benefit is the replenishment of nearby groundwater reserves and wells. The water entrapped by the dam, surface and subsurface, is primarily intended for use in irrigation during the monsoon and later during the dry season, but can also be used for livestock and domestic needs.

Check dams are built in a range of sizes using a variety of materials, including clay, stone and cement. Earthen check dams, or embankments, can easily be constructed by the farmers themselves. Masonry and reinforced cement concrete (RCC) structures, on the other hand, require some degree of advanced construction experience and monetary inputs. Earthen dams do not allow for overflow of water, in contrast to masonry and RCC structures which allow excess water to flow over the spillway.

Site Selection

Prior to the construction of any check dam, the following criteria were considered by Development Alternatives with respect to site selection.

1. The structure should be able to store a high volume of rainwater
2. The check dam should provide a long length of stored water.
3. There should be a high percentage of cropped area, or potential crop area, on either side of the length of the stored water.
4. Risk of submergence of cropped lands during flash floods should be minimal.
5. It should have a high cost-benefit ratio.

The main objectives of the check dam projects were to recharge ground water reserves and increase the availability of water for agricultural purposes. As most of the check dams were not constructed as part of any comprehensive development plan but were stand-alone activities, they were not preceded by detailed community mobilization activities. In addition, the goals and objectives of the projects were not clearly defined during the planning phase and strategies for maintaining the check dams, as well as monitoring and evaluating the projects, were not outlined. With respect to selecting villages to participate in the projects, methods varied. In some cases, district and/or block level government officials chose the villages. In other cases, Development Alternatives recruited villages for involvement in the projects or the villagers themselves requested assistance from Development Alternatives.

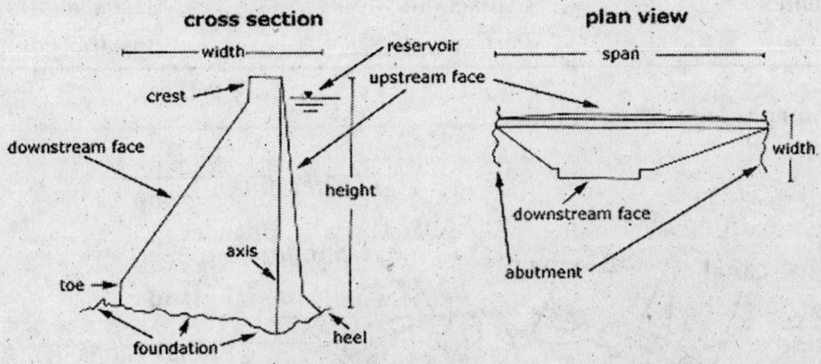

21.

Geological formation	Example	Characteristic
Aquitard	Sandy clay	Porous but less permeable
Aquifuge	Solid granite	Neither porous nor permeable
Aquiclude	Clay bed	Porous but not permeable
Aquifer	Sandy formation	Porous and permeable

22. Types of Silicates :

Major Silicate Structures

GEOMETRY OF LINKAGE OF SiO_4 TETRAHEDRA		EXAMPLE MINERAL	CHEMICAL COMPOSITION
Isolated tetrahedra: No sharing of oxygens between tetrahedra; individual tetrahedra linked to each other by bonding to cation between them		Olivine	Magnesium-iron silicate
Rings of tetrahedra: Joined by shared oxygens in three-, four-, or six-membered rings	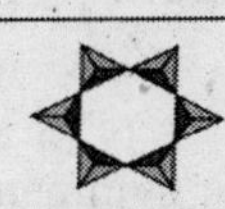	Cordierite	Magnesium-iron-aluminum silicate
Single chains: Each tetrahedron linked to two others by shared oxygens; chains bonded by cations		Pyroxene	Magnesium-iron silicate
Double chains: Two parallel chains joined by shared oxygens between every other pair of tetrahedra; the other pairs of tetrahedra bond to cations that lie between the chains	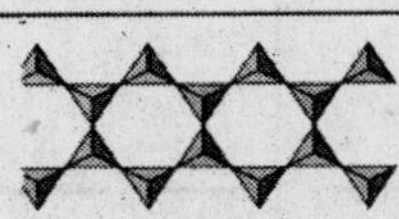	Amphibole	Calcium-magnesium-iron silicate
Sheets: Each tetrahedron linked to three others by shared oxygens; sheets bonded by cations	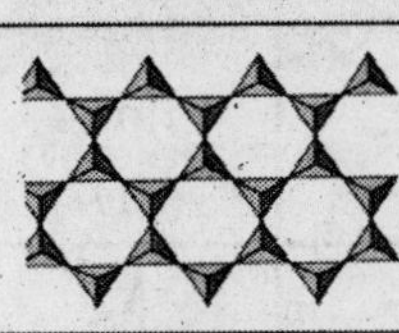	Kaolinite Mica (muscovite)	Aluminum silicate Potassium-aluminum silicate
Frameworks: Each tetrahedron shares all its oxygens with other SiO_4 tetrahedra (in quartz) or AlO_4 tetrahedra		Feldspar (orthoclase) Quartz	Potassium-aluminum silicate Silicon dioxide

23. Igneous bodies	Geological age	Remarks
Singhbhum granite	Archean (Jharkhand)	Older than Singhbhum group
Malani rhyolite	Neoproterozoic	Acid volcanic rock of Vindhyan
Deccan volcanics	Cretaceous	Basic lavaflow of central India
Panjal Traps	Permian	Basaltic composition of Kashmir

24. Functional morphology of Echinoids:

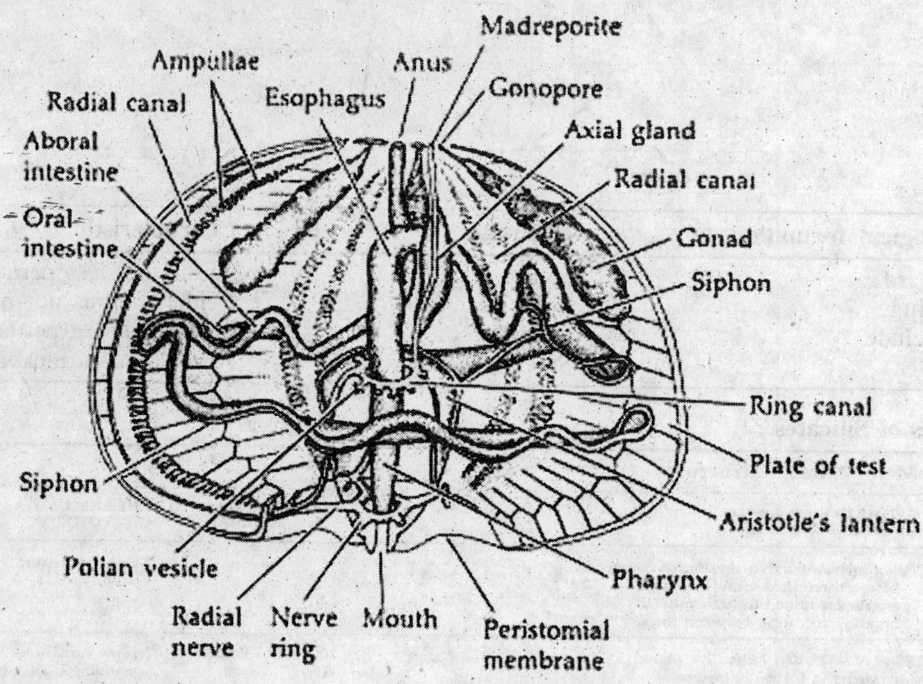

25. Point group: The 32 crystal classes, their symmetry, Hermann — Mauguin symbol and class name.

Crystal System	Crystal Class	Symmetry	Name of Class
Triclinic	1	none	Pedial
	$\bar{1}$	i	Pinacoidal
Monoclinic	2	$1A_2$	Sphenoidal
	M	1m	Domatic
	2/m	*i*, $1A_2$, 1m	Prismatic
Orthorhombic	222	$3A_2$	Rhombic-disphenoidal
	mm2 (2mm)	$1A_2$, 2m	Rhombic-pyramidal
	2/m2/m2/m	i, $3A_2$, 3m	Rhombic-dipyramidal

Crystal System	Crystal Class	Symmetry	Name of Class
Tetragonal	4	$1A_4$	Tetragonal- Pyramidal
	$\bar{4}$	$\bar{A}_4$	Tetragonal-disphenoidal
	4/m	i, $1A_4$, 1m	Tetragonal-dipyramidal
	422	$1A_4$, $4A_2$	Tetragonal-trapezohedral
	4mm	$1A_4$, 4m	Ditetragonal-pyramidal
	$\bar{4}2m$	$1\bar{A}_4$, $2A_2$, 2m	Tetragonal-scalenohedral
	4/m2/m2/m	i, $1A_4$, $4A_2$, 5m	Ditetragonal-dipyramidal
Hexagonal	3	$1A_3$	Trigonal-pyramidal
	$\bar{3}$	$1\bar{A}_3$	Rhombohedral
	32	$1A_3$, $3A_2$	Trigonal-trapezohedral
	3m	$1A_3$, 3m	Ditrigonal-pyramidal
	$\bar{3}2/m$	$1\bar{A}_3$, $3A_2$, 3m	Hexagonal-scalenohedral
	6	$1A_6$	Hexagonal-pyramidal
	$\bar{6}$	$1\bar{A}_6$	Trigonal-dipyramidal
	6/m	i, $1A_6$, 1m	Hexagonal-dipyramidal
	622	$1A_6$, $6A_2$	Hexagonal-trapezohedral
	6mm	$1A_6$, 6m	Dihexagonal-pyramidal
	$\bar{6}m2$	$1\bar{A}_6$, $3A_2$, 3m	Ditrigonal-dipyramidal
	6/m2/m2/m	i, $1A_6$, $6A_2$, 7m	Dihexagonal-dipyramidal
Isometric	23	$3A_2$, $4A_3$	Tetaroidal
	$2/m\bar{3}$	$3A_2$, 3m, $4\bar{A}_3$	Diploidal
	432	$3A_4$, $4A_3$, $6A_2$	Gyroidal
	$\bar{4}3m$	$3\bar{A}_4$, $4A_3$, 6m	Hextetrahedral
	$4/m\bar{3}2/m$	$3A_4$, $4\bar{A}_3$, $6A_2$, 9m	Hexoctahedral

26. Barrovian metamorphism: Barrovian metamorphism is the most commonly encountered. It occurs in intense tectonic conditions associated with *Volcanic Arcs*, and *Major Mountain Building*. Barrovian metamorphism is widely found across time and space on all parts of the earth, and produces the most common metamorphic rocks. It is also the most complex, and the most useful metamorphism geologically. Barrovian metamorphism is therefore of particular importance. Beginning very simply, let's examine what happens to a shale parent rock undergoing a full sequence of Barrovian metamorphism. The metamorphic sequences produces the following rocks, in order: slate, phyllite, schist and gneiss.

It can be hard to imagine at first that all these very different looking rocks can come from the same sedimentary parent, but we know that they do. This demonstrates the hidden truths that lie within the earth, and to the profound changes that earth processes can effect.

Shale >>> Slate >>> Phyllite >>> Schist >>> Gneiss

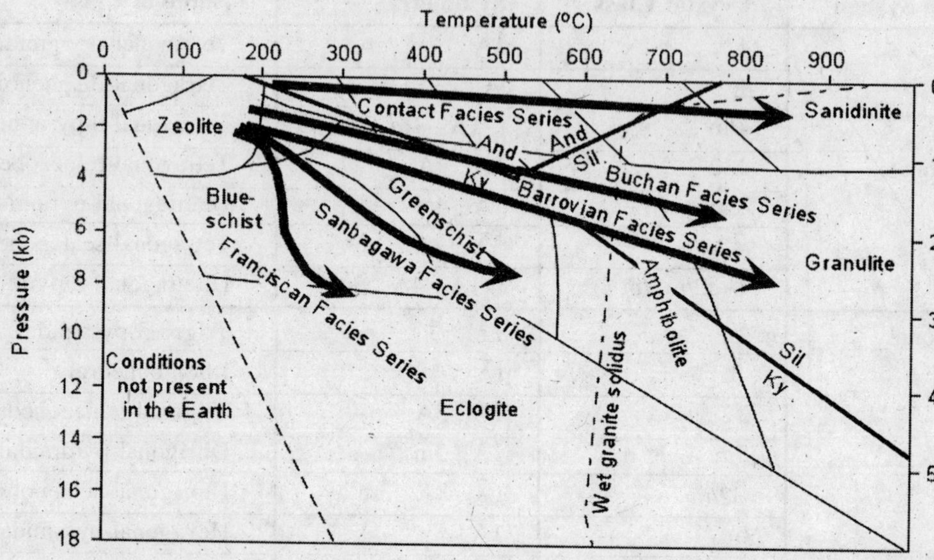

27. Metamorphic facies and Pressure — Temperature:

Metamorphic facies	Stable minerals
Granulite facies	Opx - Cpx - Plagioclase
Amphibolite facies	Hbl - plagioclase - actinolite
Eclogite facies	Pyrope - omphacite - kimberlite
Greenschist facies	Albite- epidote - actinolite

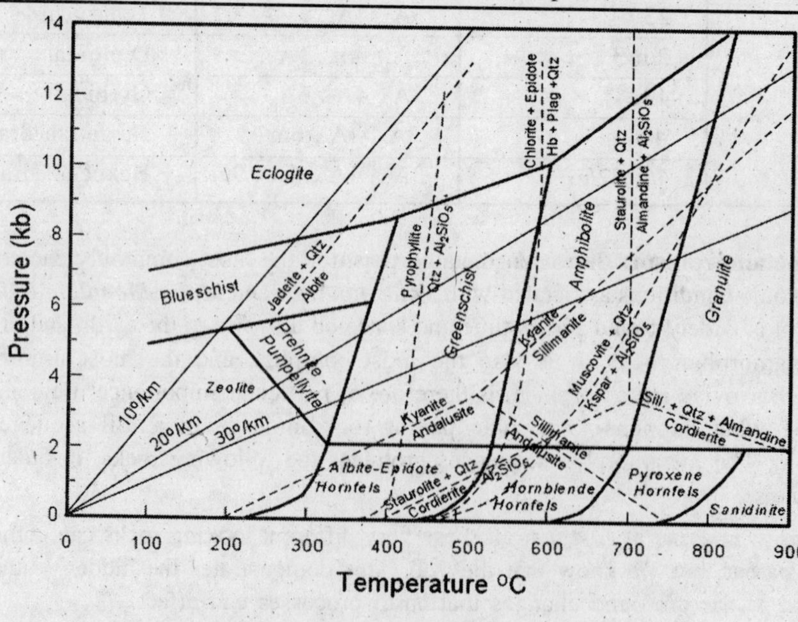

28. Classification of Sandstone:

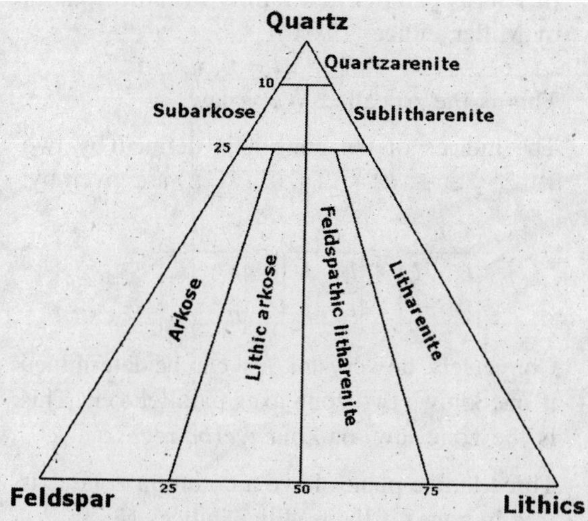

29. Interior of the Earth:

Compositional layers	Thickness (km)	Mechanical layers	Thickness (km)
Crust	40	Lithosphere	100-200
Upper mantle	40-700	Asthenosphere	200-2900
Lower mantle	700-2900	Mesosphere	2900-6371
Outer core	2900-4900		
Inner core	4900-6371		

30. Pressure and temperature of the earth:

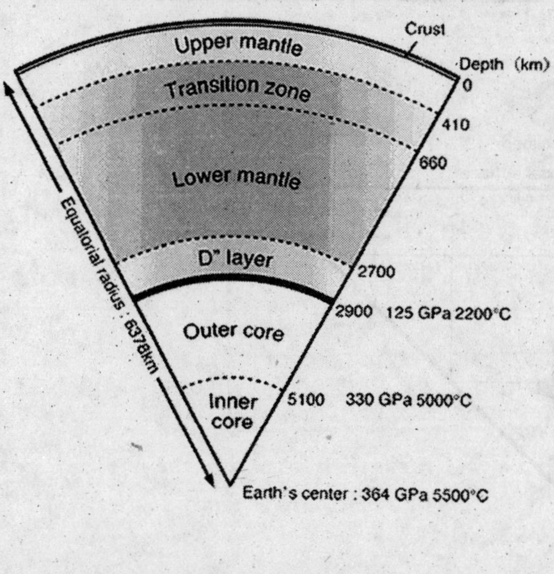

Interior Properties of Earth

Property	Crust	Mantle	Core
Fraction of Earth	<1% of mass	~70%	~30%
State	"Broken rock"	Plastic	(Semi-)liquid
Depth (kilometers)	0-30	30-3030	3030-6370
Density (grams/cubic centimeter)	2.7	3.5-5.5	10-12
Representative chemical composition	SiO_2	$(Fe,Mg)SiO_4$	Fe, Ni
Temperature (Kelvin)	300-500	500-3,000	3,000-5,300
Pressure (atmospheres)	1-1.000	10^3-10^6	10^6-10^7

31.

Landforms	Remarks
Fjords	Glacial erosion
Moraines	Glacial deposition
Drumlins	Glacial deposition
Cirques	Glacial erosion

32.

Dentition pattern	Example bivalvia
Hetrodent	*Cerastoderma*
Desmodent	*Mya*
Dysodent	*Mytilus*
Schizodont	*Trigonia*

33. From the above diagram we calculate the Lever rule:

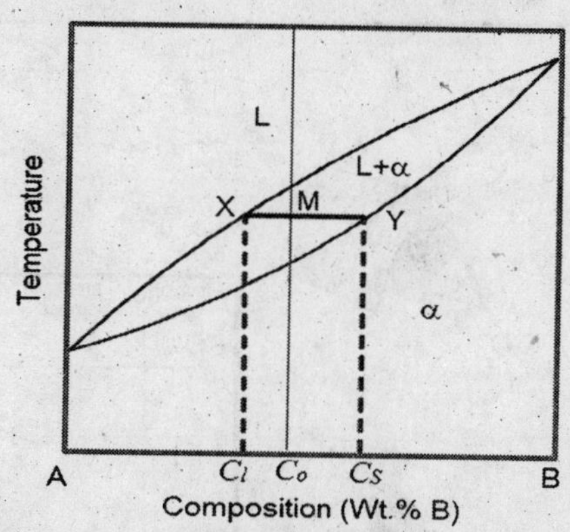

Lever rule: The relative fractions of the phases at a given temperature for an alloy composition Co is obtained by the lever rule. This rule gives the fraction of a phase by the ratio of the lengths of the tie line between Co and composition of the other phase to the total length of the tie line.

For example, fraction solid, f_s is given by :

$F_s = MX/XY = C_0 - C_l/C_s - C_l$

Similarly fraction liquid,

$f_l = MY/XY = C_s - C_0/C_s - C_l$

34. Igneous process and its texture:

Texture	Process
Graphic texture	Eutectic crystallization
Cumulate texture	Crystal settling
Porphyritic texture	Fractional crystallization
Porphyritic texture	Magma mixing

35. Zone and its axis: A **zone** is a set of crystal faces whose pairwise intersection line is parallel to a given lattice direction, called the zone axis.

A zone axis is a lattice row parallel to the intersection of two (or more) families of lattices planes. It is denoted by $[u\ v\ w]$. A zone axis $[u\ v\ w]$ is parallel to a family of lattice planes of Miller indices (hkl) if

$$uh + vk + wl = 0.$$

This is the so-called Weiss law.

The indices of the zone axis defined by two lattice planes (h_1, k_1, l_1), (h_2, k_2, l_2) are given by:

$$\frac{u}{\begin{vmatrix} k_1 & l_1 \\ k_2 & l_2 \end{vmatrix}} = \frac{v}{\begin{vmatrix} l_1 & h_1 \\ l_2 & h_2 \end{vmatrix}} = \frac{w}{\begin{vmatrix} h_1 & k_1 \\ h_2 & k_2 \end{vmatrix}}$$

Conversely, any crystal face can be determined if one knows two zone axes parallel to it. This is the zone law, or *Zonenverbandgesetz.*

Three lattice planes have a common zone axis (are in zone) if their Miller indices (h_1, k_1, l_1), (h_2, k_2, l_2), (h_3, k_3, l_3) satisfy the relation:

$$\begin{vmatrix} h_1 & k_1 & l_1 \\ h_2 & k_2 & l_2 \\ h_3 & k_3 & l_3 \end{vmatrix} = 0$$

36. Fleuty classification of the fold:

Types of folds	Character
Upright fold	Axial plane is about vertical
Recumbent fold	Axial plain is horizontal
Vertical fold	Axial plane is vertical

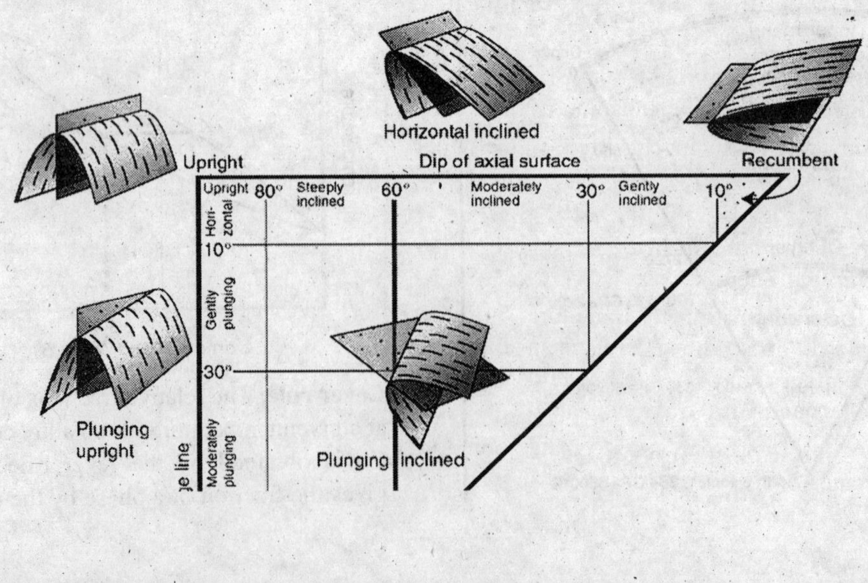

37. Formations	Economic minerals	Remarks
Barakar Formation	Coal	Gondwana supergroup
Barail Group	Lignite	Tertiary of Assam
Cuddalore Formation	Lignite	Cauvery and Godavari basins
Ariyalur Formation	Sandstone, shale and limestone	Cretaceous of Trichinopoly

39. Maturity of Sandstone: The maturity of sandstone sediments can be identified by the textures found in the sediment grains as well as the types of sandstone present.

- The compositions of the grains distinguish the source area and rock type. It also correlates with the amount of transport and erosion the sediments encounter.
- Mature sandstones can be found far from the source area. The amount of transport and weathering has formed rounded, well-sorted grain sediments.
- The minerals found in the sediment are mostly the same through out and contain very few minerals. The energy of mature sandstone is constant and is a reflection of stable depositional regions and shallow waters.
- The composition of mature sandstone is dominated with quartz. There is little to no feldspar, clastic matrix, or mica grains found within the sediment.
- It is typically bonded together with calcite or silica cement.

Immature Sandstone: Immature sandstone is equipped with diverse, unstable minerals.

- Its grains are poorlysorted and angular. The sandstone has not been transported far and is close to the source area.
- Near the source area, the sandstone has varying energy.
- The composition of immature sandstone mostly quartz, however have an abundance of other minerals, such as feldspar, mica, chert and rock fragments.
- Arkose sandstone is one type of immature sandstone. It can form as basal sandstone from granitic coastal areas or can be found in faults and low areas adjacent to granite mountains.
- It contains quartz and a great amount of feldspars, which are angular-course grains.
- Greywacke is immature sandstone that can be found deposited in unstable tectonic settings, like rising mountainous regions.
- They consist of quartz, feldspar, chert, mica and fine-grained volcanic rock particles in the matrix.

40. P and S wave shadow zone:

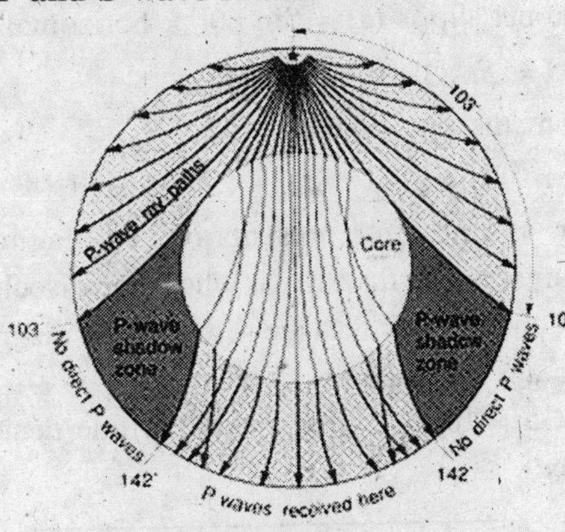

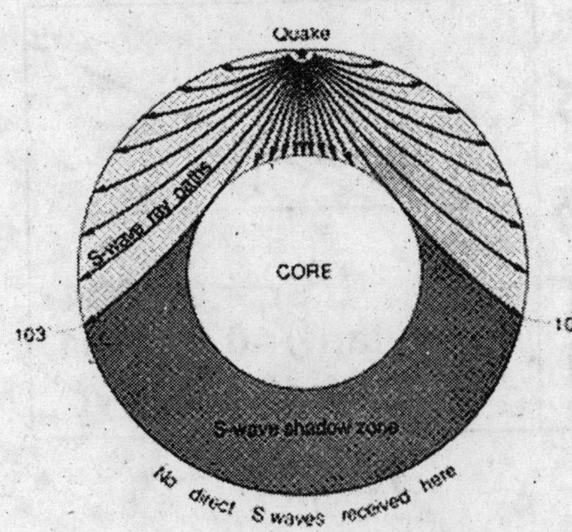

41. From the question, the bearing spotted (from the observer) = S60° E.
If position is interchanged = N60° E
The front bearing value in degree from North (measured clockwise) = 360° – 60° = 300°

43. From the question, the Weiss symbol of crystal face = $4a : 2b : c$
The Miller Index (hkl) = (124)
The value of h is = 1

46. From the question, the birefringence of a mineral thickness (t) = 30 μm
Retardation (Δ) = 0.27 μm
Retardation (Δ) = [t (birefringence)]
0.27 = 30 (birefringence)
Birefringence = 0.009

51. From the question, the pitch of the lineation = 30 degree
The horizontal slip of the fault = 1.25 m
The net slip is (x) = Sin 60° × horizontal slip
$= 2/\sqrt{3} \times 1.25 = 1.44$

55. From the equation:
$(^{87}Sr/^{86}Sr) = (^{87}Sr/^{86}Sr)_1 + (^{87}Rb/^{86}Sr)\,(e^{\lambda t} - 1)$
For a suite of representative co-magmatic granitic rocks, the Rb - Sr whole rock isochron plot and relevant data are shown in the diagram. The age of granite is calculated at --------- Ga (1 Ga = 10^9 yrs, give answer in one decimal place).

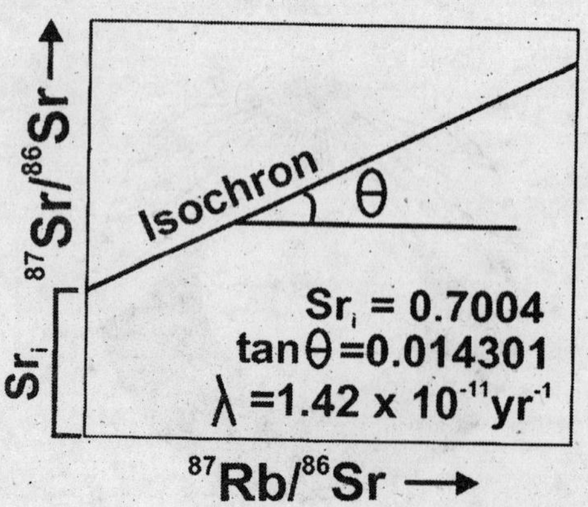

Tan θ = $(e^{\lambda t} - 1)$
Log (tan θ + 1) = $e^{\lambda t}$
1/λ log (tan θ + 1) = t
t = [1/1.42 × 10^{-11} log (0.014301 + 1)]
= 1.1 Ga

56. From the question, For granulite metamorphic facies rock:
P = 9 *k*bar
T = 850 °C
Assume a single layer crust d = 3000 kg/m^3
g = 10 m/sec^2
P = dgh
$h = 3 \times 10^4$ m
= 30 km

58. From the question, (Maximum – Minimum)
Refractive indices of P = 1.727 – 1.721
= 0.006
Refractive indices of Q = 1.553 – 1.544
= 0.009
Refractive indices of R = 1.694 – 1.672
= 0.022
Refractive indices of S = 1.658 – 1.486
= 0.172

59. In the sedimentary succession shown in the figure:

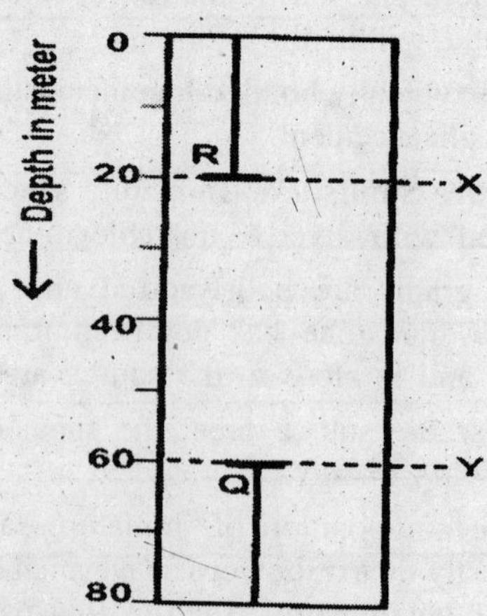

Fossil species Q = 50 Ma
Fossil species R = 30 Ma
2m = 1 Ma

YOUR SPACE

YOUR SPACE

IIT-JAM
GEOLOGY

GENERAL GEOLOGY

1. Which amongst the following planets has the highest number of known satellites?
A. Mars B. Uranus
C. Venus D. Mercury
(IIT-JAM 2015)

2. S-wave is terminated at:
A. Crust-Mantle boundary
B. Lithosphere-Asthenosphere boundary
C. Mantle-Core boundary
D. Inner and Outer core boundary
(IIT-JAM 2015)

3. Arrange the elements Fe, O, H, He, Si in decreasing order of their abundance in the solar system.
A. H > He > O > Si > Fe
B. He > H > Si > O > Fe
C. H > He > O > Fe > Si
D. Si > Fe > H > He > O
(IIT-JAM 2015)

4. In the given figure (i), (ii), (iii), and (iv) denote the plate tectonic settings: Mid Oceanic Ridge (M), Island Arc (I), Continental Arc C. and Rift Zone (R).

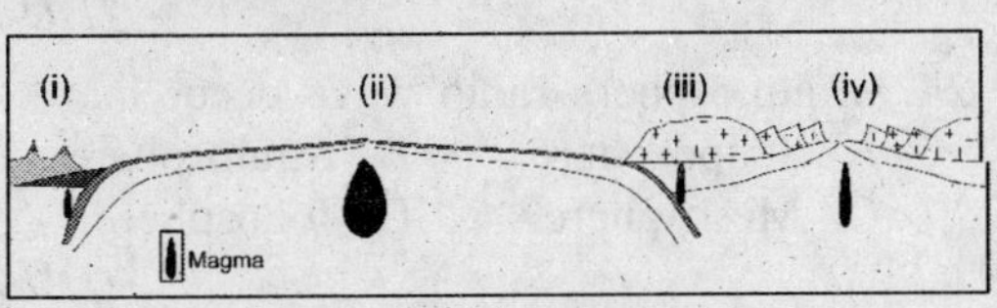

Which one of the following options is correctly matched?
A. (i) - M, (ii) - I, (iii) - C, (iv) - R
B. (i) - R, (ii) - I, (iii) - M, (iv) - C
C. (i) - I, (ii) - R, (iii) - M, (iv) - C
D. (i) - I, (ii) - M, (iii) - C, (iv) - R
(IIT-JAM 2015)

5. The mechanism of plate movement is thought to be driven by:
A. Isostasy
B. Earth's rotation
C. Thermal convection cells
D. Polar wandering *(DU-2015)*

6. Hawaii island has its origin from:
A. Subducting plate
B. Plume
C. Plate collision
D. All of the above *(AMU-2015)*

7. Most abundant element in the earth crust is:
A. Oxygen B. Silicon
C. Aluminium D. Iron *(AMU-2015)*

8. Instrument for recording earthquake wave is called:
A. Seismograph B. Seismograme
C. Seismometer D. Scintillometer
(AMU-2015)

9. The cosmic ray intensity is minimum at the magnetic:
A. Equator B. North pole
C. South pole D. Tropic of cancer
(DU-2015)

10. What powers the Earth's internal heat engine?
A. Solar energy B. Ocean tides
C. Volcanoes D. Radioactivity
(DU-2015)

11. Oxygen built up in the Earth's atmosphere is caused by:
A. Formation of ocean in early history of Earth
B. Algae and other organisms employed photosynthesis
C. Oxygen settled on Earth from planets further from the sun

D. Rocks weathered and released their oxygen.

(DU-2015)

12. Increased concentration of which of the following may cause maximum global warming?
A. Oxygen
B. Nitrogen
C. Carbon dioxide
D. Water vapour (DU-2015)

13. Global warming may lead to increased sea level as a result of:
A. Melting of ice-caps
B. Thermal expansion of sea water
C. Both A. and B.
D. Isostatic rebound (DU-2015)

14. Three major orbit ally forced cyclicities (–100 kyr, –40 kyr, –22 kyr) in the climatic changes on Earth are termed as:
A. Milankovitch cycle
B. Turner cycle
C. Glieesberg cycle
D. Wilson cycle (DU-2015)

15. Ozone layer in Earth's atmosphere, is confined to:
A. Troposphere B. Stratosphere
C. Mesosphere D. Thermosphere
(DU-2015)

16. The most abundant element in the Earth's crust is:
A. Aluminium B. Oxygen
C. Iron D. Carbon
(DU-2015)

17. The internal structure of the Earth down up to the core, has been explored with the help of:
A. P-waves
B. S-waves
C. Both P- and S-waves
D. Drilling (DU-2015)

18. We get Lehmann discontinuity at an approximate depth of:
A. 100 km B. 750 km
C. 450 km D. 220 km (DU-2015)

19. Maximum damage in the event of an earthquake is caused by:
A. P-waves
B. S-waves
C. L-waves
D. All of the above (DU-2015)

20. A negative gravity anomaly indicates:
A. An excess of mass
B. A deficiency in mass
C. A reversal of the gravitational field
D. None of these (DU-2015)

21. To qualify as a desert, a region must be:
A. Hot, with a mean annual temperature greater than 25°C (77°F)
B. Arid, with less than 15% of the ground surface vegetated
C. Both hot and arid
D. Either hot or arid (DU-2015)

22. At 30° north latitude, at the northern edge of tropical (Hadley) cell, convection takes place because:
A. Cool, dry air sinks, becoming drier as it heats up
B. Warm, moist air rises, increasing in relative humidity as it rises
C. Warm, dry air rises, becoming cooler
D. Cool moist air sinks, providing abundant rainfall
(DU-2015)

23. The Gobi of Mongolia is a desert primarily because it is located:
A. In the subtropics
B. Near the geographic South Pole
C. Near the center of a large continent
D. Next to a cold ocean current
(DU-2015)

24. In atmosphere radio wave occur in:
A. Troposphere B. Stratosphere
C. Mesosphere D. Ionosphere
(DU-2015)

25. The process by which an originally homogeneous Earth developed a dense core and a light crust is called:
A. Metamorphism
B. Differentiation
C. Accretion
D. Compression (DU-2015)

26. The Gutenberg discontinuity lies between:
A. Crust and mantle
B. Lithosphere and asthenosphere
C. Outer core and lower mantle
D. Inner core and outer core
(IIT-JAM 2014)

27. Airy's model of isostasy:
A. Requires mountains to have higher density than the oceanic crust
B. Requires mountains to have lower density than the oceanic crust
C. Requires mountains to have the same density as oceanic crust
D. Does NOT consider the densities of mountain and oceanic crust
(IIT-JAM 2014)

28. Himalaya is a product of:
A. Continent-arc collision
B. Continent-continent collision
C. Arc-arc collision
D. Continent-arc-continent collision
(BHU-2014)

29. The oldest rocks in the present ocean floor have age in range:
A. 4.0-4.6 billion years
B. 2.0-2.5 billion years
C. 185-200 million years
D. 60-65 million years *(IIT-JAM 2013)*

30. The type of plate boundary associated with deep focus earthquake?
A. Divergent B. Convergent
C. Transverse D. Both A. & B.
(IIT-JAM 2013)

31. The most abundant element in the universe is:
A. Silicon B. Helium
C. Oxygen D. Hydrogen
(BHU-2013)

32. What is the age of ocean floor?
A. 0-200 m.y. B. 0-500 m.y.
C. 0-1000 m.y. D. 0-3500 m.y.
(BHU-2013)

33. Which of the following is not one of the four inner planets?
A. Mars B. Earth
C. Neptune D. Saturn
(BHU-2013)

34. New seafloor created at a:
A. Deep sea trench
B. Mid-oceanic ridge
C. Subduction zone
D. Transform fault *(BHU-2013)*

35. By the plate tectonic processes, magma generation is highest in the following tectonic setting:
A. Convergent plate boundary
B. Continental intra-plate region
C. Oceanic intra-plate region
D. Oceanic ridges *(PU-2013)*

36. The average thickness of the continental crust is about:
A. 35-40 km B. 100-200 km
C. 1000-2000 km D. 5-10 km
(PU-2013)

37. Which one of the following features is NOT associated with an oceanic subduction?
A. Sea-mount B. Benioff zone
C. Back-arc D. Fore-arc
(IIT-JAM 2012)

38. The term isostasy refers to:
A. Gravitational equilibrium
B. Thermal equilibrium
C. Magnetic equilibrium
D. Electrical equilibrium *(IIT-JAM 2012)*

39. Deep sea trenches are associated with plate boundary:
A. Convergent B. Divergent
C. Translational D. Neutral
(BHU-2013)

40. What drive(s) the earth's internal heat engine?
A. Radioactivity B. Solar energy
C. Volcanism D. Ocean tides
(BHU-2013)

41. The 'Nebular hypothesis' was proposed by:
A. Kant
B. Laplace
C. Kant and Laplace
D. Moulten and Chamberlain *(BHU-2013)*

42 A deep valley forms where two plates pull apart:
A. Fault B. Pangea
C. Mantle D. Rift valley
(BHU-2013)

43. Mountain chain within water is:
A. Trench
B. Mid-oceanic ridge
C. Volcanic island
D. Reefs (BHU-2013)

44. A boundary where plates slide past each other in opposite directions is termed as:
A. Transform boundary
B. Tensional zone
C. Diverse boundary
D. Converging boundary (BHU-2013)

45. Wadati-Benioff zone indicates:
A. Shallow earthquake
B. Intermediate earthquake
C. Technical stable zone
D. Deep focus earthquake (BHU-2013)

46. The first wave recorded on seismograph is:
A. P-wave B. S-wave
C. P- and S-wave D. L-wave
(BHU-2013)

47. The average radius of the earth is:
A. 6378 km B. 637·1 km
C. 6372 km D. 6271 km
(BHU-2012)

48. 'Sima' is a part of:
A. Quter core B. Crust
C. Upper mantle D. Lower mantle
(BHU-2012)

49. The instrument used for recording seismic waves is:
A. Thermometer B. Seismograph
C. Barometer D. Seismogram
(BHU-2012)

50. The 'Nebular Hypothesis' was proposed by:
A. Kant
B. Laplace
C. Kant and Laplace
D. Moulten and Chamberlin (BHU-2012)

51. Himālaya is:
A. Paleozoic tectonic mountain
B. Recent folded mountain
C. Collisional mountain
D. Indian mountain (BHU-2013)

52. The Gutenberg discontinuity separates:
A. Crust and upper mantle
B. Lower to upper mantle
C. Outer and inner core
D. Outer core and lower mantle
(BHU-2013)

53. Volcanic Island is associated with:
A. Ocean-continent convergence
B. Divergent boundaries
C. Transform boundary
D. Ocean-ocean convergence (BHU-2013)

54. 'Conorad discontinuity' lies between:
A. Crust and mantle
B. Sial and sima
C. Sima and mantle
D. Mantle and core (BHU-2012)

55. The mean density of Earth is about:
A. 2650 kg m^{-3} B. 2750 kg m^{-3}
C. 4400 kg m^{-3} D. 5500 kg m^{-3}
(IIT-JAM 2011)

56. Match the scale in Group I with corresponding parameter in Group II.

Group I	Group II
P. Richter scale	1. Earthquake damage
Q. Mercalli scale	2. Hardness
R. Moh's scale	3. Particle size
S. Udden-Wentworth scale	4. Earthquake magnitude

A. P-1, Q-4, R-2, S-3
B. P-4, Q-1, R-2, S-3
C. P-3, Q-4, R-2, S-1
D. P-4, Q-3, R-2, S-1 (IIT-JAM 2011)

57. 'Wadati-Benioff zone' is well known for:
A. Deep focus earthquake
B. Shallow focus earthquake
C. Intermediate focus earthquake
D. Tectonically stable zone (BHU-2011)

58. In internal structure of earth, the outer core is composed of:
A. Solid material B. Gases
C. Liquid material D. All of these

59. Shear waves passes through:
A. Liquid only B. Solid only
C. Gas only D. All of these
(BHU-2011)

60. Asteroids are minor planetary bodies located between:
A. Earth and Mars
B. Jupiter and Saturn
C. Mars and Jupiter
D. Mercury and Venus *(IIT-JAM 2010)*

61. Which one of the following options does correctly represent the elements in order of decreasing abundance in the Earth's crust?
A. Fe > Si > O > Al
B. O > Si > Al > Fe
C. Si > O > Al > Fe
D. Si > Al > Fe > O *(IIT-JAM 2010)*

62. The average density of the earth is:
A. 4.5 gm/cm^3 B. 5.5 gm/cm^3
C. 6.5 gm/cm^3 D. 7.5 gm/cm^3
(BHU-2010)

63. Volcanic activity is usually absent along:
A. Divergent plate boundaries
B. Convergent plate boundaries
C. Transform faults
D. All of the above *(BHU-2010)*

64. What can be expected at a depth of about 35 km inside the earth?
A. Inner mantle
B. Inner core
C. Mohorovicic discontinuity
D. Gutenberg discontinuity *(BHU-2010)*

65. Sial is also known as:
A. Part of core
B. Lower continental crust
C. Mantle
D. Upper continental crust *(BHU-2009)*

66. Match the tectonic units listed in Group 1 with their geographical locations in Group 2.

Group-1	Group-2
P. Continent-oceanic lithosphere convergence	1. Himalayas
Q. Continent-continent collision	2. Andes
R. Continental rift system	3. Japanese island
S. Oceanic-oceanic lithosphere convergence	4. East Africa

A. P-2, Q-1, R-4, S-3
B. P-2, Q-3, R-4, S-1
C. P-3, Q-4, R-1, S-2
D. P-4, Q-1, R-2, S-3 *(IIT-JAM 2008)*

67. Gutenberg Discontinuity in the interior of the Earth occurs at the depth of:
A. 35 km B. 800 km
C. 2900 km D. 5200 km
(IIT-JAM 2008)

68. Isostasy involves:
A. Eustatic change
B. Gravitational balance
C. Magnetic reversal
D. Thermal balance *(IIT-JAM 2007)*

69. The upper mantle of the earth is composed largely of:
A. Peridotite B. Gabbro
C. Mafic granulite D. Eclogite
(IIT-JAM 2006)

70. The lithosphere of the earth is made up of:
A. Crust
B. Crust and upper mantle
C. Crust and part of upper mantle
D. Upper mantle *(IIT-JAM 2005)*

71. The explored part of the Moon consists dominantly of:
A. Basalt and anorthosites
B. Basalts
C. Basalts and granites
D. Basalts and andesites *(IIT-JAM 2005)*

72. Reversal of magnetic polarity in strips of ocean floor basalt parallel to ridge axis, results due to:
A. Bimodal volcanism
B. Seafloor spreading
C. Transform faults
D. Alteration of seafloor basalt
(IIT-JAM 2005)

73. The amplitude of seismic waves in an earthquake of magnitude 2 on the Richter scale is larger than that of an earthquake of magnitude 1 by:
A. 100 times B. 10 times
C. 2 times D. 1000 times
(IIT-JAM 2005)

74. Most abundant element in the earth's crust is:
A. Carbon B. Oxygen
C. Hydrogen D. Nitrogen
(BHU-2014)

75. Aseismic ridges are associated with:
A. Hotspots B. Subduction zones
C. Cratons D. Island arcs
(DU-2014)

76. The two most abundant elements in Earth's crust are:
A. Iron and Magnesium
B. Aluminium and Oxygen
C. Oxygen and Silicon
D. Silicon and Aluminium *(DU-2014)*

77. Most of the craters on moon surface are probably the result of:
A. Volcanic action
B. Meteoroid bombardment
C. Erosion
D. Deposition *(DU-2014)*

78. The most voluminous portion of the Earth is known to geologists as:
A. The crust B. The mantle
C. The core D. The lithosphere
(DU-2014)

79. Rate of Indian plate motion at present is:
A. 3 - 4 centimeters per year
B. 1 - 18 centimeters per year
C. 1 kilometer per year
D. 1,000 kilometers per year
(DU-2014)

80. The oldest seafloor on Earth is not more than:
A. 200 million years old
B. 2 billion years old
C. 20 million years old
D. 2 million years old *(DU-2014)*

81. The continental rise is:
A. A wedge of sediment at the base of the continental slope
B. An uplifted portion of the continent
C. A portion of the mid-ocean ridge
D. A flat-topped seamount *(DU-2014)*

82. Low velocity layer (LVL) within the Earth coincides with depth range of:
A. 50-100 Km B. 300-350 Km
C. 600-750 Km D. 100-200 Km
(DU-2014)

83. The surface waves are:
A. P-waves
B. S-waves
C. Both P and S waves
D. L-waves *(BHU-2015)*

84. Conrad discontinuity separates boundary between:
A. Lower crust and upper mantle
B. Upper mantle and lower mantle
C. Outer core and inner core
D. Upper crust and lower crust
(BHU-2015)

85. The new oceanic surface are created at .... plate margin.
A. Constructive B. Distractive
C. Conservative D. Convergent
(BHU-2015)

86 Match the name of scientist (Group 1) and their famous theories (Group 2):

	Group-1		**Group-2**
P.	Alfred Wegener	1.	Isostasy
Q.	Airy	2.	Sea floor spreading
R.	Herry Hess	3.	Theory of continental drift
S.	Pratt	4.	Big Bang theory

A. P-1, Q-2, R-3, S-4
B. P-3, Q-1, R-2, S-4
C. P-4, Q-1, R-2, S-3
D. P-1, Q-2, R-4, S-3 *(BHU-2015)*

87. Which of the following is supercontinents?
A. Gondwanaland B. Laurasia
C. Pangea D. Panthalassa
(BHU-2015)

88. Outer core is liquid while the inner core is solid because:
A. Outer core is heated more than the inner core
B. Melting point of material in outer core is less than the ambient temperature while it is more for the material in inner core due to high pressure

C. Outer core is in contact with liquid mantle
D. Inner core is rich in iron and nickel while outer core is not *(PU-2010)*

89. Earth's crust consists of continental and oceanic parts. Which of the following statement is not true?
A. Oceanic crust has global layering while continental crust doesn't
B. Oceanic crust is thinner compared to continental crust
C. Oceanic crust is younger than continental crust
D. Oceanic crust is more metamorphosed compared to continental crust
(PU-2010)

90. Which of the following regions would have thicker crust?
A. Mountain belt
B. Mid Oceanic ridges
C. Precambrian Shield
D. Island arc *(PU-2010)*

91. Continental crust is higher in elevation than oceanic crust because:
A. It is more buoyant than oceanic crust
B. It is older than oceanic crust
C. It is younger than oceanic crust
D. It is denser than oceanic crust
(PU-2010)

92. The average thickness of a lithospheric plate is approximately:
A. 35 km B. 8 km
C. 5 km D. 100 km *(PU-2010)*

93. The average thickness of the continental crust is about:
A. 35-40 km B. 100-200 km
C. 1000-2000 km D. 5-10 km
(PU-2013)

94. The discontinuity which represents the boundary between crust and mantle is termed:
A. Moho B. Conrad
C. Gutenberg D. Lehman
(PU-2013)

95. Benioff zone is associated with:
A. Subducted slab
B. Mid ocean ridges
C. Transform faults
D. Mountains *(PU-2013)*

96. Which of the following is seismic surface wave?
A. P wave B. S wave
C. Rayleigh wave D. PKP wave
(PU-2013)

97. The transfer of heat by the physical movement of material is called:
A. Conduction B. Convection
C. Radiation D. Scattering
(PU-2013)

98. Himalayan mountains have formed primarily as a result of:
A. Compressive deformation between Indian and Asian tectonic plates
B. Relative uplift of sediments in the Tethys Ocean in an extensional setting
C. Strike-slip deformation between Indian and Eurasian plates
D. Extrusion of volcanic material over a long period of time on the Tethys ocean floor
(PU-2013)

99. The average density of the Earth is:
A. 3.5 g/cm^3 B. 4.5 g/cm^3
C. 5.5 g/cm^3 D. 6.5 g/cm^3
(PU-2013)

100. The value of gravity becomes zero at:
A. The surface of the earth
B. The centre of the earth
C. Equator
D. Poles *(PU-2013)*

101. Earthquakes are generated due to:
A. Plastic flow of rocks and release of plastic strain energy
B. Folding of rock layers
C. Sudden release of elastic strain energy during fracture propagation
D. Sudden release of viscous strain accumulated in rocks *(PU-2013)*

102. Which of the following statements about Earth's core is true?
A. Entire core is in liquid state
B. Entire core is in solid state

C. Outer core is in solid state and inner core is in liquid state
D. Outer core is in liquid state and inner core is in solid state *(PU-2013)*

103. Outer planets in our solar system are gaseous because:
A. In rotating nebula, the gases did not move to the center
B. Gases experience more centrifugal force
C. Gases are driven away by the Sun's heat
D. Rocks were pulled inside the solar system by the Sun's gravity *(PU-2013)*

104. The three most abundant silicate minerals in the earth's continental crust are:
A. Quartz, olivine, pyroxene
B. Quartz, feldspar, pyroxene
C. Quartz, feldspar, mica
D. Feldspar, pyroxene, mica
(PU-2012)

105. The layer that occurs in between crust and core is:
A. Magma layer B. Lithosphere
C. Mantle D. Continent
(PU-2011)

106. New sea floor created as:
A. Deep sea trench
B. Mid oceanic ridge
C. Subduction zone
D. Transform fault *(PU-2011)*

107. The descent of the oceanic lithosphere into the mantle is the process of:
A. Accretion B. Subduction
C. Obduction D. Collision
(PU-2011)

108. The age of the Earth is currently considered as:
A. 6230 million years
B. 3800 million years
C. 4560 million years
D. 5540 million years *(PU-2012)*

109. The lithosphere below stable continents is approximately .... kilometers thick.
A. 30-40 B. 60-70
C. 100-150 D. 200-250
(PU-2011)

110. The continental crust is always at higher elevation than oceanic crust, because:
A. Continental crust is lower in density than Oceanic crust
B. Continents have mountain ranges
C. Oceanic crust from at Mid oceanic ridges
D. Continental crust is older compared to Oceanic crust *(PU-2011)*

111. Inner planets differ from the outer planets of the solar system having:
A. A different direction of orbit
B. More rocky material than gases material
C. Higher gravitational force
D. More number of natural satellites
(PU-2011)

112. Plate in 'Plate-tectonics' theory means:
A. Fragment of continental crust
B. Fragment of oceanic crust
C. Fragment of lithosphere
D. A part of mountain range *(PU-2011)*

113. More Tsunamis that hit Japan because:
A. Japan is in east where moon's tidal force are maximum
B. Japan's sea coast is not protected by rocks
C. Japan is along an active plae boundary where more underwater earthquakes take place
D. Average elevation of Japan is low
(PU-2011)

114. What are the most abundant elements in the Earth crust?
A. Iron and magnesium
B. Oxygen and silicon
C. Silicon and aluminium
D. Silicon and calcium *(PU-2011)*

115. The Earth's upper mantle is essentially made up of:
A. Olivine and pyroxene
B. Olivine and muscovite
C. Quartz and feldspar
D. Pyroxene and garnet *(PU-2011)*

116. The most abundant metal (by weight %) in the Earth's crust is
A. Al B. Fe
C. Na D. Mg *(IIT JAM-2016)*

117. On the Richter scale, a magnitude 8 earthquake is how much larger than a magnitude 6 earthquake
A. 2 times B. 4 times
C. 10 times D. 100 times
(*ISM-2015*)

118. What will happen to the age of the oceanic lithosphere as we move away from the mid - oceanic ridge?
A. Increases
B. First increases then decreases
C. Decreases
D. First decreases then increases (*ISM-2015*)

119. The direction of ground water-flow can be determined with the help of:
A. Depth of water level map
B. Water table contour map
C. Subsurface topographic map
D. All of the above (*ISM-2015*)

120. Why was Paleomagnetism so important in discovering plate tectonics?
A. It illustrate the location of the North pole
B. It illustrated sea floor spreading
C. It allowed measurement of mountain building rates
D. It allowed the depth of the ocean to be measured (*ISM-2015*)

121. Which of the following statements about the age of the Earth's crust is correct?
A. The oceanic crust contains the oldest rocks on earth
B. The oceanic crust is less than 200 million years old
C. The oceanic crust is oldest at the mid-oceanic ridges
D. All oceanic crust is the same age
(*ISM-2015*)

122. What is the angle of P-wave shadow zone?
A. 90 to 150 degree
B. 104 to 150 degree
C. 104 to 140 degree
D. 103 to 139 degree (*ISM-2015*)

123. The amplitude of ground motion during an earthquake of magnitude 7 in Richter scale is how many times more than that of a magnitude 5?
A. 10 B. 100
C. 1000 D. 10,000
(*IIT JAM-2016*)

124. The failed arm of a continental rift is called
A. Hot spot B. Horst
C. Decollement D. Aulacogen
(*IIT JAM-2016*)

125. How many major tectonic plates are there on the surface of the earth?
A. 4 B. 6
C. 5 D. 12 (*ISM-2014*)

126. Where are earthquakes most likely to take place?
A. Along the core-mantle boundary
B. Where the composition of the Earth tends to be uniform
C. Near the earth equator
D. Near the fault zone (*ISM-2014*)

127. At what speed does the Earth travels as it orbit on the sun?
A. 40 km/s B. 20 km/s
C. 30 km/s D. 60 km/s
(*ISM-2014*)

128. Why do we have different season on the earth?
A. Because of bio-rhythmic cycles
B. Because of the Earth's axial tilt
C. Because of rotation of the earth
D. Because of solar of the wind
(*ISM-2014*)

129. What is the most probable causes of the formation of the moon?
A. The earth's magnetic fields
B. The death of the old star
C. A meteor-like object striking the earth
D. Accretion of material already existing after formation of the earth (*ISM-2014*)

130. The earth is currently in the interglacial stage. The last glacial periods ended about:
A. 10,000 years ago
B. 20,000 years ago
C. 30,000 years ago
D. 100,000 years ago (*ISM-2014*)

131. Approximately how many years ago did the reversal of the Earth's magnetic poles occurs?
A. 400,000 B. 700,000
C. 300,000 D. 900,000 *(ISM-2014)*

132. Which statements best describe the continental and oceanic crust?
A. The continental crust is thicker and less dense than the oceanic crust
B. The continental crust is thinner and more dense than the oceanic crust
C. The continental crust is thinner and less dense than the oceanic crust
D. The continental crust is thicker and more dense than the oceanic crust *(ISM-2014)*

133. The average thickness of the earth's crust on the continents?
A. 80 km B. 100 km
C. 40 km D. 140 km *(ISM-2014)*

134. The ozone layer is located in which level of the atmosphere?
A. Thermosphere B. Stratosphere
C. Troposphere D. Mesosphere
(ISM-2014)

135. Which phenomenon is responsible for the production of about 80% of the earth's internal heat?
A. Radioactive decay
B. Magnetic decay
C. Friction between tectonic plates
D. The accelerated spinning of the inner core
(ISM-2014)

136. The rate of temperature increases below the earth's surface is greatest between depth of:
A. 3500 and 4000 km
B. 25 and 500 km
C. 1500 and 2500 km
D. 2500 and 3500 km *(ISM-2014)*

137. To get a sample material from the mantle, drilling will be done through the oceanic crust rather than through the continental crust because oceanic crust is:
A. Younger than continental crust
B. More dense than continental crust
C. Thinner than continental crust
D. Softer than continental crust *(ISM-2014)*

138. The time that an earthquake occur can be inferred by knowing the:
A. Distance between seismograph station
B. Epicentre distance and arrival time of P-waves
C. Travel time of the S-waves
D. Arrival time of P-waves *(ISM-2014)*

139. Approximately how much percentage of the Earth's crust has been uplifted?
A. 20% B. 80%
C. 60% D. 40% *(ISM-2014)*

140. Which is the best evidence that the Earth's crust has been uplifted?
A. Marine life found at high elevation above sea level
B. Younger fossils above older fossils in layers of rocks
C. Shallow-water fossils found at great ocean depths
D. The rise of sea level has occurred many times in earth's history *(ISM-2014)*

141. How many mass extinction events do scientist believe to have occurred in the earth's history?
A. 1 B. 6
C. 5 D. 3 *(ISM-2014)*

142. The inner most core of the earth is composed of what material?
A. Iron B. Nickel
C. Plutonium D. Lead *(ISM-2014)*

143. Which of the following is NOT a phenomenon that can occur as a direct result of tectonic plate movements?
A. Orogeny B. Tornodoes
C. Tsunamis D. Volcanoes
(ISM-2014)

144. Recent volcanic activity in different parts of the world supports the inference that volcanoes are located mainly in
A. The central region of the continents
B. Zone of crustal activity
C. Zones in late stages of erosion
D. The centre of landscape region
(ISM-2014)

145. The only combination of triple junction that is always stable is:
A. Ridge - Ridge - Ridge
B. Trench - Trench - Trench
C. Ridge - Ridge - Trench
D. Ridge - Trench - Fault *(ISM-2014)*

146. Which one of the following is the correct sequence of events in the Wilson cycle?
A. Breakup of continents - seafloor spreading - subduction collision
B. Seafloor spreading - break up continents - subduction collision
C. Breakup of continents - subduction collision seafloor spreading
D. Subduction collision - sea floor spreading - break up of continents *(ISM-2014)*

147. The velocity of primary waves is minimum at the:
A. Inner core B. Asthenosphere
C. Outer core D. Mantle
(SAGAR UNIVERSITY-2015)

148. Most important stage in the formation of solar system is the burning of:
A. Oxygen B. Helium
C. Hydrogen D. Carbon
(SAGAR UNIVERSITY-2015)

149. Name the main constituent elements found in the core of the earth:
A. Iron-magnesium
B. Nickel iron
C. Silicon-magnesium
D. Magnesium-nickel
(SAGAR UNIVERSITY-2015)

150. Most abundant element in the earth crust is:
A. Oxygen B. Magnesium
C. Iron D. Silicon
(SAGAR UNIVERSITY-2015)

151. The Indian ocean tsunami that caused damages was due to an earthquake off the:
A. Andaman island
B. Indonesian island of Sumatra
C. Malaysia
D. Indonesian island of Java
(SAGAR UNIVERSITY-2015)

152. The lithosphere skids over a mobile viscous layer is called:
A. The core B. The asthenosphere
C. The mesosphere D. The lithosphere
(SAGAR UNIVERSITY-2015)

153. The 250 million years ago clustering of land masses of the earth in to a single super continent were known as:
A. Tethys B. Gondwana land
C. Pangea D. Eurasia
(SAGAR UNIVERSITY-2015)

154. A linear lowland between mountain ranges created by the action of a fault is known as:
A. Rift valley B. Horst
C. Thrust belt D. Subduction zone
(SAGAR UNIVERSITY-2015)

155. Which of the following is not a tectonic plate boundary?
A. Mid-oceanic ridge
B. Subduction zone
C. Transform fault
D. Continental slope
(SAGAR UNIVERSITY-2015)

156. When plate collide, one plate goes down below another, it is known as a:
A. Collision boundary
B. Subduction boundary
C. Suture zone
D. Contact point
(SAGAR UNIVERSITY-2015)

157. Source of the Earth's heat energy is :
A. Radioactivity B. Solar energy
C. Volcanoes D. Ocean tides
(SAGAR UNIVERSITY-2015)

158. How many years ago the Big Bang took place:
A. 10 - 15 thousand years ago
B. 10 - 15 million years ago
C. 100 - 150 million years ago
D. 10 - 15 billion years ago
(SAGAR UNIVERSITY-2015)

159. The layer that lies below the crust is the:
A. Magma layer B. Lithosphere
C. Mantle D. Asthenosphere
(SAGAR UNIVERSITY-2015)

160. Volcanism is associated with which of the plate margins?
A. Convergent B. Divergent
C. Transform fault D. Transcurrent fault
(SAGAR UNIVERSITY-2015)

161. The height of the Mount Everest is:
A. 8848 m above mean sea level
B. 8800 m above mean sea level
C. 8700 m above mean sea level
D. 8650 m above mean sea level
(SAGAR UNIVERSITY-2015)

162. The Gutenberg discontinuity is located at a depth of around:
A. 35 km B. 150 km
C. 2900 km D. 5000 km
(PUNJAB UNIVERSITY-2015)

163. Thermohaline circulation in the oceans is driven by:
A. Only saline gradient
B. Both temperature and salinity gradient
C. Only temperature gradient
D. Only density difference
(PUNJAB UNIVERSITY-2015)

164. Which one of the following lakes is interpreted to be of meteoric impact origin?
A. Lunar lake B. Chilka lake
C. Kolleru lake D. Pulicut lake
(PUNJAB UNIVERSITY-2015)

165. Which of the following is located closet to the Ninety - East-Ridge?
A. Bombay high
B. Lakshadweep island
C. Andaman and Nicobar island
D. Maldives
(PUNJAB UNIVERSITY-2015)

166. Who proposed the principle "the present is the key to the past" ?
A. James Hutton B. Carl von Linnaeus
C. William Smith D. Orbigny
(PUNJAB UNIVERSITY-2015)

167. The oceanic cover about .... area of the earth surface.
A. 69% B. 21%
C. 70.8% D. 80.2%
(PUNJAB UNIVERSITY-2015)

168. Witwatersrand (South Africa) is famous for:
A. Coal deposits
B. Cooper deposits
C. Sedimentary uranium deposits
D. Sedimentary Mn and Fe deposits
(PUNJAB UNIVERSITY-2015)

169. Which of the following is true about age of ocean floor?
A. Oldest rock recorded is about 3.8 Ga age
B. Oldest rock recorded is about 4.2 Ga age
C. Youngest rock so far recorded is about 160 Ma
D. Oldest rocks are found near the trench
(AMU-2016)

170. Which of the following is true for East African rift system?
A. Continental rift
B. Active rift system
C. Active volcanic activity
D. All of the above *(AMU-2016)*

171. Transform boundaries may involve:
A. Ridge - ridge sliding
B. Ridge - trench sliding
C. Trench - trench sliding
D. All of the above *(AMU-2016)*

172. Subduction involving two plates result in:
A. Trench curved convex to subducting slab
B. Trench curved concave to subducting plate
C. Dominantly tholeiitic magmatism
D. Earthquake associated with strike-slip faults
(AMU-2016)

173. Which one of the oldest Supercontinent:
A. Rodinia B. Gondwana
C. Pangea D. Ur supercontinent
(AMU-2016)

174. Which of the following is a robust indicator of horizontal movement of plate?
A. APW paths B. Isotopic systematic
C. Ophiolites D. All the above
(AMU-2016)

175. Which of the following is an essential condition of plate tectonics?
A. Rigidity of plate

B. Buoyancy of plate
C. Convection in core
D. All the above *(AMU-2016)*

176. The 'D' layer within the Earth corresponds to a depth range between:
A. 2740 - 2890 km
B. 2010 - 2035 km
C. 2890 - 3050 km
D. 250 - 375 km
(ISM-2016)

177. The Lehmann discontinuity in the Earth is identified by:
A. 5 - 6 % decreases in P and S wave velocity
B. 3 - 4 % increases in P and S wave velocity
C. 5% decreases in density
D. 15 % increases in S wave velocity
(ISM-2016)

178. Sea-floor spreading goes through various stages. Which one of the following sequence represents the correct order of stage starting from the initial to more advanced ones?
A. Red sea - Atlantic ocean - African rift valley
B. Red sea - African rift valley - Atlantic ocean
C. Atlantic ocean - Red sea - African rift valley
D. African rift valley - Red sea - Atlantic ocean
(ISM-2016)

179. We get seasons on the Earth mainly because:
A. The Sun gets hot and cold periodically
B. The Moon gets in the way of the Sun
C. The Earth's axis is tilted with respect to its rotation ellipse
D. The Earth's gets close and for with respect to the Sun *(ISM-2016)*

180. Sea water is saline because:
A. Na^+ is the most abundant cation in sea water
B. Cl^- is the most abundant anion in sea water
C. NaCl is the first compound to precipitate from sea water
D. Na^+ has highest residence time in sea water
(ISM-2016)

181. The "ring of fire" an arcuate belt with the largest number of active volcanoes on the earth is found in:
A. Atlantic ocean
B. Pacific ocean
C. African rift valley
D. Mid-oceanic ridge *(BHU-2016)*

182. Lithosphere comprises of:
A. Crust and mantle
B. Crust, mantle and outer core
C. Crust and upper mantle
D. Crust and lower mantle *(BHU-2016)*

183. 90 degree ridge is located in:
A. Antartic ocean B. Arabian sea
C. Indian ocean D. None of the above
(BHU-2016)

ANSWER KEY

1	2	3	4	5	6	7	8	9	10
B	C	A	D	C	B	A	A	A	D
11	**12**	**13**	**14**	**15**	**16**	**17**	**18**	**19**	**20**
B	D	C	A	B	B	C	D	C	B
21	**22**	**23**	**24**	**25**	**26**	**27**	**28**	**29**	**30**
B	A	C	D	B	C	C	B	D	B
31	**32**	**33**	**34**	**35**	**36**	**37**	**38**	**39**	**40**
D	A	C,D	B	A	D	A	A	A	A
41	**42**	**43**	**44**	**45**	**46**	**47**	**48**	**49**	**50**
C	D	C	A	D	D	C	B	B	C

51	**52**	**53**	**54**	**55**	**56**	**57**	**58**	**59**	**60**
C	D	A	B	D	B	A	C	B	C
61	**62**	**63**	**64**	**65**	**66**	**67**	**68**	**69**	**70**
B	B	C	C	D	A	C	B	A	C
71	**72**	**73**	**74**	**75**	**76**	**77**	**78**	**79**	**80**
B	B	B	B	C	C	B	B	A	A
81	**82**	**83**	**84**	**85**	**86**	**87**	**88**	**89**	**90**
A	D	D	D	A	B	C	B	D	A
91	**92**	**93**	**94**	**95**	**96**	**97**	**98**	**99**	**100**
A	D	D	A	A	C	B	C	C	D
101	**102**	**103**	**104**	**105**	**106**	**107**	**108**	**109**	**110**
C	D	C	D	C	B	B	C	C	A
111	**112**	**113**	**114**	**115**	**116**	**117**	**118**	**119**	**120**
B	D	A	B	A	A	D	A	D	B
121	**122**	**123**	**124**	**125**	**126**	**127**	**128**	**129**	**130**
B	C	C	D	B	D	C	B	C	A
131	**132**	**133**	**134**	**135**	**136**	**137**	**138**	**139**	**140**
B	A	C	B	A	A	A	D	D	C
141	**142**	**143**	**144**	**145**	**146**	**147**	**148**	**149**	**150**
C	A	B	B	A	A	A	B	B	A
151	**152**	**153**	**154**	**155**	**156**	**157**	**158**	**159**	**160**
B	B	C	A	D	B	A	D	B	B
161	**162**	**163**	**164**	**165**	**166**	**167**	**168**	**169**	**170**
A	C	B	A	B	A	C	D	D	D
171	**172**	**173**	**174**	**175**	**176**	**177**	**178**	**179**	**180**
D	A	D	A	A	A	B	C	C	D
181	**182**	**183**							
B	C	C							

HINTS AND SOLUTIONS

1. Planets	**Number of satellites**
Mars	2
Uranus	14
Venus	-
Mercury	-
Jupiter	16
Neptune	2
Saturan	17
Earth	1

3. Elements in different zone	**Percentage decreasing order**
Crust	O > Si > Al > Fe > Na
Lithosphere	O > Si > Al > Fe > Na
Earth	Fe > Ni > O
Universe	H > He > Fe > O
Atmosphere	N_2 > O > CO_2 > Ar

4. From figure:

Point	Feature	Remark
1	Island Arc	It is formed in convergent plate setting. Eg. Japanese island
2	Mid Oceanic Ridge	It is formed in divergent plate where mantle plume rich the surface. Eg. Mid-Atlantic Ridge
3	Continental Arc	It is also convergent plate setting. Eg. Andes.
4	Rift Zone	It is formed in divergent plate motion. Eg. Hawaii.

6. Process	Example
Subducting plate	Alps
Plume	Hawaii Island
Plate collision	Himalayas

9. Cosmic ray intensity	Place
Minimum	Equator
Maximum	Pole
Increases towards	Pole
Decreases towards	Equator

14. Cycle	Properties
Milankovitch cycle	Cyclic climate change
Turner cycle	Basin depositional cycle
Glieesbery cycle	Rocks cycles
Wilson cycle	Opening and closing of ocean basin

26. Earth component	Disconformity
Crust and mantle	Mohorovicic
Lithosphere and asthenosphere	Repetti
Outer core and lower mantle	Gutenberg
Inner core and outer core	Lehman

29. Age of the rocks	Area
4.0-4.6 billion year	Craton
2.0-2.5 billion year	Shield area
185-200 million year	Ocean floor
60-65 million year	New Ocean floor

33. Planets	Characteristics
Mars	Inner planet
Earth	Inner planet
Neptune	Outer planet
Saturan	Outer planet

37. Features	Association
Sea-mount	Divergent plate motion
Benioff zone	Convergent plate motion.
Back-arc	Convergent plate motion
Fore-arc	Convergent plate motion

39. Features	Association
Deep sea trench	Convergent plate boundary
Mid-oceanic ridge	Divergent plate boundary
Oceanic ridge/Arcs	Translational plate boundary

47. Radius of the earth	Km
Polar radius	6356
Equitorial radius	6378
Average radius	6372

52. Earth interior division	Discontinuity
Crust and upper mantle	Mohorovicic
Lower to upper mantle	Repetti
Outer and inner core	Lehman
Outer core and lower mantle	Gutenberg

55. Component	Mean density (kg/m^3)
Crust	2600
Mantle	6400
Core	1300
Earth	5500

60. Planets	Distance from Sun (AU)	Diameter (Km)	Number of Satellite
Mercury	0.38	4800	0
Venus	0.72	12100	0
Earth	1.00	12750	1
Mars	1.52	6800	2
Asteroids Belt	2.8	-	-
Jupiter	5.2	142800	67
Saturn	9.5	120660	62
Uranus	19.2	51800	27
Neptune	30.0	49500	14

61. Elements in different zone	Percentage decreasing order
Crust	O > Si > Al > Fe > Na
Lithosphere	O > Si > Al > Fe > Na
Earth	Fe > Ni > O
Universe	H > He > Fe > O
Atmosphere	$N_2 > O > CO_2 > Ar$
Hydrosphere	$O > H_2 > Cl_2 > Na$

65. Interior of the Earth	Depth (Km)	Remarks
Core	6371	Inner and Outer core
Lower continental crust	20	Sima
Mantle	2900	Upper and Lower mantle
Upper continental crust	10	Sial

66. Group-1	Group-2	Name of the plates
Continent-oceanic lithosphere convergence	Andes	American and Nazca plate
Continent-continent collision	Himalayas	Indian and Eurasian plate
Continental rift system	East Africa	Arabian and African plate
Oceanic-oceanic lithosphere convergence	Japanese Island	Philippine and Pacific plate

67. Depth (km)	Discontinuity
35	Mohorovicic
800	Repetti
2900	Guttenberg
5200	Lehman

70. Interior of the Earth	Depth (km)	Mechanical component	Depth (km)
Crust	0-35	Lithosphere	100-200
Upper mantle	35-700	Asthenosphere	200-2900
Lower mantle	700-2900	Mesosphere	2900-6371
Outer core	2900-4900		
Inner core	4900-6371		

78. Parts of the earth	Volume (%)
The crust	1
The mantle	82
The core	17
The lithosphere	2

116. Interior of the Earth	Metal (weight%)
Crust	Aluminium
Mantle	Magnesium
Core	Iron

117. Richter Scale: The Richter scale (often shortened to Richter scale) is most common standard of measurement for earthquakes. It was invented in 1935 by Charles F. Richter of the California Institute of Technology as a mathematical device to compare the size of earthquakes. The Richter scale is used to rate the magnitude of an earthquake, that is the amount of energy released during an earthquake

Magnitude	Effect
2.5	Generally not felt, but recorded
4.5	Local damage
6.0	Can be destructive in populous region
7.0	Major earthquakes, serious damage
> 8.0	Great earthquakes

118. Mid-oceanic Ridge:

These are oceanic ridges where new oceanic lithosphere is created by upwelling mantle that melts, resulting in basaltic magmas which intrude and erupt at the oceanic ridge to create new oceanic lithosphere and crust. As new oceanic lithosphere is created, it is pushed aside in opposite directions. Thus, the age of the oceanic crust becomes progressively older in both directions away from the ridge.

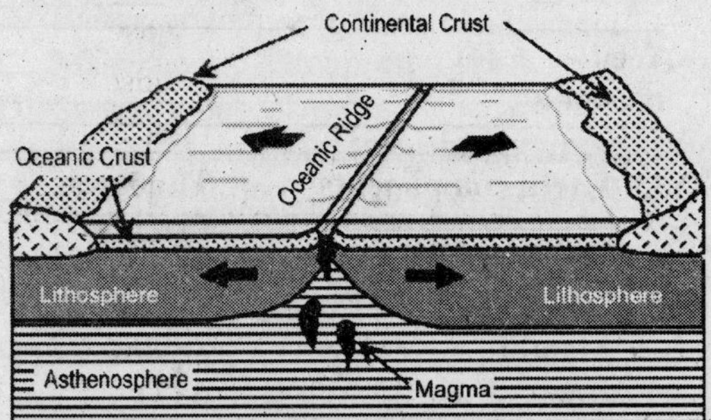

Diverging Plate Boundary
Oceanic Ridge - Spreading Center

Because oceanic lithosphere may get subducted, the age of the ocean basins is relatively young. The oldest oceanic crust occurs farthest away from a ridge. In the Atlantic Ocean, the oldest oceanic crust occurs next to the North American and African continents and is about 160 million years old (Jurassic). In the Pacific ocean, the oldest crust is also Jurassic in age, and occurs off the coast of Japan.

Because the oceanic ridges are areas of young crust, there is very little sediment accumulation on the ridges. Sediment thickness increases in both directions away of the ridge, and is thickest where the oceanic crust is the oldest.

122. The shadow zone is the area of the earth from angular distances of 104 to 140 degrees from a given earthquake that does not receive any direct P waves. The shadow zone results from S waves being stopped entirely by the liquid core and P waves being bent (refracted) by the liquid core.

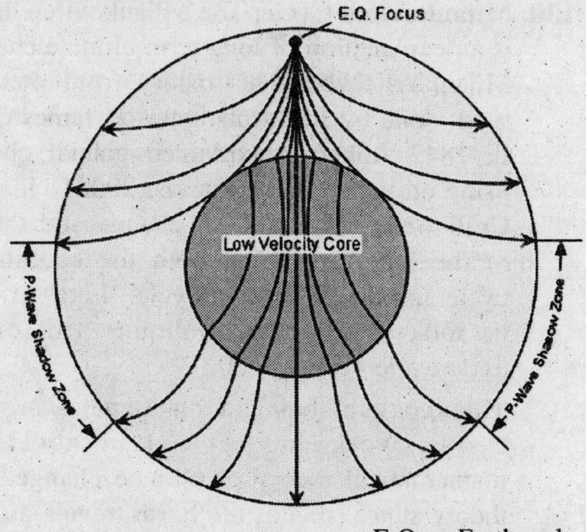

E.Q. = Earthquake

125. Plates Types	Number	Name (Example)
Major plates	7	Pacific, Indian, African, Eurasian, Australian, South America, North America
Minor plates	10	Nazca, Scotia, Arabian Caribbean plates, etc.
Micro-plates	–	Eurasian Plate (Adriatic plate, Iranian plate)

126. Earthquake Wave Propogation:

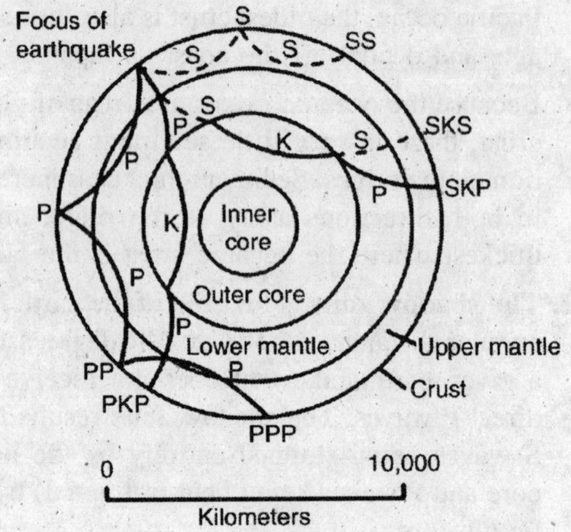

130. Milankovitch Cycle: The Milankovitch theory is an explanation of long term climate change. Milankovitch built his theory from previous work done by J.A. Adhemar and James Croll. In 1842 Adhemar explained glacial climate using only precession (Davis, 2002). In 1864 Croll wrote about orbital change and the ice of the Ice Ages using both the eccentricity cycle and the precession cycle. Later in 1875 he took into account obliquity (tilt of the axis) cycle (Davis, 2002).

Milankovitch, being a mathematician, took Croll's work and set out to develop a mathematical theory of climate change. His theory states "that as the Earth travels through space around the sun cyclical variations in three elements of Earth-Sun geometry combine to produce variations in the amount of solar energy that reaches Earth (Kaufman, 2002). These three elements that have cyclic variations are eccentricity, obliquity and precession.

Eccentricity is a term used to describe the shape of Earth's orbit around the Sun. The variation of Earth's orbit around the Sun ranges from an almost exact circle (eccentricity = 0.0005) to a slightly elongated shape (eccentricity = 0.0607) (Thomas, 2002). The time frame for the cycle is approximately 98,000 years (Davis, 2002). The impact of the variation is a change in the amount of solar energy from perihelion (around January 3) to aphelion (around July 4). Currently the Earth's eccentricity is 0.016 and there is about a 6.4 percent increase in insolation from July to January (Berger, 2001). Thomas (2002) states, "The eccentricity influences seasonal differences: when the Earth is closest to the Sun, it gets more solar radiation. If the occurs during the winter, the winter is less severe. If a hemisphere has its summer while closest to the Sun, summers are relatively warm."

132. Data on the Earth's Interior

	Thickness (km)	Density (g/cm^3) *Top*	*Bottom*	*Types of rock found*
Crust	30	2.2	–	Silicic rocks.
		–	2.9	Andesite, basalt at base.
Upper mantle	720	3.4	–	Peridotite, eclogite, olivine, spinel, garnet, pyroxene.
		–	4.4	Perovskite, oxides.
Lower mantle	2,171	4.4	–	Magnesium and
		–	5.6	silicon oxides.
Outer core	2,259	9.9	–	Iron+oxygen, sulphur,
		–	12.2	nickel alloy.
Inner core	1,221	12.8	–	Iron+oxygen, sulphur,
		–	13.1	nickel alloy.
Total thickness	**6,401**			

133. Earth components	**Thickness**
Oceanic crust	10
Continental crust	20
Mantle	2900
Core	6371

134. Atmosphere types	**Height (km)**	**Characteristics**
Exosphere	>150	Ion and gases
Ionosphere	150	Ion
Radio-wave absorbing layer	50-80	Radio wave

Mesosphere	80	Cold region
Ozone layer	30-50	Prevent ultraviolet layer
Stratosphere	50	Air rest, free of clouds
Troposphere	0-13	Densest layers

135. Four elements are responsible for the earth's internal heat:

1. U-238
2. U-235
3. U-232
4. K-41

136. Temperature distribution in Earth

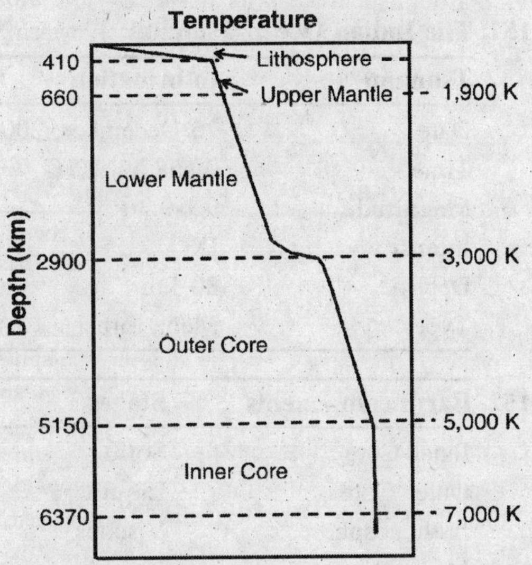

141. Mass extinction events:

There have been five mass extinction events throughout Earth's history:

1. The first great mass extinction event took place at the end of the Ordovician, when according to the fossil record, 60% of all genera of both terrestrial and marine life worldwide were exterminated.
2. 360 million years ago in the Late Devonian period, the environment that had clearly nurtured reefs for at least 13 million years turned hostile and the world plunged into the second mass extinction event.
3. The fossil record of the end Permian mass extinction reveals a staggering loss of life: perhaps 80-95% of all marine species went extinct. Reefs didn't reappear for about 10 million years, the greatest hiatus in reef building in all of Earth history.
4. The end Triassic mass extinction is estimated to have claimed about half of all marine invertebrates. Around 80% of all land quadrupeds also went extinct.
5. The end Cretaceous mass extinction 65 million years ago is famously associated with the demise of the dinosaurs. Virtually no large land animals survived. Plants were also greatly affected while tropical marine life was decimated. Global temperature was 6 to 14°C warmer than present with sea levels over 300 metres higher than current levels. At this time, the oceans flooded up to 40% of the continents.

Mass Extinction Events and Coral Reef Growth

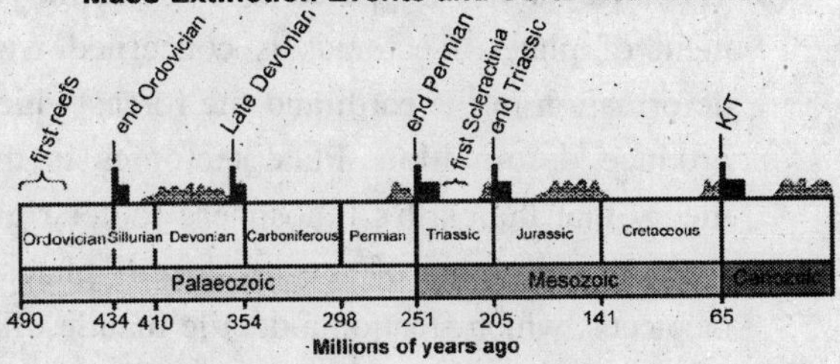

142. Earth's internal	Material
Crust	Silicon, Oxygen
Mantle	Magnesium, Iron
Core	Iron, Nickel

145. Triple junction: A triple junction is the point at which three plate boundaries meets. These boundaries can either be ridges (R), trench (T) or transform faults (F). Triple junctions were once defined simply as the meeting point of three spreading ridges. This is because this sort of triple junction is very common, usually with one of the ridges failing and the other two continuing to open a new ocean. The meeting of 4 or more plates is also theoretically possible but junctions will only exist instantaneously.

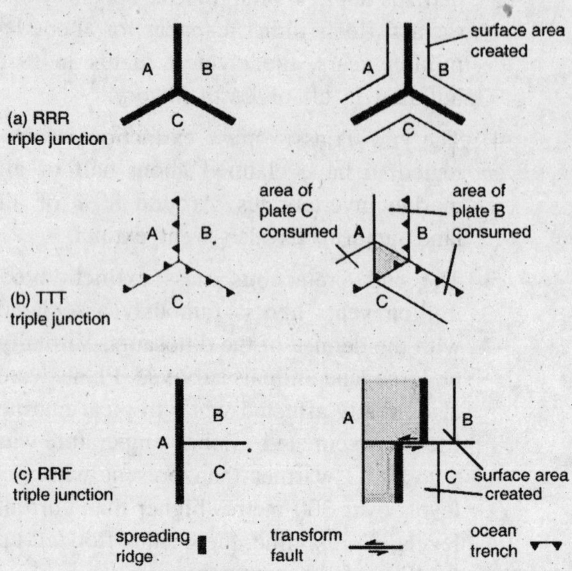

146. Wilson cycle: It is related to tectonic movement of plates, Tectonics is concerned with deformation in the earth and the forces which produce deformation. Plate tectonics is the theory that the earth's lithosphere (outer rigid shell) is composed of several dozen "plates", or pieces, which float on a ductile mantle, like slabs of ice on a pond. In plate tectonic theory earth history, at its simplest, is one of plates rifting into pieces diverging apart and new ocean basins being born, followed by motion reversal, convergence back together, plate collision, and mountain building. This cycle of opening and closing ocean basins is the Wilson Cycle.

Four events in given below in Wilson cycle:

(1) Rifting of continents by mantle diapirism

(2) Continental drift, seafloor spreading & formation of ocean basins

(3) Progressive closure of ocean basins by subduction of ocean lithosphere

(4) Continental collision and final closure of ocean basin

147. P and S-wave velocity in Earth's interior:

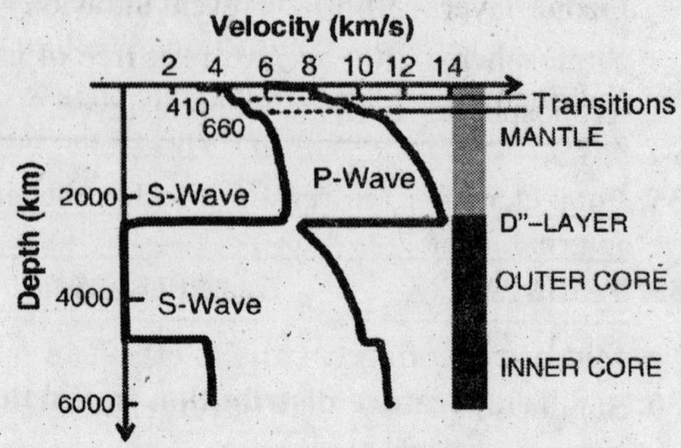

151. The Indian Ocean Tsunami:

Tsunami	Information
Date	26 December, 2004
Time	00:58:53 UTC
Magnitude	9.1-9.3
Intensity	IX
Depth	30 km
Types	Mega-thrust

152. Earth components	**Stages**
Inner Core	Solid
Outer Core	Liquid
Asthenosphere	Viscous
Mesosphere	Volatile
Lithosphere	Brittle

153. Continents	**Geological age (Ma)**
Tethys	180
Gondwana land	100
Pangea	250-280
Eurasia	15-30

154. Structure	**Definitions**
Rift valley	A linear lowland between mountain ranges created by the action of a fault is known as rift valley
Horst	When two parallel fault hade away fro another.

Thrust belt	Series of mountain foothills adjacent to orogenic belt due to contractional tectonics.
Subduction zone	The convergent plate show, one plate beneath the another plate.

155. Features	Tectonic setting
Mid-oceanic ridge	Divergent plate
Subduction Zone	Convergent plate
Transform fault	Horizontal movement of plate
Continental slope	The slope between continental shelf and deep ocean floor.

156. Plate setting	Characteristics
Collision boundary	One plate collide to another plate.
Subduction boundary	One plate goes down to another plate.
Suture zone	Locus of the collision between two plate
Contact zone	One plate meet to another

159. Layers	Depth
Sial	10 km
Sima	20
Crust	40
Lithosphere	100 - 150
Upper mantle	750
Lower mantle	2900
Outer core	4971
Inner core	6371

160. Plate setting	Features
Convergent	Trench, Back - Arc basin
Divergent	Volcanic island, MOR
Transform fault	San Andreas fault zone
Transcurrent fault	Fault zone

162. Interior Earth (Km)	Discontinuity
11	Conrad
35	Mohorovicic
950	Repetti
2900	Guttenberg
5500	Ingleman

164. Lakes	Location	Age	Composition	Origin
Lunar lake	Buldhana, Maharashtra	Pleistocene	Saline soda lake	Meteorites
Chilka lake	Puri, Odisha	Pleistocene	Brackish water	Marine
Kolleru lake	Eluru, Andhra Pradesh	Holocene	Fresh water	Fresh water
Pulicut lake	Border of A.P. and Tamilnadu	Holocene	Salty water	Brackish water

168. Witwatersrand (South Africa)	Remarks
Country	South Africa
Provenance	Vredefort meteorite impact
Age of the rocks	Archean
Rock type	Quartz, Conglomerate, Banded ironstone, Shales

173. Super continents	Age
Rodinia	750 Ma
Pannotia	600 - 540 Ma
Gondwana	100 Ma
Pangea	275 - 300 Ma
Ur supercontinent	3000 Ma

176. Different layers in the Earth:

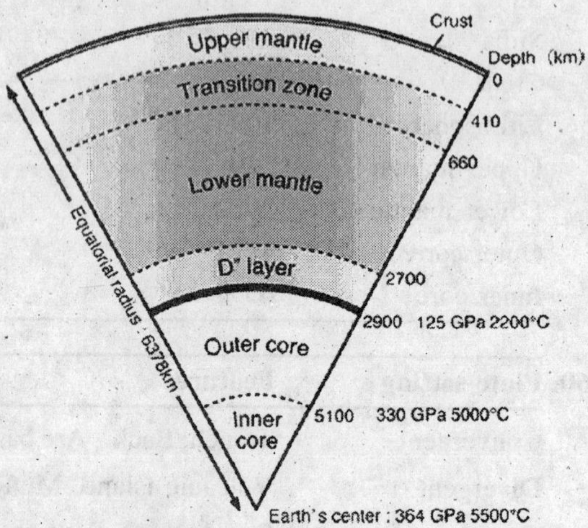

177. Different Discontinuity in the Earth:

Seismic—Discontinuities

Note that velocities increase gradually within layers (phase changes) but jump at discontinuities, which are thought mostly to be compositional changes.

LVZ ("low velocity zone") is uppermost part of asthenosphere, with low Vp due to less viscous nature (partial melts). Plates ride on the LVZ.

Discontinuities	Depth (km)	Composition
Lehman	5144	Fe solid against FeO, FeS fluid (inner/outer core boundary)
Gutenberg	2885	Fluid FeO, FeS against (Mg, Fe) silicates, velocity decrease, density increase (outer core-mantle boundary)
D"	2870	Thin, mixing of mantle and core material? (D" = D double-prime)
670 km	670	Worldwide, no earthquakes deeper, debates over whether a composition, phase, or viscosity change
400 km	400	Worldwide
LVZ	50-200	Regionally variable depth
Moho	4-55	(Abbreviation of Mo-ho-RHO-vi-chik)–sharp compositional change
Conrad	? 5-30	Mafic to felsic crust, often absent

180. Chemistry of the Sea water:

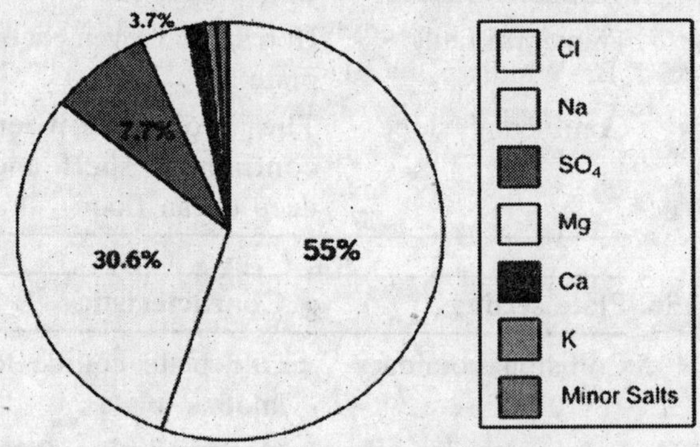

Table: Ionic Constituents of Seawater Present in Concentrations Greater than 0.001 g/kg (1 ppm)

Ionic Constituent	g/kg Seawater	Concentration (M)
Chloride, Cl^-	19.35	0.55
Sodium, Na^+	10.76	0.47
Sulphate, SO_4^{2-}	2.71	0.028
Magnesium, Mg^{2+}	1.29	0.054
Calcium, Ca^{2+}	0.412	0.010
Potassium, K^+	0.40	0.010
Carbon dioxide*	0.106	2.3×10^{-3}
Bromide, Br^-	0.067	8.3×10^{-4}
Boric acid, H_3BO_3	0.027	4.3×10^{-4}
Strontium, Sr^{2+}	0.0079	9.1×10^{-5}
Fluoride, F^-	0.0013	7.0×10^{-5}

* CO_2 is present in seawater as HCO_3^- and CO_3^{2-}

GEOMORPHOLOGY

1. Size of a kettle lake depends upon:
   A. Size of the sediment carried by the glacier
   B. Width of the glacier valley
   C. Height of the valley through which the glacier moves
   D. Size of the buried ice block within moraines *(IIT-JAM 2015)*

2. Drainage density of a basin is defined as:
   A. Sum of the number of streams of all orders divided by the area of the basin
   B. Total length of streams of all orders divided by the area of the basin
   C. Total number of the highest order streams divided by the total number of the next lower order streams
   D. Length of the largest stream divided by the area of the basin *(IIT-JAM 2015)*

3. The wind system in the equatorial areas is known as:
   A. Westerlies B. Trades
   C. Doldrums D. Monsoon *(DU-2015)*

4. The monsoon has well developed cycle in:
   A. South and South East Asia
   B. North Australia
   C. Africa
   D. East United States *(DU-2015)*

5. Yazoo stream flows through:
   A. Gorge B. Divide
   C. Backs wamp D. Water gap *(AMU-2015)*

6. The curved body of water formed when a meandering river reshapes and cuts off one of its bends is called:
   A. Essing pond B. Oxbow lake
   C. Swensen pool D. Groovy pool *(AMU-2015)*

7. Barchan are:
   A. Deflation basins
   B. Oases
   C. Crescent-shaped dunes
   D. Sand dune *(AMU-2015)*

8. Exfoliation is related to:
   A. Metamorphism
   B. Chemical weathering
   C. Carbonation
   D. None of the above *(AMU-2015)*

9. Morains are:
   A. Alluvial deposit B. Fluvial deposit
   C. Glacial deposit D. Lacustrine deposit *(AMU-2015)*

10. Elephant-skin weathering is generally found in:
    A. Shale B. Sandstone
    C. Limestone D. Conglomerate *(AMU-2015)*

11. Roche Moutonnees are formed by:
    A. Glacial B. Wind
    C. River D. Ocean wave *(AMU-2015)*

12. The geometry of sand dunes is strongly influenced by:
    A. The strength of the wind
    B. The consistency of wind direction
    C. The abundance of sand
    D. All of the above *(DU-2015)*

13. A mushroom-shaped landform consisting of a column of less resistant rock supporting a

broader extent of wind-resistant rock is termed as:

A. Blowout B. Ventifact
C. Wadies D. Yardang

(DU-2015)

14. Mars has polar ice caps that differ from the Earth's Antarctic Ice cap in that:

A. The Martian caps consist of a combination of water ice and frozen carbon dioxide (dry ice)
B. The Martian caps have a spiral fracture pattern not found in Antarctica
C. Martian caps are ablated by sublimation only; there is currently no liquid water on Mars
D. All of the above are correct

(DU-2015)

15. The first scientist who theorized the past presence of glaciers in Europe was:

A. Louis Agassiz B. Charles Darwin
C. Walter Alvarez D. Isaac Newton

(DU-2015)

16. Areas on the southern margins of the continental glaciers of the Northern Hemisphere were much .... during Plio-Pleistocene glaciations than they are today, as suggested by evidence of large .... during this time.

A. Warmer; tropical rainforests
B. Wetter; tropical rainforests
C. Drier; deserts
D. Wetter; pluvial lakes *(DU-2015)*

17. By increasing the albedo of Earth, global ice sheets produce .... conditions that are to their further advance, thus providing an example of .... feedback.

A. Detrimental; positive
B. Detrimental; negative
C. Favorable; positive
D. Favorable; negative *(DU-2015)*

18. A remnant ridge separating two bowl-shaped depressions formed by a mountain glacier is termed as:

A. Arete B. Cirque
C. Horn D. Tarn *(DU-2015)*

19. Glacial ice exhibits .... behaviour near the top, but .... behaviour beneath a depth of 60 m.

A. Brittle, ductile B. Ductile, brittle
C. Solid, liquid D. Plastic, elastic

(DU-2015)

20. Rapid chemical weathering:

A. Does not affect the rate of physical weathering in the same rock
B. Tends to reduce the rate of physical weathering in the same rock
C. Tends to increase the rate of physical weathering in the same rock
D. Does not happen in rocks those suffer physical weathering

(DU-2015)

21. Caliche is most commonly found in which type of environments?

A. Temperate forest
B. Tropical rain forests
C. Deserts
D. Grasslands *(DU-2015)*

22. The longitudinal profile of the stream is typically .... The lowest elevation below which a stream cannot cut is called its .... and occurs at the stream's ....

A. Convex-upward/water table/source
B. Convex-up/base level/mouth
C. Concave-up/base level/mouth
D. Concave-up/tributary/source *(DU-2015)*

23. The chemical reaction: $2Fe_2SiO_4 + 4H_2O + O_2 = 2Fe_2O_3 + 2H_4SiO_4$, is an example of:

A. Dissolution B. Hydration
C. Oxidation D. Reduction

(DU-2015)

24. Which of the following sequences represents increasing slope?

A. Shore-Continental shelf-continental slope
B. Continental shelf-continental slope-shore
C. Continental slope-continental shelf-shore
D. Continental shelf- shore-continental shelf

(DU-2015)

25. Earthquakes with deep foci (–700 km) are common in:

A. Mid-oceanic ridges
B. Island arcs
C. Himalayan mountain belt
D. Continental rifts *(DU-2015)*

26. Corals prefer a temperature range of:
A. 30-35°C B. 15-20°C
C. 35-40°C D. 23-29°C
(DU-2015)

27. In what type of country rock do sink-holes develop?
A. Phyllite B. Gneiss
C. Limestone D. Sandstone
(IIT-JAM 2014)

28. Match the landforms given in Group-1 with the causative processes given in Group-2.

Group-1	Group-2
P. Strath	1. Glacial deposits
Q. Drumlins	2. Glacial erosion
R. Cirque	3. Fluvial deposition
S. Point bar	4. Fluvial erosion

A. P-4, Q-1, R-2, S-3
B. P-3, Q-1, R-2, S-4
C. P-4, Q-2, R-1, S-3
D. P-1, Q-2, R-3, S-4 *(IIT-JAM 2014)*

29. Most active chemical weathering takes place:
A. At the water table
B. Below the water table
C. On the surface
D. Zone of aeration above the water table
(BHU-2014)

30 Parallel drainage patterns are found on:
A. Jointed country
B. Steep slopes with little vegetation cover
C. Granite-gneiss dome
D. Grid like joint system *(BHU-2014)*

31. When a stream or river is diverted from its own bed, and flows instead down the bed of a neighbouring stream, the phenomenon is known as:
A. Stream piracy
B. Headwater management
C. River shifting
D. Superimposed drainage *(BHU-2014)*

32. What is a 'nick point' in a river?
A. A marked change of slope along flow direction of the channel
B. A sharp bend in the river
C. A sharp meander
D. Point where the river changes its direction of flow suddenly *(BHU-2014)*

33. Which of the following is the most characteristics glacial landform:
A. Narrow 'V'-shaped valleys
B. Wide 'V'-shaped valleys
C. Hanging valleys
D. None of the above *(BHU-2014)*

34. Features developed by accumulation of annual thin layers in glacial deposits are known as:
A. Glacial lamination
B. Varves
C. Glacial beds
D. Annular rings *(BHU-2014)*

35. Frosted 'millet seed' sand grains are found in deposits formed by:
A. Rivers B. Lakes
C. Winds D. Glaciers
(BHU-2014)

36. The sand dunes with their long axes transverse to the wind direction are known as:
A. Sand sheet
B. Barchans
C. Parallel dunes
D. Longitudinal dunes *(BHU-2014)*

37. The marine deposits which form between the extreme levels of high and low tides are known as:
A. Shallow water deposits
B. Littoral deposits
C. Hadal deposits
D. Abyssal deposits *(BHU-2014)*

38. Which one of the following is a glacio-fluvial deposit?
A. Esker B. Till
C. Moraine D. Drumlin
(IIT-JAM 2013)

39. Which of the following represents tree-like pattern of flow of river?
A. Trellis B. Dendritic
C. Radial D. Rectangular
(BHU-2013)

40. The transportation through lifts and falls of materials is known as:
A. Saltation B. Traction
C. Creeping D. Attrition
(BHU-2013)

41. The highest of the Indian waterfalls is:
A. Gokak falls of the Ghatprava river
B. Dhuandhar falls of the Narmada river
C. Gerosoppa falls of Swaravati river
D. Shivasamundram falls of the Cauvery river
(BHU-2013)

42. 'Oasis' is formed by:
A. River action B. Glacier action
C. Wind action D. Sea wave action
(BHU-2013)

43. Pick the odd one out:
A. Zeugen
B. Yardang
C. Mushroom rock
D. Moraine *(BHU-2013)*

44. Meandering in river occurs during its:
A. Youth stage B. Old stage
C. Mature stage D. Rejuvenation
(BHU-2013)

45. The tendency of water to pass directly from solid state to vapour state is known as:
A. Sublimation B. Oxidation
C. Dissolution D. Dissociation
(PU-2013)

46. Which one of the following drainage patterns is typical of a doubly plunging antiformal-terrain?
A. Dendritic B. Trellis
C. Rectangular D. Radial
(IIT-JAM 2012)

47. Blind valley is found in:
A. Aeolian topography
B. Krast topography
C. Glacial topography
D. Fluvial topography *(BHU-2013)*

48. What is name of the location where volcano occurs away from the plate boundary?
A. Transform fault B. Grabbos
C. Hotspot D. Sea mount
(BHU-2013)

49. The Lagoon is:
A. Protected land behind the reef
B. Protected land ahead of the reef
C. Narrow zone of flourishing coral colonies
D. Ring shape reef unconnected with land
(BHU-2013)

50. Bhangar is the name of the:
A. Younger alluvium
B. Delataic deposit
C. Fans of river
D. Older alluvium *(BHU-2013)*

51. Mushroom, Inselberg and Zeugen are produced by:
A. Wind erosion B. Glacial erosion
C. River erosion D. Sea erosion
(BHU-2012)

52. Paternoster lake is the characteristic of:
A. Aeolian action B. Glacial action
C. Fluvial action D. Stream terraces
(BHU-2012)

53. Alluvial fans are significant:
A. Fluviallandforms
B. Glacial landforms
C. Aeolian landforms
D. Karst topography *(BHU-2012)*

54. Long, narrow and sinuous ridges of sands and gravels situated in the middle of ground moraines are:
A. Drumlins B. Crag and tail
C. Eskers D. Kames
(BHU-2012)

55 Which of the following is formed by wind erosion?
A. Yardang B. Gorges
C. Loess D. Butte *(BHU-2012)*

56. Headlands are produced by:
A. Groundwater erosion
B. River erosion

C. Marine erosion
D. Wind erosion *(BHU-2012)*

57. Stalagmites are characteristic feature of:
A. River B. Glacier
C. Groundwater D. Wind
(BHU-2012)

58. Which one among the following is a feature produced by wind?
A. Drumlins B. Loess
C. Delta D. Canyons
(BHU-2012)

59. Match the landforms in Group I with causative processes in Group II.

Group I	Group II
P. Yardang	1. Aeolian
Q. Drumlin	2. Coastal
R. Doline	3. Dissolution
S. Chenier	4. Glacial

A. P-1, Q-2, R-3, S-4
B. P-1, Q-4, R-3, S-2
C. P-4, Q-3, R-1, S-2
D. P-4, Q-3, R-2, S-1 *(IIT-JAM 2012)*

60. Glaciers are downward moving bodies of:
A. Only snow
B. Mostly ice and some snow at the top
C. Mostly snow and some ice at the top
D. Snow and ice in alternate layers
(IIT-JAM 2011)

61. The surface slope of shield volcano is gentle, but strato-volcano is steep-sided. This is due to variation in:
A. Environment of eruption
B. Duration of eruption
C. Viscosity of magma
D. Position with respect to latitude
(IIT-JAM 2011)

62. Which of the following landforms is ***NOT*** the result of strike-slip faulting?
A. Stream deflection
B. River terrace
C. Pressure ridge
D. Sag pond *(IIT-JAM 2011)*

63. Hanging valley is developed when rate of glacial erosion:
A. Is greater in the main valley as compared to the side valley
B. Is greater in the side valley as compared to the main valley
C. Is the same in both the main valley and the side valley
D. Has no effect on either the main valley or the side valley
(IIT-JAM 2011)

64. Match the geomorphic features in Group I with its corresponding details in Group II.

Group I	Group II
P. Tombol	1. Lowering of sea level
Q. Bajada	2. Sand ridge connecting island
R. Erratic	3. Coalescence of alluvial fans
S. Rejuvenated valley	4. Rock fragment carried by glacier

A. P-2, Q-4, R-1, S-3
B. P-2, Q-4, R-3, S-1
C. P-2, Q-3, R-4, S-1
D. P-4, Q-3, R-2, S-1 *(IIT-JAM 2011)*

65. Stalactite is what kind of feature?
A. Depositional feature
B. Replacement feature
C. Erosional feature
D. None of the above *(BHU-2011)*

66. Nick point is associated with:
A. Lake B. River
C. Sea D. Delta *(BHU-2014)*

67. Ventifacts are formed as a result of:
A. Attrition B. Deflation
C. Ablation D. Abrasion
(BHU-2014)

68. Varve deposits belong to:
A. Glacial deposits
B. Deltaic deposits
C. Fluvial deposits
D. Marine deposits *(BHU-2011)*

69. 'Rind of fire' the largest number of active volcanoes on earth is found in:
A. Atlantic Ocean
B. Pacific Ocean

C. Mrican rift valleys
D. The Mid-oceanic ridges *(BHU-2011)*

70. The process responsible for the lowering of mountains and derivation of sediments is:
A. Denudation B. Weathering
C. Erosion D. Decomposition
(BHU-2011)

71. Due to weathering of feldspar which of the following residual products form?
A. Limonite B. Haematite
C. Clay minerals D. Quartz grains
(BHU-2011)

72. The crescent shaped dune in which the convex side of which faces wind direction is:
A. Parabolic dune
B. Transverse dune
C. Longitudinal dune
D. Barchan dune *(BHU-2011)*

73. Antecedent streams are those which:
A. Maintain their original course across area of uplift
B. Follow the slope of the initial land surface
C. Develop on the sloping sides of stream valley
D. Flowing on exposed older rocks
(BHU-2011)

74. Loess is associated to:
A. Fluvial B. Eolian
C. Lacustrine D. Glacial
(BHU-2011)

75. In glacial environment 'eskars' are:
A. Small, smooth and elliptical hills
B. Basin like depression
C. Long winding ridges
D. Ridge or layer of tills *(BHU-2011)*

76. 'Flower structure' is associated with:
A. Transform fault boundaries
B. Divergent boundaries
C. Convergent boundaries
D. Mid-oceanic ridges *(BHU-2011)*

77. The consolidated larger fragments such as bombs or blocks of older rocks near volcanic cone are called:
A. Conglomerate B. Breccia
C. Agglomerate D. Scoriae
(BHU-2011)

78. 'Laterites' are formed as a result of:
A. Weathering
B. Chemical precipitation
C. Biogenic activity
D. Sedimentation
(BHU-2011)

79. A volcano built mainly of tephra fall deposits located immediately around the vent:
A. Cinder cone B. Strato volcano
C. Shield volcano D. Lava dome
(BHU-2011)

80. Wind-facetted pebbles are known as:
A. Lag-gravel B. Blowouts
C. Yardangs D. Ventifacts
(IIT-JAM 2010)

81. Circular reef enclosing shallow body of water are called:
A. Barrier reef B. Fringing reef
C. Spits D. Atolls
(IIT-JAM 2010)

82. The term caldera is associated with:
A. River B. Volcano
C. Glacier D. Wind *(BHU-2010)*

83. A V-shaped valley is the characteristic of:
A. Glacier B. Ocean
C. River D. Lake *(BHU-2010)*

84. Which one of the following is **not** a glacial feature?
A. Yardang B. Bergschrund
C. Arete D. Crevasse
(BHU-2010)

85. Seifs are found in which environment?
A. Glacial B. Fluvial
C. Lacustrine D. Desert *(BHU-2010)*

86. Rejuvenation in an area is suggested by which of the following?
A. Waterfalls
B. Paired river terraces
C. V-shaped valley
D. All of the above *(BHU-2010)*

87. Thermohaline circulation results due to difference in:
A. Density B. Temperature
C. Salinity D. All of these
(BHU-2010)

88. Which of the following marks the farthest advance of a glacier?
A. Terminal moraine
B. Lateral moraine
C. Medial moraine
D. Ground moraine *(BHU-2010)*

89 Which one of the following is a type of volcanic eruption?
A. Caledonian B. Strombolian
C. Venatian D. Bombolian
(BHU-2010)

90. Ventifacts are found in:
A. Deep Oceans
B. Glaciers
C. Volcanic eruptions
D. Deserts *(BHU-2010)*

91. A lagoonal circular coral reef is called:
A. Blue lagoon
B. Atoll
C. Barrier type coral reef
D. Marginal reef *(BHU-2010)*

92. The Richter scale measures:
A. Magnitude of earthquake
B. Distance of location of earthquake
C. Intensity of earthquake
D. Depth of epicentre of earthquake
(BHU-2010)

93. Tsunami is a:
A. Typhoon
B. Seismic sea wave
C. Cyclone
D. None of the above *(BHU-2010)*

94. The term 'piracy' is associated with which of the following?
A. Ocean B. Wind
C. River D. Lake *(BHU-2010)*

95. Sediment transport in which particles are moved forward in a series of short leaps and bounds is called:
A. Traction B. Suspension
C. Saltation D. None of these
(BHU-2010)

96. The raised area adjacent to a stream channel that is produced by fragment flooding and overbank deposition is:
A. Crevasse B. Point bar
C. Levee D. Oxbow lake
(IIT-JAM 2009)

97. Which of the following is a product of Aeolian processes?
A. Point bar B. Tillite
C. Calcrete D. Loess
(IIT-JAM 2009)

98. Physical weathering is not characteristics of:
A. Polar regions B. Hot desert
C. Cold desert D. Semi-arid regions
(BHU-2009)

99. Sinkhole, doline and dripstone are significant of:
A. Aeolian landform
B. Glacial landform
C. Fluvial landform
D. Karst topography *(BHU-2009)*

100. Flat-topped hills or small mountains formed by stream action are called:
A. Mesas B. Buttes
C. Cuestas D. Stream terraces
(BHU-2009)

101. An erosional ridge having a gentle slope on one side and a steep slope on the other side is:
A. Escarpment B. Gloup
C. Seifs D. Polje
(BHU-2009)

102. Which of the following rivers does not form a delta?
A. Krishna B. Godavari
C. Ganga D. Narmada
(BHU-2009)

103. Natural levee is an example of:
A. Flood basin deposit
B. Point-bar deposit
C. Flood plain deposit
D. Channel fill deposit *(BHU-2009, PU-2011)*

104. Which of the following is formed by glacial erosion?
A. Yardang B. Potholes
C. Cuesta D. Arete
(BHU-2009)

105. Richter scale is used for measuring the:
A. Relative humidity of atmosphere
B. Intensity of earth's tremor
C. Electric conductivity of water
D. Speed of wind *(BHU-2009)*

106. The mature stage feature of the river is:
A. Alluvial fan B. V-shaped valley
C. Meanders D. Ox-bow lake
(BHU-2009)

107. Yardang is the erosional feature produced by:
A. River
B. Glacier
C. Underground water
D. Wind *(BHU-2009)*

108. Which one among the following is a depositional feature produced by underground water?
A. Barchans B. Loess
C. Delta D. Stalagmites
(BHU-2009)

109. Part of the tidal flats occurring near the high water line is known as:
A. Mixedflat B. Sandflat
C. Mudflat D. Tidal flat
(BHU-2009)

110. Dripstone found rising up from the floor of a limestone cave is known as:
A. Stalactite B. Stalagmite
C. Uvala D. Poljes
(IIT-JAM 2008)

111. Which one of the following drainage pattern is observed in massive granitic terrain?
A. Trellis B. Concordant
C. Dendritic D. Annular
(IIT-JAM 2008)

112. A graded stream is characterized by:
A. Increase in erosion relative to deposition
B. Increase in deposition relative to erosion
C. Equilibrium between erosion and deposition
D. Formation of graded bedding
(IIT-JAM 2007)

113. Barchans are crescent-shaped dunes in which the ends of the crescent:
A. Point in the upwind direction
B. Are at right angles to the wind direction
C. Have no relation to the wind direction
D. Point in the downwind direction
(IIT-JAM 2007)

114. Match the following landforms (Group 1) with the corresponding geological agents (Group 2).

Group-1	Group-2
P. Point bar	1. Sea
Q. Cirque	2. River
R. Yardang	3. Glacier
S. Berm	4. Wind

A. P-2, Q-3, R-4, S-1
B. P-3, Q-1, R-4, S-2
C. P-2, Q-1, R-3, S-4
D. P-3, Q-2, R-1, S-4 *(IIT-JAM 2007)*

115. A geomorphic feature produced at divergent plate boundary is:
A. Folded mountain belt
B. Guyot
C. Island arc
D. Oceanic ridge
(IIT-JAM 2007)

116. Match the following:

Group I	Group II
P. braid bars	1. wind
Q. moraines	2. riverine
R. barchans	3. marine
S. atoll	4. glacial

A. P – 1 Q – 2 R – 3 S – 4
B. P – 2 Q – 4 R – 1 S – 3
C. P – 4 Q – 2 R – 3 S – 1
D. P – 3 Q – 4 R – 1 S – 2
(IIT-JAM 2006)

117. Mesas are common in the:
A. Eastern Ghats granulite belt
B. Himalayan mountain range
C. Deccan Traps
D. Dharwar greenstone belt
(IIT-JAM 2006)

118. Varves are typical products of:
A. Flood planes
B. Periglacial lakes
C. Coastal lagoons
D. Coal basins *(IIT-JAM 2005)*

119. A river that maintains its course by valley deepening during uplift is termed:
A. Subsequent B. Insequent
C. Consequent D. Antecedent
(IIT-JAM 2005)

120. Among the following the glacial landform is:
A. Hanging valley B. Bajada
C. Monadnock D. Blind valley
(IIT-JAM 2005)

121. Tilloids are of:
A. Glacial origin
B. Non-glacial origin
C. Fluvial origin
D. None of these *(BHU-2014)*

122. Which of the following climate regimes will have the deepest weathering profile?
A. Tundra regions
B. Steppes
C. Desserts
D. Tropical rainforests *(DU-2014)*

123. During weathering of silicate minerals, incorporation of cations from mineral into organic compounds is related to:
A. Carbonation B. Hydrolysis
C. Chelation D. Hydration
(DU-2014)

124. Which of the following volcanic landform is the largest in size?
A. Craters B. Tephra cones
C. Plug domes D. Strato volcanoes
(DU-2014)

125. The deposition of suspended and dissolved material in a soil profile is referred as:
A. Eluviation B. Illuviation
C. Humus D. Leaching
(DU-2014)

126. Laterites are:
A. Weathering product rich in Pb and Mn hydroxides
B. Weathering products rich in Fe and Al hydroxides
C. Cooling product of magma rich in Fe and Al sulphides
D. Result from high pressure metamorphism of granite
(DU-2014)

127. Which of the following is an erosional landform made by glaciers?
A. Moraine B. Eskers
C. Cirque D. Kame
(DU-2014)

128. Hydraulic action, solution, and abrasion are all examples of:
A. Stream erosion B. Stream deposition
C. Transportation D. Stream discharge
(DU-2014)

129. If CL and VL represent channel length and valley length of a stream then the sinuosity of the channel is expressed by:
A. CL +VL B. CL × VL
C. CL / VL D. VL – CL
(DU-2014)

130. The relationship between annual frequency (N) and magnitude (Ms) of earthquake is expressed as:
A. $\log N = a + bMs$
B. $\log N = a - bMs$
C. $N = a \times Ms$
D. $N = a \times bMs$ *(DU-2014)*

131. Mechanical erosion cannot takes place in the following manner:
A. Abrasion B. Attrition
C. Cavitation D. Dissolving action
(BHU-2015)

132. An isolated table-land area with steep sides called as:
A. Knick point B. Escarpment
C. Mesa D. Peneplains
(BHU-2015)

133. Which one is **not** a river pattern?
A. Antecedent B. Consequent
C. Insequent D. Trellis
(BHU-2015)

134. Which of the following term is **not** related to wind erosion?
A. Deflection B. Traction
C. Abrasion D. Attrition
(BHU-2015)

135. Blow-out is formed due to:
A. Wind B. River
C. Glacier D. Impact
(BHU-2015)

136. Fiords are formed by action of:
A. Wind B. River
C. Glacier D. Lake
(BHU-2015)

137. Which of the following is not erosional process of continental glaciers?
A. Plucking B. Avalanching
C. Rasping D. None of the above
(BHU-2015)

138. Coverns are related to:
A. Underground water
B. River
C. Glacier
D. Wind *(BHU-2015)*

139. Blind valley is formed due to:
A. Erosional B. Deposition
C. Steep slope D. Surface runoff
(BHU-2015)

140. In the following, which combination is wrong?
A. Knick point, escarpment, Mesa
B. Plucking, avalanching, rasping
C. Hamada, Yardang, sand dunes
D. Pedestal, cirques, arête
(BHU-2015)

141. Which of these environments can produce cross beds?
A. Sand dunes B. River Deltas
C. Alluvial fans D. All the above
(PU-2010)

142. In which of the following do foreset beds occur?
A. Alluvial fans B. Deltas
C. Point bars D. Natural levees
(PU-2010)

143. A drainage pattern in which streams flow in and out of lakes with irregular flow directions is:
A. Radial B. Rectangular
C. Longitudinal D. Deranged
(PU-2010)

144. Erosional remnants of floodplains that are higher than the current level of a stream are:
A. Incised meanders
B. Cut banks
C. Natural bridges
D. Stream terraces *(PU-2010)*

145. The lake beds in many deserts are:
A. Playas B. Pediments
C. Bajadas D. Inselbergs
(PU-2010)

146. In uniform flat rocks, drainage patterns tend to be:
A. Dendritic B. Radial
C. Trellis D. Braided
(PU-2010)

147. Mass wasting refers to material moved primarily by:
A. Wind B. Running water
C. Ice D. Gravity
(PU-2010)

148. The process whereby hydrogen and hydroxyl ions of water replace ions in minerals is:
A. Hydrolysis B. Oxidation
C. Carbonization D. Laterization
(PU-2010)

149. Why clay doesn't weather?
A. The grains are too large
B. It is chemically stable at the surface
C. It is doesn't occur in areas prone to weathering
D. It has too many ions *(PU-2010)*

150. Coral reefs are generally found in:
A. Polar region
B. Sub-polar region
C. Tropical region
D. All region *(PU-2010)*

151. Exogenic geomorphic processes are primarily driven by:
A. earth's internal heat
B. only gravity
C. only solar energy
D. solar energy and gravity *(PU-2013)*

152. Which of the following minerals is the most susceptible to weathering in humid climate?
A. Quartz B. K-feldspar
C. Biotite D. Olivine *(PU-2013)*

153. 'Glacial clay' is dominantly:
A. Kaolinite B. Illtite
C. Feldspar D. Quartz *(PU-2013)*

154. Which of the following is an erosional landform?
A. Barrier island B. Barchan
C. Cirque D. Floodplain *(PU-2013)*

155. Hawaii is an example of:
A. Shield volcano B. Strata volcano
C. Tephra cone D. Pyroclastic dome *(PU-2013)*

156. Shutter ridges and Sag ponds are associated with:
A. Strike slip faults B. Thrust faults
C. Grabens D. Volcanoes *(PU-2013)*

157. The movement of water from the ground surface into the soil is termed as:
A. Percolation B. Infiltration
C. Seepage D. Gravity Flow *(PU-2013)*

158. Flat topped sea mounts are termed as:
A. Guyots B. Mesa
C. Inselberg D. Monodnock *(PU-2013)*

159. The drainage pattern which signifies an area lacking structural control is:
A. radial B. rectangular
C. dendritic D. trellis *(PU-2013)*

160. Subduction is likely to happen in which of the following cases?
A. A oceanic plate moving towards a continental plate
B. A continental plate moving towards another continental plate
C. A continental plate moving past another continental plate
D. Oceanic plate moving away from another oceanic plate *(PU-2012)*

161. Most of the earthquakes originate at:
A. Plate margins
B. Plate interiors
C. Volcanic centers
D. Oceanic-continent junctions *(PU-2012)*

162. Climate change is a bigger threat to coastal countries like Bangladesh because:
A. More temperature would increase in lower latitudes
B. More carbon dioxide is emitted by coastal countries
C. Sea level rise due to melting of continental glaciers would drown the land in coastal countries
D. Fishes would die due to increase in temperature *(PU-2012)*

163. Plate tectonic like processes are not observed during recent time on any planet other than earth because:
A. No other planet has lithosphere similar to the Earth
B. No other planet's lithosphere is broken in fragments
C. No other planet has internal heat energy left to drive such processes
D. No other planet has water *(PU-2012)*

164. Earthquakes with focus deeper than 650 km are not observed because:
A. No brittle material is present deeper than that depth

B. Seismic waves travelling from deeper parts do not reach on surface
C. Liquid mantle does not allow S-waves to propagate
D. Instruments are not capable to record earthquakes from deeper parts
(PU-2012)

165. What is an irregular suture-like boundary developed in limestone, formed by pressure-controlled solution followed by immediate local redeposition called?
A. Stylolite
B. Ammoniatic suture
C. Secondary fracture
D. None of the above *(PU-2012)*

166. Dunes tend to form:
A. Parallel to the prevailing winds
B. Perpendicular to the prevailing winds
C. Either or both of the above at times
D. They have no relation to wind direction
(PU-2012)

167. The gently sloping, shallowly-submerge surface extending from the shoreline toward the deep ocean is termed:
A. Continental shelf
B. Submarine canyon
C. Continental slope
D. Ocean basin . *(PU-2012)*

168. The core of the earth is composed primarily of:
A. Iron and sulphur
B. Iron and nickel
C. Nickel and cobalt
D. Silicon and oxygen *(PU-2012)*

169. The drainage pattern which signifies an area lacking structural control is:
A. Radial B. Rectangular
C. Dendritic D. Trellis
(PU-2012)

170. Mechanical wear by rivers, wind etc. are called as:
A. Deflation B. Saltation
C. Corrosion D. Solifluction
(PU-2012)

171. Dolostone is formed by addition of .... to limestone.
A. Calcium B. Iron
C. Sodium D. Magnesium
(PU-2012)

172. Which one of the following is **not** a requirement for coral reef growth?
A. Warm water
B. Abundant sunlight
C. Shallow water
D. Abundant amount of suspended sediments
(PU-2012)

173. According to Plate Tectonics theory, most active volcanoes occur:
A. In continents
B. In large tectonic plates
C. Along plate boundaries
D. In small tectonic plates *(PU-2011)*

174. Hot desert lie between the latitude of:
A. 5° and 15° B. 20° and 45°
C. 30° and 45° D. 50° and 65°
(PU-2011)

175. In the case of wind erosion, the base level of erosion will be:
A. Mean sea level B. Sea level
C. Water level D. Lake level
(PU-2011)

176. Flat topped sea mounts are termed as:
A. Guyots B. Mesa
C. Inselberg D. Monadnock
(PU-2011)

177. The drainage pattern which signifies an area lacking structural control is:
A. Radial B. Rectangular
C. Dendritic D. Trellis
(PU-2011)

178. Volcanism is generally least along which of the following tectonic settings?
A. Subduction zone
B. Divergent plate boundary
C. Above plume head
D. Transform faults *(PU-2011)*

179. A statue made up of limestone is likely to weather faster in which of the following condition?
A. Hot humid tropical climate
B. Cold arid climate
C. Arid climate at high altitude
D. Hot desert (PU-2011)

180. Flate topped hill or small mountains formed by stream action are called:
A. Mesas B. Buttes
C. Cuestas D. River terrace
(PU-2011)

181. Streams that flow in the same direction as the original consequent but at a lower topographic level are known as:
A. Obsequent B. Subsequent
C. Insequent D. Resequent
(PU-2011)

182. Loess is an un-stratified, well-sorted Aeolian deposit composed of:
A. Clay and silt B. Sand and silt
C. Only silt D. Only clay
(PU-2011)

183. What is the most abundant biochemical precipitate in the oceans?
A. Halite B. Limestone
C. Chert D. Coal (PU-2011)

184. Which type of weathering works most effectively on limestone?
A. Dissolution B. Frost wedging
C. Hydrolysis D. Oxidation
(PU-2011)

185. While moving away from the mid oceanic ridge, the age of the mid oceanic ridge basalt:
A. Decreases
B. Increases
C. Remains constant
D. Fluctuates (PU-2011)

186. Volcanoes along the mid-ocean ridges predominantly erupt lavas of .... type?
A. Andesite B. Basalt
C. Dacite D. Trachyte
(PU-2011)

187. The tilt of the Earth's rotation axis with reference to perpendicular of the orbital plane is:
A. 12.5° B. 23.5°
C. 30° D. 45° (PU-2011)

188. Which one of the following is capable of transporting sediments against the slope?
A. River current B. Turbidity current
C. Tidal current D. Rip current
(IIT JAM-2016)

189. Which of the following rock is least effected by chemical weathering processes?
A. Shale B. Granite
C. Basalt D. Limestone
(SAGAR UNIVERSITY-2015)

190. Which of the following river is associated with the graben?
A. Ganges
B. Narmada
C. Godavari
D. Ganges, Narmada and Godavari
(SAGAR UNIVERSITY-2015)

191. Which of the following rivers does not form a delta?
A. Narmada B. Mahanadi
C. Krishna D. Godavari
(SAGAR UNIVERSITY-2015)

192. .... is an example of a shield volcano:
A. Mt.st.Helens B. Hawaii
C. Kilimanjaro D. Deccan traps
(SAGAR UNIVERSITY-2015)

193. Lonar lake is formed by which of the following?
A. Volcanic crater
B. Meteoritic impact crater
C. Karst lake
D. Glacial depression
(SAGAR UNIVERSITY-2015)

194. Dunes with the crest trending parallel to the direction of sand transport are called:
A. Linear dune B. Longitudinal dune
C. Seif dune D. Parabolic
(SAGAR UNIVERSITY-2015)

195. The number of 4th order stream(s) present in the drainage network shown below, as per hierarchical classification, is

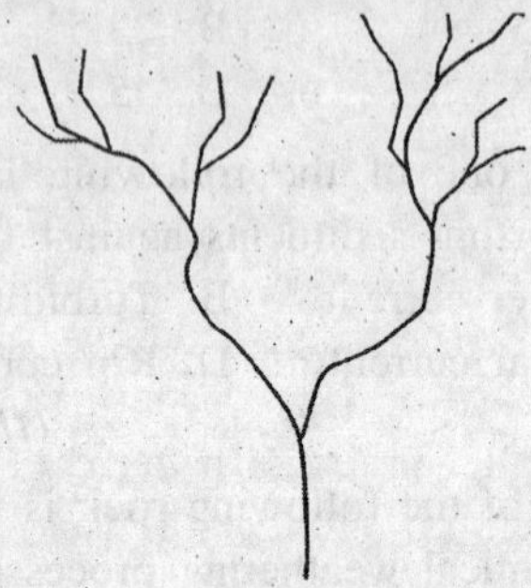

A. One B. Two
C. Three D. Four

(IIT JAM-2016)

196. Which among the following statements related to headward erosion by river is/are correct?
A. Length of tributaries increases
B. Length of the main channel increases
C. Streams get captured
D. Channel gradient increases upstream

(IIT JAM-2016)

197. Which of the following environments is/are characterized by predominant deposit of mud?
A. Barrier bar
B. Lagoon
C. Fluvial flood plain
D. Fluvial channel *(IIT JAM-2016)*

198. In which of the following are the minerals listed in order of increasing resistance to chemical weathering?
A. Hornblende quartz - muscovite - biotite
B. K-feldspar augite - biotite - olivine - quartz
C. Muscovite - hornblende - augite - quartz
D. Anorthite - albite - orthoclase - quartz

(ISM-2014)

199. Which is not a favourable condition for forming a delta where a large river enters the ocean?
A. Calm ocean condition
B. Active ocean current along the coast
C. Sluggish ocean current along the coast
D. The river is carrying a heavy load of alluvium *(ISM-2015)*

200. Knick point is not associated with:
A. A water fall
B. A sudden break in river gradient
C. A mature stage of river
D. An old stage of river *(ISM-2015)*

201. Which environment is best suited to develop cross bedding in a strata?
A. Meandering river
B. Deep marine
C. Desert
D. Swampy *(ISM-2015)*

202. Which one of the following is the not coastal landform?
A. Barrier island B. Spits
C. Tombolo D. Doline *(ISM-2015)*

203. A river developing before development of any structure is termed as?
A. Superimposed river
B. Antecedent river
C. Captured river
D. Diverted river *(ISM-2015)*

204. Retreat of glacial snout is marked by which of the following landform?
A. End moraine B. Lateral moraine
C. Drumlin D. Medial moraine

(ISM-2015)

205. The sinuous ridges of assorted and stratified sand and gravel are called as:
A. Drumlins B. Eskers
C. Moraines D. Erratic *(ISM-2015)*

206. Which of the following geomorphic features is not related to desert environment?
A. Yardang B. Bajada
C. Hamada D. Esker

(PUNJAB UNIVERSITY-2015)

207. Sinuosity of a stream is defined as:
A. Ratio of Talweg length to the valley width
B. Ratio of Talweg length to the valley length
C. Ratio of meanders length to the valley length
D. None of these

(PUNJAB UNIVERSITY-2015)

208. The deposition of suspended and dissolved material in a soil profile is referred as:
A. Eluviation B. Hiuriation
C. Humus D. Leaching
(ISM-2016)

209. The zone of sharp change in sea water temperature between 200 m and 1000 m water depth is known as:
A. Pycnocline B. Isotherm
C. Thermocline D. Compensation zone
(ISM-2016)

210. Nunatak is a:
A. Fluvial landform
B. Eolian landform
C. Glacial landform
D. Oceanic landform *(ISM-2016)*

211. The highest plateau in the world is
A. Bolvia and Peru plateau
B. Mexican plateau
C. Colorado plateau
D. Tibetan plateau *(ISM-2016)*

212. Tran lakes form
A. In meandering river
B. Due to collection of melt water in cirque
C. In limestone country
D. In damming of river by landslide
(ISM-2016)

213. In Western coastal plains of India, we observed
A. Dendritic drainage pattern
B. Trellis drainage pattern
C. Pinnate drainage pattern
D. Parallel drainage pattern *(ISM-2016)*

214. Few lakes or swamps, well develop floodplains upland divide with even slopes and with rounded to knife-edge summits with its easy view of the stream margin.
William Morrison Davis would have classified it as:
A. Adolescent B. Mature
C. Youthful D. Old stage
(ISM-2016)

ANSWER KEY

1	2	3	4	5	6	7	8	9	10
D	B	B	A	C	B	C	C	C	C
11	12	13	14	15	16	17	18	19	20
A	D	D	D	A	D	C	A	A	C
21	22	23	24	25	26	27	28	29	30
C	C	C	A	B	D	C	A	C	B
31	32	33	34	35	36	37	38	39	40
A	A	C	B	C	D	B	A	B	A
41	42	43	44	45	46	47	48	49	50
C	C	D	C	D	D	B	C	D	D
51	52	53	54	55	56	57	58	59	60
A	B	A	C	A	C	C	B	B	B
61	62	63	64	65	66	67	68	69	70
C	D	A	C	A	B	D	A	B	A
71	72	73	74	75	76	77	78	79	80
C	D	A	B	D	A	C	A	A	D
81	82	83	84	85	86	87	88	89	90
D	B	C	A	D	B	D	A	B	D
91	92	93	94	95	96	97	98	99	100
B	A	B	C	C	C	D	A	D	B

101	102	103	104	105	106	107	108	109	110
A	D	C	D	B	C	D	D	C	A
111	112	113	114	115	116	117	118	119	120
C	D	D	A	D	B	D	B	D	A
121	122	123	124	125	126	127	128	129	130
A	D	B	A	B	B	C	A	C	B
131	132	133	134	135	136	137	138	139	140
D	C	D	B	A	C	B	A	A	D
141	142	143	144	145	146	147	148	149	150
D	B	D	D	A	A	D	A	B	C
151	152	153	154	155	156	157	158	159	160
D	D	A	C	A	A	B	B	A	A
161	162	163	164	165	166	167	168	169	170
D	C	C	C	A	C	C	B	C	A
171	172	173	174	175	176	177	178	179	180
A	D	C	C	B	A	C	D	A	A
181	182	183	184	185	186	187	188	189	190
D	A	B	A	B	B	B	C	B	C
191	192	193	194	195	196	197	198	199	200
A	B	B	B	A	B,C,D	B,C	D	B	C
201	202	203	204	205	206	207	208	209	210
A	D	B	B	A	D	C	A	C	C
211	212	213	214						
D	A	B	A						

HINTS AND SOLUTIONS

2. Drainage density = Total length of stream of all orders/ Area of the basin

3. Wind system	**Properties (Area)**
Westerlies	Between 30°-60° latitude
Trades	Equatorial areas
Doldrums	Near equator where trades wind converge
Monsoon	Seasonal wind system

23. Process	**Properties (Area)**
Dissolution	$NaCl + H_2O = NaCl \cdot H_2O$
Hydration	$2Fe_2O_3 + 3H_2O = 2Fe_2O_3 \cdot H_2O$
Oxidation	$2Fe_2SiO_4 + 4H_2O + O_2 = 2Fe_2O_3 + 2H_4SiO_4$
Reduction	$2Fe_2O_3 + O_2 = 4FeO$

24. Area	**Slope (Degree)**
Shore	<1
Continental self	1
Continental slope	1-5

25. Features	**Earthquakes type**
Mid-oceanic ridge	Shallow focus
Island arc	Deep focus
Himalayan mountain belt	Intermediate focus
Continental rift	Shallow focus

27. Country rocks	**Features developed**
Phyllite	Fine grained foliated structures
Gneiss	Course grained sheet structure
Limestone	Sink-hole
Sandstone	Geode

28.

Landforms	Processes	Remarks
Strath	Fluvial erosion	Mature stage
Drumline	Glacial deposits	Glacial till
Cirque	Glacial erosion	Bowl shape feature
Point bar	Fluvial deposition	Old stage

30.

Drainage patterns	Features
Rectangular patterns	Jointed country
Parallel patterns	Steep slopes with little vegetation cover
Radial pattern	Granite-gneiss domes
Trellis	Grid like joints system

38.

Features	Agent
Esker	Glacio-fluvial deposit
Till	Glacial deposit
Moraine	Glacial deposit
Drumline	Glacial deposits

39.

River patterns	Represents pattern
Trellis	Folded structure pattern
Dendritic	Tree-like pattern
Radial	Dome-like pattern
Rectangular	Rectangular pattern

41.

Indian Falls	River	Location	Heights(Meters)
Gokak falls	Ghatprava	Balagavi of Karnataka	14
Dhaundhar falls	Narmada	Jabalpur of Madhya Pradesh	30
Gerosoppa falls	Swaravati	Simoga of Karnataka	253
Shivasamudram falls	Cauvery	Mandya of Karnataka	98

43.

Features	Product
Zeugen	Wind erosion
Yardang	Wind erosion
Mushroom rock	Wind erosion
Moraine	Glacial deposition

44.

River stage	Features
Youth stage	V-shape valley
Old stage	Delta
Mature stage	Meanderng, Fan
Rejuvenation	Flowing in down level. (Base level fall)

45.

Drainage pattern	Typical terrain type
Dendritic	Multiple set of joints
Trellis	Anticline, syncline fold, fault etc.
Rectangular	Faults, joints
Radial	Radiating form like dome basine structure

55.

Geomorphic feature	Agencies
Yardang	Wind erosion
Gorges	River erosion
Loess	Wind deposition
Butte	River erosion

58.

Geomorphic feature	Agencies
Drumline	Glacial
Loess	Wind deposition
Delta	Fluvial deposition
Canyons	Fluvial erosion

59.

Landforms	Process	Features
Yardang	Aeolian	Erosinal
Drumlin	Glacial	Depositional
Doline	Dissolution	Erosional
Chenier	Coastal	Deposotional

62.

Landform	Characteristics
Stream deflection	Strike-slip fault
River terrace	Strike-slip fault
Pressure ridge	Strike-slip faults
Sag pond	Normal faults

78.

Processes	Products
Weathering	Laterites
Chemical precipitation	Limestone
Biogenic activity	Coral reef
Sedimentation	Sandstone

80. **Landforms**	**Origin**	**Features**
Lag gravel	Wind deflation	Removal of fine material due to formation of desert pavement
Blowouts	Wind abrasion	Sandy depression due to heavy winds
Yardang	Wind abrasion	Furrod topographic feature due to wind abrasion
Ventifact	Wind abrasion	Wind facetted pebbles due to abrasion

96. **Feature**	**Characteristics**
Crevasse	Glacial fracture zone.
Point bar	It is cresentic shape deposited in the river.
Levee	The depositon parallel to river channel due to flooding or overflow of river.
Oxbow lake	Erosion due to meandering of river channel cut the outer bank and sediment deposited on inner bank

97. **Products**	**Processes**
Point bar	Fluvial deposition
Tillite	glacial deposition
Calcrete	Rain water due to climate fluctuation.
Loess	Eolian deposition

98. **Regions**	**Types of weathering**
Polar region	Chemical weathering
Hot deserts	Physical weathering
Cold desert	Physical weathering
Semi-arid region	Physical weathering

104. **Morphological feature**	**Agencies**
Yardang	Erosional feature of eolian
Potholes	Erosional feature of fluvial
Cuesta	Erosional feature of fluvial
Arete	Erosional feature of glacial

105. **Different terminology**	**Measuring scale**
Relative humidity of atmosphere	Hygrometer
Intensity of earth's tremor	Richter scale
Electric conductivity of water	Lysimeter (EC meter)
Speed of wind	Anemometer

106. **Morphological features**	**River stage**
Alluvial fan	Old stage
V-shaped valley	Youth-stage
Meanders	Youth-stage
Ox-bow lake	Mature stage

108. **Morphological feature**	**Characteristics**
Barchan	Depositional feature of eolian
Loess	Depositional feature of eolian
Delta	Depositional feature of fluvial
Stalagmites	Depositional feature of underground water

110. **Features**	**Characteristics**
Stalactite	Dripstone rising up from the floor of a limestone.
Stalagmite	Dripstone rising down from top of the limestone cave.
Uvala	Elongate depression on the floor.
Polje	Larger depression on the floor.

111. Drainage patterns	Rock type
Trellis	Shale
Concordant	Tillite
Dendritic	Massive granite
Annular	Clay

113. Wind direction	Features
Point in the upwind direction	Parabolic dune
Are at right angle to the wind direction	Transverse dune
Have no relation to the wind direction	Star dune
Point in the downward direction	Barchan dune

114. Feature	Agencies	Types of feature
Point bar	River	Deposition
Cirque	Glacial	Erosion
Yardang	Wind	Erosion
Berm	Sea	Depositional

115. Geomorphic feature	Plate boundary
Folded mountain belt	Convergent boundary
Guyot	Convergent boundary
Island arc	Convergent boundary
Oceanic ridge	Divergent boundary

116. Feature	agencies	Feature
Braid bar	river	Deposition
Moraines	glacial	Deposition
Barchans	wind	Erosion
Atoll	marine	Deposition

120. Feature	Agencies	feature
Hanging valley	Glacial	Erosion
Bajada	Fluvial	Deposition
Monadnock	Fluvial	Deposition
Blind valley	Karst topography	Erosion

127. Land forms	Characteristics
Morain	Glacial deposition
Eskers	Glacial-fluvial deposits
Cirque	Glacial erosion
Kame	Glacial-fluvial deposits

192. Example	Volcano type	Location	Composition	Elevation
Mt.st.Helens	Strato type	Skamania U.S.	Andesitic	2549 m
Hawaii	Shield	Hawaii USA	Basaltic	1143 m
Kilimanjaro	Stratovolcano	Kilimanjaro, Tanzania	Rhyolitic	5895 m
Deccan traps	Tholeiitic	West central India	Basaltic	>2000

194. Dunes type	Remarks
Linear dune	Perpendicular to the wind
Longitudinal dune	Parallel to wind direction
Seif dune	Perpendicular to wind, but single direction.
Parabolic	Perpendicular to the wind, opposite to the Barchan

198. Stability of Common Minerals under Weathering Conditions Compared with Bowen's Reaction Series

Stability of Minerals	Rate of Weathering
Most stable	**Slowest**
Iron oxides (hematite)	
Aluminum hydroxides (gibbsite)	
Quartz	
Clay minerals	
Muscovite mica	
Potassium feldspar (orthoclase)	
Biotite mica	
Sodium-rich feldspar (albite)	
Amphiboles	
Pyroxene	
Calcium-rich feldspar (anorthite)	
Olivine	
Calcite	
Halite	
Least stable	**Fastest**

Bowen's Reaction Series

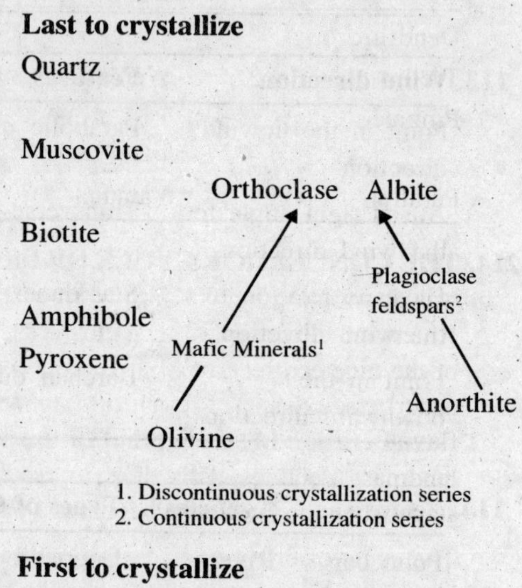

202. Landform	**Remarks**
Barrier island	Sea landform
Spits	Sea landform
Tombolos	Sea landform
Doline	Carbonate precipitation

203. River types	**Remarks**
Superimposed river	Older river overpassed on the new river
Antecedent river	River maintain its original direction
Captured river	River drainage system diverted its down bed of the neighbouring stream. It is also called Wind Gap.
Diverted river	Due to tectonic activity the river change its initial direction

206. Features	**Environments**
Yardang	Desert environment due to wind
Bajada	Fluvial landform
Hamada	Desert environment due to wind
Esker	Fluvio-Glacial environment

207. Sinuosity of a stream is defined as:
Actual path length / Shortest path length

209. Terminology	**Remarks**
Pycnocline	Density increases with depth in an ocean
Isotherm	Equal temperature
Thermocline	Change of temperature in sea water
Compensation zone	CCD, ACD

211. Plateau	**Heights**
Bolvia	3750 m
Peru	3750 m

Mexican	3600 m
Colarado	3870 m
Tibetan	5000 m

213.

Drainage pattern	Area
Dendritic	Ganga river
Trellis	Western coastal plain
Pinnate	Tributaries of the Ganga river
Parallel	Yamuna

214. THE CONCEPT OF CYCLE OF EROSION was formulated by William Morris Davis, an American Geomorphologist, towards the end of the nineteenth century. It is a concept of an orderly sequence of evolutionary stages of fluvial erosion in which relief of the available landmass declines with time to reach a late stage when the landscape becomes a peneplain.

The cycle of erosion, as envisioned by Davis, has its initial stage at a time when the landmass is rapidly elevated by internal earth forces, followed by a very long period of tectonic quiescence.

Once raised high above sea level as a landmass, streams come into existence and erosion begins to operate on the uplifted mass which is gradually worn down almost to a plain. The landmass may, at some later time, be rejuvenated and the cycle begins again and remnants of the earlier cycle of erosion are preserved at new and higher levels.

In a normal cycle three stages have been recognized as: Youth stage, Mature stage and Old stage. These follow each other in a regular sequence.

YOUTH STAGE

In this stage the river flows along an uneven surface and there is intensive bottom erosion, the gradients are steep and the erosion is rapid. The rapid deepening of the channel leads to the formation of V-shaped valleys.

Thus during the youth stage of a river, the valley form undergoes vigorous development, particularly in depth and head ward growth. Lakes, rapids, waterfalls, steep-sided valleys and gorges are of common occurrence during this stage. Besides, the phenomenon of river-capture or river piracy takes place in this stage. Youthful rivers have an irregular long profile (thalweg) from source to mouth.

RIVER CAPTURE

When one of the two rivers flowing in opposite directions from a single divide, becomes more effective in erosion due to steeper gradient (when the slopes are unequally inclined), the divide gradually recedes towards the side with the gentler slope. In other words, the river with steeper gradient extends its valley head ward thus causing a shift of the divide against the river with gentle gradient. Gradually deepening of the valley continues head ward with pronounced dissection of the ridge (divide). Sometimes this head ward migration of one river enables it to reach the river on the other side. But, as the first river has a steeper gradient than the other one, the course of the second river gets diverted and its water starts draining through the channel of the first river. This process of diversion of a river by the head ward migration of another river is known as River-Capture or River-piracy. The point where the course of the second river is diverted is known as the Elbow of capture. The captured river is known as Misfit and the deserted part of its channel through which no water flows is termed as the Wind-gap.

MATURE STAGE

In this stage rivers flow with a graded profile i.e. it attains a profile of equilibrium. The land mass is fully dissected and a well-integrated drainage system is developed. Ridges and valleys develop prominently.

Flood plains develop and river meandering takes place. The topography consists of features such as: hogbacks, cuestas, mesa, butte, meanders, oxbow lakes, natural bridge, flood plains, alluvial fans etc.

OLD STAGE

In this stage the gradients are gentle and the velocity is low. Accordingly the river lose most of its erosive power and flow in a sluggish manner. In old age a river has maximum meandering. The river at this age does little of erosion and transportation but is mostly engaged in deposition. This stage is characterised by the development of distributaries and the river flows almost at the base level of erosion. The topography consists of features like peneplains, natural levees, deltas etc. Most of the cycles of erosion do not reach the final stage, as sometime during their operation either climatic or tectonic disturbances take place, and thus results in an incomplete or partial cycle.

PALEONTOLOGY

1. Which of the following group of fossils became extinct at the Permo-Triassic boundary?
A. Graptolite B. Trilobite
C. Ammonite D. Ediacaran
(IIT-JAM 2015)

2. Which of the following Brachiopod genera is considered as "living fossil"?
A. *Lingula* B. *Productus*
C. *Atrypa* D. *Spirifer*
(IIT-JAM 2015)

3. Match the Lamellibranchia genera in Group-I with their dentition pattern in Group-II:

Group - I	Group - II
(*i*) *Glycimeris*	(P) Schizodont
(*ii*) *Trigonia*	(Q) Taxodont
(*iii*) *Lucina*	(R) Dysodont
(*iv*) *Mytilus*	(S) Heterodont

A. (*i*) - Q, (*ii*) - P, (*iii*) - S, (*iv*) - R
B. (*i*) - P, (*ii*) - Q, (*iii*) - R, (*iv*) - S
C. (*i*) - Q, (*ii*) - P, (*iii*) - R, (*iv*) - S
D. (*i*) - S, (*ii*) - R, (*iii*) - Q, (*iv*) - P
(IIT-JAM 2015)

4. Which of the following is the most suitable for biostratigraphy of Cambrian rocks?
A. Trilobite B. Ammonoid
C. Rugose coral D. Pelecypod
(AMU-2015)

5. Which of the following is **not** exclusively marine?
A. Foraminifera B. Ostracoda
C. Conodonta D. Radiolaria
(AMU-2015)

6. Which of the following is found below CCD?
A. Foraminifera B. Ostracoda
C. Radiolaria D. Coccolithophore
(AMU-2015)

7. Aristotle's lantern is found in:
A. Brachiopoda B. Gastropoda
C. Trilobita D. Echinoidea
(AMU-2015)

8. Which of the following has shortest geological range?
A. Brachiopoda B. Pelecypoda
C. Trilobita D. Conodonts
(AMU-2015)

9. Acme or peak zone is defined as the strata that contains:
A. Minimum number of species
B. Maximum number of species
C. Maximum number of individuals of one species
D. Minimum number of individuals of single species
(DU-2015)

10. Foraminifera are good index fossils for the:
A. Archaean B. Proterozoic
C. Phanerozoic. D. None of the above
(DU-2015)

11. Past seawater temperature can be reconstructed from the:
A. Oxygen isotopic composition of the calcareous foraminifera
B. Carbon isotopic composition of the calcareous foraminifera
C. Carbon isotopic composition of agglutinated foraminifera
D. All of the above *(DU-2015)*

12. 'Ediacaran' fossils have significance in determining the:
A. Archean/Proterozoic boundary
B. Permian/Triassic boundary

C. Cretaceous/Tertiary boundary
D. Precambrian/Cambrian boundary
(DU-2015)

13. First stage of horse evolution is known as:
A. *Eohippus* B. *Hyracotherium*
C. *Mesohippus* D. *Epihippus*
(DU-2015)

14. In response to the expansion of grasslands, mammals evolved:
A. Brachydont dentition
B. Bunodont dentition
C. Lophodont dentition
D. Hypsodont dentition *(DU-2015)*

15. A specimen used as diagnostic of a particular new species is known as:
A. Holotype B. Paratype
C. Lectotype D. Neotype
(DU-2015)

16. The concept of punctuated equilibrium was proposed by:
A. James Hutton & Charles Lyell
B. Niels Eldredge & Stephen Gould
C. Charles Darwin & Alfred Russell Wallace
D. William Smith & Nicholas Steno
(DU-2015)

17. Evolution above species level is known as:
A. Microevolution
B. Macroevolution
C. Iterative evolution
D. None of these
(DU-2015)

18. Origin of bird is sought among:
A. Sauropod dinosaurs
B. Dromaeosaurid theropods
C. Prosauropods
D. Omithischian dinosaurs
(DU-2015)

19. Saurischian and ornithischian dinosaurs are distinguished primarily by their:
A. Domichnia B. Fodichnia
C. Cubichnia D. Pascichnia
(DU-2015)

20. Feeding burrows made by animals are known as:
A. Domichnia B. Fodichnia
C. Cubichnia D. Pascichnia
(DU-2015)

21. Which one of the following belong to the Equidae family?
A. *Stegodon* B. *Stegoophodon*
C. *Ramapithecus* D. *Gigantopithecus*
(IIT-JAM 2014)

22. Which one of the following flora is a conifer?
A. *Glossopteris* B. *Gangamopteris*
C. *Buriadia* D. *Schizoneura*
(IIT-JAM 2014)

23. Match the genera in Group-1 with the phylum in Group-2:

Group-1	Group-2
P. *Mya*	1. Echinodermata
Q. *Micraster*	2. Mollusca
R. *Olenoides*	3. Brachiopoda
S. *Leptaena*	4. Arthopoda

A. P-3, Q-4, R-1, S-2
B. P-2, Q-1, R-3, S-4
C. P-2, Q-4, R-1, S-3
D. P-2, Q-1, R-4, S-3 *(IIT-JAM 2014)*

24. Which one of the following morphological features is absent in the class Pelecypoda?
A. Umbo B. Deltidium
C. Pallial sinus D. Resilifier
(IIT-JAM 2013)

25. The first record of Amphibians are found in:
A. Carboniferous B. Proterozoic
C. Jurassic D. Devonian
(BHU- 2013)

26. Match the morphological features described in Group I with their names listed in Group II:

Group I	Group II
P. Masticatory apparatus in Echinoidea	1. Cameral deposits
Q. Triangular cavity near umbo of Brachiopoda	2. Madreporite
R. Calcareous secretions in siphuncle of Cephalopoda	3. Delthyrium

S. Genital plate in Echinoidea — 4. Aristotle's lantern

A. P-3, Q-4, R-1, S-2
B. P-4, Q-3, R-1, S-2
C. P-3, Q-4, R-2, S-1
D. P-4, Q-2, R-3, S-1 (*IIT-JAM 2012*)

27. Which one of the following is a Primate fossil?
A. *Hipparion* B. *Ramapithecus*
C. *Equus* D. *Stegolophodon*
(*IIT-JAM 2012*)

28. Characteristics of suture are the absolute criterion for recognizing the evolution of ammonoids:
A. Early, Paleozoic period
B. Mesozoic era
C. Cenozoic era
D. Late Paleozoic period (*BHU-2013*)

29. A bivalve shell having Pallial Sinus indicates its:
A. Pelagic habitat
B. Planktonic habitat
C. Epifaunnal habitat
D. Infaunal habitat (*BHU-2013*)

30. Which man is known as Java Man?
A. *Pithecanthropus* B. *Ramapithecus*
C. *Australopithecus* D. *Sinanthropus*
(*BHU-2013*)

31. The first vertebrate reported from which geological age?
A. Devonian B. Ordovician
C. Silurian D. Cambrian
(*BHU-2013*)

32. Graptolites demonstrate which type of symmetry?
A. Radio-bilateral symmetry
B. Biradial symmetry
C. Bilateral symmetry
D. Radial symmetry (*BHU-2013*)

33. Which is known as blind trilobite?
A. *Agnostos* B. *Microdiscus*
C. *Harpetida* D. *Proeida*
(*BHU-2013*)

34. The appearance of dinosaur started:
A. Cretaceous B. Early Triassic
C. Early Jurassic D. Late Triassic
(*BHU-2013*)

35. The base of Palaeozoic Era is marked by first appearance of:
A. Corals B. Trilobites
B. Brachiopods D. Cephalopods
(*BHU-2012*)

36. Triassic begins with first appearance of:
A. *Olenus*
B. *Nautilus*
C. *Otoceras woodwardi*
D. *Macrocephalites* (*BHU-2012*)

37. The tooth like phosphatic microfossils are:
A. Conodonts B. Thecodonts
C. Bathydonts D. Coprolites
(*BHU-2012*)

38. Cephalopoda with complex suture is:
A. *Ceratites* B. *Nautilus*
C. *Goniatites* D. *Ammonites*
(*BHU-2012*)

39. The trilobite with pygidium equal to that of cephalan is called:
A. *Micropygous*
B. *Macropygous*
C. *Isopygous*
D. Either *micropygous* or *isopygous*
(*BHU-2012*)

40. Which one of the following trilobites IS characteristic of Middle Cambrian?
A. *Redlichia* B. *Olenellus*
C. *Paradoxides* D. *Olenus*
(*BHU-2012*)

41. The order Primates is for:
A. Homo sapiens B. Trilobites
C. Brachiopods D. Mollusca
(*BHU-2012*)

42. The gastropod genus *Physa* is:
A. Uncoiled
B. Dextrally coiled
C. Sinistrally coiled
D. Planispirally coiled (*BHU-2012*)

43. The trilobite having large number of lenses in the eyes is:
A. *Agnostus* B. *Microdicus*
C. *Remopleurides* D. *Olenellus*
(BHU-2012)

44. The exoskeleton of molluscs is most commonly composed of:
A. Silica B. Calcite
C. Aragonite D. Phosphate
(BHU-2012)

45. The least favourable environment for the presenration of fossils is:
A. Terrestrial B. Lacustrine
C. Fluvial D. Marine
(BHU-2012)

46. Which one is **not** a bivalve?
A. *Nautilus* B. *Lima*
C. *Nucula* D. *Trigonia*
(BHU-2012)

47. Rapidly evolving class is:
A. Bivalvia B. Gastropoda
C. Cephalopoda D. Brachiopoda
(BHU-2012)

48. The community of inter-related organisms inhabiting in area is:
A. Biotype B. Biocenose
C. Thenatocoenose D. Thenotype
(BHU-2012)

49. Trilobites got extinct towards the close of:
A. Devonian B. Permian
C. Cambrian D. Silurian
(BHU-2012)

50. In brachiopods, the pedical valve is also called:
A. Brachial valve B. Ventral valve
C. Right valve D. Dorsal valve
(BHU-2012)

51. The Corona is the part of .... shell.
A. Ammonoid B. Trilobite
C. Echinoid D. Bivalve
(BHU-2012)

52. Match the feature in Group I with appropriate taxon in Group II.

Group I	Group II
P. Monomyarian	1. Echinoidea
Q. Deltidial plates	2. Anthozoa
R. Corona	3. Pelecypoda
S. Epitheca	4. Brachiopoda

A. P-3, Q-4, R-1, S-2
B. P-4, Q-2, R-1, S-3
C. P-4, Q-3, R-2, S-1
D. P-3, Q-4, R-2, S-1 *(IIT-JAM 2011)*

53. The arrangement in which teeth are radiating from umbo is:
A. Isodont B. Desmbdont
C. Dysodont D. Schizodont
(BHU-2011)

54. The geological age of Spirifer is:
A. Orodovician to devonian
B. Triassic to present
C. Jurassic
D. Silurian to Permian *(BHU-2011)*

55. *Area*, *Ostrea*, *Cardita*, *Maya* all these belong to phylum:
A. Mollusca B. Brachiopoda
C. Arthropoda D. Echinodermata
(BHU-2011)

56. Arrange in correct order:
A. Cephalon-pygidium-thorax
B. Thorax-cephalon-pygidium
C. Cephalon-thorax-pygidium
D. Pygidium-thorax-cephalon *(BHU-2011)*

57. The form in which two adductor muscle impressions are preserved is called:
A. Isomyariyan B. Monomyariyan
C. Dimyariyan D. Anisomyariyan
(BHU-2011)

58. The aperture of the gastropod shell is covered by a plate known as:
A. Peristome B. Operculam
C. Siphon D. Umblicus
(BHU-2011)

59. The suture line with rounded saddles and angular lobes is known as:
A. Straight B. Cerratitic
C. Goniatitic D. Ammonoid
(BHU-2011)

60. The trilobites in which suture commences from posterior margin and ends at lateral margin is:
A. Protopariyan B. Hypopariyan
C. Propariyan D. Opisthopariyan
(BHU-2011)

61. Cephalon of trilobite is the part of:
A. Head B. Body
C. Tail segments D. Legs (BHU-2011)

62. Which of the following is the sinistral form of Gastropoda?
A. *Natica* B. *Murex*
C. *Physa* D. *Voluta* (BHU-2011)

63. In Brachipoda the two valves are joined at:
A. Hinge line B. Posterior
C. Anterior D. Didictor
(BHU-2011)

64. Index fossils are those which have:
A. Short geological range and short geographical range
B. Wide geological range and very wide geographical range
C. Short geological range and very wide geographical range
D. Wide geological range and short geographical range
(BHU-2011)

65. A specimen of a species from the same locality of as its holotype or syntype:
A. Lectotype B. Neotype
C. Plesitotype D. Topotype
(BHU-2011)

66. Which of the following is the phosphatic microfossil?
A. Conodont B. Diatoms
C. Dinoflagellates D. Coccoliths
(BHU-2011)

67. The organisms who are free swimmers are called:
A. Plankton B. Nekton
C. Pectin D. Taxon
(BHU-2011)

68. Which of the following is **not** the process of fossilization?
A. Petrification
B. Replacement of mineralization
C. Carbonization
D. Hydrolysis (BHU-2011)

69. The function of the lophophore in Brachiopoda is:
A. Respiration B. Gathering food
C. Movement D. Supporting organs
(BHU-2011)

70. The brachiopods having dental system are known as:
A. Articulata B. Inarticulata
C. Delthryum D. Deltidium
(BHU-2011)

71. The organism in which the last whorl shell covers all the previous whorls is:
A. *Patella* B. *Turritella*
C. *Natica* D. *Cypraea*
(BHU-2011)

72. The organism that can be divided in two equal halves is:
A. Brachiopoda B. Pelecypoda
C. Gastropoda D. Cephalopoda
(BHU-2011)

73. Lamellibranchia is classified in the following branch/branches:
A. Active B. Sessile
C. Burrowing D. All of these
(BHU-2011)

74. Match the dentition type in Group 1 with the Bivalvia genus in Group 2.

Group 1	Group 2
P. Heterodont	1. *Trigonia*
Q. Desmodont	2. *Cerastoderma*
R. Dysodont	3. *Mya*
S. Schizodont	4. *Mytilus*

A. P-2, Q-3, R-1, S-4
B. P-1, Q-3, R-2, S-4
C. P-2, Q-3, R-4, S-1
D. P-1, Q-2, R-3, S-4 (IIT-JAM 2010)

75. Which one of the following flora belongs to the Lower Gondwana?
A. *Gangamopteris* B. *Cladophlebis*
C. *Sphenopteris* D. *Ptilophyllum*
(IIT-JAM 2010)

76. Dinosaur egg shells have been found in the Lameta Formation in:
A. Kerala B. Madhya Pradesh
C. Tamil Nadu D. Punjab
(BHU-2010)

77. Trilobites became extinct at the end of:
A. Cambrian B. Devonian
C. Permian D. Cretaceous
(BHU-2010)

78. Graptolites reached their maximum development in:
A. Cambrian B. Ordovician
C. Silurian D. Carboniferous
(BHU-2010)

79. Ammonoids became extinct at the end of:
A. Tertiary B. Triassic
C. Cretaceous D. Carboniferous
(BHU-2010)

80. Foramen is found in which one of the following?
A. Trilobites B. Pelecypods
C. Graptolites D. Brachiopods
(BHU-2010)

81. Which one of the following is a coral?
A. *Motlivaltia* B. *Globigerina*
C. *Cardita* D. *Calymene*
(BHU-2010)

82. Which one is a plant fossil?
A. *Schizoneura* B. *Favosites*
C. *Productus* D. *Syringothyris*
(BHU-2010)

83. Ichnofossil is:
A. An index fossil
B. A trace fossil
C. A living fossil
D. None of these
(BHU-2010)

84. Which of the following is a middle Cambrian Trilobite?
A. *Olenus* B. *Paradoxides*
C. *Olenellus* D. *Phacops*
(IIT-JAM 2009)

85. Which one of the following is a Lower gondwana flora?
A. *Gangamopteris*
B. *Ptilophyllum*
C. *Otozamites*
D. *Nilssonia*
(IIT-JAM 2009)

86. In a pelecypod shell, if umbos curve is in such a way that the beak points in an anterior direction, the umbos are:
A. Opisthogyrate B. Prosogyrate
C. Orthogyrate D. Spirogyrate
(IIT-JAM 2009)

87. What name has been assigned to tooth like phosphatic microfossils?
A. Conodonts B. Thecodont
C. Bathydont D. Coprolites
(BHU-2009)

88. Which one is a cephalopoda with simple suture?
A. *Ceratites* B. *Nautilus*
C. *Goniatites* D. *Ammonites*
(BHU-2009)

89. Pygidium in trilobite if larger than the cephalon is called:
A. Micropygous
B. Macropygous
C. Isopygous
D. Either micropygous or isopygous
(BHU-2009)

90. Which one of these trilobites is blind?
A. *Asaphus* B. *Oleneltus*
C. *Agnostus* D. *Olenus*
(BHU-2009)

91. Homo sapiens belong to order:
A. Primates B. Rodentia
C. Chiroptera D. Mollusca
(BHU-2009)

92. The example of sinistrally coiled gastropods is:
A. *Turbo* B. *Natica*
C. *Physa* D. *Crepidula*
(BHU-2009)

93. The skeleton of entire coral colony is termed as:
A. Corralite B. Corralium
C. Columella D. Rhabdosame
(BHU-2009)

94. The exoskeleton of sponges is made up of:
A. Silica B. Calcite
C. Aragonite D. Calcium phosphate
(BHU-2009)

95. The most favourable environment for the preservation of fossils is:
A. Terrestrial B. Lacustrine
C. Fluvial D. Marine
(BHU-2009)

96. Find odd one out:
A. *Nautilus* B. *Lingula*
C. *Nucula* D. *Paradoxides*
(BHU-2009)

97. Among molluscs which class shows rapid evolution?
A. Bivalvia
B. Gastropoda
C. Cephalopods
D. Both Gastropoda and Bivalvia
(BHU-2009)

98. Terebratula is the characteristic genera of:
A. Mesozoic B. Cenozoic
C. Quaternary D. Paleozoic
(IIT-JAM 2008)

99. The occurrence of ammonites in a sequence suggests:
A. Cenozoic age and freshwater environment
B. Cenozoic age and marine environment
C. Mesozoic age and freshwater environment
D. Mesozoic age and marine environment
(IIT-JAM 2007)

100. Match the organisms listed in Group 1 with their respective symmetry elements in Group 2.

Group-1	Group-2
P. Brachiopods	1. Asymmetric
Q. Pelecypods	2. Plane of symmetry bisects the two valves
R. Echinoids	3. Plane of symmetry lies between the two valves
S. Helically coiled gastropods	4. Five-fold radial symmetry

A. P-1, Q-2, R-3, S-4
B. P-4, Q-1, R-2, S-3
C. P-3, Q-4, R-2, S-1
D. P-2, Q-3, R-4, S-1 *(IIT-JAM 2007)*

101. Match the fossil taxa listed in Group 1 with their characteristics morphological features in Group 2.

Group-1	Group-2
P. *Paradoxides*	1. Two pendent stipes
Q. *Didymograptus*	2. Sinistral coiling
R. *Physa*	3. Genal spines
S. *Ceratites*	4. Sutures with undivided saddles

A. P-3, Q-2, R-4, S-1
B. P-3, Q-1. R-2, S-4
C. P-2, Q-1, R-4, S-3
D. P-2, Q-1, R-3, S-4 *(IIT-JAM 2007)*

102. Which one of the following groups of organisms belongs to terrestrial/fresh-water environment?
A. Echinoids
B. Brachiopods
C. Corals
D. Pulmonate gastropods *(IIT-JAM 2006)*

103. Match the following:

Group I	Group II
P. *Micraster*	1. Trilobite
Q. *Phacops*	2. Dinosaur
R. *Equus*	3. Mammal
S. *Barappasaurus*	4. Echinoid

A. P - 1 Q - 2 R - 3 S - 4
B. P - 2 Q - 1 R - 4 S - 3
C. P - 3 Q - 4 R - 2 S - 1
D. P - 4 Q - 1 R - 3 S - 2 *(IIT-JAM 2006)*

104. Acme zone indicates the:
A. Range of a fossil species
B. Absence of a fossil species
C. Maximum abundance of a fossil species
D. Migration of a fossil species
(IIT-JAM 2006)

105. Vertebrate fossils are most important for the study of the:
A. Intra-trappean beds
B. Gondwana rocks
C. Siwalik rocks
D. Zewan beds *(IIT-JAM 2005)*

106. Character of sutures is the sole criterion for recognizing the evolution of ammonoids during the:
A. Early Paleozoic Periods
B. Late Paleozoic Periods
C. Mesozoic Era
D. Cenozoic Era *(IIT-JAM 2005)*

107. As index fossils, brachiopods were important during the:
A. Late Cretaceous
B. Middle Jurassic
C. Early Jurassic
D. Late Paleozoic *(IIT-JAM 2005)*

108. The presence of a pallial sinus in a bivalve shell indicates its:
A. Infaunal habitat
B. Epifaunal habitat
C. Planktonic habitat
D. Pelagic habitat *(IIT-JAM 2005)*

109. The science of the study of remains of all organisms from the past geological ages is called:
A. Neontology B. Planetology
C. Palaeontology D. Palichnology
(BHU-2014)

110. Which of the following organism's skeleton is made up entirely of calcite?
A. *Nautilus* B. *Crania*
C. *Trigonia* D. *Patella*
(BHU-2014)

111. All biological names should/may be written in:
A. Italics
B. CAPITAL LETTERS
C. Either A or B
D. None of these *(BHU-2014)*

112. The specimen selected to replace the lost holotype (from the same locality) is called:
A. Neotype B. Topotype
C. Plastotype D. Cotype
(BHU-2014)

113. Which developmental stage is commonly found during the growth of an individual after the larval stage?
A. Ephebic B. Nepionic
C. Neanic D. Gerontic
(BHU-2014)

114. Which of the following is a type of post-embryonic growth of skeletons?
A. Accretion of existing parts
B. Addition of new parts
C. Moulting
D. All of these *(BHU-2014)*

115. A modern organism that has descended from a very ancient stock with comparatively a little change, is known as:
A. Extant Organism
B. Living fossil
C. Derived fossil
D. None of these *(BHU-2014)*

116. Derived fossils:
A. Can not be used as an index fossil
B. Can be used as an index fossil
C. Are marker fossils
D. Are trace fossils *(BHU-2014)*

117. Which of the following is a cementing brachiopod?
A. *Crania* B. *Richthofenia*
C. Both A. and B. D. None of these
(BHU-2014)

118. In which of the following bivalvian genera the beaks are opisthogyrous?
A. *Trigonia* B. *Nucula*
C. Both of these D. None of these
(BHU-2014)

119. In which of the following gastropod genera the 'silt band' is present?
A. *Cypraea* B. *Bellerophon*
C. *Planorbis* D. *Patella*
(BHU-2014)

120. In an endogastric ceratocone (cephalopod), the siphuncle (si) and the hyponomic sinus (hs) are:
A. Both located ventrally
B. Both located dorsally
C. si dorsally but hs ventrally
D. si ventrally but hs dorsally
(BHU-2014)

121. The class Crinoidea is:
A. Confined in middle Cambrian
B. Confined in middle Paleozoic
C. Ranges from middle Cambrian to Holocene
D. Found in Holocene only
(BHU-2014)

122. The fresh water bryozoans are found in:
A. Late Palaeozoic only
B. Jurassic to Cretaceous beds
C. Early Tertiary times
D. Recent time only *(BHU-2014)*

123. Which of the following is a three-toed horse found in the Siwalik rocks?
A. *Hipparion* B. *Equus*
C. *Ramapithecus* D. *Bos* *(BHU-2014)*

124. Which of the following is a fossil excreta?
A. Coprolite B. Gastrolith
C. Dendrite D. Stromatolite
(BHU-2014)

125. Mold and cast are:
A. –ive impressions
B. +ive impressions
C. –ive and +ive impressions
D. +ive and –ive impressions *(BHU-2014)*

126. The sub-branch of palaeontology which deals with the impressions produced due to behavioural activities of ancient organism is called:
A. Ichnology
B. Palaeoecology
C. Taphonomy
D. Micropalaeontology *(BHU-2014)*

127. Select a pseudofossil from the following:
A. Chondrites B. Dendrites
C. Stylolites D. Uncolites
(BHU-2014)

128. In *Corbula lyrata*, the first and second names respectively stand for:
A. Genus and species
B. Species and genus
C. Genus and subgenus
D. Subgenus and species *(BHU-2014)*

129. Which of the following has agglutinated shell?
A. Conodonts B. Radiolarian
C. Foraminifera D. Pollen and spores
(BHU-2014)

130. In the bivalves the prosogyrous umboes point towards:
A. Anterior margin B. Posterior margin
C. Ventral margin D. Dorsal margin
(BHU-2014)

131. The pallial sinus in the bivalves are usually situated at:
A. Anterior side of the valve
B. Posterior side of the valve
C. Dorsal side of the valve
D. Ventral side of the valve *(BHU-2014)*

132. Select a bivalve genus from the following having taxodont type of dentition:
A. *Venus* B. *Pholadomya*
C. *Mytilus* D. *Arca*
(BHU-2014)

133. If a brachiopod shell is concavo-convex in the early growth stages but reverses to convexo-concave in maturity, it is known as:
A. Dorsibiconvex
B. Convexo-concave
C. Resupinate
D. Concavo-convex *(BHU-2014)*

134. In cephalopods, the chambered part is known as:
A. Body whorl B. Phragmocone
C. Siphuncle D. Aperture
(BHU-2014)

135. The cylindrical siphuncle with simple of straight septal neck in cephalopods is called:
A. Holochoanitic B. Ellipochoanitic
C. Orthochoanitic D. Cyrtochoanitic
(BHU-2014)

136. The biogenetic law "Ontogeny recapitulates phylogeny" is best illustrate by:
A. Gastropods B. Bivalves
C. Brachiopods D. Ammonoids
(BHU-2014)

137. Which of the following are plant fossils only of Lower Gondwana in age?
A. *Ptillophyllum, Glossopeteris*
B. *Glossopeteris, Dicrodium*
C. *Dicrodium, Ptillophyllum*
D. *Glossopeteris, Gangamopteris*
(BHU-2014)

138. $\delta^{18}O$ value of foraminifera carbonate:
A. Will remain same during interglacial & glacial stage
B. Will increase during glacial stage
C. Will increase during interglacial stage
D. Will decrease during glacial stage
(DU-2014)

139. Which of the following is associated with ancient suture zones:
A. Conglomerate B. Foraminiferal ooze
C. Radiolarian chert D. Sandstone
(DU-2014)

140. Taphonomy is the study of:
A. Environmental processes when the specimen was living
B. Post mortem (post-depositional) processes, that influenced the specimen
C. Chemical composition of the fossils
D. Internal structure of the fossils
(DU-2014)

141. Which of the following is an example of ichnofossil?
A. Fossilized Serpent trail
B. Fossilized Brachiopod shell
C. Fossilized Elephant tusk
D. None of the above *(DU-2014)*

142. Trilobite fossils are recovered from the rocks ranging in age from:
A. Pliocene to Pleistocene
B. Triassic to Jurassic
C. Cambrian to Permian
D. Pleistocene to Holocene *(DU-2014)*

143. Brachiopods have:
A. Equivalves and inequilateral shell
B. Unequal valves and equilateral shell
C. Radially symmetric shell
D. None of the above *(DU-2014)*

144. Ichthyostega represents the evolutionary stage between:
A. Reptiles and mammals
B. Fish and amphibians
C. Invertebrates and vertebrates
D. Amphibians and reptiles *(DU-2014)*

145. Which one of the following fossil groups has a property of ecdysis?
A. Trilobites B. Brachiopods
C. Foraminifera D. Connecting organs
(DU-2014)

146. Past seawater temperature can be reconstructed from the:
A. Oxygen isotopic composition of the calcareous foraminifera
B. Carbon isotopic composition of the calcareous foraminifera
C. Carbon isotopic composition agglutinated foraminifera
D. Oxygen isotopic composition of agglutinated foraminifera
(DU-2014)

147. Which is not an invertebrate fossil in the following?
A. *Productus* B. *Ptylophylum*
C. *Trigonia* D. *Murex*
(BHU-2015)

148. Which one is an index fossil?
A. *Trilobite* B. *Natiloids*
C. *Gastropods* D. *Bivalves*
(BHU-2015)

149. Match the following:

Column-A (Phylum)	**Column-B (Species)**
P. Brachiopoda	1. *Turritella*
Q. Cephalopoda	2. *Goniatite*
R. Gastropoda	3. *Sprifer*
S. Bivalvia	4. *Spondylus*

A. P-1, Q-2, R-3, S-4
B. P-1, Q-3, R-2, S-4
C. P-3, Q-1, R-4, S-2
D. P-3, Q-2, R-1, S-4 *(BHU-2015)*

150. What are the characteristics of an index fossil?
A. Wide geographic range and limited time range
B. Both time and geographic range should be wide
C. Limited geographic range and wide time range
D. None of the above *(BHU-2015)*

151. Trigonia has .... dentition pattern.
A. Isodont B. Schizodont
C. Heterodont D. Desmodont
(BHU-2015)

152. The coiling of shell in .... is sinistral.
A. *Trochus* B. *Murex*
C. *Planorobis* D. *Physa* *(BHU-2015)*

153. Which is **not** the sub-class of Cephalopoda?
A. Nautiloidea B. Ammonoidea
C. Acephala D. Dibranchia
(BHU-2015)

154. The body of Trilobite is made up of three distinct parts known as Cephalic Shield, Thorax and ....
A. Ocular plates B. Genital plates
C. Corona D. Pygidium
(BHU-2015)

155. Ammonoidea is a/an:
A. Trace fossil B. Index fossil
C. Living fossil D. Plant fossil
(BHU-2015)

156. Ceratitic suture-line have:
A. Pointed lobes and rounded saddle
B. Rounded lobes as well as rounded saddle
C. Divided lobes and rounded saddle
D. Divided lobes and divided saddles
(BHU-2015)

157. Mouth and Anus are centrally located in:
A. Regular Echinoids
B. Irregular Echinoids
C. Both regular and Irregular Echinoids
D. Neither regular Echinoids nor irregular Echinoids *(BHU-2015)*

158. Both the valves are held together by means of muscles in:
A. Articulata brachiopods
B. Inarticulata brachiopods
C. Both A and B
D. None of the above *(BHU-2015)*

159. The shell is equilateral and unequal in:
A. Acephala B. Brachiopoda
C. Gastropoda D. Ammonoidea
(BHU-2015)

160. *Pupilla* is a:
A. Discoidal Shape Gastropoda
B. Conical Shape Gastropoda
C. Cylindrical shape Gastropoda
D. Globular Shape Gastropoda
(BHU-2015)

161. Mamelon is the part of .... of the Echinoids.
A. Apical disc B. Peristome
C. Spines D. Tubercle
(BHU-2015)

162. Which group provides the fast moving invertebrate?
A. Cephalopoda B. Echinodermata
C. Gastropoda D. Brachiopoda
(PU-2010)

163. When did the Trilobite disappear from the Earth?
A. End of Ordovician
B. End of Devonian
C. End of Permian
D. End of Jurassic *(PU-2010)*

164. Petrified wood is an example of:
A. Encrustation B. Substitution
C. Altercation D. Desiccation
(PU-2010)

165. Dinosaurs are reported from the rocks of:
A. Paleozoic B. Mesozoic
C. Tertiary D. Quaternary
(PU-2010)

166. The fossil contents of elementary canal of animals are knows as:
A. Burrows B. Mould
C. Trails D. Coprolites
(PU-2010)

167. Oldest backboned animals are:
A. Pisces B. Reptilia
C. Amphibia D. Ayes *(PU-2013)*

168. *Gangamopteris* can be differentiated from *Glossopteris* by:
A. Presentation of venation
B. Absence of midrib
C. Acute leaf margin
D. Size of the leaves *(PU-2013)*

169. The bivalvia shells are jointed together at dorsal side by:
A. Adductor muscle
B. Hinge plate
C. Ligament
D. Delthyrium *(PU-2013)*

170. Which is the oldest fossil?
A. Trilobita B. Algae
C. Fish D. Graptoloidea
(PU-2013)

171. Gastropod exhibit .... symmetry.
A. Bilateral B. Radial
C. Biradial D. Lack of
(PU-2013)

172. Echinoidea are .... water animals.
A. Marine B. Brackish
C. Fresh D. Terrestrial
(PU-2013)

173. Lamellibranches with two unequal adductors are:
A. Isomyaria B. Aniosomyria
C. Monomyria D. Dimyaria
(PU-2013)

174. Of the following which group consists of the fastest moving invertebrates?
A. Cephalopoda B. Gastropoda
C. Brachiopoda D. Trilobita
(PU-2013)

175. Fossilized contents of alimentary canal of animals are known as:
A. Burrows B. Mould
C. Trails D. Coprolites
(PU-2013)

176. The two words in Binomial nomenclature of organisms designate:
A. Order and family
B. Family and genus
C. Genus and species
D. Phylum and class *(PU-2013)*

177. When did the Trilobite disappear from the Earth?
A. End of Ordovician
B. End of Devonian
C. End of Permian
D. End of Jurassic *(PU-2013)*

178. Fossils of Archean life forms is widely represented by:
A. Graphite B. Stromotolites
C. Porifera D. Echinodermata
(PU-2013)

179. The Ordovician period is known as the age of:
A. Crinoids B. Graptolites
C. Brachiopoda D. Corals
(PU-2012)

180. The most ancient ancestor of man seems to have appeared during:
A. Paleocene B. Eocene
C. Pliocene D. Pleistocene
(PU-2012)

181. Fossils Ammonites indicate .... age.
A. Cretaceous B. Tertiary
C. Carboniferous D. Cambrian
(PU-2012)

182. Most fossils are of cretures that lived in:
A. Rivers B. The Sea
C. Fresh water D. The land
(PU-2012)

183. Which group provides the fast moving invertebrate?
A. Cephalopoda B. Echinodermata
C. Gastropoda D. Brachiopoda
(PU-2012)

184. The bottom dwellers living between low tide and high tide area are termed as:
A. Vagile B. Sessile
C. Nektonic D. Littoral (PU-2012)

185. The fossil contents of elementary canal of animals are known as:
A. Burrows B. Mould
C. Trails D. Coprolites
(PU-2012)

186. Petrified wood is a example of:
A. Encrustation B. Substitution
C. Alteration D. Desication
(PU-2012)

187. Which type of coiling is rare in gastropoda?
A. Dextral B. Sinistral
C. Armestral D. Trochospiral
(PU-2011)

188. When did the Trolibite disappear from the Earth?
A. Devonian B. Carboniferous
C. End of Permian D. End of Cretaceous
(PU-2011)

189. Dinosaurs are reported from the rocks of:
A. Silurian B. Devonian
C. Triassic D. Permian
(PU-2011)

190. Find the odd out:
A. *Fusus* B. *Conus*
C. *Oliva* D. *Cardita*
(PU-2011)

191. Stromatolite structures are related to:
A. Algae B. Fungi
C. Diatom D. Foraminifera
(PU-2011)

192. The shell of foraminifera are made up of:
A. Calcium Carbonate
B. Arenaceous sediments
C. Chitrin
D. Any of the above (PU-2011)

193. The most favourable environment for the preservation of fossils is:
A. Terrestrial B. Lacustrine
C. Marine D. Fluvial
(PU-2011)

194. Each coil in a shell of Gastropoda:
A. Increse towards the base
B. Decrease towards the base
C. Increase towards the apex
D. Uniform throughout the shell
(PU-2011)

195. In Trilobites, the most primitive Suture line is described as:
A. Hypoparian B. Opisthoparian
C. Proparian D. Gonatoparian
(DU-2016)

196. The bivalvia shells are jointed together at dorsal side by:
A. Adductor muscle B. Hinge plate
C. Ligament D. Delthyrium
(PU-2011)

197. *Megalodon* belongs to phylum:
A. Brachiopod B. Pelecypod
C. Gastropoda D. Cephalopod
(PU-2011)

198. Which of the following is not matched correctly?
A. *Planorbis* - Brachiopoda
B. *Montivaltia* - Coral
C. *Physa* - Gastropoda
D. *Anisomyarian* - Gastropoda (PU-2011)

199. Match features mentioned in Group-I with the fossil types in Group-II.

Group I	Group II
i. Escutcheon	P. Brachiopoda
ii. Plastron	Q. Nautiloidea
iii. Delthyrium	R. Echinoidea
iv. Siphuncle	S. Bivalvia

A. i - S, ii - R, iii - Q, iv - P
B. i - S, ii - R, iii - P, iv - Q
C. i - S, ii - P, iii - Q, iv - R
D. i - R, ii - Q, iii - P, iv - S
(IIT JAM-2016)

200. Which one of the following is an upper Gondwana flora?
A. *Ptillophyllum* B. *Phyllotheca*
C. *Glossopteris* D. *Schizoneura*
(IIT JAM-2016)

201. Choose the burrowing form(s) of bivalve.
A. *Mya* B. *Teredo*
C. *Pecten* D. *Venus*
(IIT JAM-2016)

202. Which one of the following gastropods is planispirally coiled?
A. Turntella B. Murex
C. Pleurotona D. Bellertophon
(ISM-2014)

203. What is the common distribution range of ammonoids?
A. Jurassic to tertiary
B. Triassic to cretaceous
C. Permian to Jurassic
D. Triassic to Jurassic *(ISM-2014)*

204. What is the most commonly accepted causes of the mass extinction of the dinosaurs?
A. A viral plague
B. An immense volcanic eruption
C. A meteor striking the earth
D. Global warming *(ISM-2014)*

205. The common form of foraminifer test range in size from:
A. 0.01 to 0.1 mm
B. 0.1 to 0.3 mm
C. 0.1 to 2 mm
D. 0.1 to 5 mm
(PUNJAB UNIVERSITY-2015)

206. What is the correct sequence of the evolution of Elephant?
A. Moeritherium, Tetralophodor, Trillophodon, Philoned, Alphus and Luxodont
B. Moeritherium, Philomel, Trilophodon, Tetralophodor, Luxodont, Alphus
C. Philomel, Moeritherium, Luxodont, Tetralophodor, Trilophodon, Alphus
D. Alphus, Philomel, Moeritherium, Fletralophodon, Luxodont
(PUNJAB UNIVERSITY-2015)

207. Which of the following is not exclusively marine?
A. Pelecypoda B. Trilobite
C. Ammonoidea D. Nautiloidea
(AMU-2016)

208. Which of the following is extinct?
A. Pelecypoda B. Echinoidea
C. Ammonoidea D. Nautiloidea
(AMU-2016)

209. Which of the following is living fossils?
A. Gingko B. Latimeria
C. Limulus D. All the above
(AMU-2016)

210. The concept of punctuated equilibrium was proposed by:
A. James Huttsn & Charles Lyell
B. Niels Eldredge & Stephen Gould
C. Charles Darwin & Alfered Russell Wallace
D. William Smith & Nicholas Steno
(ISM-2016)

211. Origin of the bird is sought among:
A. Sauropod dinosaurs
B. Dromacosauridth eropods
C. Prosauropods
D. Omithischian diosaurs *(ISM-2016)*

212. Saurischian and ornithischian dinosaurs are distinguished primarily by their
A. Pelvic structure
B. Mode of locomotion
C. Food habit
D. Brain size *(ISM-2016)*

213. Steinmann's trinty refers to
A. Mid-oceanic ridge volcanics
B. Subducion zone volcanism
C. A fossilised oceanic crust
D. Subalkaline volcanics *(ISM-2016)*

214. Tephonomy is the science dealing with:
A. Study of biogeographic of fossils
B. Reconstruction of paleoenvironment
C. Mode of preservation of fossils
D. Study of fossils spore and pollens
(DU-2016)

215. Nammulites were the abundant rock building foraminifera during:
A. Silurian
B. Devonian
C. Carboniferous
D. Eocene *(DU-2016)*

216. The Shells of a brachiopoda consists of:
A. Two equal but inequilateral valve
B. Two unequal but equilateral valve
C. Two equal and equilateral valve
D. None of the above *(DU-2016)*

ANSWER KEY

1	2	3	4	5	6	7	8	9	10
B	A	A	A	B	C	D	C	C	D
11	12	13	14	15	16	17	18	19	20
A	D	A	D	A	B	B	D	A	B
21	22	23	24	25	26	27	28	29	30
B	C	D	B	D	B	B	B	C	A
31	32	33	34	35	36	37	38	39	40
A	C	A	B	B	C	A	D	C	C
41	42	43	44	45	46	47	48	49	50
A	C	C	B	B	A	C	B	B	B
51	52	53	54	55	56	57	58	59	60
C	D	C	D	A	C	C	B	B	C
61	62	63	64	65	66	67	68	69	70
A	C	A	C	A	A	A	D	A	A
71	72	73	74	75	76	77	78	79	80
D	B	D	C	A	B	C	B	D	D
81	82	83	84	85	86	87	88	89	90
A	A	B	B	A	B	A	B	B	C
91	92	93	94	95	96	97	98	99	100
A	C	B	A	D	D	C	D	D	A
101	102	103	104	105	106	107	108	109	110
B	D	D	C	C	A	D	B	A	A
111	112	113	114	115	116	117	118	119	120
A	A	B	D	B	B	A	A	C	A
121	122	123	124	125	126	127	128	129	130
C	B	B	A	C	A	A	A	C	A
131	132	133	134	135	136	137	138	139	140
A	D	A	B	C	B	D	C	C	B
141	142	143	144	145	146	147	148	149	150
A	C	A	B	A	A	B	A	D	A
151	152	153	154	155	156	157	158	159	160
B	D	C	D	B	C	A	B	B	C
161	162	163	164	165	166	167	168	169	170
D	A	C	B	B	D	A	A	C	B
171	172	173	174	175	176	177	178	179	180
D	A	D	A	A	C	C	B	B	C
181	182	183	184	185	186	187	188	189	190
A	B	A	D	D	A	B	C	C	C
191	192	193	194	195	196	197	198	199	200

A	A	C	A	A	C	B	D	B	A
201	**202**	**203**	**204**	**205**	**206**	**207**	**208**	**209**	**210**
A,D	D	B	C	A	B	A	C	D	B
211	**212**	**213**	**214**	**215**	**216**				
B	A	B	C	D	B				

HINTS AND SOLUTIONS

1.

Fossils	Origin (Geological Age)	Extinct (Geological Age)
Graptolite	Cambrian	Carboniferous
Trilobite	Early Cambrian	Late Permian
Ammonite	Devonian	Cretaceous
Ediacaran	Proterozoic (630 Ma)	Cambrian (520 Ma)

2.

Brachiopoda genera	Geological Age	Characteristics
Lingula	Cambrian to recent	Living
Productus	Carboniferous to Permian	Extinct
Atrypa	Silurian to Devonian	Extinct
Spirifer	Ordovician to Permian	Extinct

3.

Lamellibranchia genera	Dentition pattern	Geological age
Glycimeris	Texodont	Cretaceous to present
Trigonia	Schizodont	Triassic to present
Lucina	Heterodont	Cretaceous
Mytilus	Dysodont	Triassic to present

4.

Fossils	Geological age
Trilobite	Cambrian to Permian
Ammonoide	Devonian to cretaceous
Rugose coral	Ordovician to Permian
Pelecypod	Cambrian to recent

5.

Fossils	Origin	Geological age
Foraminifera	Marine	Cambrian to resent
Ostracoda	Marine/ Aquatic	Ordovician to resent
Conodont	Marine	Cambrian to Triassic
Radiolaria	Marine	Cambrian to recent

13.

Horse	Age	Characteristics
Eohippus	Eocene	First stage
Hyracotherium	Oligocene	Second stage
Epihippus	Oligocene	Third stage
Mesohippus	Oligocene	Fourth stage

21.

Example	Family
Stegodone	Elephantidae
Stegolophodon	Equidae
Ramapithecus	Hominidae
Gigantopithecus	Hominidae

22.

Flora	Characters
Glassopteris	Lower gondwana flora which is Permean age.
Gangamopteris	Lower gondwana plant fossils.
Buriadia	Vasicular plant with conifer
Schizoniura	Triassic of gondwana flora.

23. Genera	Phylum	Geological age	Indian occurrence
Mya	Mollusca	Carboniferous	Jurassic of Cutch Gujarat
Micraster	Echinoides	Late cretaceous	Cretaceous of Narmada valley
Olenoides	Arthopoda	Cambrian	Cambrian of Kashmir
Leptaena	Brachiopoda	Ordovician to permian	Cretaceous of Trichinopoly.

24. Morphological feature	Pelecypoda	Brachiopoda
Umbo	Present	Absent
Deltidium	Absent	Present
Pallial sinus	Present	Absent
Resilifier	Present	Absent

26. Morphological feature	Name	Example
Masticatory apparatus in Echinoidea	Aristotle's lentern	Regular Echinoids
Triangular cavity near umbo of Brachiopoda	Delthyrium	Orthis
Calcareous secretions in siphuncle of Cephalopoda	Cameral deposits	Ammonites
Genital plate in Echinoidea	Madreporite	Regular Echinoids

27. Fossils	Fossils type	Age	Remarks
Hipparion	Horse	Pliocene	Extincts
Ramapithecus	Human	Miocene	Extincts
Equus	Horse	Recent	Present
Stegolophodon	Elephas	Miocene-pliocene	Extinct

30. Types of man	Age
Pithecanthropus	Mid pleistocene
Ramapithecus	Upper miocene
Australopithecus	Early pleistocene
Sinanthropus	Holocene

35. Class	First appearance
Coral	Cambrian
Trilobites	Cambrian
Brachiopods	Neoproterozoic
Cephalopods	Cambrian

36. Fossils	Class
Olenus	Trilobite
Nautilus	Nautiloidea
Otoceras woodwardi	Ammonoid
Macrocephalites	Ammonoid

37. Fossils	Characteristics
Conodonts	Phosphatic tooth like microfossils
Thecodont	Dinosour fossils
Bathydont	Mollusca
Coprolite	Trace fossils

39. Trilobite types	Characteristics
Micropugous	Pygidium small than cephalon
Macropygous	Pygidium larger than cephalon
Isopygous	Pygidium equal to cephalon

40. Trilobites	Geological age
Redlichia	Lower Cambrian
Olenellus	Upper Cambrian
Paradoxides	Middle Cambrian
Olenus	Upper Cambrian to Ordovician

42. Gastropod	Genus Example
Uncoiled	*Cypraea*
Dextrally coiled	*Natica*
Sinistrally coiled	*Physa*
Planispirally coiled	*Bellerophon*

43.

Fossils	Properties
Agnostus	Trilobite eye absent
Microdiscus	Fungus
Remopleurides	Trilobite eye present with large number of lances
Olenellus	Trilobite eye present

44.

Environment	Fossils Preservation
Terrestrial	Not suitable for fossils preservation
Lacustrine	Least favourable
Fluvial	Suitable
Marine	Most suitable for fossils preservation

45.

Fossils	Class
Nautilus	Cephalopoda
Lima	Bivalve
Nucula	Bivalve
Trigonia	Bivalve

52.

Feature	Texas	Example
Monomyraian	Pelycypoda	*Spondylus*
Deltidial plates	Brachiopoda	*Rhynchonella*
Corona	Echinoidea	*Bothriocidaris*
Epitheca	Anthozoa	Corals

62.

Gastropoda	Remarks
Natica	Dextral form
Murex	Dextral form
Physa	Sinistral form
Voluta	Dextral form

66.

Microfossils	Composition
Conodont	Phosphatic
Diatoms	Siliceous
Dinoflagellates	Siliceous
Cocolith	Calcareous

74.

Dention pattern	Bivalvia Genus	Age
Hetrodent	*Cerastoderma*	Paleocene to quaternary
Desmodent	*Mya*	Tertiary to present
Dysodant	*Mytilus*	Triassic to present
Schizodont	*Trigonia*	Triassic to present

75.

Flora	Gondwana type
Gangamopteris	Lower gondwana
Cladophlebis	Upper gondwana
Sphenopteris	Upper gondwana
Ptilophyllum	Upper gondwana

81.

Name	Characteristics
Motlivaltia	Coral
Globigerina	Foraminifera
Cardita	Pelycpoda
Calymene	Trilobite

82.

Name	Characteristics
Schizoneura	Lower gondwana plant fossils
Favosites	Corals
Productus	Brachiopoda
Syringothyris	Brachiopoda

84.

Trilobite	Age
Olenus	Mid Cambrian
Paradoxides	Low Cambrian to Permian
Olenellus	Low Cambrian
Phocous	Ordovician to Devonian

85.

Flora	Gondwana type
Gangamopteris	Lower gondwana
Ptilophyllum	Upper gondwana
Otozamites	Upper gondwana
Nilsonia	Upper gondwana

86. Pelecypod	Umbos	Example
Opisthogyrate	Umbo pointly towards posterior side of the wall.	*Trigonia*
Prosogyrate	Pointly towards anterior side.	*Arca*
Orthogyrate	Towarsd each others of both.	*Pecten*
Spirogyrate	Umbo curcing towards centers	*Graphea*

87. Types of fossils	Characteristics
Conodonts	Phosphatic microfossil
Thecodont	Reptiles
Bathydont	Phosphatic microfossil
Coprolites	Burrows

88. Cephalopoda	Suture
Ceratites	Rounded suture (Complex)
Nautilus	Straight suture (Simple)
Goniatites	Angular suture (Complex)
Ammonites	Complex suture

89. Trilobites	Properties
Micropygous	Pygidium smaller than cephalon
Macropygous	Pygidium larger than cephalon
Isopygous	Pygidium and cephalon same size

90. Trilobites	Eye
Asaphus	Present
Oleneltus	Large eye
Agnostus	Blind
Olenus	Present

92. Gastropods	Forms
Turbo	Dextral
Natica	Dextral
Physa	Sinistral
Crepidula	Dextral

96. Fossils	Characteristics (Phyllum)
Nautilus	Cephalopoda
Lingula	Brachiopoda
Nucula	Pelycepoda
Paradoxides	Trilobite

98. Species	Phyllum (Genera)	Geological age
Terebratula	Brachiopoda	Paleozoic

101. Fossils taxa	Character-istics	Genus	Age
Parad-oxides	Genal spines	Trilobita	Early to middle Cambrian
Didymo-graptus	Two pendent stipes	Graptolite	Early to late Ordovician
Physa	Sinistral coiling	Gastro-poda	Jurassic to present
Ceratites	Sutures with undivided saddles	Cephalo-poda	Devonian to Triassic

102. Organism	Envionment
Echionoids	Marine
Brachiopods	Only Marine
Corals	Marine
Pulmonate gastropods	Terestrial, Marine and Fresh water

103. Fossils	Characteristics	Age
Micraster	Infaunal echinoides	Cretaceous
Phacops	Aquatic trilobite	Ordovician to Devonian
Equas	Modern horse	Pliocene
Barappa-saurus	Big leg dinosaur	Jurassic

104.

Characteristics	Zones
Range of a fossils species	Range zone
Absence of fossils species	Gap/Interval zone
Maximum abundant of a fossils species	Acme zone
Migration of fossils species	Overlap zone

105.

Stratigraphic units	Fossils Types	Example
Intratrappen beds	Invertebrate	*Natica*, *Physa*
Gondwana rocks	Vertibrate	Fishes, Amphibians
Siwalik rocks	Vertibrate	Molluscus, Ostrocods
Zewan beds	Invertebrate	Brachiopods

110.

Organism	Class	Composition
Nautilus	Nautiloidea	Aragonitic
Crania	Gastropoda	Calcium Carbonate
Trigonia	Pelycepoda	Calcareous
Patella	Cephalopoda	Aragonitic

112.

Types of species	Characteristics
Neotype	The specimen selected to replace the lost holotype.
Topotype	The specimen selected to same locality of holotype.
Plastotype	The specimen selected for the redefining.
Cotype	The multiple specimen selected for description.

123.

Fossils	Fossils Types	Age
Hipparion	Horse	Miocene
Equus	Horse	Upper Pliocene
Ramapithecus	Man	Upper Miocene
BOS	Fishes	Miocene

200.

Plant fossils	Gondwana type
Ptillophyllum	Upper Gondwana
Phyllotheca	Lower Gondwana
Glossopteris	Lower Gondwana
Schizoneura	Lower Gondwana

202.

Gastropoda	Coiled
Turntella	Conispiral
Murex	Helicoiled
Pleurotona	Conispiral
Bellertophon	Planispiral

206.

Elephants	Geological age
Moeritherium	Eocene
Philomel	Oligocene
Trilophodon	Miocene
Tetralophodor	Pliocene
Luxodont	Pleistocene
Alphus	Recent

210. Punctuated equilibrium Theory: That is certainly true in many cases, because the chances of each of those critical changing forms having been preserved as fossils are small. But in 1972, evolutionary scientists Stephen Jay Gould and Niles Eldredge proposed another explanation, which they called "punctuated equilibrium." That is, species are generally stable, changing little for millions of years. This leisurely pace is "punctuated" by a rapid burst of change that results in a new species and that leaves few fossils behind.

According to this idea, the changes leading to a new species don't usually occur in the mainstream population of an organism, where changes wouldn't endure because of so much interbreeding among like creatures. Rather, speciation is more likely at the edge of a population, where a small group can easily become separated geographically from the main body and undergo changes that can create a survival advantage and thus produce a new, non-interbreeding species.

This hypothesis predicts that the fossil record at any one site is unlikely to record the process of speciation. If a site records that the ancestral species lived there, the new species would probably be evolving somewhere else. The small size of the isolated population which is evolving into a new species reduces the odds that any of its members will be fossilized. The new species will only leave fossils at the same site as the old one if it becomes successful enough to move back into its ancestral range or different enough to exist alongside its relatives.

STRUCTURAL GEOLOGY

1. In an undulated topography, a set of uniformly inclined beds show 'V'-shaped outcrop pattern with apex pointing downstream in a valley. This indicates:

A. The dip of the beds is more than the slope of the valley
B. The dip of the beds is less than the slope of the valley
C. The beds are vertical
D. The beds are horizontal

(IIT-JAM 2015)

2. On a toposheet of 1:50,000 scale, the distance between upper and lower traces of a bed is 5 cm. The actual distance on ground in kilometers is:

A. 2 B. 2.5
C. 4 D. 5 *(IIT-JAM 2015)*

3. On a flat topography the outcrop width of a bed is 30 m. If the true dip of the bed is 30°, the actual thickness of the bed in meters is?

A. 30 B. 15
C. 45 D. 10 *(IIT-JAM 2015)*

4. An ore body is dipping 30° towards west on a flat topography. At what distance (in meters) should a borehole be placed so that it intersects the ore body perpendicularly at a vertical depth of 70 meters?

A. 16.61 B. 161.63
C. 1.63 D. 1616.3

(IIT-JAM 2015)

5. The lithostatic pressure in Mega Pascal (MPa) at a depth of 10 km in a granite batholith having density 2700 kg/m^3 is (Acceleration due to gravity = 9.8 m/s^2):

A. 26.45 B. 264.5
C. 2.64 D. 2645

(IIT-JAM 2015)

6. In an oblique slip fault having 30° dip, the net slip is 10 meters at an angle of 45° to the strike of the fault plane. The dip slip component of the fault in meters is:
(Answer to be given up to 2 decimal places).

A. 7.00 B. 7.07
C. 7.17 D. 7.70

(IIT-JAM 2015)

7. If on a geological map, contour lines run parallel to contact lines, the beds are:

A. Horizontal B. Vertical
C. Inclined D. Any of the above

(AMU-2015)

8. A thrust fault is:

A. a normal fault with fault plane at<45 degree
B. A normal fault with fault plane at >45 degree
C. A reverse fault with fault plane at < 45 degree
D. A reverse fault with fault plane at > 45 degree *(AMU-2015)*

9. Which of the following type of fold is associated with migmatite?

A. Reclind B. Suprataneous
C. Ptygmatic D. Piercing

(AMU-2015)

10. The angle between any line and its horizontal projection measured in a vertical plane is the .... of the line.

A. Pitch B. Plunge
C. Dip D. Dip and Strike

(DU-2015)

11. Maximum dip of a plane can be obtained on a line that is at .... angle with strike of the bed.

A. 25° B. 45°
C. 90° D. 65° *(DU-2015)*

12. How many components are generally required to define stress system at a point within a three-dimensional solid body?

A. Four B. Nine
C. Three D. Two *(DU-2015)*

13. If fold axis of a fold assumes near perpendicular relationship with axial trace of the fold, then the fold is interpreted as a:

A. Vertical fold B. Recumbent fold
C. Reclined fold D. Horizontal fold
(DU-2015)

14. What kind of fold will show parallelism between axial trace and fold axis in map?

A. Neutral fold B. Horizontal fold
C. Reclined fold D. None of the above
(DU-2015)

15. One can expect high angle relationship between bedding and cleavage in:

A. Fold limb
B. Any part of a fold
C. Only when the fold is overturned
D. Fold hinge *(DU-2015)*

16. Bedding fault is a special type of:

A. Strike fault
B. Strike-slip fault
C. Oblique slip fault
D. Dip-slip fault *(DU-2015)*

17. A discontinuous cleavage formed by micro folding of an existing cleavage/schistosity is called:

A. Disjunctive cleavage
B. Slaty cleavage
C. Crenulation cleavage
D. Transected cleavage *(DU-2015)*

18. In the given stereo plot, poles to three bedding planes are plotted as points (i), (ii) and (iii). Find the correct order of increasing dips of the beds.

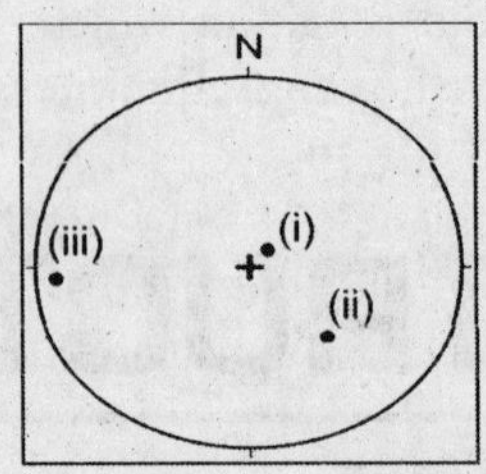

A. (i) - (ii) - (iii) B. (i) - (iii) - (ii)
C. (ii) - (i) - (iii) D. (iii) - (i) - (ii)
(IIT-JAM 2015)

19. In the given map, the true dip (in degree) of the bedding plane X-Y is ....

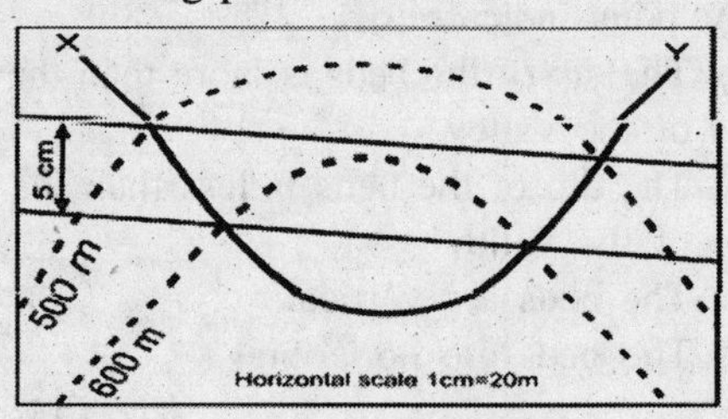

A. 40 B. 45
C. 60 D. 90 *(IIT-JAM 2015)*

20. Identify the pole of the great circle in the given stereographic projection

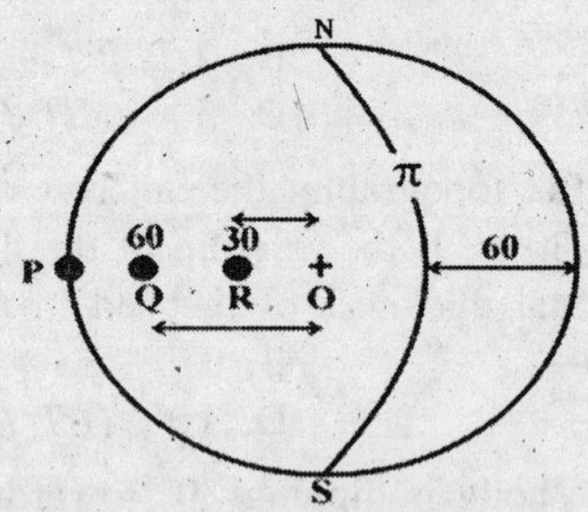

A. P B. Q
C. R D. R *(IIT-JAM 2014)*

21. Transform fault is a:

A. Strike-slip fault B. Normal fault
C. Reverse fault D. Thrust fault
(IIT-JAM 2014)

22. If the strike of the inclined bed is N15°E, the dip direction can be:

A. S75° W B. N75° E
C. S75° E D. S15° W
(IIT-JAM 2014)

23. Parallel fold is one in which:
A. The orthogonal thickness is same all around the fold.
B. The thickness is same parallel to the axial plane of the fold.
C. The outer and inner arc have same curvature.
D. The dip isogons are parallel to each other.
(IIT-JAM 2014)

24. Stress-strain deformation curves for four rocks P, Q, R and R are given below. The dots at the ends of the curves indicate the respective failure points of each rocks. Choose the rock with the highest brittle strength.

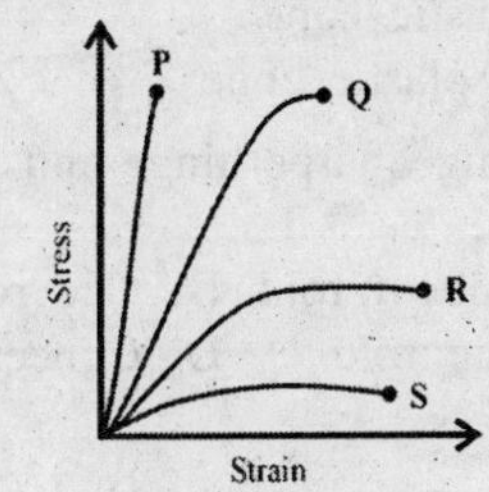

A. P B. Q
C. R D. S *(IIT-JAM 2014)*

25. Which of the following symbols indicate the Synform?

A.  B. 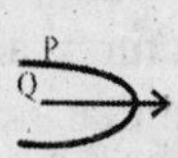

C.  D. 

(IIT-JAM 2014)

26. The type unconformity in which older and younger formation remain:
A. Non-conformity
B. Disconformity
C. Angular unconformity
D. Local unconformity
(BHU- 2014)

27. The tightness of a fold should be estimated in sections:
A. Parallel to both the fold axis and the axial plane
B. Parallel to fold axis and perpendicular to the axial plane
C. Perpendicular to the profile plane of the fold
D. Perpendicular to the fold axis
(IIT-JAM 2013)

28. Identify the primary structure:
A. Bedding plane B. Cleavage plane
C. Fault plane D. Joint plane
(BHU-2013)

29. Younger beds will always be found in the:
A. Opposite direction of the dip
B. Direction of dip
C. Direction of strike line
D. Inclined direction of dip and strike
(BHU-2013)

30. Which fold has two hinges?
A. Fan fold B. Chevron fold
C. Isoclinal fold D. Box fold
(BHU-2013)

31. If a fault plane is inclined with an angle of 35°, then the hade will be:
A. 145° B. 125°
C. 55° D. 45° *(BHU- 2013)*

32. The horizontal distance between two consecutive contours is:
A. Horizontal equivalent
B. Horizontal interval
C. Plane ground
D. Vertical surface *(BHU-2013)*

33. Match the structural processes in Group I with their products in Group II.

Group I	Group II
P. Pressure solution	1. Boudin
Q. Layer-parallel extension	2. Bending fold
R. Layer-perpendicular shortening	3. Ptygmatic fold
S. Layer-parallel shortening	4. Stylolite

A. P-4, Q-2, R-3, S-1
B. P-3, Q-2, R-4, S-1
C. P-4, Q-1, R-2, S-3
D. P-2, Q-1, R-3, S-4

34. Match the fold types in Group I with inter-limb angles of folds in Group II.

Group I	Group II
P. Tight	1. 0°
Q. Open	2. 120° – 70°
R. Isoclinal	3. 70° – 30°
S. Close	4. 30° – 0°

A. P-1, Q-2, R-3, S-4
B. P-4, Q-2, R-1, S-3
C. P-3, Q-2, R-4, S-1
D. P-3, Q-2, R-1, S-4 *(IIT-JAM 2012)*

35. The distance between any two joints may be described as:
A. Slip cleavage
B. Crenulation cleavage
C. Fracture cleavage
D. Bedding cleavage *(BHU-2012)*

36. A contour line intersect a ridge or a valley line:
A. Perpendiculary B. Parallely
C. Vertically D. Obliquily *(BHU-2013)*

37. If hanging wall is moved down, then the fault is:
A. Normal B. Reverse
C. Strike-slip D. Indeterminate *(BHU-2013)*

38. Recumbent fold is characterized by:
A. Vertical axial plane
B. Horizontal axial plane
C. Inclined axial
D. Oblique axial plane *(BHU-2013)*

39. When the minimum principle stress (σ_3) is vertical, the resultant fault is?
A. Normal fault B. Oblique slip fault
C. Thrust D. Strike slip fault *(BHU-2013)*

40. The ratio of the normal stress to strain for a rock is known as:
A. Modulus of rigidity
B. Young modulus
C. Shear modulus
D. Bulk modulus *(BHU-2013)*

41. Lateral offset in drainage line is commonly associated with:
A. Strike-slip fault
B. Reverse fault
C. Normal fault
D. Thrusting *(BHU-2013)*

42. Main boundary fault lies:
A. To the north of the Himalaya
B. To the south of the Himalaya
C. Along the river Indus
D. Along the river Ganga *(BHU-2013)*

43. Brittle material have a:
A. Large plastic zone
B. No elastic zone
C. No plastic zone
D. Small plastic zone *(BHU-2013)*

44. Fold having Sharpe hinge and smooth side is known as:
A. Recumbent fold B. Chevron fold
C. Cuspate fold D. Concentric fold *(BHU-2013)*

45. Mullions are mainly parallel to:
A. Lineation B. Foliation
C. Fold axis D. Plunge of fold *(BHU-2013)*

46. The columnar structures bounded by cylindrical structures are known as:
A. Cylinder B. Rodding
C. Mullions D. Boudinage *(BHU-2013)*

47. Which exhibit maximum strain level among the following?
A. Main boundary thrust
B. Main central thrust
C. Main Frontal thrust
D. High angle thrust *(BHU-2013)*

48. Dip of the bed is always measured in:
A. Horizontal plane
B. Vertical plane
C. Inclined plane
D. Axial plane *(BHU-2012)*

49. An unconformity with almost parallel beds overlying and underlying the surface of erosion is:

A. Non-conformity
B. Disconformity
C. Para-unconformity
D. Angular unconformity (BHU-2012)

50. The folds with thickened crests or troughs and thinner limbs are:
A. Open folds B. Closed folds
C. Isoclinal folds D. Overfolds
(BHU-2012)

51. Abrupt termination of strata marks the presence of:
A. Fold and joint B. Joint
C. Fold D. Fault (BHU-2012)

52. The folds having both the limbs overturned are:
A. Cross folds B. Conjugate folds
C. Tight folds D. Fan folds
(BHU-2012)

53. The structure having dip towards a common central point from all sides is:
A. Basin B. Fault
C. Dome D. Joint (BHU-2012)

54. A group of folds having essentially parallel axial planes is:
A. Recumbent folds
B. Isoclinal folds
C. Conjugate folds
D. Overturned folds (BHU-2012)

55. Mullions are formed under:
A. Compressive stress regime
B. Tensile stress regime
C. Shearing stress regime
D. Tensile and shearing stress regimes
(BHU-2012)

56. The behaviour of perfectly elastic body is governed by:
A. Hook's law B. Hilt's law
C. Bulk modulus D. Bode's law
(BHU-2012)

57. Compressibility can be described as the reciprocal of:
A. Bulk modulus
B. Young's modulus
C. Rigidity modulus
D. Young's and rigidity modulii
(BHU-2012)

58. Faults striking across structures like fold axes, schistosity, lineation, etc., are known as:
A. Transverse fault
B. Longitudinal fault
C. Diagonal fault
D. Bedding fault (BHU-2012)

59. A limited area of younger rocks surrounded by the older rocks is called:
A. Outlier B. Overlap
C. Inlier D. Offlap (BHU-2012)

60. A group of small size faults overlapping each other in the area of their occurrence is:
A. Parallel fault
B. Radial fault
C. Enechelon fault
D. Peripheral fault (BHU-2012)

61. Non-conformity is synonymous to:
A. Heterolithic unconformity
B. Parallel unconformity
C. Angular unconformity
D. Disconformity (BHU-2012)

62. Gravity faults are formed under:
A. Compressive stress regime
B. Tensile stress regime
C. Shear stress regime
D. Effects of torsional forces (BHU-2012)

63. Joints developed perpendicular to the fold axis are termed as:
A. Columnar joints
B. Release joints
C. Extension joints
D. Cross joints (BHU-2012)

64. The greatest principal stress axis vertical in:
A. Normal faults B. Reverse faults
C. Thrust faults D. Strike-slip faults
(BHU-2012)

65. Schuppen structures are associated with:
A. Normal faulting B. Reverse faulting
C. Thrust faults D. Recumbent folding
(BHU-2012)

66. A normal fault has a dip of 45° towards east, and a throw of 100 meters. The heave of the fault (in meters) is:
A. 50 B. 100
C. 200 D. 400 *(IIT-JAM 2011)*

67. Extensional regime in an area is characterized by the presence of:
A. Synform and antiform
B. Thrust fault
C. Normal fault
D. Strike-slip fault *(IIT-JAM 2011)*

68. In an almost flat terrain steeply dipping beds exhibit repetition as ABCDEDCBA. It is because of:
A. Erosion B. Faulting
C. Folding D. None of the above
(BHU-2014)

69. A limited area of older rock surrounded by younger rocks on a geological map is called:
A. Inlier B. Outlier
C. Offlap D. Overlap
(BHU-2014)

70. Listric fault is:
A. Steep dipping normal fault
B. Gentle dipping normal fault
C. Steep dipping fault at top and gentle dipping at bottom
D. Gentle dipping fault at top and steel dipping at bottom
(BHU-2014)

71. In a normal stratigraphic sequence, erosion can sometimes expose older rocks within younger rocks. Such exposure are called:
A. Inlier B. Window
C. Outlier D. Klippe *(BHU-2014)*

72. A major fold structure composed of many smaller anticlines and synclines and having the general form of an arch is called:
A. Geanticlines B. Anticlinorium
C. Synclinorium D. Geosynclines
(BHU-2014)

73. The principal axes of stress are:
A. Normal to surface on which no shear stress acts
B. Direction of application of force
C. Direction of maximum compression
D. Direction of maximum tension
(BHU-2014)

74. The dip of the S-surface is best measured on:
A. Horizontal B. Vertical plane
C. Inclined plane D. None of the above
(BHU-2014)

75. Which one is an example of structure having quaquaversal dip?
A. Depression B. Plateau
C. Dome D. Horse *(BHU-2014)*

76. The pole to S-surface plotted on the lower hemisphere of equal area net lies on a circle. The circle is called:
A. π-S circle B. Great circle
C. Small circle D. None of the above
(BHU-2014)

77. The quadratic elongation (λ) is:
A. (1+e) B. $(1+e)^2$
C. $(1+e)/(1+e^2)$ D. None of the above
(BHU-2014)

78. Thrust faults are developed when the maximum compressive stress (σ_1) is:
A. Vertical B. Horizontal
C. Inclined D. All of the above
(BHU-2014)

79. The fold in which the axes plunges directly down the dip of the axial surface is termed as:
A. Upright fold B. Reclined fold
C. Cascade fold D. Drag fold
(BHU-2014)

80. Horst and graben structure is mainly associated with:
A. Subduction zone
B. Divergent boundary
C. Collision zone
D. Transform boundary *(BHU-2011)*

81. Angle of inclination in a fault plane measured from vertical axis is called:
A. Heave B. Throw
C. Hade D. Rake *(BHU-2011)*

82. The angle of inclination of hing line with the horizontal is:
A. Plunge B. Dip
C. Pitch D. Strike *(BHU-2011)*

83. Thrust faults are:
A. High angle reverse fault
B. Low angle reverse fault
C. High angle normal fault
D. Low angle normal fault *(BHU-2011)*

84. In fold classification Ramsay's classified the folds on the basis of:
A. Dip isogon B. Dip amount
C. Interlimb angle D. Symmetry
(BHU-2011)

85. Foliation in a rock is defined as:
A. A planar feature with randomly oriented minerals
B. A planar feature with well-oriented minerals
C. A linear feature with randomly oriented minerals
D. A linear feature with well-oriented minerals
(BHU-2011)

86. When a bedded sedimentary rock overlying the non-bedded igneous rock, then it is:
A. Disconformity
B. Nonconformity
C. Angular unconformity
D. Blended unconformity *(BHU-2011)*

87. When younger rocks are surrounded by older rock, then this structure is called:
A. Klippe B. Window
C. Inlier D. Outlier
(BHU-2011)

88. Basalts usually comprises:
A. Mural joints B. Columnar joints
C. Sheet joints D. Bedding joints
(BHU-2011)

89. An up-fold where beds dip radially outward in all directions from the centers is:
A. Basin B. Klippe
C. Dome D. Window
(BHU-2011)

90. In a fold, the lower limb gets overturned with nearly horizontal axial plane, it is:
A. Recumbent fold
B. Overturned fold
C. Isoclinal fold
D. Fan fold *(BHU-2011)*

91. The maximum amount of slope along a line perpendicular to the strike is:
A. True dip B. Apparent dip
C. Plunge D. Pitch
(BHU-2011)

92. Sheet joints usually confined to:
A. Quartzite B. Granite
C. Shale D. Slate *(BHU-2011)*

93. In an anticline, layers of earth material converge in which direction?
A. Upward
B. Parallel to bedding planes
C. Downward
D. No convergence *(BHU-2011)*

94. In similar folds, the bed is:
A. Thinner at limb and thicker at hinge
B. Thickness does not change
C. Thicker at limb and thinner at hinge
D. The thinnest at crest
(BHU-2011)

95. In reverse fault, σ_1, stress is applied:
A. Horizontal to block
B. Perpendicular to block
C. At some angle
D. No stress applies *(BHU-2011)*

96. The transform faults are associated with the:
A. Plate boundaries B. Bedding planes
C. Veins D. Mineral deposits
(BHU-2011)

97. Fold is formed as a result of:
A. Tensional force
B. Shear force
C. Compression force
D. No force applied *(BHU-2011)*

98. In box folds, there are:
A. One hinge B. Two hinges
C. Sharp hinges D. Quaquaversal dip
(BHU-2011)

99. In isoclinals folds, the limbs are:
A. Dipping in same direction with different dip
B. Dipping towards each other
C. Dipping in same direction with same dips
D. Dipping away from each other
(BHU-2011)

100. The pitch of a fold axis on the axial plane, dipping at 45 degree is 88 degree. The fold is:
A. Vertical fold B. Inclined fold
C. Reclined fold D. Recumbent fold
(IIT-JAM 2010)

101. Shown below is a profile section of a folded sequence consisting of limestone, cross bedded sandstone and shale. The fold can be termed as:

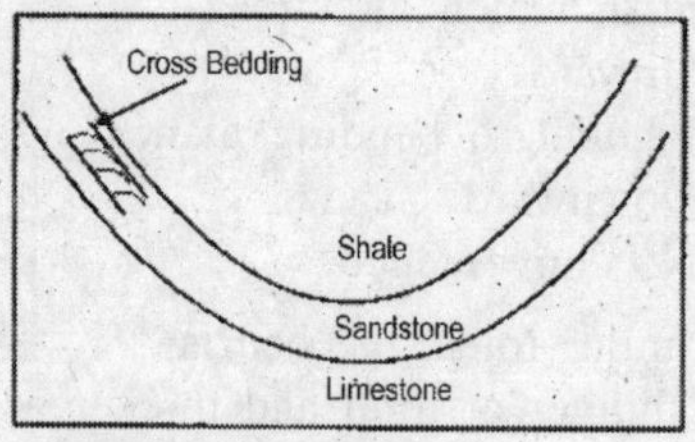

A. Synformal syncline
B. Synformal anticline
C. Antiformal syncline
D. Antiformal anticline *(IIT-JAM 2010)*

102. Amount of dip is maximum:
A. In the direction of strike
B. At 90° to the strike direction
C. At 45° to the strike direction
D. At 0° to the strike direction *(BHU-2010)*

103. In a recumbent fold the axial plane is nearly:
A. Horizontal B. Vertical
C. Inclined at 45° D. Inclined at 30°
(BHU-2010)

104. A strike line with a given value is called
A. Horizontal equivalent
B. Contour interval
C. Stratum contour
D. Contour line *(BHU-2010)*

105. Columnar joints are seen in
A. Basalts B. Sandstones
C. Gneissic rocks D. Pegmatites
(BHU-2010)

106. Boudins are formed in
A. Extensional regimes
B. Compressional regimes
C. Zero stress regimes
D. Fault regimes *(BHU-2010)*

107. Pitch is another term for
A. Hade B. Rake
C. Plunge D. Axial angle
(BHU-2010)

108. An imaginary line which joins points of equal elevation is called
A. Contour line B. Isothermal line
C. Isoseismal line D. Isograde line
(BHU-2010)

109. Folds with inter-limb angle >70° are called
A. Open folds B. Closed folds
C. Tight folds D. Isoclinal folds
(BHU-2010)

110. Gouge is associated with:
A. Joints B. Folds
C. Unconformities D. Faults *(BHU-2010)*

111. Folds having parallel limbs are:
A. Parallel folds B. Similar folds
C. Isoclinal folds D. Infinite folds
(BHU-2010)

112. The following two figure, 1 and 2, show outcrop patterns on flat grounds, What are the correct interpretations of structures in 1 and 2?

Fig. 1

Fig. 2

A. 1-fold; 2-fold
B. 1-fold; 2-fault
C. 1-fault; 2-fault
D. 1-fault; 2-fold *(IIT-JAM 2009)*

112. Which one of the following suggests the presence of an unconformity?
A. Gouge B. Mylonite
C. Cataclasite D. Basal conglomerate
(IIT-JAM 2009)

113. Which one of the following fault types is developed if the intermediate principal stress axis (σ_2) is vertical?
A. Normal B. Reverse
C. Thrust D. Strike-slip
(IIT-JAM 2009)

114. If the strike of a fault is parallel to the dip direction of the disrupted bed, the fault will be called:
A. Dip fault B. Strike fault
C. Strike-slip fault D. Dip-slip fault
(BHU-2009)

115. An unconformity in which the overlying and underlying beds are not parallel is termed as:
A. Non-conformity
B. Disconformity
C. Para-unconformity
D. Angular unconformity *(BHU-2009)*

116. An isoclinal fold has interlimb angle between:
A. 0-100 B. 10-300
C. 30-700 D. 70-1200
(BHU-2009)

117. Repetition and omission of some strata mark the presence of:
A. Fold and joint B. Joint
C. Fold D. Fault *(BHU-2009)*

118. The folds in which the thickness of rocks is not effected during the process of folding are?
A. Open folds B. Closed fold
C. Tight fold D. Gentle fold
(BHU-2009)

119. Which one is an example of a structure having quaquaversal dip?
A. Joint B. Fault
C. Dome D. Basin *(BHU-2009)*

120. Keeping other condition same, If the strain rate is increased, the behaviour of the rock will be more:
A. Ductile B. Brittle
C. Elastic D. Plastic *(BHU-2009)*

121. 'Dilation' is defined as:
A. Change in form
B. Change in volume
C. Neither change in form nor change in volume
D. Either change in form, or change in volume
(BHU-2009)

122. When a body is slightly strained, the stress opposing the change is proportional to the strain produced. This is the statement for:
A. Hook's law B. Young's modulus
C. Bulk modulus D. Bode's law
(BHU-2009)

123. Dip is the amount of inclination with respect to horizontal plane and always measured:
A. In the horizontal plane
B. In the axial plane
C. In the vertical plane
D. In any desired plane *(BHU-2009)*

124. Parallel folding is also termed as:
A. Similar folding B. Flexural folding
C. Flexural flow D. Concentric folding
(BHU-2009)

125. A limited area of older rocks surrounded by the younger rocks is called:
A. Outlier B. Overlap
C. Inlier D. Offlap *(BHU-2009)*

126. A group of faults appear emerging outward from a common central region is:
A. Parallel faults B. Radial faults
C. Enechelon faults D. Peripheral faults
(BHU-2009)

127. Heterolithic unconformity is also known as:
A. Non-conformity
B. Parallel unconformity
C. Angular unconformity
D. Disconformity *(BHU-2009)*

128. The angle between any line and its horizontal projection, measured in a vertical plane is:
A. Pitch B. Plunge
C. Dip D. Strike *(BHU-2009)*

129. Joints perpendicular to the axis of fold more common in orogenic belts are termed as:
A. Columnar joints
B. Release joints
C. Extension joints
D. Cross joints *(BHU-2009)*

130. An arrangement of elongated mineral grains along continuous lines is called:
A. Lithification B. Crenulation
C. Lineation D. Petrofabrication
(BHU-2009)

131. Slaty cleavage is best developed in the rocks rich in:
A. Arenaceous minerals
B. Micaceous minerals
C. Calcareous minerals
D. Chloritic minerals *(BHU-2009)*

132. Faults striking across structures like fold axes, schistosity, lineation etc., are known as:
A. Transverse fault B. Longitudinal fault
C. Diagonal fault D. Bedding fault
(BHU-2009)

133. In a contour map, if higher contours are closely placed, it indicates:
A. Steep slope B. Uniform slope
C. Concave slope D. Convex slope
(BHU-2009)

134. Which one of the following structure typically indicates direction of movement of fault blocks along the fault plane?
A. Slickenside and foliation
B. Slickenside and Riedel planes
C. Riedel planes and foliation
D. Foliation and joint *(IIT-JAM 2008)*

135. On an undulating ground, a sedimentary sequence of beds X, Y and Z shows repetition as Z, Y, X, Y, X, Z, Y, X. If the bed X is the oldest and Z is the youngest, then this repetition indicates the structural sequence of:
A. Syncline – anticline – fault – homocline
B. Anticline – fault – homocline – fault
C. Fault – homocline – fault – syncline
D. Homocline – fault – homocline – fault – homocline *(IIT-JAM 2008)*

136. Fold axis of a single-folded surface is located at:
A. Locus of minimum curvature of the folded surface
B. Locus of maximum curvature of the folded surface
C. Any place on the folded surface in its direction of rotation
D. Locus of the highest elevation of the folded surface *(IIT-JAM 2008)*

137. A fold in which the hinge line plunges in the direction of dip of the axial plane is:
A. Overturned fold
B. Reclined fold
C. Recumbent fold
D. Upright fold *(IIT-JAM 2007)*

138. Which of the following is a strike-slip fault?
A. Normal fault B. Reverse fault
C. Thrust fault D. Transform fault
(IIT-JAM 2007)

139. Scratches produced on the fault plane are known as:
A. Boudins
B. Mullions
C. Plumose structure
D. Slickensides *(IIT-JAM 2007)*

140. A hill face slopes at an angle 40° due east. The joint set most likely to cause landslide is the one dipping at:
A. 20° due west B. 30° due east
C. 60° due east D. 65° due south-east
(IIT-JAM 2006)

141. In a normal stratigraphic sequence, erosion can sometimes expose older rocks within younger rocks. Such exposures are called:
A. Inlier B. Window
C. Outlier D. Klippe
(IIT-JAM 2006)

142. On a Survey of India toposheet number 53 J/3, a distance of four centimeters between two points represents an actual ground distance of:
A. 1 km B. 2 km
C. 8 km D. 16 km
(IIT-JAM 2006)

143. Refer to the geological map, which shows the development of a metamorphic aureole around a granite intrusion in a sedimentary sequence. Which one of the following combinations of rocks exposed at locations X, Y and Z is correct?

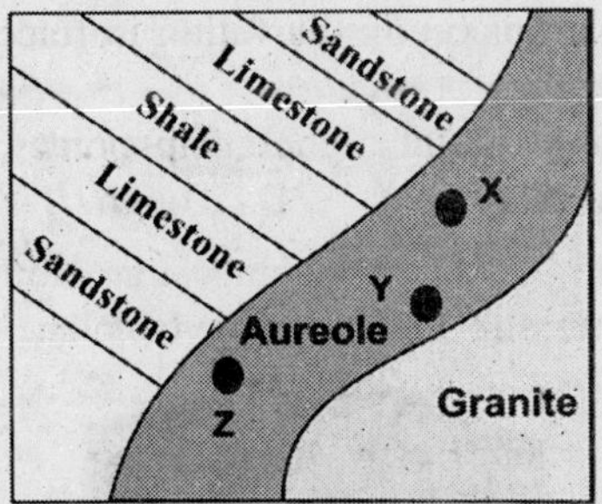

	X	Y	Z
A.	Marble	quartzite	hornfels
B.	Quartzite	hornfels	marble
C.	Hornfels	quartzite	marble
D.	Marble	hornfels	quartzite

(IIT-JAM 2006)

144. Match the following:

Group I	Group II
P. Plunging antiform	1. 
Q. Plunging synform	2. 
R. Recumbent fold	3. 
S. Reclined fold	4. 

P - 1 Q - 2 R - 3 S - 4
P - 2 Q - 1 R - 4 S - 3
P - 3 Q - 4 R - 1 S - 2
P - 4 Q - 3 R - 2 S - 1 *(IIT-JAM 2006)*

145. The continuous line on a folded surface connecting points of maximum curvature is known as the:
A. Axis B. Axial trace
C. Hinge line D. Crest
(IIT-JAM 2005)

146. Nonconformity can be recognized by occurrence of:
A. Contact effect below and above the plane of unconformity
B. Pebbles of non-sedimentary rock over the plane of unconformity
C. Paraconformable bedding planes
D. Angular discordance at the plane of unconformity
(IIT-JAM 2005)

147. When the minimum principal stress(σ_3) is vertical, the resultant fault is a:
A. Strike-slip fault B. Oblique-slip fault
C. Thrust D. Normal fault
(IIT-JAM 2005)

148. A rock made up of very finely crushed mineral grains is called:
A. Breccias B. Gouge
C. Cataclastic D. Mylonite
(DU-2014)

149. Which of the following character is typical of a transform fault?
A. Fault displacement remains same along its length
B. Fault displacement increases from one end to other end
C. Fault displacement is maximum at the central part
D. Fault displacement is minimum at the central part
(DU-2014)

150. Slickenside lineation on a fault surface indicates:
A. Direction of the net slip of the fault
B. Direction of throw of the fault
C. Direction of the heave of the fault
D. Direction of the dip of the fault
(DU-2014)

151. In a fold, the thickness of the folded layer measured perpendicular to its length remains same everywhere. Such fold is called:
A. Parallel fold
B. Supraterenous fold
C. Cuspate lobate fold
D. Similar fold *(DU-2014)*

152. An unconformity in which sedimentary rocks overlie a plutonic igneous rock body is called:
A. A discontinuity
B. A non-conformity
C. A para-conformity
D. An angular unconformity *(DU-2014)*

153. In the overturned limb of a fold, which of the following is generally true?
A. Dip of bedding is more than dip of cleavage
B. Dip of cleavage is more than dip of bedding
C. Dip of cleavage and bedding are same
D. Dip of cleavage and bedding cannot be compared. *(DU-2014)*

154. According to Anderson's theory of faulting the Earth's surface is taken as:
A. A surface with no shear stress
B. A surface with no normal stress
C. A surface with equal normal and shear stress
D. A surface with maximum shear stress *(DU-2014)*

155. A sedimentary bedding dips 40° towards N45° W. Its apparent dip towards N10° W will be around:
A. 25° B. 45°
C. 60° D. 80° *(DU-2014)*

156. Point of maximum curvature in fold is defined as:
A. Hinge point B. Infection point
C. Plunge D. Centre of curvature *(BHU-2015)*

157. Transcurrent fault is a type of:
A. Normal fault B. Reverse fault
C. Strike-slip fault D. Oblique fault *(BHU-2015)*

158. A plane is dipping 45° (True dip) towards NE, the apparent dip of plane can be:
A. >90° B. Only 0°
C. 45° to 0° D. 90° to 45° *(BHU-2015)*

159. True dip of a plane is 30°, what will be head of the plane?
A. 45° B. 60°
C. 90° D. 0° *(BHU-2015)*

160. The direction of line along which an inclined bed or plane intersects a horizontal plane is known as:
A. Fault trace B. Dip direction
C. Strike D. Plunge *(BHU-2015)*

161. Point of maximum elevation in fold is termed as:
A. Hinge point B. Tip point
C. Crest D. Tough *(BHU-2015)*

162. The net slip in following fault is:

A. ab B. ac
C. bc D. ad *(BHU-2015)*

163. Unconformity is define as:
A. Plane representing the time of non-deposition and erosion
B. Plane of asymmetric deposition
C. Plane of joint and fault on regional scale
D. Plane of faulting *(BHU-2015)*

164. Which one is **not** a deformational structure?
A. Fold B. Fault
C. Joint D. Disconformity *(BHU-2015)*

165. Match the following:

Group-1 (Terms)	**Group-2 (Definition)**
P. Cross-section	1. Dip isogones that are perpendicular to bedding through the fold
Q. Profile section	2. A vertical plane through a fold
R. Similar fold	3. Dip isogones that are parallel to each other
S. Parallel fold	4. The surface perpendicular to the hinge line

A. P-1, Q-2, R-4, S-3
B. P-4, Q-2, R-1, S-3
C. P-4, Q-2, R-3, S-1
D. P-1, Q-2, R-3, S-4 *(BHU-2015)*

166. Intermediate principle stress(σ_2) is vertical in:
A. Normal fault B. Strike-slip fault
C. Reverse fault D. Thrust fault
(BHU-2015)

167. Which of the following is a correct statement?
A. The angle between fold limbs as measured in the cross-section is called the interlimb angle
B. A fold with a horizontal axial surface by definition must have a horizontal hinge line, and is called a reclined fold
C. The high point of the hinge line in a doubly plunging fold is called the culmination
D. In an anticline, the beds are younger towards the core
(BHU-2015)

168. .... in an area of older rocks surrounded by younger rocks.
A. Inlier B. Outlier
C. Dome D. Basin *(BHU-2015)*

169. .... having net slip equal to dip-slip and rake of net slip 90°.
A. Dip fault B. Dip-slip fault
C. Strike fault D. Bedding fault
(BHU-2015)

170. .... is defined as a fracture in a rock between the sides of which there is no observable movement.
A. Fold B. Fault
C. Joint D. Shear zone
(BHU-2015)

171. A topographic contour is defined as:
A. Line joining the point of equal stratigraphic horizons
B. Line joining the point of equal elevation
C. Line joining the point of equal thickness
D. Line joining point of same age rocks
(BHU-2015)

172. If the orientation of the principal stress axes X, Y and Z does not change during the deformation, the deformation is known as:
A. Simple shear B. Pure shear
C. Triaxial stress D. Axial strain
(PU-2010)

173. A fold which is concave upward is called:
A. Synform B. Antiform
C. Syncline D. Anticline
(PU-2010)

174. A fold in which younging direction is away from the fold core is called:
A. Synform B. Antiform
C. Syncline D. Anticline
(PU-2010)

175. Fold with alternate sharp and broad hinges are called:
A. Chevron fold B. Cuspate fold
C. Box fold D. Concentric fold
(PU-2010)

176. If the axial planar cleavage is steeper than the dip of the fold limb, then limb is called:
A. Normal limb B. Overturned limb
C. Back limb D. Inverted limb
(PU-2010)

177. If rake/pitch of the net slip on the fault plane is 90°, then the fault will be:
A. Dip-slip fault
B. Strike-slip fault
C. Oblique slip fault
D. Hinge (Oblique) fault *(PU-2010)*

178. If the strike of a dip slip normal fault is N30°E, then the trend of the net slip will be:
A. N60° E B. S60° E
C. N30° W D. S30° W *(PU-2010)*

179. The linear zone separating two adjacent offset segments of a mid-oceanic ridge is:
A. Gravity fault B. Detachment zone
C. Transform fault D. Thrust fault
(PU-2013)

180. In a toposheet, a series of 'V' shaped contours represent:
A. Ridge B. Valley
C. Mountain D. Road *(PU-2013)*

181. The relative displacement between two adjacent points on either side of the fault plane is known as:
A. Offset B. Separation
C. Net slip D. Throw *(PU-2013)*

182. Hangingwall anticline structure is a feature associated with:
A. Thrust fault B. Normal fault
C. Strike-slip fault D. Scissor fault
(PU-2013)

183. Which of the following lineation cannot be used for stress analysis?
A. Mineral lineation
B. Boudin axis
C. Slickensides
D. Stretching lineation *(PU-2013)*

184. The true dip direction of an inclined plane is .... to the line of intersection of the inclined plane with an imaginary horizontal plane.
A. Parallel B. Perpendicular
C. At 45 degree D. Oblique
(PU-2013)

185. Shear strain is measured by:
A. Change in length of a line
B. Change in angle between two lines
C. Displacement of a line
D. Distortion of a plane *(PU-2013)*

186. A fault in which hanging wall moves down relative to the footwall is:
A. Reverse fault B. Thrust fault
C. Normal fault D. Transform fault
(PU-2012)

187. The angle between strike line and lineation measured on a vertical plane is:
A. Plunge B. Dip
C. Pitch D. Rake *(PU-2012)*

188. Fault is an example of:
A. Brittle Deformation
B. Ductile Deformation
C. Brittle-Ductile Deformation
D. Malleability *(PU-2012)*

189. Which of the following in the fold with parallel arrangement of limbs?
A. Inverted fold B. Isoclinal fold
C. Reclined fold D. Symmetrical fold
(PU-2012)

190. If the rake of the net slip of a fault is 900 , the fault may be:
A. Strike slip fault
B. Dip slip fault
C. Diogonal slip fault
D. Tear fault *(PU-2012)*

191. An outcrop in which older rocks surrounded by younger rocks is known as:
A. Overlap B. Offlap
C. Outlier D. Inlier *(PU-2012)*

192. The dip direction of an inclined plane is always .... to the line of intersection of the inclined plane with an imaginary horizontal plane.
A. Parallel B. Perpendicular
C. 45° D. Oblique
(PU-2012)

193. Pitch of a linear structure lying on the plane is defined as the angle between the:
A. Linear structure and the strike line of the plane
B. Horizontal projection of the linear structure and the strike line of the bed
C. Linear structure and its horizontal projection
D. Linear structure and its vertical projection
(PU-2012)

194. Cleavage is a:
A. Primary planar structure
B. Primary linear structure
C. Secondary planar structure
D. Secondary linear structure *(PU-2012)*

195. A geologic surface that seperates younger strata from older and represents a time of non-deposition, where horizontally parallel strata of sedimentary rock are deposited on tilted and eroded layers that may be vertical or at an angle to the overlying horizontal layers is named as:
A. Angular Unconformity surface
B. Paraconformity surface
C. Nonconformity surface
D. None of the above *(PU-2012)*

196. One of the following refers to Survey of India toposheet in 1:50000 scale:
A. 57 B. 57J
C. 57J/12 D. 57J/12/SW
(PU-2012)

197. The tectonic feature illustrated in the figure given below:

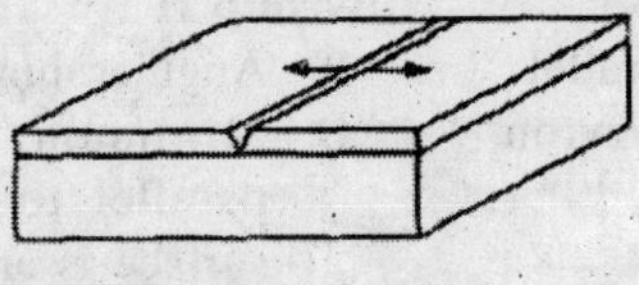

A. Transform fault
B. Divergent plate boundary
C. Convergent plate boundary
D. Thrust fault *(PU-2011)*

198. Dip direction of the vertical bed whose strike is northeast-southwest is:
A. Towards north-west
B. Towards south-west
C. Towards north-east
D. Not defined *(PU-2011)*

199. If the pitch of a linear structure on a bed of attitude N 30° E/50° SE is 90°, then the plunge of the linear structure will be:
A. 30° B. 45°
C. 50° D. 90° *(PU-2011)*

200. In the diagram, the orientation of the axial planar cleavage (dark plane) and pitch of the bedding (white band) trace on the cleavage is given. The pattern is characteristics of:

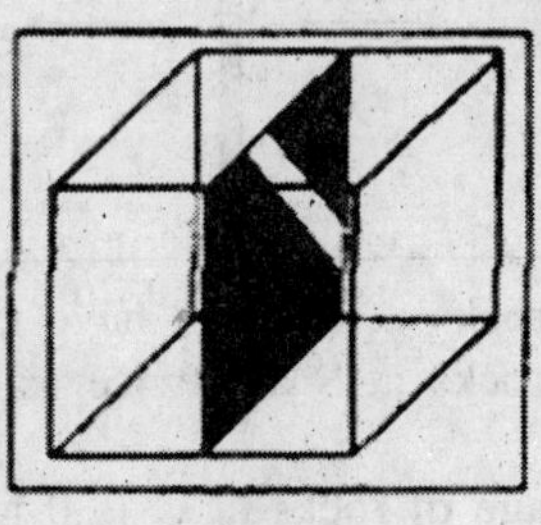

A. Upright horizontal fold
B. Upright plunging fold
C. Inclined horizontal fold
D. Inclined plunging fold *(PU-2011)*

201. Which one of the following is not a type of unconformity?
A. Nonconformity
B. Disconformity
C. Uniformity
D. Angular unconformity *(ISM-2015)*

202. Which one of the following is not related to Himalayan thrust fold system?
A. Main boundary thrust
B. Main central thrust
C. Great boundary fault
D. Indus-tsangpo suture *(ISM-2015)*

203. It is not possible to get .... fold of axis of rotation
A. Two B. Three
C. Four D. Five *(ISM-2015)*

204. Tectonic Window is an example of:
A. Rift zone
B. Diapiric intrusions
C. Overthrust sheets
D. Transform fault *(ISM-2015)*

205. In a recumbent folds the axial plane is:
A. Inclined (acute angle)
B. Inclined (obtuse angle)
C. Horizontal
D. Vertical *(ISM-2015)*

206. Attitude of linear structure can be represented by:
A. Dip and strike
B. Pitch and plunge
C. Heave and throw
D. Striation and slickensides *(ISM-2015)*

207. Find out the incorrect one:
A. Contour lines are lines of equal elevation value
B. Contour lines generally should not merge or cross
C. Closely spaced contours indicates gentle gradient
D. Contour lines are repeated to indicate reversal in gradient direction *(ISM-2015)*

208. The line of intersection of two planes pitches at 90 degree on both the planes. Which of the following intemperance is correct:
A. One plane is horizontal
B. Both plane are vertical
C. The strikes of the two mutual planes are mutually perpendicular
D. One plane is vertical *(ISM-2014)*

209. Which one of the following folds is characterized by a constant orthogonal thickness in all the parts of the folds?

A. Similar fold B. Parallel fold
C. Flow fold D. Shear fold

(ISM-2014)

210. Which of the following is correct while taking beaning of an object with a clinometer compass?

A. Pointing the N end of the magnetic needle towards the object
B. Pointing S end of the magnetic needle towards the objects
C. Pointing N marked by the crown of the dial towards the object
D. Pointing E or W marked on the dial towards the object *(ISM-2014)*

211. The stereographic projection of hinge lines of a fold be along a common great circle. What is the geometry of the folds?

A. Planar non-cylindrical
B. Planar cylindrical
C. Non-planar cylindrical
D. Non-planar non-cylindrical *(ISM-2014)*

212. Which one of the following characteristics the period of non-deposition, no erosion and no precipitable rock records?

A. Diastem B. Hiatus
C. Unconformity D. Discontinuity

(ISM-2014)

213. In the topographic map the steepest slope is

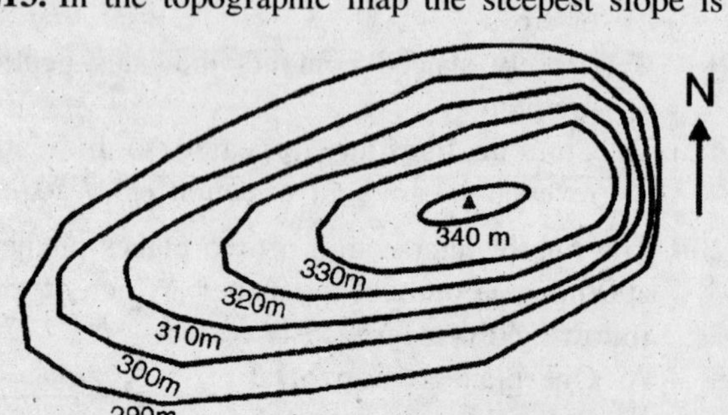

A. Northerly B. Southerly
C. Easterly D. Westerly

(IIT JAM-2016)

214. Match the folds listed in Group I with corresponding geometric characteristics in Group II.

Group I	Group II
(*i*) Parallel	P. Angular hinge
(*ii*) Chevron	Q. Minimum thickness parallel to the axial surface is at the hinge
(*iii*) Similar	R. Negative inter-limb angle
(*iv*) Mushroom	S. Constant thickness parallel to the axial surface

A. (*i*)-Q; (*ii*)-R; (*iii*)-S; (*iv*)-P
B. (*i*)-S; (*ii*)-P; (*iii*)-Q; (*iv*)-R
C. (*i*)-Q; (*ii*)-P; (*iii*)-S; (*iv*)-R
D. (*i*)-S; (*ii*)-R; (*iii*)-Q; (*iv*)-P *(IIT JAM-2016)*

215. A schematic diagram of a divergent plate boundary, with arrows indicating directions of plate movement, is given below. Which one of the following statements is NOT true for points P, Q, R and S, if the spreading rate for both the plates is uniform and same through time and space?

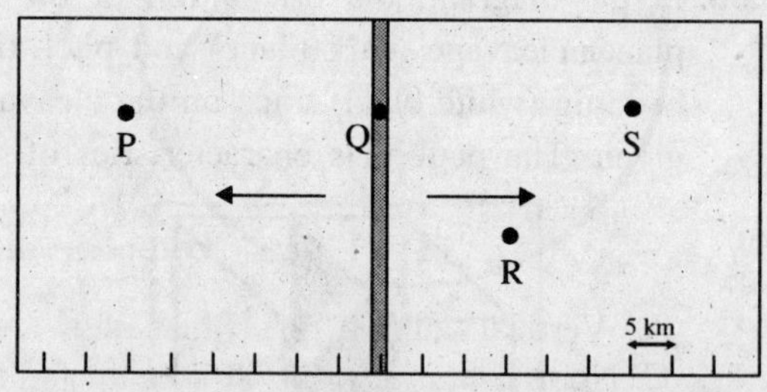

A. The rocks at P and S have the same age
B. The rocks at S are twice as old as those at R
C. The age of rocks at Q is 0 Ma
D. The age of the rocks decreases progressively from P to S *(IIT JAM-2016)*

216. Which of the following is/are found only in the brittle shear zones?

A. Cataclastic B. Mylonite
C. Psuedotachylite D. Gouge

(IIT JAM-2016)

217. A gravity dam with E-W axis is to be constructed in a narrow river valley between two N-S trending parallel ridges. The river is flowing from south to north. The lithology of the area is represented by 2 to 5 m thick metasedimentary rocks-quartzite, phyllite and schist. Which of the following geological conditions will be most suitable?

A. Beds in both the ridges strike N-S but dip towards each other

B. Beds in both the ridges strike N-S but dip in opposite direction to each other

C. Beds in both the ridges strike E-W and dip towards N (downstream)

D. Beds in both the ridges strike E-W and dip towards S (upstream) *(IIT JAM-2016)*

218. The fault for which the orientation and displacement vector are given in the following lower hemisphere stereographic projection is a

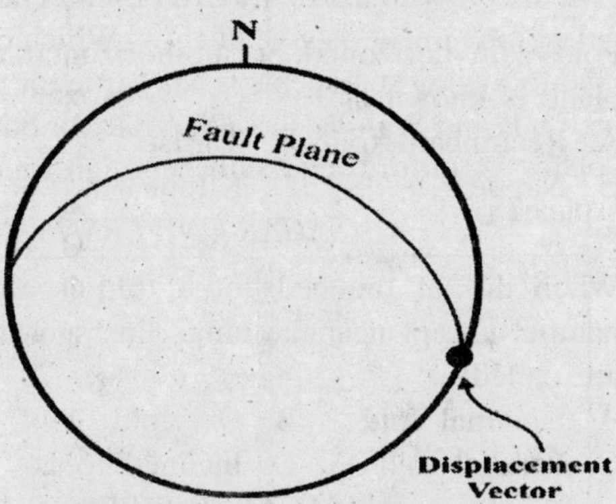

A. Vertical fault B. Thrust fault

C. Normal fault D. Strike-slip fault

(IIT JAM-2016)

219. If the elevation of a wave cut platform is 55 m above the sea level and the age of the erosional surface is 120 kilo years, the rate of rock uplift at this coastal location is .... m/kilo years (give answer in two decimal places).

A. 0.45 B. 2.45

C. 0.35 D. 20.45

(IIT JAM-2016)

220. A foliation plane has strike 25° and 60° easterly dip. A mineral lineation on this foliation plane has a rake/pitch of 90°. The plunge direction of the mineral lineation in whole circle bearing is .... degrees.

A. 115 B. 121

C. 112 D. 122 *(IIT JAM-2016)*

221. Two outcrops on a 1 : 25000 map are 12 cm apart. The ground distance between the two outcrops is .... km.

A. 3.0 B. 12.0

C. 1.0 D. 2.0 *(IIT JAM-2016)*

222. B and B′ are two points on the topographic map shown below. The distance between B and B′ along the linear traverse BB′ is 220 m. The angle of the slope along this traverse is .... degree (give answer in two decimal places).

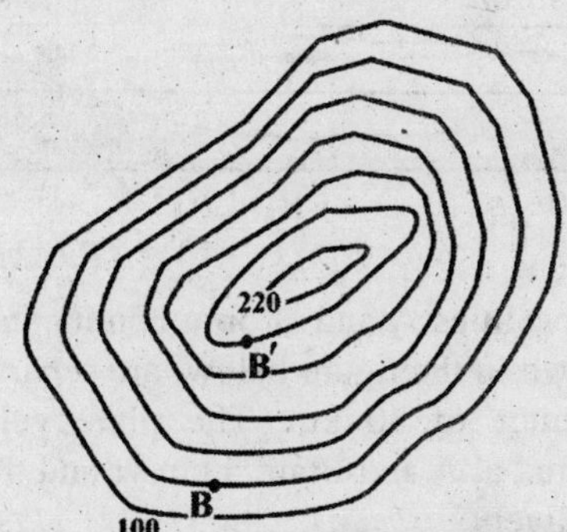

Contour Interval = 20m

A. 21.32 B. 22.23

C. 22.32 D. 23.32

(IIT JAM-2016)

223. As shown in the following figure, a vertical well intersects the top and bottom of an inclined bed at 200 m and 410 m depths, respectively. If the true dip of the bed is 60° to the north, the true thickness of the bed is .... metres.

A. 105　　B. 120
C. 122　　D. 123 *(IIT JAM-2016)*

224. In the given geologic map, the horizontal lines are stratum contours. The throw on the fault (F-F) is .... meters.

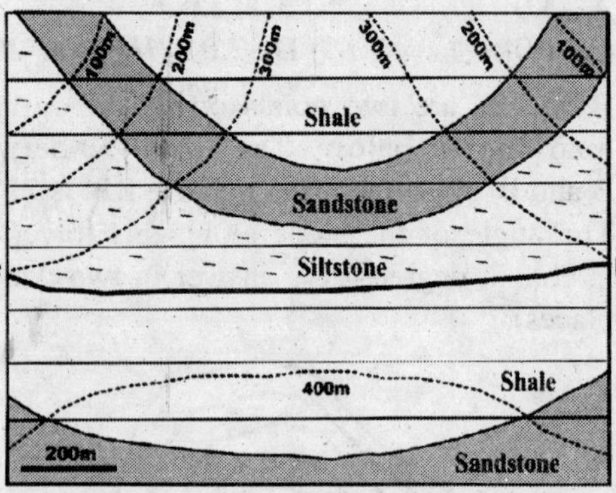

A. 400　　B. 120
C. 123　　D. 140 *(IIT JAM-2016)*

225. Two localities A and B on a continental plate as shown in the figure below, are separated by a distance of 80 km. The plate velocities measured at A and B are 4 cm/yr and 5 cm/yr, respectively.

Assuming no faulting in the area, the new distance between A and B will be .... km in one million years. Give answer in two decimal places.

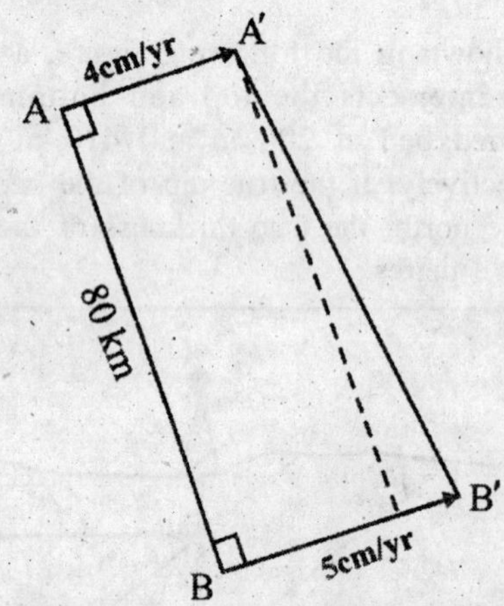

A. 80.62　　B. 28.62
C. 18.62　　D. 12.00

(IIT JAM-2016)

226. Structure formed by stretching of competent bed along bedding planes show pull apart structure, often occupied by incompetent material from either side is termed as:

A. Mylonite　　B. Breccia
C. Pukers　　D. Boudinage

(SAGAR UNIVERSITY-2015)

227. Joints which strike parallel to the direction between strike and dip of the strata is called:

A. Dip joints　　B. Oblique joints
C. Strike joints　　D. Sheer joints

(SAGAR UNIVERSITY-2015)

228. When cleavage dip is greater than the dip of the bedding plane, the beds are termed as:

A. Overturned　　B. Right side up
C. Right side down　　D. Under turned

(SAGAR UNIVERSITY-2015)

229. Fold with horizontal or sub-horizontal axial plane is known as:

A. Recumbent fold　　B. Thrust
C. Nappe　　D. Klippe

(SAGAR UNIVERSITY-2015)

230. When dip of the beds of a fold is almost parallel except near the hinge line such folds are called:

A. Isoclinal fold　　B. Overfold
C. Parallel fold　　D. Inclined folds

(SAGAR UNIVERSITY-2015)

231. Two complimentary sheer planes that in two sets which constitute?

A. Sheet structure　　B. Bedding joint
C. Shear joint　　D. Conjugate joint

(SAGAR UNIVERSITY-2015)

232. Hot springs are associated with which of the following structures?

A. Unconformity　　B. Anticlinorium
C. Faults　　D. Synclinorium

(SAGAR UNIVERSITY-2015)

233. Which of the following structure is parallel to the bedding plane?

A. Batholith　　B. Lopolith
C. Chonolith　　D. Laccolith

(SAGAR UNIVERSITY-2015)

234. The angle between the bonds joining the hydrothermal nuclei to the centre of the oxygen atom in a water molecules is:
A. 90 degree B. 96 degree
C. 105 degree D. 115 degree
(PUNJAB UNIVERSITY-2015)

235. If the folds plunge, the strike on the cleavage is:
A. Parallel to the strike on the bedding
B. Diagonal to the strike of the bedding
C. Perpendicular to the strike of the bedding
D. Relation is uncertain
(PUNJAB UNIVERSITY-2015)

236. Slickenside are defined as:
A. Foliation B. Bedding
C. Lineation D. Fault plane
(PUNJAB UNIVERSITY-2015)

237. A fault in which the displacement of limbs take place at a right angle to the surface of discontinuity (like dykes) are known as:
A. Trench fault B. Lag fault
C. Gaping fault D. Drag fault
(PUNJAB UNIVERSITY-2015)

238. Match the following :

List-A	List-B
P. Disconformities	1. Non-depositional unconformity
Q. Non-conformity	2. Between residual soil and under laying rocks
R. Local unconformity	3. Parallel unconformity
S. Blended unconformity	4. Hetrolothic unconformity

A. P-3, Q-4, R-1, S-2
B. P-1, Q-2, R-3, S-4
C. P-2, Q-3, R-4, S-1
D. P-4, Q-1, R-2, S-3
(PUNJAB UNIVERSITY-2015)

239. The columnar structures bounded by long cylindrical surface are called:
A. Boudinage B. Bedding
C. Mullions D. Cylinders
(AMU-2016)

240. Two set of joins nearly at right angle to one another, produce by the same stress system, are called
A. Joint system
B. Joint set
C. Conjugate system
D. Master joint *(AMU-2016)*

241. A strike fault separating two lithospheric plates is generally known as:
A. Slip fault B. Transform fault
C. Wrench fault D. Enechelon fault
(AMU-2016)

242. An erosional surface separates sedimentary strata dipping 60 degree towards N80 degree E from its overlaying strata dipping 5 degree towards N 45 degree E, the structure is called:
A. Disconformity
B. Nonconformity
C. Paraconformity
D. Angular unconformity *(ISM-2016)*

243. A discontinuous cleavage formed by micro-folding of an existing cleavage/schistosity in called
A. Disjunctive cleavage
B. Slaty cleavage
C. Crenulation cleavage
D. Transacted cleavage *(ISM-2016)*

244. According to the Anderson's theory of faulting, the earth's surface is assumed as a plane of:
A. Principal strain
B. Principal stress
C. Maximum shear stress
D. Maximum shear strain *(ISM-2016)*

245. The two limbs of a folds, when plotted as planes on a stereographic projection, intersect, diametrically opposite to each other. The fold is:
A. A reclined fold
B. A horizontal (non-plunging) fold
C. A steeply plunging fold
D. None of the earlier three *(ISM-2016)*

246. A coal beds dip 50° towards N45° W, its apparent dip towards N20° W will be around

A. 0° B. 45°
C. 60° D. 80° *(ISM-2016)*

247. The above diagram illustrate:

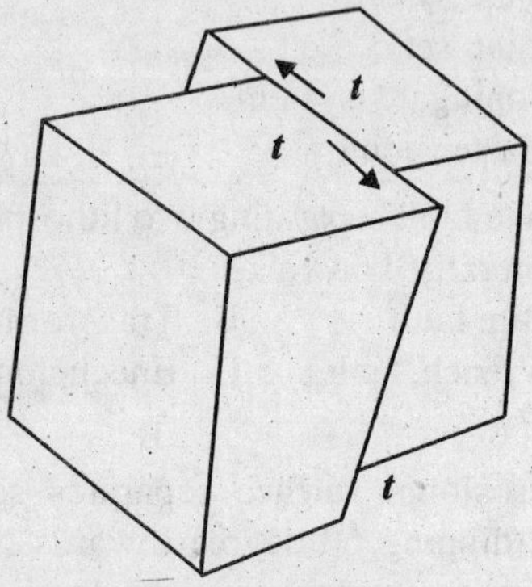

A. Dextral strike-slip fault
B. Sinistral strike-slip fault
C. Dextral dip-slip fault
D. Sinistral dip-slip fault *(ISM-2016)*

248. The Great Boundary Fault occur in the region of:
A. Alwar B. Chittorgarh
C. Jaipur D. Haridwar
(BHU-2016)

249. The fold in which the axes plunging direction shows the dip of the axial surface is termed as:
A. Normal fold B. Reclined fold
C. Concentric fold D. Drag fold
(BHU-2016)

250. When the minimum principal stress is vertical, the resulting fault is?
A. Strike-slip fault B. Oblique slip fault
C. Normal fault D. Reverse fault
(BHU-2016)

251. The R.F. of a geological map prepared on a scale of 2 cm = 1 km is:
A. 1:50000 B. 1:5000
C. 1:500 D. 1:100000
(BHU-2016)

252. Hade of the vertical fault will be:
A. 90 degree B. 45 degree
C. 0 degree D. 40 degree
(BHU-2016)

253. The point which separates a convex and concave segment of a fold is called:
A. Median point B. Inflection point
C. Hinge point D. Crest point
(BHU-2016)

254. The thrust fault will be generated when .... (where σ_1 is max. Stress):
A. σ_1 and σ_2 are horizontal
B. σ and σ_3 are horizontal
C. σ_2 and σ_3 are horizontal
D. None of the above *(BHU-2016)*

255. For obtaining true dip of the bed the minimum number of apparent dip data required is:
A. one B. two
C. three D. None of the above
(BHU-2016)

256. A synformal anticline will have:
A. upward arching beds having older rocks in core
B. downward arching beds having older rocks in core
C. upward arching beds having younger rocks in core
D. downward arching beds having younger rocks in core *(BHU-2016)*

257. According to the Anderson theory of faulting, the principal stress should be horizontal?
A. Strike-slip faulting
B. Normal faulting
C. Reverse faulting
D. Wrench faulting *(DU-2016)*

ANSWER KEY

1	2	3	4	5	6	7	8	9	10
A	B	B	B	B	B	B	C	C	B
11	12	13	14	15	16	17	18	19	20
B	B	C	B	D	A	C	B	B	B

21	22	23	24	25	26	27	28	29	30
A	C	A	A	A	D	C	A	B	D
31	32	33	34	35	36	37	38	39	40
C	D	C	D	B	A	A	B	C	A
41	42	43	44	45	46	47	48	49	50
A	B	B	B	C	D	A	A	B	D
51	52	53	54	55	56	57	58	59	60
D	D	A	A	B	A	A	A	A	C
61	62	63	64	65	66	67	68	69	70
A	B	D	A	C	B	A	C	A	C
71	72	73	74	75	76	77	78	79	80
B	C	C	B	C	B	B	B	B	C
81	82	83	84	85	86	87	88	89	90
C	A	B	A	B	B	D	B	C	A
91	92	93	94	95	96	97	98	99	100
D	B	A	A	A	A	C	B	C	C
101	102	103	104	105	106	107	108	109	110
B	B	A	C	A	A	B	A	A	D
111	112	113	114	115	116	117	118	119	120
C	D	B	A	D	A	D	A	C	B
121	122	123	124	125	126	127	128	129	130
B	A	C	D	C	B	A	B	C	C
131	132	133	134	135	136	137	138	139	140
B	C	A	B	B	B	A	D	A	C
141	142	143	144	145	146	147	148	149	150
A	B	D	A	A	B	C	B	B	A
151	152	153	154	155	156	157	158	159	160
A	B	A	A	A	A	C	C	B	C
161	162	163	164	165	166	167	168	169	170
C	B	A	D	C	B	C	A	B	C
171	172	173	174	175	176	177	178	179	180
B	B	A	D	C	A	A	B	C	B
181	182	183	184	185	186	187	188	189	190
A	A	B	B	C	C	C	A	B	B
191	192	193	194	195	196	197	198	199	200
D	B	A	C	A	B	B	A	D	C
201	202	203	204	205	206	207	208	209	210
C	C	D	C	C	A	C	C	B	A
211	212	213	214	215	216	217	218	219	220
B	C	C	C	D	A,C,D	D	D	A	A
221	222	223	224	225	226	227	228	229	230
A	A	A	A	A	D	B	C	A	A
231	232	233	234	235	236	237	238	239	240
C	C	D	C	A	D	A	A	C	C

241	242	243	244	245	246	247	248	249	250
A	D	C	B	B	B	B	B	B	D
251	252	253	254	255	256	257			
A	C	B	A	B	B	C			

HINTS AND SOLUTIONS

1.

Statement	Indicates
The dip of the bed is more than the slope of the valley.	V-shaped outcrop with apex pointing downstream in a valley.
The dip of the bed is less than the slope of the valley.	V-shaped outcrop with apex pointing upstream in valley.
The bed is vertical.	The outcrop is straight line.
The bed is horizontal.	The width of the outcrop depends on the slope of the ground.

2. The scale of the toposheet is – 1:50,000

1 cm = 500 meters
= 0.5 Km.

The actual distance of the ground is
= 5 × 0.5
= 2.5 Km.

3. The width of the bed is = 30 meters.
The dip of the bed is = 30°
The thickness of the bed is
= width × Sin θ
= 30 × ½
= 15 meters.

5. The lithostatic pressure P = *hdg*

From question, h = 10 km.
d = 2700 Kg/m^3
g = 9.8 m/s^2

Threrfore, P = 10 × 27 × 980
= 264600
= 264.6 MPa.

19. The elevation difference between in (Contours line) is = 600-500 meters
= 100 meters

The horizontal distance is = 5 cm.
= 5 × 20 m.
(Horizontal Scale 1 cm = 20 meters)
= 100 meters.

Now, True dip tan θ = Elevation difference/Horizontal difference.
= 100/100
= 1
θ = Tan^{-1} 1
= 45°.

20.

Stereographic projection points	Degree (Approximate)
O	Center (0) degree
R	30°
Q	60°
P	90° (Pole)

21.

Faults types	Characteristics
Strike-slipe fault	The movement of the block is along the strike (horizontally).
Normal fault	The hanging wall moved downward relative to the footwall.
Reverse fault	The hanging wall moved upward relative to the footwall.
Thrust fault	Low angle reverse fault is a thurst fault.

24.

Figure point	Remark
P	Minimum brittle
S	Maximum brittle

29.

Dip and strike direction	Beds position
Opposite direction of dip	Older beds
Direction of dip	Younger beds
Direction of strike	Same beds
Inclined direction of dip and strike	Older/Younger

31. The dip of the fault plane = 35°
Hade = 90-Dip = 90 – 35° = 55°

33.

Structural processes	Products	Economic minerals
Pressure solution	Stylolite	Limestone, Dolomite
Layer-parallel extension	Boudine	Gold, Copper
Layer-perpendicular shortening	Bending fold	Iron, Nickel
Layer-parallel shortening	Ptygmatic fold	Manganese

38.

Axial plane direction	Types of fold
Vertical axial plane	Symmetrical fold
Horizontal axial plane	Recumbent fold
Inclined axial plane	Overturned fold
Oblique axial plane	Isoclinal fold

40.

Terminology	Definition
Modulus of rigidity	Normal stress/Strain
Young modulus	Stress (force per unit area)/Strain (over initial length)
Shear modulus	Shear stress/Shear strain
Bulk modulus	Infinite pressure/Resultant decreasing of the volume

63.

Types of Joints	Properties	Example (Rock types)
Columnar joints	Tensional joints	Basalt
Release joints	Parallel to fold axis	Sedimentary bedding
Extension joints	Due to extension of Bedding	Granite gnesses
Cross joints	Perpendicular to fold axis	Sandstone

66. The dip of the normal fault = 45°
Throw of the fault is = 100 meters
The heave of the fault is = tan 45°
= 100/Heave
1 = 100/Heave
= 100 meters.

88.

Joints type	Rocks
Mural joints	Granite
Columnar joints	Basalt
Sheet joints	Mica
Bedding joint	Joints parallel to bedding plane

112.

Feature	Indicator
Gouge	Fault
Mylonite	Fault
Cataclasite	Fault
Basal conglomerate	Unconformity

113.

Stress (σ_1)	Stress (σ_2)	Stress (σ_3)	Types of faults
Vertical	Horizontal	Horizontal	Normal faults
Horizontal	Vertical	Horizontal	Reverse faults
Horizontal	Horizontal	Vertical	Thrust faults
Horizontal	Horizontal	Vertical	Strike-slip faults

116.

Interlimb angle (in degree)	Types of fold
0-10	Isoclinal fold
10-30	Tight fold
30-70	Closed fold
70-120	Open fold

117.

Terms	Remarks
Outlier	Younger rocks surrounded by older
Overlap	Younger strata cover of the underlying strata.
Inlier	Older rocks surrounded by younger.
Offlap	Older strata cover of the underlying strata.

138.

Types of faults	Characteristics
Normal faults	The hanging wall moved downward relative to foot-wall.
Reverse fault	The hanging wall moved upward relative to foot-wall.
Thrust fault	Low angle reverse fault is called thrust faults
Transform fault	A fault which moved along strike of the fault is called strike slip fault.

139.

Types of structures	Characteristics (Indicators)
Boudins	Scratches
Mullion	Detachment
Plumose structure	Propogation direction
Slickensides	Movement one another

141.

Types of structures	Characteristics
Inlier	The older rocks surrounded by younger rocks in normal stratigraphic order.
Outlier	The younger rocks surrounded by older in normal stratigraphy level.
Window	Erosion can some time expose older rocks within younger rocks in normal stratigraphic order.
Klippe	It is remains portion of after erosion. Example simla klippe

142. Toposheet number – 53J/3

Sensor – LISS-3

Scale – 1:50000

Resolution spatial – 23.5 m

Map scale 1 : 50000 = 1cm : 50000 cm

= 1cm : 500 m

= 1cm : 0.5 km

The given distance is 4 cm = $4 \times 0.5 = 2.0$ km

146.

Characteristics	Unconformity
Contact effect below and above and plane of unconformity	Disconformity
Pebbles of non-sedimentary rocks over the plane of unconformity	Non-conformity
Para-conformable bedding planes	Para-conformity
Angular discordance at the plane of unconformity	Angular unconformity

201.

Unconformity	Remarks
Nonconformity	An unconformity that separate igneous or metamorphic rocks from overlying sedimentary rock.
Disconformity	An erosion surface between two package of sediment, but the lower package of sediment was not tilted.
Unconformity	Unconformty are gaps in the geological record that may indicate episodes of crustal deformation, erosion and sea level variation.

Angular unconformity	Angular relation between underlying and overlaying strata.
Uniformity	A sequence of strata which is equal almost variation

202.

Thrust type	**Area**
Main boundary thrust	Between Lesser and outer Himalayan.
Main central thrust	Indian plate pushed under Eurasian plate
Great boundary fault	Between Aravalli and Vindhyan super-group
Indus-tsangpo suture	Zanskar zone and Korakoram zone (Tethys Himalayan)

205.

Axial plane	**Fold types**
Inclined	Overturned fold
Horizontal	Recumbent fold
Vertical	Symmetrical

209.

Folds type	**Characteristic**
Similar fold	Thinner in the limb and thicker near the hinge
Parallel fold	Equal thickness over all point
Flow fold	Incompetent layer show this type of fold
Shear fold	Slip folding

213.

Contour Line	**Characteristics**
Close to each other	Steepest slop
In between some distance	Gentle slop

216.

Structures	**Deformation**
Cataclastie	Brittle shear zone
Mylonite	Ductile shear zone
Pseudotachylite	Brittle shear zone
Gouge	Brittle shear zone

221. The given scale is: 1:25000

$$1 \text{ cm} = 25000 \text{ cm}$$
$$= 250 \text{ m}$$
$$= 0.25 \text{ km}$$

The scale is 1 cm = 0.25 km, then

The distance apart is = 12 cm

$$= 12 \times 0.25$$
$$= 3.0 \text{ Km}$$

223. The given figure:

The top bed = 200 m

The bottom bed = 410 m

If the true dip of the bed is = 60° N

The true thickness of the bed is:

210 = True thickness × cos 60°

True thickness = 210/2

= 105 m.

227.

Joints	**Characteristics**
Dip joints	Joint parallel to the dip of the country rocks
Oblique joints	Joints parallel to the dip and strike of the country rocks
Strike joints	Joints parallel to the strike of the country rocks
Sheet joints	Compression force develop due to this joints developed.

229.

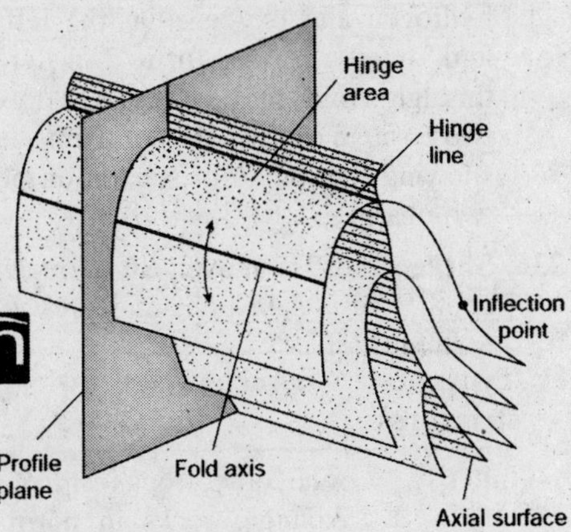

Fig. : Different part of the fold.

230. Fold types	Hinge line
Isoclinal fold	Hinge line parallel except near the hinge line
Over fold	Limbs dip in the same direction.
Parallel folds	Beds thickness equal in all point
Inclined fold	The hinge of the fold is inclined

233. Igneous structures	Characteristics
Batholith	Large igneous body cover on the earth crust
Lopolith	An igneous body which is lenticular in shape (convex downward)
Chonolith	Irregular shape of the igneous body
Laccolith	A mushroom shaped body which is convex upward

234.

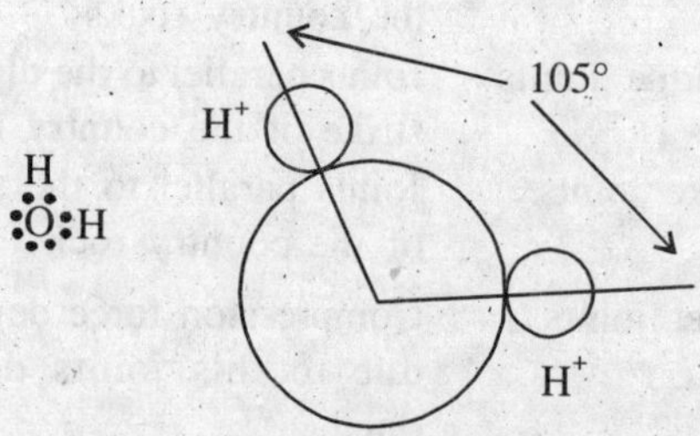

Diagram of a covalent bonding between hydrogen and oxygen. On the left hydrogen and oxygen atoms form a water molecule through covalent bonding. On the right the two positive charges (H^+) of the water molecule showing the angle of separation between the two charges.

236. Slickenside: The movement of one wall against another results in polishing and grooving of fault surface.

247. Faults	Remark
Dextral strike-slip fault	Horizontal movement of the right hand side
Sinistral strike-slip fault	Horizontal movement of the left hand side
Dextral dip fault	The movement of the dip slop with respect to right hand side
Sinistral dip-slip fault	The movement of the strike to the left hand side

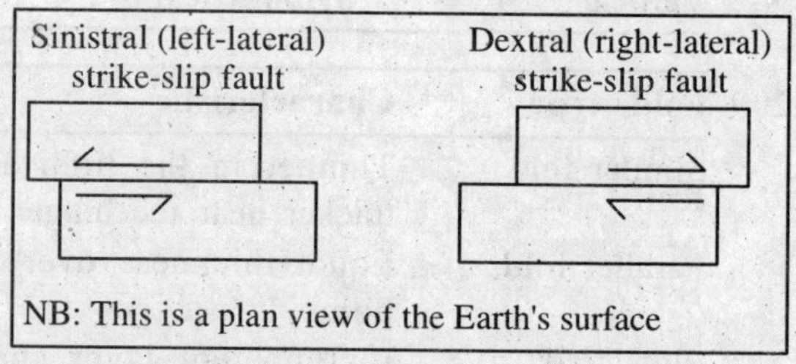

251. The given R.F. scale:

2 cm = 1 km
1 cm = 0.5 km
1 cm = 500 m
100 cm = 1 m
Then 1 cm = 500 × 100
= 50000

The scale is: 1:50,000

MINERALOGY

1. Graphic texture is indicative of:
A. Eutectic crystallization
B. Crystal settling
C. Fractional crystallization
D. Magma mixing *(IIT-JAM-2015)*

2. Which of the following is NOT a form of the isometric system?
A. Octahedron
B. Tetrahedron
C. Trapezohedron
D. Rhombohedron
(IIT-JAM 2015)

3. Silica-rich viscous magmas often produce:
A. Strato-volcano B. Shield volcano
C. Dome D. Fissure eruption
(IIT-JAM 2015)

4. Which one of the following minerals is correctly matched with its optical property?
A. Sillimanite - Isotropic
B. Orthoclase - High relief
C. Pyroxene - Nearly orthogonal cleavage
D. Garnet - Lamellar twinning
(IIT-JAM 2015)

5. The birefringence of a uniaxial mineral having refractive indices of 1.658 and 1.486 is:
A. 0.750 B. 0.172
C. 1.758 D. 0.840
(IIT-JAM 2015)

6. A crystal face has the following intercepts for the crystallographic axes: $1a_1, 1a_2, \frac{1}{2}a_3, \infty c$.
The Miller-Bravais indices for the face is:
A. (1111) B. (1120)
C. (1110) D. (1012)
(IIT-JAM 2015)

7. Total number of mirror planes in the $\frac{4}{m}\bar{3}\frac{2}{m}$ point group is:
A. 8 B. 12
C. 9 D. 10 *(IIT-JAM 2015)*

8. A radioactive isotope has 1024 atoms. How many atoms will remain after 4 half-lives?
A. 256 B. 64
C. 512 D. 32 *(IIT-JAM 2015)*

9. Waste water discharged from a coal mine has hydrogen ion $[H^+]$ concentration of 10-6 moles/liter. The pH of the water is:
A. 6 B. 5
C. 8 D. 9 *(IIT-JAM 2015)*

10. Using the phase rule, the maximum number of phases in a 3 (three) component geological system having 1 (one) degree of freedom is:
A. 3 B. 4
C. 2 D. 1 *(IIT-JAM 2015)*

11. Calculate the average atomic weight (answer to be given up to 3 decimal places) of Rubidium using the given data.

Isotope	Abundance (%)	Atomic weight
^{85}Rb	72.17	84.912
^{87}Rb	27.83	86.909

A. 85.000 B. 85.465
C. 8.545 D. 854.600
(IIT-JAM 2015)

12. A mineral has a weight of 4.5 grams in air and 3.2 grams in water. Assuming density of water to be $1 g/cm^3$, the specific gravity (up to 2 decimal places) of the mineral is:
A. 3.45 B. 3.46
C. 3.49 D. 3.50
(IIT-JAM 2015)

13. What do you mean by low pH?
A. High H^+ concentration
B. Low H^+ concentration
C. High basicity
D. High CO_2 *(DU-2015)*

14. Which of the following is the gem variety of olivine?
A. Amethyst B. Sapphire
C. Aquamarine D. Peridote
(AMU-2015)

15. Which of the following mineral has no cleavage?
A. Orthoclase B. Muscovite
C. Quartz D. Calcite
(AMU-2015)

16. A-type granites are associated with:
A. Subduction zone
B. Island arc
C. Continental rift zone
D. Continent-continent collision
(AMU-2015)

17. Plagioclases of Lunar Anorthosite are:
A. Calcic B. Sodic
C. Albitic D. both B&C
(AMU-2015)

18. Clay mineral analysis is carried out by:
A. XRD B. XRF
C. ICP-AES D. Flamephotometer
(AMU-2015)

19. Talc is formed from the alteration of olivine by:
A. Oxidation B. Hydration
C. Carbonation D. None of the above
(AMU-2015)

20. Which of the following is the hardest mineral?
A. Topaz B. Tourmaline
C. Beryl D. Sapphire
(AMU-2015)

21. Among the different kinds of flood basalt, the SiO_2 content is highest in:
A. Continental tholeiites
B. Oceanic tholeiites
C. Alkaline olivine basalts
D. Spilites *(AMU-2015)*

22. The variation diagram in which the weight percentage of SiO_2 is plotted on the X-axis and the concentration of other oxides on the Y-axis is called:
A. Cartesian diagram
B. AFM diagram
C. QAF diagram
D. Harker diagram *(AMU-2015)*

23. Texture having intergrowth of pyroxene and plagioclase is called:
A. Porphyritic B. Poikilitic
C. Ophitic D. Myrmekite
(AMU-2015)

24. Diamond crystallizes in:
A. Cubic system
B. Hexagonal system
C. Monoclinic system
D. Triclinic system *(AMU-2015)*

25. Amber is a:
A. Mineral
B. Fossilized natural resin
C. Gemstone
D. Both B&C *(AMU-2015)*

26. A mineral having one set of perfect basal cleavage, pearly lusture and hardness less than 3:
A. Quartz B. Calcite
C. Muscovite D. Plagioclase
(AMU-2015)

27. Which of the following mineral show lamellar twining under microscope?
A. Orthoclase B. Plagioclase
C. Microcline D. Quartz
(AMU-2015)

28. Which of the following methods is best suited to date the marine carbonates up to the last 30,000 yr BP?
A. Thermoluminescence dating
B. Radiocarbon dating
C. ^{210}Pb dating
D. U/Th dating *(DU-2015)*

29. The correct statement in respect of cleavage refraction is:
A. Layered rock does not show cleavage refraction
B. Large cleavage-bedding angle is characterized by less competent layers

C. Smaller cleavage-bedding angle is characterized by more competent layer
D. None of the above *(DU-2015)*

30. Aragonite and calcite have same chemical composition. These represent:
A. Polytypism
B. Polymorphism
C. Pseudomorphism
D. None of the above *(DU-2015)*

31. Tectosilicates represent sharing of:
A. Cations for joining with the adjacent units
B. One oxygen ion for joining with adjacent units
C. All oxygen ions for joining with adjacent units
D. Three oxygen ions for joining with adjacent units
(DU-2015)

32. This mineral lacks good cleavage:
A. Fluorite B. Calcite
C. Galena D. Quartz *(DU-2015)*

33. Fibrolite is:
A. Phyllosilicate
B. A clay mineral
C. Has composition Al_2SiO_5
D. An asbestos bearing rock *(DU-2015)*

34. The ratio of a mineral's weight to the weight of an equal volume of water is its:
A. Specific gravity
B. Atomic mass number
C. Equivalent mass
D. Equivalent weight *(DU-2015)*

35. In the given phase diagram, the sequence of minerals crystallizing from a magma of composition "X" under equilibrium conditions is:

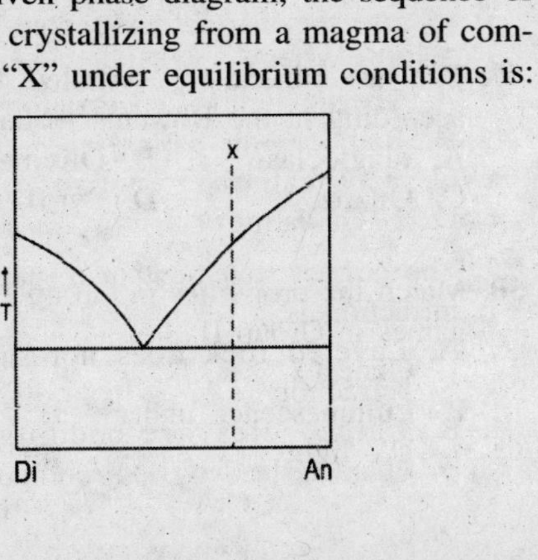

A. An → Di B. An → An + Di
C. Di → An D. Di → Di + An
(IIT-JAM 2015)

36. The geochemical trends illustrated in the diagrams for a suite of cogenetic and coeval igneous rocks indicate:

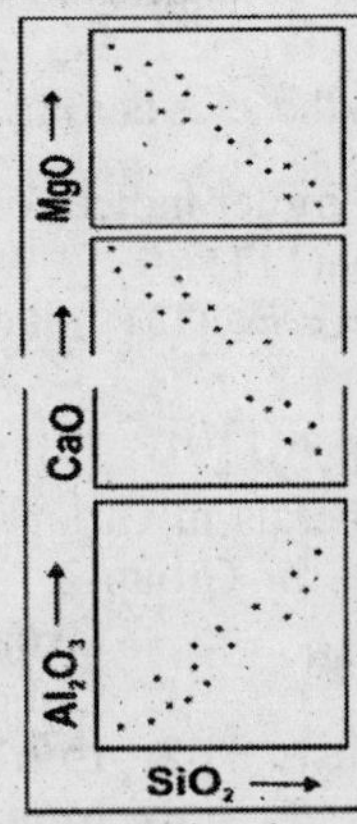

A. Olivine crystallization
B. Clinopyroxene crystallization
C. Plagioclase crystallization
D. Alkali feldspar crystallization
(IIT-JAM 2015)

37. In the given diagram, the percentage of Plagioclase in a rock of composition 'X' is ....

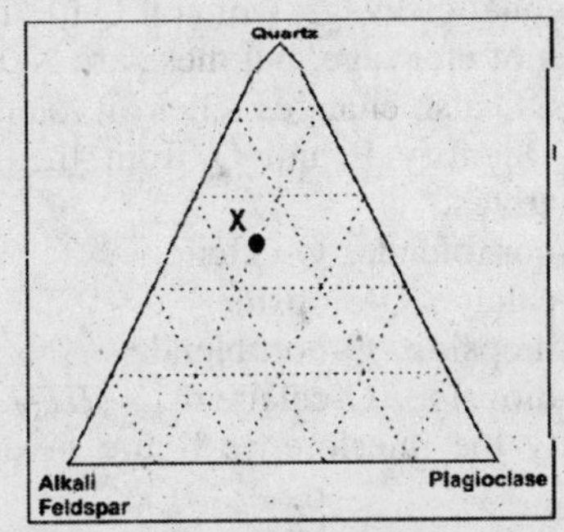

A. 45 B. 20
C. 80 D. 70 *(IIT-JAM 2015)*

38. Which of the following minerals belong to the cubic system?
A. Orthoclase B. Quartz
C. Garnet D. Apatite
(IIT-JAM 2014)

39. Which one of the following minerals is commonly used for determination of vibration direction of the polarizer in a petrological microscope?

A. Muscovite B. Hornblende
C. Augite D. Tourmaline
(IIT-JAM 2014)

40. A closed form belonging to crystal class $\frac{4}{m}\bar{3}\frac{2}{m}$ intercept the axes at a = 1, b = 1 and c = ∞. The name and miller indices of the form are:

A. Octahedron(111)
B. Hexoctahedron(321)
C. Cube(100)
D. Dodcahedron(110) (IIT-JAM 2014)

41. Match the minerals in Group-1 with their optical properties in Group-2:

Group-1	Group-2
P. Quartz	1. Oblique extinction
Q. Hornblende	2. Isotropic
R. Muscovite	3. Mottled extinction
S. Garnet	4. Uniaxial positive

A. P-3, Q-2, R-1, S-4
B. P-4, Q-3, R-3, S-1
C. P-4, Q-1, R-3, S-2
D. P-3, Q-4, R-1, S-2 (IIT-JAM 2014)

42. Two minerals P and Q, were studied under the microscope. Mineral P is characterized by two orthogonal cleavage. Mineral Q has more than one set of cleavage, but these are NOT at right angles, Q also changes relief on rotation of the stage. Identify P and Q from the following alternatives:

A. P-hornblende, Q-calcite
B. P-calcite, Q-diopside
C. P-diopside, Q-hornblende
D. P-diopside, Q-calcite (IIT-JAM 2014)

43. Identify the interference figure given below.

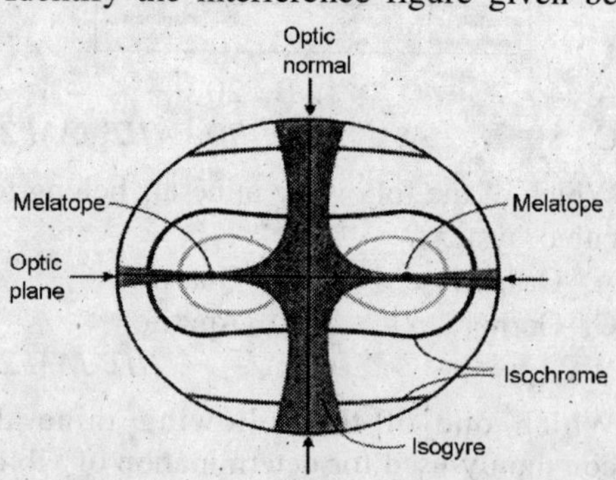

A. Biaxial optic normal figure
B. Biaxial acute bisectrix figure
C. Uniaxial optic axis figure
D. Uniaxial flash figure (IIT-JAM 2014)

44. Which one of the following represents a Closed form?

A. Tetragonal prism
B. Dihexagonal pyramid
C. Orthodome
D. Trigonal dipyramid (IIT-JAM 2013)

45. Match the minerals listed in Group-1 with their corresponding twin type in Group-2:

Group-1	Group-2
P. Staurolite	1. Japan twin
Q. Quartz	2. Swallow-tail twin
R. Gypsum	3. Carlsbad twin
S. Orthoclase	4. Cruciform twin

A. P-2, Q-4,R-1, S-3
B. P-4, Q-2, R-1, S-3
C. P-2, Q-3, R-4, S-1
D. P-4,Q-1, R-2, S-3 (IIT-JAM 2013)

46. The hardest silicate mineral in Mohs scale of hardness is:

A. Feldspar B. Topaz
C. Quartz D. Corundum
(IIT-JAM 2013)

47. The strongly pleochroic mineral is:

A. Quartz B. Hornblende
C. Orthoclase D. Muscovite
(BHU- 2013)

48. Which of the following is an iso-axial mineral?

A. Hornblende B. Garnet
C. Calcite D. Biotite
(BHU-2013)

49. The last mineral to crystallize from magma, according to the Bowen's reaction series is:

A. Plagioclase B. Olivine
C. Quartz D. Pyroxene
(PU-2013)

50. Match the properties in Group I with mineral names in Group II.

Group I	Group II
P. Luminescence under UV light	1. Talc

Q. Pisolitic structure — 2. Galena
R. Soapy feel — 3. Scheelite
S. High specific gravity — 4. Bauxite

A. P-3, Q-2, R-1, S-4
B. P-2, Q-4, R-1, S-3
C. P-3, Q-4, R-1, S-2
D. P-4, Q-3, R-2, S-1 *(IIT-JAM 2012)*

51. Arrange the following silicate minerals in order of increasing sharing of $(SiO_4)^{4-}$ tetrahedra.
A. Olivine, Augite, Hornblende, Muscovite, Orthoclase
B. Orthoclase, Muscovite, Hornblende, Augite, Olivine
C. Olivine, Hornblende, Augite, Muscovite, Orthoclase
D. Orthoclase, Augite, Hornblende, Muscovite, Olivine

(IIT-JAM 2012)

52. Match the crystallographic axes and their angular orientations in Group I with corresponding crystal systems in Group II:

Group I	Group II
P. $a \neq b \neq c$, $\alpha = \beta = \gamma = 90°$	1. Monoclinic
Q. $a = b \neq c$, $\alpha = \beta = \gamma = 90°$	2. Isometric
R. $a \neq b \neq c$, $\alpha = \gamma = 90°$, $\beta \neq 90°$	3. Tetragonal
S. $a = b = c$,	4. Orthorhombic

A. P-1, Q-3, R-4, S-2
B. P-4, Q-2, R-1, S-3
C. P-3, Q-2, R-1, S-4
D. P-4, Q-3, R-1, S-2 *(IIT-JAM 2012)*

53. A snowball garnet is an example of:
A. Inter-kinematic mineral growth
B. Syn-kinematic mineral growth
C. Post-kinematic mineral growth
D. Pre-kinematic mineral growth

(IIT-JAM 2012)

54. The number of crystal faces in a rhombohedron is:
A. 4 B. 6
C. 12 D. 16 *(IIT-JAM 2012)*

55. Match mineral textures in Group I with processes in Group II.

Group I	Group II
P. Replacement of coesite by quartz	1. Polymorphic transformation
Q. Crystallographically oriented lamellae of albite in orthoclase	2. Exsolution
R. Graphic intergrowth of orthoclase and quartz	3. Eutectic crystallization
S. Cross-hatched twinning in microcline	4. Cation ordering

A. P-1, Q-2, R-3, S-4
B. P-2, Q-4, R-3, S-1
C. P-4, Q-2, R-3, S-1
D. P-2, Q-1, R-4, S-3 *(IIT-JAM 2012)*

56. Decay of which one of the following isotopes can be used for dating Archean rocks?
A. ^{14}C B. ^{10}Be
C. ^{147}Sm D. ^{210}Pb

(IIT-JAM 2012)

57. The total number of crystal system is:
A. 7 B. 5
C. 6 D. 4 *(BHU-2013)*

58. Class 'm3m' belong to:
A. Trigonal system
B. Isometric system
C. Hexagonal system
D. Triclinic system *(BHU- 2013)*

59. Galena crystallizes in:
A. Isometric system
B. Triclinic system
C. Trigonal system
D. Monoclinic system *(BHU-2013)*

60. The total number of Bravais lattices is:
A. 12 B. 13
C. 14 D. 16 *(BHU-2013)*

61. The magnetic ferrous mineral is:
A. Hematite B. Magnetite
C. Fe-pyrite D. Sphalerite

(BHU-2013)

62. Based on the Moh's scale, Topaz has a hardness of:
A. 2 B. 3
C. 6 D. 8 *(BHU-2013)*

63. The silica deficient mineral is:
A. Olivine B. Feldspar
C. Feldspathoid D. Amphibole
(BHU-2013)

64. Which of the following minerals would crystallize early from cooling silicate magma?
A. Quartz B. Olivine
C. Biotite D. Hornblende
(BHU-2013)

65. Amphibole usually crystallize in:
A. Triclinic system
B. Trigonal system
C. Hexagonal system
D. Monoclinic system *(BHU-2013)*

66. The Coordination number(CN) of octahedron is:
A. 8 B. 6
C. 12 D. 4 *(BHU-2013)*

67. Looking down the optic axis, the mineral has:
A. Uncertain birefringence
B. Average birefringence
C. Maximum birefringence
D. Zero birefringence *(BHU-2013)*

68. Quartz is:
A. Uniaxial negative
B. Uniaxial positive
C. Biaxial positive
D. Biaxial negative *(BHU-2013)*

69. Which of these is **not** a solid solution?
A. Forsterite-Fayalite
B. Albite-Anorthite
C. Pyrite-Hedenbergite
D. Enstatie-Ferrosilite *(BHU-2013)*

70. Flint, Chert, and Jasper are microcrystalline varieties of:
A. Hematite B. Halite
C. Quartz D. Calcite *(BHU-2013)*

71. The gem mineral which is an oxide of aluminum is:
A. Diamond B. Topaz
C. Sapphire D. Zircon *(BHU-2013)*

72. Among the following, which is not associated with pegmatite?
A. Biotite B. Tourmaline
C. Emerald D. Kyanite
(BHU-2013)

73. The ratio of mineral weight of an equal volume of water is its:
A. Density B. Specific gravity
C. Atomic mass D. Porosity
(BHU-2013)

74. Which mineral show dual hardness?
A. Fluorite B. Epidote
C. Felspar D. Kyanite
(BHU-2013)

75. Which mineral have prismatic cleavage of 56 degree and 124 degree:
A. Olivine B. Plagioclage
C. Calcite D. Amphibole
(BHU-2013)

76. Which is sodium rich mica?
A. Biotite B. Muscovite
C. Paragonite D. Lepidolite
(BHU-2013)

77. Which one is the polymorph of Aragonite?
A. Garnet B. Nephaline
C. K-feldspar D. Calcite
(BHU-2013)

78. Andalusite and Sillimanite crystallize in which system?
A. Hexagonal B. Monoclinic
C. Triclinic D. Orthorhombic
(BHU-2013)

79. Lepidolite mica is rich of:
A. Cr bearing B. Fe bearing
C. Li bearing D. Fl bearing
(BHU-2013)

80. Kaolinite is an alteration product of:
A. Pyroxene B. Feldspar
C. Muscovite D. Calcite
(BHU-2013)

81. Plaster of Paris is obtained from:
A. Bauxite B. Gypsum
C. Kaolin D. Limestone
(BHU-2012)

82. The simplest of all the silicate structures is:
A. Orthosilicates B. Sorosilicates
C Metasilicates D. Inosilicates
(BHU-2012)

83. Which one is **not** a potash felspar?
A. Orthoclase B. Oligoclage
C. Sanidine D. Microcline
(BHU-2012)

84. The main difference between graphite and diamond is:
A. Composition B. Crystal structure
C. Density D. Colour
(BHU-2012)

85. Which of the following minerals is having the chemical formula (Fe, Mg) SiO_3?
A. Andalusite B. Topaz
C. Hypersthene D. Barite *(BHU-2012)*

86. Which one of the following is **not** an example of triclinic system?
A. Kyanite B. Albite
C. Andalusite D. Microcline
(BHU-2012)

87. Mineral diamond crystallizes in:
A. Orthorhombic system
B. Tetragonal system
C. Cubic system
D. Monoclinic system *(BHU-2012)*

88. Which of the following systems has all closed forms?
A. Triclinic B. Cubic
C. Trigonal D. Monoclinic
(BHU-2012)

89. Which of the following has 3 axes of 4-fold symmetry?
A. Baryte B. Gypsum
C. Galena D. Rutile
(BHU-2012)

90. Which one is isotropic mineral?
A. Quartz B. Garnet
C. Tourmaline D. Gypsum
(BHU-2012)

91. In an uniaxial positive mineral, the velocity of ordinary ray is:
A. Greater than that of extraordinary ray
B. Equal to that of extraordinary ray
C. Less than that of extraordinary ray
D. Neither less nor equal to that of extraordinary ray
(BHU-2012)

92. A typical monomineralic rock is:
A. Syenite B. Granite
C. Anorthosite D. Dolerite
(BHU-2012)

93. A texture in which phenocrysts are embedded in fine grained ground mass is:
A. Perthite B. Porphyritic
C. Graphic texture D. Seriate texture
(BHU-2012)

94. The chemical composition of jaedite is:
A. $MgSiO_3$ B. $NaFe(SiO_3)$
C. $NaAl(SiO_3)$ D. $LiAl(SiO_3)$
(BHU-2012)

95. The hardness of Orthoclase on Moh's scale is:
A. 7 B. 6
C. 5 D. 8 *(BHU-2012)*

96. The end members of orthorhombic pyroxene are:
A. Enstatite-diopside
B. Enstatite-hypresthene
C. Diopside-augite
D. Augite-jaedite *(BHU-2012)*

97. Fibrous variety of quartz is:
A. Flint B. Chalcedony
C. Chert D. Amethyst
(BHU-2012)

98. Which of the following is **not** a magnetic mineral?
A. Pyrrhotite B. Hematite
C. Orthoclase D. Magnetite
(BHU-2012)

99. Which one is a light mica?
A. Phlogopite B. Biotite
C. Paragonite D. Zinwaldite
(BHU-2012)

100. Olivine belongs to:
A. Neosilicate B. Inosilicate
C. Sorosilicate D. Cyclosilicate
(BHU-2012)

101. What is mineral 'wool'?
A. Albite B. Andalusite
C. Asbestos D. Kyanite
(BHU-2012)

102. The native mineral having hackly fracture is:
A. Sulphur B. Copper
C. Gold D. Borax
(BHU-2012)

103. Which of the following properties is **not** observed under ordinary light?
A. Colour B. Inclusions
C. Pleochroism D. Refractive Index (RI)
(BHU-2012)

104. Augite shows:
A. 1st order interference colours
B. 2nd order interference colours
C. 3rd order interference colours
D. 4th order interference colours
(BHU-2012)

105. The indicatrix of a uniaxial positive crystal is a/an:
A. Sphere
B. Ellipsoid
C. Oblate spheroid
D. Prolate spheroid *(BHU-2012)*

106. Vertical section of a uniaxial mineral shows:
A. Dichroism B. Pleochroism
C. Isotropism D. Equal illumination
(BHU-2012)

107. Refregence is the ability to produce:
A. Reflection B. Extinction
C. Refraction D. Interference
(BHU-2012)

108. The most fundamental and common form of pyroxene is:
A. Pyramid B. Prism
C. Basal pinacoid D. Side pinacoid
(BHU-2012)

109. Find the odd one out:
A. Biotite B. Phlogopite
C. Enstatite D. Muscovite
(BHU-2012)

110. The silicate class has .... subclasses.
A. 4 B. 6
C. 8 D. 10 *(BHU-2012)*

111. Leucoxene is a variety of:
A. Tourmaline B. Talc
C. Sphene D. Topaz *(BHU-2012)*

112. The basal section of amphibole is six-sided while the same of pyroxene is eight-sided. This is because of the:
A. Crystal face
B. Crystal axis
C. Both A & B
D. None of these *(IIT-JAM 2011)*

113. Match the mineral in Group I with corresponding property in Group II.

Group I	Group II
P. Fayalite	1. Purple colour
Q. Calcite	2. Specific gravity-4.4
R. Autunite	3. Radioactive
S. Amethyst	4. Schalenohedron

A. P-3, Q-4, R-2, S-1
B. P-2, Q-1, R-4, S-2
C. P-4, Q-2, R-3, S-1
D. P-2, Q-4, R-3, S-1 *(IIT-JAM 2011)*

114. An apatite crystal cut perpendicular to c-axis shows outward movement of isochromes in NE and SW quadrants on insertion of quartz wedge from SE quadrant because apatite is:
A. Uniaxial positive
B. Biaxial negative
C. Uniaxial negative
D. Biaxial positive *(IIT-JAM 2011)*

115. Which of the given minerals is a product of reaction between siliceous magmatic fluid and limestone?
A. Andalusite B. Fayalite
C. Muscovite D. Wollastonite
(IIT-JAM 2011)

116. Prism and Pinacoids are type of:
A. Open form
B. Closed form
C. Symmetry operator
D. All of the above *(BHU-2014)*

117. Miller indices for octahedral plane in cubic crystal:
A. (100) B. (110)
C. (111) D. None of the above *(BHU-2014)*

118. Repeatable entity of a crystal structure is known as:
A. Crystal B. Lattice
C. Unit cell D. Miller indices *(BHU-2014)*

119. A family of direction is represented by:
A. (hkl) B. <uvw>
C. {hkl} D. [uvw] *(BHU-2014)*

120. The atomic diameter of an BBC crystal (if a is lattice parameter):
A. a B. a/2
C. $a/(4/\sqrt{3})$ D. $a/(4/\sqrt{2})$ *(BHU-2014)*

121. In diamond, the coordination no. of carbon is:
A. 4 and its unit cell has 8 carbons atoms
B. 4 and its unit cell has 6 carbon atoms
C. 6 with 4 carbon atoms in unit cell
D. 4 with 4 carbon atoms in unit cell *(BHU-2014)*

122. A compound formed by elements A and B crystallizes in cubic structure, in which atoms of A are at the corners while that of B are the face center. The formula of the compound is:
A. AB B. AB_3
C. A_3B D. None of the above *(BHU-2014)*

123. Which of the following mineral belongs to orthorhombic crystal system?
A. Diopside B. Tourmaline
C. Topaz D. Benitoite *(BHU-2014)*

124. The angle between [111] and [112] directions in a cubic crystal is (in degree):
A. 0 B. 45
C. 90 D. 180 *(BHU-2014)*

125. Swallow Tail Twins are commonly observed in the mineral:
A. Gypsum B. Orthoclase
C. Albite D. All of these *(BHU-2014)*

126. 'Butterfly twinning' is found in:
A. Rutile B. Gypsum
C. Aragonite D. Pyrite *(BHU-2014)*

127. Stereogram of a crystal shows primarily:
A. The interfacial angle projected on the stereogram in the forms of arcs and straight lines
B. The distribution of faces in the zones
C. A symmetry in the distribution of face
D. All the above *(BHU-2014)*

128. The penetrative twin is shown by:
A. Staurolite B. Pyrite
C. Fluorite D. All the above *(BHU-2014)*

129. The classification of the crystal system is based on:
A. Angular relationship existing between the crystallographic axes
B. Number of crystallographic axes
C. Relative length of the crystallographic axes
D. All the above *(BHU-2014)*

130. Double reflection phenomenon shown by:
A. Isotropic substances only
B. Anisotropic substance only
C. Both isotropic and anisotropic substances
D. None of the above *(BHU-2014)*

131. Uniaxial crystal are positive if:
A. Ordinary ray has the greater velocity then the Extraordinary
B. Extraordinary ray has the greater velocity then the Ordinary ray
C. Ordinary and Extraordinary rays have same velocity in all direction
D. None of the above *(BHU-2014)*

132. Mark the correct one regarding Uniaxial minerals:
A. Basal section are isotropic
B. Only prismatic sections show pleochroism
C. Dichroic
D. All the above (BHU-2014)

133. Which of the following produces 147nm retardation in the interference colour?
A. Quartz wedge B. Mica wedge
C. Gypsum plate D. None of these
(BHU-2014)

134. Which of the following is **not** an isotropic mineral?
A. Halite B. Fluorite
C. Garnet D. Gypsum
(BHU-2014)

135. The inclined Extinction is sown by:
A. Hypersthene B. Hornblende
C. Biotite D. All of these
(BHU-2014)

136. 'Tartan plaid' twinning is shown by:
A. Albite B. Oligoclage
C. Microcline D. All of these
(BHU-2014)

137. The Bertrand lens is used to:
A. Determine the interference colour
B. Determine the optical sign
C. Analysis of plane of vibration of light
D. None of the above (BHU-2014)

138. In Hypersthene, a well-marked pleochroism which shows:
A. X-pink, Y-green, Z-yellow
B. X-yellow, Y-pink, Z-green
C. X-green, Y-yellow, Z-pink
D. X-pink, Y-yellow, Z-green (BHU-2014)

139. Which type of extinction is often shown by Quartz mineral?
A. Straight B. Oblique
C. Symmetrical D. Wavy (BHU-2014)

140. Crystals which do not possess plane and center of symmetry, contain a form that occurs in two positions and which are mirror image of each other is called:
A. Holohedral form
B. Enantiomorphic form
C. Hemihedral form
D. Hemimorphic form (BHU-2011)

141. Which of the following minerals has a low solubility and therefore is least susceptible to chemical weathering at the earth's surface?
A. Calcite B. Plagioclase
C. Olivine D. Quartz
(BHU-2011)

142. For a given mineral, the physical property which displays the greatest variation is:
A. Color B. Hardness
C. Streak D. Luster
(BHU-2011)

143. Which of the following common items is **not** a mineral?
A. Salt B. Ice
C. Talcum powder D. Asphalt
(BHU-2011)

144. 'Olivine' commonly alters to:
A. Serpentine B. Clay minerals
C. Feldspar D. Amphiboles
(BHU-2011)

145. Which one of the following is fibrous variety of amphibole?
A. Gaucophane B. Crocidolite
C. Pargasite D. Tschemakite
(BHU-2011)

146. In feldspar group sanidine is:
A. A high temperature orthoclase
B. A low temperature orthoclase
C. Gem variety of albite
D. Bright green microcline (BHU-2011)

147. Uvarovite belongs to which garnet group?
A. Ca-Al bearing B. Mg-Al bearing
C. Ca-Cr bearing D. Mn-Al bearing
(BHU-2011)

148. Which one of the following is Li-bearing pyroxene?
A. Hedenbergite B. Augite
C. Spodumene D. Johannsenite
(BHU-2011)

149. In Scalenohedron, there are:
A. 12 triangular faces
B. 6 rhomb-shaped faces
C. 24 triangular faces
D. 8 equilateral triangular faces
(BHU-2011)

150. In double chain silicate, the Si : O ratio is:
A. 1 : 3 B. 4 : 11
C. 4 : 10 D. 2 : 7 *(BHU-2011)*

151. Which of the following minerals is having dual hardness?
A. Talc B. Epidote
C. Kyanite D. Gypsum
(BHU-2011)

152. 'Pericline twinning' is associated with:
A. Plagioclase B. Orthoclase
C. Rutile D. Aragonite
(BHU-2011)

153. Domes are example of:
A. Open form
B. Closed form
C. Combination form
D. General form *(BHU-2011)*

154. Which system has maximum number of classes?
A. Tetragonal B. Hexagonal
C. Isometric D. Triclinic
(BHU-2011)

155. In a phase diagram an eutectic point represents the point where?
A. Only liquid phase exists
B. Only solid phase exists
C. All the phases coexist
D. No phase exists *(BHU-2011)*

156. Which of the following rock does not contain alkali feldspar?
A. Andesite B. Trachyte
C. Adamellite D. Syenite
(BHU-2011)

157. K-feldspar decomposes in lucite and quartz on melting. It is an example of:
A. Congruent melting
B. Incongruent melting
C. Transformation
D. None of the above *(BHU-2011)*

158. Silicon has a co-ordination number six in:
A. Coesite B. Stishovite
C. Tridymite D. Cristobalite
(IIT-JAM 2010)

159. An eight-sided basal section of a mineral containing two sets of clevages at an angle of 90 degree is:
A. Hornblende B. Diopside
C. Calcite D. Sanidine
(IIT-JAM 2010)

160. The crystal class which is not a planar point group is?
A. 3m B. 4 mm
C. 3 D. 6/m
(IIT-JAM 2010)

161. Addition of a centre of symmetry converts the crystal class 622 to:
A. 6 mm B. 6/m 2/m 2/m
C. 6 bar m2 D. 6 bar 2m
(IIT-JAM 2010)

162. Fibrous form is characteristic for which mineral?
A. Chrysotile B. Hornblende
C. Actinolite D. Tremolite
(BHU-2010)

163. The streak of a mineral is:
A. The colour of its powder
B. The cleavage in monochromatic light
C. The breaking due to pressure
D. Hardness in different directions
(BHU-2010)

164. In which mineral there is no cleavage?
A. Garnet B. Spinel
C. Quartz D. All of these
(BHU-2010)

165. Aquamarine is a variety of:
A. Corrundum B. Microcline
C. Tourmaline D. Beryl *(BHU-2010)*

166. 'Cyclic twinning' is commonly seen in:
A. Aragonite B. Staurolite
C. Plagioclase D. Gypsum
(BHU-2010)

167. Staurolite is characteristic of which rock type?
A. Igneous B. Sedimentary
C. Metamorphic D. None of these
(BHU-2010)

168. Which mineral has no silicate structure?
A. Quartz B. Calcite
C. Augite D. Muscovite
(BHU-2010)

169. Which one of the following is an orthopyroxene?
A. Enstatite B. Aegirine
C. Tremolite D. Diopside
(BHU-2010)

170. Of the following, the **first** mineral to form, according to Bowen's reaction series:
A. Hornblende B. Albite
C. Quartz D. Olivine
(BHU-2010)

171. Which one of the following contains manganese?
A. Spessartite B. Almandine
C. Andradite D. Grossularite
(BHU-2010)

172. Which of the following is an incongruently melting compound?
A. Orthoclase B. Olivine
C. Albite D. Tourmaline
(BHU-2010)

173. Which of the following is **not** a variety of gypsum?
A. Alabaster B. Selenite
C. Satin spar D. Iceland spar
(BHU-2010)

174. Which one of the following is a high temperature potassium feldspar?
A. Orthoclase B. Albite
C. Sanidine D. Anorthite
(BHU-2010)

175. Orthoclase shows which type of twinning?
A. Carlsbad B. Baveno
C. Manebach D. All of these
(BHU-2010)

176. The mineral fuchsite belongs to which group?
A. Mica B. Amphibole
C. Pyroxene D. Garnet
(BHU-2010)

177. Which of the following feldspars does not contain potassium?
A. Oligoclase B. Sandine
C. Orthoclase D. Amazonstone
(BHU-2010)

178. Calcium is **not** present in which one of the following minerals?
A. Apatite B. Anorthite
C. Acmite D. Actinolite
(BHU-2010)

179. Which one of the following is radioactive?
A. Microcline B. Talc
C. Monazite D. Asbestos
(BHU-2010)

180. Which one of the following is the softest mineral?
A. Chlorapatite B. Satin spar
C. Bort D. Iceland spar
(BHU-2010)

181. The mineral wollastonite crystallises in which crystal system?
A. Orthorhombic B. Monoclinic
C. Triclinic D. Hexagonal
(BHU-2010)

182. Which of the following is used in crystallography?
A. Weiss symbol
B. Hermann Maugin notation
C. Miller indices
D. All of the above *(BHU-2010)*

183. Except for a centre of symmetry the other axial elements are absent in:
A. Tourmaline B. Microcline
C. Gypsum D. Orthoclase
(BHU-2010)

184. What will be the Miller symbol of a face whose intercepts on the axes are $\infty a : 5b : 4c$?
A. ∞ 45 B. 045
C. 054 D. 154 *(BHU-2010)*

185. The general symbol for a pyritohedron is:
A. (hOO) B. (hkO)
C. (hkl) D. (001) *(BHU-2010)*

186. A five-fold axis of symmetry is present in:
A. Quartz B. Riebeckite
C. Tourmaline D. None of these
(BHU-2010)

187. Of the feldspars, Orthoclase, Microcline, Anorthoclase, Albite, Andesine, Bytownite, Anorthite, Hyalophane and Celsian, how many are monoclinic?
A. Two B. Three
C. Four D. Five *(BHU-2010)*

188. In how many crystal systems the *c*-axis makes an angle with the *b*-axis?
A. One B. Two
C. Three D. Four *(BHU-2010)*

189. The angle between the *a*-axis and *c*-axis is called:
A. α B. β
C. γ D. δ *(BHU-2010)*

190. For which crystal system the Miller symbol for the basal pinacoid is not 001?
A. Tetragonal B. Hexagonal
C. Monoclinic D. Triclinic
(BHU-2010)

191. Of the thirty-two classes of symmetry, which of the following system has three classes?
A. Tetragonal B. Hexagonal
C. Monoclinic D. Triclinic
(BHU-2010)

192. The number of crystallographic axes in uniaxial minerals is:
A. Three B. Three or four
C. Four D. One
(BHU-2010)

193. 2V is the angle between the:
A. *a* and *c* crystallographic axes
B. Biaxial optic axes
C. Fastest and slowest direction
D. None of the above *(BHU-2010)*

194. A mineral 'X' is invisible in Canada Balsam. The refractive index of 'X' is:
A. 1·54 B. 1.44
C. 1.64 D. 1·00 *(BHU-2010)*

195. The highest relief will be shown by which one of the following minerals?
A. 'A', RI = 1.42 B. 'B' RI =1.44
C. 'C' RI =1·59 D. 'D' RI =1·62
(BHU-2010)

196. A mineral is in extinction position when its vibration plane is parallel to:
A. Vibration direction of polarizer
B. Vibration direction of analyser
C. Both A. and B.
D. None of the above *(BHU-2010)*

197. A mineral has three values for its refractive indices. The mineral crystallises in:
A. Cubic system
B. Tetragonal system
C. Hexagonal system
D. Orthorhombic system *(BHU-2010)*

198. In uniaxial minerals, there is no double refraction along which crystallographic axis?
A. *a*-axis B. *b*-axis
C. *c*-axis D. All of these
(BHU-2010)

199. The optic axes in biaxial minerals always lie in:
A. X-Y plane
B. X-Z plane
C. Y-Z plane
D. Any of the above combinations
(BHU-2010)

200. Which mineral shows one set of cleavage in thin sections?
A. Muscovite B. Calcite
C. Wollastonite D. Microdine
(BHU-2010)

201. Which mineral commonly shows euhedral shape?
A. Quartz B. Zircon
C. Microcline D. Muscovite
(BHU-2010)

202. Inclined extinctions are shown by which of the following mineral sets?

A. Andesine, Hornblende, Kyanite
B. Augite, Muscovite, Beryl
C. Biotite, Tourmaline, Quartz
D. Garnet, Microcline, Hypersthene
(BHU-2010)

203. Which mineral is isotropic under the microscope?
A. Almandine B. Biotite
C. Tourmaline D. All of these
(BHU-2010)

204. Isochromatic lines are seen as a part of the interference figure in:
A. Tetragonal minerals
B. Hexagonal minerals
C. Monoclinic minerals
D. All of the above *(BHU-2010)*

205. Calcite is:
A. Uniaxial positive B. Uniaxial negative
C. Biaxial positive D. Biaxial negative
(BHU-2010)

206. Inclined extinction is shown by:
A. Calcite
B. Basal section of augite
C. Basal section of hornblende
D. None of the above *(BHU-2010)*

207. Elements of B-subgroups (in periodic table) with 18 electrons in outermost shell are:
A. Atmophile B. Lithophile
C. Chalcophile D. Siderophile
(BHU-2010)

208. Rubidium Strontium Dating is useful for:
A. Relatively young sediments
B. Relatively old geological material
C. A hundred to thousand year old rocks
D. Non-geological material
(BHU-2010)

209. Element of which atomic number is absent in the earth?
A. 41 B. 42
C. 43 D. 44 *(BHU-2010)*

210. The weight percent oxygen in the earth crust is about:
A. 42 B. 46
C. 49 D. 53 *(BHU-2010)*

211. Which of the following are lithophile elements?
A. Ag, Zn, Pb B. Li, Zr, V
C. Ru, Os, Au D. S, Se, Te
(BHU-2010)

212. An igneous rock contains SiO_2, Al_2O_3, Na_2O, K_2O, FeO, MnO, CaO and MgO. Assuming that the degree of freedom is 2, the number of minerals present in the rock is:
A. 4 B. 8
C. 10 D. 1 *(IIT-JAM 2009)*

213. Following two figure (1 & 2) show the optical indicatrix of uniaxial crystals. Which one of the following interpretations is correct?

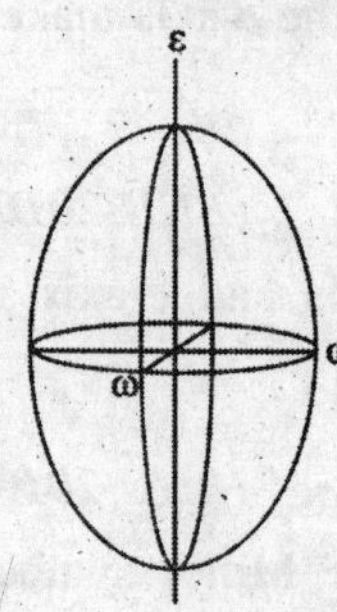

Fig. 1

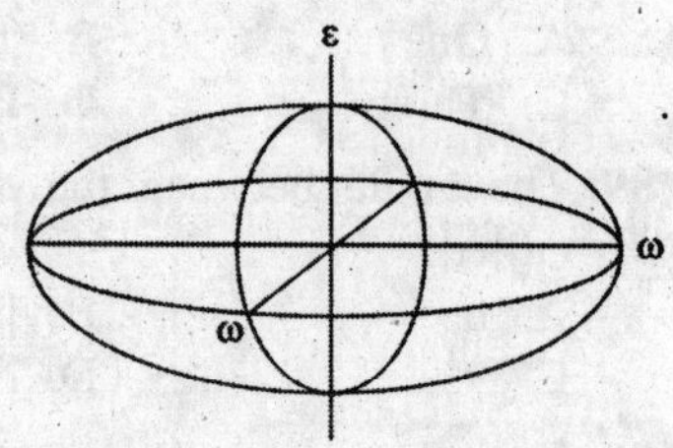

Fig. 2

A. 1-positive, 2-negative
B. 1-negative, 2- positive
C. 1-positive, 2-positive
D. 1-negative, 2-negative *(IIT-JAM 2009)*

214. Match the following:

Group 1 (Silicate structure)	Group 2 (Minerals)
P. Nesosilicate	1. Muscovite
Q. Cyclosilicate	2. Quartz
R. Phyllosilicate	3. Forsterite
S. Tectosilicate	4. Beryl

A. P-1, Q-2, R-3, S-4
B. P-4, Q-3, R-2, S-1
C. P-3, Q-4, R-1,S-2
D. P-2, Q-1, R-4, S-3 *(IIT-JAM 2009)*

215. In igneous rocks graphic texture forms at:
A. Peritectic condition
B. Eutectic condition
C. Triple point
D. Critical point *(IIT-JAM 2009)*

216. The chemical composition of orthoclase is:
A. K Al Si_2O_4 B. K Al Si_3O_8
C. K Al Si_3O_6 D. K Al Si_4O_8
(BHU-2009)

217. Normal abundance of an element in barren earth material is called:
A. Anomaly B. Background value
C. Threshold D. Contrast
(BHU-2009)

218. Two set of rhombic cleavage are shown by the mineral:
A. Hornblende B. Enstatite
C. Microcline D. Gypsum
(BHU-2009)

219. Which of the following minerals is having the chemical formula Al_2SiO_5?
A. Andalusite B. Topaz
C. Flourite D. Barite *(BHU-2009)*

220. Crystal face with miller indices (111) is known as:
A. Parametral face B. Unit face
C. Solid face D. Inclined face
(BHU-2009)

221. Mineral diamond crystallizes in:
A. Orthorhombic system
B. Tetragonal system
C. Cubic system
D. Monoclinic system *(BHU-2009)*

222. The symmetry elements present in monoclinic system crystal are:
A. One diad
B. One diad and center of symmetry
C. One diad, one mirror plane and center of symmetry
D. Four triads *(BHU-2009)*

223. If polarizing microscope the angle between polarizer and analyzer is:
A. 120° B. 180°
C. 90° D. 45° *(BHU-2009)*

224. Calcite is:
A. Uniaxial positive
B. Uniaxial negative
C. Biaxial positive
D. Biaxial negative *(BHU-2009)*

225. Augite shows:
A. 1st order interference colours
B. 2nd order interference colours
C. 3rd order interference colour
D. 4th order interference colour
(BHU-2009)

226. Concordant intrusive igneous plutons found in folded terrains as:
A. Phacolith B. Batholith
C. Laccolith D. Plug
(BHU-2009)

227. With the increase in load pressure, melting point of silicates:
A. Increases
B. Decreases
C. Remains the same
D. Sometimes increases, sometimes decreases
(BHU-2009)

228. Which of the following minerals displays twinkling?
A. Calcite B. Gypsum
C. Augite D. Talc
(BHU-2009)

229. Quartz has the following hardness on Moh's scale:
A. 7 B. 6
C. 5 D. 8 *(BHU-2009)*

230. Which of the following minerals exhibits admantine luster?
A. Diamond B. Calcite
C. Quartz D. Nepheline
(BHU-2009)

231. Which of the following is the precious variety of beryl?
A. Emerald B. Ruby
C. Sapphire D. Citrine *(BHU-2009)*

232. Which of the following minerals exhibits the property of magnetism?
A. Pyrrhotite B. Biotite
C. Orthoclase D. Albite *(BHU-2009)*

233. Muscovite exhibits .... luster
A. Silky B. Vitreous
C. Pearly D. Dull *(BHU-2009)*

234. Quartz exhibits .... fracture.
A. Conchoidal B. Hackly
C. Uneven D. Even *(BHU-2009)*

235. Olivine belongs to:
A. Neosilicate B. Inosilicate
C. Sorosilicate D. Cyclosilicate
(BHU-2009)

236. The property of lanthanide contraction is present in:
A. REE
B. Transition elements
C. Alkali elements
D. Alkaline earth elements *(BHU-2009)*

237. Members of inosilicate group exhibit:
A. Prismatic form B. Botryoidal form
C. Granular form D. Tabular form
(BHU-2009)

238. Plagioclase group of minerals exhibits:
A. Isomorphism B. Polymorphism
C. Polytypism D. Heteromorphism
(BHU-2009)

239. Which one is non-pleochroic mineral?
A. Biotite B. Olivine
C. Garnet D. Tourmaline
(BHU-2009)

240. Irregular or asymmetrical dispersion is found in:
A. Monoclinic system
B. Orthoclinic system
C. Triclinic system
D. Tetragonal system *(BHU-2009)*

241. Which of the following minerals will have the maximum tendency to concentrate $O^{18}\delta$?
A. Magnetite B. Biotite
C. Quartz D. Muscovite
(BHU-2009)

242. Rutile is source of:
A. Tin B. Tungsten
C. Titanium D. Iron
(BHU-2009)

243. Plaster of Paris obtained from:
A. Bauxite B. Gypsum
C. Talc D. Limestone
(BHU-2009)

244. Lamprophyres exhibit the following texture:
A. Panidiomorphic B. Allotriomorphic
C. Hypidiomorphic D. Polymorphic
(BHU-2009)

245. The highest symmetry elements of any crystal system is shown by:
A. Hemihedral form
B. Holohedral form
C. Tetartohedral form
D. Hemimorphic form *(IIT-JAM 2008)*

246. Ruby is a gem variety of:
A. Tourmaline B. Corundum
C. Garnet D. Beryl
(IIT-JAM 2008)

247. The normal class of Triclinic system is:
A. Barite type B. Gypsum type
C. Axinite type D. Calcite type
(IIT-JAM 2008)

248. Obtuse bisectrix figure is obtained if the crystal section is oriented:
A. Normal to acute bisectrix
B. Normal to obtuse bisectrix
C. Parallel to basal section
D. Parallel to obtuse bisectrix
(IIT-JAM 2008)

249. Which one of the following minerals shows four sets of cleavage?
A. Calcite B. Albite
C. Barite D. Flourite
(IIT-JAM 2008)

250. A mineral with patchy blue colour and variable hardness is:
A. Azurite B. Kaolinite
C. Kyanite D. Sillimanite
(IIT-JAM 2007)

251. Group 1 contains some minerals and Group 2 gives a list of crystal systems. Find the correct match.

Group 1	Group 2
P. Gypsum	1. Hexagonal
Q. Beryl	2. Triclinic
R. Galena	3. Monoclinic
S. Axinite	4. Cubic

A. P-3, Q-2, R-4, S-1
B. P-1, Q-3, R-4, S-2
C. P-2, Q-1, R-4, S-3
D. P-3, Q-1, R-4, S-2 *(IIT-JAM 2007)*

252. Minerals in which of the following pairs do NOT form a natural association?
A. Corundum - Sillimanite
B. Nepheline - Quartz
C. Olivine - Clinopyroxene
D. Olivine - Orthopyroxene *(IIT-JAM 2007)*

253. In phyllosilicates, the tetrahedral cation to oxygen ratio is:
A. 1:2 B. 1:3
C. 2:5 D. 4:11
(IIT-JAM 2006)

254. The optical indicatrix of garnet is:
A. Prolate B. Oblate
C. Sphere D. Triaxial ellipsoid
(IIT-JAM 2006)

255. The feldspathoid present in phonolite is:
A. Nepheline B. Sanidine
C. Kalsilite D. Albite
(IIT-JAM 2006)

256. The angle between the "Y" vibration direction and the optic axis of a monoclinic mineral is:
A. 0° B. 45°
C. 90° D. 102°
(IIT-JAM 2006)

257. The incorrectly matched pair among the following is:
A. Forsterite- orthosilicate
B. Augite- double chain silicate
C. Biotite- sheet silicate
D. Orthoclase- framework silicate
(IIT-JAM 2005)

258. If a crystal face has Weiss parameters of 1a, 2b, 1/2c, its Miller indices will be:
A. 124 B. 214
C. 421 D. 412
(IIT-JAM 2005)

259. The 010 section of a monoclinic mineral necessarily shows:
A. Inclined extinction
B. Straight extinction
C. Symmetric extinction
D. Undulose extinction *(IIT-JAM 2005)*

260. Which of the following is a mineral?
A. Water B. Steel
C. Mercury D. Halite *(BHU-2014)*

261. The fundamental building block of a mineral is its:
A. Unit cell B. Crystal system
C. Size D. Shape
(BHU-2014)

262. Iron-rich olivine is:
A. Forsterite B. Monticellite
C. Fayalite D. Sepentine
(BHU-2014)

263. Which of the following is a typical metamorphic mineral?
A. Cordierite B. Chert
C. Jasper D. Chalcedony
(BHU-2014)

264. Garnet present in eclogite is:
A. Pyrope B. Almandine
C. Uvaravite D. Spessartite
(BHU-2014)

265. A crystal form having only one face is:
A. Basal pinacoid B. Pedion
C. Pyramid D. Prism
(BHU-2014)

266. High-pressure pyroxene is:
A. Jadiete B. Johannsenite
C. Jasper D. Zoisite
(BHU-2014)

267. Calcite and aragonite are:
A. Polymorphs B. Isomorphs
C. Paramorphs D. Metamicts
(BHU-2014)

268. Which of the following contains water in its structure?
A. Lepidolite B. Hypersthene
C. Anorthoclase D. Fayalite
(BHU-2014)

269. Glaucophane is considered to be a characteristic mineral of:
A. Collision zones B. Subduction zones
C. Mantle plumes D. Alteration zones
(BHU-2014)

270. Asterism in a property displayed by which group of minerals?
A. Alumino silicates B. Olivine
C. Mica D. Feldspar
(BHU-2014)

271. Which of the following minerals can co-exist with free silica?
A. Forsterite B. Leucite
C. Nepheline D. Fayalite
(BHU-2014)

272. Cleavage is absent in:
A. Orthoclase B. Garnet
C. Enstatite D. Actinolite
(BHU-2014)

273. The most abundant of pyroxene group of minerals is:
A. Diopside B. Augite
C. Enstatite D. Aegerine
(BHU-2014)

274. In view of relative susceptibility to chemical weathering, which of the following is the correct sequence of increasing stability?
A. Albite – Forsterite – Muscovite – Kaolinite
B. Muscovite – Enstatite – Nephaline – Wollastonite
C. Forsterite – Diopside – Microcline – Quartz
D. Microcline – Gibbsite – Diopside – Anorthosite
(DU-2014)

275. Which of the following minerals crystallizes first from a silicate melt?
A. Diopside B. Enstatite
C. Quartz D. Hornblende
(DU-2014)

276. A Mid Oceanic Ridge is characterized by?
A. Normal fault, shallow earthquake and low-K tholeitiic lava
B. Reverse fault, shallow earthquake and low-K tholeitiic lava
C. Revarse fault, deep earthquake and alkaline lava
D. Normal fault, shallow earthquake and andesitic lava
(DU-2014)

277. A silicate minerals will remain dark on full rotation of microscope stage under cross-nicol if:
A. The mineral crystallized in orthorhombic system
B. The mineral section is cut parallel to optic axis
C. The mineral section is cut perpendicular to optic axis
D. The mineral crystallized in triclinic system
(DU-2014)

278. Polymorphic transformations represent ________ type of reactions.
A. Continuous B. Discontinuous
C. Dehydration D. Decarbonation
(DU-2014)

279. The difference between orthorhombic and monoclinic crystal is:
A. All crystallographic axes are unequal in orthorhombic and two axes are equal in monoclinic
B. All crystallographic axes are unequal in both and inter axial angles are 90 degrees
C. All crystallographic axes are unequal in both and all inter-axial angles are 90 degree in orthorhombic and one angle is not equal to 90 degree in monoclinic.
D. All crystallographic axes are unequal in both and all inter axial angles are 90 degree in monoclinic and one angle is not equal to 90 degree in orthorhombic.
(DU-2014)

280. Symplectite texture is commonly inferred as an indicator of:
A. Burial
B. Decompression
C. Contact metamorphism
D. Cataclasis *(DU-2014)*

281. A pinacoid is made up of .... faces.
A. 1 B. 2
C. 3 D. 4 *(DU-2014)*

282. Isotropic crystals have .... optic axes.
A. 0 B. 1
C. 2 D. Infinite *(DU-2014)*

283. Spinel + Quartz = Corundum + ...
A. Garnet B. Sillimanite
C. Ilmenite D. Plagioclase
(DU-2014)

284. There are .... lattices, ... classes and .... space groups.
A. 14,32,230 B. 32,14,230
C. 230,14,32 D. 14,230,32
(DU-2014)

285. Which of the following symmetry elements is **not** present in any crystal system?
A. Axis of 3 fold symmetry
B. Axis of 5 fold symmetry
C. Axis of 4 fold symmetry
D. Axis of 6 fold symmetry *(DU-2014)*

286. Becke Line, appears to move:
A. Into the mineral with the higher refractive index as the objective of the microscope is raised
B. Away from the mineral with the higher refractive index as the objective of the microscope is raised
C. Into the mineral with the higher refractive index as the objective of the microscope is lowered
D. Into the mineral with the lower refractive index as the objective of the microscope is raised *(DU-2014)*

287. Melting points of minerals:
A. Increase with pressure
B. Decrease with pressure
C. Do not change with change in pressure
D. Increases with increase in water content
(DU-2014)

288. Silicon: Oxygen ratio is highest in:
A. Tectosilicates B. Inosilicates
C. Nesosilicates D. Sorosilicates
(DU-2014)

289. In high grade rocks of granulite facies where muscovite is unstable, sillimanite coexists with:
A. Na-feldspar B. K-feldspar
C. Plagioclase D. Quartz
(DU-2014)

290. The incorrectly matched pair among the following:
A. Nesosilicate - forsterite
B. Cyclosilicate - talc
C. Phyllosilicate - muscovite
D. Tectosilicate - orthoclase *(BHU-2015)*

291. Which of the following is **not** a mineral?
A. Olivine B. Talc
C. Calcite D. Quartzite
(BHU-2015)

292. Which is not a polymorph of Al_2SiO_5?
A. Kyanite B. Calcite
C. Sillimanite D. Andalusite
(BHU-2015)

293. Which mineral shows double hardness?
A. Magnesite B. Quartz
C. Feldspar D. Kyanite
(BHU-2015)

294. Hardness of fluorite is:
A. 3 B. 4
C. 7 D. 5 *(BHU-2015)*

295. Augite crystallized in which crystal system?
A. Isometric system
B. Monoclinic
C. Triclinic
D. Orthorhombic *(BHU-2015)*

296. The maximum number of axes of symmetry .... found in crystal system.
A. 13 B. 10
C. 22 D. 9 *(BHU-2015)*

297. Match the crystal system with correct number of classes:

Group-1 (Crystal system)	Group-2 (Classes)
P. Isometric	1. 3
Q. Orthorhombic	2. 12
R. Hexagonal	3. 7
S. Tetragonal	4. 5

A. P-1, Q-2, R-3, S-4
B. P-4, Q-1, R-3, S-2
C. P-3, Q-4, R-2, S-1
D. P-4, Q-1, R-2, S-3 *(BHU-2015)*

298. Total number of classes in crystal system is:
A. 22 B. 6
C. 32 D. 13 *(BHU-2015)*

299. .... is an imaginary plane which passes through the centre of crystal and divides it into two equal parts where one part is the mirror image of the other.
A. Axial plane
B. Reference plane
C. Plane of symmetry
D. Twin plane *(BHU-2015)*

300. Which one is **not** a physical property of a mineral?
A. Hardness B. Streak
C. Form D. Extinction
(BHU-2015)

301. Which crystal is the non-pleochroic system?
A. Hexagonal B. Triclinic
C. Cubic D. Tetragonal
(BHU-2015)

302. Biaxial minerals are not belong to .... crystal system.
A. Triclinic B. Tetragonal
C. Monoclinic D. Orthorhombic
(BHU-2015)

303. Which mineral shows the double refraction?
A. Calcite B. Quartz
C. Feldspar D. Galena
(BHU-2015)

304. Streak of hematite is:
A. Black colour B. Yellow
C. Cherry red D. None of these
(BHU-2015)

305. A crystal appears unchanged when rotated 180°. Such an axis of rotation is axis of:
A. 1-fold symmetry
B. 2-fold symmetry
C. 3-fold symmetry
D. 4-fold symmetry *(PU-2010)*

306. In a crystal 4 non-parallel face intersect in a point. Form of these faces is:
A. Prism B. Pinacoid
C. Pyramid D. Dome
(PU-2010)

307. Electronegativity of elements increases:
A. From left to right of the Periodic table
B. From right to left of the Periodic table
C. In atoms that have partially filled outer shell
D. With decrease in neutrons *(PU-2010)*

308. Structurally, quartz is a:
A. Framework silicate
B. Ring silicate
C. Chain silicate
D. Isolated silicate *(PU-2010)*

309. A mineral is scratchable by quartz but not by apatite. Hardness of this mineral on Mohs scale is:
A. Less than apatite
B. Less than 7
C. More than 7
D. Less than 5 *(PU-2010)*

310. Which of the following mineral is a chain silicate?
A. Augite B. Biotite
C. Anorthite D. Olivine *(PU-2010)*

311. Which of the following is a K-feldspar?
A. Oligoclase B. Orthoclase
C. Andesine D. Anorthite
(PU-2010)

312. If a set of equivalent crystal faces are related to each other by a three fold axis of rotation, what will be angle between these faces?
A. 30° B. 60°
C. 90° D. 120° *(PU-2010)*

313. Which one of the following minerals crystallize in Orthorhombic system?
A. Calcite
B. Aragonite
C. Rhodochrosite
D. Dolerite *(PU-2010)*

314. Which one of the following minerals crystallize in cubic system?
A. Spinel B. Siderite
C. Hematite D. Corundum
(PU-2010)

315. Which one of the following mineral belongs to spinet group?
A. Zircon B. Ilmenite
C. Hematite D. Magnetite
(PU-2010)

316. The axial ratio of a crystal is determined as 0.82: 1.94. Which crystal system this may belong to?
A. Isometric or Tetragonal
B. Hexagonal or Tetragonal
C. Orthorhombic or Monoclinic
D. Isometric or Hexagonal *(PU-2010)*

317. Which one of the following is an open form?
A. Prism B. Octahedron
C. Tetrahedron D. Pyramid
(PU-2010)

318. Which one of the following is permitted in the Miller's indices?
A. Zero B. Infinity
C. Real numbers D. Fractions
(PU-2010)

319. Which one of the following forms belongs to the isometric system?
A. Pyramid B. Prism
C. Octahedral D. Sphenoid
(PU-2010)

320. A Pyroxine mineral having equal amount of CaO and FeO molecules is known as:
A. Hedenbergite B. Pigeonite
C. Bronzite D. Diopside
(PU-2010)

321. Plagioclase having 30 to 50 % anorthite content is known as:
A. Anorthoclase B. Andesine
C. Oligoclase D. Labradorite
(PU-2010)

322. To observe interference colour of a mineral under petrological microscope, the correct settings are:
A. Polarizers crossed and rotate 90° off extinction
B. Parallel polarizers and rotate 45° off extinction
C. Polarizers crossed and rotate 45° off extinction
D. Parallel polarizers and rotate 90° off extinction *(PU-2013)*

323. High pressure polymorph of quartz is:
A. Tridymite B. Cristloblite
C. Coesite D. β-quartz
(PU-2013)

324. Muscovite and biotite are distinguished based on:
A. Lustre B. Cleavage
C. Habit D. Colour
(PU-2013)

325. Intersection of two sets of cleavages at 87° and 93° in basal section of crystals is observed in:
A. Calcite B. Pyroxene
C. Hornblende D. Kyanite
(PU-2013)

326. Three fold axis of rotational symmetry is present in:
A. Only in hexagonal system
B. Only in cubic system
C. Hexagonal and cubic systems
D. Tetragonal, hexagonal and cubic systems
(PU-2013)

327. Classification of minerals into various groups is based on:
A. Anion or anionic complex
B. Cation or cationic complex
C. Abundance in the crust
D. Hardness *(PU-2013)*

328. A crystal that has only a center of symmetry belongs to point group:
A. $\bar{1}$ B. 1
C. m D. m/1 *(PU-2013)*

329. The crystal with the point group 222 belongs to:
A. Triclinic system
B. Monoclinic system

C. Orthorhombic system
D. Cubic system (PU-2013)

330. A Miller index of a face, whose Weiss symbol is 2a : 2b : 1c, is:
A. (221) B. (212)
C. (112) D. (236) (PU-2013)

331. Two non-parallel face in a crystal that can be related to each other by a mirror plane constitute the form:
A. Pinacoid B. Prism
C. Pyramid D. Dome (PU-2013)

332. The optical indicatrix of an isotropic mineral is:
A. Prolate ellipsoid
B. Oblate ellipsoid
C. Spheroid
D. Irregular in shape (PU-2013)

333. Which of the following is **not** a silicate mineral?
A. Garnet B. Magnetite
C. Mica D. Olivine
(PU-2013)

334. Which of the following minerals is a chain silicate?
A. Feldspar B. Olivine
C. Pyroxene D. Quartz (PU-2013)

335. In a monoclinic crystal:
A. All the three crystallographic axes are inclined to each other
B. All the three crystallographic axes are mutually perpendicular to each other
C. One of the crystallographic axes is inclined to other two mutually perpendicular axes
D. There are three horizontal and one vertical crystallographic axes
(PU-2012)

336. Tetragonal system is characterized by the presence of:
A. 4-fold axis of symmetry
B. 6-fold axis of symmetry
C. 3-fold axis of symmetry
D. At least two improper axes of symmetry
(PU-2012)

337. The total number of crystal classes are:
A. 32 B. 7
C. 364 D. 30000 (PU-2012)

338. A radioactive isotope with half life of 10 hours would decay to one-fourth of its original amount in:
A. 40 hours B. 20 hours
C. 2.5 hours D. 5 hours
(PU-2012)

339. A radioactive isotopes with atomic number 19 and mass 40 decays by beta decay. The atomic number and mass of the daughter isotope would be:
A. 20, 40 B. 18, 40
C. 20, 41 D. 18, 41 (PU-2012)

340. Which of the following is NOT an example of thermodynamic state function?
A. Enthalpy
B. Entropy
C. Gibbs free energy
D. Work done (PU-2012)

341. Crystal belonging to the which of the following crystal system have mutually perpendicular but unequal crystallographic axes:
A. Cubic B. Monoclinic
C. Orthorhombic D. Triclinic
(PU-2011)

342. Which of the following pairs of minerals is not an example of polymorphs?
A. Kyanite and Andalusite
B. Calcite and Aragonite
C. Quartz and Tridymite
D. Diopside and Augite (PU-2011)

343. Which one of the following choices contains minerals that do NOT show pleochroism?
A. Rutile, Hypersthene and tourmaline
B. Wollastonite, muscovite and tremolite
C. Sphene, augite and epidote
D. Staurolite, chlorite and glaucophane
(PU-2011)

344. Which of the following is not a mineral?
A. Olivine B. Limestone
C. Calcite D. Quartz
(PU-2011)

345. Which of the following minerals has a framework silicate structure?
A. Feldspar B. Mica
C. Olivine D. Pyroxene
(PU-2011)

346. Which one of the following is **not** a silicate minerals?
A. Mica B. Olivine
C. Galena D. Hornblende
(PU-2011)

347. Select sequence of the minerals according to increasing melting point:
A. Quartz, Forsterite, Fayalite
B. Hornblende, Fayalite, Forsterite
C. Fayalite, Forsterite, Diopside
D. Forsterite, Fayalite, Quartz (PU-2011)

348. Which of the following minerals will not leave a streak on streak plate?
A. Magnetite B. Talc
C. Calcite D. Topaz (PU-2011)

349. Presence of gypsum in a rock indicates following conditions during its formation:
A. High precipitation
B. Evaporative condition
C. Humid condition
D. Cold condition (PU-2011)

350. Which one of the following statement about the petrological microscope is true?
A. A circular stage and a polarizer are present
B. Two polarizers, one above and another below the circular stage are present
C. Both the polarizers are fitted below the circular stage
D. Both the polarizer are fitted above the circular stage (PU-2011)

351. Spinal group of minerals are classified as:
A. Oxides
B. Carbonates
C. Island silicates
D. Framework silicates (PU-2011)

352. At an invariant point (number of phase component) is equal to:
A. C B. C + 3
C. C + 2 D. C + 1 (PU-2011)

353. The reaction $Mg_3SiO_4O_{10}$ $(OH)_2$ (Talc) + $4MgSiO_3$ (Enstatite) = $Mg_2SiO_{22}(OH)_2$ (Anthophyllite) is a type of:
A. Devolatilization reaction
B. Ion-exchange reaction
C. Solid-solid net transfer reaction
D. Polymorphic transformation reaction
(PU-2011)

354. The correct order of increasing stability of minerals during chemical weathering is
A. Olivine, pyroxene, biotite, quartz
B. Olivine, biotite, pyroxene, quartz
C. Quartz, biotite, pyroxene, olivine
D. Pyroxene, olivine, biotite, quartz
(IIT JAM-2016)

355. The hardest mineral (with the exception of diamond) in the Moh's scale of hardness is
A. An oxide
B. A silicate
C. A phosphate
D. A carbonate
(IIT JAM-2016)

356. Match minerals in Group I with the corresponding silicate structure in Group II.

Group I	**Group II**
(i) Anthophyllite	P. Cyclosilicate
(ii) Hedenbergite	Q. Tectosilicate
(iii) Cordierite	R. Double chain silicate
(iv) Zeolite	S. Single chain silicate

A. (*i*) R; (*ii*) S; (*iii*) Q; (*iv*) P
B. (*i*) S; (*ii*) R; (*iii*) Q; (*iv*) P
C. (*i*) R; (*ii*) S; (*iii*) P; (*iv*) Q
D. (*i*) P; (*ii*) Q; (*iii*) R; (*iv*) S
(IIT JAM-2016)

357. Which one of the following optical properties of minerals is NOT observed under crossed nicols?
A. Extinction
B. Interference colour
C. Interference figure
D. Pleochroism (IIT JAM-2016)

358. Choose the correct stereographic projection among the following that represents 222 crystal symmetry.

A. 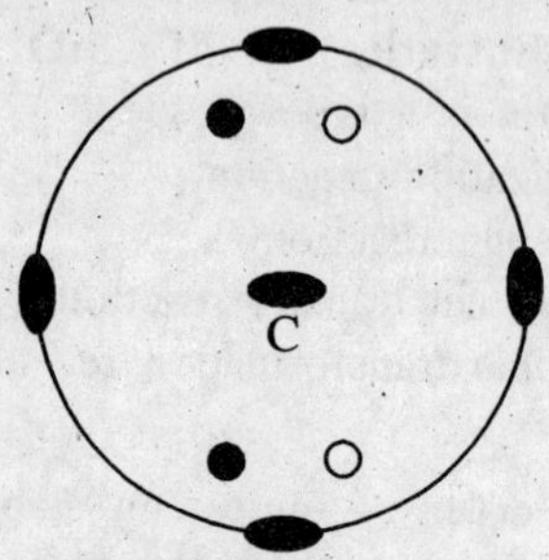

B. 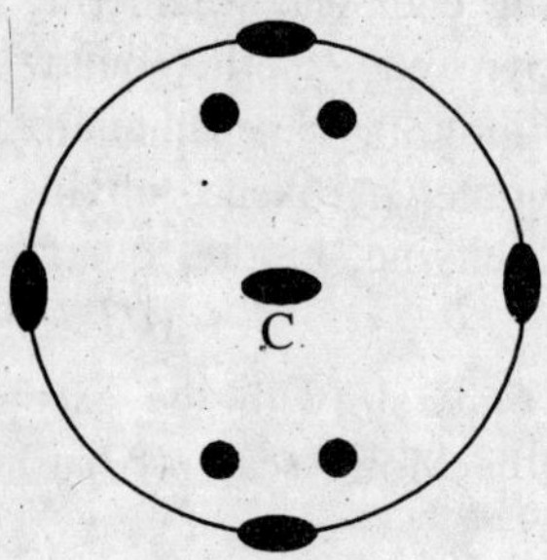

C. 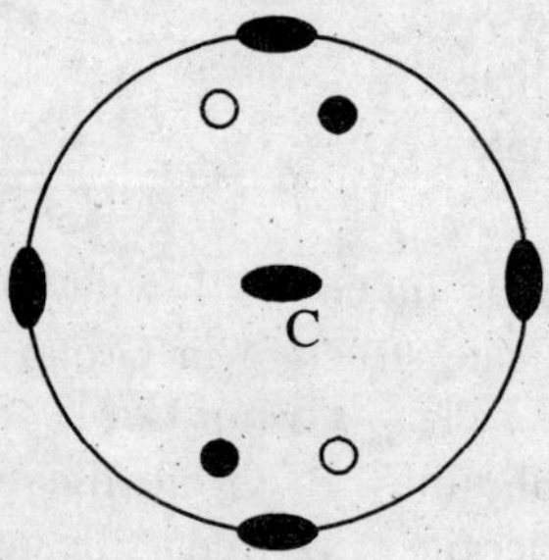

D. 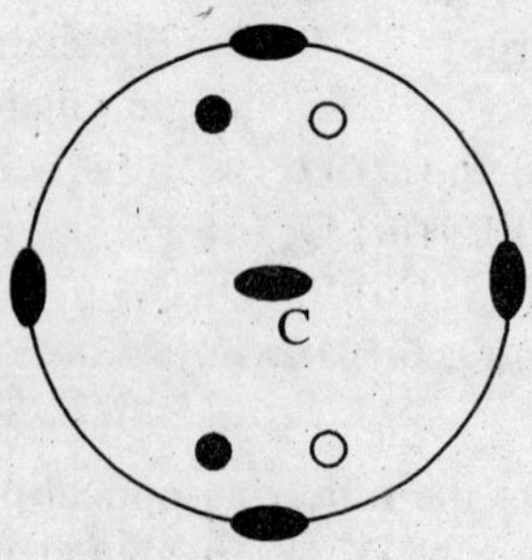

(IIT JAM-2016)

359. The change in coordination number of silicon (Si) when α-quartz transforms to stishovite, which has octahedral coordination, is

A. 2 B. 4
C. 6 D. 8 *(IIT JAM-2016)*

360. In a thin section of 30 μm thickness, the R.I. of a mineral are: $\varepsilon = 1.565$ and $\omega = 1.468$. Choose the correct statement(s) about its optical properties.

A. Uniaxial +ve
B. Uniaxial –ve
C. 1st order interference colour
D. Low to moderate relief *(IIT JAM-2016)*

361. In the following figure, the exterior angle measured between (001) ∧ (021) with a goniometer in a crystal is 40°. The interior angle between (010) ∧ (021) in degrees is ....

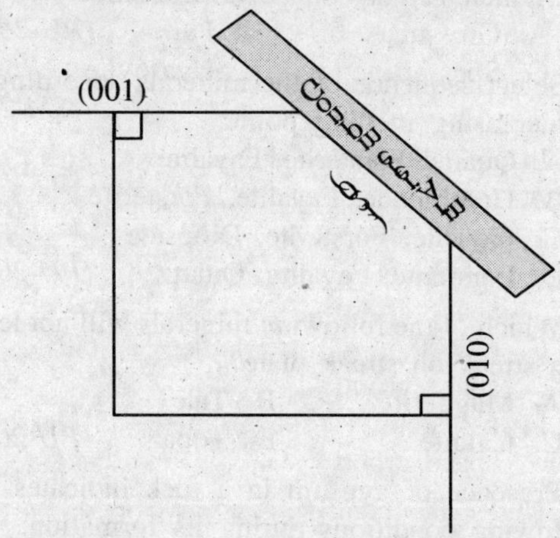

A. 140 B. 120
C. 125 D. 129 *(IIT JAM-2016)*

362. In the following schematic diagram, the aluminosilicate triple point is located at the pressure of 4.50 kbar and temperature of 823°K. If the reaction andalusite = sillimanite has negative slope with a value of –18.22 bar/°K, the pressure of the reaction at 923°K is .... kbar (give answer in two decimal places).

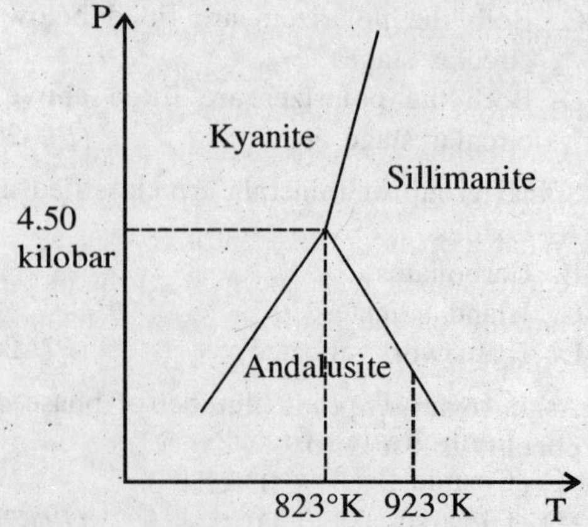

(Figure not to scale)

A. 2.67 B. 12.67
C. 1.67 D. 1.70 *(IIT JAM-2016)*

363. Which of the following combinations characterise cubic system?
A. 9 planes, 13 axes, and center
B. 13 planes, 9 axes, and center
C. 3 planes, 7 axes, and center
D. 6 planes, 7 axes, no center *(ISM-2014)*

364. Which garnet contain chromium?
A. Grossular B. Pyrope
C. Spessartine D. Uvarovite
(ISM-2014)

365. In crystallographic there are
A. 6-system, 14 space lattice, 12 class, 210 space group
B. 6 system, 32 space lattice, 14 classes, 160 space group
C. 6 system, 8 space lattice, 20 classes, 200 space group
D. 6-system, 12 space lattice, 32 classes, 14 space group *(ISM-2014)*

366. Under the polarizing microscope a minerals grain under crossed nicol appears dark at all positions of the stage. The minerals will be isotropic if the grain
A. Remain dark under microscopic observation
B. Remains dark in plane polarized light
C. Shows isogyres under conoscopic observation
D. Shows small spaces of light with the aperture closed *(ISM-2014)*

367. Which of the following represents composition of stony-iron meteorites?
A. Nickel, iron and sulphates
B. Iron, nickel and silicates
C. Platinum, iron and carbonates
D. Nickel, iron and phosphates *(ISM-2014)*

368. What term refers to meteorites that centre the earth's atmosphere and pass out again without landing on the earth or getting disintegrated completely?
A. Close calls
B. Earth-grazing fireballs
C. Cosmic flesh wounds
D. Glow streaks *(ISM-2014)*

369. What is tektites:
A. Glassy meteorites
B. Microcrystalline glassy product of meteoritic impact on crustal rocks
C. Glossy igneous rocks
D. Tectonically deformed ultra-mylonites
(ISM-2014)

370. Approximately how many meteoroids pass through the earth's atmosphere every day?
A. 10 B. 100
C. 1000 D. 100000 *(ISM-2014)*

371. In which of the following situations a biaxial minerals is positive?
A. X is acute bisectrix
B. Y is acute isectrix
C. Z is acute bisectrix
D. Z is obtuse bisectrix *(ISM-2014)*

372. The minerals Glaucophane and jadcite are indicate of:
A. High T, Low P
B. High P, Low T
C. High P, High T
D. Low P, Low T *(ISM-2014)*

373. The isogyre would show which of the following position in case of large 2V
A. Highly curved
B. Touching each other
C. At right angles to each other
D. Virtually straight *(ISM-2014)*

374. In which of the following minerals sequence are the minerals arrange the order of increasing hardness?
A. Hematite-nephaline-galena-orthoclase
B. Epidote-garnet-zircon-dolomite
C. Graphite-fluorite-topaz-kyanite
D. Chlorite-barite-sphene-tourmaline
(ISM-2014)

375. Diaphaneity is a property of crystal which?
A. Allow water to penetrate through it
B. Can transmitting light through it
C. Shows diamagnetism
D. Exhibit two poles in it *(ISM-2015)*

376. Give the term that describe different crystalline modification of the same chemical substance:

A. Isomorphism B. Polymorphism
C. Tetrahedral D. Perovskite
(ISM-2015)

377. Which one of the following minerals is biaxial positive?
A. Biotite B. Hexagonal
C. Glauconite D. Topoz *(ISM-2015)*

378. The chemical composition of the Enstatite is:
A. $CaMgSi_2O_6$ B. $MgSiO_3$
C. $CaFeSi_2O_6$ D. $FeSiO_3$ *(ISM-2015)*

379. Which of the following twinning is most common for orthoclase Feldspar?
A. Pericline B. Albite
C. Manebach D. Carlsbad
(ISM-2015)

380. When stage of the microscope is rotated the change in the colour or intensity or both of a minerals is called?
A. Polarization colours
B. Anisotropism
C. Birefringence
D. Pleochroism
(SAGAR UNIVERSITY-2015)

381. Albite and anorthite end members belong to which of the following minerals?
A. Orthoclase B. Plagioclase
C. Microcline D. Nepheline
(SAGAR UNIVERSITY-2015)

382. Ferro-magnesium minerals alter into which of the following secondary minerals:
A. Sphene B. Rutile
C. Anatase D. Chlorite
(SAGAR UNIVERSITY-2015)

383. Leucite and Nephaline can be distinguished by which of the most important character?
A. Relief
B. Polarization colours
C. Pleochroism
D. Isotropism
(SAGAR UNIVERSITY-2015)

384. Which is the most important optical property of the biotite?
A. High relief
B. High order polarization
C. Strong pleochroism
D. Straight extinction
(SAGAR UNIVERSITY-2015)

385. Which of the following minerals is highly pleochroic?
A. Augite B. Biotite
C. Hornblende D. Hypersthene
(SAGAR UNIVERSITY-2015)

386. Calcite is optically:
A. Uniaxial negative
B. Biaxial negative
C. Uniaxial positive
D. Biaxial positive
(SAGAR UNIVERSITY-2015)

387. Bytownite minerals belong to which of the following :
A. Biotite B. Muscovite
C. Plagioclase D. Orthoclase
(SAGAR UNIVERSITY-2015)

388. The bond which is responsible for the coherence of inert gases when condensed to solids at low temperature is:
A. Polar covenant bond
B. Polar bond
C. Residual bond
D. Ionic bond
(SAGAR UNIVERSITY-2015)

389. Mg garnet is termed as:
A. Pyrope B. Almandine
C. Spessartite D. Grossularite
(SAGAR UNIVERSITY-2015)

390. When original minerals and texture are preserved after the recrystallization, the stricter is known as?
A. Xenoblast B. Maculose
C. Blastopsammitic D. Palimpsest
(SAGAR UNIVERSITY-2015)

391. Presence of tourmaline in granite is indicative of:
A. Sulphur metasomatism
B. Fluorine
C. Boran metasomatism
D. Water vapour metasomatism
(SAGAR UNIVERSITY-2015)

392. Siderolites are:
A. Iron meteorites
B. Stony iron meteorites
C. Stony meteorite
D. Silicate meteorites
(PUNJAB UNIVERSITY-2015)

393. Sensitive High Resolution Ion Probe (SHRIP) is a/an:
A. Age ending method by Zircon crystal
B. Type of remote sensing survey
C. Type of film used for remote sensing survey
D. Types of minerals exploration method
(PUNJAB UNIVERSITY-2015)

394. Which method is used for dating relatively recent geological event?
A. K-Ar method
B. U-Pb method
C. Carbon-14 method
D. Rb-Sr method
(PUNJAB UNIVERSITY-2015)

395. Elements which readily from ion with an outermost 8-electron shell are:
A. Siderophile B. Chalcophile
C. Lithophile D. Atmophile
(PUNJAB UNIVERSITY-2015)

396. The stability series of minerals of igneous rocks towards weathering is given by:
A. Bowmen (1919) B. Goldich (1938)
C. Imbrie (1959) D. Goldschmidt (1940)
(PUNJAB UNIVERSITY-2015)

397. The term pH stands for:
A. Pure hydrogen
B. Pusissance de hydrogen
C. Potential hydrogen
D. Per hydrogen ion
(PUNJAB UNIVERSITY-2015)

398. A zoned plagioclase crystal is typically zoned from:
A. Calcic core to calcic outershell
B. Sodic core to sodic outershell
C. Calcic core to sodic outershell
D. Sodic core to calcic outershell
(AMU-2016)

399. Europium anomaly is chiefly controlled by:
A. Plagioclase B. Olivine
C. Orthopyroxene D. Clinopyroxene
(AMU-2016)

400. The processes where by a solid phase directly goes to a vapour phase is called:
A. Sublimation B. Assimilation
C. Fractionation D. Partial melting
(AMU-2016)

401. Maximum amount of substitution of Al for Si is found in:
A. Orthoclase B. Microcline
C. Albite D. Anorthite
(AMU-2016)

402. Refractive Index (RI) of Canada balsam is
A. 1.597 B. 0.1537
C. 1.552 D. 1.536 *(AMU-2016)*

403. While going down to discontinues series of Bowen's reaction, replacement of Si by Al:
A. Always increases
B. Always decreases
C. Remain the same
D. Decreases and increases *(AMU-2016)*

404. Complete extinction occur only when the section is:
A. Parallel to optic axis
B. Perpendicular to optic axis symmetry
C. Parallel to optic axial plane
D. Perpendicular to optic axial plane
(AMU-2016)

405. Fibrous microcrystalline quartz varieties include:
A. Agate B. Onyx
C. Sard D. All the above
(AMU-2016)

406. Tsavorite is:
A. A green variety of beryl
B. A brilliant green garnet
C. A red variety of corundum
D. A orange variety of topaz *(AMU-2016)*

407. The following cation content in orthopyroxene is a reliable thermometer in high temperature rock

A. Al B. Si
C. Mg D. Fe *(ISM-2016)*

408. Tectosilicates represent sharing of:
A. Cations for joining with the adjacent units
B. One oxygen ion for joining with adjacent units
C. All oxygen ions for joining with adjacent units
D. Three oxygen ions for joining with adjacent units *(ISM-2016)*

409. Cleavage angle of Hornblende minerals is:
A. 56 & 122 degree
B. 56 & 124 degree
C. 86 & 124 degree
D. 89 & 124 degree *(ISM-2016)*

410. A polar wandering curve:
A. Shows wandering of magnetic poles relative to fixed continents
B. Shows wandering of rotational poles relative to fixed continents
C. Shows wandering of continents relative to generally fixed pole positions
D. Is a graph of the Mercalli index *(ISM-2016)*

411. A single isogyre in a biaxial interference figure indicates
A. Optic normal figure
B. Optic axis figure
C. Acute bisectrix figure
D. Obtuse bisectrix figure *(ISM-2016)*

412. Optical axial plane is the plane containing
A. Two of the three vibration directions (*x* & *z*)
B. Two vibration direction (*x* & *z*) and optic axis
C. Y-vibration direction and two optic axes
D. Two optic axes only *(ISM-2016)*

413. The 010 section of a monoclinic minerals necessarily shows
A. Inclined extinction
B. Straight extinction
C. Symmetric extinction
D. Undulose extinction *(ISM-2016)*

414. Which of the following is not a minerals:
A. Hedenbergite B. Fayalite
C. Pseudotachylite D. Monazite *(ISM-2016)*

415. In the following diagram where the assemblage Gros + Qtz + Wo will be possible?

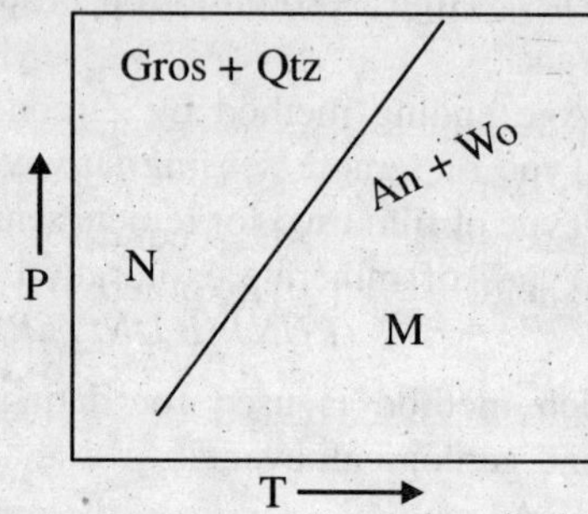

A. In field N
B. In field M
C. On the univariant line separating the two fields
D. Nowhere *(ISM-2016)*

416. Match the following:

Group-A (Silicate structure)	**Group-B (Minerals)**
P. Nesosilicate	1. Biotite
Q. Cyclosilicate	2. Quartz
R. Phyllosilicate	3. Zircon
S. Tektosilicate	4. Beryl

A. P-1, Q-2, R-3, S-4
B. P-4, Q-3, R-2, S-1
C. P-3, Q-4, R-1, S-2
D. P-2, Q-1, R-4, S-3 *(ISM-2016)*

417. Staurolite crystallizes in:
A. Monoclinic B. Orthorhombic
C. Hexagonal D. Triclinic *(BHU-2016)*

418. Symbol of Trapezohedron:
A. hhl B. hll
C. hkl D. hol *(BHU-2016)*

419. Skew twin observed in:
A. staurolite B. pyrite
C. plagioclase D. gypsum *(BHU-2016)*

420. In pyrite, twinning is present:
A. geniculate B. iron cross
C. mimetic D. swallo *(BHU-2016)*

421. It represented by only one face:
A. pinacoid B. pedion
C. prism D. pyramids
(BHU-2016)

422. Which crystal system has one asymmetric class?
A. Isometric B. Trigonal
C. Monoclinic D. Triclinic
(BHU-2016)

423. Which of the following denotes a class that has no symmetry at all?
A. Pinacoidal B. Pedial
C. Domatic D. Spheroidal
(BHU-2016)

424. The Bravais lattice of sodium chlorite structure is:
A. Primitive cell
B. Body centered cube
C. Face centred cube
D. Base centra *(BHU-2016)*

425. In the skew twins the twin plain is:
A. Dome B. Bipyramid
C. Prism D. Side pinacoid
(BHU-2016)

426. Mineral which crystallize in orthorhombic form:
A. Topaz B. Beryl
C. Quartz D. Tourmaline
(BHU-2016)

427. Enantiomorphic crystal forms possess:
A. Only a centre of symmetry
B. Only a plane of symmetry
C. Neither a plane nor a centre of symmetry
D. Both a plane and a centre of symmery
(BHU-2016)

428. Open form of intermediate between a prism and a pyramid whose faces cut the vertical axes and one of the horizontal axes
A. Dome B. Pyramid
C. Prism D. Diametral Prism
(BHU-2016)

429. The base centred arrangement of a unit cell is found in:
A. Orthorhombic crystal
B. Triclinic crystal
C. Cubic crystal
D. Hexagonal crystal *(BHU-2016)*

430. Zircon crystallise in
A. Isometric B. Hexagonal
C. Tetragonal D. Triclinic
(BHU-2016)

431. An interfacial angle:
A. An external angle between a given face and the extension of the adjoining face
B. An internal angle between two adjoining faces
C. An angle between their faces on the same side of a crystal
D. None of the above *(BHU-2016)*

432. The twin plane is:
A. A plane in a twin crystal which is common to both the halves of the crystal and is essentially a plane of symmetry of one of the twins
B. A plain in a twin crystal which is common to both the halves of the crystal and is never a plane of symmetry of one of the twins
C. A plane in a twin crystal having no definite bearing with plane of symmetry
D. A and B are correct *(BHU-2016)*

433. Which of the following properties can be used to distinguished between plagioclase feldspar and Quartz?
A. Lustre B. Streak
C. Colour D. Cleavage
(BHU-2016)

434. Choose the diamagnetic minerals:
A. Pyrite B. Ilemenite
C. Calcite D. Apatite *(BHU-2016)*

435. Blue Beryl is:
A. Aquamarine B. Agate
C. Topaz D. Both A and C
(BHU-2016)

436. Which of the following minerals does not contain silica?
A. Quartz B. Muscovite
C. Halite D. Plagioclage
(BHU-2016)

437. A minerals has pearly lustre:
A. Diamond B. Muscovite
C. Opal D. Asbestos
(BHU-2016)

438. Scheelite shows:
A. Phosphorescence
B. Fluorescence
C. Triboluminescence
D. Thermoluminescenc *(BHU-2016)*

439. Optically Negative and has low extinction angles:
A. Quartz B. Chlorite
C. Hornblende D. Acmite *(BHU-2016)*

440. Quartz is:
A. Positive uniaxial
B. Negative uniaxial
C. Negative biaxial
D. Positive biaxial *(BHU-2016)*

441. Property seen when several grain are vieved collectively under microscope in polarized light:
A. Play of colours B. Twinkling
C. Pleochroism D. Interference colour
(BHU-2016)

442. The mineralogical phase rule is given by the formula:
A. $P + F = C - 2$ B. $P - F = C + 2$
C. $F = C + P - 2$ D. $P + F = C + 2$
(BHU-2016)

443. High temperature polymorph of quartz:
A. Coesite B. Cristobalite
C. Tridymite D. Opal *(BHU-2016)*

444. Which of the following is used to measure retardation?
A. Bertland lens
B. Condenser
C. Break compensator
D. Iris diaphragm *(BHU-2016)*

445. Diaphaneity is the property related to:
A. Dispersion of light
B. Transmission of light
C. Reflection of light
D. Refraction of light *(BHU-2016)*

446. Diamond is exposed to sun, it exhibit:
A. Phosphorescence B. Fluorescence
C. Opalescence D. Pleochroism
(BHU-2016)

447. Tantalite and Brookite are example of:
A. Polymorphs B. Isomorphs
C. Pseudomorphs D. Paramorphs
(BHU-2016)

448. Highest Birefringence in:
A. Plagioclase B. Tourmaline
C. Zircon D. Calcite *(BHU-2016)*

449. Which of the following is a homopolar bond?
A. Ionic bond B. Covalent bond
C. Metallic bond D. van der Waal's bond
(BHU-2016)

450. We can know internal structure of minerals from their study by:
A. X-rays B. UV rays
C. Cosmic rays D. Visible rays
(BHU-2016)

451. Which of the following is a high pressure variety of silica?
A. Chert B. Chalcedony
C. Cristobalite D. Coesite *(BHU-2016)*

452. Fluorite is often characterised by how many sets of cleavage?
A. 5 B. 1
C. 0 D. 4 *(BHU-2016)*

453. The streak of Barites is:
A. Blue B. Colourless
C. White D. Yellow *(BHU-2016)*

454. Which of the following minerals display florescence?
A. Actinolite B. Feldspar
C. Scheelite D. Pyrope *(BHU-2016)*

455. Which amongst the following has largest ionic radius?
A. Oxygen B. Aluminium
C. Silicon D. Magnesium
(BHU-2016)

456. Which of the following is an orthosilicate?
A. Axinite B. Enstatite
C. Quartz D. Olivine *(BHU-2016)*

457. The co-ordination number of Na^+ in NaCl is:
A. 2 B. 6
C. 4 D. 8 *(BHU-2016)*

458. Ca-bearing olivine is:
A. Fayalite B. Monticellite
C. Tephroite D. Knebelite *(BHU-2016)*

459. Which of the following is not a metamorphic minerals?
A. Staurolite B. Jasper
C. Olivine D. Garnet *(BHU-2016)*

460. Most common amphibole is:
A. Hornblende B. Pigeonite
C. Anthophyllite D. Glaucophane *(BHU-2016)*

461. Which of the following garnets is used as pathfinder minerals in diamond (kimberlite) exploration?
A. Almandine B. Spessartine
C. Pyrope D. Uvaravite *(BHU-2016)*

462. Chiastolite is a variety of:
A. Anorthite B. Anorthoclase
C. Andalusite D. Andradite *(BHU-2016)*

463. Which amongst the following minerals crystallize in triclinic system?
A. Sillimanite B. Fayalite
C. Kyanite D. Tremolite *(BHU-2016)*

464. Which is not a minerals?
A. Hedenburgite B. Fayalite
C. Pseudotachylite D. Monazite *(DU-2016)*

465. The minerals having highest hardness amongst in following:
A. Graphite B. K-feldspar
C. Calcite D. Quartz *(DU-2016)*

466. Which mineral does not form when muscovite and quartz react at high temperature?
A. K-feldspar B. Oligoclase
C. Sillimanite D. Andalusite *(DU-2016)*

467. Kyanite-staurolite schist represent :
A. A mafic protolith
B. A pellitic protolith
C. A calcic protolith
D. An arenite protolith *(DU-2016)*

468. Olivine and Quartz can not together because:
A. One is plutonic and other is volcanic minerals
B. One is a metamorphic and other magmatic
C. They are chemically immiscible
D. One is siliceous and other are calcareous *(DU-2016)*

469. Coesite is a high temperature polymorphs of:
A. Garnet B. Orthoclase
C. Apatite D. Quartz *(DU-2016)*

470. Chief elements founds as alloy in the core of the earth are:
A. Iron-magnesium
B. Nickel-iron
C. Silica-magnesium
D. Magnesium-nickel *(DU-2016)*

471. Which of the following pair of elements are diadochic in the lattice structures of dolomite?
A. Fe – Mg B. Ca – Na
C. Fe – Al D. Na – K *(DU-2016)*

ANSWER KEY

1	2	3	4	5	6	7	8	9	10
A	D	A	C	B	B	C	B	A	B
11	12	13	14	15	16	17	18	19	20
B	B	A	D	C	A	A	A	B	D
21	22	23	24	25	26	27	28	29	30
A	D	C	A	D	C	B	B	D	B

31	32	33	34	35	36	37	38	39	40
C	D	C	A	B	A,B	B	C	D	D
41	42	43	44	45	46	47	48	49	50
C	D	B	C	D	B	B	B	C	C
51	52	53	54	55	56	57	58	59	60
A	D	B	B	B	C	C	B	A	C
61	62	63	64	65	66	67	68	69	70
B	D	C	B	D	B	D	B	C	C
71	72	73	74	75	76	77	78	79	80
C	A	B	D	D	C	D	D	C	B
81	82	83	84	85	86	87	88	89	90
B	A	B	B	C	C	C	B	C	B
91	92	93	94	95	96	97	98	99	100
A	C	B	C	B	B	B	C	A	A
101	102	103	104	105	106	107	108	109	110
C	B	C	A	D	C	D	B	C	B
111	112	113	114	115	116	117	118	119	120
C	A	D	C	D	A	C	C	A	C
121	122	123	124	125	126	127	128	129	130
A	B	C	A	A	B	D	D	D	B
131	132	133	134	135	136	137	138	139	140
A	D	B	D	B	A	A	D	D	A
141	142	143	144	145	146	147	148	149	150
D	A	D	A	B	A	C	C	A	B
151	152	153	154	155	156	157	158	159	160
C	A	A	B	C	A	B	B	D	D
161	162	163	164	165	166	167	168	169	170
A	A	A	D	D	A	C	B	A	D
171	172	173	174	175	176	177	178	179	180
A	A	D	C	D	A	A	C	C	B
181	182	183	184	185	186	187	188	189	190
C	D	B	B	B	D	B	A	B	B
191	192	193	194	195	196	197	198	199	200
C	B	B	A	A	C	D	C	B	A
201	202	203	204	205	206	207	208	209	210
B	A	A	D	B	D	C	B	C	B
211	212	213	214	215	216	217	218	219	220
B	B	A	C	B	B	B	A	A	B
221	222	223	224	225	226	227	228	229	230
C	C	C	B	B	A	A	A	A	A

231	232	233	234	235	236	237	238	239	240
A	A	C	A	A	A	A	A	C	C
241	**242**	**243**	**244**	**245**	**246**	**247**	**248**	**249**	**250**
C	C	B	A	B	B	C	B	A	C
251	**252**	**253**	**254**	**255**	**256**	**257**	**258**	**259**	**260**
D	B	C	C	B	A	B	B	A	D
261	**262**	**263**	**264**	**265**	**266**	**267**	**268**	**269**	**270**
A	C	A	A	B	A	A	A	B	A
271	**272**	**273**	**274**	**275**	**276**	**277**	**278**	**279**	**280**
A	B	B	C	B	A	B	A	C	B
281	**282**	**283**	**284**	**285**	**286**	**287**	**288**	**289**	**290**
B	B	A	A	B	A	A	B	C	B
291	**292**	**293**	**294**	**295**	**296**	**297**	**298**	**299**	**300**
D	B	D	B	B	A	D	C	C	D
301	**302**	**303**	**304**	**305**	**306**	**307**	**308**	**309**	**310**
C	B	A	C	A	A	A	A	B	A
311	**312**	**313**	**314**	**315**	**316**	**317**	**318**	**319**	**320**
B	D	B	C	A	C	A	B	C	A
321	**322**	**323**	**324**	**325**	**326**	**327**	**328**	**329**	**330**
B	C	C	D	B	B	A	A	C	C
331	**332**	**333**	**334**	**335**	**336**	**337**	**338**	**339**	**340**
C	C	B	C	C	A	A	C	A	D
341	**342**	**343**	**344**	**345**	**346**	**347**	**348**	**349**	**350**
C	D	B	B	A	C	B	D	B	B
351	**352**	**353**	**354**	**355**	**356**	**357**	**358**	**359**	**360**
A	C	C	A	A	C	D	C	A, D	A, D
361	**362**	**363**	**364**	**365**	**366**	**367**	**368**	**369**	**370**
A	A	A	D	D	B	B	C	B	D
371	**372**	**373**	**374**	**375**	**376**	**377**	**378**	**379**	**380**
A	B	C	D	B	B	D	B	D	D
381	**382**	**383**	**384**	**385**	**386**	**387**	**388**	**389**	**390**
B	D	D	C	D	A	C	A	A	A
391	**392**	**393**	**394**	**395**	**396**	**397**	**398**	**399**	**400**
C	B	A	C	D	B	C	C	A	A
401	**402**	**403**	**404**	**405**	**406**	**407**	**408**	**409**	**410**
D	C	A	A	D	A	D	C	B	A
411	**412**	**413**	**414**	**415**	**416**	**417**	**418**	**419**	**420**
D	B	B	C	A	C	B	B	A	B

421	422	423	424	425	426	427	428	429	430
B	D	B	D	B	A	C	A	A	C
431	**432**	**433**	**434**	**435**	**436**	**437**	**438**	**439**	**440**
B	C	D	C	A	C	B	B	D	A
441	**442**	**443**	**444**	**445**	**446**	**447**	**448**	**449**	**450**
B	D	B	C	B	A	B	D	B	A
451	**452**	**453**	**454**	**455**	**456**	**457**	**458**	**459**	**460**
D	D	C	C	A	D	B	B	B	A
461	**462**	**463**	**464**	**465**	**466**	**467**	**468**	**469**	**470**
C	C	C	C	D	B	B	C	D	B
471									
A									

HINTS AND SOLUTIONS

1.

Process	**Texture**
Eutectic crystallization	Graphic Texture
Crystal settling	Cumulate texture.
Fractional crystallization	Porphyritic texture
Magma mixing	Porphyritic texture

2.

Forms	**Crystal system**	**Mineral example**
Octahedron	Isometric system	Diamond
Tetrahedron	Isometric system	Tetrahedrite
Trapezohedron	Isometric system	Phosgenite
Rhombohedron	Hexagonal system	Siderite

3.

Volcanoes	**Characteristics**	**Example**
Strato volcano	Rhyolitic and Andesitic composition.	Mount Fujj Japan
Shield volcano	Basic lava composition.	Hawaiian island
Dome	Basaltic lava types.	Mount Unzen Japan
Fissure eruption	Basic lava.	Deccan traps

4.

Minerals	**Isotropic**	**Relief**	**Cleavage**
Sillimanite	Anisotropic	Moderate	Orthorhombic
Orthoclase	Aniso tropic	Low	Intersect at 90°
Pyroxeme	Aniso tropic	Moderate	87° and 92°
Garnet	Isotropic	High	Absent

5. Birefringence: The difference between the refractive indices of the minerals is called birefringence.

The refractive indices of the first uniaxial mineral is

$$= 1.658$$

The refractive indices of the second uniaxial mineral is

$$= 1.486$$

Therefore, Birefringence = 1.658 – 1.486

$$= 0.172.$$

6. A crystal face intercepts for the crystallographic axes is – $1a_1$, $1a_2$, $\frac{1}{2}a_3$, ∞c.

$$= \frac{1}{1}, \frac{1}{1}, \frac{2}{1}, \frac{1}{\infty}$$

= 1, 1, 2, 0

The face is = (1120)

7. The point group is = $\frac{4}{m}\bar{3}\frac{2}{m}$

Crystal symmetry = Isometric system

Class = Hexoctahedral

Symmetry elements = I, $3A_4$, $4A_3$, $6A_2$

Number of plane = 9

Minerals example = Fluorite.

8. The radioactive isotope atoms = 1024

The remains of four half lives- $(½)^4 = 1024$

1/16 = 1024

= 64

Then, 64 atoms.

OR

The remains atoms of first half life = 512

The remains atoms of second half life = 256

The remains atoms of third half life = 128

The remains of the fourth half life = 64.

9. The concentration of Hydrogen ion (H^+)

= 10^{-6} mol/liter.

The pH of the water = $-\log_{10}(H^+)$

= $-\log_{10}(10^{-6})$

= 6 (Basic in nature).

11. The average atomic weight percentage is

= 72.17/100 × 84.912

= 61.280

= 27.83/100 × 86.902

= 24.184

Therefore, = 61.280 + 24.184

= 84.465

12. A mineral weight in air is = 4.5 gm.

A mineral weight in water is = 3.2 gm.

The density of the water is = 1 gm/cm^3

Specific gravity of the mineral = weight in air/ weight in air-weight in water.

= 4.5/ 4.5 – 3.2

= 3.46.

14.

Gem variety	Minerals
Amethyst	Quartz
Sapphire	Corundum
Aquamarine	Beryl
Peridote	Olivine

15.

Minerals	Minerals class	Colour	Cleavage
Orthoclase	Tectosilicate	White	Perfect
Muscovite	Phyllosilicate	Colourless	Perfect
Quartz	Tectosilicate	Colourless	Absent
Calcite	Carbonate	White	Perfect

16.

Association	Types of Granite
Subduction zone	M- type
Island arc	M- type
Continental rift zone	A- type
Continent-continent collision	S- type

18.

Technique	Minerals
XRD	Clay minerals
XRF	Ores minerals
ICP-AES	Isotope minerals
Flame photometer	Metal ions minerals

20.

Minerals	Hardness	Colour
Topaz	8	Colourless
Tourmaline	6-7	Black
Beryl	7-8	White
Sapphire	9	Blue

21.

Types of Basalts	SiO_2% (Designation)
Continental tholeiites	Saturated
Oceanic tholeiites	Saturated
Alkaline olivine basalt	Undersaturated
Spilites	Saturated

24.

Minerals	Crystal system
Diamond	Cubic system
Corundum	Hexagonal system
Topaz	Orthorhombic system

Quartz	Hexagonal system
Feldspar	Orthorhombic system
Apatite	Hexagonal system
Fluorite	Cubic system
Calcite	Hexagonal system
Gypsum	Monoclinic system
TalcMonoclinic system	

25.

Minerals	Cleavage	Set of cleavage	Lusture	Hardness
Quartz	absent	–	vitreous	7
Calcite	present	3-set	vitreous	3
Muscovite	Present	1-set	pearly	2
Plagioclase	present	2-set	vitreous	6

27.

Minerals	Twinning (Microscope)
Orthoclase	Simple
Plagioclase	Lamellar
Microcline	Cross-hatch twinning
Quartz	Simple

32.

Minerals	Cleavage
Fluorite	Perfect, 2-set (octahedral)
Calcite	Perfect, 3-set (rhombohedral)
Galena	Perfect, 2-set
Quartz	Absent

38.

Minerals	Crystal system
Orthoclase	Orthorhombic System
Garnet	Isometric System
Quartz	Hexagonal System
Apatite	Hexagonal System

40. The crystal class is 4/m 3bar 2/m-
Crystal system-Isometric system
Crystal class-Hexoctahedron (321)
Plane-9

Crystal forms	Crystal system	Number of face
Octahedron(111)	Isometric system	8
Hexaoctahedron(321)	Isometric system	48
Cube(100)	Isometric system	6
Dodecahedron(110)	Isometric system	12

41.

Minerals	Optical properties	Microscope stage
Quartz	Uniaxial positive	Under Cross-nicol
Hornblede	Oblique extinction	Under Cross- nicol
Muscovite	Mottled extinction	Under Cross- nicol
Garnet	Isotropic	Under Cross- nicol

42.

Properties Minerals	Calcite (carbonate)	Hornblande (amphibole)	Diopside (Pyroxene)
Colour	Colourless	Brown	Grey
Cleavage	Oblique angle	Perfect on 56 degree and 124 degree.	Perfect at 90 degree.
Set of cleavage	3-set perfect	2-set	2-set
Relief	Moderate	Moderate	Moderate
Diagnostic feature			
Hardness	3	5-6	6
Crystal system	Trigonal	Monoclinic	Monoclinic

44.

Forms	Designation	Crystal system	No. of face
Tetragonal prism	Open form	Tetragonal system	4
Dihexagonal pyramid	Open form	Hexagonal system	12
Orthodome	Close form	Monoclinic system	2
Trigonal dipyramid	Open form	Tetragonal system	6

45.

Minerals	Twin types	Crystal crystal
Staurolite	Cruciform	Orthorhombic
Quartz	Japan twin	Trigonal
Gypsum	Shallow twin	Monoclinic
Orthoclase	Carlsbad twin	Triclinic

46. Moh's scale	Minerals	Class
1	Talc	Phyllosilicate
2	Gypsum	Sulphate
3	Calcite	Carbonate
4	Fluorite	Halides
5	Apatite	Phosphate
6	Orthoclase	Tectosilicate
7	Quartz	Tectosilicate
8	Topaz	Nesosilicate
9	Corundum	Oxide
10	Diamond	Native element

48. Minerals	Properties	Crystal system
Hornblende	Anisotropic	Monoclinic
Garnet	Isotropic	Isometric
Calcite	Anisotropic	Trigonal
Biotite	Anisotropic	Monoclinic

49. Bowen's Reaction Series

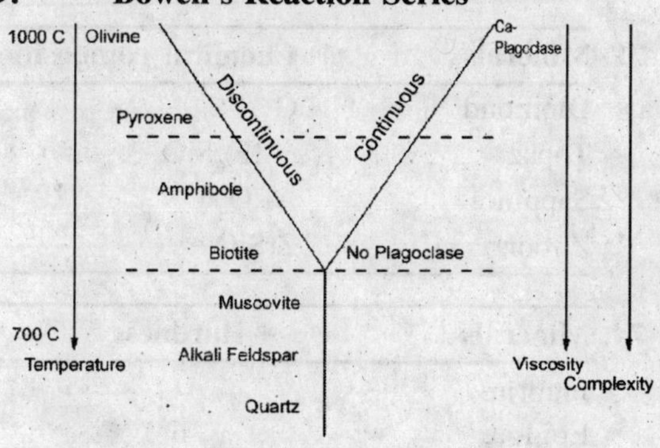

50. Minerals	Hardness	Specific gravity	Characteristics
Talc 1 ($Mg_3Si_4O_{10}(OH)_2$)	2-8	2.7-2.8	White colour, Soapy feel.
Galena (Pbs)	2-3	7.60	Gray colour, High specific gravity.
Bauxite ($Al_2O_3H_2O$)	1-3	2-3	Pisolitic shape, — grey colour.
Scheelite ($CaWo_4$)	4-5	6-7	Tabular crystal, yellowish colour.

51. Minerals	Silicate class	Si:O
Olivine	Nesosilicate	1:4
Augite	Single chain silicate	1:3
Hornblende	Double chain silicate	4:11
Muscovite	Sheet silicate	4:10
Orthoclase	Tectosilicate	1:2

52. Crystallographic feature	Crystal system	Example
$a \neq b \neq c, \alpha = \beta = \gamma = 90°$	Orthorhombic	Enargite
$a = b \neq c, \alpha = \beta = \gamma = 90°$	Tetragonal	Rutile
$a \neq b \neq c, \alpha = \gamma = 90°, \beta \neq 90°$	Monoclinic	Tremolite
$a = b = c, \beta = \beta = \gamma = 90°$	Isometric	Smaltite

54. Number of Crystal face	Examples	Crystal system
4	Tetrahedron	Isometric
6	Rhombohedron	Hexagonal
12	Scalenohedron	Hexagonal
16	Ditetragonal dipyramid	Tetragonal

55. Mineral texture	Processes
Replacement of coesite by quartz	Polymorphic transformation
Crystallographically oriented lamellae of albite in orthoclase	Exsolution
Graphic intergrowth of orthoclase and quartz	Eutectic crystallization

Crass-hatched twinning in microcline	Cation ordering

56.

Isotopes	Datting rocks type
C-14	Recent
Be-10	Mesozoic
Sm-147	Archean
Pb-210	Proterozoic

58.

Crystal system	Class
Trigonal system	3 m
Isometric system	m3m
Hexagonal system	6 mm
Triclinic system	1

59.

Crystal system	Minerals
Isometric system	Galena
Triclinic system	Axinite
Trigonal system	Smithsonite
Monoclinic system	Trona

61.

Minerals	Properties
Hematite	Magnetic mineral
Magnetite	Ferromagnetic mineral
Fe-pyrite	Diamagnetic mineral
Sphalerite	Non-magnetic minerals

62.

Minerals	Moh's scale
Gypsum	2
Calcite	3
Feldspar	6
Topaz	8

63.

Minerals	Designation
Olivine	Silica-oversaturated
Feldspar	Silica-oversaturated
Feldspathoid	Silica-undersaturated
Amphibole	Silica saturated

65.

Crystal system	Amphibole Minerals
Orthorhombic system	Anthrophyllite
Triclinic system	Cossyrite
Monoclinic system	Hornblende, Actinolite

66.

Coordination number	Forms
8	Cube
6	Octahedron
12	Cube octahedron
4	Tetrahedron

68.

Optical properties	Forms
Uniaxial negative	Calcite
Uniaxial positive	Quartz
Biaxial positive	Forsterite(Olivine)
Biaxial negative	Fayalite(Olivine)

71.

Minerals	Chemical composition
Diomond	C
Topaz	$(AlF)_2SiO_4$
Sapphire	Al_2O_3
Zircon	$ZrSiO_4$

74.

Minerals	Hardness
Fluorite	4
Epidote	6
Felspar	6
Kyanite	5-7

75.

Minerals	Cleavage
Olivine	Absent
Plagioclase	Two-set, 90 degree
Calcite	Three-set, rhombohedral
Amphibole	Two-set, 56 degree-124 degree

76.

Minerals	Mica rich
Biotite	Mg rich/Black mica
Muscovite	K- rich/ White mica
Paragonite	Na-rich mica
Lepidolite	Lithium mica

79. Mica	Rich
Cr bearing	Fuschite
Fe bearing	Biotite
Li bearing	Lepidolite
Fl bearing	Zinwaldite

83. Felspar	Type
Orthoclase	Potash felspar
Oligoclage	Plagioclase felspar
Sanidine	Potash felspar
Microcline	Potash feldspar

85. Minerals	Chemical formula
Andalusite	Al_2SiO_5
Topaz	$Al_2SiO_4(OH,F)_2$
Hypersthene	$(Fe,Mg) SiO_3$
Barite	$BaSO_4$

86. Minerals	Crystal system
Kyanite	Triclinic
Albite	Triclinic
Andalusite	Orthorhombic
Microcline	Triclinic

89. Minerals	Symmetry elements
Baryte	3 axes of 2-fold symmetry
Gypsum	1 axes of 2-fold symmetry
Galena	3 axes of 4-fold symmetry
Rutile	1 axes of 4-fold symmetry

90. Minerals	Crystal system
Quartz	Hexagonal system
Garnet	Isometric system
Tourmaline	Hexagonal system
Gypsum	Monoclinic system

92. Rocks	Minerals
Syenite	Alkali feldspar
Granite	Quartz+Feldspar
Anorthosite	Anorthite
Dolerite	Labrodorite+Augite

94. Chemical composition	Minerals
$MgSiO_3$	Enstatite
$NaFe(SiO_3)_3$	Aegirine
$NaAl(SiO_3)_2$	Jaedite
$LiAl(SiO_3)_2$	Spodumene

98. Minerals	Characteristics
Pyrrhotite	Magnetic pyrite (Magnetic)
Hematite	Iron oxide (Magnetic)
Orthoclase	K-feldspar (Nonmagnetic)
Magnetite	Iron oxide (Magnetic)

99. Types of Mica	Remarks
Phlogopite	White/Light mica
Biotite	Black mica
Paragonite	Pale yellow mica
Zinwaldite	Lithium mica

100. Silicate class	Minerals
Nesosilicate	Olivine
Inosilicate	Pyroxene/Amphibole
Sorosilicate	Melilite
Cyclosilicate	Tourmaline

103. Properties	Observed
Colour	Ordinary light
Inclusions	Ordinary light
Pleochroism	Polarized light
Refractive Index(RI)	Ordinary light

113. Properties	Fayalite	Calcite	Amethyst	Autonite
Colour	Brown	White	Purple colour	Yellow
Cleavage	Absent	Perfect	Absent	Present
Form	Crystalline	Crystalline	Amorphous	Crystalline
Radioactivity	Not radioactive	Not radioactive	Not radioactive	Radioactive
hardness	6-7	3	7	2-3
Specific gravity	4.5	2.75 - 3.0	2.6 - 2.7	3.0-3.75
Crystal class	Orthorhombic	Trigonal	-	Tetragonal
Mineral class	Nesosilicate	Carbonate	Tectosilicate	Phosphates

116. Types of form	Forms (example)
Open form	Prism, Pinacoids
Closed form	Cube, Octahedron

117. Miller indices	Plane(Cubic crystal)
(100)	Cube
(110)	Dodecahedron
(111)	Octahedron

123. Minerals	Crystal system
Diopside	Monoclinic
Tourmaline	Trigonal
Topaz	Orthorhombic
Benitoite	Hexagonal

125. Minerals	Twins type
Gypsum	Swallow Tail Twins
Orthoclase	Carlsbad
Albite	Albite twin

126. Minerals	Twinning
Rutile	Geniculate
Gypsum	Butterfly
Aragonite	Cyclic twin
Pyrite	Iron-cross twin

134. Minerals	Crystal system
Halite	Isometric
Fluorite	Isometric
Garnet	Isometric
Gypsum	Monoclinic

135. Minerals	Extinction
Hypersthene	Parallel
Hornblende	Inclined
Biotite	Parallel

139. Extinction	Minerals
Straight	Orthopyroxene
Oblique	Clinopyroxene
Symmetrical	Amphibole
Wavy	Quartz

147. Minerals (Garnet)	Garnet minerals
Ca-Al bearing	Grossular
Mg-Al bearing	Pyrope
Ca-Cr bearing	Uvarovite
Mn-Al bearing	Spessertite

148. Pyroxene	Chemical formula
Hedenbergite	$CaFeSi_2O_6$
Augite	$CaMgSi_2O_6$
Spodumene	$LiAlSi_2O_6$
Johansenite	$CaMnSi_2O_6$

150.

Si:O	Silicate type	Mineral example
1:3	Cyclosilicate	Beryl
4:11	Double chain silicate	Amphibole
4:10	Phyllosilicate	Mica
2:7	Sorosilicate	Epidote

151.

Minerals	Hardness
Talc 1	
Epidote	6
Kyanite	5-7
Gypsum	2

152.

Minerals	Twinning
Plagioclase	Pericline
Orthoclase	Carlsbad
Rutile	Simple
Aragonite	Polysynthetic parallel

153.

Forms	Example
Open form	Domes
Closed form	Dipyramid
Combination form	Pinacoid
General form	Cube

154.

Crystal system	Classes
Tetragonal system	7
Hexagonal system	12
Isometric system	5
Triclinic system	2
Monoclinic system	3
Orthorhombic system	3

156.

Rocks type	Minerals
Andesite	Plagioclase dominant
Trachyte	Alkali feldspar dominant
Adamellite	Alkali feldspar
Syenite	Alkali feldspar

158.

Minerals	Co-ordination number
Coesite	4
Stishovite	6
Tridymite	4
Cristobalite	4

159.

Minerals	Minerals Group	Cleavage angle
Hornblande	Amphibole	56-124 degree
Diopside	Pyroxene	87-93 degree
Calcite	Carbonate	Rhombohedral
Sanidine	Feldspar	90 degree

160.

Crystal class	Crystal system
3 m	Trigonal system
4 mm	Tetragonal system
3	Trigonal system
6/m	Hexagonal system

162.

Minerals	Minerals Group	Forms
Chrysotile	Serpentine	Fibrous
Hornblende	Amphibole	Prismatic
Actinolite	Asbestos	Fibrous
Tremolite	Amphibole	Prismatic

165.

Minerals	Variety
Corrundum	Emery
Microcline	Amazonite
Tourmaline	Rubellite
Beryl	Aquamarine

166.

Minerals	Twinning
Aragonite	Cyclic twining
Staurolite	Lamellar
Plagioclase	Lamellar
Gypsum	Simple

168. Minerals	Silicate structure
Quartz	Tectosilicate
Calcite	Carbonate
Augite	Single chain silicate
Muscovite	Sheet silicate/Phyllosilicate

169. Minerals	Pyroxene
Enstatite	Orthopyroxene
Aegirine	Alkali pyroxene
Tremolite	Amphibole
Diopside	Clinopyroxene

171. Minerals	Contains
Spessartite	$Mn_3Al_2(SiO_4)_3$
Almandine	$Fe_3Al_2(SiO_4)_3$
Andradite	$Ca_2Fe_2(SiO_4)_3$
Grossularite	$Ca_3Al_2(SiO_4)_3$

173. Minerals	Variety
Alabaster	Gypsum
Selenite	Gypsum
Satinspar	Gypsum
Iceland spar	Calcite

174. Minerals	P/T
Orthoclase	Low temperature potassium feldspar
Albite	Low temperature plagioclase feldspar
Sanidine	High temperature potassium feldspar
Anorthite	High temperature plagioclase feldapar

180. Variety	Minerals	Hardness
Chlorapatite	Apatite	5
Satin spar	Gypsum	2
Bort	Diamond	10
Iceland spar	Calcite	3

187. Minerals	Crystal system
Orthoclase	Monoclinic system
Microcline	Triclinic system
Anorthoclase	Triclinic system
Albite	Triclinic system
Andesine	Triclinic system
Bytownite	Triclinic system
Anorthite	Triclinic system
Hyalophane	Monoclinic system
Celsian	Monoclinic system

200. Minerals	Set of cleavage (Thin section)
Muscovite	1-set
Calcite	3-set
Wollastonite	2-set
Microcline	2-set

201. Minerals	Shape (Crystal)
Quartz	Anhedral
Zircon	Euhedral
Microcline	Anhedral
Muscovite	Subhedral

202. Minerals	Extinction
Andesine	Inclined
Hornblende	Inclined
Kyanite	Inclined
Augite	Inclined
Muscovite	Parallel
Beryl	Parallel
Biotite	Parallel
Tourmaline	Straight
Quartz	Wavy Extinction
Garnet	Isotropic
Microcline	Inclined
Hypersthene	Parallel

203. Minerals	Crystal system	Remarks
Almandine	Isometric system	Isotropic
Biotite	Monoclinic system	Anisotropic
Tourmaline	Tetragonal system	Anisotropic

205. Optical properties	Minerals example
Uniaxial positive	Quartz
Uniaxial negative	Calcite, Cristobalite
Biaxial positive	Olivine, Chlorite
Biaxial negative	Biotite, Glauconite

211. Elements	Classification
Ag, Zn, Pb	Chalcophiles elements
Li, Zr, V	Lithophiles elements
Ru, Os, Au	Siderophiles elements
S, Se, Te	Chalcophiles elements

214. Silicate structure	Example	Si:O
Nesosilicate	Forsterite	1:4
Cyclosilicate	Beryl	1:3
Phyllosilicate	Muscovite	4:10
Tectosilicate	Quartz	1:2

218. Minerals	Cleavage
Hornblende	Two set rhombic cleavage
Enstatite	Two set prismatic cleavage
Microcline	Two set prismatic
Gypsum	Two set perfect

219. Minerals	Formula
Andalusite	Al_2SiO_5
Topaz	$Al_2SiO_4(OH,F)_2$
Flourite	CaF_2
Barite	$BaSO_4$

226. Igneous bodies	Characteristics
Phacolith	Concordant in folded terrain
Batholith	Discordant, large dimension
Laccolith	Concordant, dome shape
Plug	Discordant, volcanic vent

229. Moh's scale	Minerals
7	Quartz
6	Feldspar
5	Apatite
8	Topaz

230. Minerals	Luster
Diomond	Admantine
Calcite	Vitreous
Quartz	Vitreous
Nepheline	Vitreous

231. Variety	Minerals
Emerald	Beryl
Ruby	Corundum
Sapphire	Corundum
Citrine	Quartz

233. Types of luster	Minerals
Silky	Gypsum
Vitreous	Quartz
Pearly	Calcite
Dull	Kaolin

235. Silicate	Example
Nesosilicate	Olivine
Inosilicate	Pyroxene, Amphibole
Sorosilicate	Epidote
Cyclosilicate	Beryl

246. Minerals	Gem variety
Tourmaline	Rubellite, Sapphire
Corundum	Ruby, Emerald
Garnet	Demantoid, Melanite
Beryl	Emerald, Aquamarine

247. Normal class	Crystal system
Barite type	Orthorhombic system
Gypsum type	Monoclinic system
Axinite type	Triclinic system
Calcite type	Hexagonal system

249. Minerals	Set of cleavage
Calcite	4- set
Albite	2- set
Barite	3- set
Fluorite	3- set

250. Minerals	Colour	Hardness	Cleavage	Specific gravity	Remarks
Azurite	Blue	3-4	Perfect	3.50-4.00	Copper ores
Kaolinite	White	2-3	Perfect	2.50-3.00	Clay mineral
Kyanite	Patchy blue	5-7	Good	3.50-3.75	Polymorph of Al_2SiO_5
Sillimanite	Brown	6-7	good	3.25-3.50	Polymorph of Al_2SiO_5

251. Minerals	Crystal system	Colour
Gypsum	Monoclinic	Colourless
Beryl	Hexagonal	Green
Galena	Isometric	Grey
Axinite	Triclinic	Brown

253. Cation to oxygen ratio (Si:O)	Silicate class	Example
1:2	Framework silicate/ Tectosilicate	Feldspar
1:3	Cyclosilicate/ Ring silicate	Beryl
2:5	Phyllosilicate/ Phyllosilicate	Biotite
4:11	Double chain silicate	Amphibole
1:4	Nesosilicate	Garnet
2:7	Sorosilicate	Epidote
1:3	Single chain silicate	Pyroxene

254. Optical indicatrix (shape)	Characteristics
Prolate	Uniaxial positive
Oblate	Uniaxial negative
Sphere	Isotropic mineral
Triaxial ellipsoid	Hexagonal system

255. Minerals	Characteristics
Nephaline	Feldspathoid, silica undre saturated volcanic rocks
Sanidine	Feldspar, felsic volcanic rocks
Kalsilite	Feldspathoid minerals
Albite	Feldspar

257. Silicates	Minerals
Orthosilicate	Forsterite
Double chain silicate	Amphibole
Sheet silicate	Biotite, muscovite
Framework silicate	Orthoclase, quartz
Single chain silicate	Pyroxene

258. The miller indices will be:

Weiss parameter = 1a, 2b, 1/2c

Written as = 1, 2, 1/2

Reciprocal of the parameter = 1/1, 1/2, 2/1

Clear the friction = 2, 1, 4

The Miller indices is = (214)

262. Olivine minerals	Chemical formula
Forsterite	Mg_2SiO_4
Monticellite	$CaMgSiO_4$
Fayalite	Fe_2SiO_4
Serpentine	$Mg_6[Si_4O_{10}]\ (OH)_8$

265. Crystal form	Number of face
Basal pinacoid	2
Pedion	1
Pyramid	3
Prism	4

268. Minerals	Groups
Lepidolite	Mica
Hypersthene	Pyroxene
Anorthoclase	Feldspar
Fayalite	Olivine

272. Minerals	Cleavage
Orthoclase	2-set
Garnet	Absent
Enstatite	2-set
Actinolite	2-set

285. Symmetry elements	Crystal system
Axis of 3 fold symmetry	Isometric system
Axis of 5 fold symmetry	Not present
Axis of 4 fold symmetry	Isometric system
Axis of 6 fold symmetry	Hexagonal system

290. Minerals	Silicate class
Forsterite	Nesosilicate
Talc Phyllosilicate	
Muscovite	Phyllosilicate
Orthoclase	Tectosilicate

294. Hardness	Minerals
3	Calcite
4	Fluorite
7	Quartz
5	Apatite

323. Polymorph of Quartz	P/T
Tridymite	High temperature low pressure
Cristloblite	High temperature intermediate pressure
Coesite	Low temperature high pressure
b-Quartz	Low temperature intermediate pressure

324. Property	Muscovite	Biotite
Luster	Pearly	Pearly
Cleavage	One-set	Perfect
Habit	Platy	Prismatic
Colour	White	Black

330. The Wiess parameter is = 2a, 2b, 1c

The miller indices is = 1a, 1b, 1/2c

= 1/1, 1/1, 2/1

= (112)

333. Minerals	Class
Garnet	Nesosilicate
Magnetite	Iron oxide
Mica	Sheet silicate
Olivine	Nesosilicate

334. Minerals	Silicate class
Feldspar	Framework silicate
Olivine	Nesosilicate
Pyroxene	Single chain silicate
Quartz	Framework silicate

336. Crystal symmetry	Crystal system
4-fold axis of symmetry	Tetragonal system
6-fold axis of symmetry	Hexagonal system
3-fold axis of symmetry	Isometric system
At least two improper axis of symmetry	Monoclinic system

337.

Crystal system	No of crystal class
Isometric system	5
Tetragonal system	7
Hexagonal system	12
Monoclinic system	3
Triclinic system	2
Orthorhombic system	3

343.

Minerals	Crystal system	Pleochroism
Rutile	Tetragonal system	Present
Hypersthene	Orthorhombic system	Present
Tourmaline	Trigonal system	Present
Wollastonite	Triclinic system	Absent
Muscovite	Monoclinic system	Absent
Tremolite	Monoclinic system	Absent
Sphene	Monoclinic system	Present
Augite	Monoclinic system	Absent
Epidote	Monoclinic system	Present
Staurolite	Orthorhombic system	Present
Chlorite	Monoclinic system	Present
Glaucophane	Monoclinic system	Present

345.

Minerals	Silicate class
Feldspar	Framework silicate
Mica	Sheet silicate
Olivine	Nesosilicate
Pyroxene	Single chain silicate

346.

Minerals	Groups
Mica	Silicate
Olivine	Silicate
Galena	Sulphide
Hornblende	Silicate

347.

Minerals	Melting point
Quartz	1660°C
Forsterite	1890°C
Fayalite	1200°C
Hornblende	1060°C
Diopside	1390°C

348.

Minerals	Hardness
Magnetite	6
Talc 1	
Calcite	3
Topaz	8

354. Order of Minerals Stability

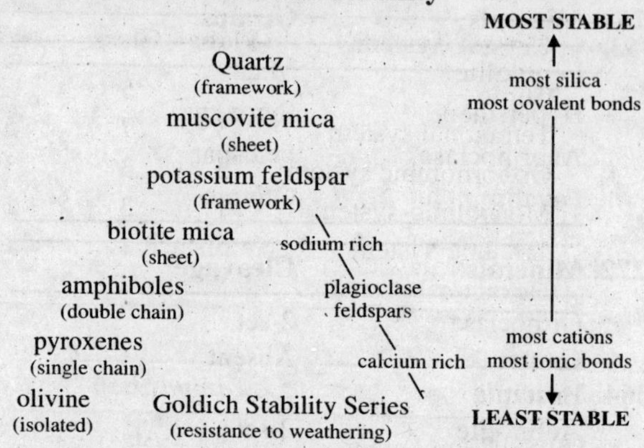

355.

Hardness	Minerals	Minerals Group
1	Talc	Silicate
2	Gypsum	Sulphate
3	Calcite	Carbonate
4	Fluorite	Halides
5	Apatite	Phosphate
6	Orthoclase	Silicate
7	Quartz	Silicate
8	Topaz	Silicate
9	Corundum	Oxide
10	Diamond	Native element

356.

Minerals	Minerals Group	Silicate Structure	Si:O	Example
Anthophyllite	Amphibole	Double chain structure	4:11	Amphibole
Hedenbergite	Pyroxene	Single chain	1:3	Augite
Cordierite	Cordierite	Cyclosilicate	1:3	Beryl
Zeolite	Zeolite	Framework silicate	1:2	Quartz

357. Optical properties	Observed	Terminology	Example
Extinction	Under cross-Nicol	Inclined	Microcline
Interference colour	Under cross-Nicol	Gray to white	Quartz
Interference figure	Under cross-Nicol	Fosterite	+ve
Pleochroism	Plane polarized light	Yellow, Green	Hornblende

363. Crystal System	Plane	Class	Axes	Example	Remarks
Cubic	9	5	13	Sphalerite	Centre of symmetry present
Tetragonal system	5	7	5	Pyrites	Centre of symmetry present
Orthorhombic system	3	3	3	Boehmite	Centre of symmetry present
Monoclinic system	1	3	1	Kernite	Centre of symmetry present
Triclinic system	No plane	2	No axes	Chalcanthite	Centre of symmetry present
Hexagonal system	7	12	7	Niccolite	Centre of symmetry present

364. *Minerals*	*Composition*
Grossular	Ca-Al
Pyrope	Mg-Al
Spessartine	Mn-Al
Uvarovite	Ca-Cr

365. *Crystallographic Terminology*	*Remarks*
System	6
Lattice	14
Classes	32
Space group	230

371. Acute Bisectrix Figure (BX_A)

Looking down the acute bisectrix, (the γ-direction perpendicular to the stage if the crystal is optically positive, or the γ-direction perpendicular to the stage if the crystal is negative), at 45° off extinction in conoscope mode, one would see the interference figure shown in the left-hand diagram below.

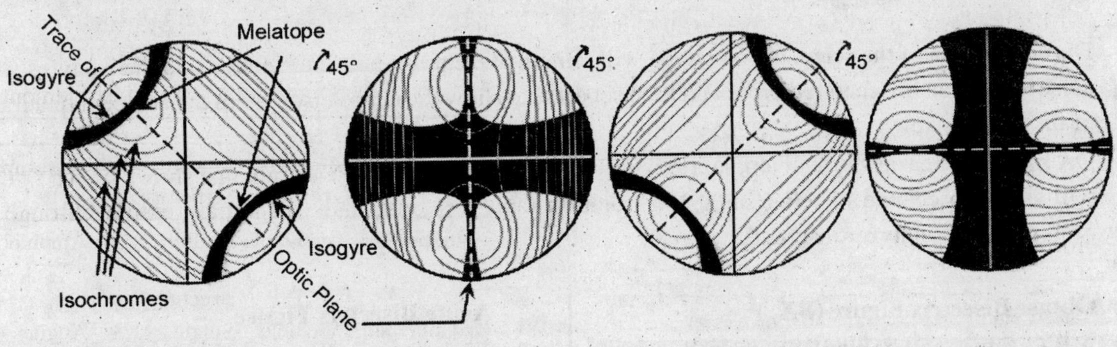

- The dark isogyres mark the positions where light vibrating parallel to the polarizer has passed through the crystal.
- At the points of maximum curvature of the isogyres are the two melatopes that mark the positions where rays that travelled along the optic axis emerge from the field of view.
- Note that the distance between the two melatopes is proportional to the angle 2V between the optic axes.

- Also seen are isochromes, which show increasing interference colours in all directions away from the melatopes. The number of isochromes and maximum order of the interference colours seen will increase with increasing thickness and absolute birefringence of the crystal.
- Shown in the figure is the trace of the optic axial plane which includes the two optic axes.

As the stage is rotated 45° from this initial position, the isogyres will close to produce a cross. In this position the crystal would be extinct in orthoscope mode. The melatopes will be rotated so that both lie along the N-S cross hair. Rotation by an additional 45° will result in the isogyres then separating again to show the interference figure in the third diagram. Another 45° rotation will again cause the isogyres to close into a cross, this time with the OAP lying parallel to the polarizing direction of the microscope. The crystal would again be extinct in orthoscope mode. Another 45° rotation would return the view to the first diagram in the series.

Optic Axis Figure (OA)

If one of the optic axes is oriented perpendicular to the stage, placing the microscope in conoscope mode will result in an optic axis interference figure. This is similar to the BX_A figure, except one of the isogyres and melatopes will be outside of the field of view (unless the 2V angle is very small).

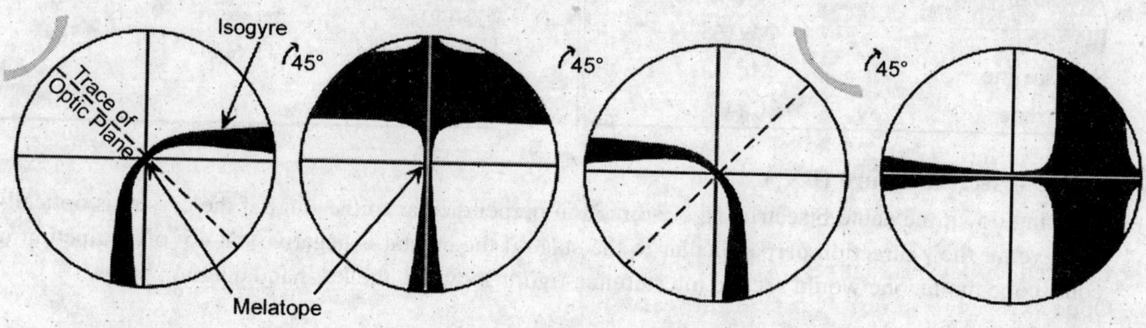

During rotation of the stage, the melatope will remain at the cross-hair intersection and the isogyres will close to form an off-centered cross and then separate to show the curved isogyre in the adjacent quadrant of the field of view.

OA figures are easiest to find among randomly oriented grains, because a grain that shows such a figure will show no change in relief on a 360° rotation (analyzer out), and will remain extinct through a 360° rotation (analyzer inserted).

Obtuse Bisectrix Figure (BX_O)

A BX_O figure will be similar to the BX_A figure, except that the melatopes will be outside of the field of view most of the time during a 360° rotation. Still, every 90° the broad cross will form as the OAP becomes parallel to one of the cross hairs.

Acute Bisectrix Figure

To determine the optic sign of a biaxial mineral from a BX_A figure, position the isogyres so that the melatopes are in the NE and SW quadrants. There should be an area near the melatopes that shows a 1° gray interference colour. Observe this area as you insert the 550 nm or 1° red compensator. If the 1° gray area in region between the two isogyres turns yellow, the mineral is biaxial positive. If the 1° gray area inside of both the isogyres turns yellow the mineral is biaxial negative. Note that it is easiest to remember this if you imagine the area inside the isogyres turns yellow and a line drawn inside the isogyres crosses the slow direction of the compensator like the vertical stroke on a plus sign. For a negative mineral, the line connecting the two yellow areas is parallel to the slow direction in the compensator, forming the minus sign.

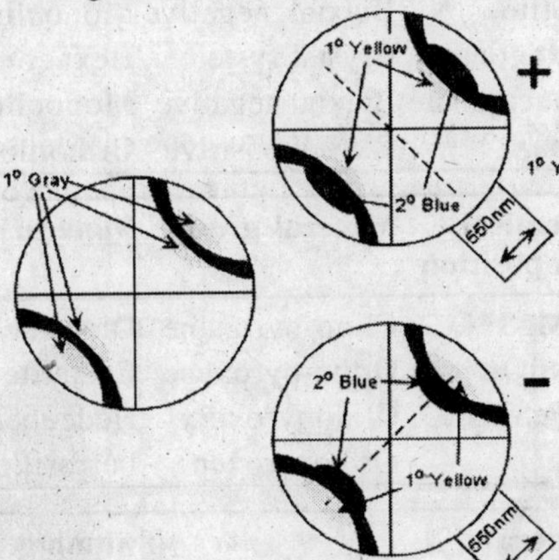

Centred Optic Axis Figure

Optic axis figures probably provide the easiest method for determination of optic sign because grains with an orientation that would produce an OA figure are perhaps the easiest to find.

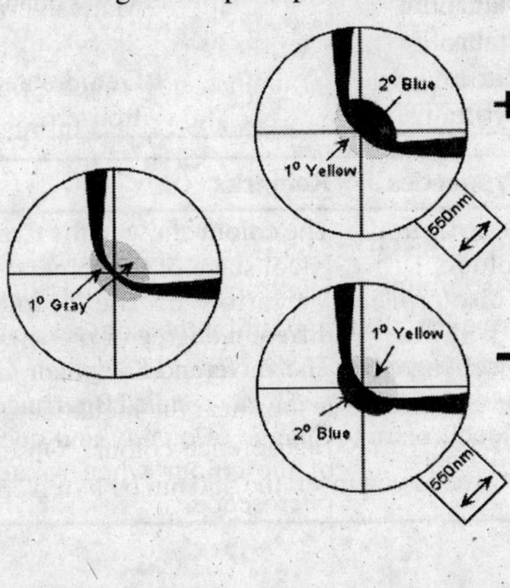

The method is similar to the B_{XA} figure, except you will be looking at only one of the isogyres. Again place the isogyre so that the inside of the isogyre is in the NW quadrant. Find the area that shows 1° gray close to the melatope. Observe this area as the 550 nm compensator plate is inserted. If the area outside of the isogyre turns yellow, the mineral is biaxial positive. If the area inside the isogyre turns yellow, the mineral is biaxial negative.

372. Metamorphic Facies Diagram:

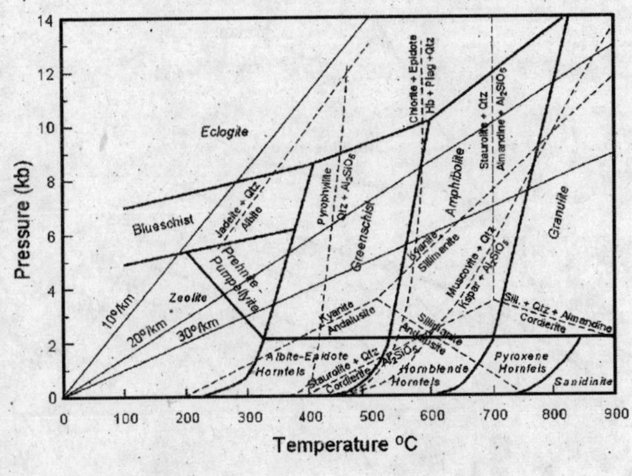

374. Minerals	Hardness
Hematite	5.5-6.5
Nephaline	6
Galena	2.5-3
Orthoclase	6
Epidote	6-7
Garnet	6.5-7.5
Zircon	7.5
Dolomite	3.5-4
Graphite	1-2
Fluorite	4
Topaz	8
Kyanite	6-7.5
Chlorite	2-2.5
Barite	3-3.5
Sphene	5-5.5
Tourmaline	7-7.5

376. Isomorphism

The word derives from the Greek iso, meaning "equal," and morphosis, meaning "to form" or "to shape." Formally, an isomorphism is bijective morphism. Informally, an isomorphism is a map that preserves sets and relations among elements.

Example: Albite – Anorthite System

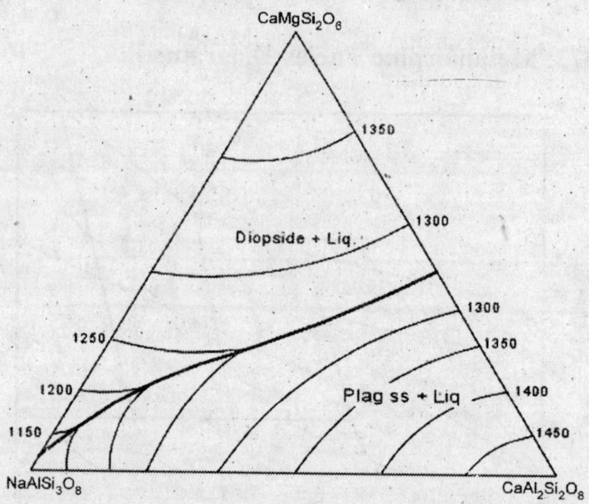

Polymorphism

Polymorphism means "many forms". In mineralogy it means that a single chemical composition can exist with two or more different crystal structures.

Example:

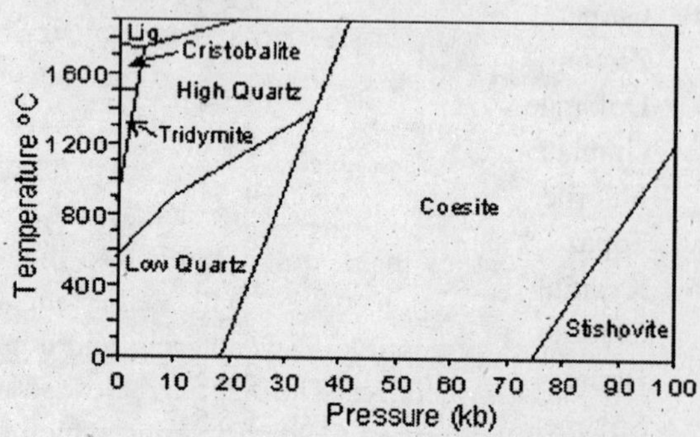

Fig. : Polymorph of the SiO_2

Modification	Crystalline system	Density g/cm³	Formation conditions
Quartz	Trigonal	2.65	T < 573°C
Quartz	Hexagonal	2.53	T > 573°C
Tridimite	Monoclinic	2.27	
Tridimite	Hexagonal	2.26	T > 870°C
Cristobalite	Tetragonal	2.32	
Cristobalite	Cubic	2.20	T > 1470°C
Coesite	Monoclinic	3.01	P > 20 kbar
Stishovite	Tetragonal	4.35	P > 80 kbar

377. Minerals	Optical properties	Crystal system
Biotite	Biaxial negative	Monoclinic
Hexagonal	Crystal system	Hexagonal
Glaucophane	Biaxial negative	Monoclinic
Topaz	Biaxial positive	Orthorhombic

378. Chemical composition	Mineral group	Mineral name
$CaMgSi_2O_6$	Clino-pyroxene	Diopside
$MgSiO_3$	Ortho-pyroxene	Enstatite
$CaFeSi_2O_6$	Clinopyroxene	Hedenbergite
$FeSiO_3$	Orthopyroxene	Ferrosillite

379. Minerals	Twinning
Plagioclase	Pericline
Plagioclase	Albite
Orthoclase and Plagioclase	Manebach
Orthoclase and Plagioclase	Carlsbad
Aragonite	Cyclic
Staurolite	Maitese-cross
Staurolite	Skew
Rutile	Geniculate
Pyrite	Iron-cross

380. Properties	Remarks
Polarization colours	The colour show in the Crossed Nicol stage of the minerals.
Anisotropism	Variation of the colour in different degree of pleochroic.
Birefringence	The difference between O-ray and E-ray is called Birefringence.
Pleochroism	Change of quality and quantity of the colour when rotate the microscope.

381. Classification of Feldspar:

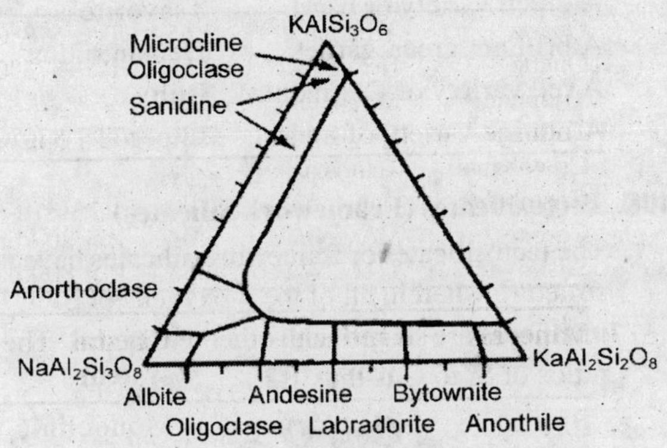

383. Properties	Nephaline	Leucite
Chemical composition	Na_3 (Na, Al) $[Al_4Si_4O_{16}]$	$KAlSi_2O_6$
Crystal system	Hexagonal	Tetragonal
Minerals group	Feldspathoid	Feldspathoid
Hardness	5.5-6	5.5-6
Specific gravity	2.56-2.66	2.47-2.50
Uniaxial	Negative	Positive
Relief	Low	Low
Polarization colours	Colourless	Colourless
Habit	Anhedral	Euhedral crystal

384. Optical property of Biotite:

Properties	Remarks
Colour	Brownish
Pleochroic	Strong
Habit	Platy crystal
Relief	Moderate
Alteration	Chlorite
Birefringence	High
2V	–Ve

386. Properties	Minerals example
Uniaxial negative	Calcite
Biaxial negative	Faylite
Uniaxial positive	Quartz
Biaxial positive	Fosterite

388. Bond type	Example
Polar covenant bond	Inert gases
Polar bond	HCl
Residual bond	Liquid
Ionic bond	Salt

389. Garnet minerals	Rich minerals
Pyrope	Mg-Al rich
Almandine	Fe-Al
Spessartite	Mn-Al
Grossularite	Ca-Al
Andradite	Ca-Fe
Uvarovite	Ca-Cr

394. Dating in Geological Event

Methods	Half-life	Event time
K-Ar method	11900 Ma	Archean to not less than 0.1 Ma
U-Pb method	4500 Ma	Archean to recent
Carbon-14 method	5700 years	Recent
Rb-Sr method	50000	Archean to recent

396. Goldich (1938) series for stability: The Goldich dissolution series is a way of predicting the relative stability or weathering rate of various minerals on the Earth's surface. S. S. Goldich came up with the series in 1938 after studying soil profiles. He found that minerals that form at higher temperatures and pressures are less stable on the surface than minerals that form at lower temperatures and pressures. This pattern follows the same pattern of the Bowen's reaction series, with the minerals that are first to crystallize also the first the undergo chemical weathering.

In 1938, Samuel Goldich published a study of the weathering of various igneous rocks in which he concluded that igneous silicate minerals weather in an order much like that of Bowens Reactions Series, with mafic silicates the most susceptible to weathering and quartz the least susceptible. The diagram below shows a modified form of Goldich's series, which has subsequently been confirmed experimentally.

Zircon and several non-silicates have been added below in the apparent order of their susceptibility of weathering (the position of calcite relative to olivine and anorthite is debatable).

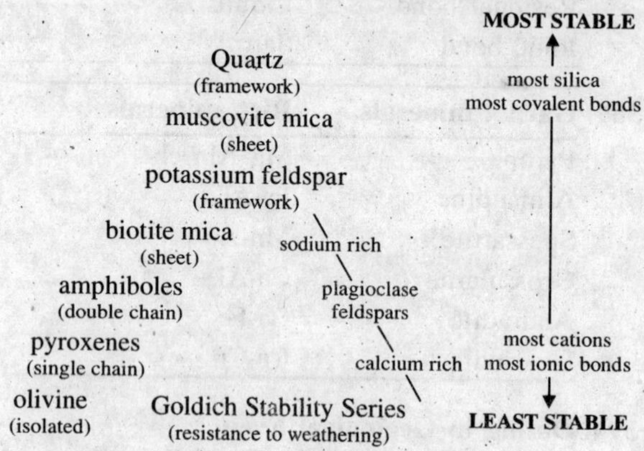

402. Refractive Index	Minerals
1.40 – 1.50	Moderate relief
1.50 – 1.58	Low relief
1.552	Canada balsam
1.58 – 1.67	Moderate
1.67 – 1.76	High
>1.76	Very high

406. Properties	Mineral name
A green variety of beryl	Tsavorite
A brilliant green garnet	Demantoid
A red variety of Corundum	Ruby
A orange variety of topaz	Imperial Topaz

408. Tectosilicates (Framework Silicates)

The tectosilicates or framework silicates have a structure wherein all of the 4 oxygens of SiO_4^{-4} tetrahedra are shared with other tetrahedra. The ratios of Si to O is thus 1:2.

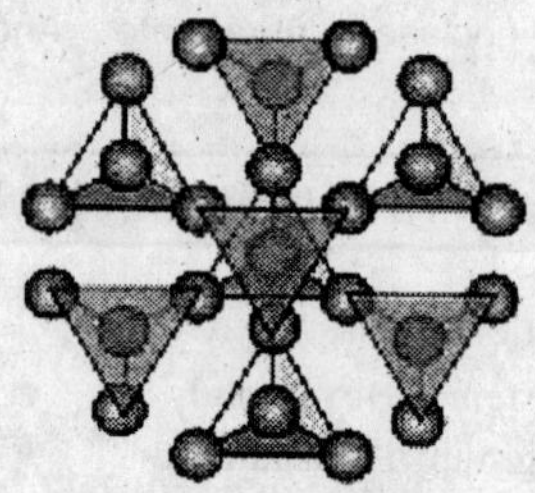

SiO_2 Minerals

Since the Si– O bonds are strong covalent bonds and since the structure is interlocking, the tectosilicate minerals tend to have a high hardness.

PETROLOGY

1. A sandstone with less than 15% matrix content is known as:

A. Siltstone B. Mudstone
C. Greywacke D. Arenite

(IIT-JAM 2015)

2. The mineral coesite is expected to be stable in which of the following metamorphic facies?

A. Greenschist B. Blueschist
C. Eclogite D. Granulite

(IIT-JAM 2015)

3. Match the sedimentary rocks given in Group-I with their characteristics in Group-II:

Group - I	Group - II
(*i*) Sandstone	(P) Chemical sediment
(*ii*) Arkose	(Q) Grain size 1/16 - 2 mm
(*iii*) Limestone	(R) Feldspar-rich
(*iv*) Shale	(S) Grain size < 1/16 mm

A. (*i*) - R, (*ii*) - P, (*iii*) - S, (*iv*) - Q
B. (*i*) - Q, (*ii*) - R, (*iii*) - P, (*iv*) - S
C. (*i*) - Q, (*ii*) - R, (*iii*) - S, (*iv*) - P
D. (*i*) - P, (*ii*) - S, (*iii*) - R, (*iv*) - Q

(IIT-JAM 2015)

4. A dolerite dyke metamorphosed under amphibolite facies condition is expected to have the mineral assemblage:

A. Chlorite + Actinolite + Albite
B. Lawsonite + Glaucophane + Epidote
C. Orthopyroxene + Clinopyroxene + Plagioclase
D. Hornblende + Plagioclase

(IIT-JAM 2015)

5. Which of the following statements related to depositional environments is/are correct?

A. Herringbone cross-stratification indicates glacio-fluvial environment
B. Dune is characterized by the presence of finer-grains at the top and coarser-grains at the bottom
C. Dropstone is of glacial origin
D. Bouma sequence indicates turbidite deposit

(IIT-JAM 2015)

6. Which of the following sedimentary structure could be used for determination top and bottom?

A. Planer cross bedding
B. Asymetrical ripple marks
C. Symmetrical ripple marks
D. None of the above *(AMU-2015)*

7. Which of the following sedimentary environments oscillation ripples are formed?

A. Alluvial B. Beach
C. Deep sea D. Desert *(AMU-2015)*

8. Which of the following is in decreasing order of particle size?

A. Sandstone, siltstone, conglomerate
B. Sandstone, conglomerate, siltstone
C. Conglomerate, sandstone, siltstone
D. Siltstone, sandstone, conglomerate

(AMU-2015)

9. Graded bedding is the result of deposition by:

A. River B. Blowing wind
C. Moving ice D. Turbidity currents

(AMU-2015)

10. Choose the correct sequence of the deformation and metamorphism:

A. Shale, slate, schist, phyllite, gneiss
B. Shale, slate, phyllite, schist, gneiss
C. Slate, shale, phyllite, schist, gneiss
D. Slate, shale, schist, phyllite, gneiss

(AMU-2015)

11. Andesites on an average are intermediate in composition between:

A. Syenites and gabbros
B. Basalt and gabbros
C. Rhyolites and Trachytes
D. Anorthosite and Ijolites *(AMU-2015)*

12. TTGs may form due to:

A. Subducting slab melting
B. Mantle wedge melting
C. Melting of mafic crust
D. Both A&C *(AMU-2015)*

13. Which of the following pair of rock and eruptive setting is correctly matched?

A. Schoshonite – Active continental margin
B. Carbonatite – Ocean island
C. Boninite – Greenstone belt
D. Trachyte – Ocean-continent collision *(AMU-2015)*

14. Which of the following Transition Element is incompatible in basalt?

A. Chromium B. Vanadium
C. Manganese D. Titanium *(AMU-2015)*

15. Dacite is a volcanic equivalent of:

A. Syenite B. Lherzolite
C. Gabbro D. Granodiorite *(AMU-2015*

16. A rocks derived from the primary basaltic magma is characterized by:

A. Occurrence of olivine and clinopyroxene
B. Absence of olivine and feldspathoid
C. Presence of clinopyroxene and quartz
D. Presence of feldspar and quartz *(AMU-2015)*

17. Palaeoproterozoic rocks found in the Himalayan fold belt are suggestive of:

A. Continental Fragmentation
B. Suture zone
C. Continent-continent collision
D. Sea-floor spreading *(AMU-2015)*

18. Coarse grained gneissose rock rich in hypersthene is called:

A. Charnockite B. Khondalite
C. Kodurite D. Gondite *(AMU-2015)*

19. Kimberlite is a:

A. Mineral
B. Igneous rocks
C. Sedimentary rock
D. Metamorphic rock *(AMU-2015)*

20. As compared with metamorphism, diagenesis is........

A. Means exactly the same thing
B. Takes place at lower temperature and pressures
C. Takes place at higher temperature and pressures
D. Takes place at greater depth that are well within the mantle *(DU-2015)*

21. If water is the transport medium of sediment, the grain size of sedimentary deposits most closely indicates the:

A. Geographic extent of the weathering source rock at outcrop
B. Average velocity of the water from the time of erosion until deposition
C. Velocity of the water at the moment the sediment settled to the bottom
D. Velocity of water when it keeps sediment in suspension *(DU-2015)*

22. A lenticular clast-supported conglomerate with a(t) b(i) clast imbrication is interpreted as the product of:

A. Debris flow
B. Turbidity current
C. Channel lag
D. Hyperconcentrated flow *(DU-2015)*

23. One can expect precipitation of .... as dominant carbonate mineralogy from saturated sea water when CO_2 pressure in atmosphere becomes high.

A. Low-Mg calcite and aragonite

B. High-Mg calcite
C. High-Mg calcite and aragonite
D. High-Mg calcite and dolomite

(DU-2015)

24. Syneresis cracks are:
A. A form of desiccation crack
B. A type of trace fossil
C. Subaquous shrinkage cracks
D. A type of penecontemporaneous deformation feature

(DU-2015)

25. Occurrence of ferroan calcite cement in a carbonate rock is indicator of:
A. Sea floor diagenesis
B. Vadose diagenesis
C. Phreatic diagenesis
D. Deep-burial diagenesis *(DU-2015)*

26. A cross-stratified sandstone bed is a:
A. Isochronous unit
B. Time-diachronous unit
C. Suspension deposit
D. None of the above *(DU-2015)*

27. Syn-sedimentary deformation structures are a result of:
A. Low sedimentation
B. High sedimentation
C. Marine transgression
D. Marine regression *(DU-2015)*

28. A sedimentary rock included fossils of two species that are known to have lived between 480-430 and 450-250 million years ago. What could be the age of the rock?
A. 445 million years
B. 480 million years
C. 250 million years
D. 410 million years *(DU-2015)*

29. Cementation before compaction in sediments will result in:
A. Increase in porosity
D. Decrease in grain to grain contact
C. Increase in grain to grain contact
D. Increase in permeability *(DU-2015)*

30. Most robust radiomatric dating method that is used to date a Precambrian granite is:
A. C14 B. U-Pb
C. Rb-Sr D. K-Ar *(DU-2015)*

31. The rock characterized by pronounced negative 'Eu' anomaly is:
A. Basalt
B. Granite
C. TTG suite of rocks
D. Lunar basalt *(DU-2015)*

32. Andesitic rocks form in:
A. Continent-Continent collision
B. Ocean-Continent collision
C. Ocean-Ocean collision
D. Continental rift *(DU-2015)*

33. Pyrolite is:
A. Three part peridotite and one part basalt
B. A rock found in continental crust
C. A model rock composition which can occur any where
D. A rock found on ocean floor

(DU-2015)

34. Relicts of original minerals and textures found in metamorphic rocks are described:
A. Palimpsest textures
B. Nematoblastic textures
C. Lepidoblastic textures
D. Decussate textures *(DU-2015)*

35. Mylonite represents:
A. Regional metamorphism
B. Dynamic metamorphism
C. Burial metamorphism
D. Contact metamorphism *(DU-2015)*

36. Which represents (in parts per million) the nearest concentration of Ni in rhyolite-andesite-basalt- kimberlite?
A. 6-18-134-1050 B. 20-55-213-1200
C. 1-51-104-250 D. 13-23-59-234

(DU-2015)

37. Which one of the following rocks has maximum compressive strength?
A. Marble B. Salate
C. Sandstone D. Granite

(IIT-JAM 2014)

38. From the sedimentary rocks listed below, selected most fine grained:

A. Sandstone B. Conglomerate
C. Siltstone D. Claystone

(IIT-JAM 2014)

39. A mafic rock in the eclogite facies may contain the assemblage:

A. Chlorite-epidote-albite
B. Garnet-clinopyroxene-hornblende-plagioclase
C. Garnet-clinopyroxene
D. Garnet-clinopyroxene-orthopyroxene-plagioclase *(IIT-JAM 2014)*

40. Choose the correct statement about the igneous forms, lopolith and laccolith.

A. Both are concordant, lopolith is convex upward while laccolith is concave upward.
B. Both are discordant, both are convex upward.
C. Both are concordant, laccolith is convex upward, while lopolith is convex downward.
D. Both are discordant, both are convex downward. *(IIT-JAM 2014)*

41. Match sedimentary features in Group-1 with the processes in Group-2:

Group-1	Group-2
P. Flute cast	1. Deformation of sediments
Q. Convolute lamination	2. Erosion
R. Cross- stratification	3. Fluctuating current
S. Flaser bedding	4. Migration of bed forms

A. P-2, Q-1, R-3, S-4
B. P-2, Q-1, R-4, S-3
C. P-1, Q-2, R-4, S-3
D. P-2, Q-3, R-4, S-1 *(IIT-JAM 2014)*

42. A potash-rich ultrabasic rock containing phlogopite both as megacryst and groundmass is:

A. Leucite syenite B. K-rich granite
C. Carbonatite D. Kimberlite

(IIT-JAM 2014)

43. Match the geological feature listed in Group-1 with their corresponding processes in Group-2.

Group-1	Group-2
P. Cumulate	1. Sub-aqueous eruption
Q. Migmatite	2. Magmatic segregation
R. Fenitized rock	3. Anatexis
S. Pillow basalt	4. Volatile transport

A. P-2, Q-1, R-4, S-3
B. P-4, Q-3, R-1, S-2
C. P-2, Q-3, R-4, S-1
D. P-4, Q-3, R-2, S-1 *(IIT-JAM 2013)*

44. Which of the following is NOT a depositional sedimentary structure?

A. Trough cross-bedding
B. Massive bedding
C. Convolute bedding
D. Graded bedding *(IIT-JAM 2013)*

45. Rudite is a rock comprising of:

A. Pebble B. Silt
C. Sand D. Clay

(BHU- 2013)

46. Pick the odd one out:

A. Poikilitic B. Pseudotachylite
C. Porphyroclast D. Ribbon texture

(BHU- 2013)

47. Which of the following is foliated?

A. Gouge B. Breccia
C. Mylonite D. Cataclasite

(BHU- 2013)

48. Which of the following is post-kinematic metamorphic texture?

A. Helicitic fold B. Pressure shadow
C. Kink bands D. Deformation twin

(BHU- 2013)

49. The green schist facies assemblage does not include:

A. Chlorite B. Epidote
C. Actinolite D. Glaucophane

(BHU- 2013)

50. Sanidinite facies developed under:

A. High temperature and high pressure
B. Low temperature and high pressure
C. High temperature and low pressure
D. Low temperature and low pressure

(BHU- 2013)

51. Prehnite-pumpellyte facies develops under:
A. High temperature and high pressure
B. Low temperature and high pressure
C. High temperature and low pressure
D. Low temperature and low pressure
(BHU- 2013)

52. Mechanical fragmentation, sliding and rotation of fragments are characteristics of:
A. Cataclastic flow
B. Pressure solution
C. Intracrystalline deformation
D. Recovery *(BHU- 2013)*

53. Match the following:

1. Green schist	A. Monomineralic
2. Blue schist	B. Chlorite rich
3. Quartzite	C. High pressure
4. Serpentinite	D. Ultramafic

A. 1-A, 2-D, 3-C, 4-B
B. 1-B, 2-C, 3-A, 4-D
C. 1-D, 2-B, 3-C, 4-A
D. 1-A, 2-B, 3-C, 4-D *(BHU-2013)*

54. In general, igneous rocks:
A. Form at relatively low temperatures
B. Form by crystallization of liquid to solid
C. Form at low temperature
D. Form at high temperature
(BHU-2013)

55. The plutonic equivalent of andesite is:
A. Granite B. Gabbro
C. Diorite D. Peridotite
(PU- 2013)

56. Ophitic texture is the key characteristics of which one of the following rocks?
A. Gabbro B. Basalt
C. Dolerite D. Peridotite
(PU- 2013)

57. Which one of the following facies represent highest grade of metamorphism?
A. Zeolite facies B. Granulite facies
C. Green schist facies
D. Amphibolite facies *(PU- 2013)*

58. Which of the following index minerals represent the highest grade of regional metamorphism?
A. Chlorite B. Sillimanite
C. Biotite D. Garnet
(PU- 2013)

59. Which of the following sedimentary structure represents the highest flow energy condition?
A. Trough cross bed
B. Planar cross bed
C. Wavy bedding
D. Plane bedding with parting lineation
(PU- 2013)

60. A silica undersaturated plutonic igneous rock is:
A. Nepheline syenite
B. Granodiorite
C. Anorthosite
D. Syenite *(IIT-JAM 2012)*

61. The metamorphic facies series that best characterizes a subduction-zone tectonic setting is:
A. Prehnite-Pumpellyite → Blueschist → Eclogite
B. Albite-Epidote hornfels → Hornblende hornfels → Pyroxene hornfels → Sanidinite
C. Greenschist → Amphibolite → Granulite
D. Pyroxene hornfels → Granulite → Eclogite
(IIT-JAM 2012)

62. Match the definitions of magmatic bodies in Group I with their nomenclature in Group II.

Group I	**Group II**
P. Concordant intrusive body with a flat base and domed roof	1. Phacolith
Q. Plutonic igneous body with an aerial extent > 100 km^2	2. Lopolith
R. Large, conformable, saucer-shaped layered intrusive body	3. Batholith
S. Lens-shaped pluton that occupies the crest of an antiform or trough of a synform	4. Laccolith

A. P-1, Q-2, R-3, S-4
B. P-2, Q-3, R-1, S-4
C. P-4, Q-1, R-3, S-2
D. P-4, Q-3, R-2, S-1 *(IIT-JAM 2012)*

63. Match igneous rocks in Group I with their most common textures in Group II.

Group I	Group II
P. Komatiite	1. Porphyritic
Q. Dolerite	2. Spinifex
R. Lamprophyre	3. Ophitic
S. Andesite	4. Panidiomorphic

A. P-3, Q-1, R-2, S-4
B. P-2, Q-1, R-4, S-3
C. P-4, Q-3, R-1, S-2
D. P-2, Q-3, R-4, S-1 *(IIT-JAM 2012)*

64. Match the sedimentary structures in Group I with the processes of their formation in Group II.

Group I	Group II
P. Flute cast	1. Fluctuating current
Q. Convolute lamination	2. Exposure
R. Rain print	3. Erosion
S. Flaser bedding	4. Syn-depositional deformation

A. P-3, Q-4, R-1, S-2
B. P-1, Q-3, R-2, S-4
C. P-4, Q-3, R-1, S-2
D. P-3, Q-4, R-2, S-1 *(IIT-JAM 2012)*

65. In which depositional environment are the sand grains best sorted and rounded?
A. Glacial B. Aeolian
C. Fluvial D. Deep marine
(IIT-JAM 2012)

66. Why basalt is fine grained that Gabbro?
A. Basalt formed from quick cooling of magma
B. Basalt formed from slow cooling of magma
C. Gabbro formed from quick cooling of magma
D. Formed at high temperature
(BHU-2013)

67. Mafic igneous rocks contain more amounts of .... as compared to felsic igneous rock.
A. Quartz B. Aluminum
C. Pink feldspar D. Iron
(BHU-2013)

68. Reynolds Number(N) is a ratio of:
A. Inertial force to viscous force
B. Viscous force to inertial force
C. Horizontal to vertical flow
D. Density of fluid to solid *(BHU-2013)*

69. What is Psephite?
A. Medium fragmented rock
B. Finely fragmented rock
C. Lutites
D. Coarsely fragmented rock *(BHU-2013)*

70. Find odd one out:
A. Granite B. Basalt
C. Slate D. Diorite
(BHU-2012)

71. Which one is an argillaceous rock?
A. Sandstone B. Limestone
C. Conglomerate D. Shale
(BHU-2012)

72. Indicate non-feldspar bearing lamprophyre
A. Monchiquite B. Sovite
C. Komatites D. Kersanite
(BHU-2013)

73. Which rocks contains least silica?
A. Phonolite B. Rhyolite
C. Nepheline syanite D. Trachyte
(BHU-2013)

74. If a limestone is dominantly composed of organic detritus, called:
A. Oolite B. Coquina
C. Encrinite D. Spergenite
(BHU-2013)

75. If sediment is dominantly composed of organic detritus, called:
A. No skewness
B. Negative skewness
C. Positive skewness
D. Complex skewness *(BHU-2013)*

76. Which metamorphic facies represent transition phase between diagenesis and regional metamorphism?
A. Green schist facies
B. Hornfels facies
C. Zeolite facies
D. Sanidine facies *(BHU-2013)*

77. Khondalite mainly comprises of the following minerals:
A. Sillimanite+Quartz+Garnet+Graphite
B. Quartz+Perthite+K-feldspar
C. Hornblende+Plagioclase+Soda Amphibole
D. Pyroxene+Amphibole+Quartz+Halite
(BHU-2013)

78. Which sedimentary rock is most likely to changed to slate during regional metamorphism?
A. Breccia B. Conglomerate
C. Dolostone D. Shale *(BHU-2013)*

79. Sphene is mainly:
A. Magnesium silicate
B. Calcium silicate
C. Calcium titanosilicate
D. Iron titanosilicate *(BHU-2013)*

80. Myrmekitic structures are generally observed in:
A. Dunite rock B. Diorite rock
C. Granite rock D. Basaltic rock
(BHU-2013)

81. Cataclastic cleavage is a type of:
A. Slip cleavage
B. Slaty cleavage
C. Crenulation cleavage
D. Shear cleavage *(BHU-2013)*

82. Newly deposited clays have porosity:
A. Up to 5% B. Up to 100%
C. Up to 70% D. Up to 30%
(BHU-2012)

83. Antidunes develop under:
A. Transitional flow regime
B. Upper flow regime
C. Lower flow regime
D. Both in lower and upper flow regime
(BHU-2012)

84. Large-scale cross-beds have the minimum thickness of:
A. 10 cm B. 50 cm
C. 66 cm D. 5 cm *(BHU-2012)*

85. Bauma sequence forms in:
A. Shallow sea B. Rivers
C. Desert D. Deep sea
(BHU-2012)

86. Average depth of fair weather wave base is:
A. 40-50 m B. 8-15 m
C. 70-80 m D. 100 m
(BHU-2012)

87. Debris flow produce:
A. Clast supported texture
B. Grain supported texture
C. Matrix supported texture
D. Cement supported texture *(BHU-2012)*

88. Presence of mineral glauconite suggests:
A. Desert environment
B. Glacial environment
C. Fluvial environment
D. Marine environment *(BHU-2012)*

89. Facies association of a prograding delta is:
A. Coarsening upward
B. Fining upward
C. Disorganized
D. Both fining upward and disorganized
(BHU-2012)

90. In shallow marine environment, palaeocurrent patterns are:
A. Unimodal
B. Bimodal
C. Polymodal
D. Both unimodal and bimodal *(BHU-2012)*

91. Braided River Channels develop due to:
A. Low surface gradient and stable banks
B. High suspension load
C. Steep surface gradient and unstable banks
D. Point bars *(BHU-2012)*

92. Greywacke sandstone indicates:
A. Active provenance tectonics and prolonged transport
B. Stable provenance and prolonged transport
C. Active provenance tectonics and less transport
D. Stable provenance and less transport
(BHU-2012)

93. Which one is the completely unfoliated rock?
A. Slate B. Schist
C. Phyllite D. Hornfels
(BHU-2012)

94. Pyrometamorphism generally includes:
A. High pressure changes
B. High temperature changes
C. Low pressure and temperature changes
D. High pressure and temperature changes
(BHU-2012)

95. Migmatites are the result of:
A. Retrograde metamorphism
B. Ultrametamorphism
C. Palingenesis
D. Metasomatism *(BHU-2012)*

96. Khondalites are characteristic rocks of:
A. Amphibolite facies
B. Granulite facies
C. Eclogite facies
D. Green-schist facies *(BHU-2012)*

97. Find odd one out:
A. Marble B. Slate
C. Granite D. Phyllite
(BHU-2012)

98. Tormalinisation is a combined effect of:
A. Water and fluorine
B. Water, barone and fluorine
C. Water, carbon dioxide and chlorine
D. Water, carbon dioxide and hydrogen
(BHU-2012)

99. Thermal metamorphism of shales produces:
A. Hornfels B. Novaculite
C. Phyllite D. Schist
(BHU-2012)

100. The metamorphic rock essentially composed of hornblende and plagioclase is:
A. Amphibolite B. Hornblendite
C. Blue schist D. Hornfels
(BHU-2012)

101. The most common accessory mineral in eclogites is:
A. Ilmenite B. Zoisite
C. Rutile D. Sphene
(BHU-2012)

102. Which one is a metamorphic texture?
A. Ophitic B. Clastic
C. Granoblastic D. Subophitic
(BHU-2012)

103. The metamorphic rock with maculose structure is:
A. Granulose B. Hornfels
C. Cataclastic D. Schistose
(BHU-2012)

104. The gabbroic rock without pyroxenes containing mainly feldspars and olivine is:
A. Norite B. Troctolite
C. Dunite D. Eucrite
(BHU-2012)

105. The volcanic rocks containing the highest percentage of silica is:
A. Rhyolites B. Trachytes
C. Andesites D. Basalts
(BHU-2012)

106. Granophyres are hypabyssal equivalent of:
A. Basalt B. Granite
C. Gabbro D. Diorite *(BHU-2012)*

107. An example of a discordant igneous intrusion is:
A. Chonolith B. Laccolith
C. Lopolith D. Bysmalith
(BHU-2012)

108. Pegmatite rock contains one of the following in abundance:
A. Tourmaline B. Spinel
C. Forsterite D. Andesine
(BHU-2012)

109. Find odd one out:
A. Idioblastic B. Non-clastic
C. Poikilitic D. Blastoporphyritic
(BHU-2012)

110. Orthophyric texture is a type of:
A. Intergrowth texture
B. Inequigranular texture
C. Equigranular texture
D. Intergranular texture *(BHU-2012)*

111. Peridotile is:
A. An amphibole
B. A pyroxene
C. An acid igneous rock
D. An ultra mafic rock *(BHU-2012)*

112. Blue-granite is also known as:
A. Monozonite B. Diorite
C. Larvikite D. Nordmarkite
(BHU-2012)

113. Lavas containing numerous gas cavities of irregular shape are:
A. Scona B. Pumice
C. Amygdales D. Ignimbrites
(BHU-2012)

114. The variety of peridotite, in which olivine is altered to serpentine, is:
A. Pyroxenite B. Kimberlite
C. Dunite D. Bronzitites
(BHU-2012)

115. Which of the following is an oxide of titanium and iron?
A. Rutile B. Anatase
C. Ilmenite D. Brookite
(BHU-2012)

116. Determine the correctness or otherwise of the following Assertion [a] and Reason [r].
Assertion [a]: Blueschist can form in the subduction zone.
Reason [r]: Geothermal gradient in the subduction zone is low compared to that in stable cratons.
A. Both [a] and [r] are true and [r] is the correct reason for [a]
B. Both [a] and [r] are true but [r] is NOT the correct reason for [a]
C. Both [a] and [r] are false
D. [a] is true but [r] is false
(IIT-JAM 2011)

117. A plutonic rock consisting only of plagioclase, pyroxene and hornblende as major minerals is known as:
A. Dunite B. Lherzolite
C. Granite D. Diorite
(IIT-JAM 2011)

118. Consider the following statements regarding way-up criteria in sets of trough cross-beds.
P: Cross-beds are concave upwards
Q: Cross-beds are truncated towards the top
R: Cross-beds have tangential contact towards the bottom
A. P is true, Q is false and R is true
B. P is false, Q is true and R is true
C. P is true, Q is true and R is true
D. P is false, Q is false and R is false
(IIT-JAM 2011)

119. The correct sequence of rocks from top to bottom in an ophiolite is:
A. Radiolarian chert - pillow basalt - dolerite dyke - gabbro - peridotite
B. Peridotite - pillow basalt - gabbro - dolerite dyke - radiolarian chert
C. Pillow basalt - radiolarian chert - dolerite dyke - gabbro - peridotite
D. Gabbro - pillow basalt - dolerite dyke - radiolarian chert - peridotite
(IIT-JAM 2011)

120. Match the item in Group I with corresponding item in Group II.

Group I	Group II
P. Compositional zoning	1. High grade metamorphism
Q. Perthitic texture	2. Low grade metamorphism
R. Segregation banding	3. Disequilibrium crystallization
S. Slaty cleavage	4. Exsolution

A. P-2, Q-4, R-1, S-3
B. P-2, Q-1, R-4, S-2
C. P-4, Q-3, R-2, S-1
D. P-3, Q-4, R-1, S-2
(IIT-JAM 2011)

121. Schist can split into number of thin slabs because of its:
A. Lineation B. Foliation
C. Fracture D. Joints
(BHU-2014)

122. .... is a coarse grained, igneous plutonic rock which is composed of 90% olivine.
A. Peridotite B. Monzonite
C. Dunite D. Kimberlite
(BHU-2011)

123. In Bowen's reaction series, from olivine to quartz, the temperature:
A. Increases
B. Decreases
C. First increases then decreases
D. First decreases then increases
(BHU-2011)

124. The process of differentiation is favoured by:
I. Rate of cooling
II. Settling of crystallized heavy minerals
III. Escaping of gases
A. I only B. II only
C. I and II only D. II and III only
(BHU-2011)

125. Lamprophyres are:
A. Metamorphic rock
B. Plutonic rock
C. Hypabyssal rock
D. Volcanic rock
(BHU-2011)

126. In anorthosite, the feldspar dominantly present is:
A. Anorthite B. Labradorite
C. Sandine D. Microcline
(BHU-2011)

127. The crescent shaped bodies of igneous rocks that occupy crests and toughs of folded strata is:
A. Phacolith B. Lopolith
C. Laccolith D. Batholith
(BHU-2011)

128. Igneous rocks containing partly digested xenoliths are called:
A. Hybrid rocks B. Acid rocks
C. Basic rocks D. Ultrabasic rocks
(BHU-2011)

129. Which of the following magma is having high viscosity?
A. Andesitic B. Basaltic
C. Rhyolitic D. None of the above
(BHU-2011)

130. Volcahic equivalent of nepheline syenite is:
A. Ijolite B. Rhyodacite
C. Phonolite D. Trachyandesite
(BHU-2011)

131. An intergrowth of two feldspars—a host grain of K-feldspar, $KAlSi_3O_8$ includes exsolved lamellae or irregular intergrowths of albite $NaAlSi_3O_8$ is:
A. Perthitic texture
B. Anti-perthitic texture
C. Graphic texture
D. Myrmekitic texture *(BHU-2011)*

132. Khondalite is:
A. Graphite bearing-rock
B. Hypersthene bearing-rock
C. Dominantly pyroxene bearing-rock
D. Mainly composed of actinolite
(BHU-2011)

133. In a phase diagram, the solidus curve represents the:
A. Temperature above which substance is stable in solid state
B. Temperature below which substance is stable in solid state
C. Pressure below which substance is stable in solid state
D. Pressure above which substance is stable in solid state
(BHU-2011)

134. Which one of the following rocks does not contain olivine?
A. Dunite B. Basanite
C. Basalt D. Andesite
(BHU-2011)

135. Pegmatites are associated with:
A. Basalt B. Gabbro
C. Granite D. Kimberlite
(BHU-2011)

136. Devitrification is the process in which:
A. Glassy material converts into crystallized state
B. Crystallized material converts into glassy material
C. Glass forms directly from the magma
D. None of the above *(BHU-2011)*

137. Arenaceous rocks are composed of:
A. Pebbles B. Clay
C. Boulders D. Sand
(BHU-2011)

138. Dolomites are .... rocks.
A. Clastic
B. Non-clastic
C. Organically formed
D. Scoriae (BHU-2011)

139. A well-sorted rock will be defined as the rock composed of:
A. Well rounded grains
B. Angular grains
C. Oblate grain
D. Prolate grains (BHU-2011)

140. Salt rocks are:
A. Sedimentary rocks
B. Igneous rocks
C. Metamorphic rocks
D. None of the above (BHU-2011)

141. The tendency for variations in current velocity to segregate sediments on the basis of particle size is called:
A. Lithification B. Compaction
C. Metamorphism D. Sorting
(BHU-2011)

142. Particles move into the open ocean or lake from a river and settle by:
A. Size and density
B. Only size
C. Only density
D. Specific gravity (BHU-2011)

143. What is the main difference between a conglomerate and a sedimentary breccia?
A. Breccia clasts are angular; conglomerate clasts are rounded
B. A breccia is well stratified; a conglomerate is poorly stratified
C. Breccia clasts are the size of baseballs; conglomerate clasts are larger
D. Breccia has a compacted, clay-rich matrix; conglomerate has no matrix
(BHU-2011)

144. Coarse clastic material can be transported into a deep marine environment by:
A. Turbidity currents
B. Wind
C. River
D. All of these (BHU-2011)

145. In which of the following environments would you expect to find synunetrical ripples?
A. Alluvial B. Beach
C. Deep sea D. Desert
(BHU-2011)

146. What is the porosity of newly deposited mud?
A. <5% B. 5%-25%
C. 25%-50% D. >50%
(BHU-2011)

147. Shale refers to a rock formed from:
A. Sand size material
B. Plant remains
C. Clay minerals
D. Carbonates (BHU-2011)

148. The sedimentary rock dark gray in colour, composed of mainly the rock fragments with more feldspar than quartz is known as:
A. Arkose B. Sub-arkose
C. Gray-wacke D. Quartz arenite
(BHU-2011)

149. Which of the following structures is best indicator of the flow direction?
A. Ripple marks B. Mud cracks
C. Graded bedding D. Rain imprint
(BHU-2011)

150. Mylonite is a/an:
A. Igneous rock
B. Sedimentary rock
C. Metamorphic rock
D. Structure (BHU-2011)

151. The mineral found in silica-undersaturated rock is:
A. Perovskite B. Tchermekite
C. Fuschite D. Coesite
(BHU-2011)

152. The igneous rock, Obsidian possesses .... texture.
A. Porphyritic B. Sub-ophitic
C. Aphanitic D. Ophitic
(BHU-2011)

153. The kimberlites pipes are holding diamond as:
A. Xenocrysts B. Porphyroblasts
C. Porphyroclasts D. Clasts
(BHU-2011)

154. Which one of the following constituents of limestone is characterized by concentric structure?

A. Peloid B. Ooid
C. Sparry calcite D. Intraclast

(IIT-JAM 2010)

155. The mineral assemblage which does not correctly define a metamorphic facies is:

A. Albite + Epidote + Actinolite + Chlorite
B. Hornblende + Plagioclase + Ilmenite
C. Omphacite + Garnet + Plagioclase
D. Orthopyroxene + Clinopyroxene + Plagioclase + Hornblende

(IIT-JAM 2010)

156. Match texture in Group 1 with its corresponding rock types in Group 2.

Group-1	Group 2
P. Spinifex texture	1. Dolerite
Q. Graphic texture	2. Lamprophyre
R. Ophitic texture	3. Pegmatite
S. Panidiomorphic texture	4. Komatiite

A. P-1, Q-4, R-3, S-2
B. P-4, Q-3, R-1, S-2
C. P-2, Q-3, R-1, S-4
D. P-4, Q-1, R-2, S-3 *(IIT-JAM 2010)*

157. Which one of the following sedimentary structures DOES NOT mark the top surface of a bed?

A. Flute cast B. Current ripple
C. Wave ripple D. Rain print

(IIT-JAM 2010)

158. Match the mode of occurrence if igneous rocks in Group 1 with their corresponding rock types in Group 2.

Group 1	Group 2
P. Batholith	1. Kimberlite
Q. Lopolith	2. Granite
R. Pahoehoe lava	3. Basalt
S. Pipe	4. Gabbro-norite

A. P-4, Q-2, R-3, S-1
B. P-2, Q-1, R-3, S-4
C. P-2, Q-4, R-3, S-1
D. P-4, Q-3, R-1, S-2 *(IIT-JAM 2010)*

159. A pillowed metabasalt containing the mineral assemblage glaucophane + lowsonite + epidote+ chlorite indicate:

A. Ocean floor metamorphism
B. Continent-continent collision metamorphism
C. Contact metamorphism
D. Subduction zone metamorphism

(IIT-JAM 2010)

160. The rock, which DOES NOT have a volcanic equivalent is:

A. Anorthosite B. Granite
C. Syenite D. Gabbro

(IIT-JAM 2010)

161. Spinifex texture is found in which rock?

A. Lamprophyre B. Komatiite
C. Tholeiite D. Rhyolite

(BHU-2010)

162. Dolerite shows which texture?

A. Ophitic B. Panidiomorphic
C. Allotriomorphic D. Seriate

(BHU-2010)

163. Eutectic crystallisation of quartz and K-feldspar give rise to:

A. Hypidiomorphic texture
B. Porphyritic texture
C. Porphyritic texture
D. All of !he above *(BHU-2010)*

164. In the IUGS classification, the field of diorite falls near which end of the triangle?

A. Quartz (Q)
B. Alkali feldspar A.
C. Plagioclase (P)
D. None of the above *(BHU-2010)*

165. The periclase-silica phase diagram exhibits:

A. Eutectic points
B. Peritectic point
C. Liquid immiscibility
D. All of the above *(BHU-2010)*

166. As per the Bowen's reaction series the correct order of crystallisation is shown by:

A. Albite - Anorthite - Labradorite
B. Anorthite - Bytownite - Labradorite

C. Albite - Oligoclase - Andesine
D. Anorthite - Labradorite - Bytownite
(BHU-2010)

167. Which one is a type of meteorite?
A. Anthracite B. Anticilite
C. Aubrite D. Adulite
(BHU-2010)

168. Sideromelane is a synonym for:
A. Tachylyte B. Basinite
C. Boninite D. Trachyte
(BHU-2010)

169. A komatiite should contain:
A. CaO > 16% B. $Na_2O + K_2O > 22\%$
C. $SiO_2 > 54\%$ D. MgO > 18%
(BHU-2010)

170. Filter pressing is associated with:
A. Magmatic differentiation
B. Texture
C. Volatiles
D. Viscosity *(BHU-2010)*

171. Pillow structure is shown by:
A. Granites B. Rhyolites
C. Lamprophyres D. Basaltic lava
(BHU-2010)

172. Ankaramite and picrite are types of:
A. Basalts B. Granite
C. Fossils D. Lamprophyres
(BHU-2010)

173. The condensed phase rule is given by the formula:
A. $P + C = F - 1$ B. $1 - P = F - C$
C. $F + C = P - 2$ D. $P + F = C + 2$
(BHU-2010)

174. Viscosity of a magma increases with:
A. Increase in Mg
B. Increase in Al
C. Increase in H_2O
D. All of the above *(BHU-2010)*

175. A magmatic association with felsic and mafic members predominating is termed as:
A. Bioomial B. Bimagmatic
C. Bimodal D. Bi-petrotectonic
(BHU-2010)

176. Of the following which one is **not** a meteorite?
A. Chondrite B. Impactite
C. Achondrite D. Siderolite
(BHU-2010)

177. Dacite is the volcanic equivalent of:
A. Granite B. Syenite
C. Diorite D. Granodiorite
(BHU-2010)

178. In a phase diagram, the phase rule at an invariant point is defined by:
A. F = 0 B. F = 1
C. F = 2 D. F = 3 *(BHU-2010)*

179. Which sedimentary structures can be used to decipher top and bottom of beds?
A. Graded bedding
B. Current ripples
C. Deformational structures
D. All of the above *(BHU-2010)*

180. The composition of an arkose would be:
A. Quartz + Lithics + (Matrix > 15%)
B. Quartz + Feldspars + (Matrix > 15%)
C. Quartz + Feldspars + (Matrix < 15%)
D. Quartz + Lithics + (Matrix < 15%)
(BHU-2010)

181. An authigenic growth forms during:
A. Sedimentation B. Diagenesis
C. Palingenesis D. Anatexis
(BHU-2010)

182. A turbidite deposit may show:
A. Ripple marks and good sorting
B. Current lamination and ripple marks
C. Rain prints and flute marks
D. Graded bedding and poor sorting
(BHU-2010)

183. As per Wentworth's scale, clay size is defined as:
A. Less that 1/256 mm
B. 1 mm to 1/256 mm
C. More than 256 mm
D. 0.5 mm to 1.5 mm *(BHU-2010)*

184. Micrite is:
A. A fine grained carbonate
B. A type of stalactite
C. Similar to myrmekite
D. A hexagonal iron carbonate *(BHU-2010)*

185. Rudite is a rock comprising:
A. Clay B. Silt
C. Sand D. Pebbles
(BHU-2010)

186. A rock composed mainly of quartz with matrix less than 15% is:
A. Quartz arenite B. Greywacke
C. Litharenite D. Mudstone
(BHU-2010)

187. Sabkha is a:
A. Superatidal deposit
B. Desert deposit
C. Deep sea deposit
D. River terrace deposit *(BHU-2010)*

188. A sedimentary rock may form by weathering and deposition of which rock type?
A. Igneous/Sedimentary/Metamorphic
B. Only Sedimentary/Metamorphic
C. Only Igneous/Metamorphic
D. Only Igneous/Sedimentary
(BHU-2010)

189. Omphacite; characteristically found in eclogite is a:
A. Sodic hornblende
B. Sodic pyroxene
C. Sodic garnet
D. Sodic plagioclase *(BHU-2010)*

190. Texture formed by parallel arrangement of acicular minerals in a metamorphic rock is called:
A. Crystalloblastic B. Nematoblastic
C. Lepidoplastic D. Porphyroblastic
(BHU-2010)

191. Saccharoidal texture is seen in which rock?
A. Syenite B. Hornfels
C. Marble D. Biotite schist
(BHU-2010)

192. An AFM diagram depicts the metamorphic mineral assemblage of:
A. Pelitic rocks B. Charnockites
C. Hornfels D. Greenstones
(BHU-2010)

193. The beginning of metamorphism is indicated by the first appearance of:
A. Staurolite B. Lawsonite
C. Quartz D. Kyanite
(BHU-2010)

194. Laumontite is stable at water pressure of:
A. >7 kb B. 7 to 5kb
C. 5 to 3kb D. < 3 kb
(BHU-2010)

195. Vesuvianite is formed in:
A. Metamorphosed arenites
B. Metamorphosed pelites
C. Metamorphosed marls
D. Metamorphosed carbonates *(BHU-2010)*

196. Texture of charnockite is:
A. Granoblastic B. Porphyritic
C. Hypidiomorphic D. Lepidoblastic
(BHU-2010)

197. With increasing metamorphism, sequence of index minerals will be:
A. Biotite-Chlorite-Kyanite
B. Kyanite-Sillmanite-Almandine
C. Almandine-Staurolite-Kyanite
D. Biotite-Sillimanite-Kyanite
(BHU-2010)

198. Muscovite + Quartz = K-feldspar + Al_2SiO_8 + H_2O (water pressure <3·5 kb) defines:
A. Beginning of very low grade metamorphism
B. Beginning of low grade metamorphism
C. Beginning of medium grade metamorphism
D. Beginning of high grade metamorphism
(BHU-2010)

199. If the Reynolds Number is >2000, the flow regime is:
A. Laminar B. Turbulent
C. Transitional D. Critical
(IIT-JAM 2009)

200. A rock containing equal proportions of plagioclase, pyroxene and olivine is:
A. Anorthosite B. Dunite
C. Lherzolite D. Norite
(IIT-JAM 2009)

201. Which of the following is formed by storm wave?
A. Herring-bone cross beds
B. Antidunes
C. Linguoid ripples
D. Hummocky cross beds *(IIT-JAM 2009)*

202. Which one of the following assemblages is characteristic of the eclogite facies?
A. Hornblende + plagioclase + staurolite
B. Orthopyroxene + clinopyroxene + plagioclase (no staurolite, no muscovite)
C. Omphacite + garnet (no plagioclase)
D. Actinolite + chlorite + epidote + albite *(IIT-JAM 2009)*

203. Which one is a rudaceous rocks?
A. Sandstone B. Limestone
C. Conglomerate D. Shale *(BHU-2009)*

204. Bauma sequence is the product of:
A. Debris flow process
B. Turbidity flow process
C. Mass flow process
D. Rock flow process *(BHU-2009)*

205. The environment between the high tide and low tide levels of the sea is known as:
A. Neritic B. Littoral
C. Bathyal D. Abyssal *(BHU-2009)*

206. The sedimentary rock without stratification is:
A. Tillite B. Sandstone
C. Limestone D. Shale *(BHU-2009)*

207. Coastal lakes connected to sea are known as:
A. Freshwater lakes
B. Sabkha
C. Lagoons
D. Embayment *(BHU-2009)*

208. Rivers experiencing tidal action are known as:
A. Floodplains
B. Estuary
C. Tidal flats
D. Tidal flats and floodplains *(BHU-2009)*

209. Tidal bundies form in:
A. Rivers B. Glaciers
C. Mountains D. Shallow sea *(BHU-2009)*

210. Black shale facies is characterized by the presence of:
A. Large volume of stagnant water
B. Running water
C. Salty water
D. Frozen water *(BHU-2009)*

211. Rocks having particles of gravel size is known as:
A. Arenaceous B. Rudaceous
C. Argillaceous D. Calcareous *(BHU-2009)*

212. When top and bottom sets of a cross bedding meet at an acute such cross bedding is known as:
A. Trough B. Planer
C. Torrantial D. Wedge trough *(BHU-2009)*

213. The smallest megascopic layer of particles in a sedimentary sequence, made up of sediments is known as:
A. Varve B. Stratum
C. Lamina D. Bed *(BHU-2009)*

214. When lacustrine environment is affected by marine conditions, the environment is known as:
A. Paludal B. Paralic
C. Limnic D. Estuarine *(BHU-2009)*

215. Sediment precipitated from solutions within the basin of deposition and having later moved within the basin are:
A. Orthochemical B. Allochemical
C. Epiclastic D. Euxinic *(BHU-2009)*

216. Silt and clay deposit formed by the sediment carried in suspension by air current is:
A. Marlites B. Novaculites
C. Loess D. Delta *(BHU-2009)*

217. During metamorphism the mineral graphite is changed to:
A. Pure carbon B. Pure Iron
C. Pure copper D. Pure lead *(BHU-2009)*

218. A line joining the points where the rocks have the same grade of metamorphism is called:
A. Isogyre B. Isograde
C. Isopach D. Isochore
(BHU-2009)

219. Complete destruction of original texture is due to:
A. Regional metamorphism
B. Contact metamorphism
C. Dynamothermal metamorphism
D. Metasomatism *(BHU-2009)*

220. Which of the following rocks is produced due to dynamothermal metamorphism?
A. Slate B. Schist
C. Phyllite D. Quartzite
(BHU-2009)

221. The usual parent rock of marble is:
A. Limestone B. Shale
C. Granite D. Sandstone
(BHU-2009)

222. The typical product of contact metamorphism with maculose structure is:
A. Granulose B. Cataclastic
C. Hornfels D. Schistose
(BHU-2009)

223. The mineral of highest zone of metamorphism is:
A. Almandine B. Kyanite
C. Sillimanite D. Biotite *(BHU-2009)*

224. Isograd is a line on a geological map of metamorphic rock joining the points of;
A. Same thickness B. Same altitude
C. Same grade D. Same pressure
(BHU-2009)

225. Green schist are metamorphic rocks of which igneous composition?
A. Acidic B. Basic
C. Intermediate D. Ultrabasic
(BHU-2009)

226. One is not a metamorphic texture?
A. Lepidoblastic B. Clastic
C. Granoblastic D. Crystalloblastic
(BHU-2009)

227. Migmatite is formed due to:
A. Anatexis B. Metasomatism
C. Palingenesis D. Pneumatolysis
(BHU-2009)

228. Ophitic texture is commonly exhibited by:
A. Lamprophyres B. Andesite
C. Dolerite D. Trachyte
(BHU-2009)

229. The process of complete melting of rock is known as:
A. Antexis B. Assimilation
C. Palingenesis D. Metasomatism
(BHU-2009)

230. Khondalite is a .... rock.
A. Metamorphic
B. Igneous
C. Sedimentary
D. Igneous and sedimentary *(BHU-2009)*

231. Find Odd one out:
A. Schist B. Shale
C. Sandstone D. Limestone
(BHU-2009)

232. The facies which is developed in the deepest part of geosynclinal environment is:
A. Amphibolite facies
B. Granulite facies
C. Green schist facies
D. Sanidinite facies *(BHU-2009)*

233. Which one of the following rocks has the highest percentage of silica?
A. Rhyolite B. Andesite
C. Dacite D. Basalt *(BHU-2009)*

234. Ophitic texture contains laths of plagioclase enclosed in the crystal of:
A. Augite B. Hypersthene
C. Hornblende D. Calcite
(BHU-2009)

235. Granite is:
A. An amphibole
B. A pyroxene
C. An acidic igneous rock
D. An ultra-basic rock *(BHU-2009)*

236. Which one of the common minerals is present in granite?
A. Calcite B. Dolomite
C. Hematite D. Quartz *(BHU-2009)*

237. Spherulitic texture is associated with:
A. Plutonic rock
B. Volcanic rock
C. Hypabyssal rock
D. Metamorphic rock *(BHU-2009)*

238. Pegmatite rock contains one of the following in abundance:
A. Tourmaline B. Spinel
C. Forsterite D. Andesine
(BHU-2009)

239. Charnikite rock contains which of the following assemblage?
A. Quartz + Orthoclase + Hypersthene
B. Plagioclase + Orthoclase + Hypersthene
C. Quartz + Calcite + Hypersthene
D. Quartz + Haematite + Hypersthene
(BHU-2009)

240. The rock without feldspars essentially having olivine and pyroxenes in abundance is:
A. Eucrite B. Troctolite
C. Dunite D. Picrite *(BHU-2009)*

241. The gabbroic rock without pyroxenes containing mainly feldspars and olivine is:
A. Troctolite B. Andesite
C. Dacite D. Basalt *(BHU-2009)*

242. A typical monomineralic rock containing labradorite is:
A. Anorthosite B. Gabbro
C. Granite D. Norite *(BHU-2009)*

243. Batholiths are usually associated with:
A. Earthquake zone B. Island arcs
C. Orogenic belt D. Fold and faults
(BHU-2009)

244. The following paleocurrent directions were determined from asymmetrical ripple marks on a sandstone outcrop:

N 10°, N 355°, N 15°, N 345°, N 5°, N 2°, N 350°, N 358°, N 340°, N 20°

The mean paleocurrent direction is:
A. N 360° B. N 180°
C. N 10° D. N 340°
(IIT-JAM 2008)

245. Match the magmatic processes listed in Group-1 with the resultant products in Group-2.

	Group-1	Group-2
P.	Fractional crystallization	1. Exsolution lammelle
Q.	Liquid immiscibility	2. Cumulate
R.	Filter pressing	3. Anorthite-Albite solid solution
S.	Gravitational Settling	4. Pegmatite

A. P-2, Q-1, R-3, S-4
B. P-1, Q-3, R-4, S-2
C. P-3, Q-1, R-4, S-2
D. P-3, Q-2, R-4, S-1 *(IIT-JAM 2008)*

246. Relationship between the grain size measurements in ϕ scale and d (mm) scale is:
A. $\phi = -\log_2 d$ B. $\phi = -\log_{10} d$
C. $\phi = \log_{10} d$ D. $\phi = -\ln d$
(IIT-JAM 2008)

247. The glaucophane and lowsonite minerals assemblage is diagnostic of:
A. Eclogite facies
B. Amphibolite facies
C. Blueschist facies
D. Greenschist facies *(IIT-JAM 2008)*

248. The basaltic lava flow having ropy and wavy surface is known as:
A. Aa
B. Pahoehoe
C. Tuff
D. Volcanic breccia *(IIT-JAM 2007)*

249. Continued crystallization of augite after the formation of plagioclase can result in:
A. Hyalopilitic texture
B. Ophitic texture
C. Porphyritic texture
D. Trachyte texture
(IIT-JAM 2007)

250. Match the plutonic rocks listed in Group 1 with their corresponding volcanic equivalents in Group 2.

Group-1	**Group-2**
P. Granite	1. Dacite
Q. Granodiorite	2. Trachyte
R. Diorite	3. Rhyolite
S. Syenite	4. Andesite

A. P-1, Q-2, R-4, S-3
B. P-2, Q-1, R-3, S-4
C. P-3, Q-1, R-4, S-2
D. P-3, Q-2, R-4, S-1 *(IIT-JAM 2007)*

251. Staurolite, andalusite and sillimanite from during progressive regional metamorphism of:
A. Arenaceous rocks
B. Siliceous carbonate rocks
C. Basic igneous rocks
D. Argillaceous rocks *(IIT-JAM 2007)*

252. The pressure- temperature ranges of the metamorphic facies:
P: Granulite, Q: Zeolite,
R: Amphibolite, S: Greenschist
Increasing in the order:
A. $Q < R < S < P$
B. $Q < P < R < S$
C. $Q < S < R < P$
D. $Q < R < P < S$ *(IIT-JAM 2007)*

253. Which of the following sedimentary rocks is/are formed by chemical precipitation?
P : Limestone Q : Chert
R : Sandstone S : Shale
A. P only B. Both P and Q
C. Both P and R D. Both Q and S
(IIT-JAM 2007)

254. Which of the following sedimentary structures is erosional in nature?
A. Cross-stratification
B. Flute casts
C. Graded bedding
D. Ripple marks *(IIT-JAM 2007)*

255. The presence of reef corals in a sedimentary sequence indicates that deposition took place:
A. Near the equator
B. Near the Poles
C. In fluvial environment
D. In deep sea environment
(IIT-JAM 2007)

256. The orthopyroxene - clinopyroxene - plagioclase mineral assemblage in metabasic rocks is diagnostic of:
A. Granulite facies
B. Amphibolite facies
C. Eclogite facies
D. Greenschist facies *(IIT-JAM 2006)*

257. In clastic sedimentary rocks, grains ranging in size from 0.0625 mm to 2.0 mm are called:
A. Granule B. Sand
C. Silt D. Clay
(IIT-JAM 2006)

258. Which one of the following sedimentary features can be used to determine the unidirectional palaeocurrent?
A. Asymmetrical ripple marks
B. Symmetrical ripple marks
C. Parting lineation
D. Load cast *(IIT-JAM 2006)*

259. Match the following:

Group I	**Group II**
(P) plutonic rocks	1. sanidine
(Q) subduction zone	2. stishovite
(R) impact crater	3. microcline
(S) volcanic rocks	4. jadeite

A. P - 2 Q - 3 R - 4 S - 1
B. P - 3 Q - 1 R - 2 S - 4
C. P - 4 Q - 1 R - 2 S - 3
D. P - 3 Q - 4 R - 2 S - 1 *(IIT-JAM 2006)*

260. The presence of pillow structures in igneous rocks is indicative of:
A. Layered basic intrusion
B. Subaerial volcanism
C. Submarine volcanism
D. Shield volcanism *(IIT-JAM 2006)*

261. Ophiolitic mélange is characteristic of:
A. Lesser Himalaya
B. Higher Himalaya
C. Outer Himalaya
D. Indus Suture Zone *(IIT-JAM 2006)*

262. Exsolution texture results from:
A. Eutectic crystallization
B. Phase separation in the solid phase
C. Crystallization following phase separation
D. Replacement *(IIT-JAM 2005)*

263. When crystals are in equilibrium with their coexisting melt in the bi-component albite-anorthite system, the Ca/Na ratio of the crystals will at all times be:
A. Higher than that of coexisting melt
B. Lower than that of coexisting melt
C. Same as that of coexisting melt
D. Same as that of the initial melt
(IIT-JAM 2005)

264. Match the following:

	Group-1	Group-2
P.	Myrmekite	1. Augite grains enclosed by laths of plagioclase
Q.	Perthite	2. Intergrowth of quartz and K-feldspar
R.	Graphic	3. Vermicular quartz in plagioclase
S.	Ophitic	4. Intergrowth of albite in K-feldspar

A. P-3, Q-4, R-1, S-2
B. P-3, Q-4, R-2, S-1
C. P-3, Q-2, R-4, S-1
D. P-2, Q-4, R-3, S-1 *(IIT-JAM 2005)*

265. A metamorphic rock has garnet and omphacite. The metamorphic facies of the rock is:
A. Greenschist B. Granulite
C. Amphibolite D. Eclogite
(IIT-JAM 2005)

266. Of the following features, one that is not useful in determination of the order of superposition is:
A. Truncated crossbedding
B. Graded bedding
C. Mud-cracks
D. Current ripples *(IIT-JAM 2005)*

267. In a sequence of layered rock, the bedding plane indicates a single minor period of:
A. Erosion
B. Sublevation
C. Non-deposition or scour
D. Peneplanation *(IIT-JAM 2005)*

268. What is primitive magma?
A. It is a speculative magma derived from a mantle source
B. It is magma as it exists immediately after separation from its source region
C. It is derived magma
D. It is a fractionated magma
(BHU-2014)

269. Which of the following systems explains peritectic reaction mechanism?
A. Albite-Anorthite system
B. Forsterite-Silica system
C. Diopside-Anorthite system
D. Forsterite-Olivine system *(BHU-2014)*

270. Eutectic in any binary system is:
A. Trivariant point B. Bi-variant point
C. Uni-variant point D. Invariant point
(BHU-2014)

271. What is liquid immiscibility?
A. Homogeneous mixing of two liquids of different composition
B. The unmixing of two liquids of different composition
C. The mixing of two liquids of same composition
D. The unmixing of two liquids of same composition *(BHU-2014)*

272. In general magma flow is laminar and depends on:
A. Viscosity
B. Shape of the conduit
C. Pressure gradient
D. All of the above *(BHU-2014)*

273. Glomeroporphyritic texture is:
A. Distinct clusters of phenocrysts in fine-grained groundmass
B. Equigranular in size
C. Phenocrysts are absent
D. A single big crystal embedded in fine grained groundmass
(BHU-2014)

274. Ophitic texture may be explained as:
A. Large pyroxene grains enclose small random plagioclase laths
B. Small pyroxene grains enclose large random plagioclase laths
C. Large hornblende grains enclose small random biotite laths
D. Large pyroxene grains enclose small random pyroxene laths
(BHU-2014)

275. Micrographic intergrowth of quartz and alkali feldspar is known as:
A. Myrmekite texture
B. Granophyre texture
C. Variolitic texture
D. Symplectite texture *(BHU-2014)*

276. Ultramafic igneous rocks contain:
A. 30-50% mafic minerals
B. >90% mafic minerals
C. >90% felsic minerals
D. <90% mafic minerals *(BHU-2014)*

277. Mineral composition of Norite is:
A. Hypersthene and plagioclase feldspar
B. Augite and plagioclase feldspar
C. Olivine and plagioclase feldspar
D. Hornblende and plagioclase feldspar
(BHU-2014)

278. Phonolite is volcanic equivalent of:
A. Granite B. Gabbro
C. Syenite D. Nepheline syenite
(BHU-2014)

279. Which of the following textures is typical observed in komatiite?
A. Cumulate texture
B. Poikilitic texture
C. Spinifex texture
D. Spherulitic texture *(BHU-2014)*

280. Which of the following is not a mono-mineralic igneous rocks?
A. Granite B. Dunite
C. Anorthosite D. Pyroxenite
(BHU-2014)

281. Obsidian is composed of:
A. Entirely of crystals
B. Entirely of glassy material
C. Mixture of glass and crystals
D. All sizes of crystals *(BHU-2014)*

282. Intergranular and intersertal textures are commonly observed in:
A. Basalt B. Gabbro
C. Pyroxenite D. Trachyte
(BHU-2014)

283. Coarse-grained sediments are transported by:
A. Traction process
B. Saltation process
C. Suspension process
D. None *(BHU-2014)*

284. Ripple marks occur on the:
A. Lower surface B. Upper surface
C. Internal structure D. None
(BHU-2014)

285. Ripple marks with bifurcated crest and ripple index from 2-5 are generated by:
A. Wave B. Current
C. Tide D. Storm *(BHU-2014)*

286. Graded beds from in marine environments:
A. By traction currents
B. By turbidity currents
C. By suspension fall out
D. By debris flow *(BHU-2014)*

287. Arkoses are sandstones which contain:
A. Appreciable amount of feldspar
B. Low percentage of feldspar
C. Negligible amount of feldspar
D. No feldspar *(BHU-2014)*

288. Lithic arenites are:
A. Quartz rich sandstones
B. Feldspar rich sandstones
C. Rock fragments rich sandstones
D. Clay rich sandstones *(BHU-2014)*

289. Argillaceous sediments containing clay between 33-66 % are known as:
A. Mudstone B. Siltstone
C. Claystone D. Dropstone
(BHU-2014)

290. Oolites from in:
A. Agitated B. Calm water
C. Cool water D. None
(BHU-2014)

291. Mineralogically mature sandstones are rich in:
A. Quartz B. Feldspar
C. Mica D. Rock fragments
(BHU-2014)

292. Convolute bedding forms along:
A. Coast
B. In rivers
C. Slopes
D. Deep marine part *(BHU-2014)*

293. Cross-beds are:
A. Lower surface
B. Upper surface
C. Internal structure
D. None of these *(BHU-2014)*

294. Heavy minerals are used for deciphering:
A. Provenance B. Environment
C. Diagenesis D. None of these
(BHU-2014)

295. Conversion of chlorite into illite is known as:
A. Chloritization B. Illitization
C. Fluidization D. None of these
(BHU-2014)

296. Foliation is an important characteristic of most metamorphic rocks and it indicates that the rocks formed in an environment of:
A. High water content in the minerals
B. Very high temperature and low pressure
C. Directed pressure
D. Partial molten rocks *(BHU-2014)*

297. Low temperature high pressure metamorphic rocks are the characteristic of:
A. Metamorphic zones abound granitic intrusions
B. Mountain building events at continental collision
C. Volcanic eruptions
D. Area adjacent to subduction zone
(BHU-2014)

298. Which of the following does not occur during metamorphism?
A. Mineral orients themselves in a parallel alignment
B. Water flower through the rocks and helps transport elements
C. Radical changes in mineral composition
D. Minerals breakdown and the element recombine to form new minerals
(BHU-2014)

299. Grain growth and size in metamorphic rock are a function of:
A. Nucleation kinetics
B. Material transfer processes
C. Growth rate and P-T crystallization
D. All the above *(BHU-2014)*

300. Retrograde metamorphism is also known as:
A. Diapthoresis
B. Additive metamorphism
C. Injective metamorphism
D. None of these *(BHU-2014)*

301. The predominant agents in contact metamorphism is:
A. Temperature B. Pressure
C. Chemical fluids D. All the above
(BHU-2014)

302. Staurolite forms by the reaction of:
A. Chlorite and sillimanite
B. Muscovite and chloritoid
C. Chlorite and Muscovite
D. Chloritoid and muscovite
(BHU-2014)

303. Quartz microcline hypersthene is the assemblage of:
A. Khondalite B. Charnockite
C. Gondite D. Kudurite
(BHU-2014)

304. What minerals cannot be shown in AKF diagram?
A. Na-bearing minerals
B. K-bearing minerals
C. Ca-bearing minerals
D. None of these *(BHU-2014)*

305. Migmatites are the rock which are characterized by:
A. Granitic nature
B. High grade regional metamorphism
C. Low grade regional metamorphism
D. Both A and B *(BHU-2014)*

306. The correct sequence of metamorphic zones is:
A. Chlorite, Biotite, Garnet, Staurolite, Kyanite and Sillimanite
B. Biotite, Chlorite, Staurolite, Garnet, Kyanite and Sillimanite
C. Chlorite, Biotite, Garnet, Staurolite, Sillimanite and Kyanite
D. Chlorite, Biotite, Staurolite, Garnet, Sillimanite and Kyanite
(BHU-2014)

307. Eskola's ACF and AKF diagrams are used only for the rocks with:
A. Excess SiO_2 B. Low SiO_2
C. Excess $CaCO_3$ D. Low $CaCO_3$
(BHU-2014)

308. In enderbite the dominant feldspar is:
A. Plagioclase B. Microcline
C. Orthoclase D. Bytownite
(BHU-2014)

309. The greenschist facies include the:
A. Chlorite zone
B. Chlorite and Biotite zone
C. Biotite and Garnet zone
D. Garnet and Kyanite zone *(BHU-2014)*

310. Shallow igneous Plano-convex lens shaped bodies are called:
A. Batholith B. Lopolith
C. Laccoliths D. Tropical Domes
(DU-2014)

311. Which of the following volcanic activity is related to little or no explosive activity?
A. Rhyolitic B. Dacitic
C. Andesitic D. Basaltic *(DU-2014)*

312. Metamorphic rock that forms under high pressure and low to moderate temperature in subduction zones are called?
A. Granulites
B. Skarns
C. Blueschists
D. Amphibolites *(DU-2014)*

313. Which of the following cannot be used for determining sediment provenance?
A. Geochemistry of sediments
B. Thickness of the sediment column
C. Grain size of sediment
D. Sorting of sediment *(DU-2014)*

314. From Visser plot of grain size distribution we infer about:
A. Mode of transport of sediment
B. Sorting of sediments
C. Skewness of sediment
D. Spread of sediment size *(DU-2014)*

315. A conglomerate in which long axis (a) of pebbles is oriented perpendicular to direction of flow and intermediate axis (b) sloping in upcurrent direction can be inferred as product of:
A. Mass flow
B. Lag deposit
C. Bed load deposit
D. Scree deposit *(DU-2014)*

316. We different an arkose from a granite on the basis of:
A. Mineralogy B. Geochemistry
C. Grain size D. Texture
(DU-2014)

317. Antiperthite is a lamellar intergrowth of:
A. Sodic plagioclase feldspar within K-feldspar
B. K-feldspar within sodic plagioclase
C. K-feldspar within calcic plagioclase
D. Calcic plagioclase within sodic plagioclase
(DU-2014)

318. Obsidian is a:
A. Silica rich glassy volcanic rock
B. Silica deficient volcanic rock
C. Mafic plutonic rock
D. Basic hypabyssal rock *(DU-2014)*

319. In AFM diagrams projected through muscovite, A is computed as:
A. Al_2O_3
B. $(Al_2O_3 -3K_2O)/(Al_2O_3 -3K_2O +FeO + MgO)$
C. $(Al_2O_3 -K_2O)/(Al_2O_3 -3K_2O +FeO + MgO)$
D. $Al_2O_3 /(FeO + MgO)$
(DU-2014)

320. A Barrovian metamorphic sequence represents:
A. Intermediate pressure facies
B. Low pressure facies
C. High pressure facies
D. Contact metamorphism *(DU-2014)*

321. A mafic rock metamorphosed into a garnetiferous amphibolite indicates:
A. Mg enriched mafic protolith
B. Fe enriched mafic protolith
C. Al enriched mafic protolith
D. Ca enriched mafic protolith *(DU-2014)*

322. A reaction isograd in ACF diagram will be reflected as:
A. Rotated tie line
B. Tie line flip
C. Change of slope of tie line
D. None of the above *(DU-2014)*

323. In a three component system, the maximum number of phases at invariance will be:
A. 3 B. 4
C. 5 D. 6 *(DU-2014)*

324. Graphic texture signifies:
A. Peritectic crystallization
B. Eutectic crystallization
C. Fractional crystallization
D. Partial melting *(DU-2014)*

325. In paired metamorphic belts high pressure metamorphic belt is located:
A. At the mid oceanic ridges
B. Towards continental plate side
C. Towards oceanic plate side
D. In obduction zone *(DU-2014)*

326. Identify the correct sequence in ascending order of metamorphism:
A. Biotite, chlorite, kyanite, garnet
B. Chlorite,biotite, garnet, kyanite
C. Chlorite, garnet, biotite, kyanite
D. Chlorite, biotite, kyanite, garnet.
(DU-2014)

327. Which of the following facies is **not** present in low pressure metamorphic series?
A. Green schist facies
B. Amphibolites facies
C. Epidote amphibolite facies
D. Granulitic facies *(DU-2014)*

328. Metasomatism represents:
A. Closed system B. Open system
C. Isolated system D. Adiabatic system
(DU-2014)

329. Which of the following is contact metamorphic facies?
A. Zeolite facies B. Sanidinite facies
C. Granulite facies D. Greenschist facies
(DU-2014)

330. Presence of volcanic glass of terrestrial origin, in marine sediments, is an example of:
A. Autochthonous component
B. Allochthonous component
C. Piedmont component
D. Primary component *(DU-2014)*

331. Fine grained igneous rocks are:
A. Intrusive body
B. Plutonic
C. Extrusive body
D. Sedimentary body *(BHU-2015)*

332. Hard, massive, compact with interlocking grains and absence of fossil and bedding is the essential feature of:
A. Igneous rocks
B. Sedimentary rocks
C. Metamorphic rocks
D. Limestone *(BHU-2015)*

333. Volcanic equivalent of granite is:
A. Basalt B. Gabbro
C. Rhyolite D. Diorite
(BHU-2015)

334. Gabbro is a .... type of igneous rocks.
A. Volcanic
B. Plutonic
C. Hypabyssal
D. Metasedimentary
(BHU-2015)

335. Which one is **not** concordant body?
A. Laccolith B. Lopoliths
C. Phacoliths D. Batholiths
(BHU-2015)

336. Which mineral crystallized last from the magma?
A. Olivine B. Quartz
C. Hornblende D. Labrodorite
(BHU-2015)

337. The 'aa structure' is also known as:
A. Pahoehoe B. Ropy lava
C. Blocky lava D. Flow structure
(BHU-2015)

338. The occurrence of foreign materials within igneous rocks are formed a structure that is:
A. Xenolithic structure
B. Myrmekite structure
C. Sheet structure
D. Amygdaloidal structure *(BHU-2015)*

339. When the igneous rocks made of only glassy materials known as:
A. Holocrystalline B. Hemicrystalline
C. Holohyaline D. Semihyaline
(BHU-2015)

340. Perlitic cracks in crystal represent the conversion from:
A. Basic magma B. Acidic magma
C. Glassy condition D. Plutonic condition
(BHU-2015)

341. Eutectic point is a:
A. Univariant B. Invariant
C. Di-variant D. Tri-varient
(BHU-2015)

342. Silica content in basic igneous rock is:
A. 44 to 55% B. More than 65%
C. 55 to 65% D. More than 80%
(BHU-2015)

343. Gem quality olivine is termed as:
A. Norite B. Satin-spar
C. Peridot D. Ruby
(BHU-2015)

344. Intergrowth of quartz and feldspar resulting:
A. Perthite B. Orbicular structure
C. Graphic texture D. Myrmekite structure
(BHU-2015)

345. Correctly match the following:

Column-A
P. Laths of plagioclase are totally surrounded in a course grained matrix of pyroxene crystals
Q. Smaller grains of one minerals are completely enclosed in large
R. Plagioclase grains show a preferred orientation due to flowage
S. Spherical intergrowths of radiating quartz and feldspar replace glass as a result devitrification

Column-B
1. Spherulitic texture
2. Ophitic texture
3. Poikilitic texture
4. Trachytic texture

A. P-2, Q-3, R-4, S-1
B. P-1, Q-2, R-3, S-4
C. P-2, Q-4, R-3, S-1
D. P-1, Q-3, R-2, S-4 *(BHU-2015)*

346. Principal of Uniformitarianism is proposed by:
A. William Smith B. James Hutton
C. Gottlob Worner D. Harry Hess
(BHU-2015)

347. Which one is the most stable mineral?
A. Quartz B. Feldspar
C. Zircon D. Olivine *(BHU-2015)*

348. The phi-scale is given by:
A. $\phi = -\log_2 d$ B. $\phi = -\log_5 d$
C. $\phi = -\log_{10} d$ D. $\phi = -\log_2 (-d)$
(BHU-2015)

349. What is the sand particle size?
A. 2 to 0.092 mm B. 4 to 2 mm
C. 2 to 0.0625 mm D. 2 to 0.0825 mm
(BHU-2015)

350. Grain size analysis of clastic rock cannot interpret:
A. Environment
B. Source rock
C. Energy flow
D. Nature of transportation *(BHU-2015)*

351. The river sands are distinguished by:
A. Positive skewness
B. Negative skewness
C. High porosity
D. Sand dunes *(BHU-2015)*

352. The ratio of number of grain to grain contact to the total number of grains counted in the traverse is:
A. Packing density B. Packing proximity
C. True porosity D. True density
(BHU-2015)

353. The diamictic conglomerate are characterizes by:
A. High porosity B. High matrix
C. Low matrix D. Zero matrix
(BHU-2015)

354. Endogenetic rocks are deposited due to:
A. Precipitation
B. River deposition
C. Volcanic ash
D. Mechanical weathering *(BHU-2015)*

355. Graywacke is a type of rock:
A. Volcanic rock
B. Non-clastic sedimentary
C. Rudaceous sedimentary rock
D. Arenaceous sedimentary rock
(BHU-2015)

356. What is mineralogical maturity of sandstone?
A. Ratio of quartz to olivine
B. Ratio of feldspar to the quartz
C. Ratio of quartz to feldspar
D. Ratio of flaky minerals to feldspar
(BHU-2015)

357. At 200 temperature, what will critical Reynolds number for laminar to turbulent flow?
A. 500 B. 1200
C. 700 D. 1000 *(BHU-2015)*

358. Calcium carbonate could not found in:
A. Above CCD B. Below CCD
C. Below ACD D. Above ACD
(BHU-2015)

359. Which of the following original components bound together at time deposition?
A. Mudstone B. Wackestone
C. Packstone D. Boundstone
(BHU-2015)

360. Which of the following statement is false?
A. Graded bedding shows the grain size variation within a bed
B. Convolute bedding a deformed bedding structure
C. Secondary sedimentary structure form before the diagenesis of a sedimentary rocks
D. Hummocky cross-stratification is primary sedimentary structure
(BHU-2015)

361. Which one among the following is **not** a metamorphic rock?
A. Hornfels B. Orthoquartzite
C. Metagranite D. Biotite gneiss
(BHU-2015)

362. Which one of the following belongs to high grade metamorphism?
A. Epizone B. Acme zone
C. Mesozone D. Katazone
(BHU-2015)

363. .... structure is found in the rocks composed of equidimensional minerals.
A. Gneissose B. Maculose
C. Granulose D. Schistose
(BHU-2015)

364. Which rock is a parametamorphic?
A. Metacharnokite B. Metaqurtzite
C. Biotite gneiss D. Khondalite
(BHU-2015)

365. Which one from the following is non-foliated metamorphic rock?
A. Gneiss B. Phyllite
C. Slate D. Quartzite
(BHU-2015)

366. .... is the high temperature and high pressure metamorphic facies.

A. Blueschist B. Greenschist
C. Eclogite D. Amphibolite

(BHU-2015)

367. The term 'blast or blastic' used as prefix in the case of:

A. Palimpset texture
B. Aphanitic texture
C. Phanitic texture
D. Glassy texture *(BHU-2015)*

368. In the Al_2SiO_5 polymorphs, which shows highest grade of metamorphism?

A. Andalusite B. Kyanite
C. Sillimanite D. All the above

(BHU-2015)

369. In the following, which indicates the ascending order of temperature of formation?

A. Eclogite, Greenschist, Zeolite, Blueschist
B. Greenschist, Blueschist, Eclogite, Granulite
C. Zeolite, Amphibolite, Granulite, Eclogite
D. Blueschist, Granulite, Amphibolite, Eclogite

(BHU-2015)

370. Omphacite and Pyrope Garnet are the representative minerals for:

A. Zeolite facies
B. Greenschist facies
C. Glaucophane-schist facies
D. Eclogite facies *(BHU-2015)*

371. Which is **not** related to metamorphism?

A. Wind
B. Pressure
C. Temperature
D. Chemically active fluid *(BHU-2015)*

372. An uniform pressure is also called:

A. Direct pressure B. Hydrostatic pressure
C. Compression D. Tension

(BHU-2015)

373. .... metamorphism occurs around larger intrusives at comparatively low temperature:

A. Pyrometamorphism
B. Contact metamorphism
C. Metasomatism
D. Auto-metamorphism *(BHU-2015)*

374. Match the following columns and give the correct answer:

	Column-A		Column-B
P.	Blueschist facies	1.	Chlorite
Q.	Greenschist	2.	Plagioclase
R.	Granulite	3.	Mg-Garnet
S.	Eclogite	4.	Glaucophane

A. P-4, Q-1, R-2, S-3
B. P-1, Q-4, R-3, S-2
C. P-2, Q-3, R-1, S-4
D. P-3, Q-2, R-4, S-1 *(BHU-2015)*

375. Correctly match Column A and B:

	Column-A		Column-B
P.	Shale	1.	Quartzite
Q.	Sandstone	2.	Amphibolite
R.	Granite	3.	Slate
S.	Basalt	4.	Gneiss

A. P-1, Q-2, R-4, S-3
B. P-2, Q-1, R-4, S-3
C. P-3, Q-1, R-4, S-2
D. P-4, Q-3, R-2, S-1 *(BHU-2015)*

376. Why basalt is fine grained than gabbro?

A. Basalt formed from quick cooling of magma
B. Gabbro formed from quick cooling of magma
C. Basalt has a mafic composition
D. Gabbro has a mafic composition

(BHU-2015)

377. Rocks are formed when magma:

A. Erodes
B. Crystallizes
C. Undergoes radioactive decay
D. Weathers *(BHU-2015)*

378. Basic source of magnetism is:

A. Charged particles alone
B. Magnetic domain
C. Magnetic dipoles
D. Movement of charged particles

(BHU-2015)

379. Obsidian is a/an:
A. Metamorphic rock which contains high mafic mineral
B. Igneous rock which contains high Fe and Mg
C. Volcanic glass containing high silica
D. Sedimentary rock formed by weathering of basaltic rock
(BHU-2015)

380. According to Plate Tectonics Theory, most active volcanoes occur:
A. On continents
B. In large tectonic plate
C. Along plate boundaries
D. Randomly over continents *(BHU-2015)*

381. Match each items and choose the correct option:

Column-A	Column-B
P. Felsic	1. Magma that flow out Earth's surface
Q. Kimberlite	2. Dark-coloured rock such as gabbro that rich in Fe and Mg.
R. Lava	3. Rock that is rich in silica
S. Mafic	4. Ultramafic rock that can contain diamonds

A. P-1, Q-3, R-4, S-2
B. P-3, Q-4, R-1, S-2
C. P-2, Q-1, R-4, S-3
D. P-4, Q-1, R-3, S-2 *(BHU-2015)*

382. Ores near Earth's surface are generally obtained from:
A. Waste-removal facilities
B. Underground mines
C. Open-pit mines
D. Bodies of water with high concentration of dissolved minerals
(BHU-2015)

383. What is the main difference between a conglomerate and breccia?
A. Breccia clasts are angular, conglomerate clasts are rounded
B. A breccia is well stratified; a conglomerate is poorly stratified
C. Breccia clasts are the size of baseballa; conglomerate clasts are larger
D. Breccia has a compacted, clay-rich matrix; conglomerate has no matrix
(BHU-2015)

384. Detrital sedimentary rocks are classified based on the:
A. Colours of the cementing materials
B. Grain sizes of the detrital particles
C. Composition of soluble minerals
D. Degree of compaction and lithification
(BHU-2015)

385. Which of the following sedimentary rocks indicate long distance transportation of the sediments?
A. Quartz arenite
B. Breccia
C. Arkose (sandstone with lots of feldspar particles)
D. Felspathic wacke *(BHU-2015)*

386. A course grained rock mainly composed of orthopyroxene and plagioclase is called as:
A. Norite B. Gabbro
C. Lherzolite D. Harzbugite
(PU-2010)

387. Which of the following is a concordant intrusive body?
A. Cone sheet B. Ring dyke
C. Stock D. Lopolith *(PU-2010)*

388. When a single pyroxene envelops several well-developed plagioclase laths, the texture is called:
A. Ophitic texture B. Poikiloblast
C. Porphyroblast D. Granoblast
(PU-2010)

389. Pyroclastic deposits are usually found associated with:
A. Explosive volcano
B. Shield volcano
C. Pahoehoe flow
D. Ropy lava *(PU-2010)*

390. Grains that crystallize to a significantly larger size than those of the matrix, in an igneous rocks are called:
A. Phenocrysts B. Porphyroblasts
C. Xenoliths D. Relic intrusion
(PU-2010)

391. Silica of oceanic crust that have been emplaced (Obducted) onto continental crust:
A. Mylonite B. Migmatite
C. Ophiolite D. Granulite
(PU-2010)

392. Which one of the following statements about Andalusite, Kyanite and Siliimanite is WRONG?
A. They have same chemical formula
B. Commonly one of them is present in Metamorphosed peptic rocks
C. If all the three are present the pressure and temperature conditions can be determined.
D. They have same crystal structure
(PU-2010)

393. A rock containing Phenocrysts of Olivine set in a fine grained matrix of pyroxerie and plagioclase is known as:
A. Gabbro B. Basalt
C. Norite D. Diorite *(PU-2010)*

394. The major minerals present in granite are:
A. Microcline, quartz and albite
B. Sanidine, quartz and labradorite
C. Orthoclase, quartz and nepheline
D. Nepheline, quartz and labradorite
(PU-2010)

395. Volcanic equivalent of the plutonic rock granodiorite is:
A. Rhyolite B. Rhodacite
C. Dacite D. Andesite *(PU-2010)*

396. At the binary eutectic, which of the following phases coexist?
A. A crystal phase and a liquid
B. Two different crystal phases and a liquid
C. Two different crystal phases
D. Three different crystal phases *(PU-2010)*

397. Complete solid solution is exhibited at 1 atmospheric pressure by:
A. Albite - Anorthite
B. Diopside - Enstatite
C. Enstatite - Ferrosilite
D. Actinolite - Tremolite *(PU-2010)*

398. Komatites are typically found in:
A. Archean greenstone belts
B. Proterozoic mobile belts
C. Proterozoic sedimentary basins
D. Tertiary volcanic provinces *(PU-2010)*

399. Predominant magma that erupted during the Deccan Volcanism is:
A. Alkali basalt B. Tholeritic basalt
C. Andesite D. Trachyte
(PU-2010)

400. Metamorphic facies are defined by the:
A. Critical mineral assemblages
B. Texture and structure of the rock types
C. Particular rock type
D. Pressure-Temperature condition
(PU-2010)

401. Metamorphic grade refers to:
A. Particular rock type
B. Intensity of metamorphism
C. Texture and structure of the rock types
D. Critical mineral assemblages
(PU-2010)

402. Which of the following metamorphic facies is characterized by maximum temperature and minimum pressures?
A. Blueschist facies
B. Greenschist facies
C. Hornblende-hornfels facies
D. Sanidinite facies *(PU-2010)*

403. The mineral omphacite is characteristic of:
A. Greenschist facies
B. Granulite facies
C. Blueschist facies
D. Eclogite facies *(PU-2010)*

404. In case of Barrovian metamorphism, the lowest grade of metamorphism is marked by the mineral:
A. Biotite B. Garnet
C. Chlorite D. Kyanite
(PU-2010)

405. Anhydrous rocks are characteristically present in rocks belonging to:
A. Eclogite facies
B. Greenschist facies
C. Amphibolite facies
D. Granulite facies *(PU-2010)*

406. The reaction Ferrosilite + Diopside = Enstatite + Hedenbergite, is a kind of:
A. Solid-solid net-transfer reaction
B. Devolatilization reaction
C. Ion-exchange reaction
D. Oxidation reaction *(PU-2010)*

407. In the triangular ACF diagrams used to designate the mineralogical and chemical composition of metamorphic facies, the 'A' apex represents:
A. Al_2O_3
B. Al_2O_3+ Fe_2O_3 - Na_2O- K_2O
C. Al_2 O_3 + $Fe_2O_3K_2O$
D. Al_2O_3 + Na_2 + K_2O *(PU-2010)*

408. Paired metamorphic belt is a characteristic feature of:
A. Continental collision zone
B. Subduction zone
C. Sea floor spreading zone
D. Continental rift zone *(PU-2010)*

409. Which of the metamorphic zone of Barrow is similar to the green hilt facies?
A. Biotite zone B. Garnet zone
C. Staurolite zone D. Sillimanite zone *(PU-2010)*

410. The high pressure polymorph of silica, coesite, stishovite, may be associated with:
A. Hydrothermal metamorphism
B. Burrial metamorphism
C. Contact metamorphism
D. Impact metamorphism *(PU-2010)*

411. Sedimentary breccia is a rare rock type because:
A. Gravel is rounded quickly during transport
B. Clay is less abundant than other sedimentary particles
C. Feldspars are chemically unstable
D. Sand deposits are typically well-sorted *(PU-2010)*

412. Which of the following is detrital sediment?
A. Broken sea shells
B. Ions in solution
C. Graded bedding
D. Quartz sand *(PU-2010)*

413. If an aggregate of sediment consists of particles that are all about the same size, it is said to be:
A. Well-sorted B. Sandstone
C. Poorly rounded D. Lithified *(PU-2010)*

414. Which of the following can be used to determine paleocurrent direction?
A. Mud cracks
B. Turbidity currents
C. Graded bedding
D. Cross-bedding *(PU-2010)*

415. Bedding or stratification in a sedimentary rock:
A. Is due to directed pressure from earth forces
B. Results from heat
C. Is always exactly horizontal
D. Represents variations which took place during deposition *(PU-2010)*

416. The least stable material in elastic detritus is:
A. Quartz B. Amphibole
C. Olivine D. Hornblende *(PU-2010)*

417. Graded bedding usually signifies:
A. Fast-moving water that gradually slowed down
B. Slow-moving water that gradually speeded up
C. Stagnant water
D. Deposition on dry land *(PU-2010)*

418. Mud cracks are most likely to form by:
A. Rapidly changing patterns of erosion and deposition
B. Gently oscillating waves
C. Periodic exposure to the air and drying out
D. Fast-moving water that gradually slowed down *(PU-2010)*

419. Which would be least likely to indicate a desert environment?
A. Large cross-beds
B. Limestone
C. Evaporite
D. Mud cracks
(PU-2010)

420. By the plate-tectonic processes, magma generation is highest in the following tectonic setting:
A. Convergent plate boundaries
B. Continental intra-plate region
C. Oceanic intra-plate region
D. Oceanic ridges *(PU-2013)*

421. The last mineral to crystallize from magma, according to the Bowen's reaction series is:
A. Plagioclase
B. Olivine
C. Quartz
D. Pyroxene *(PU-2013)*

422. The plutonic equivalent of andesite is:
A. Granite B. Gabbro
C. Diorite D. Tonalite *(PU-2013)*

423. Ophitic texture is the key characteristics of which one of following rocks?
A. Gabbro B. Basalt
C. Dolerite D. Peridotite
(PU-2013)

424. Which one of the following facies represent highest grade of metamorphism?
A. Zeolite facies
B. Granulite facies
C. Greenschist facies
D. Amphibolite facies *(PU-2013)*

425. Which one of the following index minerals represents the highest grade of regional metamorphism?
A. Chlorite B. Sillimanite
C. Biotite D. Garnet
(PU-2013)

426. Which of the following sedimentary structures represents the highest flow energy condition?
A. Trough cross bed
B. Planar Cross bed
C. Wavy bedding
D. Plane bedding with parting lineation
(PU-2013)

427. Complete solid solution is observed between the minerals:
A. Tremolite cummingtonite
B. Calcile - magnesite
C. Forsterite fayalite
D. Diopside - enstatite *(PU-2013)*

428. Basalt is mainly made up of:
A. Pyroxene, calcic plagioclase and olivine
B. Amphibole, orthoclase and leucite
C. Albite, orthoclase and quartz
D. Albite, hornblende and nepheline
(PU-2013)

429. An igneous rock is fully crystalline and made up of equidimensional crystals of about 3 mm in size. What are the conditions of its emplacement and crystallization?
A. Surface of continental crust, sudden cooling
B. Shallow depth in crust, fast cooling
C. Moderate depth in crust, slow cooling
D. On ocean floor, sudden-cooling
(PU-2013)

430. A Composition of magma erupted along the mid-Ocean ridges is:
A. Andesite B. Basalt
C. Syenite D. Granoiorite
(PU-2013)

431. A suite of igneous rocks with gradual changes in composition, found at Island arc is known as:
A. Calc-alkaline series
B. Alkaline series
C. Tholeiitic series
D. Syenitic series *(PU-2013)*

432. Where do komatiites usually occur?
A. Recent island-arc belts
B. Cretaceous basins
C. Gondwana basins
D. Archean greenstone belts *(PU-2013)*

Study the phase diagram and answer the following four Questions

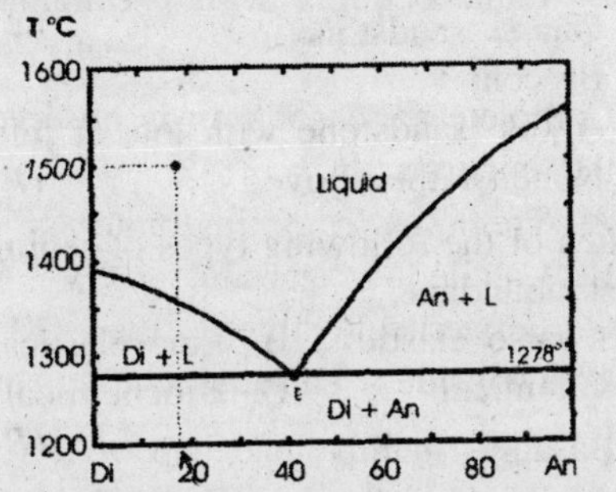

433. What was the degree of freedom when magma marked (o) reached the liquidus curve?

A. 0 B. 1
C. 2 D. 3 *(PU-2013)*

434. The magma cooled further and reached eutectic then it will crystallize:

A. Only anorthite
B. Only diopside
C. Diopside and anorthite
D. Quartz, anorthite and diopside *(PU-2013)*

435. At what temperature the last drop of magma will crystallize?

A. 1350 B. 1278
C. 1500 D. 1225 *(PU-2013)*

436. On complete crystallization of the magma, rack formed will be made up of:

A. 40 % diopside and 60 % anorthite
B. 60 % diopside and 40 % anorthite
C. 18 % diopside and 82 % anorthite
D. 82 % diopside and 18 % anorthite *(PU-2013)*

437. Which one of the following minerals occurs as a phenocryst in basalt?

A. Quartz B. Biotite
C. Microcline D. Olivine *(PU-2012)*

438. Rocks exhibiting hypidiomorphic texture were formed as:

A. Metamorphic rocks
B. Plutonic rocks
C. Continental volcanic rocks
D. Submarine volcanic rocks *(PU-2012)*

439. Phenocrysts of quartz are found in:

A. Andesite B. Basalt
C. Rhyolite D. Dacite *(PU-2012)*

440. In an igneous rock .... cannot occur along with quartz under equilibrium condition.

A. Hypersthene
B. Nepheline
C. Hornblende
D. Magnetite *(PU-2012)*

441. Syenites are essentially made up of:

A. Plagioclase feldspar
B. Alkali feldspar
C. Plagioclase feldspar and augite
D. Diopside and quartz *(PU-2012)*

442. The discontinuous series in the Bowen's reaction series consists of minerals:

A. Olivine, pyroxene, anorthite, albite
B. Olivine, amphibole, anorthite, albite
C. Olivine, pyroxene, albite, quartz
D. Olivine, pyroxene, amphibole, biotite *(PU-2012)*

443. The minerals of amygdules usually the composition of the host rock, for example those basalts bearing nepheline commonly contain zeolites, what mineral in the source rock can causes the formation of Chalcodony?

A. Chalcocite
B. Calcite
C. Sodalite
D. Silica *(PU-2012)*

444. Quartz grains embedded in large single crystals of the feldspar have shapes that resemble the characters of cuneiform writing. This texture is known as:

A. Corona B. Symplectite
C. Graphic D. Ophitic *(PU-2012)*

445. In metamorphosed calc-silicate rocks, one of the following sequence of the minerals appears in the mineral assemblage with increasing metamorphic grade:

A. Tremolite - Diopside - Tacl - Forsterite
B. Talc - Forsterite - Diopside - Tremolite
C. Talc - Tremolite - Diopside - Forsterite
D. Forsterite - Tremolite - Talc - Diopside
(PU-2012)

446. Sapphirine + Quartz assemblage indicates .... metamorphism condition.
A. High pressure
B. Ultra-high pressure
C. Ultra-high temperature
D. Low pressure *(PU-2012)*

447. In case of the dynamic metamorphism, the principal agent of metamorphism is:
A. Temperature
B. Fluid
C. Deviatoric stress
D. Both temperature and deviatoric stress
(PU-2012)

448. Metamorphic grade refers to:
A. Intensity of metamorphism
B. Collection of minerals assemblage from rocks of various bulk composition that crystallized at same P, T condition
C. Particular mineral observed at a specific P, T condition
D. Specific P, T condition of metamorphism
(PU-2012)

449. A clastic rock is:
A. A rocks formed from the cementation of transported grains
B. A rocks formed from evaporation of sea water
C. Transformed by heat into limestone
D. Transformed by pressure into limestone
(PU-2012)

450. What is probably the single most important, original, depositional feature in sedimentary rocks?
A. Sizes of the sand grain
B. Degree of lithification
C. Bedding or stratification
D. Compaction of the mud and clay
(PU-2012)

451. Which of the following sedimentary rocks indicate long-distance transportation of the sediments?
A. Quartz sandstone
B. Breccia
C. Arkos (sandstone with lots of feldspar)
D. None of the above *(PU-2012)*

452. Which of the following types of sediments are most abundant?
A. Coarse elastics B. Fine elastics
C. Chemical D. Biochemical
(PU-2012)

453. Which of the following types of currents can transported sand grains?
A. Rivers B. Wind
C. Ocean waves D. All of these
(PU-2012)

454. Which of the following lists is written in order of decreasing particle size?
A. Sandstone, siltstone, conglomerate
B. Sandstone, conglomerate, siltstone
C. Conglomerate, sandstone, siltstone
D. Siltstone, sandstone, conglomerate
(PU-2012)

455. What is the difference between a breccia and a conglomerate?
A. Breccias are course grained and conglomerates are fine grained
B. Conglomerates are course grained and breccias are fine
C. Breccias have rounded fragments and conglomerats have angular fragments
D. Breccias have angular fragments and conglomerats have rounded fragments
(PU-2012)

456. Megascopically identifiable quartz and potassium feldspar grains are the main constituents of:
A. Granite B. Gabbro
C. Basalt D. Rhyolite
(PU-2011)

457. Which of the following minerals would crystallize early from a high-Mg basalt magma?
A. Biotite B. Plagioclase
C. Olivine D. Spinel
(PU-2011)

458. Flint, chert and Jasper are microcrystalline forms of:
A. Quartz B. Hematite
C. Halite D. Calcite *(PU-2011)*

459. A granitic body intrudes into a sedimentary formation. Which of the following geological process would likely take place in such settings?
A. Sedimentary rocks would undergo thermal metamorphism along the contact of granitic intrusion
B. Granitic body would from a high mountain chain
C. Sedimentary rocks would get weathered faster
D. Intrusion would cause basaltic volcanism on surface
(PU-2011)

460. Which of the following structures can help to determine the direction of current that deposited the rock?
A. Asymmetric ripple marks
B. Symmetric ripple marks
C. Mud cracks
D. Graded bedding *(PU-2011)*

461. Which of the following rock types would have finest size particles?
A. Conglomerate B. Sandstone
C. Mudstone D. Greywacke
(PU-2011)

462. Shale refers to a rock formed from:
A. Sand size material
B. Plant remains
C. Clay minerals
D. Carbonate *(PU-2011)*

463. The tendency for variations in current velocity to segregate sediments on the basis of particle size is called:
A. Lithification B. Compaction
C. Metamorphism D. Sorting *(PU-2011)*

464. Which of the following lists is written in order of decreasing particle size?
A. Sandstone, siltstone, conglomerate
B. Sandstone, conglomerate, siltstone
C. Conglomerate, sandstone, siltstone
D. Siltstone, sandstone, conglomerate
(PU-2011)

465. What is the difference between a breccia and a conglomerate?
A. Breccias are course grained and conglomerate are fine grained
B. Conglomerate are course grained and breccias are fine grained
C. Breccias have rounded fragments and conglomerate have angular fragments
D. Breccias have angular fragments and conglomerate have rounded fragments
(PU-2011)

466. A feldspar rich sandstone is called:
A. Arkose
B. Litharenite
C. Quartz arenite
D. Shale *(PU-2011)*

467. Which of the following lists is written in the order of increasing temperature?
A. Sedimentation, metamorphism, diagenesis
B. Diagenesis, sedimentation, metamorphism
C. Sedimentation, diagenesis, metamorphism
D. Metamorphism, diagenesis, sedimentation
(PU-2011)

468. In which of the following sedimentary environments would you least expect to find gravel?
A. Active margin beach
B. Alluvial fans
C. Glacial
D. Deep marine *(PU-2011)*

469. Which one of the following represents increasing order of metamorphism?
A. Slate, phyllite, schist, gneiss
B. Slate, schist, phyllite, gneiss
C. Phyllite, slate, schist, gneiss
D. Phyllite, schist, slate, gneiss *(PU-2011)*

470. Rock formed by consolidation of fine volcanic ash is known as:
A. Obsidian B. Tephra
C. Rhyolite D. Tuff *(PU-2011)*

471. Which of the following statements is/are NOT true for equilibrium crystallization of plagioclase from a melt in the binary system $NaAlSi_3O_8$–$CaAlSi_2O_8$ at constant pressure?

A. Composition of the first-formed solid depends on the initial composition of the melt.

B. Solid composition is always more anorthitic than the coexisting melt.

C. Composition of the final solid is the same as the initial composition of the melt.

D. Equilibrium crystallization leads to the formation of zoned crystals.

(IIT JAM-2016)

472. Gabbro contains much more alumina than peridotite because it is richer in

A. Orthopyroxene B. Orthoamphibole
C. Olivine D. Plagioclase

(IIT JAM-2016)

473. The correct name for the well sorted sandstone, whose framework composition plots at Y in the following QFR diagram, is

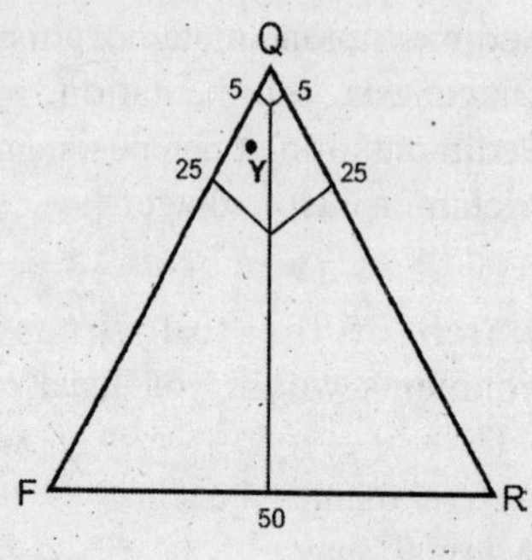

A. Arkose B. Subarkose
C. Lithic arkose D. Quartz arenite

(IIT JAM-2016)

474. During metamorphism, temperature can increase at constant pressure in the case of

A. Exhumation by erosion
B. Burial by subduction
C. Burial by underthrusting
D. Intrusion of batholith

(IIT JAM-2016)

475. Identify the rock types at X and Y in the following QAP diagram of IUGS.

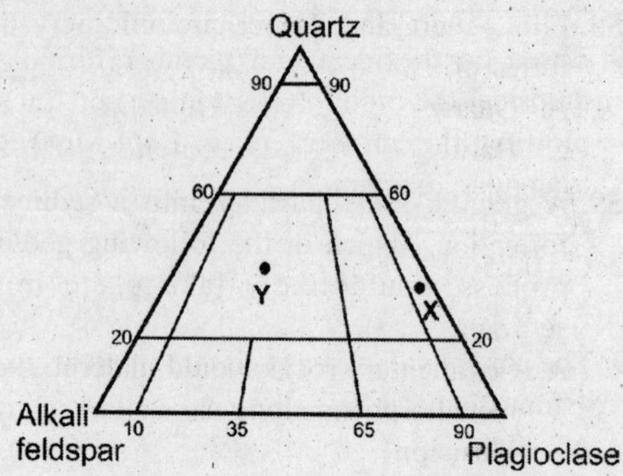

A. X is granodiorite and Y is granite
B. X is granodiorite and Y is alkali feldspar granite
C. X is tonalite and Y is granite
D. X is tonalite and Y is alkali feldspar granite

(IIT JAM-2016)

476. Which one of the following sedimentary structures is related to gravitational instability?

A. Groove cast B. Load cast
C. Gutter cast D. Flute cast

(IIT JAM-2016)

477. Choose the correct match of items in Group I with the items in Group II.

Group-I	Group-II
(i) Eutectic crystallisation	(P) Perthitic texture
(ii) Fractional crystallisation	(Q) Graphic texture
(iii) Exsolution	(R) Pillow structure
(iv) Submarine eruption	(S) Crystal zoning

A. i - Q, ii - S, iii - R, iv - P
B. i - Q, ii - S, iii - P, iv - R
C. i - S, ii - Q, iii - P, iv - R
D. i - S, ii - Q, iii - R, iv - P

(IIT JAM-2016)

478. Migmatite is a rock

A. In which mafic-rich parts are intermixed with pods or layers of granitic composition
B. With melanosome and leucosome
C. With a solid residue and partial melt
D. Which forms at high grade metamorphic conditions

(IIT JAM-2016)

479. Based on the ideal end member formula of diopside, the mole proportion (%) of CaO for plotting the mineral on a CaO-MgO-SiO_2 triangular diagram is________.

A. 25.0 B. 2.25

C. 0.25 D. 12.25

(IIT JAM-2016)

480. The phi (ϕ) value of a sediment particle having 4 mm diameter is ________.

A. −2 B. +2

C. 0 D. 1 *(IIT JAM-2016)*

481. Calcite, quartz, wollastonite and CO_2 fluid were present in equilibrium during the formation of a calc-silicate rock. In the chemical system CaO–SiO_2–CO_2, the degree of freedom of this assemblage is ________.

A. 1.0 B. 2.0

C. 3.0 D. 4.0 *(IIT JAM-2016)*

482. A melt in the binary system MgO–SiO_2 contains 89.92 wt% SiO_2. If all the magnesium is consumed to form enstatite ($MgSiO_3$), how many moles of this mineral will crystallize from 100 grams of the melt? Give answer in two decimal places. *Molecular weight* : MgO = 40.3, SiO_2 = 60.1

A. 0.25 B. 0.35

C. 0.45 D. 1.25 *(IIT JAM-2016)*

483. A Lherzolite xenolith from the mantle contains 50 volume % olivine, the rest being equal proportions of orthopyroxene and clinopyroxene. If the densities of the minerals are (in g/cc) olivine = 3.42, orthopyroxene = 3.28 and clinopyroxene = 3.46, the bulk density of the xenoliths in g/cc is_______ (give answer in two decimal places).

A. 3.39 B. 4.39

C. 1.39 D. 12.39 *(IIT JAM-2016)*

484. The SiO_2 value, recalculated on volatile free basis, of the rock whose major oxide (wt%) composition given below is _______. Give answer in two decimal places.

Oxide	wt%
SiO_2	45.58
Al_2O_3	12.49
FeO (T)	12.44
MgO	11.36
CaO	09.93
Na_2O	02.25
K_2O	01.67
Total	95.72

A. 47.61 B. 48.61

C. 49.61 D. 50.79

(IIT JAM-2016)

485. Diorite and Gabbro can best be distinguished on which basis?

A. Gabbro contains pyroxene but diorite contain hornblende

B. Gabbro does not contain quartz but diorite contain quartz

C. Diorite contain both orthoclase and plagioclase feldspar but Gabbro contains only plagioclase

D. Plagioclase in diorites is more Na-rich but in gabbro it is more Ca rich

(ISM-2014)

486. Epsilon cross bedding in fluvial channels originate due to:

A. Migration of ripple along the flow direction of the river

B. Scouring of the channel base and late filling it

C. Crossing over transportation to deposition field due to slowing down of current flow

D. Lateral shifting of a meandering channel

(ISM-2014)

487. Red Shales are characteristics features of:

A. Intertidal zone B. Supratidal zone

C. Subtidal zone D. Swash zone

(ISM-2014)

488. As the depth of burial increases the progressive transformation of clay rich sediments to shale and then to metamorphic rocks of successive higher grade is characterized by which of the following ?

A. The density of the rock increases while the water content and the grain size decreases

B. H_2O and CO_2 are driven out if metamorphism has taken place in a closed system

C. The mineral components change progressively to new forms that are stable at higher temperature and pressure
D. If the rock is a part of the closed system its minerals composition remains the same but the packing of grains become tighter *(ISM-2014)*

489. Consider the following igneous ultramafic rocks
1. Peridotite 2. Alkali Peridotite
3. Dunite 4. Pyroxenite

What is the correct sequence of these in the order of decreasing MgO and increasing CaO?
A. 3-2-1-4 B. 4-1-2-3
C. 3-1-2-4 D. 4-2-1-3 *(ISM-2014)*

490. Which of the following pairs of minerals are characteristics of eclogites?
A. Clinopyroxene + Orthopyroxene
B. Sapphirine + Garnet
C. Jadeite + Plagioclase
D. Omphacite + Pyrope *(ISM-2014)*

491. Match List-A and select List-B:
A. Websterite 1. 60% Ol, 14% Opx, 25% Cpx
B. Lherzolite 2. 60% Ol, 10% Opx, 30 Cpx
C. Harzbugite 3. 60% Ol, 30% Opx, 10% Cpx
D. Wehrlite 4. 10% Ol, 50%Opx, 40% Cpx
A. A-2, B-3, C-1, D-4
B. A-4, B-3, C-1, D-2
C. A-2, B-1, C-3, D-4
D. A-4, B-1, C-3, D-2 *(ISM-2014)*

492. Which one of the following can be used as away up criteria?
A. Pillow lava B. Blocky lava
C. Ropy lava D. Pa-hoe-hoe lava *(ISM-2014)*

493. Which of the following minerals assemblage represent granulite facies?
A. Quartz-albite-epidote-muscovite
B. Quartz-oligoclase-muscovite-andalusite-cordierite
C. Calcite-quartz-epidote-tremolite
D. Quartz-orthoclase-sillimanite- almandine *(ISM-2014)*

494. During field excursion greyish rock is exposed which composed of pearl like miniature spheres and react with dilute hydrochloric acid. The rock is:
A. Conglomerate B. Micrite-limestone
C. Oolitic limestone D. Breccia *(ISM-2015)*

495. Which are the plutonic rocks of intermediate chemical composition in total alkali silica plot?
A. Synite and gabbro
B. Synite and ijolite
C. Synite and diorite
D. Granite and granodiorote *(ISM- 2015)*

496. Pisolite is a:
A. Sedimentary B. Igneous
C. Metamorphic D. Meta-sedimentary *(ISM-2015)*

497. A rock texture that has a bimodal size distribution, with large crystal surrounded by numerours small grain or glass, is described as:
A. Ophitic B. Graphic
C. Sub-ophitic D. Porphyritic *(ISM-2015)*

498. What is the texture of schist having large garnets in fine grained biotite, muscovite and quartz matrix?
A. Granoblastic B. Porphyroblastic
C. Poikiloblastic D. Porphyritic *(ISM-2015)*

499. Crystalloblastic texture will be exhibited when rock is:
A. Crystallized under conditions of low viscosity with low uneven pressure
B. Crystallized under conditions of high viscosity with high uneven pressure
C. Recrystallization under conditions of low viscosity and directed pressure
D. Recrystallization under conditions of high viscosity and directed pressure *(ISM-2015)*

500. When orthoclase is rimmed with sodic plagioclase the texture is known as?
A. Rapakivi B. Orbicular
C. Graphic D. Ophitic
(ISM-2015)

501. Which of the following sedimentary structure in not useful in analysis of paleo-current direction?
A. Cross bedding
B. Oscillation ripple mark
C. Groove cast
D. Sole mark
(ISM-2015)

502. The graphic texture is generally found in:
A. Basalt B. Granite
C. Diorite D. Dunite
(ISM-2015)

503. ———is a type of rocks formation that rises from the floor of a cave found in limestone deposited area:
A. Stalagmite B. Stalagtite
C. Doline D. Barrier
(ISM-2015)

504. Ophilites are:
A. Cratonic rock B. Continental rock
C. Oceanic crust D. Volcanic rock
(ISM-2015)

505. The blue colour of the rocks of blue schist facies comes from the presence of:
A. Glaucophane B. Omphacite
C. Almandine D. Pyrope
(ISM-2015)

506. Gabbro with more orthopyroxene is called:
A. Mugerite B. Limburgite
C. Norite D. Oceanite
(SAGAR UNIVERSITY-2015)

507. Low viscosity thin, glassy lavas from tongues and lobes overlapping one another is called:
A. Lava tubes
B. Aa and pa-hoe-hoe lavas
C. Aa and blocky lavas
D. Pollow lavas
(SAGAR UNIVERSITY-2015)

508. The crystallization of An-Ab-Kf components in ternary systems involved:
A. Three binary system
B. Three binary eutectic
C. Three peritectic
D. Incongruent melting
(SAGAR UNIVERSITY-2015)

509. Volcanic breccias is known as:
A. Lapilli B. Conglomerate
C. Agglomerate D. Volcanoclast
(SAGAR UNIVERSITY-2015)

510. Komatiite contains high concentration of:
A. MgO B. FeO/Fe_2O_3
C. MnO D. CaO
(SAGAR UNIVERSITY-2015)

511. Pegmatite rock contain which of the following minerals in abundance?
A. Tourmaline B. Spinel
C. Forsterite D. Andesite
(SAGAR UNIVERSITY-2015)

512. Largest accumulation of the tholeiitic rocks in India is known by which of the following name?
A. Rajamahal traps B. Sylhet traps
C. Bhavali traps D. Deccan traps
(SAGAR UNIVERSITY-2015)

513. The plagioclase + hypersthene + biotite minerals assemblage represent :
A. Amphibolite facies
B. Green schist facies
C. Epidote facies
D. Pyroxene – hornfels facies
(SAGAR UNIVERSITY-2015)

514. The mineral which represents highest zone of metamorphism is:
A. Biotite B. Sillimanite
C. Kyanite D. Almandine
(SAGAR UNIVERSITY-2015)

515. Which of the following is associated with the eugeosyncline deposit?
A. Molasses B. Arkose
C. Flysch D. Graywackes
(SAGAR UNIVERSITY-2015)

516. Sole marks having longitudinal appearance is called:

A. Flute cast
B. Load cast
C. Sole marks
D. Dendrites *(SAGAR UNIVERSITY-2015)*

517. A gap in the deposition observed between parallel horizontally disposed two sedimentary units is called:

A. Angular unconformity
B. Unconformity
C. Disconformity
D. Non-conformity

(SAGAR UNIVERSITY-2015)

518. A sandstone which contains abundance decomposed feldspar is known as:

A. Grit B. Greywacke
C. Marl D. Arkose

(SAGAR UNIVERSITY-2015)

519. Conglomerate is a _____ type of sedimentary rock:

A. Arenaceous B. Argillaceous
C. Rudaceous D. Calcareous

(SAGAR UNIVERSITY-2015)

520. Current bedding is indicative of:

A. Shallow lake environment
B. River water environment
C. Deep water marine environment
D. Shelf water environment

(SAGAR UNIVERSITY-2015)

521. Oolites and pisolites are distinguished by:

A. Shape B. Size
C. Composition D. Sphericity

(SAGAR UNIVERSITY-2015)

522. Limestone and dolostone is distinguished by:

A. Alumina content
B. Lime content
C. Magnesia content
D. Soda content

(SAGAR UNIVERSITY-2015)

523. Gondites is a:

A. Metamorphosed manganiferous sediment
B. Magniferous sediment intruded by igneous rocks
C. Maganiferous beds resemble the Archean complex
D. Maganniferous beds resemble the pre-cambrian iron formation

(SAGAR UNIVERSITY-2015)

524. Contacts between parallel layers formed by periods of non deposition is called:

A. Paraconformity
B. Angular unconformity
C. Non-conformity
D. Disconformity

(SAGAR UNIVERSITY-2015)

525. Large intrusive body with steep walls lackes in visible floor is called as:

A. Conolith B. Lopoliths
C. Phacoliths D. Batholith

(SAGAR UNIVERSITY-2015)

526. With time, sediment particles transported by rivers become:

A. Platy and smaller
B. Platy and larger
C. Rounded and larger
D. Rounded and smaller

(SAGAR UNIVERSITY-2015)

527. Karst topography is associated with which of the following rocks?

A. Silicate B. Felsic
C. Carbonate D. Clay

(SAGAR UNIVERSITY-2015)

528. Cavities in lava flows formed by volatile escape and filled with secondary minerals is called:

A. Pumice B. Amygdules
C. Oolites D. Pisolites

(SAGAR UNIVERSITY-2015)

529. Processes of metamorphism dominates:

A. Below the zone of weathering
B. Below the zone of deposition
C. Below the base level
D. Above the zone of weathering

(SAGAR UNIVERSITY-2015)

530. Mylonite is associated with which type of metamorphism?
A. Plutonic
B. Cataclastic
C. Thermal
D. Dynamothermal
(SAGAR UNIVERSITY-2015)

531. Metamorphism of wood produces:
A. Lignite B. Sapropal
C. Graphite D. Charcoal
(SAGAR UNIVERSITY-2015)

532. Current bedding is a type of:
A. Inclined bedding
B. Discordant bedding
C. Planar bedding
D. Concordant bedding
(SAGAR UNIVERSITY-2015)

533. Extrusive igneous rocks are characterized by:
A. Rapid cooling and coarse grain
B. Rapid cooling and fine grain
C. Slow cooling and fine grain
D. Slow cooling and coarse grains
(SAGAR UNIVERSITY-2015)

534. Carbonate rocks after metamorphosed gives rise to:
A. Mylonite B. Migmatite
C. Marble D. Marl
(SAGAR UNIVERSITY-2015)

535. Metamorphic aureole is mostly associated with which magma types?
A. Basaltic magma
B. Granitic magma
C. Dacite magma
D. Acid magma
(SAGAR UNIVERSITY-2015)

536. Palingenesis starts at a temperature is about:
A. 1400 °C B. 1500 °C
C. 1000 °C D. 800 °C
(SAGAR UNIVERSITY-2015)

537. Eclogites is a:
A. Olivine rock
B. Olivine and pyroxene
C. Garnet and pyroxene rocks
D. Garnet and olivine rocks
(PUNJAB UNIVERSITY-2015)

538. Geochemical data on the Dhanjori volcanic of the Singhbhum craton indicate:
A. Basaltic in nature
B. Rhyolitic in nature
C. Ranging from basalt to andesite
D. Ranging from rhyolitic to rachitic
(PUNJAB UNIVERSITY-2015)

539. The processes of conversion of glassy materials to crystallised material is:
A. Crystallisation
B. Devitrification
C. Eutectic processes
D. Assimilation
(PUNJAB UNIVERSITY-2015)

540. A rock made up of euhedral crystal, the texture is described as:
A. Allotrimorphic B. Hypidiomorphic
C. Panidomorphic D. None of these
(PUNJAB UNIVERSITY-2015)

541. Ophitic texture is common shown by:
A. Trachyte B. Lamprophyres
C. Andesite D. Dolerite
(PUNJAB UNIVERSITY-2015)

542. A common decomposition in basaltic rocks involve:
A. Oxidation of Mg
B. Oxidation of Fe
C. Hydration of Fe
D. Hydration of K
(PUNJAB UNIVERSITY-2015)

543. Match the following:

List-A	***List-B***
P. A course grained light coloured rock	1. Rhyolite
Q. A course grained dark coloured rock	2. Gabbro
R. A fine grained light coloured rocks	3. Granite
S. A fine grained dark coloured rocks	4. Basalt

A. P-4, Q-2, R-3, S-1
B. P-3, Q-2, R-1, S-4
C. P-2, Q-1, R-4, S-3
D. P-3, Q-2, R-4, S-1
(PUNJAB UNIVERSITY-2015)

544. The first minerals to crystallize from a completely or almost anhydrous melt at high temperature are called:
A. Hydatogenetic minerals
B. Pyrogenetic minerals
C. Pegmatitic minerals
D. Orthomagmatic minerals
(PUNJAB UNIVERSITY-2015)

545. Match the following:

List-A	*List-B*
P. Leucocratic	1. Dark coloured rocks
Q. Mesocratic	2. Light coloured rocks
R. Melanocratic	3. Extremely dark coloured rocks
S. Hypermelan-ocratic	4. Intermediate in colour

A. P-2, Q-4, R-1, S-3
B. P-3, Q-2, R-4, S-1
C. P-1, Q-2, R-3, S-4
D. P-4, Q-3, R-2, S-1
(PUNJAB UNIVERSITY-2015)

546. The temperature range for which the Bowen's reaction series has been worked out is?
A. 1200 °C to 635 °C
B. 1100 °C to 450 °C
C. 1100 °C to 573 °C
D. 900 °C to 573 °C
(PUNJAB UNIVERSITY-2015)

547. Match the following:

List-A	*List-B*
P. Laccolith	1. Saucer shaped intrusive body
Q. Lopolith	2. Cresentic shaped igneous body
R. Bysmalith	3. Convex top and flat bottom
S. Phacolith	4. Broken intrusive body

A. P-3, Q-1, R-4, S-2
B. P-2, Q-1, R-4, S-3
C. P-1, Q-2, R-3, S-4
D. P-4, Q-3, R-2, S-1
(PUNJAB UNIVERSITY-2015)

548. Current bedding are formed due to:
A. Frequent changes are formed velocity of the flow of the stream
B. Frequent change in direction of the flow of the stream
C. Frequent change in velocity and direction of the flow of the stream
D. None of these
(PUNJAB UNIVERSITY-2015)

549. Variation of the paleocurrent direction is minimum in the deposit formed by:
A. Glacial processes
B. Fluvial processes
C. Aeolian processes
D. Beach processes
(PUNJAB UNIVERSITY-2015)

550. A thickness map which show drilled intervals, irrespective of the amount of dip, is called:
A. Isopach map B. Isochore map
C. Isobar map D. Isolith map
(PUNJAB UNIVERSITY-2015)

551. Quartz sandstone is said mature when it is having :
A. Well rounded, poor sorted quartz grain
B. Poor-rounded, poor sorted quartz grain
C. Well rounded, well sorted quartz grain
D. Poor rounded, well sorted quartz grain
(PUNJAB UNIVERSITY-2015)

552. If a body of sea water is totally evaporated, what sequence of rock types result ?
A. Calcite, gypsum, anhydrite and halite
B. Gypsum, calcite, halite and anhydrite
C. Halite, gypsum, anhydrite and calcite
D. Gypsum, anhydrite, halite and calcite
(PUNJAB UNIVERSITY-2015)

553. Greywacke is dark in colour and rich in clay rock fragment and contain less of quartz sand grains, which are:
A. Rounded fine and well sorted
B. Angular fine and well sorted
C. Rounded, course and unsorted
D. Angular, course and unsorted
(PUNJAB UNIVERSITY-2015)

554. Nappe structure can be formed by:
A. Overthrusting
B. Recumbent folding
C. Normal faulting
D. Both A & B *(PUNJAB UNIVERSITY-2015)*

555. Which term will be more appropriate for the slate rock?
A. Slaty cleavage
B. Schistosity
C. Continues cleavage
D. None of these
(PUNJAB UNIVERSITY-2015)

556. Choose the correct statement regarding current ripple marks:
A. The rounded trough is convex toward the older beds
B. The rounded trough is convex toward the younger beds
C. The upper portion of the ripple mark indicate younger beds
D. The current marks cannot be used to determine top and bottom
(PUNJAB UNIVERSITY-2015)

557. Which type of cross-bedding cannot used to tell top and bottom?
A. Tabular cross-bedding
B. Planar cross bedding
C. Lenticular cross bedding
D. Wedge shape crossbedding
(PUNJAB UNIVERSITY-2015)

558. Ocean Island Basalt (IOB) form at:
A. Divergent oceanic plate
B. Convergent oceanic plate
C. Intra-oceanic plate
D. Transform oceanic plate *(AMU-2016)*

559. Calk-alkaline andesite are related to:
A. Melting at great depth
B. Melting at shallow depth
C. Melting at intermediate depth
D. None of the above *(AMU-2016)*

560. Hawaii Island is related to:
A. Subducting plate B. Plume
C. Plate collision D. All the above
(AMU-2016)

561. Among the different kinds of flood basalt, the SiO_2 content is highest in:
A. Continental tholeiite
B. Oceanic tholeiite
C. Alkaline olivine tholeiite
D. Spilites *(AMU-2016)*

562. A-type granites are typically associated with:
A. Subduction zone
B. Island Arc
C. Continental rift zone
D. Continent–continent collision
(AMU-2016)

563. The variation diagram in which the weight percentage of SiO_2 is ploted on the X-axis and the concentration of other oxides on the Y-axis is describe as:
A. Cartesian diagram
B. AFM diagram
C. QAP diagram
D. Harker diagram *(AMU-2016)*

564. Texture in which intergrowth of pyroxene and plagioclase occurs:
A. Porthyritic B. Poikilitic
C. Ophitic D. Myrmekite
(AMU-2016)

565. Match the correct solid solution series of plagioclase:
A. Anorthite – Bytownite – Labradorite – Andesine – Oligoclase – Albite
B. Anorthite – Bytownite – Andesine – Labradorite – Oligoclase – Albite
C. Albite – Bytownite – Andesine – Labradorite – Oligoclase – Anorthite
D. Anorthite – Bytownite – Andesite – Oligoclases – Labradorite – Albite
(AMU-2016)

566. TTGs are
A. Highly fractionated, Na_2O rich, LILE enriched HREE depleted rocks
B. Highly fractionated, Na_2O poor, LILE enriched HREE depleted rocks
C. Highly fractionated, Na_2O rich, LILE depleted HREE enriched rocks
D. Highly fractioned, Na_2O rich, LILE enriched HREE enriched rocks
(AMU-2016)

567. The most characteristic rocks of the Island – Arc system are
A. Andesites B. Granodiorite
C. Blueschist D. Basalts
(AMU-2016)

568. Which of the following characteristics oceanic trenches?
A. High heat flow
B. Positive gravity anomaly
C. Reverse magnetic polarity
D. Low heat flow *(AMU-2016)*

569. Which of the following is the correct sequence of ophiolites from top to bottom?
A. Layered gabbro – pillow basalt – sheeted dykes – peridotite
B. Peridotite –sheeted dyke – pillow basalt – layered gabbro
C. Pillow basalt – sheeted dyke – layered gabbro – peridotite
D. Pillow basalt – layered gabbro – sheetted dyke – peridotite *(AMU-2016)*

570. Ovoid shaped K-feldspar surrounded by overgrowth of plagioclase is:
A. Spinifex texture
B. Ophitic texture
C. Rapakivi texture
D. Porphyritic texture *(AMU-2016)*

571. Hornfels is characteristics rock of:
A. Cataclasic metamorphism
B. Dynamic metamorphism
C. Contact metamorphism
D. Regional metamorphism *(AMU-2016)*

572. Which one does not belong to Barrovian Sequence?
A. Chlorite Zone B. Cordierite Zone
C. Staurolite Zone D. Sillimanite Zone *(AMU-2016)*

573. Which one of the contact metamorphic facies?
A. Blue schist facies
B. Green schist facies
C. Eclogite facies
D. Sanidinite facies *(AMU-2016)*

574. Omphacite + Pyrope are the characteristics minerals of:
A. Amphibolite facies
B. Granulite facies
C. Eclogite facies
D. Serpentinite facies *(AMU-2016)*

575. Hypersthene bearing high grade metamorphic rocks is:
A. Charnokite B. Khondalite
C. Granulite D. Eclogite *(AMU-2016)*

576. A rock which does not show the primarily cohesion
A. Mylonite B. Cataclasite
C. Pseudotachylite D. Fault Gouge *(AMU-2016)*

577. High pressure facies:
A. Green schist facies
B. Granulite facies
C. Eclogite facies
D. Zeolite facies *(AMU-2016)*

578. Glaucophane + Lawsonite/Epidote + Albite ± Chlorite is a characteristic mineral aasemblage of which of the following metamorphic facies:
A. Eclogite B. Greenschist
C. Blueschist D. Amphibolite *(AMU-2016)*

579. Allochems in limestone consist of:
A. Micrite B. Speroid
C. Ooids only D. Biochemical ooids *(AMU-2016)*

580. Which type of cross bedding is definite indicator of fluvial environments?
A. Epsilon cross bedding
B. Herring bone bedding
C. Hummocky cross bedding
D. Trough cross bedding *(AMU-2016)*

581. A salt dome is characterised by:
A. Low velocity and low density
B. Low velocity and high density
C. High velocity and high density
D. High velocity and low density *(AMU -2016)*

582. If the diameter of a sand particles is 1mm, the equivalent phi is
A. –1 B. 0
C. 1 D. 2 *(AMU-2016)*

583. Lower limit of temperature of metamorphism is:
A. < 70 °C B. 100 – 150 °C
C. 400 – 500 °C D. > 500 °C *(AMU-2016)*

584. Which of the following can be used to determine paleo-current direction?
A. Wave ripple B. Mud cracks
C. Current ripple D. Parting lineation
(ISM-2016)

585. Rock melt at high temperature because with increasing temperature
A. Entropy of rock increase more than the entropy of melt
B. Entropy of melt increase more than the entropy of rock
C. Entropy of rock and melt both decreases
D. Temperature has no relation with the G of the two phase
(ISM-2016)

586. A cumulative grain size analysis data of a well sorted sandstone will plot in a probality paper as:
A. Bell shaped curve
B. Sigmoidal curve
C. Straight line with positive slope
D. Straight line with negative slope
(ISM-2016)

587. In froud No. (F) > 1, we get
A. Out of phase flow
B. Rapid flow
C. Sluggish flow
D. In-phase flow *(ISM-2016)*

588. Relicts the following, the highest porosity value can be expected from:
A. Palimpsest texture
B. Nematoblastic texture
C. Lepidoblastic texture
D. Decaussate texture *(ISM-2016)*

589. ———— of metamorphism is formed at high temperature and low pressure:
A. Eclogite facies
B. Greenschist facies
C. Zeolite facies
D. Sanidinite facies *(ISM-2016)*

590. Large garnet with numerous inclusion giving it a sieve like appearance occurs in a schist made up of matrix of fine grained biotite, muscovite and quartz. This texture is known as:
A. Granoblastic B. Poikiloblastic
C. Porphyroblastic D. Porphyritic
(ISM-2016)

591. The sequence of crystalline in a cooling calc alkaline basaltic magma is likely to be
A. Forsterite – diopside – enstatite
B. Enstatite – forsterite – enstatite
C. Enstatite – diopside – forsterite
D. Diopside – forsterite – enstatite
(ISM-2016)

592. Granitic magma produced by partial melting becomes mobile if the melt volume is approximately
A. 1% B. 5%
C. 10% D. 30% *(ISM-2016)*

593. The igneous rock filling in the shaded field of the figure below is:

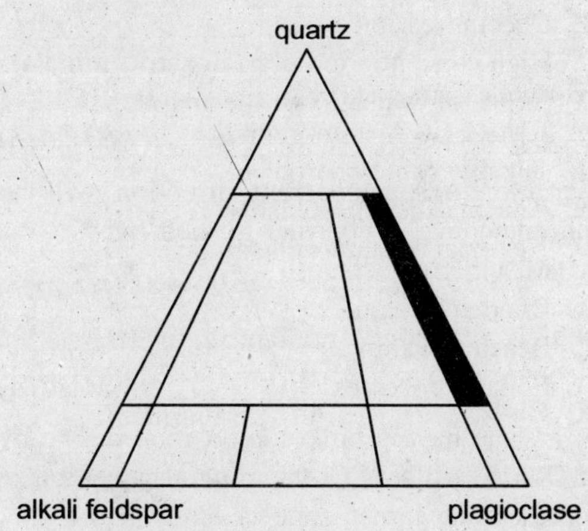

A. Monzonite
B. Plagioclase Gabbro
C. Tonalite
D. Syenite *(ISM-2016)*

594. Out of the following, which one is the characteristic minerals of Blueschist facies metamorphism?
A. Omphacite B. Plagioclase
C. Muscovite D. Glaucophane
(ISM-2016)

595. Goldschmidt mineralogical phase rule states that number of phase is equal to number of components. The underlying assumption is
A. It is degenerating system
B. It is adiabatic system
C. Degree of freedom in the system is 2
D. It is valid only in one components system *(ISM-2016)*

596. Staurolite > cordierite > andalusite > sillimanite > ....
A. Represent a Barrovian metamorphic sequence
B. Represent Buchan zone
C. Represent a typical mafic protolith
D. Representing decreasing metamorphic grade *(ISM-2016)*

597. Partial melting of peridotite beneath mid oceanic ridge is caused by
A. Frictional heating
B. Decompression
C. High concentration of radioactive minerals
D. Influx of water in to the mantle *(ISM-2016)*

598. Variation of paleocurrent direction will be minimum in the deposits formed by:
A. Meandering stream
B. Braided stream
C. Beach stream
D. Eolian processes *(ISM-2016)*

599. In a shale having bulk composition define by K_2O-Al_2O_3-SiO_2-H_2O, the minerals assemblage representing garnet zone is likely to be
A. Garnet – muscovite – quartz
B. Kyanite – muscovite – quartz
C. Garnet – kyanite – muscovite – quartz
D. Biotite – kyanite – muscovite – quartz *(ISM-2016)*

600. This facies is absent in low-pressure series metamorphic rocks:
A. Pyroxene hornfels facies
B. Sanidinite facies
C. Epidote-amphibolite facies
D. Amphibolite facies *(ISM-2016)*

601. In term of increasing flow energy, primarily sedimentary structures can be arranged as
A. Flute marks – cross stratification – parting lineation
B. Parting lineation – flute marks – cross stratification
C. Cross stratification – parting lineation – flute marks
D. Flute marks – parting lineation – cross stratification *(ISM -2016)*

602. A flow with a velocity 1 cm/sec will turn turbulent if the flow depth is:
A. 2 cm B. 20 cm
C. 200 cm D. 2000 cm *(ISM-2016)*

603. Lithification is the primary process in which the formation of one of the following
A. Gneiss B. Granite
C. Marble D. Conglomerate *(ISM-2016)*

604. Schist represents a material which has mechanical properties as:
A. Homogeneous and isotropic
B. Homogeneous and anisotropic
C. Inhomogeneous and anisotropic
D. Inhomogeneous and isotropic *(BHU-2016)*

605. Where do the ophiolite suit of rocks occur?
A. Indus suture zone
B. Lesser Himalaya
C. Brahmaputra valley
D. Himalayan foot hill zone *(BHU-2016)*

606. Primary magma:
A. Is formed by magmatic differentiation
B. Is a magma as it exist immediately after separation from its source region
C. Is formed by contamination
D. Is a fractionated magma *(BHU-2016)*

607. Norite is essentially composed of:
A. Clinopyroxene and plagioclage feldspar
B. Orthopyroxene and olivine
C. Only orthopyroxene
D. Orthopyroxene and plagioclase feldspar *(BHU-2016)*

608. Alkali granite is essentially composed of:
A. Alkali feldspar and quartz only
B. Alkali and plagioclase feldspar and quartz
C. Alkali and plagioclase feldspar, quartz and alkali mafic minerals
D. Plagioclase feldspar, quartz and hornblende
(BHU-2016)

609. Plutonic equivalent of phonolite is:
A. Mela nepheline syenite
B. Granite
C. Syenite
D. Gabbro *(BHU-2016)*

610. Alkali feldspar syenite is essentially composed of:
A. Both alkali and plagioclase feldspar
B. Minerals need microscope to see
C. Only alkali feldspar
D. Plagioclase feldspar and biotite
(BHU-2016)

611. What is phaneritic texture in igneous rocks?
A. Minerals are tool large to see with naked eye
B. Minerals need microscope to see
C. Glassy in nature
D. Very fine-grained groundmass
(BHU-2016)

612. Mafic minerals in igneous rocks are:
A. Light in colour due to present of Al and Si
B. Dark in colour due to presence of Fe and Mg
C. Dark in colour due to presence of Al and Si
D. Dark in colour due to presence of Si and Fe
(BHU-2016)

613. Colour Index (CI) of a melanocratic igneous rock varies between:
A. 0 and 50 B. 35 and 65
C. 20 and 65 D. 65 and 100
(BHU-2016)

614. Silica contents in a felsic igneous rock is:
A. Always less than 55%
B. Always more than 66%
C. Varies between 55% and 66%
D. Varies between 45% and 52%
(BHU-2016)

615. In a peraluminous igneous rocks:
A. $Al_2O_3 > CaO + N_2O + K_2O$
B. $Al_2O_3 < CaO + Na_2O + K_2O$
C. $Al_2O_3 = CaO + Na_2O + K_2O$
D. $Al_2O_3 > CaO + K_2O$
(BHU-2016)

616. Which of the following is not an ultramafic igneous rocks?
A. Troctolite B. Komatiite
C. Peridotite D. Dunite
(BHU-2016)

617. Which of the following is a best criterion to classify plutonic igneous rocks?
A. Texture
B. Essential mineral constituents
C. Nature of groundmass
D. Mode of occurrence *(BHU-2016)*

618. Minerals composition of websterite is:
A. Orthopyroxene and olivine
B. Clinopyroxene and olivine
C. Orthopyroxene and clinopyroxene
D. Hornblende and biotite *(BHU-2016)*

619. Following oxides are used in TAS classification:
A. Al_2O_3 and SiO_2
B. MgO, CaO and SiO_2
C. Na_2O, K_2O and MgO
D. SiO_2, Na_2O and K_2O *(BHU-2016)*

620. Granularity in an igneous rock defines:
A. Degree of crystallization
B. Absolute and relative sizes of crystals
C. Relationship between crystal and groundmass
D. Shape of crystal *(BHU-2016)*

621. Course grained sediments are transported by:
A. Traction processes
B. Saltation process
C. Suspension process
D. None of these *(BHU-2016)*

622. Particle size range of arenaceous rocks is:
A. 2 – 4 mm B. 1/16 – 2 mm
C. 1/256 – 1/16 mm D. <1/256 mm
(BHU-2016)

623. Arkoses are sandstones which contain?
A. Appreciable amount of feldspar
B. Low percentage of feldspar
C. Negligible amount of feldspar
D. No feldspar
(BHU-2016)

624. Sandstone containing > 15% matrix and < 75% quartz are classified as:
A. Arkose B. Greywacke
C. Lithic arenite D. Quartz arenite
(BHU-2016)

625. Lithic arenites are:
A. Quartz rich sandstone
B. Feldspar rich sandstone
C. Rock fragments rich sandstone
D. Clay rich sandstone *(BHU-2016)*

626. Claystones contain:
A. < 33% clay B. 33 – 66% clay
C. > 66% clay D. > 100% clay
(BHU-2016)

627. Argillaceous sediments containing clay between 33 – 66% are know as:
A. Mudstone B. Siltstone
C. Claystone D. Dropstone
(BHU-2016)

628. Micrites are orthochems containing crystals:
A. < 4 μm B. 4 – 10 μm
C. > 10 μm D. > 100 μm
(BHU-2016)

629. Oolites form in:
A. Agitated water B. Calm water
C. Cool water D. None of these
(BHU-2016)

630. Large scale metamorphism associated with increases in both temperature and pressure best defines:
A. Contact metamorphism
B. Burial metamorphism
C. Parting melting
D. Regional metamorphism *(BHU-2016)*

631. Alkali amphibole :
A. Glaucophane B. Albite
C. Muscovite D. Jadeite *(BHU-2016)*

632. Which one of the following is not an agent of metamorphism?
A. Fluid
B. Time
C. Rock type
D. Heat-temperature *(BHU-2016)*

633. What types of metamorphism is associated with the zeolite metamorphic facies?
A. High grade
B. Low grade
C. Rock melt
D. Intermediate grade *(BHU-2016)*

634. Which of the following metamorphic facies is characterized by the pyrope rich garnet + omphacite assemblage:
A. Blue schist B. Granulite
C. Eclogite D. Greenschist
(BHU-2016)

635. Migmatite is an example of:
A. Non foliated rock
B. Weakly foliated rock
C. Strongly foliated
D. None of above *(BHU-2016)*

636. Ca-poor pyroxene is:
A. Pigeonite B. Omphacite
C. Enstatitte D. Jadeite
(BHU-2016)

637. A amphibole which is characteristic of high temperature?
A. Edenite B. Hornblende
C. Tremolite D. Paragasite
(BHU-2016)

638. A phase diagram with a specific bulk composition is known as:
A. Isograde diagram
B. AFM diagram
C. ACF diagram
D. Pseudosection *(BHU-2016)*

639. Characteristic minerals of charnockites:
A. Hypersthene B. Pyroxene
C. Garnet D. Sillimanite
(BHU-2016)

640. Which is the typical contact zone metamorphic reaction?
A. Chlorite + muscovite = Garnet + biotite + quartz + H_2O
B. Chlorite + K-feldspar = Biotite + quartz + H_2O
C. Chlorite + Muscovite = Cordierite + quartz + H_2O
D. Kyanite = Sillimanite *(DU-2016)*

641. Magmas move upward because of:
A. Lower density and higher volatile content than the adjacent solid rocks
B. Higher silica content than the surrounding rocks
C. Higher mafic minerals content than the surrounding rocks
D. Higher the viscosity than the surrounding rocks
(DU-2016)

642. ______ is a common ore of uranium:
A. Azurite B. Celadonite
C. Pitchblend D. Cubinite *(DU-2016)*

643. Plutonic and their volcanic equivalent are given respectively, match the following:

List-A	***List-B***
1. Gabbro	P. Dacite
2. Granite	Q. Tholeiitic basalt
3. Monzosyenite	R. Tapherite
4. Granodiorite	S. Phonlite
5. Monzodiorite	T. Rholite

A. 1-Q, 2-T, 3-S, 4-P, 5-R
B. 1-R, 2-S, 3-T, 4-P, 5-Q
C. 1-S, 2-T, 3-Q, 4-R, 5-P
D. 1-Q, 2-T, 3-S, 4-P, 5-R
(DU-2016)

644. Bowl shaped, large shallow mafic intrusion with inward gentle slippling called?
A. Laccolith B. Phacolith
C. Chonolith D. Lopolith
(DU-2016)

645. Sole of the rocks exposed in a orogenic lithosphere is emplaced as continental minerals as:
A. Spilite B. Bonzite
C. Sphalerite D. Cipholite
(DU-2016)

646. In binary system, two crystallization curves joining a point where simultaneous crystallization of two minerals components makes place is called as:
A. Peritectic point
B. Eutectic point
C. Cotectic point
D. None of the above *(DU-2016)*

647. Rf factor is associated with which of the following?
A. Chromatography
B. AAS
C. XRD
D. XRF *(DU-2016)*

648. Pyrolite is
A. Is a rock found in mantle
B. Is a rocks found in continental crust
C. Is a model composition
D. In a rock found on ocean floor
(DU-2016)

649. U-Pb zircon datting is very useful to date:
A. Basaltic rocks
B. Granitic rocks
C. Ultramafic rocks
D. All the above *(DU-2016)*

650. Positive Eu anomaly in a rock indicaters:
A. Fractionation of plagioclase
B. Accumulation of plagioclase
C. Fractionation of olivine
D. Accumulation of olivine + pyroxene
(DU-2016)

651. Spinifex texture indicates:
A. Very low cooling of magma in plutonic condition
B. Rapid cooling of magma as lava flows
C. Intermediate cooling in a dyke
D. None of the above *(DU-2016)*

ANSWER KEY

1	**2**	**3**	**4**	**5**	**6**	**7**	**8**	**9**	**10**
D	C	B	D	C,D	C	B	C	D	B
11	**12**	**13**	**14**	**15**	**16**	**17**	**18**	**19**	**20**
B	D	A	D	D	A	A	A	B	B

21	**22**	**23**	**24**	**25**	**26**	**27**	**28**	**29**	**30**
C	C	A	C	D	B	B	A	B	B
31	**32**	**33**	**34**	**35**	**36**	**37**	**38**	**39**	**40**
B	C	A	A	B	A	D	D	C	C
41	**42**	**43**	**44**	**45**	**46**	**47**	**48**	**49**	**50**
B	D	C	C	A	C	C	B	D	C
51	**52**	**53**	**54**	**55**	**56**	**57**	**58**	**59**	**60**
D	A	B	B	C	C	B	B	D	A
61	**62**	**63**	**64**	**65**	**66**	**67**	**68**	**69**	**70**
A	D	D	B	D	A	D	A	D	C
71	**72**	**73**	**74**	**75**	**76**	**77**	**78**	**79**	**80**
D	A	C	B	B	C	A	D	C	B
81	**82**	**83**	**84**	**85**	**86**	**87**	**88**	**89**	**90**
D	B	C	D	D	B	C	D	A	B
91	**92**	**93**	**94**	**95**	**96**	**97**	**98**	**99**	**100**
B	D	D	B	B	B	C	B	A	B
101	**102**	**103**	**104**	**105**	**106**	**107**	**108**	**109**	**110**
C	C	B	B	A	B	A	A	B	B
111	**112**	**113**	**114**	**115**	**116**	**117**	**118**	**119**	**120**
D	A	C	C	C	B	D	A	A	D
121	**122**	**123**	**124**	**125**	**126**	**127**	**128**	**129**	**130**
B	C	B	D	C	B	A	A	C	C
131	**132**	**133**	**134**	**135**	**136**	**137**	**138**	**139**	**140**
A	A	B	B	C	B	D	C	A	A
141	**142**	**143**	**144**	**145**	**146**	**147**	**148**	**149**	**150**
D	A	A	A	D	D	C	A	A	C
151	**152**	**153**	**154**	**155**	**156**	**157**	**158**	**159**	**160**
A	C	B	B	D	B	A	C	A	A
161	**162**	**163**	**164**	**165**	**166**	**167**	**168**	**169**	**170**
B	A	C	C	D	B	C	A	D	A
171	**172**	**173**	**174**	**175**	**176**	**177**	**178**	**179**	**180**
D	A	B	B	C	B	D	A	A	C
181	**182**	**183**	**184**	**185**	**186**	**187**	**188**	**189**	**190**
B	D	A	A	D	A	A	A	B	B
191	**192**	**193**	**194**	**195**	**196**	**197**	**198**	**199**	**200**
C	A	B	D	C	A	C	D	B	D
201	**202**	**203**	**204**	**205**	**206**	**207**	**208**	**209**	**210**
D	C	C	B	B	A	C	B	D	A
211	**212**	**213**	**214**	**215**	**216**	**217**	**218**	**219**	**220**
B	D	C	B	B	C	A	B	A	C

221	222	223	224	225	226	227	228	229	230
A	C	C	C	B	B	A	C	C	A
231	**232**	**233**	**234**	**235**	**236**	**237**	**238**	**239**	**240**
A	A	A	A	C	D	B	A	A	C
241	**242**	**243**	**244**	**245**	**246**	**247**	**248**	**249**	**250**
A	A	C	A	A	A	C	B	B	C
251	**252**	**253**	**254**	**255**	**256**	**257**	**258**	**259**	**260**
D	C	B	B	A	A	B	A	D	C
261	**262**	**263**	**264**	**265**	**266**	**267**	**268**	**269**	**270**
D	C	C	B	D	D	C	A	B	D
271	**272**	**273**	**274**	**275**	**276**	**277**	**278**	**279**	**280**
B	D	A	A	A	B	A	D	C	A
281	**282**	**283**	**284**	**285**	**286**	**287**	**288**	**289**	**290**
B	A	A	B	B	B	A	A	A	A
291	**292**	**293**	**294**	**295**	**296**	**297**	**298**	**299**	**300**
A	C	C	B	B	C	D	B	D	A
301	**302**	**303**	**304**	**305**	**306**	**307**	**308**	**309**	**310**
A	D	B	C	D	A	A	A	B	C
311	**312**	**313**	**314**	**315**	**316**	**317**	**318**	**319**	**320**
A	C	B	A	B	A	A	A	B	A
321	**322**	**323**	**324**	**325**	**326**	**327**	**328**	**329**	**330**
A	B	C	B	C	B	C	A	B	B
331	**332**	**333**	**334**	**335**	**336**	**337**	**338**	**339**	**340**
C	A	C	D	D	B	C	A	C	C
341	**342**	**343**	**344**	**345**	**346**	**347**	**348**	**349**	**350**
B	A	C	C	A	B	C	A	C	B
351	**352**	**353**	**354**	**355**	**356**	**357**	**358**	**359**	**360**
A	B	B	A	D	C	A	B	D	C
361	**362**	**363**	**364**	**365**	**366**	**367**	**368**	**369**	**370**
B	D	C	B	D	C	A	C	C	D
371	**372**	**373**	**374**	**375**	**376**	**377**	**378**	**379**	**380**
A	B	B	A	C	A	B	D	C	C
381	**382**	**383**	**384**	**385**	**386**	**387**	**388**	**389**	**390**
B	C	A	B	A	A	D	A	A	A
391	**392**	**393**	**394**	**395**	**396**	**397**	**398**	**399**	**400**
C	D	B	A	C	C	A	A	B	A
401	**402**	**403**	**404**	**405**	**406**	**407**	**408**	**409**	**410**
B	D	D	C	D	C	B	B	A	B
411	**412**	**413**	**414**	**415**	**416**	**417**	**418**	**419**	**420**
A	D	A	D	D	C	B	C	D	A
421	**422**	**423**	**424**	**425**	**426**	**427**	**428**	**429**	**430**
C	C	C	B	B	B	C	A	C	B

431	432	433	434	435	436	437	438	439	440
B	D	C	C	B	B	D	B	C	C
441	**442**	**443**	**444**	**445**	**446**	**447**	**448**	**449**	**450**
B	D	D	C	C	C	C	A	A	A
451	**452**	**453**	**454**	**455**	**456**	**457**	**458**	**459**	**460**
A	C	D	C	D	A	C	A	A	A
461	**462**	**463**	**464**	**465**	**466**	**467**	**468**	**469**	**470**
C	C	D	C	D	A	B	D	A	D
471	**472**	**473**	**474**	**475**	**476**	**477**	**478**	**479**	**480**
D	D	B	D	C	B	B	B,C,D	A	A
481	**482**	**483**	**484**	**485**	**486**	**487**	**488**	**489**	**490**
A	A	A	A	C	D	C	C	C	D
491	**492**	**493**	**494**	**495**	**496**	**497**	**498**	**499**	**500**
C	A	D	C	B	A	D	B	A	A
501	**502**	**503**	**504**	**505**	**506**	**507**	**508**	**509**	**510**
C	D	B	C	A	C	B	B	A	A
511	**512**	**513**	**514**	**515**	**516**	**517**	**518**	**519**	**520**
A	D	D	B	D	C	C	D	C	D
521	**522**	**523**	**524**	**525**	**526**	**527**	**528**	**529**	**530**
B	C	C	D	B	D	C	B	D	D
531	**532**	**533**	**534**	**535**	**536**	**537**	**538**	**539**	**540**
B	C	B	C	B	D	C	B	B	C
541	**542**	**543**	**544**	**545**	**546**	**547**	**548**	**549**	**550**
D	A	B	B	A	C	A	C	A	B
551	**552**	**553**	**554**	**555**	**556**	**557**	**558**	**559**	**560**
C	A	D	A	A	A	C	C	C	B
561	**562**	**563**	**564**	**565**	**566**	**567**	**568**	**569**	**570**
A	C	D	C	A	A	A	D	C	C
571	**572**	**573**	**574**	**575**	**576**	**577**	**578**	**579**	**580**
C	B	D	C	A	D	C	C	D	B
581	**582**	**583**	**584**	**585**	**586**	**587**	**588**	**589**	**590**
D	B	B	C	A	C	B	A	D	B
591	**592**	**593**	**594**	**595**	**596**	**597**	**598**	**599**	**600**
C	C	C	D	C	D	B	C	A	D
601	**602**	**603**	**604**	**605**	**606**	**607**	**608**	**609**	**610**
D	D	D	B	A	B	D	C	A	B
611	**612**	**613**	**614**	**615**	**616**	**617**	**618**	**619**	**620**
A	B	D	B	A	A	B	C	D	B
621	**622**	**623**	**624**	**625**	**626**	**627**	**628**	**629**	**630**
A	B	A	B	C	C	A	A	A	D
631	**632**	**633**	**634**	**635**	**636**	**637**	**638**	**639**	**640**
A	C	B	C	B	A	B	D	A	C
641	**642**	**643**	**644**	**645**	**646**	**647**	**648**	**649**	**650**
A	C	A	D	D	B	A	C	B	B
651									
B									

HINTS AND SOLUTIONS

1. Sandstone	Matrix%
Siltstone	Absent
Mudstone	<10%
Greywacke	15-75%
Arenite	<15%

2. Metamorphic facies	Stable minerals	P/T
Greenschist facies	Epidote, Chalorite, Albite	LowT/ LowP
Blueschist facies	Hornblande, Plagioclage	LowT/ ModerateP
Eclogite facies	Coesite, Garnet, Omphacite	HighT/ HighP
Granulite facies	Opx-Cpx-Plagioclage	HighT/ ModerateP

3. Sedimentary rocks	Characteristics	Remarks
Sandstone	Formed by Sand	1/16-2 mm.
Arkose	Rich in feldspar	>25% feldspar.
Limestone	Formed by chemical process.	Chemical sediment.
Shale	Formed by clay.	Clay size <1/16 mm.

4. Mineral assemblage	Metamorphic facies	P/T
Chalorite + Actinolite + Albite	Green schist facies.	LowP/ LowT
Lowsonite + Glaucophane + Epidote	Glacophane. lowsonite facies	LowP/ LowT
Orthopyroxene + Clinopyroxene + Plagioclase	Granulite facies.	HighT/ ModerateP
Hornblende + Plagioclase	Amphibolite facies.	ModT/ ModP

6. Sedimentary structure	Application
Planer cross bedding	Flow direction
Asymmetrical ripple marks	Flow direction
Symmetrical ripple marks	Top and bottom
Oscillation ripple	Top and bottom

8. Rocks type	Particle size
Conglomerate	>2 mm
Sandstone	2-1/16 mm
Siltstone	1/16- 1/256 mm

13. Rocks type	Origin setting	Composition
Shoshonite	Active continental margin	Olivine-augite-plagioclase
Carbonatite	Continental rift	> 50% carbonate minerals
Boninite	Subduction zone	Pyroxene-olivine-andesite
Trachyte	Continental rifting	Alkali-feldspar-quartz-feldspar

15. Plutonic rocks	Volcanic equivalent
Syenite	Trachyte
Lherzolite	Ijolite
Gabbro	Basalt
Granodiorite	Dacite

18. Rock type	Characteristics
Charnokites	Course grained gneissose rock rich in hypersthene
Khondolite	Garnet-sillimnite-schist rocks
Kodurite	Manganese rich gneiss rocks
Gondite	Manganese bearing gneiss rocks

25. Diagenesis	Example (cement)
Seafloor spreading	Red cement
Vadose diagensis	Dolomitic cement
Phreatic diagenesis	Dolomitic cement
Deep burial diagenesis	Ferroan calcite cement

31. Rocks	Example (Anomaly)
Basalt	Large negative Eu anomaly
Granite	Negative Eu anomaly
TTG suit of rocks	Absent Eu anomaly
Lunar basalt	Large negative Eu anomaly

32. Processes	Rocks type
Continent-continent collision	Granite
Ocean-continent collision	Basalt
Ocean-ocean collision	Andesite
Continental rift	Carbonatites

35. Types of Metamorphism	Represent (Texture/Structure)
Regional metamorphism	Gneissose structure
Dynamic metamorphism	Mylonite
Burial metamorphism	Leptynites
Contact metamorphism	Hornfels

37. Rocks types	Compressive strength (psi)	Tensile strength (psi)
Sandstone	9000	300
Slate	14000	500
Marble	16000	600
Granite	19000	700

38. Sedimentary rocks	Grain size
Sandstone	2 – 1/16 mm
Conglomerate	>2 mm
Siltstone	1/16 – 1/256 mm
Claystone	<1/256 mm

39. Mafic rocks assemblage	Metamorphic facies
Chalorite-epidotealbite	Green schist facies
Garnet cpx-hornblende -plag.	Amphibolite facies
Garnet-Cpx	Eclogite facies
Garnet-cpx-opx-plag.	Granulite facies

40. Igneous forms	Explanations
Concordant bodies	An igneous body which parallel to the rocks strats.
Disconcordant bodies	An igneous body which cross cut relation to the country rocks.
Lopolith	An igneous body which is lenticular in shape (Convex downward).
Lacolith	A mushroom shaped body which is convex upward.

42. Rocks	Megacrysts and Groundmass minerals	Rocks type
Leucite syenite	Alkali feldspar and leucite	Basic
K-rich granite	K-feldspar and quartz	Acidic
Carbonatite	Granular texture (Intrusive relations)	Ultrabasic
Kimberlite	K-feldspar and olivine, phlogopite	Ultrabasic

43. Geological feature	Process	Example
Cumulate	Magmatic segregation	Chromite deposit
Migmatite	Anatexis	Andalusite, sillimanite deposit
Fenitized rocks	Volatile transportation	Gold deposit
Pillow basalt	Sub-aques eruption	Iron deposit

44. Sedimentary structure	Features	Application
Trough cross-bedding	Depositional	Flow direction
Massive bedding	Deposition	Top and bottom
Convolute bedding	Erosion	Deformation
Graded bedding	Depositional	Top and bottom

45. Grain size	Rocks
Pebble	Rudaceous rocks
Silt	Argillaceous rocks
Sand	Arenaceous rocks
Clay Argillaceous rocks	

47. Feature	Characteristics
Gouge	Powderly material
Breccia	Angular fragment
Mylonite	Foliated rocks due to ductile deformation
Cataclasite	Cataclastic structure

49. Minerals	Metamorphic facies
Chlorite+Epidote+ Actinolite	Greenschist facies
Glaucophane+ Lowsonite+Zeolite	Glaucophane-Lowsonite facies

50. Pressure-Temperature	Metamorphic facies
High temperature/High pressure	Granulite Facies
Low temperature/High pressure	Blueschist facies
High temperature/Low pressure	Sanidinite facies
Low temperature/Low pressure	Zeolite facies

51. Metamorphic facies	Minerals	P/T
Prehnite-pumpellyite facies	Zeolite + laumontite + albite + pumpellyite	LowT/ LowP

55. Plutonic rocks	Volcanic equivalent
Granite	Rhylotie
Gabbro	Basalt
Diorite	Andesite
Tonalite	Dacite

56. Rocks	Texture
Gabbro	Porphyritic
Basalt	Porphyritic
Dolerite	Ophitic
Peridotite	Cumulate

57. Metamorphic facies	Grade of metamorphism
Zeolite facies	Lowest grade metamorphism
Granulite facies	Highest grade metamorphism
Greenschist facies	Low grade metamorphism
Amphibolite facies	High grade of metamorphism

60. Plutonic Igneous Rocks	Volcanic Equivalent	Silica disig-nation	% SiO_2
Nephaline Syenite	Phonolite	Under-saturated	55
Granodiorite	Dacite	Over-saturated	65
Anorthosite	Limburgite	Saturated	45
Synite	Trachyte	Saturated	60

61. Metamorphic Facies	Characteristics Environments
Prehnite-Pumpellyite-Blueschist-Eclogite	Subduction zone
Albite-Epidote hornfels-Hornblende hornfels-Pyroxene hornfels-Sanidinite	Trench
Greenschist-Amphibolite-Granulite	Subduction margin
Pyroxene hornfels-Granulite- Eclogite	Continental margin

62. Magmatic bodies	Names	Example
Concordant intrusive body with a flate base and dome roof.	Lacco-lith	Comoro Island
Plutonic igneous body with an areal extent> 100 km^2	Batho-lith	Chhota nagpur platue
Large, conformable, sausor-shaped layerd intrusive body.	Lopo-lith	In Dharwari-an rocks
Lens-shaped pluton that occupies the crest of an antiform or trough of synform.	Phaco-lith	In Baster craton

63. Igneous rocks	Texture	Essential minerals
Komatiite	Spinifex	Olivine, pyxn, amphibole- Ultramafic
Dolerite	Ophitic	Olivine, augite, plagi.- mafic
Lampro-phyre	Panidio-morphic	Mica, cpx, plagi.- mafic
Andesine	Porphyritic	Andesine, pyxn, Intermediate

64. Sedimentary structure	Origin	Applications
Flute cast	Fluctuating current	Top and bottom
Convolute lamination	Erosion	Deformation history
Rain print	Exposure	Top and bottom
Flaser bedding	Syn-deposi-tional deformation	Paleo-environ-ments

65. Depositional Enviornment	Sand grain characteristics
Glacial	Unsorted and rounded
Aeolian	Sorted and rounded
Fluvial	Unsorted and unrounded
Deep marine	Best sorted and rounded

68. Reynolds number	Types of flow
<2000	Laminar flow
2000-4000	Transitional flow
>4000	Turbulant flow

70. Rocks	Types
Granite	Igneous Plutonic rocks
Basalt	Igneous Volcanic rocks
Slate	Metamorphic rocks
Diorite	Igneous Plutonic rocks

71. Rocks	Types
Sandstone	Arneceous rocks
Limestone	Argilaceous rocks
Conglomerate	Rudaceouse rocks
Shale	Argilaceous rocks

73. Rocks	Designations
Phonolite	Undersaturated volcanic rocks
Rhyolite	Oversaturated volcanic rocks
Nepheline syenite	Undersaturated plutonic rocks
Trachyte	Saturated volcanic rocks

74. Terms	Characteristics
Oolite	Spherical grain of organic or inorganic material
Coquina	Fragment of invertebrate
Encrinite	Bioclast
Spergenite	Bioclast with debris of quartz

75. Types of Skewness	Material
No skewness	Grain size not defined
Negative skewness	Course size particles
Positive skewness	Fine particles
Complex skewness	Course and fine particles

88. Environments	Minerals presence
Desert environment	Red deposit
Glacial environment	Till
Fluvial environment	Heavy minerals
Marine environment	Glauconite

90. Palaeocurrent patterns	Environment
Unimodal	Fluvial
Bimodal	Marine
Polymodal	Wind

94. Metamorphism	Temperature/Pressure
Dynamic metamorphism	High pressure changes
Pyrometamorphism	High temperature changes
Contact metamorphism	Low pressure and temperature changes
Dynamothermal metamorphism	High pressure and temperature changes

95. Processes	Minerals
Retrograde metamorphism	Talc
Ultrametamorphism	Migmatites
Palingenesis	Olivine, Pyroxene
Metasomatism	Skarn

96. Metamorphic facies	Rocks type
Amphibolite facies	Gondite
Granulite facies	Khondalite
Eclogite facies	Kodirites
Greenschist facies	Gondites

97. Rocks type	Designations
Marble	Metamorphic rock
Slate	Metamorphic rock
Granite	Igneous rock
Phyllite	Metamorphic rock

102. Texture	Classification
Ophitic texture	Igneous texture
Clastic texture	Sedimentary texture
Granobalastic texture	Metamorphic texture
Subophitic texture	Igneous texture

103. Metamorphic rocks	Structure
Granulose	Granulose structure
Hornfels	Maculose structure
Cataclastic	Cataclastic structure
Schistose	Schistose structure

104. Rocks types	Minerals
Norite	Opx-labrodorite-olivine
Troctolite	Olivine-plagioclase
Dunite	Olivine-pyroxene
Eucrite	Achondrite meteorites

105. Volcanic rocks	Characteristics
Rhyolite	Acid
Trachytes	Intermediate
Andesite	Intermediate
Basalt	Basics

107. Igneous bodies	Remark
Chonolith	Discordant
Laccolith	Concordant
Lopolith	Concordant
Bysmalith	Concordant

117. Rocks	Essential Minerals
Dunite	Olivine, Pyxn,Charomite
Lherzolite	Olivine,Opx,Cpx,Plagi.
Granite	Potassium feldspar, plagioclase, quartz, biotite
Diorite	Andesine, biotite, hornblende, pyroxene

137. Particles	Rock type
Pebbles	Rudaceous rock
Clay	Argillaceous rocks
Boulders	Rudaceous rocks
Sand	Arenaceous rocks

151. Minerals	Silica designations
Perovskite	Undersaturated rocks
Tchermekite	Saturated rocks
Fuschite	Saturated rocks
Coesite	Oversaturated rocks

155.

Minerals assemblage	Metamorphic facies
Albite+Epidote+ Actinolite+Chlorite	Greenschist facies
Hornblende+ Plagioclage+Ilmenite	Amphibolite facies
Omphacite+Garnet +Plagioclage	Eclogit facies
Opx+Cpx+Plagi +Hbl	Not defined

158.

Igneous structure	Rocks type	Example
Batholith	Granite	Singhbhum craton
Lopolith	Gabbro-norite	
Pa-hoe-hoe lava	Basalt	Deccan basalt
Pipe	Kimberlite	Clospet granite in Dharwar supergroup

159.

Metamorphic Processes	Minerals assemblage
Ocean floor metamorphism	Serpentine, Chlorite, Epidote, Zeolite
Continent-continent collision metamorphism	Eclogite series minerals
Contact metamorphism	Hornfels series minerals
Subduction zone metamorphism	Blueschist series minerals

160.

Rocks	Volcanic Equivalent
Anorthosite	No volcanic equivalent
Gabbro	Basalt
Syenite	Trachyte
Granite	Rhyolite

167.

Terms	Characteristics
Anthracite	High rank coal
Anticilite	Igneous rock
Aubrite	Meteorite type
Adulite	Gemstone

177.

Plutonic rocks	Volcanic rocks
Granite	Rhyolite
Syenite	Trachyte
Diorite	Andesite
Granodiorite	Dacite

202.

Minerals assemblage	Facies
Hornblende+plagioclage +Staurolite	Amphibolite Facies
Opsthopyroxene+ Clinoptroxene+plagioclage	Granulite facies
Omphacite+garnet	Eclogite facies
Actinolite+chlorite+Epidote	Greenschist facies

203.

Rocks type	Classification
Sandstone	Arenaceous rock
Limestone	Argillaceous rock
Conglomerate	Rudaceous rock
Shale	Argillaceous rock

210.

Characteristics environment	Facies type
Large volume of stagnant water	Black shale facies
Running water	Sand facies
Salty water	Mud facies
Frozen water	Carbnate water

211.

Rocks	Particle size
Arenaceous rock	Sand size
Rudaceous rock	Gravel size
Argillaceous rock	Clay size
Calcareous rock	Particle size not defined

218. Terms	Characteristics
Isogyre	Anisotropic minerals show figure in microscope
Isograde	A line which show same grade of metamorphism
Isopach	A line which show equal true thickness
Isochore	A line which show equal vertical thickness

220. Rock type	Products
Slate	Cataclastic metamorphism
Schist	Regional metamorphism
Phyllite	Dynamothermal metamorphism
Quartzite	Thermal metamorphism

221. Rock type	Parent rocks
Marble	Limestone
Slate	Shale
Granite	Granite
Quartzite	Sandstone

222. Typical products	Types of metamorphism
Granulose	Thermal metamorphism
Cataclastic	Cataclastic metamorphism
Hornfels	Contact metamorphism
Schistose	Dynamothermal metamorphism

223. Stage	Zone of metamorphism (Increasing order)
Lowest	Chlorite Zone
-	Biotite Zone
-	Garnet Zone
-	Staurolite Zone
-	Kyanite Zone
Highest	Sillimanite Zone

228. Rocks type	Textures
Lamprophyres	Porphyritic texture
Andesite	Porphyritic texture
Dolerite	Ophitic texture
Trachyte	Porphyritic texture

232. Metamorphic facies	Environment
Amphibolite facies	Depth
Granulite facies	Subduction zone environment
Greenschist facies	Shallow geosynclinals environment
Sanidinite facies	High temperature and low pressure environment

233. Igneous rocks	Silica %	Designation
Rhyolite	>65	Acid igneous rocks
Andesite	55-65	Intermediate igneous rocks
Dacite	>65	Acid igneous rocks
Basalt	45-55	Basic igneous rocks

240. Rocks	Abundance minerals
Eucrite	Poor pyroxene, Ca-rich plagioclase
Trocrolite	Olivine, Anorthite with pyroxene
Dunite	Olivine, pyroxene
Picrite	Olivine, pyroxene, minor plagioclase

247. Metamorphic Facies	Mineral assemblage
Eclogite facies	Omphacite- Garnet-kimberlite-pipe
Amphibolite facies	Hbl- Plagi- Almandine-Sillimanite
Blueschist facies	Glaucophane-Lawsonite-Jadeite
Green schist facies	Epidote- Chalorite-Albite.

250. Plutonic rocks	Volcanic equivalent	Character-istics	SiO_2%
Granite	Rhyolite	oversaturated	72
Granodi-orite	Dacite	oversaturated	65
Diorite	Andesite	Saturated	58
Syenite	Trachyte	Saturated	60

254. Sedimentary structure	Nature of structure	Characteristics
Cross-stratifi-cation	Deposition	Flow direction
Flute casts	Erosion	Top and bottom
Graded bedding	Deposition	Top and bottom
Ripple marks	Deposition	Flow direction

256. Metamorphic facies	Mineral assemblage
Granulite facies	Orthopyroxene-Clinopyroxene-plagioclase
Amphibolite facies	Hbl-plagiclase-actinolite
Eclogite facies	Pyrope-omphacite-kimberlite
Greenschist facies	Albite- epidote-chalorite

257. Clastic Sedimentry rocks	Size(mm)
Granule	2-4
Sand	2-0.625
Silt 0.625-0.0039	
Clay	Less than 0.0039

258. Sedimentary feature	Uses
Asymmetrical ripple marks	Unidirectional palaeocurrent
Symmetrical ripple marks	Flow direction
Parting lineation direction	Paleocurrent
Load cast	Top and bottom

262. Process	Texture
Eutectic crystallization	Graphic
Phase separation of solid state	Intergrowth
Crystallization following phase separation	Exsolution
Replacement	Intergrowth

265. Metamorphic facies	Minerals
Greenschist facies	Chlorite, epidote, albite
Granulite facies	Orthopyroxene, clinopyroxene, plagioclase
Amphibolite facies	Hornblende, plagioclase, almandine
Eclogite facies	Pyrope rich garnet, omphacite,kimberlite

269. System	Characteristics
Albite-Anorthite system	Solid solution
Forsterite-Silica system	Peritectic reaction
Diopside-Anorthite system	Eutectic reaction
Forsterite-Olivine system	Eutectic reaction

278. Plutonic rocks	Volcanic equivalent
Granite	Rhyolite
Gabbro	Basalt
Syenite	Trachyte
Nepheline syenite	Phonolite

279. Textures	Example
Cumulate texture	Anorthite
Poikilitic texture	Gabbro
Spinifex texture	Komatiite
Spherulitic texture	Obsidian

289. Rocks	Clay%
Mudstone	33-66%
Siltstone	<33 %
Claystone	>90 %
Dropstone	<10 %

295. Process	Characteristics
Chloritization	Conversion of pyroxene into Chlorite
Illitization	Conversion of chlorite into illite
Fluidization	Conversion of solid minerals into fluid forms

304. AKF components	Characteristics
A	$[Al_2O_3 + Fe_2O_3] - [Na_2O + K_2O + CaO]$
K	$[K_2O]$
F	$[FeO + MgO + MnO]$

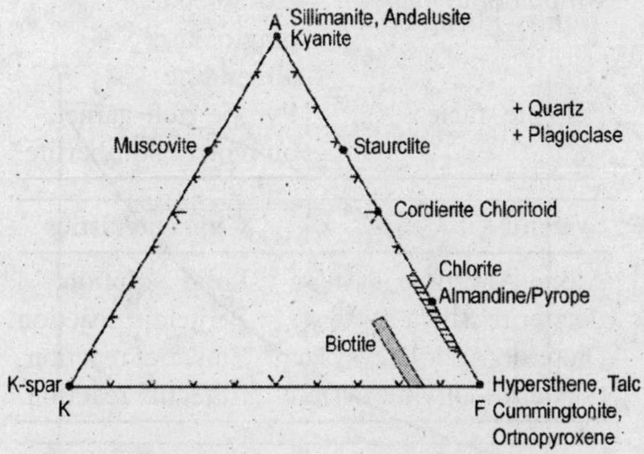

319. AFM components	Characteristics
A	$[Al_2O_3]$
F	$[FeO]$
M	$[MgO]$

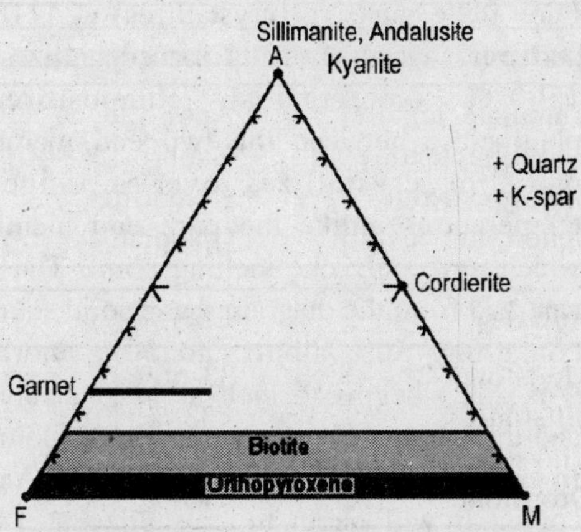

322. ACF components	Characteristics
A	Al_2O_3
C	CaO
F	$FeO + MgO$

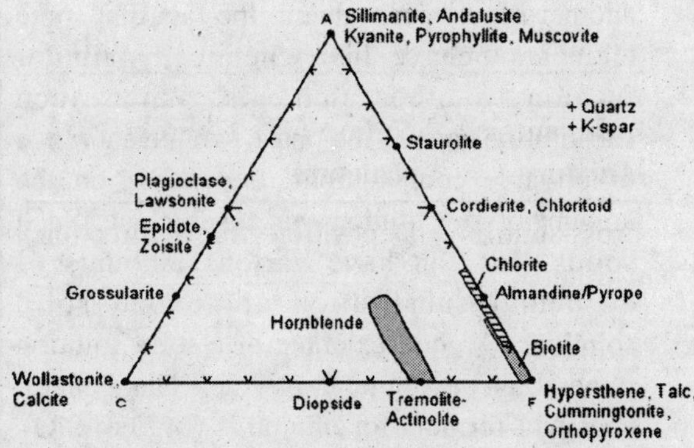

323. Number of components C. $= 3$

$$F = C + 2 - \phi$$

$$0 = 3 + 2 - \phi$$

$$\phi = 5 \quad \text{Ans.}$$

329. Facies	Characteristics
Zeolite facies	Regional Metamorphism
Sanidine facies	Contact metamorphism
Granulite facies	High grade regional metamorphism
Greenschist facies	Regional metamorphism

424. Facies	Grade of metamorphism
Zeolite facies	Lowest grade of metamorphism
Granulite facies	Highest grade of metamorphism
Greenschist facies	Low grade of metamorphism
Amphibolite facies	High grade of metamorphism

471. Albite – Anorthite system:

In the systems we've discussed so far, all of the mineral or solid phases have been pure phases, that is they have one and only one possible composition. This is not usually the case in nature, since substitution of one element for another often occurs due to the fact that some elements behave in a chemically similar fashion to other elements. When such substitutions occur, the phase can have a range of possible compositions, depending on the amount of substitution that takes place. Such solids that can have various amounts of elemental substitution are called solid solutions. A good example of a solid solution mineral is the mineral olivine. The general chemical formula for olivine is $(Mg,Fe)_2SiO_4$. Since Mg^{+2} and Fe^{+2} are about the same size and have the same charge, they may substitute for one another in the crystal structure of olivine. Thus olivine may have a composition anywhere between the pure Mg end member, forsterite (Mg_2SiO_4), and the pure Fe end member, fayalite (Fe_2SiO_4). When all compositions between two end members are possible, the solid solution is said to be a complete solid solution.

Another good example of a complete solid solution is displayed in the plagioclase feldspars. In this case the solid solution is between the end members albite ($NaAlSi_3O_8$) and anorthite ($CaAl_2Si_2O_8$). In order to maintain charge balance we cannot simply substitute Na^+ for Ca^{+2}, so this solid solution is what is called a coupled solid solution. In this case Na^+Si^{+4} is substituted for $Ca^{+2}Al^{+3}$ in the plagioclase structure to produce intermediate compositions of plagioclase.

Because the elements that substitute are not exactly the same size (they are similar in size) the amount of substitution is dependent on temperature and pressure and the solid solutions behave in a somewhat orderly fashion as illustrated below.

Since plagioclase is one of the most common minerals in the earth's crust, we will discuss the phase diagram for the plagioclase system. The phase relations in the plagioclase system are shown in Figure at constant pressure equal to that of the atmosphere (atmospheric pressure is 1 bar). In Figure the upper curve is called the liquidus and the lower curve is called the solidus. At temperatures above the liquidus everything is liquid, below the solidus everything is solid (crystals of plagioclase solid solution). At temperatures between the solidus and liquidus crystals of plagioclase solid solution coexist in equilibrium with liquid.

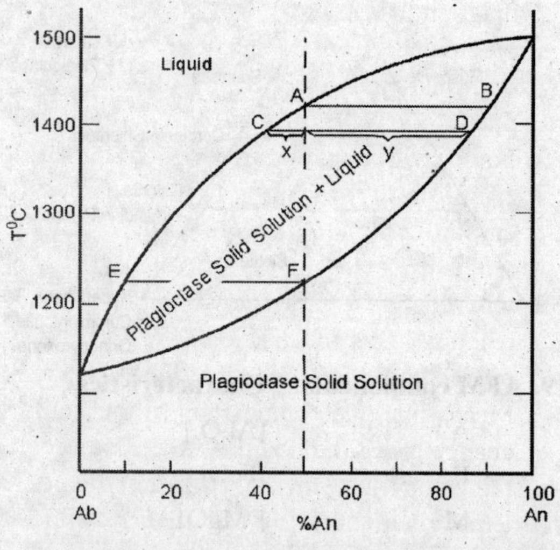

Pure albite melts (or crystallizes) at 1118°C, and pure anorthite melts (or crystallizes) at 1500°C. Note that any composition of plagioclase between the two end members melts or crystallizes over a range of temperatures unlike the pure end members which have only one melting point. Thus we can read from the diagram that a solid solution containing 50% albite and 50% anorthite ($Ab_{50}An_{50}$) begins to melt at 1220°, point F, and the melting is complete at 1410°, point A. Inversely, if a melt of composition $Ab_{50}An_{50}$ is cooled it will begin to crystallize at 1410°

and will be completely crystalline at 1220°. We will now trace the crystallization history of composition X, which is $Ab_{50}An_{50}$. Composition X is completely liquid above the liquidus (above 1410°). Cooling to the liquidus at point A results in the crystallization of a small amount of plagioclase solid solution. The composition of this plagioclase can be found by drawing an isotherm (line of constant temperature, a horizontal line in this diagram) through the temperature 1410°. Where this isotherm intersects the solidus (at point B), the composition of the solid can be found by drawing a vertical line to the base of the diagram. Thus it is seen that the first crystals precipitated from composition X will have the composition $Ab_{10}An_{90}$. Note that in this diagram crystals that are in equilibrium with liquid will always be enriched in anorthite component relative to the liquid. As crystallization continues with lowering of temperature the composition of the plagioclase will change along the solidus, continually reacting with the liquid to produce crystals more enriched in the Ab component. Meanwhile, the composition of the liquid will change along the liquidus, thus also becoming more enriched in the Ab component. At a temperature of 1395° the liquid composition will be at point C, while the solid composition will be at point D. Crystallization proceeds until a temperature of about 1220°, at which point the last remaining liquid will have a composition at E, and the solid will have a composition equal to the original starting composition at point F.

At this point all of the liquid will be consumed and the final crystalline product will have the composition $Ab_{50}An_{50}$.

During crystallization the proportion of the solid continually increases while that of the liquid continually decreases. Thus as the composition of the liquid becomes more sodic, approaching E, its volume steadily decreases. Thus it can be seen that the amount of liquid in equilibrium with the solid of composition F will be extremely small.

473. Classification of Sandstone:

Sandstone is the indurated equivalent of unconsolidated sand containing clasts between 2 and 1/16 mm. They make up between 10% and 20% of sedimentary rocks. Interesting to sedimentologists for the information they contain about the provenance of sediment and depositional environments. For the practical person, they are important as fluid reservoirs (aquifers and petroleum reservoir rock). Indeed, sandstones hold over 1/2 of the worlds oil reserves.

Sandstone classification: There are MANY different classification schemes for sandstones, but the two defining parameters are:

- the relative abundance of framework (sand) matrix (mud)
- the composition of sand framework grains.

These are refined by consideration of texture, including:

- grain size and distribution
- sorting
- rounding
- fabric

Taken together, these factors can be encompassed in terms of maturity (see below).

Framework composition: Framework grains can consist of any rock or mineral, but for practical purposes, we reduce these to:

- **Crystalline quartz:** Generally abundant because of its resistance to weathering.
- **Feldspar:** Both K-feldspar and plagioclase are more susceptible to chemical and physical weathering, so they are less abundant in sediments, even if they are common in source rock. Their presence implies minimal chemical weathering. This can be due to:

* Environmental conditions that minimize chemical weathering such as aridity or extreme cold.
* High topographic relief in the source region, so that it is eroded before it is chemically transformed.

- **Rock (lithic) fragments:** Fragments of whole rock containing more than one mineral grain. These provide information on sediment provenance.

Matrix: Fine grained matrix (< 30 μm), can be detrital or products of *in situ* feldspar diagenesis. Depending on the percentage of matrix, a sandstone can be a:

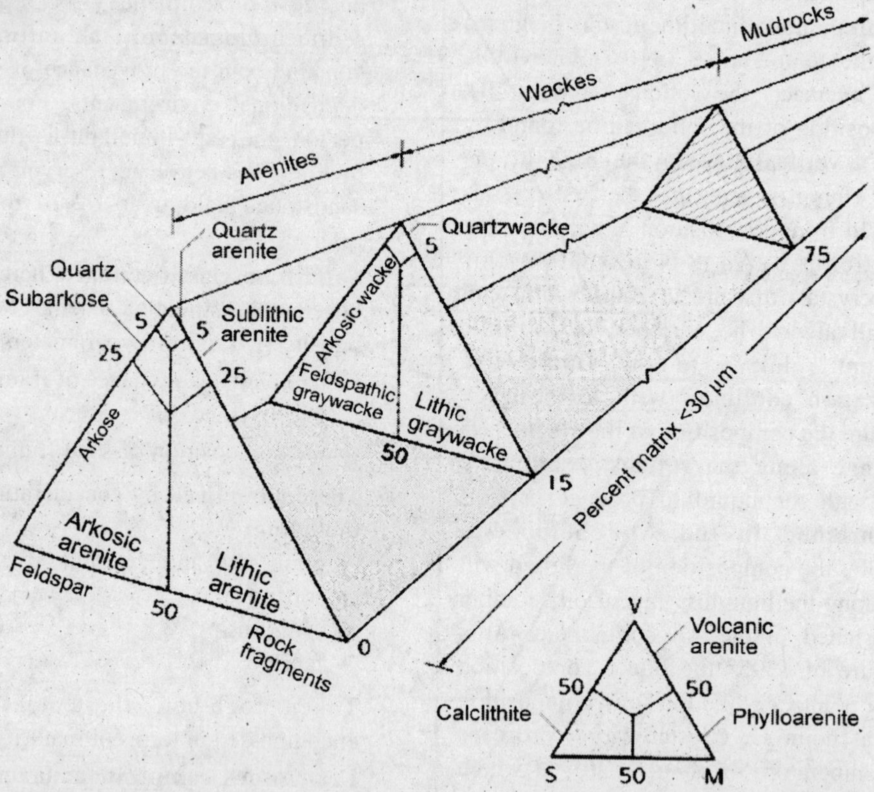

- Arenite: [Latin arena = sand]. 0% - 15% matrix.
- Wacke: (A.K.A. "graywacke") 15% - 75% matrix. i.e. significant matrix by volume.

Any more matrix and the rock is regarded as a mudrock.

We use the classification system of Dott (1964) that can nicely be summarized by a ternary diagram showing relative volume of quartz, feldspar, and rock fragments, with an extra dimension for matrix volume.

Depending on the predominant framework composition, arenites are broken down into:

- Quartzarenite
- Feldspathic arenite (A.k.a. arkose)
- Lithic arenite

Wackes, similarly break down into:

- Quartzwacke
- Feldspathic graywacke
- Lithic graywacke

475. QAP Diagram:

A QAPF diagram is a double triangle diagram which is used to classify igneous rocks based on mineralogic composition. The acronym, QAPF, stands for "Quartz, Alkali feldspar, Plagioclase, Feldspathoid (Foid)". These are the mineral groups used for classification in QAPF diagram.

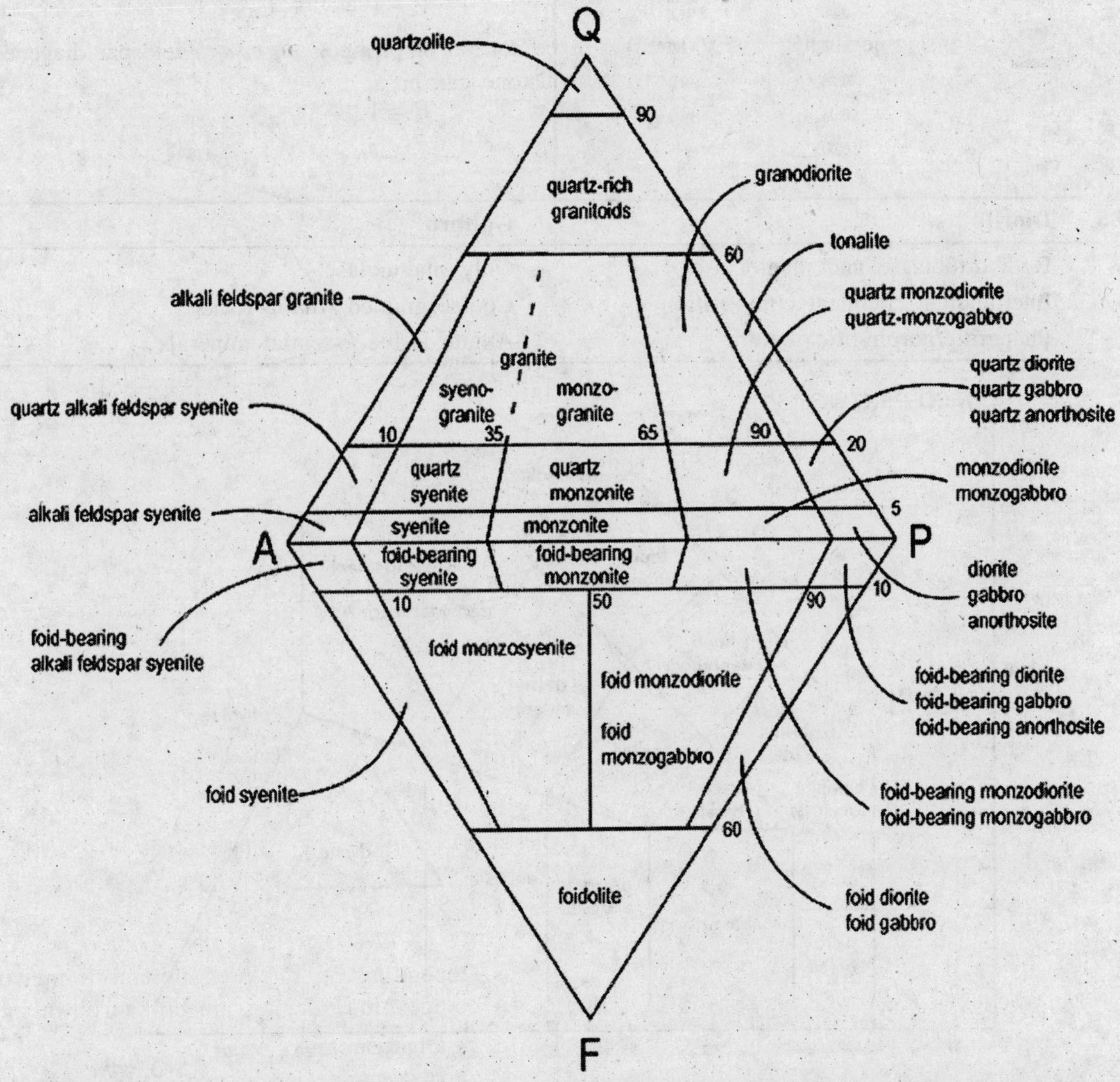

477. Processes	Texture	Example
Eutectic crystallization	Graphic	Graphic granite
Fractional crystallization	Crystal zoning	Basaltic rock
Exsolution	Perthite texture	Granitic rock
Submarine eruption	Pillow structure	Basaltic rock

480. Phi value of the sediment particles:

Wentworth Scale of Grain Size

	Wentworth Scale	Millimeters	Phi
GRAVEL	Cobble/Boulder	64.0	-6.0
	Pebble	4.00	-2.0
	Granule	2.00	-1.0
SAND	Coars	0.50	+1.0
	Medium	0.25	+2.0
	Fine	0.06	+4.0
MUD	Sil	0.002	+9.0
	Clay		

481. Number of phase = 2

Number of component is = 2

Then

$$P + F = C + 1$$

$$2 + F = 2 + 1$$

$$F = 1$$

485. **Diorite**	**Gabbro**
Both orthoclase and plagioclase	Only plagioclase
Intermediate chemical composition	Course grained mafic rocks
Phaneritic/porphyritic texture	Augite is the essential minerals

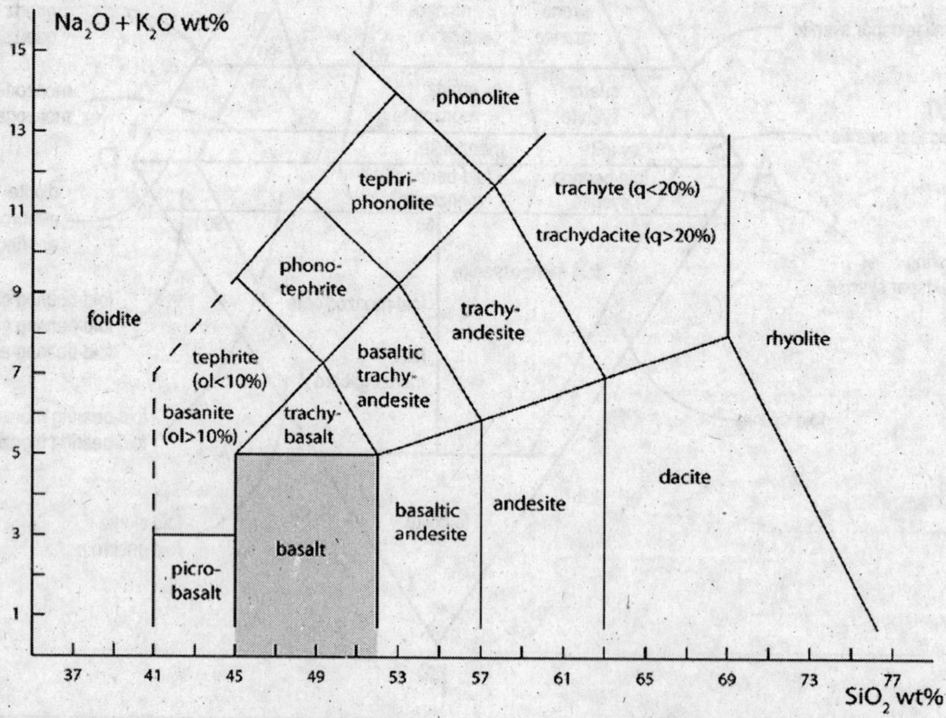

489. Igneous rocks	**Remark**	**Minerals**
Peridotite	Course grained, ultramafic rocks	Olivine and pyroxene
Alkali Peridotite	Do	Alkali minerals present
Dunite	Course grained, ultramafic rocks	> 90 % olivine
Pyroxenite	Course grained, ultramafic rocks	> 90 % pyroxene

490. Minerals assemblage	Metamorphic facies
Omphacite + Pyrope	Eclogite facies
Chlorite + Actinolite + Albite	Green schist facies
Opx + Cpx + Plagioclase	Granulite facies
Hornblende + Plagioclase	Amphibolite facies

491.

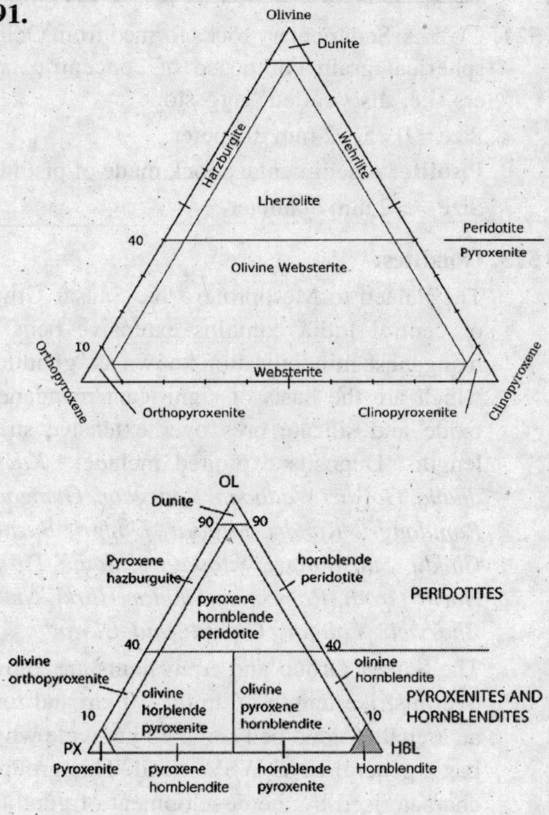

495. TAS Diagram

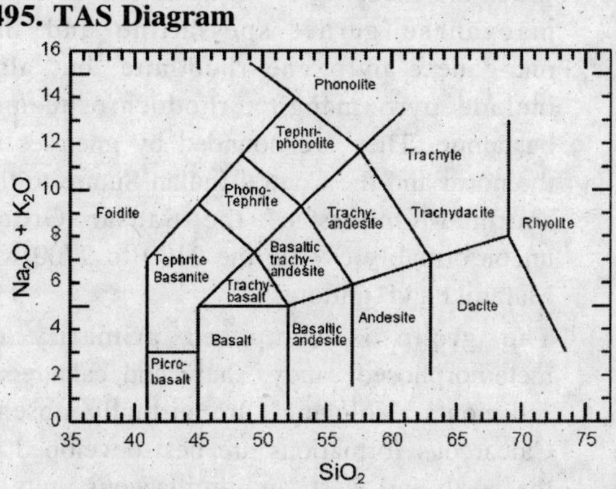

497. Texture	Characteristic	Remarks
Ophitic	Bigger crystal of augite enclosed in lath of feldspar	Dolerite
Graphic	Intergrowth of quartz and feldspar	Granite
Sub-ophitic	Bigger crystal of augite enclosed in lath of feldspar	Dolerite
Porphyritic	Large crystal surrounded by the small crystal	Gabbro

502. Igneous rocks	Texture
Basalt	Porphyritic
Granite	Porphyritic
Diorite	Orbicular/porphyritic
Dunite	Graphic

503. Figure of the limestone feature:

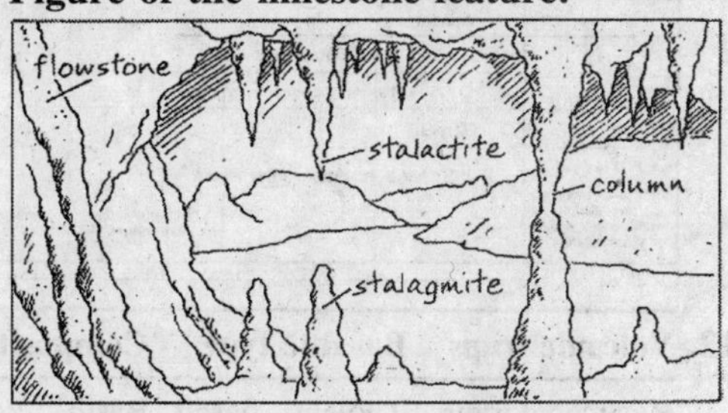

506. Classification of the Gabbro:

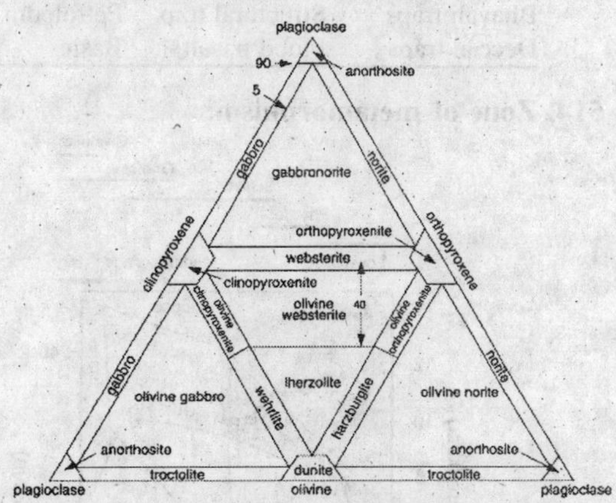

507. Lava Types	Composition
Lava tubes	Basic composition/Acidic composition
Aa lavas	Basic composition
Pa-hoe-hoe lava	Basic composition
Blocky lava	Acidic composition
Pillow lava	Basic composition

510. Rocks and its minerals contain:

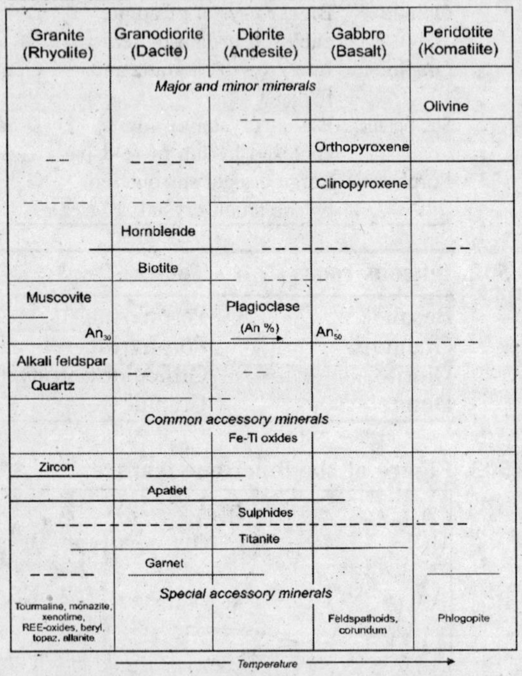

512. Volcanic traps	Basaltic type	Composition
Rajamahal traps	Tholeiitic basalt	Basic
Sylhet traps	Alkali basalt	Basic
Bhavali traps	Structural trap	Petroleum basin
Deccan traps	Flood basalt	Basic

514. Zone of metamorphism:

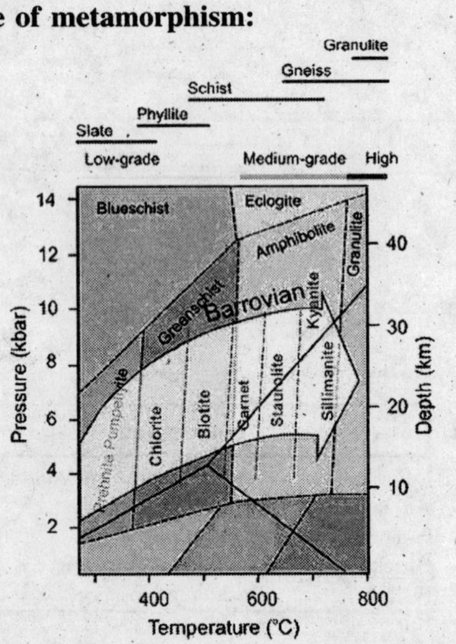

519. Classification of Sedimentary Rocks:

Classification of the sediment	Particle size	Particle name	Rock types
Arenaceous	2 – 1/16 mm	Sand	Sandstone
Argillaceous	1/16 – 1/256 mm	Clay	Shale
Rudaceous	>2 mm	Pebble	Conglomerate

521. Oolites: Sedimentary rocks formed from Ooids, spherical grain composed of concentric layers. i.e. also called "egg stone".

Size = 0.25 – 2 mm diameter.

Pisolite: A sedimentary rock made of pisoid.

Size = > 2 mm diameter.

523. Gondites:

The Palaeo to Mesoproterozoic Sausar Group of central India contains extensive beds of manganese mineralisation known as "gondites" which are the basis of significant manganese oxide and silicate ores over extensive strike lengths. Deposits exploited include: *Kachidhana, Gowari Wadhona, Kodegaon, Gumgaon, Ramdongri, Kandri, Mansar, Dongri, Buzurg, Chikla, Sitasaongri, Sitapatore, South Tirodi, North Tirodi, Ramrama, Shodan Hurki, Netra, Bharweli, Laugor, Ghondi and Ukwa.*

The Sausar Group and equivalents are known over a strike length of up to 210 km and form an arcuate folded belt some 75 km wide which has a general ENE-WSW trend. This group is characterised by the development of gondites, which are made up of manganese oxides, the manganese garnet spessartine and the manganese pyroxene rhodonite, but also include pyroxmangite, rhodochrosite and bustamite. They are bounded by gneisses to the north and the Central Indian Suture to the south. Members of the Sausar Group unconformably overlie the 2100 to 2400 Ma Malanjkhand pluton.

The group is composed primarily of metamorphosed sandy, shaly and calcareous sediments. Volcanics are virtually absent. Calcareous formations are best developed to the north and west, and argillaceous units to the south and east. In general they have been metamorphosed to upper amphibolite facies, but have also undergone anatexis to produce migmatites and gneisses. In general the metamorphism increases to the north and west. Near the Balaghat Mn mine they are of greenschist facies, while farthest to the north-west rocks with a granulite assemblage are encountered. Deformation is intense and is coeval with the metamorphism, producing shearing, thrusting and nappe development (Mahmood Naqvi & Rogers, 1987; Jain, et al., 1995).

540. Texture	**Crystal**
Allotrimorphic	Anhedral
Hypidiomorphic	Subhedral
Panidiomorphic	Euhedral

543. Rocks and its texture and minerals contain:

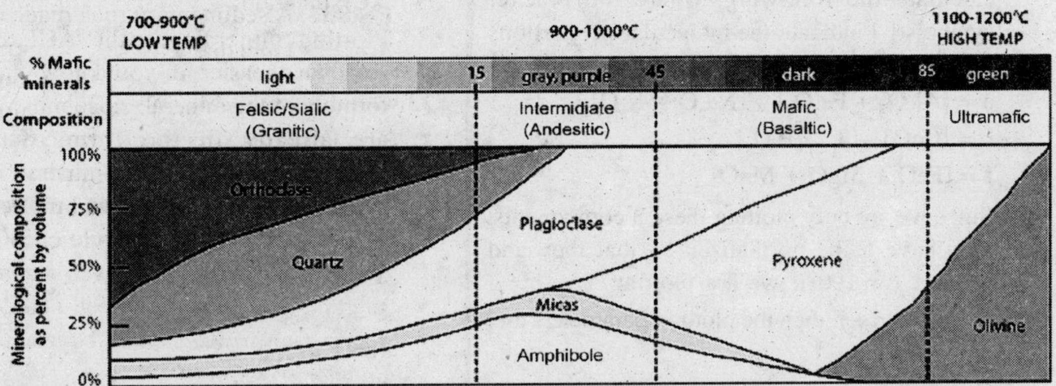

Texture	Rock names			
Coarse-grained	Granite	Diorite	Gabbro	Peridotite
Fine-grained	Rhyolite	Andesite	Basalt	Uncommon (rarely encountered)
Pyroclastic	Rhyolite Tuff	Andesite Tuff	Basalt Tuff	
Glassy	Obsidian			
Vesicular	Pumice (like meringue)		Scoria	

550. Terminology	**Remarks**
Isopach map	A map show equal thickness
Isochore map	A map show drilled interval, irrespective amont of dip
Isobar map	A map show the equal pressure
Isolith map	A map show the equal thickness of the strata

563. ACF Diagrams

One of the first uses of these types of diagrams was by Eskola (1915) who employed a diagram known as the ACF diagram in his study of metamorphic rocks. To plot a rock on the ACF diagram, the chemical analysis of the rock is first recalculated to molecular proportions by dividing the molecular weight of each oxide constituent by the molecular weight of that oxide. Nominally, the ACF diagram plots the following components:

$A = Al_2O_3$

$C = CaO$, and

$F = FeO + MgO$

However, the A value we want is the value of excess Al_2O_3 left after allotting Na_2O and K_2O to form alkali feldspar. The CaO value we want is the excess CaO after allotting P_2O_5 to form apatite, assuming that any P_2O_5 in the rock will suck up CaO to form apatite. We will assume then that all mineral assemblages plotted may also contain alkali feldspar and quartz (and apatite).

So, to obtain the plotting parameters, we calculate the following, where the bracket symbols [ ] indicate the molecular proportions of the oxides.

$a = [Al_2O_3 + Fe_2O_3] - [Na_2O + K_2O]$

$c = [CaO] - 3.33[P_2O_5]$

$f = [FeO + MgO + MnO]$

Since we are only plotting these 3 components, they have to be normalized so that they add up to 1 (or 100 if we are plotting %).

if $t = a + c + f$, then the plotting parameters are:

A = 100 * a/t

C = 100 * c/t

F = 100 * f/t

When these calculations are done for a wide variety of rock compositions and grouped as pelitic, quartzo-feldspathic, basic, and calcareous, the fields are as shown here.

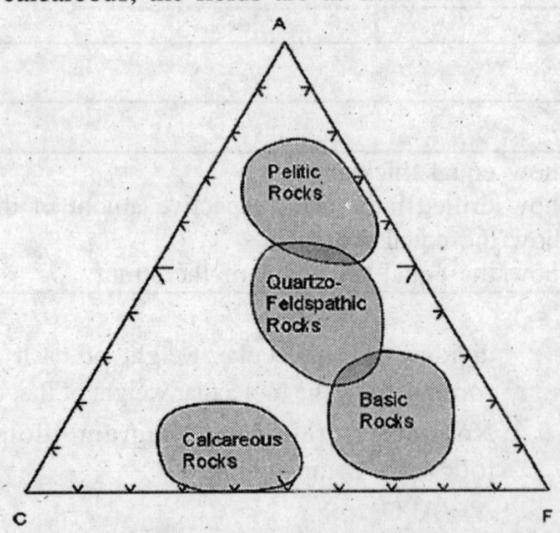

Most shales will plot in the field of Pelitic Rocks. Quartzo-feldspathic rocks like feldspathic sandstones, granites, and rhyolites will plot in the Quartzo-Feldspathic field. Basic igneous rocks, like basalts and gabbros will plot in the field of Basic Rocks, and siliceous limestones and dolomites will plot in the field of Calcareous Rocks.

This diagram and the fields shown will become an important part of later discussions, so it is wise to know approximately where the fields of these different chemical types occur on the ACF diagram.

Plotting minerals on the ACF diagram is somewhat easier if you know the chemical formula of the mineral, since mineral formulae are already in the form of molecular proportions. Thus for a mineral like hypersthene, $(Mg,Fe)SiO_3$, we have 1 molecule of (FeO + MgO) for every 1 molecule of SiO_2. Thus:

a = 0

c = 0

f = 1

t =1

so, the plotting parameters become

A = 100 * 0/1 = 0

C = 100 * 0/1 = 0

F = 100 * 1/1 = 100%

and we see that hypersthene would plot at the F corner of the ACF diagram.

As a second example, look at the formula for tremolite - $Ca_2(Mg,Fe)_5Si_8O_{22}(OH)_2$

We can rewrite this formula as 2CaO 5(Mg,Fe)O $8SiO_2H_2O$. Then:

a = 0

c = 2

f = 5

t = 7

so the plotting parameters become:

A = 100 * 0/7 = 0

C = 100 * 2/7 = 28.57%

F = 100 * 5/7 = 71.43%

Other minerals are more complicated. For example, the formula for chlorite is (Mg,Al,

$Fe)_{12}(Si,Al)_8O_{20}(OH)_{16}$. Thus we have several possibilities for writing the formula for chlorite, and depending on which formula we use, chlorite will plot at different locations on the diagrams. You will explore the possible solid solution ranges in a lab on this topic.

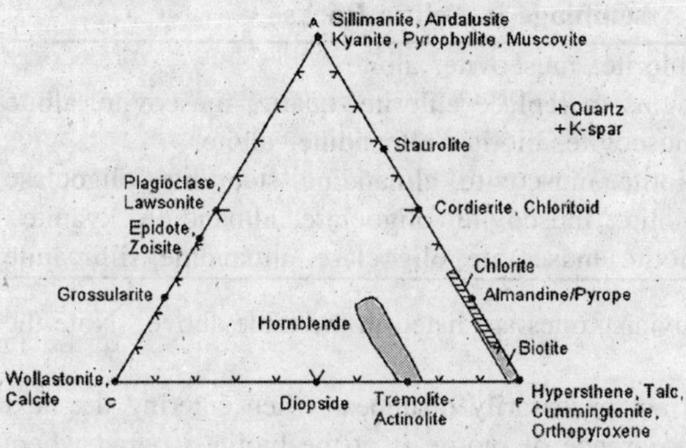

Although you will calculate the plotting positions of some rocks and a wide variety of minerals in lab, the diagram above shows the plotting positions of some of the more common minerals that occur in metamorphic rocks. Note that this diagram is for reference only, it does not show mineral assemblages in rocks (there are no tie lines). Note that alkali feldspars do not plot in this diagram, but are assumed to be present because of the way we calculate the A component.

AKF Diagrams

In AKF diagrams we assume that both alkali feldspar and plagioclase feldspar can be present, thus the amount of Al_2O_3 that we use is the excess Al_2O_3 left after allotting it to all of the feldspars. To obtain the plotting parameters for ACF diagrams, calculate the following:

$a = [Al_2O_3 + Fe_2O_3] - [Na_2O + K_2O + CaO]$

$k = [K_2O]$

$f = [FeO + MgO + MnO]$

Let $t = a + k + f$, then the plotting parameters in % are:

$A = 100 * a/t$

$K = 100 * k/t$

$F = 100 * f/t$

Minerals are plotted in the same way as was done for the ACF diagrams, and an example AKF diagram showing the plotting positions of common metamorphic minerals is shown below.

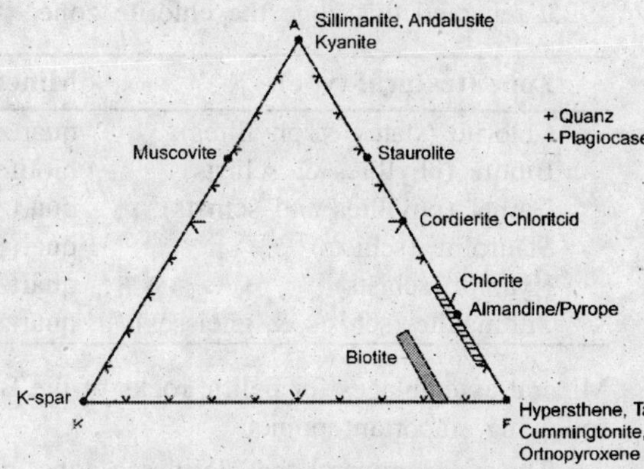

Note that K-feldspar plots in the lower right hand corner. To see why, we first take the chemical formula of K-feldspar $KAlSi_3O_8$ and rewrite it in oxide form as $1/2K_2O$ $1/2Al_2O_3$ $3SiO_2$. Then:

$a = \frac{1}{2} - \frac{1}{2} = 0$

k = ½

f = 0

t = ½

So,

A = 100 * 0/½ = 0%

K = 100 * ½ /½ = 100%

F = 100 * 0/½ = 0%

Another example is muscovite $KAl_3Si_3O_{10}(OH)_2$ or $1/2K_2O$ $3/2Al_2O_3$ $3SiO_2$ H_2O. For muscovite:

a = 3/2 – 1/2 = 1

k = 1/2

f = 0

$t = 1\frac{1}{2} = 1.5$

So,

A = 100 * 1/1.5 = 66.7%

K = 100 * 0.5/1.5 = 33.33%

F = 100 * 0 = 0%

Note that AKF diagrams are used for CaO-poor, K_2O-rich rocks, whereas ACF diagrams should be used for Al_2O_3 and CaO - rich rocks.

572. Barrovian Sequence: The boundaries for his zones were based on the first appearance of a particular mineral, called an index mineral, which is characteristic of the zone. These boundaries were later called isograds (equal grade) and likely represent surfaces in a three dimensional sense. He called the zone of lowest grade rocks the "zone of digested clastic mica," but Tilley, mapping the area in 1925, renamed this zone the chlorite zone.

Zone (textural type)	Mineral Assemblage in Pelitic Rocks
Chlorite (slates & phyllites)	quartz, chlorite, muscovite, albite
Biotite (phyllites & schists)	biotite begins to replace chlorite, quartz, muscovite, albite
Garnet (phyllites and schists)	quartz, muscovite, biotite, almandine, albite
Staurolite (schists)	quartz, biotite, muscovite, almandine, staurolite, oligoclase
Kyanite (schists)	quartz, biotite, muscovite, oligoclase, almandine, kyanite
Sillimanite (schists & gneisses)	quartz, biotite, muscovite, oligoclase, almandine, sillimanite

Mineral assemblages for pelitic rocks of the Barrovian Zones are listed in the table above. Note the following important points:

- The index mineral that defines a zone, does not necessarily disappear when entering the next higher grade zone. For example the first appearance of biotite is at the biotite isograd where chlorite is seen to be reacting to produce biotite. Biotite does not disappear at the garnet isograd, and, in fact continues to be seen though the garnet, sillimanite, staurolite, kyanite, and sillimanite zones.
- Staurolite, kyanite, and sillimanite only occur in the staurolite, kyanite, and sillimanite zones, respectively.
- The composition of the plagioclase changes with increasing grade of metamorphism. It is nearly pure albite in the chlorite and biotite zones, and becomes somewhat more calcic (oligoclase) in higher grade zones.
- The texture of the rocks changes from slates and phyllites in the chlorite zone to schists in the staurolite and kyanite zones, to schists and gneisses in the sillimanite zone.

As mentioned above, Tilley mapped the area in 1925, and extended the zones across the area between the Highland Boundary Fault and into the Moinian Series, and renamed the chlorite zone. Later, Wiseman (1934), mapped the metabasic rocks and Kennedy (1949) mapped the meta calcareous sediments. Mineral assemblages in these rocks and their correlation with the metamorphic zones in the pelitic rocks are shown in the table below.

<table>
<tr><th>Index Mineral (Pelitic Rocks)</th><th>Basic Rocks</th><th>Calcareous Rocks</th><th>Facies</th></tr>
<tr><td>Chlorite</td><td rowspan="2">Chlorite, albite, epidote, sphene, ± calcite ± actinolite</td><td>qtz, muscovite, biotite, calcite</td><td rowspan="2">Greenschist</td></tr>
<tr><td>Biotite</td><td rowspan="2">garnet, zoisite, sodic plagioclase, biotite or hornblende</td></tr>
<tr><td>Garnet</td><td rowspan="4">Hornblende, plagioclase, ± epidote, ± almandine, ± diopside</td><td rowspan="4">Amphibolite</td></tr>
<tr><td>Staurolite</td><td rowspan="2">garnet, anorthite or bytownite, hornblende</td></tr>
<tr><td>Kyanite</td></tr>
<tr><td>Sillimanite</td><td>garnet, anorthite or bytownite, pyroxene</td></tr>
</table>

577. P-T diagram of Metamorphic Facies:

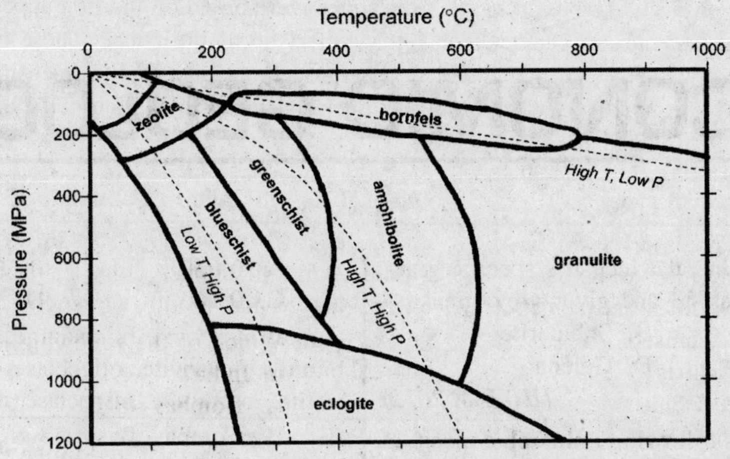

596. Buchan zone:

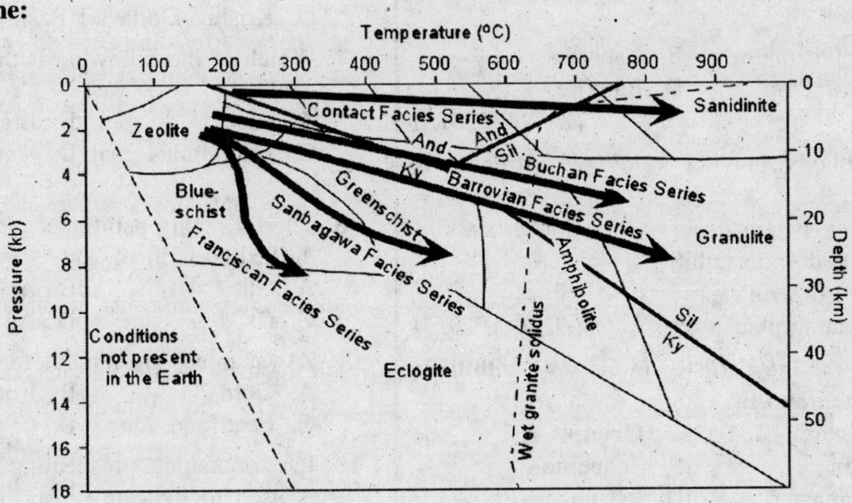

599. AKF diagram:

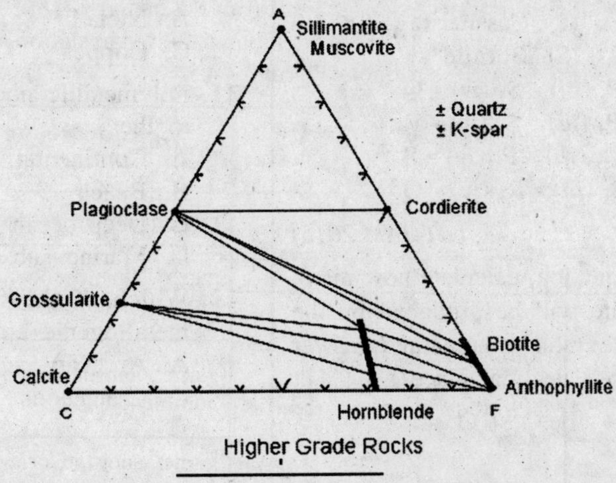

ECONOMIC GEOLOGY

1. Identify the mineral which has specific gravity >5, hardness <3 and gives grey streak.
   A. Barite B. Sphalerite
   C. Graphite D. Galena
   (IIT-JAM 2015)

2. Which of the following mineral deposit is formed exclusively by surface geological processes?
   A. Wollastonite B. Asbestos
   C. Corundum D. Bauxite
   (IIT-JAM 2015)

3. An ore body showing 'saddle reef' structure is formed by:
   A. Early magmatic crystallization process
   B. Liquid immiscibility
   C. Hydrothermal process
   D. Metamorphic process *(IIT-JAM 2015)*

4. Match the metals listed in Group-I with their ores in Group-II:

Group - I	Group - II
(*i*) Tin	(P) Scheelite
(*ii*) Tungsten	(Q) Pyrolusite
(*iii*) Copper	(R) Cassiterite
(*iv*) Manganese	(S) Bornite

   A. (*i*) - R, (*ii*) - P, (*iii*) - S, (*iv*) - Q
   B. (*i*) - Q, (*ii*) - P, (*iii*) - S, (*iv*) - R
   C. (*i*) - R, (*ii*) - Q, (*iii*) - P, (*iv*) - S
   D. (*i*) - P, (*ii*) - S, (*iii*) - R, (*iv*) - Q
   (IIT-JAM 2015)

5. From the given equation, calculate how many moles of magnetite will be produced by the reduction of 1 (one) mole of hematite (Answer to be given up to 3 decimal places).

$$6Fe_2O_3 \rightarrow 4Fe_3O_4 + O_2$$

   A. 0.660 B. 0.666
   C. 0.160 D. 1.60 *(IIT-JAM 2015)*

6. Which of the following is the oldest oil-field in India?
   A. Bombay high
   B. Combay Basin
   C. Digboi
   D. Krishna-Godavari Basin *(AMU-2015)*

7. Which of the following is the highest rank of coal?
   A. Peat B. Lignite
   C. Bituminous coal D. Anthracite
   (AMU-2015)

8. The following pathfinder element is useful in the exploration of Au:
   A. Cu B. As
   C. Cr D. Ag *(AMU-2015)*

9. Zawar mines are famous for:
   A. Gold B. Iron
   C. Lead and Zinc D. Copper *(AMU-2015)*

10. In geobotanical prospecting Douglas Fir plant is used to indicate:
    A. Nickel B. Arsenic
    C. Copper D. Silver *(AMU-2015)*

11. Polymetallic nodules are generally restricted to the:
    A. Continental shelf
    B. Beach
    C. Deep ocean floor
    D. Marine subsoil *(AMU-2015)*

12. Which of the following is the major gold metallogenic epoch?
    A. Archean B. Proterozoic
    C. Cretaceous D. Paleozoic
    (DU-2015)

13. Major Al-deposits in India are associated with:
A. Gondwana rocks
B. Archean rocks in Singhbhum
C. Vindhyan rocks
D. Khondalites along east coast of India *(DU-2015)*

14. Following is also associated with brecciated Iron Ore Copper-Gold (IOCG) deposits:
A. Uranium B. Molybdenum
C. Fluorite D. Zinc *(DU-2015)*

15. Economic concentration of this metal is commonly ascribed to liquid immiscibility:
A. Au B. Pt
C. Cu D. Pb *(DU-2015)*

16. Which of these is an ore mineral of lead?
A. Galena B. Pyrite
C. Wolframite D. Cubanite *(DU-2015)*

17. Placer deposits are formed by:
A. Fluid boiling
B. Magma segregation
C. Gravitational separation
D. Wall-rock alteration *(DU-2015)*

18. In a vein body comprising 'galena - pyrite - sphalerite - quartz', which is/are the gangue mineral(s) ?
A. Quartz and pyrite
B. Sphalerite and quartz
C. Galena and sphalerite
D. Only quartz *(DU-2015)*

19. Which of these is a magnesium carbonate ore?
A. Pyrite B. Malachite
C. Rhodochrosite D. Pyrolusite *(DU-2015)*

20. Which of these is **not** an ore mineral of copper?
A. Cuprite B. Chalcopyrite
C. Bornite D. Molybdenite *(DU-2015)*

21. Which of these rocks could be an important source of diamonds?
A. Komatiite B. Gabbro
C. Eclogite D. Kimberlite *(DU-2015)*

22. Among the following ores, of which is gossan a characteristics surface indicator?
A. Pyrite – Chalcopyrite
B. Chromite – mabnetite
C. Pyrolusite – psilomelane
D. Clumbite – tantalite *(IIT-JAM 2014)*

23. Which of the following physical properties characterize galena?
A. Prismatic form B. Cherry red streak
C. Yellow colour D. High specific gravity *(IIT-JAM 2014)*

24. 'Comb structure' with 'symmetrical banding' is formed by:
A. Early magmatic process
B. Hydrothermal process
C. Chemical sedimentary process
D. Residual concentration process *(IIT-JAM 2014)*

25. Match the items of Group-I with appropriate items in Group-II.

Group-I	Group-II
P. Banded Iron Formation	1. Copper
Q. Andesite porphyry	2. Diamond
R. Lamproit	3. Rare metals
S. Pegmatite	4. Blue dust

A. P-4, Q-1, R-2, S-3
B. P-3, Q-2, R-4, S-1
C. P-4, Q-3, R-1, S-2
D. P-4, Q-1, R-3, S-2 *(IIT-JAM 2014)*

26. Match the mineral deposits in Group-1 with the localities of their occurrences in Group-2.

Group-1	Group-2
P. Lignite	1. Dariba
Q. Fluorite	2. Noamundi
R. Chromite	3. Panandhro
S. Sphalerite-Galena	4. Amba Dongar
	5. Naushani

A. P-2, Q-3, R-4, S-1
B. P-3, Q-4, R-2, S-1
C. P-3, Q-4, R-2, S-1
D. P-4, Q-5, R-1, S-2 *(IIT-JAM 2014)*

27. Malachite and Azurite are characteristics ore minerals of:
A. Gossan zone
B. Zone of oxidation

C. Zone of supergene enrichment
D. Protore *(IIT-JAM 2013)*

28. Match the economic deposits in Group I with their Indian occurrences in Group II.

Group I	Group II
P. Coal	1. Balaghat
Q. Manganese	2. Koderma
R. Magnesite	3. Talcher
S. Mica	4. Salem

A. P-4, Q-1, R-3, S-2
B. P-2, Q-1, R-4, S-3
C. P-4, Q-3, R-2, S-1
D. P-3, Q-1, R-4, S-2 *(IIT-JAM 2012)*

29. Extensive hydrothermal alteration is generally associated with:
A. Stratiform chromite deposit
B. Quartz-pebble conglomerate-hosted gold deposit
C. Superior-type iron deposit
D. Porphyry copper deposit *(IIT-JAM 2012)*

30. Geiger Müller counter is commonly used for the exploration of:
A. Bauxite deposit B. Pb-Zn deposit
C. Uranium deposit D. Iron ore deposit
(IIT-JAM 2012)

31. Mineral often referred to as 'Fool's gold' is:
A. Pyrite B. Magnetite
C. Chromite D. Ilmenite
(BHU-2013)

32. Sukinda in Odisha is well-known for deposits of:
A. Chromite B. Iron
C. Cassiterite D. Pyrrhotite
(BHU-2013)

33. Economically recoverable quantities of each resources are called:
A. Reservoir B. Reserves
C. Resource D. Stuff *(BHU-2013)*

34. Iron ore of Goa State is mainly of:
A. Magnetite B. Hematite
C. Blue dust D. BHJ *(BHU-2013)*

35. Supergene sulphides enrichment occurs:
A. Above the water table
B. A depth independent of water table
C. Below the water table
D. Both above and below the groundwater table
(BHU-2013)

36. Which one is **not** an ore of manganese?
A. Braunite B. Bixybite
C. Jacobite D. Hemimorphite
(BHU-2013)

37. Rampura-Agucha mine is unique due to:
A. Type of ore minerals
B. Grade of metamorphism
C. Presence of rock assemblage
D. Very high grade of ores *(BHU-2013)*

38. Which one of the largest Kyanite deposit of the World?
A. Bhandara region
B. Lapsa Buru
C. Nongstoin area
D. Pipra *(BHU-2013)*

39. What is chemical composition of Siderite?
A. $FeCO_3$ B. $2Fe_2O_3 \cdot 3H_2O$
C. Fe_5S_6 D. Fe_3O_3
(BHU-2013)

40. Which is known as Cobalt Bloom?
A. Asbolite B. Erythrite
C. Smaltite D. Cobalatite
(BHU-2013)

41. Which state among the following has maximum gypsum deposit?
A. Mizoram B. Chhattisgarh
C. Uttar Pradesh D. Jammu & Kashmir
(BHU-2013)

42. Which important mineral is found in Subathu/Krol group of rocks?
A. Graphite B. Talc
C. Pyrophyllite D. Rock salt
(BHU-2013)

43. The pathfinder element of gold is:
A. As B. Se
C. Mo D. SO_4 *(BHU-2013)*

44. Panchpatmali hills are famous for:
A. Bauxite deposits
B. Chromite deposits
C. Coal deposits
D. Iron ore deposits *(BHU-2013)*

45. Which one of the following contains tin ore?
A. Spessartite B. Andradite
C. Cassiterite D. Pitch blend
(BHU-2013)

46. Chalcopyrite is ore mineral of:
A. Aluminium B. Copper
C. Iron D. Silver *(BHU-2012)*

47. A vertical dyke showing transverse veins is known as:
A. Stock work B. Saddle reef
C. Ladder vein D. Vug *(BHU-2012)*

48. The chemical composition of haematite is:
A. Fe_2O_3 B. Fe_3O_4
C. Fe_3O D. Fe_2O_4 *(BHU-2012)*

49. Placer gold deposits are mostly:
A. Elluvial B. Colluvial
C. Aeolian D. Alluvial
(BHU-2012)

50. Banded manganese ores are generally:
A. Epigenetic
B. Syngenetic
C. Paragenetic
D. Both epigenetic and paragenetic
(BHU-2012)

51. Chromite deposits are mostly of:
A. Igneous origin
B. Metamorphic origin
C. Sedimentary origin
D. Both metamorphic and sedimentary origin
(BHU-2012)

52. The chief ore of aluminium is:
A. Pyrolucite B. Sphalerite
C. Bauxite D. Chalcopyrite
(BHU-2012)

53. The most important ore of lead is:
A. Rutile B. Psilomelane
C. Sphalerite D. Galena *(BHU-2012)*

54. The Kolar Gold Field is located in:
A. Bihar B. Karnataka
C. Andhra Predesh D. Tamil Nadu
(BHU-2012)

55. 'Smarskite' is an ore mineral of:
A. Thorium B. Uranium
C. Cobalt D. Copper *(BHU-2012)*

56. The purest form of iron is:
A. Native iron B. Pig iron
C. Wrought Iron D. Steel
(BHU-2012)

57. Uranium deposits of Jaduguda are of:
A. Metamorphic origin
B. Sedimentary origin
C. Magmatic origin
D. Hydrothermal origin *(BHU-2012)*

58. Coal seams are often found to be associated with:
A. China clay B. Fire clay
C. Pottery clay D. Bentonites
(BHU-2012)

59. Match the item in Group-I with appropriate item in Group-II.

Group I	Group II
P. Beach placers	1. Submarine hydrothermal system
Q. Sulphide chimney	2. Mechanical concentration
R. Bauxite	3. Biogenic
S. Phosphorite	4. Residual Concentration

A. P-4, Q-3, R-2, S-1
B. P-4, Q-1, R-3, S-2
C. P-2, Q-1, R-4, S-3
D. P-4, Q-2, R-3, S-1 *(IIT-JAM 2011)*

60. Choose the correct sequence of ore minerals arranged in the increasing order of hardness.
A. Pyrite < Galena < Sphalerite < Magnetite
B. Galena < Sphalerite < Magnetite < Pyrite
C. Pyrite < Magnetite < Sphalerite < Galena
D. Galena < Sphalerite < Pyrite < Magnetite
(IIT-JAM 2011)

61. Which of the given mineral species of gold is rare to form in nature?
A. Au-sulphide B. Au-telluride
C. Native Au D. Au-Ag alloy
(IIT-JAM 2011)

62. Which of the following minerals is used as the source of phosphate in fertilizer?
A. Dolomite B. Fluorite
C. Apatite D. Corundum
(BHU-2011)

63. The metal content of an ore is called:

A. Gangue B. Tenor
C. Carat D. Ore mineral

(BHU-2011)

64. Epigenetic ore deposits are those which formed:

A. Before the host rock formation
B. Same time as the host rock
C. Later than the host rock formation
D. No relation with host rock *(BHU-2011)*

65. Which one of the following does not belong to magmatic deposits?

A. Segregation deposits
B. Immiscible liquid injection
C. Dissemination deposits
D. Sublimation deposits *(BHU-2011)*

66. The placer deposits that occur along hill slopes are:

A. Beach placer B. Eolian placer
C. Stram placer D. Eluvial placer

(BHU-2011)

67. Supergene enrichment deposit has the process except:

A. Oxidation and solution in the zone of oxidation
B. Deposition in the zone of oxidation
C. Supergene sulphide deposition
D. Precipitation from carbonate rich water

(BHU-2011)

68. Which one of the following is **not** the ore of manganese?

A. Braunite B. Bornite
C. Manganite D. Psilomelane

(BHU-2011)

69. Which one of the following is an ore mineral of iron?

A. Manganite B. Magnesite
C. Magnetite D. Malachite

(BHU-2011)

70. The diaspore, aluminium bearing mineral has the .... chemical composition.

A. AlO(OH) B. $Al_2O_3 . H_2O$
C. $Al_2O_3 . 2H_2O$ D. $Al_2O_3 . 3H_2O$

(BHU-2011)

71. Sulphide ore of mercury is:

A. Galena B. Sphalerite
C. Chalcopyrite D. Cinnabar

(BHU-2011)

72. Bauxite is occurring in .... mode.

A. Blanket deposits
B. Pocket deposits
C. Detrial deposits
D. All of these *(BHU-2011)*

73. Rampura-Agucha mines are famous for:

A. Hg-Sn B. Fe-Cu
C. Pb-Zn D. Co-Ni *(BHU-2011)*

74. Azurite and malachite are the .... deposits of copper.

A. Sulphide B. Carbonate
C. Oxide D. Silicate *(BHU-2011)*

75. Which one of the following is **not** a gold field?

A. Kolar Field B. Hutti Field
C. Koderma Field D. Wyanad Field

(BHU-2011)

76. Which one of the following is known as 'brown coal'?

A. Peat B. Lignite
C. Bituminous D. Anthracite

(BHU-2011)

77. Magnesite is formed by action of CO_2 and H_2O on:

A. Ambhibolite B. Serpentine
C. Dacite D. Komattite

(BHU-2011)

78. Match the following:

	Group-1		**Group-2**
P.	Lead	1.	Magmatic
Q.	Aluminium	2.	Pegmatitic
R.	Chromite	3.	Residual
S.	Muscovite	4.	Hydrothermal

A. P-2, Q-1, R-3, S-4
B. P-4, Q-3, R-1, S-2
C. P-3, Q-1, R-4, S-2
D. P-1, Q-2, R-3, S-4 *(BHU-2011)*

79. The placer deposites are formed by the process of:
A. Replacement
B. Mechanical concentration
C. Segregation
D. Dissemination *(BHU-2011)*

80. Gondite and kodurites are:
A. Mg deposits B. Mn deposits
C. Mo deposits D. Cu deposits
(BHU-2011)

81. Smithsonite is an ore of:
A. Zn B. Pb
C. Hg D. Tn *(BHU-2011)*

82. Which one of the following is an INCORRECT statement?
A. Siderite is an iron ore of carbonate composition
B. Hematite is an iron ore with ferromagnetic character
C. Limonite is hydrated iron ore
D. Magnetite is an iron ore that crystallizes in an isometric system *(IIT-JAM 2010)*

83. Commercial magnesite deposit can be hosted in:
A. Serpentinite B. Potassic Granite
C. Gabbro D. Nepheline syenite
(IIT-JAM 2010)

84. Match the economic deposits in Group 1 with their corresponding occurrences in Group 2.

Group 1	Group 2
P. Lignite	1. Jadugoda
Q. Petroleum	2. Malanjkhand
R. Copper	3. Neyveli
S. Uranium	4. Rugrasagar

A. P-3, Q-4, R-1, S-2
B. P-2, Q-4, R-1, S-3
C. P-3, Q-4, R-2, S-1
D. P-4, Q-3, R-2, S-1 *(IIT-JAM 2010)*

85. A copper-nickel-sulphide ore deposit occurring in the basal part of a layered gabbro can originate by the following process:
A. Early magmatic segregation
B. Late magmatic residual liquid segregation
C. Early magmatic injection
D. Late magmatic immiscible liquid segregation
(IIT-JAM 2010)

86. In India, the iron ore deposits are mainly of which type?
A. Magnetite B. Hematite
C. Limonite D. Siderite
(BHU-2010)

87. 'Blue Dust' is a type of:
A. Iron deposit in Goa
B. Copper deposit in Khetri
C. Manganese deposit in Madhya Pradesh
D. Cobalt deposit *(BHU-2010)*

88. Sphalerite is generally associated with:
A. Braunite B. Magnetite
C. Cuprite D. Galena
(BHU-2010)

89. Lead and zinc is mainly mined from:
A. Rajasthan B. Kerala
C. Madhya Pradesh D. Uttar Pradesh
(BHU-2010)

90. Jaduguda is famous for:
A. Uranium B. Copper
C. Manganese D. Tungsten
(BHU-2010)

91. Kimberlites are important because:
A. They contain crustal zenoliths
B. They exhibit fenitisation
C. They host diamonds
D. All of the above *(BHU-2010)*

92. Which of the following is (are) used in the steel industry?
A. Coal
B. Manganese ore
C. Limestone, dolomite, quartz
D. All of the above *(BHU-2010)*

93. The largest iron ore deposits of igneous origin are found in:
A. India B. Denmark
C. Sweden D. Australia
(BHU-2010)

94. Hydrothermal ore deposits of mesothermal type are formed at which temperatures?
A. 0 to 100°C B. 100 to 200°C
C. 200 to 300°C D. 300 to 500°C
(BHU-2010)

95. The chromite deposits of Orissa are found in:
A. Sukinda B. Sitampundi
C. Ratnagiri D. Byrapur Area
(BHU-2010)

96. The bauxite deposits of Ranchi have a composition of about:
A. 20 to 30% Al_2O_3
B. 40 to 50% Al_2O_3
C. 50 to 60% Al_2O_3
D. 10 to 20% Al_2O_3 *(BHU-2010)*

97. Which mineral is used for the separation of aluminium metal by electrolysis?
A. Chrysolite B. Chrysotile
C. Cryolite D. Chiastolite
(BHU-2010)

98. Which one of the following can be used to recover gold by amalgamation?
A. Zinc B. Mercury
C. Copper D. Magnesite
(BHU-2010)

99. Which one of the following is a Neutral Refractory Mineral?
A. Chromite B. Kyanite
C. Sillimanite D. Magnesite
(BHU-2010)

100. Which one of the following is used in the cement industry?
A. Zinc B. Pyrite
C. Gypsum D. Kyanite
(BHU-2010)

101. Iron-ore group is known from:
A. Uttar Pradesh B. Maharashtra
C. Arunachal D. Jharkhand-Orissa
(BHU-2010)

102. Which of the following is an ore mineral of uranium?
A. Sphalerite B. Pitchblende
C. Cassiterite D. Wolframite
(IIT-JAM 2009)

103. Which one of the following minerals is characterized by black streak?
A. Hematite B. Sphalerite
C. Magnetite D. Malachite
(IIT-JAM 2009)

104. Tin and Tungsten mineralization is commonly associated with:
A. Basalt B. Gabbro
C. Dunite D. Granite
(IIT-JAM 2009)

105. Ore mineral of copper most common in the zone of supergene sulphide enrichment is:
A. Chalcopyrite B. Chalcocite
C. Cuprite D. Covelite
(BHU-2009)

106. Epigenetic minerals are generally formed by:
A. Mountain building process
B. Crustal movement
C. Dislocation
D. Igneous intrusion *(BHU-2009)*

107. The position of graphite in India is:
A. Strategic
B. Critical
C. Essential
D. Both strategic and essential
(BHU-2009)

108. Chromite is a member mineral of:
A. Spinel group B. Epidote group
C. Olivine group D. Melilite group
(BHU-2009)

109. Cu and Pb of Agnigundala belong to which group of Cuddapah supergroup?
A. Papaghani B. Cheyair
C. Nallamalai D. Kistna
(BHU-2009)

110. In Rajasthan, copper is confined to:
A. Raialo group B. Bhilwara group
C. Delhi group D. Aravalli group
(BHU-2009)

111. Kolar Gold deposit is an example of:
A. Fissure vein deposit
B. Shear zone deposit
C. Ladder vein deposit
D. Stock work *(BHU-2009)*

112. Which one of the following does **not** belong to mica group?
A. Oxide B. Silicate
C. Carbonate D. Phosphate
(BHU-2009)

113. The most common impurity in iron ore is:
A. Muscovite B. Hornblende
C. Biotite D. Lepidolite
(BHU-2009)

114. Chalk Hill of Salem Tamil Nadu is famous for:
A. Limestone B. Baryte
C. China clay D. Magnesite
(BHU-2009)

115. Which of the following is quick silver?
A. Native silver
B. Native mercury
C. Native amalgam
D. Mixture of lead and silver *(BHU-2009)*

116. Which of the following minerals is known as heavy spar?
A. Baryte B. Calcite
C. Flourite D. Apatite *(BHU-2009)*

117. What type of metamorphism is responsible for the formation of hornfelsic rocks?
A. Burial B. Regional
C. Contact D. Cataclastic
(BHU-2009)

118. Ajabgarh formation is associated with:
A. Iron B. Zinc
C. Manganese D. Copper *(BHU-2009)*

119. Lower Gondwana sediments are of great economic significance because of presence of:
A. Iron B. Petroleum
C. Coal D. Plant fossils
(BHU-2009)

120. Cherry red streak is shown by the mineral:
A. Magnetite B. Limonite
C. Hematite D. Siderite *(BHU-2009)*

121. Mineral bauxite is formed by:
A. Magmatic process
B. Sedimentary process
C. Pegmatitic process
D. Post-magmatic process *(BHU-2009)*

122. The ascending ore-forming fluids within the earth's crust give rise to:
A. Hypogene deposit B.Supergene deposit
C. Syngenetic deposit D. Stratiform deposit
(IIT-JAM 2008)

123. Match the coalfields listed in Group 1 with their occurrences in State listed in Group 2.

Group-1	Group-2
P. Talchir	1. Uttar Pradesh
Q. Nayveli	2. Orissa
R. Singrauli	3. Tamil Nadu
S. Jhingurda	4. Madhya Pradesh

A. P-1, Q-2, R-3, S-4
B. P-2, Q-3, R-4, S-1
C. P-3, Q-4, R-1, S-2
D. P-4, Q-1, R-2, S-3 *(IIT-JAM 2008)*

124. Match the mineral deposits listed in Group 1 with their occurrences listed in Group 2.

Group-1	Group-2
P. Gold	1. Majhgawan
Q. Bauxite	2. Zawarmala
R. Lead-Zinc	3. Palamau
S. Diamond	4. Hutti

A. P-2, Q-1, R-3, S-4
B. P-4, Q-3, R-4, S-1
C. P-3, Q-1, R-2, S-3
D. P-1, Q-3, R-4, S-2 *(IIT-JAM 2008)*

125. A heavy mineral assemblage in a rock consists of rounded grains of tourmaline, zircon, rutile, kyanite and staurolite. The provenance of such assemblage is:
A. Hydrothermal deposits
B. Reworked sedimentary rock
C. Basic igneous rock
D. Acidic igneous rock *(IIT-JAM 2008)*

126. A greyish brown coloured mineral of metallic luster gives cherry red streak. The minerals is:
A. Wolframite
B. Hematite
C. Chromite
D. Magnetite
(IIT-JAM 2008)

127. Hydrothermal deposits formed at a temperature range of 50 – 200° C are known as:
A. Mesothermal
B. Hypothermal
C. Epithermal
D. Telethermal *(IIT-JAM 2008)*

128. Which of the following processes of ore formation produces gold deposits in the form of reef?
A. Hydrothermal
B. Magmatic injection
C. Magmatic segregation
D. Supergene enrichment *(IIT-JAM 2007)*

129. Beach placers in India are an important source of:
A. Copper B. Lead
C. Thorium D. Uranium
(IIT-JAM 2007)

130. Chromite deposits result from:
A. Early magmatic dissemination
B. Early magmatic segregation
C. Early magmatic injection
D. Late magmatic residual fluid injection
(IIT-JAM 2006)

131. The largest known kyanite deposit in the world is in:
A. Jharkhand B. Bihar
C. Assam D. Tamil Nadu
(IIT-JAM 2006)

132. Of the following, pick the mineral that is **not** associated with pegmatites:
A. Mica B. Emerald
C. Feldspar D. Kyanite
(IIT-JAM 2005)

133. Supergene sulphide enrichment occurs:
A. Below the water table
B. Above the water table
C. At depths independent of water table
D. Both above and below the water table
(IIT-JAM 2005)

134. A mineral has steel-grey colour, metallic lusture, two sets of perfect octahedral cleavage and very high specific gravity. The mineral is:
A. Galena B. Magnetite
C. Molybdenite D. Barite
(IIT-JAM 2005)

135. Among the following, commercial asbestos is:
A. Cristobalite B. Chrysotile
C. Cryolite D. Chaistolite
(IIT-JAM 2005)

136. Graphite deposits from largely by:
A. Magmatic concentration
B. Sedimentation
C. Metamorphism
D. Residual concentration *(BHU-2014)*

137. Chromite deposits of Sukinda is formed by:
A. Magmatic concentration
B. Contact metasomatism
C. Hydrothermal processes
D. Metamorphism *(BHU-2014)*

138. 'Bort' is the term used for:
A. Gem diamond
B. Industrial diamond
C. Variety of corundum
D. Variety of garnet *(BHU-2014)*

139. Rhodochrosite is an ore of:
A. Copper B. Iron
C. Maganese D. Lead *(BHU-2014)*

140. Muscovite is associated with:
A. Magmatic deposits
B. Pegmatitic deposits
C. Placer deposits
D. Sedimentary deposits *(BHU-2014)*

141. India is deficient in:
A. Iron deposits
B. Bauxite deposits
C. Mica deposits
D. Tungsten deposits *(BHU-2014)*

142. Which of the following groups of minerals are prominently associated with the beach sand of Kerala?
A. Monazite-magnetite-wolframite
B. Monazite-cassiterite-rutile
C. Monazite-chromite-zircon
D. Monazite-ilmenite+rutile *(BHU-2014)*

143. Which of the following economic mineral is used in the refractory industry?
A. Orthoclase B. Tourmaline
C. Chromite D. Aragonite
(BHU-2014)

144. Match the list-1 and list-2 and select the correct answer using the codes given below the lists:

List-1	List-2
I. Abrasive	1. Kynaite
II. Ceramic	2. Sulphur

III. Refractory 3. Orthoclase
IV. Chemical 4. Corundum

	I	II	III	IV
A.	1	3	4	2
B.	4	3	1	2
C.	4	2	1	3
D.	3	1	2	4

(BHU-2014)

145. Gypusm to be used as fertilizers should have:
A. 70% $CaSO_4.2H_2O$
B. 75% $CaSO_4.2H_2O$
C. 80% $CaSO_4.2H_2O$
D. 85% $CaSO_4.2H_2O$ *(BHU-2014)*

146. Guano is:
A. An organic calcareous rock
B. An organic phosphatic rock
C. An organic ferruginous rock
D. None of the above *(BHU-2014)*

147. Which amongst the following is the main constituent mineral of a true Gossan?
A. Hematite B. Magnesite
C. Limonite D. Pyrite

148. Which one of the following is single largest deposit of iron ores in India?
A. Chiria B. Noamundi
C. Kiriburu D. Bailadila
(BHU-2014)

149. Major Iron ore deposits of the world belongs to:
A. Phanerozoic
B. Cenozoic
C. Neoproterozoic
D. Paleoproterozoic and Archean *(DU-2014)*

150. Archean greenstone belts are associated with economic deposits of:
A. Gold B. Phosphorus
C. Lead D. Rare earth elements
(DU-2014)

151. India's largest aluminium deposits are associated with:
A. Gondawana Supergroup
B. Eastern Ghats mobile belt
C. Lesser Himalayas
D. Mahakoshal group *(DU-2014)*

152. World's major platinum group of element (PGE) deposits located in:
A. Australia
B. South Africa
C. United states of America
D. Ethiopia *(DU-2014)*

153. Economic deposit of Uranium is most likely to be found in:
A. Gabbro B. Rhyolite
C. Sandstone D. Peridotite *(DU-2014)*

154. The metal content in an ore is called as:
A. Grade B. Tenor
C. Rank D. Gangue *(BHU-2015)*

155. .... deposits comprise an assemblage of high temperature metamorphic minerals.
A. Hypothermal B. Mesothermal
C. Skarn D. Residual
(BHU-2015)

156. Among the following which one does not belong to the factor affecting the residual concentration?
A. Climate
B. Local relief
C. Existence of proper drainage
D. Gravity *(BHU-2015)*

157. .... is a hard mantle which is produced by useless residual materials and desiccated product of oxidation on the surface.
A. Gossan B. overburden
C. Gangue D. Placer *(BHU-2015)*

158. Goethite is a/an:
A. Silicate B. Sulphide
C. Carbonate D. Oxide *(BHU-2015)*

159. Tenorite is an ore mineral of:
A. Iron B. Manganese
C. Copper D. Lead *(BHU-2015)*

160. Which of the following is the ore mineral of manganese?
A. Smithsonite B. Hematite
C. Bornite D. Braunite
(BHU-2015)

161. Manganite ore mineral has been found in the rock of:
A. Gondite B. Charnockite
C. Orthoquarzite D. Marl
(BHU-2015)

162. Bauxite deposits occur as:
A. Disseminated
B. Loades
C. Blanket
D. Hydrothermal solution *(BHU-2015)*

163. Chromite deposits occur in:
A. Granite B. Limestone
C. Acidic rocks D. Ultrabasic rocks *(BHU-2015)*

164. Kudramukh is famous for:
A. Banded Hematite Quartz (BHQ)
B. Magnetite
C. Chromite
D. Psilomelane *(BHU-2015)*

165. Ruby is a gem variety of:
A. Olivine B. Labradorite
C. Quartz D. Corundum *(BHU-2015)*

166. .... is an imperfectly crystallized diamond with inclusions.
A. Bort B. Ballas
C. Carborando D. Moonstone *(BHU-2015)*

167. Secondary product of Diaspore and Gibbsite is:
A. Bohemite B. Bauxite
C. Hydrogillite D. Amazon stone *(BHU-2015)*

168. Match the following and choose the correct option:

Column-A	Column-B
P. Muscovite	1. Residual concentration
Q. Bauxite	2. Magmatic segregation
R. Copper	3. Pegmatitic deposits
S. Chromite	4. Hydrothermal solution

A. P-4, Q-2, R-1, S-3
B. P-3, Q-1, R-2, S-4
C. P-3, Q-1, R-4, S-2
D. P-4, Q-2, R-3, S-1 *(BHU-2015)*

169. Chromite deposits result from:
A. Early magmatic dissemination
B. Early magmatic segregation
C. Early magmatic injection
D. Late magmatic segregation *(BHU-2015)*

170. One of the following is the rarest metal in the earth crust:
A. Gold B. Platinum
C. Rhenium D. Niobium *(PU-201)*

171. Wall rock alteration is a characteristic feature of:
A. Magmatic deposits
B. Hydrothermal deposits
C. Metamorphic deposits
D. Volcano-sedimentary deposits *(PU-2010)*

172. One of the following is an ore mineral of copper:
A. Bornite B. Braunite
C. Bixbyite D. Boehmite *(PU-2010)*

173. Talc deposit can be formed by metasomatism of one of the following rocks.
A. Limestone B. Shale
C. Ultramafic rock D. Granite *(PU-2010)*

174. The reason for the correct answer to the previous question is that this rock contains:
A. Mg silicates B. Fe silicates
C. Al silicates D. Ca silicates *(PU-2010)*

175. Identify the odd type of ore deposit among the following, by considering the age of known ore deposits.
A. Banded iron formation
B. Ferromanganese nodules
C. Lateritic bauxite
D. Phosphatic nodules *(PU-2010)*

176. In the previous question, the odd type of ore deposit has formed in the following geological time.
A. Precambrian B. Paleozoic
C. Mesozoic D. Tertiary *(PU-2010)*

177. What is called "green marble"?
A. Amphibolite B. Pyroxenite
C. Peridotite D. Serpentinite *(PU-2010)*

178. "Green marble" is mined as building stone from:
A. Amphibolite of Dharwar schist belts
B. Pyroxenite of Nausahi ultramafic complex
C. Peridotite of Sukinda ultramafic complex
D. Serpentinite of Rishabdev ultramafic belt *(PU-2010)*

179. One of the following sulphide minerals can be transluscent or transparent.
A. Pyrite
B. Chalcopyrite
C. Sphalerite
D. Galena (PU-2010)

180. One of the following oxide minerals can be transluscent or transparent:
A. Chromite B. Pyrolusite
C. Wolframite D. Cassiterite (PU-2010)

181. Gold deposits are NOT associated with one of the following:
A. Laterite
B. Banded iron formation
C. Conglomerate
D. Shale (PU-2010)

182. Uranium deposits have NOT formed by one of the following processes:
A. Detrital sedimentary
B. Circulation of groundwater
C. Hydrothermal
D. Magmatic (PU-2010)

183. Emerald is a gem variety of:
A. Corundum B. Beryl
C. Topaz D. Tourmaline (PU-2010)

184. Which one of the following pairs of minerals is NOT dimorphic?
A. Graphite-diamond
B. Calcite-aragonite
C. Magnetite-hematite
D. Kyanite-sillimanite (PU-2010)

185. Contact metasomatic deposits are more common in:
A. Shale B. Greywacke
C. Sandstone D. Limestone (PU-2010)

186. One of the following ore minerals does **not** contain copper:
A. Covellite B. Pyrrhotite
C. Bornite D. Chalcocite (PU-2010)

187. The largest zinc deposit in India is located at:
A. Zawar, Rajasthan
B. Agucha, Rajasthan
C. Agnigundla, Andhra Pradesh
D. Sargipalle, Orissa (PU-2010)

188. The largest bauxite mines in India is located at:
A. Panchpatmali, Orissa
B. Korba, Madhya Pradesh
C. Yercaud, Tamil Nadu
D. Kolli Hills, Tamil Nadu (PU-2010)

189. Majority of world's coal resources are restricted to the following geological time period:
A. Triassic
B. Permo-Carboniferous
C. Cambro-Ordovician
D. Eocene (PU-2013)

190. Which one of the followings is a sulphide ore mineral?
A. Scheelite B. Siderite
C. Sphalerite D. Celestite (PU-2013)

191. Which one of the following minerals has the highest specific gravity?
A. Galena B. Diamond
C. Native silver D. Native gold (PU-2013)

192. Which one of the following economic minerals does **not** belong to the category of magmatic deposits?
A. Chromite
B. Titaniferous magnetite
C. Cu-Ni sulphide
D. Magnesite (PU-2013)

193. Porphyry copper deposits are formed by which one of the following ore forming process?
A. Magmatic
B. Magmatic hydrothermal
C. Sedimentary
D. Metamorphic (PU-2013)

194. In India, which one of the following ore minerals is produced mostly from beach sand?
A. Cassiterite B. Hematite
C. Ilmenite D. Magnetite (PU-2013)

195. Which one of the following metals occurs as native metal, oxide and sulphide in the earth's crust?

A. Iron B. Copper
C. Uranium D. Gold *(PU-2013)*

196. Identify the odd. pair of minerals among the following.

A. Goethite - siderite
B. Pyrite - marcasite
C. Pyrite - pyrrhotite
D. Hematite - magnetite *(PU-2013)*

197. In which one of the following lists, minerals are arranged in order of increasing temperature of crystallization?

A. Galena - gibbsite - chromite
B. Gibbsite - galena - chromite
C. Gibbsite - chromite - galena
D. Galena - chromite gibbsite *(PU-2013)*

198. Which one of the following isotopes is the naturally occurring fissile atom of interest in nuclear reactors?

A. ^{238}U B. ^{235}U
C. ^{234}U D. ^{232}Th *(PU-2013)*

199. Petroleum and Natural gas are chiefly composed of:

A. Hydrogen
B. Hydrogen and Carbon
C. Nitrogen
D. Oxygen *(PU-2013)*

200. Which of the following minerals would be most likely to form a clay mineral during chemical weathering?

A. Iron oxide B. Mica
C. Calcite D. Quartz *(PU-2012)*

201. What are the two most abundant elements in the Earth's crust?

A. Iron and magnesium
B. Oxygen and silicon
C. Nitrogen and oxygen
D. Silicon and calcium *(PU-2012)*

202. Gold deposits are NOT associated with one of the following:

A. Quartz lode
B. Banded iron formation
C. Conglomerate
D. Shale *(PU-2012)*

203. Uranium deposits have NOT formed by one of the following processes:

A. Detrital sedimentary
B. Circulation of groundwater
C. Hydrothermal
D. Magmatic *(PU-2012)*

204. India is a leading producer of one of the following metals:

A. Gold B. Aluminium
C. Copper D. Uranium *(PU-2012)*

205. One of the following sulphide minerals can be transluscent or transparent:

A. Pyrite B. Chalcopyrite
C. Sphalerite D. Galena *(PU-2012)*

206. One of the following oxide minerals can be transluscent or transparent:

A. Chromite B. Pyrolusite
C. Wolframite D. Cassiterite *(PU-2012)*

207. Identify the odd type of ore deposit among the following by considering the environment of ore formation.

A. Banded iron formation
B. Ferromanganese nodules
C. Lateritic bauxite
D. Phosphatic nodules *(PU-2012)*

208. In the previous question, the odd type of the ore deposit has formed in the following environment.

A. Shallow marine B. Deep sea
C. Lake D. Terrestrial *(PU-2012)*

209. Ore deposits of the following metals is formed by magmatic process:

A. Aluminium B. Iron
C. Lead D. Zinc *(PU-2012)*

210. One of the following locations does not have a copper mines.

A. Rakha B. Malanjkhand
C. Kolihan D. Byrapur *(PU-2012)*

211. The correct answer to previous question is a location where there is:
A. Lead-zinc mines
B. Bauxite mines
C. Chromite mines
D. Iron ore mines *(PU-2012)*

212. One of the following defines the cut-off grade of an ore:
A. Minimum metal content of an ore
B. Average metal content of an ore
C. Minimum thickness of an ore body
D. Average thickness of an ore body *(PU-2012)*

213. In a lateritic bauxite deposit, the uppermost litho-unit is:
A. Bauxite
B. Laterite
C. Lithomarge
D. Partially weathered bedrock *(PU-2012)*

214. One of the following defines the hydrothermal ore forming process:
A. Metals are transported and precipitated from carbonic fluid
B. Metals are transported and precipitated from hot carbonic fluid
C. Metals are transported and precipitated from aqueous fluid
D. Metals are transported and precipitated from hot aqueous fluid *(PU-2012)*

215. One of the following minerals deposits is **not** associated with granite pegmatites:
A. Chalcopyrite B. Cassiterite
C. Muscovite D. Beryl *(PU-2012)*

216. One of the following represents the chemical composition of pyrrhotite
A. $Fe_{1-X}S$ B. FeS_{1-X}
C. $Fe_{1-X}S_2$ D. Fe_2S_{1-X} *(PU-2012)*

217. One of the following represents the chemical composition of magnetite:
A. $Fe + 2Fe_3 + 2O_4$
B. $Fe + 3Fe_2 + 2O_4$
C. $Fe_2 + Fe_3 + O_3$
D. $Fe_3 + 2O_3$ *(PU-2012)*

218. The bands names Leica, Olympus, Nikon, Zeiss refer to manufacturers of:
A. Geological maps
B. Polarizing microscope
C. Spectrometers
D. Geological field kit *(PU-2012)*

219. The pocket lens used in geological field work has a magnification of:
A. 5X B. 10X
C. 50X D. 100X *(PU-2012)*

220. One of the following minerals does **not** from placer deposit:
A. Magnetite B. Ilmenite
C. Diomond D. Sphalerite *(PU-2012)*

221. Nickel sulfide ore deposits are example of
A. Early magmatic deposit
B. Late magmatic deposit
C. Contact metamorphic deposit
D. Residual sedimentary deposit *(PU-2011)*

222. Which of the following is the example of Residual sedimentary ore deposit?
A. Chromite
B. Gold
C. Banded Iron Formation
D. Bauxite *(PU-2011)*

223. Gossans are indicative of mineral occurrences. They are formed due to:
A. Contact metamorphism because of magma intrusion
B. Hydrothermal alteration of country rock
C. Intensely oxidized, weathering or decomposed rock of an ore deposit
D. Sedimentary exhalative processes *(PU-2011)*

224. Coking coal in India is found in
A. Neyvelli, Tamil Nadu
B. Jharia, Jharkhand
C. Palana, Rajasthan
D. Garampani, Meghalya *(IIT JAM-2016)*

225. Match List-A and List-B and select the correct answer:

List-A (Processes)	List-B (Mechanism)
A. Magmatic	1. Cavity filling
B. Hydrothermal	2. Dissemination
C. Contact metasomatic	3. Recombination
D. Residual concentration	4. Weathering

A. A-4, B-3, C-1, D-2
B. A-4, B-1, C-3, D-2
C. A-2, B-1, C-3, D-4
D. A-4, B-3, C-1, D-2 *(ISM-2014)*

226. Match the ore deposits in Group I with the localities in Group II

Group-I	Group-II
(i) Copper	(P) Balaghat, M.P.
(ii) Lead-Zinc	(Q) Panchpatmali, Odisha
(iii) Manganese	(R) Rampura-Agucha, Rajasthan
(iv) Bauxite	(S) Khetri, Rajasthan

A. (i) - S, (ii) - R, (iii) - Q, (iv) - P
B. (i) - S, (ii) - R, (iii) - P, (iv) - Q
C. (i) - P, (ii) - R, (iii) - Q, (iv) - S
D. (i) - S, (ii) - Q, (iii) - R, (iv) - P
(IIT JAM-2016)

227. Which of the following ore deposit(s) is/are formed only by hydrothermal process?
A. 'Sn-W' ore associated with greissenised rock
B. Layered type chromite ore associated with dunite-peridotite-pyroxenite
C. Vein type gold ore associated with green-stone belt
D. Ni-Cu sulphide ore associated with gab-broic rocks *(IIT JAM-2016)*

228. Gossan zones on the surface would probably indicate of:
A. Hydrothermal sulphide enrichment
B. Supergene sulphide enrichment
C. Development of skarn zones with sulphide enrichment
D. Development of bauxite *(ISM-2014)*

229. Which one of the following is a tertiary coalfield?
A. Korba (Chattisgarh)
B. Jhan (Jharkhand)
C. Godavari (A.P.)
D. Palana (Rajasthan) *(ISM-2014)*

230. Match List-1 with List-2 and select the correct answer:

List-1	List-2
A. Copper	1. Chandapathar
B. Tungusten	2. Ambadongar
C. Gold	3. Agnigundala
D. Fluorite	4. Kundrekocha

A. A-2, B-4, C-1, D-3
B. A-4, B-1, C-3, D-2
C. A-1, B-3, C-2, D-4
D. A-3, B-1, C-4, D-2 *(ISM-2014)*

231. The reservoir rocks at Bombay High hydrocarbon field is predominantly:
A. Biomicrites B. Biosparite
C. Oomicrites D. Oosparites
(ISM-2014)

232. Which of the following has the least volatile content and highest carbon?
A. Bituminous coal B. Lignite
C. Anthracite D. Peat *(ISM-2015)*

233. Best quality of Indian coking coal is reported from:
A. Jharia coal field
B. Bokaro coal field
C. Raniganj coal field
D. Jabalpur coal field *(ISM-2015)*

234. Lateritic bauxite is also known as:
A. Carbonate bauxite
B. Silicate bauxite
C. Gibbsite
D. Karst bauxite *(ISM-2015)*

235. The most important oil producing horizon in India, Bombay High is of:
A. Proterozoic age B. Miocene
C. Paleocene D. Mesozoic
(ISM-2015)

236. East coast bauxite deposit is associated which of the following rocks:
A. Gondite
B. Basalt
C. Kodurite
D. Khondalite *(SAGAR UNIVERSITY-2015)*

237. Ore mineral that are diagnostic of an oxidizing environment are?
A. Native gold, copper and silver
B. Limonite and pyrolusite
C. Manganese minerals
D. Carbonate, sulphates and silicates of Cu, Zn
(SAGAR UNIVERSITY-2015/

238. Example of hydrothermal deposits:
A. Mica
B. Molybdenite
C. Copper
D. Lead-zinc
(SAGAR UNIVERSITY-2015)

239. Lake Superior copper is a example of which type of deposit?
A. Pore-space filling
B. Solution cavity filling
C. Vesicular filling
D. Gas veins *(SAGAR UNIVERSITY-2015)*

240. What is the composition of charnockite?
A. Hypersthene + K-feldspar + Plagioclase + Quartz
B. Hypersthene + Quartz
C. Feldspar + Plagioclase + Quartz
D. Hypersthene + Plagioclase + K-feldspar + Quartz
(SAGAR UNIVERSITY-2015)

241. Which important ore is formed by magmatic concentration as stratified?
A. Chromite
B. Gold
C. A & B
D. Copper
(SAGAR UNIVERSITY-2015)

242. Limestone and dolomite is distinguished by:
A. Alumina content
B. Lime content
C. Magnesia content
D. Soda content
(SAGAR UNIVERSITY-2015)

243. Gossans indicates subsurface:
A. Groundwater
B. Oil deposit
C. Oxide deposit
D. Sulphide deposit
(SAGAR UNIVERSITY-2015)

244. Sargipalli is known for:
A. Chromite deposit
B. Mn-ore deposit
C. Fe-ore deposit
D. Base metal deposit
(SAGAR UNIVERSITY-2015)

245. The Proterozoic Mn-deposit of India:
A. They are not associated with iron-formation and black shells
B. They were all deposited in shallow water basin margin
C. They are associated with iron-formation only
D. They are associated with shallow marine water basins
(SAGAR UNIVERSITY-2015)

246. Chrysolite asbestos result from the:
A. Magmatic liquid
B. Alteration of serpentine
C. Alteration of olivine to serpentine
D. Hydrothermal solutions
(SAGAR UNIVERSITY-2015)

247. Sukinda in Orissa is known for which of the following deposits:
A. Molybdenite deposits
B. Chromite deposits
C. Pyrite deposits
D. Fluorite deposits
(SAGAR UNIVERSITY-2015)

248. Host rock in the Malanjkhand deposit is:
A. Granite
B. Quartz reef
C. Dolomitic limestone
D. Dolerite *(SAGAR UNIVERSITY-2015)*

249. Bauxite is an aggregate of:
A. Gibbsite and diaspora
B. Gibbsite and bonhomie
C. Diaspora and bonhomie
D. Gibbsite and diaspora and boehmite
(PUNJAB UNIVERSITY-2015)

250. Chrysolite asbestos of Cuddapah district (A.P.) is found in:
A. Srisailam quartzite
B. Kolamanala shells
C. Contact of a bed of Cuddapah limestone with a dolerite still
D. Tadpatri shales
(PUNJAB UNIVERSITY-2015)

251. Match the following :

List-A	List-B
P. Bauxite deposits of central and western India	1. Khondalites
Q. Bauxite deposits of Andhra Pradesh and Orissa	2. Detrital orogin
R. Bauxite deposits of Tamil Nadu	3. Deccan traps
S. Bauxite deposits of Kati (M.P)	4. Charnokites

A. P-3, Q-1, R-2, S-4
B. P-3, Q-4, R-2, S-1
C. P-3, Q-1, R-4, S-2
D. P-2, Q-4, R-3, S-1
(PUNJAB UNIVERSITY-2015)

252. Chromite belongs to which of the minerals group:
A. Olivine B. Spinel
C. Pyroxene D. Garnet
(PUNJAB UNIVERSITY-2015)

253. Mark the correct statements about the Mosabani copper deposit:
A. The sulphide assemblage of Mosabani situated with the in Singhbhum shear zone
B. Quartz chlorite-biotite schist is the major host rock
C. Chalcopyrite , pyrite, pyrrhotite constitute the main ore minerals
D. All the correct
(PUNJAB UNIVERSITY-2015)

254. Choose the incorrect pair from the following:
A. Azurite – Copper
B. Goethite – Iron
C. Rhodonite – Manganese
D. Saxonite – Chromium *(AMU-2016)*

255. Which of the following is sulphide minerals?
A. Cinnabar B. Sylvite
C. Sulphur D. Rhodochrosite
(AMU-2016)

256. Which of the following is a copper ore?
A. Pyrolusite B. Psilomelane
C. Bornite D. Galena *(AMU-2016)*

257. Mineral deposits that form in oceanic basins near ridge-related hot spring are called?
A. Evaporates B. Magmatic
C. Pegmatite D. Hydrothermal
(AMU-2016)

258. The major elements in Mn-nodules are:
A. Fe and Mn
B. Fe, Mn and Ni
C. Fe, Mn, Ni and Cu
D. Fe, Mn, Ni, Cu and Co *(AMU-2016)*

259. The formation temperature of different magmatic deposits varies from:
A. 2000 °C to 500 °C
B. 1500 °C to 300 °C
C. 1000 °C to 200 °C
D. 800 °C to 100 °C *(AMU-2016)*

260. The metasomatic deposits are usually resulted in:
A. Calcareous rocks
B. Arenaceous rocks
C. Ferrogenous rocks
D. Siliceous rocks *(AMU-2016)*

261. Economic concentration of the following metal is commonly ascribed to liquid immiscibility
A. Au B. Pt
C. Cu D. Pb *(ISM-2016)*

262. The following minerals is also associated with brecciated Iron Ore-Copper–Gold(IOCG) deposits
A. Uranium B. Molybdenum
C. Fluorite D. Zinc
(ISM-2016)

263. A metallogenic epoch is
A. A geographic area comprising a number of ore deposit formed by similar or related processes
B. The specific interval of geological time during which ore formation was most intense
C. The barren area between various ore zone in a region comprising multiple ore deposit
D. The specific interval of geological time during which ore formation was inactive
(ISM-2016)

264. Match the following in List-A with List-B

List-A (Mineral)	***List-B (Area)***
P. Gold	1. Jaduguda
Q. Hematite	2. Quilon
R. Ilmenite	3. Noamundi
S. Uraninite	4. Hutti

A. P-1, Q-2, R-3, S-4
B. P-4, Q-3, R-2, S-1
C. P-4, Q-3, R-1, S-2
D. P-3, Q-2, R-1, S-2 *(ISM-2016)*

265. In a recent placer deposit olivine is not found but monazite is present. The reason is at the surface condition:

A. $G_{monazite} < G_{olivine}$
B. $G_{monazite} > G_{olivine}$
C. $G_{monazite} = G_{olivine}$
D. $G_{monazite} \geq G_{olivine}$ *(ISM-2016)*

266. Which copper ore is showing peacock colour?

A. Bornite B. Braunite
C. Malachite D. Azurite *(BHU-2016)*

267. The minerals constituting an example of ______ solid solution is:

A. Pyrolusite B. Pyrite
C. Pyrrhotite D. Pigeonite *(BHU-2016)*

268. The concentration of the heavy minerals by stream or wave action leading to an economic deposits is called:

A. Hydrothermal deposits
B. Placer deposits
C. Skarn deposits
D. Stratiform deposits *(DU-2016)*

269. Which of the following elements is associated with the Crude oil?

A. Cd B. Pb
C. Ag D. V *(DU-2016)*

270. Hydrocarbon reservoir in Bombay High is:

A. Well sorted sandstone
B. Fossiliferous sandstone
C. Fractured granitic basement
D. Salt layer *(DU-2016)*

271. Which of the following represent an ore minerals of tin?

A. Wolframite B. Rutile
C. Cassiterite D. Molybdenite *(DU-2016)*

272. Chromite deposits are associated with _____ rocks:

A. Pigmatitic B. Felsic
C. Ultamafic D. Leaucocratic *(DU-2016)*

273. India in the largest production ________ of world:

A. Mica B. Gold
C. Copper D. Bauxite *(DU-2016)*

274. Ore deposits formed at shallow deposit in the homogeneous range of the 50 °C to 200 °C may range be:

A. Hypothermal
B. Mesothermal
C. Xenothermal
D. Epithermal *(DU-2016)*

275. The term BHQ, BMQ and BHJ generally refer is ______ ores:

A. Iron B. Copper
C. Bauxite D. Manganese *(DU-2016)*

276. A greisen deposits exposes the ______ part of the body:

A. Lower B. Uppermost
C. Middle D. Basal *(DU-2016)*

277. Match the following:

List-A	**List-B**
1. Scheelite	A. ZnS
2. Pyrite	B. FeS_1
3. Galena	C. $CaWO_4$
4. Sphalerite	D. PbS

A. 1-B, 2-A, 3-C, 4-D
B. 1-C, 2-B, 3-D, 4-A
C. 1-C, 2-B, 3-A, 4-D
D. 1-B, 2-A, 3-D, 4-C *(DU-2016)*

ANSWER KEY

1	2	3	4	5	6	7	8	9	10
D	D	C	A	B	C	D	B	C	D
11	**12**	**13**	**14**	**15**	**16**	**17**	**18**	**19**	**20**
C	A	D	A	B	A	C	A	C	D
21	**22**	**23**	**24**	**25**	**26**	**27**	**28**	**29**	**30**
D	A	D	B	A	B	B	D	D	C
31	**32**	**33**	**34**	**35**	**36**	**37**	**38**	**39**	**40**
A	A	B	C	C	D	A	B	A	B
41	**42**	**43**	**44**	**45**	**46**	**47**	**48**	**49**	**50**
D	D	A	A	C	B	C	A	D	B
51	**52**	**53**	**54**	**55**	**56**	**57**	**58**	**59**	**60**
A	C	D	B	B	C	C	D	C	C
61	**62**	**63**	**64**	**65**	**66**	**67**	**68**	**69**	**70**
B	C	B	C	D	D	D	B	C	A
71	**72**	**73**	**74**	**75**	**76**	**77**	**78**	**79**	**80**
D	D	C	A	C	B	B	B	B	B
81	**82**	**83**	**84**	**85**	**86**	**87**	**88**	**89**	**90**
A	B	A	C	C	B	A	D	A	A
91	**92**	**93**	**94**	**95**	**96**	**97**	**98**	**99**	**100**
C	D	C	C	A	C	C	B	A	C
101	**102**	**103**	**104**	**105**	**106**	**107**	**108**	**109**	**110**
D	B	C	D	B	B	B	A	C	C
111	**112**	**113**	**114**	**115**	**116**	**117**	**118**	**119**	**120**
A	B	B	D	B	A	C	D	C	C
121	**122**	**123**	**124**	**125**	**126**	**127**	**128**	**129**	**130**
B	A	B	B	B	B	C	A	C	B
131	**132**	**133**	**134**	**135**	**136**	**137**	**138**	**139**	**140**
A	C	A	A	B	C	A	B	C	B
141	**142**	**143**	**144**	**145**	**146**	**147**	**148**	**149**	**150**
D	D	A	B	B	B	C	B	D	A
151	**152**	**153**	**154**	**155**	**156**	**157**	**158**	**159**	**160**
D	B	C	B	C	D	A	D	D	D
161	**162**	**163**	**164**	**165**	**166**	**167**	**168**	**169**	**170**
A	C	D	B	D	A	B	C	B	D
171	**172**	**173**	**174**	**175**	**176**	**177**	**178**	**179**	**180**
B	A	C	A	D	D	D	D	C	D

181	182	183	184	185	186	187	188	189	190
D	B	B	C	A	B	A	A	C	C
191	**192**	**193**	**194**	**195**	**196**	**197**	**198**	**199**	**200**
D	D	B	A	A	C	B	B	B	A
201	**202**	**203**	**204**	**205**	**206**	**207**	**208**	**209**	**210**
B	D	B	B	C	D	C	D	B	D
211	**212**	**213**	**214**	**215**	**216**	**217**	**218**	**219**	**220**
C	A	D	D	A	A	A	B	B	D
221	**222**	**223**	**224**	**225**	**226**	**227**	**228**	**229**	**230**
B	D	C	B	C	B	A, C	B	D	D
231	**232**	**233**	**234**	**235**	**236**	**237**	**238**	**239**	**240**
A	C	A	C	B	D	B	C	D	A
241	**242**	**243**	**244**	**245**	**246**	**247**	**248**	**249**	**250**
C	C	C	A	D	C	B	A	C	C
251	**252**	**253**	**254**	**255**	**256**	**257**	**258**	**259**	**260**
C	D	D	D	A	C	D	D	B	A
261	**262**	**263**	**264**	**265**	**266**	**267**	**268**	**269**	**270**
D	A	B	B	B	A	C	B	D	B
271	**272**	**273**	**274**	**275**	**276**	**277**			
C	C	A	D	A	B	B			

HINTS AND SOLUTIONS

1.

Minerals	**Specific gravity**	**Hardness**	**Streak**	**Colours**
Barite ($BaSO_4$)	4.3-4.6	3-3.5	White	White
Sphalerite (ZnS)	3.9	3.5-4	Pale	Brown
Graphite (C)	2.1	1-2	Black	Black
Galena (PbS)	7.6	2-3	Grey	Grey

2.

Mineral deposits	**Processes**	**Deposits**
Wallostonite	Sub surface	Metamorphic deposit
Asbestus	Sub surface	Magmatic deposit
Corundum	Sub surface	Pegmatitic deposit
Bauxite	Surface	Mechanical concentration deposit

3.

Ores processes minerals	**Structure**	**Economic**
Early magmatic crystallization	Diamond pipe	Diamond, chromite
Liquide immiscibility	Sills, dykes	Iron, nickel, copper
Hydrothermal processes	Saddle reef, vuge	Gold, silver, lead
Metamorphic processes	Fold, fault	Talc, kyanite, graphite

4.

Metals	**Ores**	**Distributions**	**Source rocks**
Tin	Cassiterite	Salem, Koraput in Orissa.	Pegmatite vein.
Tungsten	Scheelite	East Godavri basin of Andhra Pradesh	Pegmatite.
Cooper	Bornite	Singhbhum Copper Belt, Jharkhand.	Quartz schist.
Manganese	Pyrolusite	Sausor Group of rocks in MH&MP.	Gondites, Quartzite

6. Oil fields	Working time
Bombay high	1974
Cambay basin	1958
Digboi oil field	1931
Krishna-Godavari basin	1983

7. Types of coal	% of Carbon	% of Oxygen	% of Hydrogen
Peat	55	35	6
Lignite	72	20	5
Bituminous	85	8	5
Anthracite	93	2	2

9. Ores mineral	Famous mine	State
Gold	Kolar	Karnataka
Iron	Singhbhum	Jharkhand
Lead and Zinc	Zawer	Rajasthan
Copper	Khetri	Rajasthan

13. Rocks unit	Ore deposits
Gondwana rocks	Coal
Archean rocks of Singhbhum	Iron ores
Vindhayan rocks	Diamond
Khondalite along east coast of India	Major Al-deposits

16. Minerals	Ores
Galena (PbS)	Lead
Pyrite (Fe_2S)	Iron
Wolframite (Fe, Mn)WO_4	Tunguston
Cubanite ($CuFe_2S_3$)	Copper-iron sulphide

19. Ores mineral	Properties	Formula
Pyrite	Iron sulphides	FeS_2
Malachite	Copper carbonate	$CuCO_3(OH)_2$
Rhodochrosite	Magnesium carbonate	$MnCO_3$
Pyrolusite	Magnisium oxide	MnO_3

20. Minerals	Ores
Cuprite	Oxide of Copper
Chalcopyrite	Sulphide of Copper
Bornite	Sulphide of Copper
Molybdenite	Sulphide of Molybdenum

21. Rocks	Source rocks
Komatiite	Magnisium
Gabbro	Iron
Eclogite	Garnet
Kimberlite	Diamond

22. Ores	Surface indicator	Minerals
Pyrite-chalcopyrite	Gossan Zone	Copper
Chromite-magnetite	Enrichment zone.	Chromium, Iron
Pyrolusite-psilomelane	Oxidized zone	Magnesium
Columbite-tantelite	Protolith	Tantalum, Niobium

23. Physical properties of Galena.	
Galena	PbS
Crystal system	Isometric system
Form	Granular
Colour	Grey
Streak	Grey
Cleavage	Three set
Fracture	Concoidal
Hardness	2-3
Specific gravity	High (7.6)
Lustre	Metallic

24. Process	Structure formed
Early magmatic process	Sills, Dykes
Hydrothermal process	Comb structure, Vugs
Chemical sedimentary process	Bedding plane, Ripplemarks
Residual concentration	Tidal flow

25. Formation	Ore minerals	Indian example
Banded Iron	Blue dust formation	BIF belt of Jharkhand
Andesite prphyry	Copper	Malanjkhand copper deposit of Madhaya Predesh.
Lamproite	Diamond	Panna diamond of Madhya Pradesh.
Pigmatite	Rare metal	Pigmatite deposits of Jharkhand.

26. Mineral deposits	Localities	Area (State)	Source rocks
Lignite	Panandhro	Gujarat	Shale, Sandstone
Fluorite	Ambadongar	Gujarat	Carbonatite
Chromite	Naushahi	Orissa	Schistose rocks
Sphalerite-Galena	Rampura - Dariba	Rajasthan	Dolomite

27. Zones	Ore minerals
Gossan Zone	Pyrite
Zone of oxidation	Cuprite, malachite, azurite
Zone of supergene enrichment	Chalcopyrite, covelite
Protore(primary zone)	Pyrite, bornite

28. Economic deposits	Indian occurrence	State	Source rocks
Mangnese	Mansar formation	Maharastra	Mica-schist
Mica	Koderma mica belt	Jharkhand	Mica-schist.
Coal	Talchir coal field	Odisha	Boulder bed, Sandstone
Magnesite	Salem magnesite deposits	Andhra Pradesh	Serpentinite

29. Types deposits	Associations
Stratiform chromite deposit	Banded deposits
Quartz-pebble conglomerate-hosted gold deposit	Lodes deposits
Superior-type iron deposit	Extensive iron deposits in transgressive sea.
Porphyry copper deposit	Extensional hydrothermal alteration

30. Mineral deposits	Exploration technique	Exploration Instrument
Bauxite deposit	Gravity technique	Gravimeter
Pb-Zn deposit	Gravity technique	Gravimeter
Uranium deposit	Radiometric technique	Geiger Mueller counter
Iron ore deposit	Magnetic technique	Magnetometer

31. Mineral deposits	Location	Source rocks
Chromite	Sukinda, Odisha	Pyroxenite, Peridotite
Iron	Singhbhum, Jharkhand	Granodiorite, Granite
Cassiterite	Koraput Orissa	Pegmatite veins
Pyrrhotite	Mosaboni Jharkhand	Schistose rocks

34. Iron ores	State type
Magnetite	Tamil Nadu
Hematite	Jharkhand
Blue dust	Goa
BHJ	Orissa

36. Minerals	Ores
Braunite	Mn-ores
Bixybite	Mn-ores
Jacobite	Mn-ores
Hemimorphite	Zn-ores

37. Chemical composition	Iron Ores minerals
$FeCO_3$	Siderite
$2Fe_2O_3.3H_2O$	Limonite
$Fe_5 S_6$	Pyrite
Fe_3O_3	Hematite

40. Minerals	Characteristics
Asbolite	Mn, Cu, Fe ores of oxide and sulphide form
Erythrite	Cobalt and nickel ores
Smaltite	Tin-white cobalt
Cobalatite	Copper-Iron sulphide

41. State	Maximum ores deposit
Mizoram	Oil and Gas
Chhattisgarh	Coal
Uttar Pradesh	Silica Sand
Jammu & Kashmir	Gypsum

42. Minerals	Groups of rocks
Graphite	Dharwarian rocks
TalcKornool group	
Pyrophyllite	Schistose rocks of Singhbhum
Rock salt	Subathu/Krol group

44. Types of deposits	Ores minerals	Source rocks
Panchpatmali Hills of Orissa	Bauxite	Khondalite
Sukinda chromite	Chromite	
Coal deposits of Jharia	Bituminous	Shale
Iron ores of Orissa	BHQ & BHJ	Quartzite

45. Minerals	Contain
Spessartite	Mn-Garnet
Andradite	Ca-Fe-Garnet
Cassiterite	Tin ores
Pitch blend	Uranium ore

48. Chemical composition	Names
Fe_2O_3	Hematite
Fe_3O_4	Magnetite
Fe_3O_2	Ferrous oxide
Fe_2O_4	Ferrous oxide

49. Types of placer	Economic minerals
Elluvial	Tunguston
Colluvial	Magnetite
Aeolian	Chromite
Alluvial	Gold

52. Minerals	Ores
Pyrolucite	Mn
Sphalerite	Zinc
Bauxite	Aluminium
Chalcopyrite	Copper

53. Minerals	Ores
Rutile	Titanium
Psilomelane	Manganese
Sphelerite	Zinc
Galena	Lead

59. Deposits	Processes	Economic minerals
Beach placers	Mechanical concentration	Gold, Platinum, Silver
Sulphide chimney	Submarine hydrothermal system	Copper, zinc
Bausite	Residual concentration	Alluminium
Phosphorite	Biogenic	Phosphate

60. Minerals	Ore minerals	Hardness
Pyrite(Fe_2S)	Iron	6-6.5
Sphelerite	Zinc	3.5-4
Magnesite	Magnesium	6
Galena	Lead	2-2.5

62. Minerals	Uses
Dolomite	Refractory
Fluorite	Chemical industry
Apatite	Fertilizer
Corundum	Abrasive

68. Ores	Minerals
Braunite	Manganese
Bornite	Copper
Manganite	Manganese
Psilomelane	Manganese

69. Ores	Minerals
Manganite	Manganese
Magnesite	Magnesiem
Magmetite	Iron
Malachite	Copper

70. Chemical composition	Minerals
(Fe,Mn,Al)O(OH)	Diaspore
(Fe,Mn,Al)O(OH)	Boehmite
$Al(OH)_3$	Bauxite
$Al(OH)_3$	Gibbsite

71. Ores	Minerals
Galena	Sulphide ore of Lead
Sphalerite	Sulphide ore of Zinc
Chalcopyrite	Sulphide ore of Copper
Cinnabar	Sulphide ore of Mercury

74. Types of Copper ores	Example
Sulphide	Bornite, Covellite, Chalcocite
Carbonate	Azurite, Malachite
Oxide	Cuprite, Tenorite
Silicate	Chrysocolla

75. Economic deposits	Occur-rence	State	Source rocks
Kolar field	Gold	Karnataka	Quartz reefs
Hutti field	Gold	Karnataka	Quartz-reefs
Koderma field	Mica	Jharkhand	Mica-schist
Wyanad field	Gold	Tamilnadu	Quartz-reefs with gneiss

82. Ores	Chemical composition	Ores-Minerals
Hematite	Fe_2O_3	Iron oxidė
Siderite	$FeCO_3$	Iron carbonate
Limonite	$Fe_2O_3\ 3H_2O$	Iron hydrate
Magnetite	$MgCO_3$	Iron carbonate

84. Deposits	Area	Source rocks
Nayeli lignite deposit	Tamil Nadu	Limestone
Rudrasagar oil field	Assam	Brail Sandstone
Malanjkhand copper deposit	Madhya Pradesh	Biotite granite
Jaduguda uranium deposit	Jharkhand	Schist gneisses

85. Process	Ores minerals
Early magmatic segregation	Nickle, Chromium, Platinum
Late magmatic residual magmatic injection	Ilmenite, Magnetite
Late magmatic immiscible liquid segregation	Nickle, Platinum
Early magmatic injection	Copper, Chromium

94. Hydrothermal deposits	Temperature
Telethermal type	0 to 100ºC
Epithermal type	100 to 200ºC
Mesothermal type	200 to 300ºC
Hypothermal type	300 to 500ºC

100. Minerals	Uses
Zinc	Fertilizer Industry
Pyrite	Iron Industry
Gypsum	Cement Industry
Kyanite	Refractory Industry

102. Ore mineral	Minerals
Hematite(Fe_2O_3)	Iron oxide
Pitchblende($2UO_3UO_2$)	Uranium oxide
Cassiterite(SnO_2)	Tin oxide
Wolframite(Fe,Mn)WO_4	Tungusten

103. Mineral	Streak
Hematite	Red brown(cherry red)
Sphelerite	Light blue
Magnetite	Black
Malachite	Green

104. Rocks	Associated minerals
Basalt	Gold, Copper
Gabbro	Magnetite, Ilmenite
Dunite	Magnesite, Quartz,
Granite	Quartz, Feldspar, mica.

105. Copper ores	Zone
Chalcopyrite	Primary zone
Chalcocite	Supergene enrichment zone
Cuprite	Oxidized zone
Covelite	Oxidized zone

107. Classification of mineral	Example mineral
Strategic	Aluminium
Critical	Uranium, Thorium
Essential	Iron, Copper
Strategic & essential	Diamond, Gold, Copper

108. Minerals group	Example mineral
Spinel group	Chromite, Magnetite
Epidote group	Zoisite, Clinozoisite
Olivine group	Forsterite, Fayalite
Mellilite group	Gehlenite, Akermanite

109. Cuddapah supergroup	Economic minerals
Papaghani group	Asbestos, Barytes
Cheyair group	Jasper
Nallamalai group	Lead(Pb)& Copper(Cu)
Kistna	--------

110. Stratigraphy of Rajasthan	Economic minerals
Raialo Group	Marble
Bhilwara Group	Manganese
Delhi Group	Copper
Aravalli Group	Lead & Zinc

111. Types of cavity filling	Example
Fissure vein deposits	Kolar gold mine Karnataka
Shear zone deposits	Singhbhum shear zone
Ladder vein deposits	Magnesite deposit of Karnataka
Stock work	Zawar mine of Rajasthan

116. Minerals	Properties (Name)
Baryte	Heavy Spar
Calcite	Calc Spar
Flourite	Fluorspar
Apatite	Fluorapatite

117. Types of metamorphism	Rocks-formation
Burial metamorphism	Granulites
Regional metamorphism	Marble
Contact metamorphism	Hornfels
Cataclastic metamorphism	Mylonite

118. Economic mineral	Formations
Iron	Chaibasa formation
Zinc	Nallamallai formation
Manganese	Mansar formation
Copper	Ajabgarh formation

120. Minerals	Streak
Magnetite	Black
Limonite	Brown
Hematite	Cherry red
Siderite	White

121. Process	Minerals deposits
Magmatic process	Iron, Diamond
Sedimentary process	Bauxite, Copper, Iron
Pegmatitic process	Mica, Uranium, Tin
Post-magmatic process	Manganese

123. Coal fields	State	Coal type
Talchir	Orissa	Bituminous
Nayveli	Tamil Nadu	Lignite
Singrauli	Madhya Pradesh	Bituminous
Jhingurda	Uttar Pradesh	Bituminous

124. Mineral deposits	Occurrences	Source rocks
Gold	Hutti, Karnataka	Hornblende schist
Bauxite	Palamau, Madhya Pradesh	Laterite
Lead-Zinc	Zawarmala Rajasthan	Dolomite
Diamond	Majhgawan	Kimberlite pipe

125. Provenance	Minerals assemblage
Hydrothermal deposits	Tin, Mercury
Reworked sedimentary rock	Tourmaline, zircon, rutile, kyanite
Basic igneous rock	Pyroxene, olivine
Acidic igneous rock	Biotite, quartz, feldspar

126. Colour	Lusture	Strek	Hardness
Wolframite	Submetallic	Reddish brown	4 - 5
Hematite	Submetallic	Cherry red	5 - 6
Chromite	Submetallic	Brown	5 - 6
Magnetite	Metallic	Black	6

127. Hydrothermal deposits	Temperature range
Epithermal deposits	50-200°C
Mesothermal deposits	200-300°C
Hypothemal deposits	300-600°C
Telethermal deposits	50°C - near surface.

128. Processes	Economic minerals
Hydrothermal	Gold(reef)
Magmatic injection	Magnetite(vein)
Magmatic segregation	Magnetite, Chromite
Supergene enrichment	Covellite, Chalcocite

129. Minerals	Types of placers
Copper	Eluvial placers
Lead	Stream placers
Thorium	Beach placers
Uranium	Stream placers

130. Magmatic process	Economic mineral
Early magmatic dissemination	Diomond
Early magmatic segregation	Chromite
Early magmatic injection	Iron
Late magmatic residual	Gold, Copper
Fluid injection	

132. Minerals	Associations
Mica	Mica associated with pegmatites in Precambrian rocks of Jharkhand.

Emerald	Emerald is gem verity of corundum which associated pigmatites rocks of Udaipur Rajasthan.
Feldspar	Feldspar is not associated with pegmatites.
Kyanite	Kyanite is also associated with pegmatites world largest deposits of Jharkhand.

134. Proper-ties	Galena	Magne-tite	Molyb-denite	Barite
Formula	PbS	Fe_3O_4	MoS_2	$BaSO_4$
Colour	Grey	Black	Grey	White
Lusture	Metallic	Metallic	Metallic	Vitreous
Cleavage	Perfect	Perfect	Parallel	Perfect
S.G.	7.4-7.6	5.2	4.7-4.8	4.50

135. Minerals	Formula	Properties
Cristobalite	SiO_2	Polymorph of Quartz
Chrysotile	$Mg_3Si_2O_5(OH)_4$	Commercial asbestos
Cryolite	Na_3AlF_6	Aluminum ore
Chaistolite	Al_2SiO_5	Variety of Andalusite

140. Types of deposits	Associated minerals
Magmatic deposits	Iron, Diamond
Pegmatitic deposits	Muscovite
Placer deposits	Gold, Thorium, Platinum
Sedimentary deposits	Aluminium, Silver, Gold

143. Minerals	Uses
Orthoclase	Refractory minerals
Tourmaline	Gemstone
Chromite	Metals
Aragonite	Abrasive

190. Ores Minerals	Characteristics
Scheelite	Calcium molybdates
Siderite	Iron carbonate
Sphalerite	Zinc sulphide
Celestite	Strontium sulphates

191. Minerals	Specific gravity
Galena	7.5- 7.6
Diamond	3.5
Native silver	10-11
Native gold	19-20

196. Minerals	Specific Characteristics
Goethite	Iron oxide
Siderite	Iron Carbonate
Pyrite	Iron sulphide
Marcasite	Iron sulphide
Pyrrhotite	Iron with nickel ore
Hematite	Iron sulphide
Magnetite	Iron oxide

197. Minerals	Crystallization temperature (°C)
Gibsite	60-70
Galena	330
Chromite	1100

224. Coking Coal: Coking coal is an essential ingredient in steel production. It is different to thermal coal which is used to generate power. Coking coal, also known as metallurgic coal, is heated in a coke oven which forces out impurities to produce coke, which is almost pure carbon.

- Very low ash levels (less than 3%) - thus reduced waste material
- Very low phosphorus (less than 0.01%) - a major positive for steel strength
- Very good coke characteristics - it performs well in the furnace
- Moderate sulphur (around 1%) - reduced by coal processing

The quality of the coking coal our mine is able to produce will be very important as the higher quality, the greater demand from steel producers looking to maximise their productivity. Tests from the area lead us to believe that we can extract high quality 'hard' coking coal, which is the most desirable.

225. Processes	Mechanism	Example
Magmatic	Dissemination	Diamond
Hydrothermal	Cavity filling	Gold, Silver, Lead, Zinc
Contact metasomatic	Recombination	Copper, Mn
Residual concentration	Weathering	Bauxite, Iron, Kyanite

226. Minerals	Area	Host rocks
Copper	Khetri, Rajasthan	Dolomite
Lead-Zinc	Rampura-Agucha, Rajasthan	Mica schist dolomite
Manganese	Balaghat, M.P.	Granulite
Bauxite	Panchpatmali, Odisha	Khondalite

229. Coal field	Coal type	Geological age
Korba, Chattisgarh	Bituminous	Gondawana age
Jhan, Jharkhand	Bituminous	Gondawana age
Godavari, A.P.	Bituminous	Gondawana age
Palana, Rajasthan	Lignite	Tertiary

228. Gosssan zone and water table:

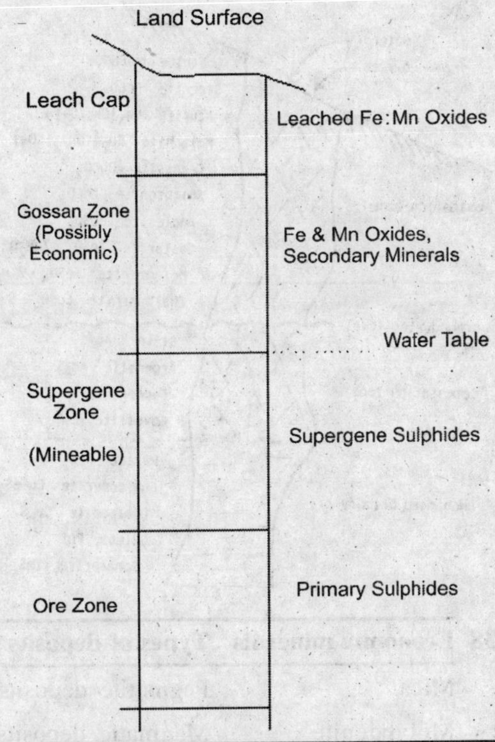

236. Rock types	Remarks
Gondite	Mn bearing rocks
Basalt	Ca rich
Kodurite	Mn rich
Khondalite	Garnet-silliminite schist rocks

232. Fuel type	Higher calorific value (Mj/kg, dry)	Volatile matter	Fixed carbon	Ash	C	H	O	N	S
		(wt%, dry)			(wt%, dry (ash-free)				
biomass	19.7	76.4	20.4	3.2	49.8	6.3	43.5	0.4	0.01
Lignite	26.5	36.8	35.2	28.0	70.7	4.9	23.1	1.0	0.3
sub-bituminous	26.9	44.5	41.4	14.1	76.3	6.4	15.3	1.3	0.7
sub-bituminous	27.2	42.5	49.0	8.5	69.1	5.4	23.8	1.1	0.6
hv-bituminous	32.4	37.6	55.1	7.3	83.6	5.5	7.7	1.6	1.5
hv-bituminous	29.7	32.5	53.2	14.3	83.4	5.4	8.8	1.9	0.5
mv-bituminous	30.9	21.8	70.8	7.4	87.6	4.7	4.9	2.1	0.8
lv-bituminous	29.3	14.3	68.8	16.9	89.0	4.3	4.6	1.6	0.4
petroleum coke	36.8	11.8	85.9	2.5	86.3	3.9	1.3	2.9	5.7

237. Weathering of An Ore Body

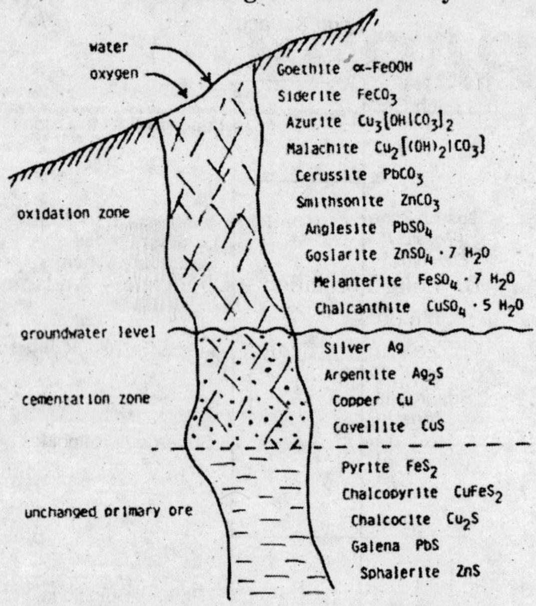

238. Economic minerals	Types of deposits
Mica	Pegmatite deposits
Molybdenite	Magmatic deposits
Copper	Hydrothermal deposits
Lead-Zinc	Cavity filling deposits

244. Deposits	Distribution
Chromite deposits	Keonjhar deposits of Orissa.
Mn-ore deposits	Mansar formation of the Sausor group
Fe-ore deposits	Singhbhum iron deposits
Base metal deposits	Archean rocks of the Bihar and Odisha

248. Host rocks	Deposits
Granite	Malanjkhand copper deposit
Quartz reef	Kolar gold field
Dolomitic limestone	Carbonate minerals in India
Dolerite	Phosphatic noudules of the Utatur stage.

254. Ore minerals	Composition	Ores
Azurite	$Cu_3(CO_3)2(OH)_2$	Copper
Goethite	FeO(OH)	Iron
Rhodonite	(Mn, Ca) SiO_3	Manganese
Saxonite	Cr-rich dunite	Chromium

256.	Minerals	Ores
	Pyrolusite	Mn
	Psilomelane	Mn
	Bornite	Cu
	Galena	Pb

STRATIGRAPHY

1. Arrange the following granitoid bodies in order of their decreasing age.
   P. Erinpura Granite
   Q. Closepet Granite
   R. Singhbhum Granite
   S. Ladakh Granitoids
   A. R-P-Q-S B. S-Q-R-P
   C. R-Q-P-S D. Q-R-S-P
   (IIT-JAM 2015)

2. Arrange the following formations of the Gondwana Supergroup in order of their decreasing age:
   (*i*) Raniganj Formation
   (*ii*) Barakar Formation
   (*iii*) Kulti Formation
   (*iv*) Karharbari Formation
   A. (*i*) - (*iii*) - (*ii*) - (*iv*)
   B. (*i*) - (*ii*) - (*iii*) - (*iv*)
   C. (*iv*) - (*iii*) - (*ii*) - (*i*)
   D. (*iv*) - (*ii*) - (*iii*) - (*i*) *(IIT-JAM 2015)*

3. Which of the following chronostratigraphic unit(s) is/are correctly matched with the corresponding geochronologic unit(s)?
   A. System-period B. Stage-era
   C. Series-epoch D. Chronozone-age
   (IIT-JAM 2015)

4. Which of the following formation(s) was/were deposited in glacial and/or fluvial environment?
   A. Subathu Formation
   B. Karewa Formation
   C. Indus Formation
   D. Blaini Formation *(IIT-JAM 2015)*

5. Which of the following stratigraphic units are correctly matched with their ages and geographical localities?
   A. Niniyur Formation - Cretaceous - Cauvery basin
   B. Long Formation - Oligocene - Andaman-Nicobar
   C. Zewan Formation - Permian - Kashmir-Lidar valley
   D. Bhander Group - Early Proterozoic - Vindhyan basin
   (IIT-JAM 2015)

6. The duration of the Proterozoic Eon in Ma is:
   A. 2000 B. 2500
   C. 630 D. 230 *(IIT-JAM 2015)*

7. Which of the following is the youngest?
   A. Muth Quartzite B. Lameta Beds
   C. Golden Oolite D. Karewas
   (AMU-2015)

8. Ordovician period was named after a:
   A. Place B. Tribe
   C. Rock type D. Geologist
   (AMU-2015)

9. Which of the following is the type area of Jarassic rocks in India?
   A. Kutch B. Trichinappally
   C. Spiti D. Jaisalmer
   (AMU-2015)

10. The smallest division of geological time scale is:
    A. Eon B. Era
    C. Period D. Epoch *(AMU-2015)*

11. Siwalik hills are famous for:
    A. Invertebrate fossils
    B. Plant fossils
    C. Dinosaur fossils
    D. Mammal fossils *(AMU-2015)*

12. Dinosaur were living during:
A. Proterozoic B. Palaeozoic
C. Mesozoic D. Cenozoic
(AMU-2015)

13. The age of Bagh Beds is:
A. Cretaceous B. Jurassic
C. Triassic D. Permian
(AMU-2015)

14. The age of Fenestella Shales is:
A. Ordovician B. Devonian
C. Carboniferous D. Permian
(AMU-2015)

15. Which of the following fossils are evidence for Gondwanaland continental assembly?
A. *Mesosaurus* B. *Lystrosaurus*
C. *Glossopteris* D. All the above
(AMU-2015)

16. The basic unit in Biostratigraphy is:
A. Subzone B. Zone
C. Chron D. System *(DU-2015)*

17. Glossopteris is a plant fossil known from the rocks of:
A. Upper Gondwana
B. Middle Gondwana
C. Lower Gondwana
D. Rajmahal lnter-trappean Beds *(DU-2015)*

18. The age of Muthquartzite is:
A. Silurian B. Ordovician
C. Devonian D. Archean *(DU-2015)*

19. Permian is represented in Spiti by:
A. Kanawar Group
B. Kuling Group
C. Muth Quartzite
D. Agglomerate Shale *(DU-2015)*

20. Age of the oldest rocks in the ocean basin is:
A. Devonian B. Eocene
C. Pennian D. Jurassic *(DU-2015)*

21. The sills and dykes of basaltic composition are found at different places of Gondwana succession. They found in Satpura and Rewa are thought to be ____ of age, while of Damodar Valley and Assam of ____ age, respectively.
A. Rajmahal - Deccan Trap
B. Deccan Trap - Rajmahal
C. Both are Rajmahal Trap
D. Both are Deccan Trap *(DU-2015)*

22. Syringothyris Cuspidata is the characteristic fossil of:
A. Silurian
B. Lower Devonian
C. Lower Carboniferous
D. Upper Carboniferous *(DU-2015)*

23. Out of these, which is intrusive into Delhi system:
A. Bijli Rhyolite
B. Erinpura Granite
C. Bundelkhand Granite
D. Singhbhum Granite *(DU-2015)*

24. Which combination of geological entities occurs in the same craton?
A. Chitradurga schist belt-Closepet Granite-Hatti gold mine
B. Bhilwara Supergroup-Sitapundi anorthosite - Zawer Pb-Zn deposit
C. Kolar schist belt-Malani rhyolite-Malanjkhand copper deposit
D. Mansar formation – Makrana marble – Sukinda chromite deposit
(IIT-JAM 2014)

25. Choose the correct statement from the following:
A. Vindhayan supergroup is devoid of limestone
B. The top of the Cuddapah Supergroup is Phanerozoic
C. Kaladgi basin is situated in the Baster Craton
D. The Mesozoic Bhuj Formation consist predominantly of sandstone
(IIT-JAM 2014)

26. Identify the correct chronostratigraphic sequence of the Siwalik Group from the following (Oldest to youngest):
A. Pinjor-Nagri-Dhok Pathan-Tatrot
B. Nagri-Dhok Pathan-Tatrot-Pinjor
C. Dhok Pathan-Tatrot-Nagri-Pinjor
D. Tatrot-Pinjor-Nagri-Dhok Pathan
(IIT-JAM 2014)

27. Match the lithostratigraphic units in Group-1 with their corresponding geological period in Group-2.

Group-1	**Group-2**
P. Mammal Formation	1. Triassic
Q. Fenestella Shale	2. Permian
R. Maleri Formation	3. Cretaceous
S. Lameta Formation	4. Carboniferous

A. P-3, Q-4, R-1, S-2
B. P-4, Q-2, R-1, S-3
C. P-2, Q-3, R-4, S-1
D. P-4, Q-1, R-2, S-3 *(IIT-JAM 2013)*

28. The concept of 'uniformitarianism' are found in:

A. J. Hutton B. Wegner
C. Dutton D. Kant *(BHU-2013)*

29. Match the following volcanic stratigraphic units in Group I with their given ages in Group II.

Group I	**Group II**
P. Malani Rhyolite	1. Meso- to Paleoproterozoic
Q. Dalma Volcanics	2. Paleozoic
R. Panjal Trap	3. Mesozoic
S. Rajmahal Trap	4. Neoproterozoic

A. P-4, Q-1, R-2, S-3
B. P-1, Q-4, R-3, S-2
C. P-2, Q-4, R-1, S-3
D. P-4, Q-3, R-2, S-1 *(IIT-JAM 2012)*

30. Which one of the following fossils is found in rocks of Cambrian age?

A. *Redlichia* B. *Fenestella*
C. *Syringothyris* D. *Otoceras*
(IIT-JAM 2012)

31. Which one of the following flora represents Upper Gondwana?

A. *Noeggerathiopsis* B. *Gangamopteris*
C. *Dicroidium* D. *Vertebraria*
(IIT-JAM 2012)

32. Which is the youngest era in the geological time scale?

A. Precambrian B. Mesozoic
C. Cenozoic D. Paleozoic
(BHU-2013)

33. The characteristics of index fossils are?

A. Big and easy to see in field
B. Easily preserved hard shell
C. Widespread geographically and limited short span of time
D. Widespread in time span and limited geographically
(BHU-2013)

34. Which system is very rich of vertebrate fossils?

A. Gondwana B. Zewan beds
C. Vindhyan D. Siwalik rock
(BHU-2013)

35. Devonian marker horizon in Himalaya is known as:

A. Po formation
B. Lipak formation
C. Muth quartzite
D. Syringothyris limestone *(BHU-2013)*

36. A well-preserved evidence of stromatolites in Semari group is abundant:

A. Kagrahat limestone
B. Fawn limestone
C. Bleaching shale
D. Suket shale *(BHU-2013)*

37. Nagthat formation belong to which geological age?

A. Mesozoic B. Paleozoic
C. Proterozoic D. Archean
(BHU-2013)

38. Near Jabalpur, Lameta beds belong to:

A. Fluvio-lacustrine facies
B. Volcanic facies
C. Marine transgression facies
D. Marine facies *(BHU-2013)*

39. Youngest succession of Kutch Basin rock represented by:

A. Madh formation
B. Porabandar formation
C. Umia formation
D. Bhuj formation *(BHU-2013)*

40. The main boundary fault seperates:

A. Siwalik and Archean
B. Siwalik and Tertiary

C. Siwalik and Delhi group
D. Vindhyan and Aravalli *(BHU-2013)*

41. The Permian rock represented in the Sikkim region belongs to:
A. Lipak series B. Lachi series
C. Laki series D. Lathi formation *(BHU-2013)*

42. Division of Himalaya from South to North is:
A. Trans-Siwalik-Lesser-Higher
B. Siwalik-Lesser-Higher-Trans
C. Higher-Lesser-Siwalik-Trans
D. Lesser-Siwalik-Higher-Trans *(BHU-2013)*

43. Most fossilliferous bed of Siwalik Himalaya is:
A. Nagri B. Kamlial
C. Dhokpathan D. Chingi *(BHU-2013)*

44. Zewan beds are located in:
A. Meghaiaya B. Sikkim
C. Mizoram D. Kashmir *(BHU-2013)*

45. Which one does **not** belong to Brail series?
A. Tura sandstone stage
B. Tikak Parbat stage
C. Naogaon stage
D. Bragolai stage *(BHU-2013)*

46. The Mn-ores is associated in India as:
A. Chorbaoli formation
B. Mansar formation
C. Bochum formation
D. Lohangi formation *(BHU-2013)*

47. The Dalma and Dhanjori lava belongs to:
A. Cuddapah B. Vindhyan
C. Dharwar D. Singhbhum *(BHU-2013)*

48. Match the following:

A. Alwar series	1. Railo series
B. Bhagwanpura limestone	2. Aravalli
C. Binota shale	3. Delhi system
D. Semri series	4. Vindhyan

A. A-2, B-4, C-1, D-3
B. A-1, B-4, C-3, D-2
C. A-3, B-1, C-2, D-4
D. A-4, B-1, C-3, D-2 *(BHU-2013)*

49. The fundamental unit of chronostratigraphic classification is:
A. Erathem B. System
C. Stage D. Series *(BHU-2012)*

50. Find odd one out:
A. Period B. Zone
C. Age D. Epoch *(BHU-2012)*

51. The Geological Time Scale (2004) includes:
A. Two eons B. Three eons
C. Five eons D. Seven eons *(BHU-2012)*

52. Which one is **not** a chronostratigraphic unit?
A. System B. Formation
C. Stage D. Series *(BHU-2012)*

53. The Chari Formation is best developed in:
A. Spiti basin B. Kachchh basin
C. Jaisalmer basin D. Godavari basin *(BHU-2012)*

54. Isopach maps are used for analysis of:
A. Stratigraphic thickness
B. Depositional environments
C. Sedimentary structures
D. Structural features *(BHU-2012)*

55. The Himalayan Neogene succession is represented by:
A. Dagshai Formation
B. Siwalik Group and Karewa Formation
C. Subathu Formation
D. Kasauli Formation *(BHU-2012)*

56. The law of 'order of superposition of beds' was proposed by:
A. Stena B. Smith
C. Kiev D. Phillip *(BHU-2012)*

57. The Palaeozoic/Mesozoic boundary lies at:
A. 270 ma B. 251 ma
C. 240 ma D. 255 ma *(BHU-2012)*

58. The close of Cretaceous marks the extinction of:
A. Bivalves B. Trilobites
C. Corals D. Dinosaurs *(BHU-2012)*

59. Nallamalai Group is famous for:

A. Iron B. Pyrite

C. Copper and lead D. Mica *(BHU-2012)*

60. The strike of Aravalli is:

A. NNW-SSE B. NNE-SSW

C. NE-SW D. SW-NE *(BHU-2012)*

61. Find odd one out

A. Lathi formation

B. Chari formation

C. Jaisalmer formation

D. Bhadasar formation *(BHU-2012)*

62. The type locality for Triassic is in:

A. England B. Germany

C. Canada D. France *(BHU-2012)*

63. The lower Gondwana rocks are of age:

A. Cambrian B. Permian

C. Jurassic D. Triassic *(BHU-2012)*

64. The characteristic flora of Middle Gondwana is:

A. *Glossopteris* B. *Ptilophyllum*

C. *Dichroidium* D. *Gangamopteris*

(BHU-2012)

65. Katrul formation belongs to:

A. Triassic of Spiti

B. Jurassic of Jaisalmer

C. Jurassic of Kachchh

D. Cretaceous of Jabalpur *(BHU-2012)*

66. Match the fossil group in Group-I with corresponding stratigraphic unit in Group-II.

Group-I	Group-II
P. *Equidae*	1. Triassic of Spiti
Q. *Stromatolites*	2. Gondwana Super-group
R. Dinosaurs	3. Siwalik Group
S. Brachiopoda	4. Vindhyan Super-group

A. P-4, Q-3, R-2, S-1

B. P-3, Q-4, R-2, S-1

C. P-3, Q-1, R-4, S-2

D. P-3, Q-2, R-4, S-1 *(IIT-JAM 2011)*

67. Ammonoid genus Macrocephalites is found in:

A. Talchir Formation of Satpura Basin

B. Bijori Formation of Satpura Basin

C. Patcham Formation of Kachchh

D. Bhuj Formation of Kachchh

(IIT-JAM 2011)

68. Which of the following sequences of litho-stratigraphic units is arranged in the correct order from older to younger age?

A. Kamlial - Nagri - Chinji - Dhokpathan - Pinjor - Tatrot

B. Papaghni - Cheyair - Nallamalai - Kistna

C. Talchir - Barakar - Pachmarhi - Bijori - Motur - Bagra

D. Banded Gneissic Complex - Raialo - Aravalli - Delhi

(IIT-JAM 2011)

69. Match the lithostratigraphic unit in Group-I with corresponding time unit in Group-II.

Group-I	Group-II
P. Kolhan Group	1. Triassic
Q. Kaharbari Formation	2. Proterozoic
R. Kota Formation	3. Permian
S. Kamthi Formation	4. Jurassic

A. P-2, Q-3, R-4, S-1

B. P-3, Q-1, R-4, S-2

C. P-2, Q-4, R-1, S-2

D. P-3, Q-2, R-4, S-1 *(IIT-JAM 2011)*

70. Which of the following formations is rich in mammalian fauna?

A. Siwalik

B. Lower Himalaya

C. Tethys Himalaya

D. Nagri *(BHU-2011)*

71. Syringothyris limestone of permo-carboniferous is associated with:

A. Kashmir B. Kutch

C. Cudappah D. Dharwar

(BHU-2011)

72. Age of Deccan traps is:

A. Paleocene B. Tertiary

C. Jurassic D. Triassic

(BHU-2011)

73. The _______ was an era dominated by the dinosaurs.

A. Precambrian B. Paleozoic

C. Mesozoic D. Cenozoic

(BHU-2011)

74. The boundaries between .... seem to coincide with major changes in the life forms present on earth.
A. Precambrians B. Systems
C. Eras D. Epochs
(BHU-2011)

75. What is the correct sequence of Kutch?
A. Patcham-Chari-Katrol-Umia-Bhuj
B. Chari-Patcham-Umia-Katrol-Bhuj
C. Umia-Patcham-Katrol-Chari-Bhuj
D. Katrol-Patcham-Chari-Umia-Bhuj
(BHU-2011)

76. Which of the following radioactive isotopes is most useful for dating a very young sample (< 20000 years) of wood?
A. Rubidium-87 B. Uranium-238
C. Carbon-14 D. Potassium-40
(BHU-2011)

77. Which of the following represents the longest time period?
A. Precambrian B. Paleozoic
C. Mesozoic D. Cenozoic
(BHU-2011)

78. Trilobites are index fossil of:
A. Paleozoic B. Mesozoic
C. Cenozoic D. Archean
(BHU-2011)

79. Tadpatri shales belong to:
A. Papaghani group
B. Cheyair group
C. Nallamlai group
D. Kistna group *(BHU-2011)*

80. Purple sandstone and neobolus shale belong to:
A. Salt range B. Kashmir
C. Spiti D. Trichnapally
(BHU-2011)

81. The Vindhyan is delineated from the Siwalik by:
A. Main central thrust
B. Main boundary thrust
C. Main frontal thrust
D. Tsangpo Suture zone *(BHU-2011)*

82. The principal of original horizontality states that:
A. Most rocks in the earth's crust are layered horizontally
B. Igneous rocks form essentially horizontal layers
C. Metamorphic gradients are essentially horizontal before deformation
D. Sediments are deposited as essentially horizontal layers
(BHU-2011)

83. What scientific avenue of investigation gave scientists the best estimate of the age of the earth?
A. Dating fossils
B. Archaeological dating
C. Radiometric dating
D. Carbon dating *(BHU-2011)*

84. Which one of the following successions does represent the correct stratigraphic sequence from bottom to top?
A. Uttatur Group-Trichinapally Group-Ariyalur Group-Niniyur Group
B. Trichinapally Group-Uttatur Group-Ariyalur Group-Niniyur Group
C. Uttatur Group-Trichinapally Group-Niniyur Group-Ariyalur Group
D. Niniyur Group-Trichinapally Group-Uttatur Group- Ariyalur Group
(IIT-JAM 2010)

85. Which one of the following stratigraphic units does occur outside the Aravalli Craton?
A. Berach Granite
B. Untala Granite
C. Gingla Granite
D. Closepet Granite
(IIT-JAM 2010)

86. The Fawn Limestone contains which one of the following?
A. *Collenia columnaris*
B. *Conophyton cylindricus*
C. *Collenia* sp.
D. All of the above *(BHU-2010)*

87. The Jodhpur Sandstone is unconformably underlain by:
A. Malani Volcanics
B. Aravalli Group
C. Banded Gneissic Complex
D. Jhiri Shales *(BHU-2010)*

88. Which of the following represents the correct chronostratigraphic sequence?
A. Zone-Stage-Series-System-Erathem
B. Series-Zone-Stage-System-Erathem
C. Series-System-Erathem-Zone-Stage
D. Zone-Stage-System-Series-Erathem
(BHU-2010)

89. The boundary between Mesozoic and Cenozoic is approximately at:
A. 55 Ma B. 65 Ma
C. 75 Ma D. 125 Ma
(BHU-2010)

90. Diamondiferous conglomerate occurs in which one of the following?
A. Rewa Group B. Semri Group
C. Bhander Group D. Kaimur Group
(BHU-2010)

91. The Salkhala Group is overlain by:
A. Talchir formation
B. Dogra slates
C. Mandhali formation
D. Gneissic rocks *(BHU-2010)*

92. In Kashmir, the Muth Quartzite is conformably overlain by:
A. Fenestella shales
B. Zewan formation
C. Syringothyris limestone
D. None of the above *(BHU-2010)*

93. Which one is an Upper Gondwana Flora?
A. Gangamopteris B. Ptilophyllum
C. Vertebraria D. Glossopteris
(BHU-2010)

94. Stegodon is a characteristic fossil of which formation?
A. Kamalial formation
B. Pinjor formation
C. Nagri formation
D. Dhokpathan formation *(BHU-2010)*

95. When did the first plants appear on land?
A. Cambrian B. Ordovician
C. Silurian D. Devonian
(BHU-2010)

96. The Sargur Schist Complex is:
A. Older than Dharwar Group
B. Younger than Dharwar Group
C. Equivalent to Closepet Granite
D. Younger than Papaghani Group
(BHU-2010)

97. Which one of the following stratigraphic units is marine?
A. Umaria Beds
B. Kamthi Beds
C. Pali Beds
D. Bareen Measures *(IIT-JAM 2009)*

98. Match the stratigraphic units listed in Group-1 with their respective ages mentioned in Group-2.

Group 1	Group 2
P. Neobolus Beds	1. Devonian
Q. Muth Quartzites	2. Miocene
R. Katrol Group	3. Jurassic
S. Tipam Sandstone	4. Cambrian

A. P-1, Q-2, R-3, S-4
B. P-2, Q-3, R-1, S-4
C. P-4, Q-1, R-3, S-2
D. P-2, Q-1, R-4, S-3 *(IIT-JAM 2009)*

99. What is the most commonly agreed age of the Archean-Proterozoic boundary?
A. 1500 Ma B. 2500 Ma
C. 570 Ma D. 3500 Ma
(IIT-JAM 2009)

100. Otoceras woodwardi marks the beginning of:
A. Jurassic B. Triassic
C. Cretaceous D. Paleogene
(BHU-2009)

101. The fundamental unit of lithostratigraphic classification is:
A. Group B. Formation
C. Member D. Bed *(BHU-2009)*

102. Which one is **not** a time-unit?
A. Period B. Zone
C. Age D. Epoch
(BHU-2009)

103. Mount Abu is situated on:
A. Kailash granite B. Erinpura granite
C. Jalor granite D. Closepet granite
(BHU-2009)

104. Find odd one out:
A. System B. Formation
C. Stage D. Series *(BHU-2009)*

105. The Jaisalmer formation of Rajasthan is equivalent to .... in Kachchh.
A. Katrol formation
B. Chari formation
C. Patcham formation
D. Umia formation *(BHU-2009)*

106. Analysis of stratigraphic thickness is carried out with the help of:
A. Isopach maps
B. Lithostratigraphic maps
C. Geologic maps
D. Structural maps *(BHU-2009)*

107. The Miocene of Assam is represented by:
A. Surma and Tipam group
B. Barail group
C. Barail group Jaintia group
D. Dihing group *(BHU-2009)*

108. The law of floral and faunal succession was proposed by:
A. Steno B. Smith
C. Kiew D. Phillip *(BHU-2009)*

109. The Precambrian/Cambrian boundary lies at:
A. 542 ma B. 550 ma
C. 900 ma D. 670 ma *(BHU-2009)*

110. Dinosaurs got extinct towards the close of:
A. Jurassic B. Triassic
C. Cretaceous D. Tertiary *(BHU-2009)*

111. Bijagarh shale is famous for:
A. Iron B. Pyrite
C. Copper D. Lead *(BHU-2009)*

112. The strike of Dharwar supergroup is:
A. NNW-SSE B. NNE-SSW
C. NE-SW D. SE-NW
(BHU-2009)

113. Select the correct stratigraphic sequence of the Vindhyan Supergroup:
A. Semri - Kaimur - Rewa - Bhander
B. Rewa - Bhander - Kaimur - Semri
C. Semri - Rewa - Kaimur - Bhander
D. Semri - Bhander - Rewa - Kaimur
(IIT-JAM 2008)

114. Globotruncana-bearing sequence indicates a shallow marine sedimentary succession deposited during:
A. Cretaceous B. Eocene
C. Jurassic D. Oligocene
(IIT-JAM 2008)

115. Find odd one out:
A. Lathi formation
B. Chari formation
C. Katrol formation
D. Umia formation *(BHU-2009)*

116. The age of Gondwana coal deposits is:
A. Late Triassic B. Permian
C. Early Jurassic D. Cretaceous
(BHU-2009)

117. Match the stratigraphic units listed in Group-1 with their respective ages mentioned in Group-2.

Group-1	Group-2
P. Deccan Trap	1. Jurassic
Q. Panjal Trap	2. Cretaceous
R. Malani Rhyolite	3. Permian
S. Rajmahal Trap	4. Proterozoic

A. P-1, Q-2, R-3, S-4
B. P-2, Q-3, R-4, S-1
C. P-2, Q-4, R-3, S-1
D. P-1, Q-3, R-4, S-2 *(IIT-JAM 2008)*

118. Which of the following stratigraphic units contains dinosaur fossils?
A. Lameta B. Muth Quartzite
C. Siwalik D. Subathu
(IIT-JAM 2007)

119. Match sedimentary formations in Group-1 with their corresponding ages in Group-2.

Group-1	Group-2
P. Maleri	1. Oligocene

Q. Karewa 2. Plio-Pleistocene
R. Ariyalur 3. Cretaceous
S. Barail 4. Triassic
A. P-1, Q-2, R-3, S-4
B. P-2, Q-1, R-4, S-3
C. P-4, Q-3, R-2, S-1
D. P-4, Q-2, R-3, S-1 *(IIT-JAM 2007)*

120. Which of the following series of stratigraphic units is arranged in the order from older to younger age?
A. Cheyair - Rewa - Sausar - Gondwana
B. Sausar - Cheyair - Rewa - Gondwana
C. Gondwana - Rewa - Cheyair - Sausar
D. Rewa - Gondwana - Sausar - Cheyair
(IIT-JAM 2007)

121. Match the following:

Group I	Group II
P. Lameta Formation	1. *Elephas*
Q. Raniganj Formation	2. *Glossopteris*
R. Pinjor Formation	3. *Titanosaurus*
S. Umaria Marine Beds	4. *Productus*

A. P - 4 Q - 1 R - 2 S - 3
B. P - 1 Q - 3 R - 4 S - 2
C. P - 2 Q - 1 R - 3 S - 4
D. P - 3 Q - 2 R - 1 S - 4 *(IIT-JAM 2006)*

122. Identify the incorrect pair:
A. Lithostratigraphy - Formation
B. Biostratigraphy - Interval Zone
C. Chronostratigraphy - Series
D. Geochronology - System
(IIT-JAM 2006)

123. Which one of the following series of rock formations is arranged in the correct stratigraphic order (i.e. from older to younger)?
A. Malani volcanics - Panjal volcanics - Rajmahal volcanics - Deccan volcanics
B. Dhokpathan - Tatrot - Boulder Conglomerate - Pinjor
C. Uttatur - Ariyalur - Niniyur - Trichinapally
D. Talchir - Panchet - Damuda - Mahadeva
(IIT-JAM 2006)

124. The method applied for isotope dating of the Holocene rocks involves:
A. K – Ar B. C – 14
C. U – Pb D. Rb – Sr
(IIT-JAM 2006)

125. The Devonian marker horizon in the Himalaya is known as:
A. Lipak Formation
B. Muth Quartzite
C. Po Formation
D. Syringothyris Limestone
(IIT-JAM 2005)

126. The Principle of Uniformitarianism was given by:
A. James Hutton
B. Nicholas Steno
C. William Smeeth
D. Walther *(DU-2014)*

127. The stratigraphic equivalent of Salkhala and Jutogh Group of western Himalaya in eastern Himalaya is:
A. Darjeeling Gneisses
B. Daling Group
C. Tejam Group
D. Kurnool Group *(DU-2014)*

128. Unusually large traces of rare element iridium (Ir) are found in sediments deposited at the end of:
A. Cenozoic B. Mesozoic
C. Cambrian D. Jurassic *(DU-2014)*

129. Which of the following stratigraphic unit underwent major age revision?
A. Siwalik Super group
B. Dharwar Super group
C. Krol Group
D. Singbhum Group *(DU-2014)*

130. Standard subdivision of each interval of geological time scale is based on some group of fossils. Which of the following fossil groups form the basis of subdivision of Mesozoic?
A. Graptolites B. Gastropods
C. Ammonites D. Trilobites
(DU-2014)

131. Caledonian Orogeny occurred during:
A. Tertiary
B. Jurassic
C. Early Palaeozoic
D. Cretaceous *(DU-2014)*

132. Which of the following stratigraphic boundaries is **not** defined on paleontological criterion?
A. Permian-Triassic boundary
B. Miocene-Pliocene boundary
C. Pleistocene-Holocene boundary
D. Cretaceous-Tertiary boundary
(DU-2014)

133. Which of the following is **not** a chronostratigraphic unit?
A. System B. Erathem
C. Formation D. Series
(DU-2014)

134. Which of the following is correct sequence of lithostratigraphic units in Siwalik super group in ascending stratigraphic order?
A. Kamalial Formation, Chinji Formation, Nagri Formation, Dhok Pathan Formation, Tatrot Formation, Pinjor Formation.
B. Chinji Formation, Tatrot Formation, Kamalial Formation, Pinjor Formation, Nagri Formation
C. Tatrol Formation, Kamalial Formation, Pinjor Formation, Nagri Formation, Dhok Pathan Formation.
D. Pinjor Formation, Dhok Pathan Formation, Kamalial Formation, Nagri Formation, Tatrot Formation. *(DU-2014)*

135. First vertebrate (Fish) appeared in:
A. Triassic B. Ordovician
C. Cretaceous D. Permian *(DU-2014)*

136. Which one of the following is litho-stratigraphic unit?
A. Eratham B. Acme zone
C. Series D. Formation
(BHU-2015)

137. The Paleozoic era is restricted between:
A. 65 to 225 Ma
B. 1.8 to 65 Ma
C. 570 to 2500 Ma
D. 225 to 570 Ma *(BHU-2015)*

138. Which era is called as 'Age of Reptiles'?
A. Palaeozoic era B. Mesozoic era
C. Cenozoic D. Proterozoic era
(BHU-2015)

139. Which formation belongs to the lower Gondwana Sequence?
A. Zawar B. Poladpur
C. Rohtas D. Iron stone
(BHU-2015)

140. Semri Group is related to the:
A. Upper Vindhyan B. Lower Gondwana
C. Lower Vindhyan D. Upper Gondwana
(BHU-2015)

141. Which is **not** related to coal field?
A. Chandrapur B. Sohagpur
C. Ramgarh D. Panna *(BHU-2015)*

142. Arranged in correct sequence from older to younger.
A. Sargurschist complex, Peninsula gneissic complex, Dharwar super group
B. Patcham, Umia, Katrol, Chari
C. Karharbari, Raniganj, Barakar, Kulti
D. Papaghani, Nallamalai, Cheyair, Kistna
(BHU-2015)

143. Which one of the following separate the Eastern Dharwar to Western Dharwar?
A. Peninsula gneissic complex
B. Bababudan
C. Rani Bennur
D. Closepet granite *(BHU-2015)*

144. The maximum Genera and Species in Upper Gondwana Sequence found in:
A. Mahadeva B. Jabalpur
C. Rajmahal D. Maleri *(BHU-2015)*

145. Kaladgi group belongs to:
A. Archaean era
B. Proterozoic era
C. Palaeozoic era
D. Mesozoic era *(BHU-2015)*

146. The general trend of Aravalli Group is:
A. NW-SE B. E-W
C. ESE-WNW D. NE-SW *(BHU-2015)*

147. Dhokpathan formation is overlain by .... formation.
A. Pinjor B. Tatrot
C. Nagri D. Chinji *(BHU-2015)*

148. Match the following:

	Column-A		**Column-B**
P.	Dhosa Oolite	1.	Gondwana Sequence
Q.	Mansar	2.	Dharwar Super group
R.	Umaria Marine Bed	3.	Sausar Group
S.	Hutti Goldfield	4.	Jurassic of Kutch

A. P-3, Q-1, R-4, S-2
B. P-2, Q-4, R-1, S-3
C. P-1, Q-2, R-3, S-4
D. P-4, Q-3, R-1, S-2 *(BHU-2015)*

149. Barail Series overlain by:
A. Tipam Series B. Dupitila Series
C. Surma Series D. Jaintia Series
(BHU-2015)

150. Dinosaur commonly found in which formation?
A. Lameta Bed B. Bagh Bed
C. Deccan Trap D. Rajmahal Trap
(BHU-2015)

151. The Olive Series is developed in:
A. Simla B. Kashmir
C. Kumaon D. Salt range
(PU-2010)

152. What is the age of Muth quartzite?
A. Ordovician B. Silurian
C. Devonian D. Carboniferous
(PU-2010)

153. Which combination represents the Phanerozoic?
A. Cenozoic, Precambrian, Paleozoic
B. Eocene, Cretaceous, Archean
C. Precambrian, Cenozoic, Eocene
D. Cenozoic, Mesozoic, Paleozoic
(PU-2013)

154. Copper Belt Thrust (CBT) also known as:
A. Singhbhum Shear Zone
B. Great Boundary Fault
C. Older Metamorphic Group
D. Main Boundary Fault *(PU-2013)*

155. Archaean rocks in Kashmir-Hazara area are known as:
A. Salkhala Series B. Champian Gneiss
C. Buxa Series D. Darjeeling Series
(PU-2013)

156. Who proposed the Principle "the present is the key to the past"?
A. William Smith
B. James Hutton
C. d'Orbingy
D. Carl von Linnaeus *(PU-2013)*

157. The first fundamental generalization of stratigraphy is:
A. Order of superposition
B. Principle of uniformitarianism
C. Stratigraphic analysis
D. Law of faunal succession *(PU-2013)*

158. Which one of the following dinosaur fossils is a theropod?
A. Kotasaurus
B. Titanosaurus
C. Rajasaurus
D. Barapasaurus *(IIT-2023)*

159. Neogene period includes:
A. Paleogene and Eocene
B. Paleogene, Eocene and Oligocene
C. Oligocene, Miocene and Pliocene
D. Miocene and Pliocene *(PU-2013)*

160. The general trend of the Aravalli mountain range is:
A. North South
B. East-West
C. Northeast-Southwest
D. Northwest-Southeast *(PU-2013)*

161. Hercynian or Variscan orogeny took place during:
A. Silurian
B. Devonian
C. Permo carboniferous
D. Jurassic *(PU-2013)*

162. Which of the following is **not** a lithostratigraphic unit?
A. Group B. Formation
C. Series D. Member *(PU-2012)*

163. The age of the Cuddapah Super group is approximately:
A. 1000 Ma B. 1200 Ma
C. 1600 Ma D. 2000 Ma *(PU-2012)*

164. Shield and craton refer to some parts of:
A. Earth
B. Earth crust
C. Continental crust
D. Oceanic crust *(PU-2012)*

165. One of the following is an example of mobile belt:
A. Vindhyan basin
B. Deccan trap
C. Himalaya
D. None of the above *(PU-2012)*

166. Which of the following represents the longest time period?
A. Precambrian B. Paleozoic
C. Mesozoic D. Cenozoic
(PU-2011)

167. The Main Central Thrust separates:
A. Higher Himalaya and Tethyan Himalaya
B. Higher Himalaya from Lesser Himalaya
C. Siwalik from Higher Himalaya
D. Siwalik and Lesser Himalaya *(PU-2011)*

168. Oldest rocks from India are reported from:
A. Dharwar craton B. Bastar craton
C. Aravalli craton D. Eastern Ghats
(PU-2011)

169. Indian manganese and iron deposits are mostly found in:
A. Hadean
B. Upper Proterozoic
C. Lower Proterozoic
D. Paleozoic *(PU-2011)*

170. Choose correct sequence from the following:
A. Blaini-Krol-Tal
B. Subathu-Kasauli-Dagshai
C. Tipam-Surma-Barail
D. Nari-Gaj-Kirthar *(PU-2011)*

171. The main unconformity separating the Archaean and Proterozoic succession occurs at:
A. 3200 Million years
B. 2700 Million years
C. 600 Million years
D. 2500 Million years *(PU-2011)*

172. Which of the following is **not** a time rock unit?
A. System B. Series
C. Epoch D. Stage *(PU-2011)*

173. The first fundamental generalization of stratigraphy is:
A. Order of superposition
B. Principle of uniformitarianism
C. Stratighaphic analysis
D. Law of faunal succession *(PU-2011)*

174. The first land plant appeared during:
A. Cambrian B. Ordovician
C. Silurian D. Carboniferous
(PU-2011)

175. Oldest rocks from India are reported from:
A. Dharwar craton B. Baster craton
C. Aravalli craton D. Eastern ghats
(PU-2011)

176. Choose the correct chronological order from oldest to youngest for the following stratigraphic units
P - Chitradurga Group;
Q - Sargur Group;
R - Ajabgarh Group;
S - Udaipur Group
A. Q, P, S, R B. P, Q, S, R
C. Q, R, P, S D. Q, R, S, P
(IIT JAM-2016)

177. Sudden death of Dinosaurs occurred at:
A. End of cretaceous
B. End of Permian
C. End of Jurassic
D. End of Triassic
(SAGAR UNIVERSITY-2015)

178. The main coal-formation epoch in India is:
A. Cretaceous
B. Permian
C. Tertiary
D. Eocene *(SAGAR UNIVERSITY-2015)*

179. Diomond bearing conglomerate underlies which of the following:
A. Semri B. Kaimur
C. Bhander D. Rewa
(SAGAR UNIVERSITY-2015)

180. The term era is used in:
A. Lithostratigraphy
B. Biostratigraphy
C. Chronostratigraphy
D. Magnetostratigraphy
(SAGAR UNIVERSITY-2015)

181. Delhi ridge is an extension of:
A. Aravallis B. Vindhyans
C. Nilgiris D. Satpuras
(SAGAR UNIVERSITY-2015)

182. Neyveli lignite deposit lies in which of the following states:
A. Karnataka B. Tripura
C. Tamil Nadu D. Andhra Pradesh
(SAGAR UNIVERSITY-2015)

183. When Human being evolved on the Earth:
A. Cenozoic B. Mesozoic
C. Paleozoic D. Precambrian
(SAGAR UNIVERSITY-2015)

184. Appearance of land plants is associated with which of the following Era:
A. Cenozoic B. Paleozoic
C. Mesozoic D. Proterozoic
(SAGAR UNIVERSITY-2015)

185. Greenstone belt is associated with which of the following:
A. Bundelkhand massif
B. Baster craton
C. Mahakaoshal Group of rocks
D. Sargur group of rocks
(SAGAR UNIVERSITY-2015)

186. Gondite is associated with which of the following stratigraphic succession:
A. Saussars B. Sakolis
C. Chilpis D. Bijawars
(SAGAR UNIVERSITY-2015)

187. Late maastrichtian fresh water facies of sediments in India is called:
A. Bagh beds
B. Baripada beds
C. Cuddalore sandstone beds
D. Lameta beds
(SAGAR UNIVERSITY-2015)

188. Graptolires are confined to:
A. Upper Paleozoic rocks
B. Middle Paleozoic rocks
C. Lower Paleozoic rocks
D. Lower Triassic rocks
(SAGAR UNIVERSITY-2015)

189. The Mansar formation belongs to:
A. Amgoan group B. Chilpi group
C. Sakoli group D. Saucer group
(SAGAR UNIVERSITY-2015)

190. Patcham, Chari, Katrol and Umia are the stages of which of the following :
A. Triassic of Spiti
B. Jurassic of Kutch
C. Cretaceous of Trichinapalli
D. Tertiary of Assam
(SAGAR UNIVERSITY-2015)

191. Gwalior system is equivalent to:
A. Lower Vindhyan
B. Cuddapah system
C. Dharwar system
D. Upper Vindhyan
(SAGAR UNIVERSITY-2015)

192. The age of the Rajmahal Traps in geological time scale is:
A. Cretaceous B. Proterozoic
C. Mesozoic D. Cambrian
(ISM-2015)

193. Which of the following is the smallest unit of standard stratigraphic scale?
A. Stage B. Series
C. Zone D. System *(ISM-2015)*

194. *Glossopteris indica* is characteristic fossil of:
A. Upper Gondwana
B. Lower Gondwana
C. Both of the above
D. None of the above *(ISM-2015)*

195. Which is the boundary known as the Mother of all mass extinction?
A. Permian – Triassic
B. Precambrian – Cambrian
C. Cretaceous – Tertiary
D. Jurassic – Cretaceous *(ISM-2015)*

196. Mississippian and Pennsylvanian belong to which period?

A. Permian
B. Devonian
C. Carboniferous
D. Silurian *(IIT JAM-2016)*

197. Match the following stratigraphic units (Group-I) with their ages (Group-II)

Group-I	**Group-II**
(i) Cumbum Formation	(P) Cenozoic
(ii) Baisakhi Formation	(Q) Mesozoic
(iii) Kopili Formation	(R) Palaeozoic
(iv) Barakar Formation	(S) Proterozoic

A. (i) - R, (ii) - P, (iii) - Q, (iv) – S
B. (i) - S, (ii) - Q, (iii) - P, (iv) - R
C. (i) - S, (ii) - R, (iii) - P, (iv) – Q
D. (i) - S, (ii) - Q, (iii) - R, (iv) – P *(IIT JAM-2016)*

198. Which of the following stratigraphic units belong(s) to Cretaceous?

A. Bhuj Formation
B. Ariyalur Group
C. Patcham Formation
D. Katrol Formation *(ISM-2014)*

199. Which one of the following is a counter example of Low of Uniformitarianism?

A. Development of sedimentary basin
B. Development of deformed terrain
C. Development of glacial terrain
D. Development of banded iron formation *(ISM-2014)*

200. Life appeared on the planet Earth within approximately how many years of its formation?

A. 1 billion years
B. 100 million years
C. 500 million years
D. 1 million years *(ISM -2014)*

201. Which periods marks the disappearance of Rugose Corals?

A. Permian B. Triassic
C. Jurassic D. Cretaceous *(ISM-2014)*

202. When did the most scientist predict that the Earth will be largely destroyed by the warming and expansion of the sun as it reaches the Red Giant Stage?

A. 100 billion years
B. 500 billions years
C. 20 billion years
D. 5 billion years *(ISM-2014)*

203. Which one of the following represent the longest time period?

A. Precambrian B. Mesozoic
C. Paleozoic D. Cenozoic *(ISM-2015)*

204. The boundaries between ________ seem to coincide with major changes in the life forms represent on earth.

A. System B. Eras
C. Epoch D. Beds *(ISM-2015)*

205. Geological processes operating at the present time are the same processes that have operated in the past is a statement of the principle of:

A. Cross-cutting relationship
B. Uniformitarianism
C. Original horizontality
D. Faunal succession *(ISM-2015)*

206. What is the age of Brail Series?

A. Jurassic
B. Paleocene
C. Oligocene
D. Miocene *(PUNJAB UNIVERSITY-2015)*

207. Which one of the following is a typical Lower Gondawana plant fossils assemblage?

A. *Glossopteris, Ptilophyllus, Nilssonia, Bucklnadia*
B. *Glossopteris, Gangamopteris, Schizoneura, Sphenophyllum*
C. *Gangmopteris, Lycopodites, Brachyphyllum, Nilsonia*
D. *Vertibraria, Alerthopteris, Otozamites, Glossopteris* *(PUNJAB UNIVERSITY-2015)*

208. The Sulekere series of Shimoga belt is equivalent to:
A. Upper Dharwar
B. Middle Dharwar
C. Lower Dharwar
D. None of these
(PUNJAB UNIVERSITY-2015)

209. The Bijlli Rhyolites belongs to:
A. Kairagarh group
B. Nandgaon group
C. Sakoli group
D. Sausor group
(PUNJAB UNIVERSITY-2015)

210. Match the correct one:

	List-A		List-B
P.	Alwar series	1.	Raialo series
Q.	Bhagawanpura limestone	2.	Aravalli
R.	Binota shale	3.	Delhi super group
S.	Semri series	4.	Vindhyan

A. P-2, Q-4, R-1, S-3
B. P-1, Q-4, R-3, S-2
C. P-3, Q-1, R-2, S-4
D. None of them
(PUNJAB UNIVERSITY-2015)

211. Graptolites are very useful index fossils of:
A. Cretaceous
B. Permian to Triassic
C. Carboniferous
D. Ordovician to Silurian
(PUNJAB UNIVERSITY-2015)

212. Hyperion was a:
A. Upper Pliocene horse
B. Lower Pliocene horse
C. Lower Cambrian horse
D. Upper Permian
(PUNJAB UNIVERSITY-2015)

213. The early Middle and late Proterozoic with time boundaries at:
A. 2500 – 1500 and 900 m.y. respectively
B. 3000 – 1500 and 1000 m.y. respectively
C. 3000 – 2000 and 1000 m.y. respectively
D. 2800 – 2200 and 1000 m.y. respectively
(PUNJAB UNIVERSITY-2015)

214. In the Bhandra craton manganese ores are restricted to:
A. Sausar group
B. Sakoli group
C. Dongargarh group
D. Basement gneiss and granites
(AMU-2016)

215. Which of the following is a Paleoproterozoic supercontinents?
A. Ur
B. Pangea
C. Laurasia
D. None of the above *(AMU-2016)*

216. Which of the following is most suitable for the biostratigraphy of Cambrian rocks?
A. Pelecypoda
B. Trilobite
C. Ammonoidea
D. Nautiloidea *(AMU-2016)*

217. Deccan flood basalts are considered to be related to:
A. Hawaii hot spot
B. Iceland hot spot
C. Easter hot spot
D. Reunion hot spot *(AMU-2016)*

218. The oldest rocks of Aravalli craton is:
A. Gneiss
B. Supracrustal sequence
C. Undeformed granitoids
D. Mafic dyke *(AMU-2016)*

219. Age of the Aravalli Supergroup is:
A. Paleoproterozoic to Mesoproterozoic
B. Mesoproterozoic to Neoproterozoic
C. Neoproterozoic
D. Neoarchean *(AMU-2016)*

220. Age of the Delhi supergroup
A. Mesoarchean
B. Neoarchean
C. Mesoproterozoic to Neoproterozoic
D. Paleoproterozoic *(AMU-2016)*

221. A sedimentary rock includes fossils of two species that are known to have lived between 480–430 and 450–250 million years ago. What could be the age of the rock?

A. 445 million years
B. 480 million years
C. 250 million years
D. 410 million years *(ISM-2016)*

222. Acme or peak zone is defined as the strata that contains
A. Minimum number of species
B. Maximum number of species
C. Maximum number of individual of one species
D. Minimum number of individual of single species
(ISM-2016)

223. The age of the Muthquartzite is:
A. Silurian B. Ordovician
C. Devonian D. Archean
(ISM-2016)

224. Permian is represented in Spiti by:
A. Kanawar group
B. Kulling group
C. Muth quartzite
D. Agglomerate shale *(ISM-2016)*

225. The sills and dykes of basaltic composition are found in Gondwana succession at different places. Those in Satpura and Rewa region are through to be of _____ age, while those in Damodar Valley and Assam are of _____ age, respectively
A. Rajmahal, Deccan trap
B. Deccan trap, Rajmahal
C. Both are Rajmahal
D. Both are Deccan trap *(ISM-2016)*

226. The basic unit of the lithostratigraphic classification is:
A. Super group B. Group
C. Formation D. Bed *(BHU-2016)*

227. In stratigraphic classification which of the following units is used for igneous and sedimentary rocks?
A. System B. Period
C. Eon D. Lithodeme
(BHU-2016)

228. In a transgressive sequence which of the following type of sediment accumulation exist?
A. Coarsening upward
B. Homogeneous sized sediment
C. Alternating coarse and fine sediment
D. Fining upward sequence
(BHU-2016)

229. In which part of India full marine Mesozoic succession is exposed?
A. Eastern Ghat
B. Western Ghat
C. Outer Himalaya
D. Tethyan Himalaya *(BHU-2016)*

230. The total geological span of Mesozoic Era is:
A. 175 Ma B. 185 Ma
C. 195 Ma D. 205 Ma
(BHU-2016)

231. During which of the following eras, the largest number of paleogeography reconstruction tool place?
A. Paleozoic B. Cenozoic
C. Mesozoic D. Proterozoic
(BHU-2016)

232. The poor record of marine fauna during Early Triassic is mainly attributed to:
A. Sudden marine transgression
B. Abnormal salinity
C. Magnetic reversal
D. Unusual tectonism *(BHU-2016)*

233. During which of the following times the cretaceous sea was on its peak of transgression as well as having maximum temperature?
A. Maastrichtian
B. Albian
C. Campanian
D. Turonian *(BHU-2016)*

234. The "cyclotheme" sedimentation is best observed in:
A. Jurassic sediments
B. Triassic sediments
C. Cretaceous sediments
D. Gondwana sediments *(BHU-2016)*

235. Rajmahal Formation consists of:
A. Sedimentary rocks only
B. Volcanic rocks only
C. Both sedimentary and volcanic rocks
D. Volcanic and Plutonic rocks *(BHU-2016)*

236. Which of the following represents a Lower Gondwana Marine intercalation?
A. Umaria marine beds
B. Athgarh Sandstone beds
C. Umia beds
D. Raghavpuram shale beds *(BHU-2016)*

237. The complete Triassic succession of Spiti region is known as:
A. Lilang Group
B. Spiti Shales
C. Kioto Limestone
D. Tagling Stage *(BHU-2016)*

238. The modern concept of stratigraphy is applicable to:
A. Sedimentary rocks
B. Sedimentary and volcanic igneous rocks
C. Sedimentary and low grade metamorphic rocks
D. Sedimentary igneous and metamorphic rocks *(BHU-2016)*

239. A stratum is a:
A. Smallest layered units of sedimentary rocks
B. Two layered unit of sedimentary rocks
C. The whole sedimentary units of a basin
D. Can be a sedimentary rock of any dimension
(BHU-2016)

240. The oldest paleozoic sediments of Kashmir Himalaya belongs to:
A. Dogra slate B. Salkhala group
C. Haimanta group D. Thango formation
(BHU-2016)

241. The igneous activity during palaeozoic era in Kashmir Himalaya is stratigraphically known as
A. Deccan traps B. Panjal trap
C. Rajmahal traps D. Bhimtal volcanics
(BHU-2016)

242. The youngest Palaeozoic sediments in Kashmir Himalaya is grouped as:
A. Syringothyris limestone
B. Fenestella shale
C. Zewan formation
D. Lipak formation *(BHU-2016)*

243. Umaria Marine beds are marine interrelation in:
A. Karewa group
B. Siwalik group
C. Lower gondwana
D. Middle gondwana *(BHU-2016)*

244. The beginning of Gondwana sedimentation is characterised by deposition of:
A. Lacustrine sediments
B. Marine sediments
C. Deltaic sediments
D. Glacial sediments *(BHU-2016)*

245. Sedimentary magnesite deposit is found :
A. Pithoragarh B. Almora
C. Chamoli D. All the above
(BHU-2016)

246. Important polymetallic sulphide deposit is found to occur in:
A. Rajpura – Dariba B. Sukindha
C. Khetri D. None of the above
(BHU-2016)

247. Economic point of view which type of pegmatite is very important:
A. Complex pegmatite
B. Simple pegmatite
C. Both A. and B.
D. None of the above *(BHU-2016)*

248. Which of the following is not a chronostratigraphic unit?
A. System B. Erathem
C. Group D. Series *(DU-2016)*

249. Seismic profile help to:
A. Interpret stratigraphic which is not exposed
B. Record subsurface rocks
C. Find focus of the earthquake
D. Find fossils *(DU-2016)*

250. Syringothyris is characteristic fossils of:
A. Silurian
B. Ordovician
C. Carboniferous
D. Cambrian *(DU-2016)*

251. Which of the following sediment types is expected to the marine stable?
A. Globigerina ooze
B. Pteropod ooze
C. Melange
D. Fossiliferous sandstone *(DU-2016)*

252. Which one of the following stratigraphic boundary marks maximum events?
A. Cretaceous – Tertiary boundary
B. Precambrian – Cambrian boundary
C. Jurassic – Cretaceous boundary
D. Permian – Triassic boundary *(DU-2016)*

253. First stage of the horse evolution is known as:
A. Eohippus B. Hyracodium
C. Mesohippus D. Epihippus
(DU-2016)

254. In response to the expansion of grasslands, mammals evolved:
A. Brachiydent dentition
B. Bonodent dentition
C. Lophodont dentition
D. Hypsodont dentition *(DU-2016)*

255. A specimen used as diagnostic of a particular new species is known as:
A. Holotype
B. Paratype
C. Lectotype
D. Neotype *(DU-2016)*

256. The concept of punctuated equilibrium was proposed by:
A. James Hutton & Charles Lyell
B. Neils Eldredge & Stephen Gourd
C. Charles Darwin and Alferd Wegner
D. William Smith and Nicholus
(DU-2016)

257. Evolution above species level is known as:
A. Microevolution
B. Macroevolution
C. Iterative evolution
D. None of these *(DU-2016)*

258. Origin of the Bird is sought among
A. Sauropod dinosaur
B. Omithician dinosaur
C. Prosauropods
D. Dromacosaurid theropods *(DU-2016)*

259. Glassopteris plant fossils
A. Upper Gondwana
B. Middle Gondwana
C. Lower Gondwana
D. Rajmahal *(DU-2016)*

260. Corals prefer to grow in:
A. Polar ocean
B. Cool subtropical ocean
C. Stagnant water
D. Tropical water *(DU-2016)*

261. Graphic correlation is used in Paleontology is used:
A. Synchronically of FADs & LADs
B. Morphological variation
C. Paleobiogeographic variation
D. None of the above *(DU-2016)*

262. Gungko is a
A. Index fossils
B. Living fossils
C. Invertebrate fossils
D. Vertebrate fossils *(DU-2016)*

263. Body of strata corresponding to the overlaying to the more specific fossils taxa is known as:
A. Taxon range biozone
B. Lineage biozone
C. Assemblage biozone
D. Concurrent range biozone *(DU-2016)*

264. Angiosperms first fossils record of:
A. Late Cretaceous
B. Middle Cretaceous
C. Early Cretaceous
D. Late Jurassic *(DU-2016)*

ANSWER KEY

1	2	3	4	5	6	7	8	9	10
C	D	A,C	A,D	A,C	B	D	B	A	D
11	**12**	**13**	**14**	**15**	**16**	**17**	**18**	**19**	**20**
D	C	A	C	D	B	C	C	B	D
21	**22**	**23**	**24**	**25**	**26**	**27**	**28**	**29**	**30**
B	C	B	A	D	B	A	A	B	A
31	**32**	**33**	**34**	**35**	**36**	**37**	**38**	**39**	**40**
C	C	C	D	C	B	B	D	B	B
41	**42**	**43**	**44**	**45**	**46**	**47**	**48**	**49**	**50**
A	B	C	D	A	B	D	C	C	B
51	**52**	**53**	**54**	**55**	**56**	**57**	**58**	**59**	**60**
A	B	B	A	B	A	B	D	C	C
61	**62**	**63**	**64**	**65**	**66**	**67**	**68**	**69**	**70**
D	B	B	C	C	B	C	B	A	A
71	**72**	**73**	**74**	**75**	**76**	**77**	**78**	**79**	**80**
A	B	C	A	A	C	A	A	B	A
81	**82**	**83**	**84**	**85**	**86**	**87**	**88**	**89**	**90**
B	D	C	A	D	D	A	A	B	A
91	**92**	**93**	**94**	**95**	**96**	**97**	**98**	**99**	**100**
B	C	B	D	C	A	A	C	B	B
101	**102**	**103**	**104**	**105**	**106**	**107**	**108**	**109**	**110**
B	B	B	B	B	A	A	B	A	C
111	**112**	**113**	**114**	**115**	**116**	**117**	**118**	**119**	**120**
B	A	A	A	A	B	B	A	D	B
121	**122**	**123**	**124**	**125**	**126**	**127**	**128**	**129**	**130**
D	D	A	B	B	A	C	B	C	C
131	**132**	**133**	**134**	**135**	**136**	**137**	**138**	**139**	**140**
C	D	C	A	B	D	D	B	D	C
141	**142**	**143**	**144**	**145**	**146**	**147**	**148**	**149**	**150**
D	A	D	C	B	D	B	D	D	A
151	**152**	**153**	**154**	**155**	**156**	**157**	**158**	**159**	**160**
D	C	D	A	A	B	A	C	D	C
161	**162**	**163**	**164**	**165**	**166**	**167**	**168**	**169**	**170**
B	C	D	B	C	A	B	A	C	C

171	172	173	174	175	176	177	178	179	180
D	C	A	C	A	A	D	B	D	C
181	**182**	**183**	**184**	**185**	**186**	**187**	**188**	**189**	**190**
A	D	A	C	D	A	D	C	D	B
191	**192**	**193**	**194**	**195**	**196**	**197**	**198**	**199**	**200**
A	A	C	B	C	C	B	B	A	C
201	**202**	**203**	**204**	**205**	**206**	**207**	**208**	**209**	**210**
D	B	A	C	B	C	B	A	A	C
211	**212**	**213**	**214**	**215**	**216**	**217**	**218**	**219**	**220**
C	A	A	A	D	B	D	A	A	C
221	**222**	**223**	**224**	**225**	**226**	**227**	**228**	**229**	**230**
B	C	C	B	B	C	D	D	D	B
231	**232**	**233**	**234**	**235**	**236**	**237**	**238**	**239**	**240**
C	B	C	D	C	A	A	D	A	A
241	**242**	**243**	**244**	**245**	**246**	**247**	**248**	**249**	**250**
B	C	C	D	D	A	A	C	C	C
251	**252**	**253**	**254**	**255**	**256**	**257**	**258**	**259**	**260**
C	D	A	D	A	B	B	D	C	D
261	**262**	**263**	**264**						
A	B	D	C						

HINTS AND SOLUTIONS

1.

Granitoid bodies	Geological age	Area
Erinpura Granite	Phanerozoic	Delhi Super group
Closepet Granite	Archean	Dharwar Super group
Singhbhum Granite	Archean	Singhbhum Super group
Ladakh Granitoid	Mesozoic	Jammu Kashmir

2.

Formations	Geological age	Economic significant	Area
Raniganj Formation	Upper Permian	Coal	Jharkhand
Barakar Formation	Lower Permian	Coal	Jharkhand
Kulti Formation	Lower Trissic	Coal	West Bengal
Karharbari Formation	Lower Permian	Coal	Jharkhand

3.

Chronostratigraphic units	Geochronologic units	Example
System	Period	Neogene
Stage	Age	Globorotalia
Series	Epoch	Miocene
Chronozone	Phase	Dinosaur

4. Formation	Deposition environment	Geological age	Chief rocks type	Fossils
Subathu Formation	Fluvial	Eocene	Sandstone, Limestone	Foraminifera, Ostrocoda, Molluscus
Karewa Formation	Fluvial and Lacustrine	Pleistocene	Gravels, Lignite	Ostrocode, Vertibrate
Indus Formation	Fluvial	Cretaceous	Shale, Siltstone.	Gastopodes, Ostocode
Blaini Formation	Glacial	Precambrian	Shale, Quartzite	Gondwana Flora

5. Formation	Geological age	Geographical localities	Remarks/Fossils
Niniyur Formation	Cretaceous	Trichinapally	*Foraminifera, Gastopode.*
Long Formation	Miocene	Andaman-Nicobar	*Foraminifera.*
Zewan Formation	Permian	Vihi ditrict.	*Lammelibranchia.*
Bhander Group	Upper purana.	Madhya Pradesh	*Gastropode*

6. The starting time of the Proterozoic Eon is = 2500 Ma.
The end time of the Proterozoic Eon is = 580 Ma.
The total Proterozoic Eon in Ma is = 2500 – 580 = 1920 Ma.

7. Stratigraphic unit	Geological age	Important fossils
Muth Quartzite	Devonian	Unfossilliferous
Lameta bed	Cretaceous	*Titaniferous*
Golden Oolite	Jurassic	Ammonoides
Karewas	Pleistocene	Molluscus

9. Type area	Geological age	Fossils	Rocks type
Kutch	Jurassic	Ammonoides	Sandstone, limestone
Trichinapally	Cretaceous	Brachiopoda	Sandstone, limestone
Spiti	Triassic	Gastropoda	Shale, limestone
Jaisalmer	Jurassic	Ammonoidea	Shale

10. Geological time scale	Example
Eon	Proterozoic
Era	Cenozoic
Period	Tertiary
Epoch	Miocene

11. Fossils	Famous area
Invertibrate fossils	Spiti
Plant fossils	Gondwana
Dinosaur fossils	Lameta beds
Mammals fossils	Siwalik hills

16. Units	Characteristics
Subzone	Biostratigraphic unit
Zone	Biostratigraphic unit
Chron	Chronostratigraphic unit
System	Chronostratigraphic unit

17. Gondwana type	Plant fossils
Upper gondwana	Ptilophyllum
Middle gondwana	Pacopteris
Lower gondwana	Glossopteris
Rajmahal intertrapean bed	Dictyozamites

24. Geological entites	Craton	Age
Chitradurga schist belt	Dharwar	Archean

Clospet Granite	Dharwar	Archean
Hutti gold mine	Dharwar	Archean
Bhilwara super group	Bundelkhand	Archean
Sitapundi anorthosite	Dharwar	Archean
Zawer Pb-Zn deposits	Aravalli	Proterozoic
Kolar schist belt	Dharwar	Archean
Malani rhyolite	Aravalli	Protetozoic
Malankhajng copper deposits	Dharwar	Archean
Mansar formation	Baster	Archean
Makrana marble	Aravalli	Proterozoic
Sukinda cromite deposits	Singhbhum	Archean

25.

Statements	Characteristics
A	The chief rocks of Vindhyan Super group is Sandstone, limestone, Shale.
B	The age of the Cuddapah is Proterozoic. The top of the Cuddapah is Vidhyan, which is not a Proterozoic
C	The Kaladgi basin is situated in Bijapur district of Maharashtra of archean rocks. It is part of Darwar craton
D	The Mesozoic Bhuj formation predominently of Sandstone

26.

Chronostratigraphic sequence	Rocks types	Fossils
Boulder Conglomerate	Boulder of Granite	Equus
Pinjor Formation	Sandstone, Conglomerate	Elephas
Tatrot Formation	Sandstone, Clay	
Dhokpathan Formation	Sandstone, Clay, Shale	Giraffoid
Nagri Formation	Sandstone, Shale	Hipparian
Chinji Formation	Clays, Shale, Sandstone	Mammalian
Kamlial Formation	Shale, Sandstone	Anthropoids

27.

Lithostratigraphic unit	Geological Period	Important Fossils
Mammal formation	Miocene	Mammals, birds
Fenestella shale	Corboniferous	Bivalvia, Corals
Maleri formation	Triassic	Reptiles
Lameta formation	Cretaceous	Dinosaurs

28.

Scientist	Concept
J. Hutton	Uniformitarianism
Wegener	Continental drift
Dutton	Isostasy
Kant	Nebular hypothesis

29.

Volcanoes	Age	Compositions
Malani Rhyolite	Meso-paleoproterozoic	Acidic
Dalma Volcanics	Neoproterozoic	Basic
Panjal Trap	Mesozoic	Ultrabasic
Rajmahal Trap	Peolozoic	Basic

30.

Fossils	Age	Indian example
Redlichia	Lower cambrian	Saline series
Fenestella	Carboniferous to permean	Kashmir valley
Syringothyris	Carboniferous to Permean	Zewan formation
Otoceras	Triassic	Spiti

31. Flora	Gondwana type	Age
Noeggerathiopsis	Lower gondwana	Carboniferous to Triassic
Gangamopteris	Lower gondwana	Carboniferous to Triassic
Dicroidium	Upper gondwana	Triassic to Cretaceous
Vertibraria	Lower gondwana	Carboniferous to Triassic

34. System	Important fossils
Gondwana	Invertibrate
Zewan beds	Brachiopoda, Corals
Vindhyan	Stromatolites, Brachiopoda
Siwalik	Vertibrate

35. Stratigraphic units	Marker horizons
Po formation	Carboniferous of Spiti
Lipok formation	Carboniferous of Spiti
Muth Quartzite	Devonian of Himalaya
Syrinngothysis Limestone	Carboniferous of Kashmir

38. Facies (environment)	Example (Stratigraphic unit)
Fluvio-lacustrine facies	Ladakh Himalaya
Volcanic facies	Dalma Volcanoes
Marine transgression facies	Gondwana Basin
Marine facies	Lameta bed

39. Formations	Geological age
Madh formation	Palaeocene
Porbandar formation	Pleistocene
Umia formation	Jurassic
Bhuj formation	Jurassic

41. Series	Area
Lipak series	Permian of Sikkim
Lachi series	Triassic of Spiti
Laki series	Tertiary of Sind
Lathi formation	Mesozoic of Rajasthan

43. Siwalik Himalaya	Fossils
Nagri	Hipparion
Kamlial	Hypoboops
Dhokpathan	Stegodon
Chingi	Ramapithicus

45. Stage	Series	Age
Tura sandstone	Jaintia series	Eocene of Assam
Tikak Parbat	Barail series stage	Oligocene of Assam
Naogaon stage	Brail series	Oligocene of Assam
Bragolai stage	Brail series	Oligocene of Assam

46. Formation	Economic minerals
Chorbaoli formation	Quartzite
Mansar formation	Mn-Ores
Bochum formation	Marble
Lohangi formation	Mica

50. Stratigraphic units	Classifications
Period	Time unit
Zone	Chrono unit
Age	Time unit
Epoch	Time unit

55. Himalaya formation	Succession
Dagshai formation	Himalayan Paleogene
Siwalik group of Karewa formation	Himalayan Neogeme
Subathu formation	Himalayan Paleogene
Kasauli formation	Himalayan Paleogene

59. Cuddapah super group	Ore minerals
Kistna Group	Iron
Nallamalai group	Copper and Lead
Cheyair group	Jasper, Mica
Papaghani group	Barytes, Asbestos

60. Strike	Succession
NNW-SSE	Dharwar super group
NNE-SSW	Singhbhum Super group
NE-SW	Aravalli Super group
SW-NE	Eastern ghat

61. Formations	Successions
Lathi formation	Mesozoic of Barmer of Western Rajasthan
Chari formation	Mesozoic of Kutch
Jaisalmer formation	Mesozoic of Western Rajasthan
Bhadasar formation	Mesozoic of Western Rajasthan

64. Gondwana flora	Class
Glossopteris	Lower gondwana
Ptilophyllum	Upper gondwana
Dichroidium	Middle gondwana
Gangamopteris	Lower gondwana

66. Fossils	stratigraphic unit	important rocks type
Equidae	Siwalik group	Sandstone, conglomerate
Stramarolite	Vindhyan supergroup	Sandstone, Shale
Dinosaurs	Gondwana super group	Boulder beds, Shale
Brachiopoda	Triassic of Spiti	Shale, Limestone

67. Formations	Characteristics Fossils
Talchir formation of Satpura basine	Brachiopoda, Pelycypoda, Glossopteris Indica
Bijori formation of Satpura basine	Glosspteris flora
Patcham formation of Cutch	Macrocephalites
Bhuj formation of Cutch	Ammonoides

68. Lithostratigraphy sequence	Characteristics
Kamlial-chinji-Nagri-Dhokpathan-Tatrot-Pinjor	Siwalik group
Papaghani-Cheyair-Nallamalai-Kistna	Cuddapah super group
Talchir-Damuda-Panchet-Mahadewa-Rajmahal-Jabalpur	Gondwana super group
BGC-Aravalli-Railo-Delhi	Archean of Rajasthan

69. Lithostratigraphy units	Time units	Important character
Kolhan formation	Proterozoic	Singhbhum super group
Khorabari formation	Permian	Lower Gondwana super group
Kathi formation	Triassic	Upper Gondwana super group
Kota formation	Jurassic	Upper Gondwana super group

79. Cuddapah super group	Shales
Papaghani group	Vempalle shale
Cheyair group	Tadpatri shale
Nallamalai group	Cumbum shale
Kistna group	Kolamnala shale

84. Stratigraphy succession of cretaceous of Trichinapally.

Succession	Chief rocks	Fossils
Niniyar group	Sandstone, shale, limestone	Gastropods, Foraminifera
Ariyalur group	Sandstone, shale	Pelycpoda, echinoids, ammonites
Trichinapally group	Sandstone, limestone	Lamellibranchia, gastropoda
Uttatur group	Shale, limestone	Cephalopods, gastropoda

85.

Stratigraphic Unit	Associated Craton	Age
Berach Granite	Aravalli craton of Rajasthan	Archean
Untala Granite	Aravalli craton	Archean
Gingla Granite	Aravalli craton	Archean
Closepet Granite	Dharwar craton of Karnataka	Archean

94.

Formations	Fossils
Kamalial formation	Tetrabelodon
Pinjor formation	Elephas
Nagri formation	Mastodon
Dhokphathan formation	Stegodon

97.

Stratigraphic unit	Origin	Age	Area
Umaria beds	Marine	Cambrian	Maharashtra
Kamthi beds	Marine	Lower Triassic	Orissa
Pali beds	Marine	Upper Triassic	Son Valley
Bareen Measures	Glacial	Mid Permian	Damodar Valley

98.

Stratigraphic unit	Age	Important fossils
Neobolus Beds	Cambrian	Brachiopods, Trilobites.
Muth Quartizite	Devonian	Unfossiliferous.
Katrol Group	Jurassic	Ammonites.
Tipam sandstone	Miocene	Plant fossils.

99.

Age (Ma)	Boundary
1500	Paleoproterozoic - Mesoproterozoic
2500	Archean - Proterozoic
570	Proterozoic - Cambrian
3500	Eoarchean - Paleoarchean

101.

Lithostratigraphic units	Example
Group	Sausor group
Formation	Mansar formation
Member	Syringothyris limestone
Bed	Limestone bed

102.

Time units	Example
Period	Quaternary
Zone	Chronostratigraphic unit
Age	Miocene
Epoch	Neogene

104.

Terminology	Characteristics
System	Chronostratigraphic units
Formation	Lithostratigraphic units
Stage	Chronostratigraphic units
Series	Chronostratigraphic units

107.

Tertiary of Assam	Geological age
Dihing group	Pleistocene
Dupitila group	Pliocene
Tipam group	Miocene
Surma group	Miocene
Barail group	Oligocene
Jaintia group	Palaeocene

112.

Strike direction	Example
NNW-SSE	Dharwar super group
NNE-SSW	Iron ore group
NE-SW	Eastern Ghat
SE-NW	Baster Craton

115. Formations	Remarks
Lathi formation	Jurassic of Jaisalmer
Chari formation	Jurassic of Cutch
Katrol formation	Jurassic of Cutch
Umia formation	Jurassic of Cutch

117. Stratigraphic unit	Age	Composition
Deccan Trap	Cretaceous	Basic
Panjal Trap	Permian	Basic
Malani Rhyolite	Proterozoic	Acid
Rajmahal Trap	Jurassic	Basic

118. Stratigraphic units	Fossils
Lameta beds	Dinosaur
Muth quartzite	Unfossileferous
Siwalik	Mammals
Subathu formation	Ostrocods, molluscas, Foraminifera

119. Formation	Age	Area
Maleri	Triassic	Gondwana
Karewa	Plio-Pleistocene	Kashmir Valley
Ariyalur	Cretaceous	Trichinapally
Barail	Oligocene	Assam

120. Stratigraphic units of India	Characteristics
Sausar group	Archean rocks of Maharashtra and Madhya Pradesh.
Cheyair	Proterozoic rocks of Cuddapah of Andhra Pradesh.
Rewa	Proterozoic rocks of Vindhyan of Madhya Pradesh.
Gondwana	Paleozoic rocks of Gondwana.

121. Formation	Fossils	Family	Age (Fossils)
Lameta formation	*Titanosaurus*	Dinasaur	Cretaceous
Raniganj formation	*Glossopteris*	Gondwana flora	Permian
Pinjor formation	*Elephas*	Elephant	Pleistocene
Umaria Marine beds	*Productus*	Brachiopoda	Carboniferous to Permian

122. Stratigraphic classifications	Units
Lithostratigraphy	Supergroup-Group- Member-Formation-Beds
Biostratigraphy	Interval zone- biozone
Chronostratigraphy	Erathem- system- Series-Stage - Zone
Geochronology	It is determined age of the rocks.

Chronostratigraphic	Time Units	Example
Erathem	Era	Mesozoic
System	Period	Quaternary
Series	Epoch	Miocene
Stage	Age	Panonian
Zone	Phase	Globorotalia tumida

123. Rocks formation	Age
Malani volcano	Precambrian
Panjal volcano	Permian
Rajmahal volcano	early cretaceous
Deccan volcano	late cretaceous

Rocks formation	Characteristics
Dhokpathan-tatrot-pinjor-boulder congumerate	Siwalik succession of Tertiary rocks.
Uttatur-Trichinapally-Ariyallur- Niniyar	Cretaceous of Trichinapally
Talchir-Damuda-Panchet- Mahadewa	Gondwana succession of India

124. Methods	Uses
K-Ar	Archean to recent
C-14	Holocene
U-Pb	Archean to resent
Rb-Sr	Archean to recent

125.

Formation	Area	Age
Lipak formation	Spiti	Carboniferous
Muth quartzite	Spiti	Devonian
Po formation	Spiti	Corboniferous
Syringothyria Limestone	Kashmir	Carboniferous

126.

Name (scientist)	Statements (Principal)
James Hutton	"The present is key to the past"
Nicholas Steno	Low of order of superposition.
William Smeeth	Principal of Uniformitarianism.
Walther	Walther facies variation low.

127.

Stratigraphic units	Area
Darjeeling Gneisses	Eastern Himalaya
Daling Group	Eastern Himalaya
Tejam Group	Kumaun Himalaya
Kurnool Group	Kurnool district of Andhra Pradesh

155.

Formation	Area	Geological age
Salkhala Series	Kashmir	Paleozoic
Champion Gneiss	Mysore	Archean
Buxa Series	Eastern Himalaya	Precambrian
Darjeeling Series	Darjeeling and Sikkim region	Precambrian

160.

Trend	Stratigraphic units
North-South	Vaikrita group
East-West	Narmada Valley
Northeast-Southwest	Aravalli mountain
Northwest-Southeast	Pandian Mobile belt

176.

Stratigraphic unit	Geological age (Ma)
Chitradurga Group	2100-2380
Sargur schist	>3200
Ajabgarh Group	1600
Udaipur Group	2500

179.

Stratigaphic Units (Vindhyan Supergroup)	Economic minerals
Semri	Shale
Kaimur	Jaspars, Pyrites
Bhander	Building stone
Rewa	Diamond

180.

Stratigraphic Units	Term
Lithostratigraphy	Group, Formation, Beds
Biostratigraphy	Zone, Biozone
Chronostratigraphy	Erathem, Era, System
Magnetostratigraphy	Magnetic behaviour of the Earth

183. Evolution of Human:

Era	Years Ago	Period		Epoch	Characterized by
CENOZOIC	70,000,000-60,000,000 60,000,000-40,000,000 40,000,000-25,000,000	Paleogene	Tertiary	Paleocene Eocene	advent of birds, mammals presence of modern mammals Oligocene saber-toothed cats
	25,000,000-10,000,000 10,000,000-1,000,000	Neocene		Miocene Pliocene	grazing mammals growth of mountains; increase in size and numbers of mammals; gradual cooling of climate
	1,000,000-10,000 10,000-present	Quaternary		Pleistocene Recent	widespread glacial ice, development of human

184. Evolution of Plants:

Era	Years Ago	Period	Characterized by
MESOZOIC	220,000,000-180,000,000	Triassic	volcanic activity; marine reptiles, dinosaurs
	180,000,000-135,000,000	Jurassic	dinosaurs, conifers
	135,000,000-70,000,000	Cretaceous	extinction of giant reptiles; advent of modern insects; flowering plants

187. Stratigraphic units	Origin
Bagh beds	Marine Origin
Baripada beds	Miocene of the Marine
Cuddalore sandstone beds	Marine origin
Lameta beds	Fresh water

190. Stratigraphic units	Stages
Triassic of Spiti	Lower Triassic-Middle Triassic-Upper Triassic
Jurassic of Kutch	Patcham-Chari-Katrol-Umia
Cretaceous of Trichinapalli	Uttatur-Trichinopoly-Ariyalur-Niniyur
Tertiary of Assam	Barail-Surma-Tipam-Dupitila-Dihing

198. Stratigraphic Units	Location	Geological Age
Bhuj Formation	Kutch	Jurassic
Ariyalur Formation	Trichinapally	Cretaceous
Patcham formation	Kutch	Jurassic
Katrol formation	Kutch	Jurassic

224. Stratigraphic Units	Geological age
Kanawar group	Neoproterozoic
Kulling group	Eocene
Muth quartzite	Devonian
Agglomerate shale	Mesozoic

203. **PHANEROZOIC AND PRECAMBRIAN CHRONOSTRATIGRAPHY:**

International Stratigraphic Chart

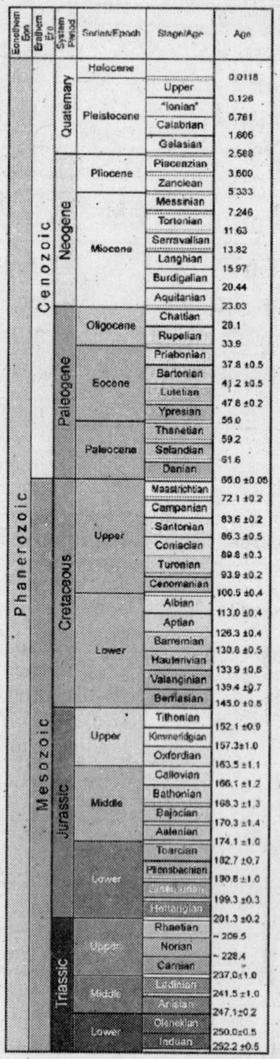

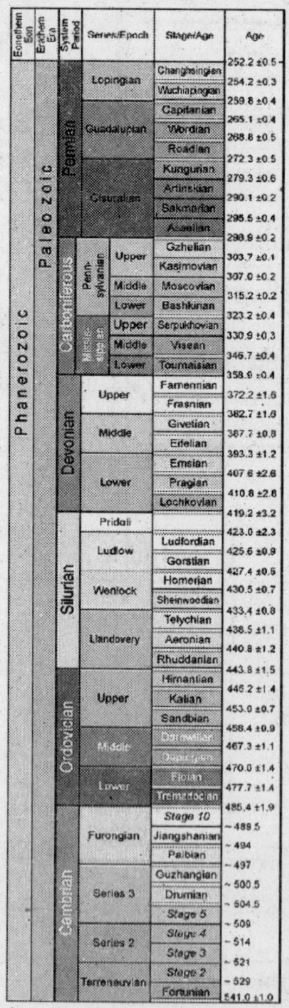

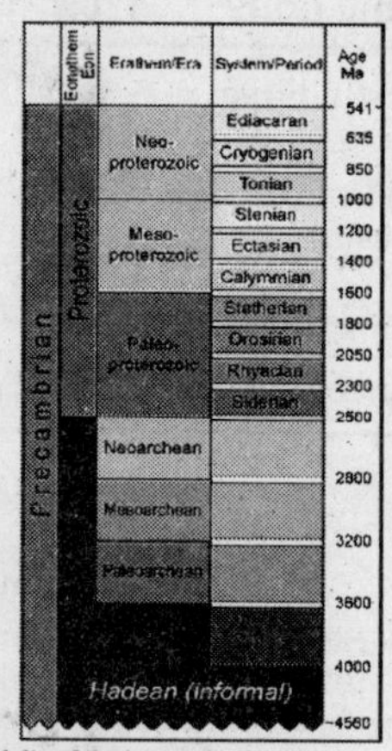

Units of the international chronostratigraphic scale with estimated numerical ages from the GTS2012 age model.

Colors are according to the Commission for the Geological Map of the World.

Subdivisions of the Phanerozoic are formally defined by a Global boundary Stratotype Section and Point (GSSP) at each lower boundary. Thick yellow lines between stages on this diagram denote GSSPs approved by the International Commission on Stratigraphy (ICS) and ratified by the International Union of Geological Sciences (IUGS).

Precambrian units are formally defined by absolute age (Global Standard Stratigraphic Age — GSSA), with the exception of the Ediacaran System defined by a basal GSSP.

Numerical ages assigned to unit boundaries are subject to revision upon formal decision or revision of GSSPs and when enhanced radio-isotopic and cyclostratigraphy studies enable improvements to the age models.

Stratigraphic information and details on international and regional geologic units can be found on the websites of the ICS (*www.stratigraphy.org*) and the Geologic TimeScale Foundation (*https://engineering.purdue.edu/stratigraphy*).

APPLIED GEOLOGY

1. A unit which is semi-permeable and does **not** yield any significant quantity of groundwater is called:

A. Aquiclude B. Aquifer
C. Aquitard D. Artesian Aquifer

(IIT-JAM 2015)

2. A mine excavation made along the strike of a 2 meter thick tabular ore body dipping 30° is called:

A. Crosscut B. Raise
C. Drive D. Shaft

(IIT-JAM 2015)

3. Petroleum producing basins of India belong to:

A. Cambrian B. Cretaceous
C. Tertiary D. Quaternary

(IIT-JAM 2015)

4. A 10 cm^3 sandstone block has a mass of 20 grams. If the average density of sand grains is 2.5 g/cm^3, the porosity (in percent) of the sandstone is

A. 25 B. 20
C. 28 D. 30 *(IIT-JAM 2015)*

5. Field capacity represents the maximum quantity of water held:

A. In the soil against the force of gravity
B. In the ground water zone
C. In the intermediate zone
D. Below the water table *(AMU-2015)*

6. Which of the following parameter has the least influence on porosity?

A. Grain size
B. Sorting
C. Degree of cementation
D. Dissolution *(AMU-2015)*

7. The stream which receive groundwater discharge is called:

A. Influent B. Effluent
C. Ephemeral D. Perennial

(AMU-2015)

8. Clay is example of:

A. Aquifer B. Aquiclude
C. Aquitard D. Aquifuge

(AMU-2015)

9. Which of the following has secondary porosity?

A. Vesicular basalt B. Sand
C. Limestone D. Gravel

(AMU-2015)

10. A measure of the amount of calcium, magnesium and iron dissolved in water is called:

A. TDS B. E.C.
C. Hardness D. SAR *(AMU-2015)*

11. The recommended permissible limit of fluoride in drinking water as per Indian Standard is:

A. 1 B. 1.5
C. <1 D. >1 *(AMU-2015)*

12. Excess of nitrate in drinking water causes a disease in infants known as:

A. Fluorosis B. Silicosis
C. Asbestosis D. Blue-baby syndrome

(AMU-2015)

13. Which of the following is the most suitable reservoir rocks?

A. Sandstone with poorly sorted angular grains
B. Sandstone with well sorted rounded grain
C. Shale
D. Black shale *(AMU-2015)*

14. The principal difference between a debris slide and a debris avalanche is in:

A. Shape of the path taken by the moving mass

B. Grain size of the moving mass

C. Water content; former mass contains abundant water whereas the latter is dry

D. Character of source region; former mass contains pyroclastic debris from a volcanic eruption *(DU-2015)*

15. Which soil horizon is the zone of accumulation, so named because dissolved matter, leached from other parts of the soil, precipitates to from new minerals?

A. A-horizon B. B-horizon

C. C-horizon D. O-horizon *(DU-2015)*

16. Among the following, highest porosity value can be obtained from:

A. Consolidated mud

B. Unconsolidated mud

C. Well sorted sandstone

D. Poorly sorted sandstone *(DU-2015)*

17. Sea water is saline because:

A. Na^+ is most abundant cation in sea water

B. Cl^- is most abundant anion in sea water

C. NaCl is the first compound to precipitate from sea water

D. Na^+ has highest residence time in sea water *(DU-2015)*

18. Which statement is correct for fair weather wave base?

A. It is situated below storm wave base

B. Fair weather wave- base coincides with the base of shoreface zone

C. Fair weather wave base laterally grades to storm wave base

D. Fair weather wave base coincided with the base of beach-foreshore zone *(DU-2015)*

19. Hydraulic conductivity is:

A. Transmissivity/saturated aquifer thickness

B. Storativity aquifer thickness

C. Extension of the aquifer

D. Specific capacity *(DU-2015)*

20. In steady state/equilibrium ground-water flow situation, the water table head during pumping:

A. Changes with time

B. Changes without time

C. Does not change with time

D. Remains constant *(DU-2015)*

21. In unconfined aquifer, Storage coefficient is approximately equal to:

A. Conductivity

B. Specific yield

C. Intrinsic permeability

D. Specific retention *(DU-2015)*

22. Flownets in hydrogeology can be made using:

A. Depth to water level map

B. Water level fluctuation map

C. Water table contour map

D. Topographic map *(DU-2015)*

23. Porosity of a formation is summation of:

A. Hydraulic conductivity and Transmissivity

B. Specific storage and Specific retention.

C. Specific retention and Specific yield

D. None of the above *(DU-2015)*

24. In resistivity sounding of groundwater exploration:

A. Variation of resistivity with depth at a point is measured

B. Horizontal variation in resistivity is measured

C. Electrical conductivity is measured

D. TDS of groundwater is measured *(DU-2015)*

25. This equation for unsteady state radial flow to a tubewell in confined aquifer can be used to estimate:

A. Transmissivity only

B. Storativity only

C. Specific yield only

D. Both Transmissivity and Storativity *(DU-2015)*

26. The dimension of Storativity is:

A. LT^1 B. UT^{-1}

C. UT D. None of the above

(DU-2015)

27. Aquifer anisotropy relates to:

A. Temporal variation in aquifer parameters

B. Both spatial and temporal variation in aquifer parameters

C. Directional variation in aquifer parameters

D. Spatial variation in aquifer parameters

(DU-2015)

28. Suitability of groundwater for irrigation uses is commonly studied by measuring:

A. Sodium Adsorption Ratio

B. Hydraulic conductivity

C. Grade of metamorphism

D. One of the above *(DU-2015)*

29.

Group I	Group II
(*i*) Creep	1. Tounge-shaped mass movement
(*ii*) Earth flow	2. Curved tree trunks
(*iii*) Slump	3. Scree formation at the base
	4. Curved surface of rupture

Correct match between the nature of mass movement listed in Group I with the evidences listed in Group II is:

A. (*i*)-2, (*ii*)-1, (*iii*)-4

B. (*i*)-1, (*ii*)-3, (*iii*)-4

C. (*i*)-2, (*ii*)-4, (*iii*)-3

D. (*i*)-4, (*ii*)-1, (*iii*)-3 *(DU-2015)*

30. Which of the following conditions is the most suitable for construction of a tunnel ?

A. Tunnel along the strike of a strata

B. Tunnel across the strike of strata

C. Tunnel along the axis of a syncline

D. Tunnel along the axis of an Anticline

(DU-2015)

31. Elevation contours of ground surface (values in parenthesis) and groundwater table (values in normal font) are given in the figure below.

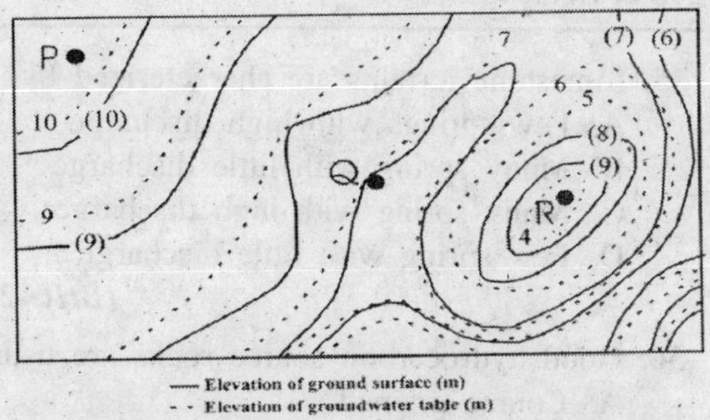

What do the points P, Q and R represent?

A. P-recharge area, Q-spring, R-discharge area

B. P- discharge area, Q-spring, R-recharge area

C. P-spring, Q-recharge area, R-discharge area

D. P-discharge area, Q-recharge area, R-spring

(IIT-JAM 2014)

32. In the given block diagram, select the most suitable site amongst (i), (ii), (iii), and (iv) for constructing a tunnel.

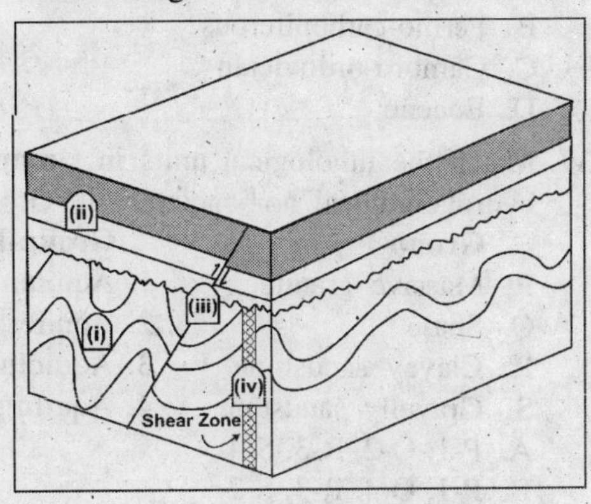

A. (*i*) B. (*ii*)

C. (*iii*) D. (*iv*)

(IIT-JAM 2015)

33. In which of the following rocks does ground water flow obey Darcy's law?

A. Shale B. Sandstone

C. Marble D. Granite

(IIT-JAM 2014)

34. Most of the groundwater in hard rocks is present in the:

A. Pores

B. Cracks and joints

C. Intergranular space

D. Intragranular space *(BHU-2014)*

35. Limestone terrains are characterized by:
A. Few springs with high discharge
B. Many spring with little discharge
C. Many spring with high discharge
D. Few spring with little discharge
(BHU-2014)

36. Good hydrocarbon source rocks are usually:
A. Course grained
B. Fine grained
C. Medium grained
D. Very course grain *(BHU-2013)*

37. The most sluggish process of mass wasting is:
A. Creep B. Flow
C. Slide D. Fall *(BHU-2013)*

38. Majority of world's coal resources are restricted to the following geological time period:
A. Triassic
B. Permo-carboniferous
C. Cambro-ordovician
D. Eocene *(PU- 2013)*

39. Match the lithological units in Group-I with hydrogeological nomenclature in Group-II.

Group-I	Group-II
P. Massive granite	1. Aquitard
Q. Shale	2. Aquifer
R. Clayey sandstone	3. Aquiclude
S. Gravelly sandstone	4. Aquifuge

A. P-1, Q-2, R-3, S-4
B. P-1, Q-4, R-3, S-2
C. P-4, Q-3, R-1, S-2
D. P-2, Q-1, R-4, S-3 *(IIT-JAM 2012)*

40. The mass movement process in which cohesive blocks of earth move on a failure plane with concave-up geometry, is known as:
A. Debris flow
B. Creep
C. Rotational slide
D. Translational slide *(IIT-JAM 2012)*

41. Which one of the following tunnel alignments is considered as geologically favourable?
A. Tunnel through the core of a synform with parallel tunnel- and fold axes
B. Tunnel through the core of an antiform with parallel tunnel- and fold axes
C. Tunnel through a synform with tunnel- and fold axes perpendicular to each other
D. Tunnel through an antiform with tunnel- and fold axes perpendicular to each other
(IIT-JAM 2012)

42. Which of the following may be a good reservoir rocks?
A. Granite B. Porcilenite
C. Massive basalt D. Sandstone
(BHU-2013)

43. When an artesian condition developed?
A. Water table is above groundwater level
B. Water table is perched
C. Aquifer is unconfined
D. Piezometric surface is above groundwater table *(BHU-2013)*

44. Which building stone possesses higher crushing strength?
A. Kota stone
B. Vindhyan sandstone
C. Makarana marble
D. Bundelkhand granite *(BHU-2013)*

45. The underground water that originates from molten magma, known as:
A. Connate water
B. Juvenile water
C. Hygroscopic water
D. Fluid inclusion *(BHU-2012)*

46. Which type of clay is used as drilling mud?
A. China clay B. Bentonite clay
C. Fire clay D. Fuller clay
(BHU-2013)

47. The RF of the geological map prepared on a scale of 5 cm = 1 km is:
A. 1 : 20000 B. 1 : 2000
C. 1 : 10000 D. 1 : 1000
(BHU-2013)

48. Which geo-material has higher porosity?
A. Gravel B. Sandstone
C. Silt D. Clay
(BHU-2013)

49. Deep confined aquifer can be recharged:
A. Water spreading B. Inverted well
C. Shaft recharge D. Induced recharge
(BHU-2013)

50. The thickest coal seam belongs to:
A. Bishrampur, SECL
B. Talchir, MCL
C. Jingurdah, NCL
D. Chandrapur, WCL *(BHU-2013)*

51. What type of river is most suitable for construction of dam?
A. Effluent B. Influent
C. Insulated D. Low gradient
(BHU-2013)

52. Coal breaks along joints are called:
A. Columnar joints B. Vertical joints
C. Mural joints D. Cleats
(BHU-2013)

53. Ball coal consists of materials:
A. Calcium carbonate
B. Magnesium carbonate
C. Iron carbonate
D. Coal *(BHU-2013)*

54. The capacity of a rock to transmits fluid is known as:
A. Porosity B. Permeability
C. Transmitivity D. Water absorption
(BHU-2013)

55. An artesian well is characterized by:
A. Development of cone of depression around the well
B. Located in specific geographic locations
C. That is developed in an aquitard
D. That flow to the surface without pumping
(BHU-2013)

56. The reservoir rock of Bombay High oil field is:
A. Sandstone B. Limestone
C. Shale D. Clay
(BHU-2012)

57. The presence of coal deposits impart great economic significance to:
A. Middle Gondwana sediments
B. Lower Gondwana sediments
C. Upper Gondwana sediments
D. Late Triassic sediments *(BHU-2012)*

58. Palana legnite deposit is assigned to age:
A. Late Triassic B. Tertiary
C. Early Jurassic D. Cretaceous
(BHU-2012)

59. Leakage through rock fractures below a dam can be reduced significantly by:
A. Bolting B. Grouting
C. Drilling D. Blasting
(IIT-JAM 2011)

60. Adit is an underground excavation, which is:
A. A horizontal opening giving access to main tunnel
B. A horizontal opening used for transporting water to powerhouse
C. A vertical opening used for multiple purposes
D. An inclined opening providing access to main tunnel *(IIT-JAM 2011)*

61. The ratio of volume of underground water released under gravity to the total volume of saturated aquifer is called:
A. Transmissivity
B. Permeability
C. Storage coefficient
D. Specific yield *(IIT-JAM 2011)*

62. A petroleum system does **not** contain:
A. Source rock B. Reservoir rock
C. Traps D. Boreholes
(BHU-2011)

63. Which one of the following is **not** a type of oil traps?
A. Anticlines B. Salt domes
C. Unconformity D. Mesa *(BHU-2011)*

64. The oil-bearing formation in Digboi Oil Field is:
A. Tipam B. Kalol
C. Panna D. Barail *(BHU-2011)*

65. The process of formation of coal is:
A. Preservation of plant material
B. Biochemical changes
C. Carbonization and metamorphism
D. All of the above *(BHU-2011)*

66. The reservoir rocks in Bombay High are:
A. Sandstone B. Limestone
C. Shale D. Clay *(BHU-2011)*

67. The marble used in Taj Mahal had brought from:

A. Reewa B. Sultanpur
C. Jabalpur D. Makrana

(BHU-2011)

68. Match the hydrological terms in Group-1 with those in Group-2.

Group 1	Group 2
P. Aquifer	1. Non-porous, non-permeable
Q. Aquiclude	2. Porous, non-permeable
R. Aquitard	3. Poorly porous and poorly permeable
S. Aquifuge	4. Porous and permeable

A. P-1, Q-2, R-3, S-4
B. P-2, Q-3, R-4, S-1
C. P-4, Q-2, R-3, S-1
D. P-4, Q-3, R-1, S-2 *(IIT-JAM 2010)*

69. What is the porosity of sandstone with solid rock volume of 70%?

A. 0.7 B. 0.4
C. 0.3 D. 0.07 *(IIT-JAM 2009)*

70. Which part of the arch dam experiences maximum stress?

A. Heel B. Toe
C. Abutment D. Spillway

(IIT-JAM 2008)

71. In hydrological context, a porous and impermeable earth mass is known as:

A. Aquifer B. Aquitard
C. Aquiclude D. Aquifuge

(IIT-JAM 2008)

72. In an underground mine, shaft and ore body are connected by:

A. Bench B. Raise
C. Adit D. Crosscut

(IIT-JAM 2008)

73. Rapid mass movement of water-saturated regolith is called:

A. Landslide B. Creep
C. Solifluction D. Earth flow

(IIT-JAM 2008)

74. The reservoir rock for petroleum in Bombay High is:

A. Feldspathic sandstone
B. Graywacke
C. Limestone
D. Orthoquartzite *(IIT-JAM 2007)*

75. Which of the following rocks can form good aquifers?

A. Sandstone and highly jointed granite
B. Sandstone and massive granite
C. Sandstone and shale
D. Shale and massive granite

(IIT-JAM 2007)

76. Of the following alternatives, the most suitable condition for a road cut in a hilly terrain is where the foliations:

A. Are vertical
B. Dip away from the slope
C. Dip towards the slope at high angle
D. Dip towards the slope at low angle

(IIT-JAM 2007)

77. A rock that is porous but not permeable is:

A. Sandstone B. Siltstone
C. Shale D. Marble

(IIT-JAM 2006)

78. Match the following:

Group I	Group II
P. Oil	1. Bokaro
Q. Coal	2. Zawar
R. Lead - zinc	3. Jamar Kotra
S. Phosphate	4. Ankleshwar

A. P - 1 Q - 3 R - 2 S - 4
B. P - 2 Q - 4 R - 3 S - 1
C. P - 4 Q - 1 R - 3 S - 2
D. P - 4 Q - 1 R - 2 S - 3 *(IIT-JAM 2006)*

79. Artesion condition results when the:

A. Water table is perched
B. Aquifer is unconfined
C. Piezometric surface is above ground level
D. Water table is above ground level

(IIT-JAM 2005)

80. For the construction of a gravity dam choose the most favorable geological condition of the following:

A. Strata dipping downstream in narrow valley
B. Strata dipping upstream in wide valley
C. Strata dipping downstream in wide valley
D. Strata dipping upstream in narrow valley

(IIT-JAM 2005)

81. Mass wasting related to barely perceptible and non-accelerating downslope movement is called:
A. Earth flow B. Creep
C. Rock Topple D. Rock fall
(DU-2014)

82. Which of the following hill slope component is related to pedogenic process?
A. Colluvial foot slope
B. Fall face
C. Convex creep slope
D. Interfluve *(DU-2014)*

83. The unconformity that marks end of rift sedimentation phase in Bombay high belong to:
A. Early to middle Miocene
B. Middle Oligocene
C. Middle to late Paleocene
D. Late Miocene *(DU-2014)*

84. Lithosphere subjected to local and regional loads react by phenomenon that is known as:
A. Rifting B. Subduction
C. Obduction D. Flexure *(DU-2014)*

85. In resistivity survey the lowest resistivity can be expected from:
A. A sandstone with freshwater in pore space
B. A sandstone with oil in pore space
C. A sandstone with seawater in pore space
D. Well-cemented sandstone
(DU-2014)

86. Sea water is saline because:
A. Na^+ and Cl^- are most abundant cation and anion in seawater
B. Na^{+2} is abundant but Cl^- is less abundant
C. Ca^+ is less abundant in sea water than Na^{+2}
D. Na^{+2} has higher residency time compared to Ca^{+2} *(DU-2014)*

87. Water table is found at the top of the
A. Zone of alteration
B. Zone of recharge
C. Zone of discharge
D. Zone of saturation *(DU-2014)*

88. Out of the following which has maximum salinity?
A. Atlantic ocean B. Pacific ocean
C. Indian ocean D. Southern ocean
(DU-2014)

89. Which of the following is a mechanically strained rock?
A. Marble B. Mylonite
C. Granulite D. Hornfels *(DU-2014)*

90. Capillary fringe:
A. Overlies the true water table
B. Underlies the true water table
C. May occur above or under the water table
D. None of the above *(DU-2014)*

91. A positive gravity anomaly indicates:
A. An excess of mass
B. A deficiency in mass
C. A reversal of the gravitational field
D. None of these *(DU-2014)*

92. Which of the following is characteristic of a tsunami?
A. Very long wavelength
B. Very fast moving wave
C. Very low wave amplitude in the open ocean
D. All of the above *(DU-2014)*

93. Rock Salt is found in which place of India?
A. Rajban B. Rampur
C. Mandi D. Manali *(BHU-2015)*

94. A rock that is porous but not permeable:
A. Marble B. Sandstone
C. Siltstone D. Granite
(BHU-2015)

95. A rock-cut slope having inclination 400 due East. The joint set most likely to cause failure is the one dipping at:
A. 200 due 0 B. 300 due E
C. 600 due E D. 650 due E
(BHU-2015)

96. Petroleum and Natural gas are chiefly composed of:
A. Hydrogen B. Hydrogen & Carbon
C. Nitrogen D. Oxygen
(PU-2010)

97. The tendency of water to pass directly from solid state to vapour state is known as:
A. Sublimation B. Oxidation
C. Dissolution D. Dissociation
(PU-2013)

98. The average amount of time that a water molecule resides in a particular reservoir before transferring to another reservoir is known as:
A. Residence time B. Infiltration
C. Ion exchange D. Base exchange
(PU-2013)

99. The measure of suspended and colloidal matter in water such as clay, silt, organic matter and microscopic organisms is known as:
A. Conductivity
B. Turbidity
C. Biochemical oxygen demand
D. Total dissolved load *(PU-2013)*

100. Urbanisation can cause:
A. Increase in net infiltration
B. Decrease in base flow
C. Decrease in amplitude of stream hydrograph
D. Decrease in overland flow
(PU-2013)

101. In a black and white aerial photograph, a totally reflecting object on earth (example dry sand) will appear:
A. Black B. White
C. Dark gray D. Medium gray
(PU-2013)

102. Blue colour of sky is due to light by the aerosols:
A. Absorption B. Adsorption
C. Scattering D. Emission
(PU-2013)

103. Which of the following rock has high porosity and high permeability?
A. Shale B. Greywacke
C. Sandstone D. Quartzite
(PU-2013)

104. A dike measures 10 cm in length on a geological map prepared in 1: 50,000 scale. What is the actual length of the dike?
A. 10 cm B. 50 m
C. 500 m D. 5 km *(PU-2013)*

105. GPS is acronym for:
A. Global Positioning System
B. Global Polar System
C. Global Position Satellite
D. Global Polar Satellite *(PU-2013)*

106. The ultraviolet rays do not reach earth's surface. This is because absorption of the rays by:
A. Ozone B. Carbon monoxide
C. Water vapour D. Oxygen
(PU-2013)

107. Chandrayan-1 mission of the India was a mission to:
A. Launch a remote sensing satellite in earth's orbit
B. Send an orbiter in Moon's orbit
C. Detect life on moon
D. Send a living being on Moon
(PU-2012)

108. A local water table positioned above the regional water table is said to be:
A. Stranded B. Perched
C. Displaced D. Depressed
(PU-2012)

109. The ability of an Earth material to transmit water is a measure of its:
A. Porosity
B. Aquifer characteristics
C. Chemical cement
D. Permeability *(PU-2012)*

110. The percentage of a rock's total volume that is taken up by pore space is called the:
A. Permeability B. Recharge
C. Aquifer D. Porosity
(PU-2012)

111. Which of the following rocks possess higher amount of primary porosity?
A. Claystone B. Sandstone
C. Limestone D. Siltstone
(PU-2011)

112. Which is the most abundant ion in the normal seawater?
A. Cl^- B. SO_4^-
C. Na^+ D. K^+ *(IIT JAM-2016)*

113. The value of gravity at the Earth's surface is dependent on
A. Latitude
B. Altitude
C. Composition of underlying material
D. Relative position of Sun–Earth
(IIT JAM-2016)

114. Weight of a 10 cm^3 medium grained sandstone block with 20% (v/v) porosity, in dry state is 26g.
The density of the block when fully saturated with water is ________ g/cm^3.
A. 2.8 B. 3.8
C. 4.8 D. 1.8 *(IIT JAM-2016)*

115. Fine muds are deposited at a rate of 1 cm per 1000 y. Assuming constant sedimentation rate and absence of compaction, a 1 km thick sequence would be deposited in ________ million years.
A. 100 B. 1000
C. 10 D. 10000
(IIT JAM-2016)

116. The half life of a radionuclide A is double that of a radionuclide B. The fraction of A remaining when B is reduced to 1/64 is ________.
Give answer in three decimal places.
A. 0.125 B. 0.325
C. 0.425 D. 0.525
(IIT JAM-2016)

117. The total metal content of a mineable 40 m × 40 m × 3 m ore block having bulk density 2.75 g/cm^3 and assay value 1.5 wt% Cu is ________ metric tonnes.
A. 198 B. 298
C. 398 D. 498
(IIT JAM-2016)

118. The temperature at the Earth's surface is 25 °C. The temperature at the base of the Earth's crust (30 km thick), if the geothermal gradients are 25 °C/km up to 15 km depth and 15 °C/km further down, is ______ °C.
A. 625 B. 525
C. 245 D. 0.25
(IIT JAM-2016)

119. Unsaturated zone above the water table is known as:
A. Vadose zone
B. Effluent zone
C. Dry zone
D. None-porous zone *(ISM-2015)*

120. The bridge across the Hooghly river Calcutta is a:
A. Suspension bridge
B. Cable stayed bridge
C. Balanced cantilever bridge
D. Level bridge
(PUNJAB UNIVERSITY-2015)

121. The logging tool for the direct determination of permeability is:
A. Induction B. Litho-density
C. Sonic D. NMR
(PUNJAB UNIVERSITY-2015

122. Crude oil density, in degree API (American Petroleum Institute) is a measure of viscosity. The value of 10 API is of:
A. Water B. Heavy oil
C. Average oil D. Light oil
(PUNJAB UNIVERSITY-2015)

123. Which type of clay is used as a drilling mud?
A. China clay B. Fullers clay
C. Bentonite clay D. Fire clay
(PUNJAB UNIVERSITY-2015)

124. High magnitude of anomalous elements in a particular area is called:
A. Anomaly
B. Background value
C. Threshold value
D. Excessive value
(PUNJAB UNIVERSITY-2015)

125. Artificial porosity and permeability can be achieved by:
A. Bore ballast technique
B. Hydraulic fracturing
C. Piston fracturing
D. All the above
(PUNJAB UNIVERSITY-2015)

126. When groundwater body is separated from the main ground water by a relative impermeable stratum of small extent it is called:
A. Unconfined ground water
B. Confined ground water
C. Perched ground water
D. Pressure ground water
(PUNJAB UNIVERSITY-2015)

127. Land slide are:
A. Downward movement of slope material
B. Outward movement of slope material
C. Downward and outward movement of slope material
D. Downward and inward movement of slope material
(PUNJAB UNIVERSITY-2015)

128. Long-term changes in the geochemical cycles are known as:
A. Periodic change
B. Permanent change
C. Secular change
D. Permial change
(PUNJAB UNIVERSITY-2015)

129. Boggier correction in gravity prospecting account for:
A. Genocidal shape of the earth
B. Attraction of material between the station and the datum plane
C. Change in elevation between stations
D. Surface irregularities in vicinity of the station
(PUNJAB UNIVERSITY-2015)

130. From the following which minerals has a high dielectric capacity:
A. Quartz B. Mica
C. Feldspar D. Calcite
(PUNJAB UNIVERSITY-2015)

131. The processes of making image data conform to a map projection system is known as:
A. Refrencing image B. Rectification
C. Raster data D. Vector data
(AMU-2016)

132. The resolution (m) of LISS-IV resourcesat-I is
A. 5.8 m B. 23.5 m
C. 56 m D. 70.5 m *(AMU-2016)*

133. Which of the following is not a method of energy scattering in the atmosphere?
A. Raleigh scattering
B. Non-selective
C. Mie scattering
D. Amalgamated scattering *(AMU-2016)*

134. The art and science of manipulation of digital data including enhancement, classification and rectification operation etc. is known as:
A. Orthorectification
B. Image processing
C. Rectification
D. Geometric correction *(AMU-2016)*

135. Intermittent sliding of rock fragments, rock blocks or soil down slope along a curved plane caused by rotational movement is called:
A. Debris flow B. Slumping
C. Mudflow D. None of them
(AMU-2016)

136. A saturated but poorly permeable stratum that imposed groundwater movement is called as:
A. Aquitard B. Aquifuge
C. Aquiclude D. Aquifer
(AMU-2016)

137. Thickness of capillary zone increases with:
A. Increases in pore size
B. Decreases in pore size
C. Pore size has no influence
D. None of the above *(AMU-2016)*

138. Amongst the following the highest the porosity value can be expected from
A. Consolidated mud
B. Unconsolidated mud
C. Well sorted sandstone
D. Poorly sorted sandstone *(ISM-2016)*

139. When sea level rise, the sea invades a river valley, producing a nearshore body of water with mixed and variability salinity is termed as
A. An atoll B. A lagoon
C. An estuary D. A jetty *(ISM-2016)*

140. A matrix supported conglomerate with planar, non-erosional base and clast imbrication is interpreted as a product of:
A. Channel lag deposition
B. Debris flow deposition
C. Glacial outwash deposition
D. Residual lag deposition *(ISM-2016)*

141. The value of Poisson's ratio lies between
A. −1 and −2 B. −1/2 and +1
C. +1 and +1/2 D. −1 and 0.5
(ISM-2016)

142. Effective precipitation in a drainage basin refers to:
A. Precipitation that ultimately reach the water table
B. Precipitation that flow overland
C. Precipitation that is stored in water reservoir
D. Precipitation that ultimately reaches the river channel
(ISM-2016)

143. Permeability is a measure of the ability of the fluid to :
A. Contract when pressure is applied
B. Flow through a porous rock
C. Produce energy when burned
D. None of the above *(BHU-2016)*

144. Find the correct pair:

List-A	List-B
P. Gwalior basin	1. Well sorted
Q. Tadpatri sill	2. Paleoproterozoic
R. Marwar supergroup	3. Cuddapah
S. Limestone	4. Neoproterozoic-cambrian
T. Beach sandstone	5. Chemically deposited

A. P-2, Q-3, R-4, S-5, T-1
B. P-3, Q-1, R-4, S-5, T-2
C. P-4, Q-5, R-2, S-1, T-3
D. P-5, Q-4, R-3, S-2, T-1
(DU-2016)

145. Hydraulic conductivity of a function of:
A. Medium only
B. Fluid only
C. Both fluid and medium
D. Either fluid or medium *(DU-2016)*

146. In unconfined aquifer strativity is appropriate called
A. Conductivity
B. Intrinsic permeability
C. Specific retention
D. None of the above *(DU-2016)*

147. Intrinsic permeability is dependent on:
A. Property of the fluid alone
B. Proper of the medium alone
C. Property of both fluid and medium
D. None of the above *(DU-2016)*

148. The dimension of the specific storage is:
A. LT^{-1} B. L^2T^{-1}
C. L^3T^{-1} D. L^{-1} *(DU-2016)*

149. In an Isotopic aquifer
A. Aquifer parameter are same in all direction
B. Aquifer parameters do not vary spatially
C. Aquifer parameter vary in space and dimension both
D. None of the above *(DU-2016)*

150. Transmissivity is:
A. Hydraulic conductivity * saturated aquifer thickness
B. S * b
C. Extension of the aquifer
D. Specific capacity *(DU-2016)*

151. Which of the following maps can give direction of the groundwater?
A. Depth of the water table map
B. Water level fluctuating map
C. Water table contour map
D. None of the above
(DU-2016)

152. The dimension of the hydraulic conductivity is;
A. LT^{-1} B. $L^3 T^{-1}$
C. $L^2 T^{-1}$ D. L^{-1} *(DU-2016)*

153. Aquifer homogenetly related to
A. Temporal variation in aquifer parameters
B. Both spatial and temporal variation in aquifer parameter
C. Directional variation in aquifer parameter
D. Spatial variation in aquifer parameter
(DU-2016)

154. Sodium Absorption ratio in groundwater quality is used for studying
A. Hydrochemical facies of water facies
B. Metamorphic facies of rock sample
C. Suitability of groundwater for irrigation uses
D. Portability of the water *(DU-2016)*

155. Electrical conductivity of groundwater is a direct function of the
A. Intrinsic permeability of the formation
B. Stortivity of the formation
C. Total dissolved solids in the ground water
D. None of the above *(DU-2016)*

156. Ganga-Brahmaputra province includes
A. The Vindhyans
B. The Deccan basin
C. The Aravalli
D. Bhabar and Tarai belts *(DU-2016)*

157. The D layer with in the Earth correspond to the depth range between:
A. 2740–2890 km
B. 2010–2035 km
C. 2890–3050 km
D. 250–375 km *(DU-2016)*

158. A matrix supported conglomerate is found with characteristic concerned clast are found in bedding perpendicular orientation as well as the deposit by
A. Debris flow
B. Liquefied flow
C. Turbidity flow
D. Water flow on a granular substrata *(DU-2016)*

159. A current crescent structure can be used for
A. Finding top and bottom of the beds
B. Determination of the paleocurrent
C. Both stratigraphic and paleocurrent
D. None of the above *(DU-2016)*

160. Govindnagar is the name of a reservoir associated with which dam:
A. Rana Pratap Sagar
B. Sardar Sarovar
C. Bhakra Dam
D. Koyna Dam *(DU-2016)*

161. The Tehri Dam has its foundation on:
A. Slate B. Phyllite
C. Gneiss D. Schist *(DU-2016)*

ANSWER KEY

1	2	3	4	5	6	7	8	9	10
C	C	B,C	B	A	A	B	B	A	C
11	**12**	**13**	**14**	**15**	**16**	**17**	**18**	**19**	**20**
B	D	B	B	B	B	D	B	A	C
21	**22**	**23**	**24**	**25**	**26**	**27**	**28**	**29**	**30**
B	C	C	A	D	D	C	A	A	D
31	**32**	**33**	**34**	**35**	**36**	**37**	**38**	**39**	**40**
A	B	B	C	A	B	A	C	C	D
41	**42**	**43**	**44**	**45**	**46**	**47**	**48**	**49**	**50**
B	D	D	D	B	B	A	A	D	B
51	**52**	**53**	**54**	**55**	**56**	**57**	**58**	**59**	**60**
A	D	A	B	D	B	B	B	A	A
61	**62**	**63**	**64**	**65**	**66**	**67**	**68**	**69**	**70**
D	D	D	A	D	B	D	C	A	C
71	**72**	**73**	**74**	**75**	**76**	**77**	**78**	**79**	**80**
C	C	A	C	A		C	D	C	D
81	**82**	**83**	**84**	**85**	**86**	**87**	**88**	**89**	**90**
B	A	A	A	C	D	D	A	A	A

91	92	93	94	95	96	97	98	99	100
B	D	C	B	B	B	A	A	B	B
101	102	103	104	105	106	107	108	109	110
B	C	C	D	A	A	B	B	D	D
111	112	113	114	115	116	117	118	119	120
B	A	A,B,C	A	A	A	A	A	D	A
121	122	123	124	125	126	127	128	129	130
D	A	C	C	D	C	A	C	D	D
131	132	133	134	135	136	137	138	139	140
B	A	D	B	B	A	B	C	D	B
141	142	143	144	145	146	147	148	149	150
D	D	B	A	C	D	B	D	A	A
151	152	153	154	155	156	157	158	159	160
C	A	D	C	C	D	A	A	C	C
161									
B									

HINTS AND SOLUTIONS

1. Units	Characters
Aquiclude	Porous but not permeable.
Aquifer	Porus and permeable.
Aquitard	Porous and poorly permeable.
Artesian aquifer	Porous with good permeable.

5. Mine excavation	Designation
Crosscut	A mine that cut across the geological structure.
Raise	A mine which developed vertical or inclined for live the place.
Drive	A mine excavated in horizontally. (Along the strike).
Shaft	A mine which is vertical opening to great depth.

7. Types of stream	Designation
Influent	Which receive ground water recharge.
Effluent	Which receive ground water discharge.
Ephemeral	Water not available throughout the year.
Perennial	Water available throughout the year.

8. Hydrological properties	Example
Aquifer	Sandstone formation
Aquiclude	Clay bed
Aquitard	Sandy clay
Aquifuge	Solid basalt

10. TDS (Total Dissolved Solution) (mg/l)	Water quality
<500	Very good
500 - 1000	Good
1000 - 2000	Very poor
>2000	Not acceptable (Toxic)

E.C. = Electrical conductivity of water at 25°C.

Hardness = [Ca* combining weight of $CaCO_3$/Combing weight of Ca + Mg* combining weight of $CaCO_3$/Combining weight of Mg]

Hardness	Water quality
<50	Soft
50-100	Moderately hard
>200	Hard

$$SAR = \frac{Na^+}{\sqrt{\frac{Ca^{++} + Mg^{++}}{2}}}$$

SAR	Water quality (Irrigation water)
<10	Very sensitive
10-18	Sensitive
18-26	Moderately tolerant
>26	Tolerant

11. Elements	Indian recommended (mg/l)	WHO recommended (mg/l)
Fluoride	1.5	1.0
Arsenic	0.05	0.01
Lead	0.1	0.01
Zinc	5.0	3
Mercury	0.001	0.001
Iron	0.3	0.2

12. Diseases	Excess elements in drinking water
Fluorosis	Fluoride
Silicosis	Silica
Asbestosis	Asbestos
Blue-baby syndrome	Nitrate

32. Tunnel site	Designations
1	It is not suitable because due to folding of the strata.
2	It is most suitable because stable and near the surface.
3	It is not suitable due to fault plane it is unstable zone.
4	It is also not suitable because due to shear zone.

33. Condition of Darcy Law

1. The medium must be a Porous.
2. Flow through channel/pipe medium.
3. Laminar flow.
4. Reynolds number is less than 2000.

Rocks types	Hydrological properties
Shale	Porosity but not permeable.
Sandstone	Porosity and permeability.
Marble	Neither porous not permeable.
Granite	Neither porous not permeable.

37. Terms	Remarks
Creep	Sluggish mass wasting
Flow	Very fast movement
Slide	Slow movement with surface
Fall	Sudden goes down

39. Lithological unit	Hydrological unit	Character
Massive granite	Aquifuge	No porosity and no permeability
Shale	Aquiclud	Porosity but not permeability
Clayey sandstone	Aquitard	Porosity but less permeability
Gravelly sandstone	Aquifer	Porosity and permeability

44. Types of stone	Crushing strength (psi)
Kota stone	1000
Vindhyan sandstone	1500
Makarana marble	1400
Bundelkhand granite	2000

45. Underground water	Origin
Connate water	Fossils
Juvenile water	Magmatic
Hygroscopic water	Rain water

46. Types of clay	**Uses**
China clay	Spark pluge insulater
Bentonite clay	Drilling mud
Fire clay	Refractories
Fuller earth	Petroleum refinery

47. The scale of the geological map-

5 cm = 1 km.

1 cm = 200 m

1 cm = 20000 cm

RF = 1:20000

68. Hydrological terms	**Example**	**Characteristics**
Aquifer	Sandy formation	Porous and permeable
Aquiclude	Clay bed	Porous but not permeable
Aquitard	Sandy clay	Porous but less permeable
Aquifuse	Solid granite	Neither porous nor permeable

69. Porosity of sandstone = volume of pore spaces/ total volume of aquifer*100

72. Mine (Terms)	**Characteristics**
Bench	Open pit excavations
Raise	Vertical or inclined excavations
Adit	Horizontal excavations
Crosscut	Box shape excavations

75. Statements	**Characteristics**
A	Sandstone and highly jointed granite can good aquifers.
B	Sandstone is good aquifer but massive granite is aquifuse.
C	Sandstone is aquifer and shale is aquicluid.
D	Shale is aquicluid and massive granite is aquifuse.

77. Rocks	**Hydrological properties**
Sandstone	porous and permeable
Siltstone	porous and permeable
Shale	porous but not permeable
Marble	neither porous nor permeable

78. Group-1	**Locations**	**Source rocks**	**Age**
Oil	Ankleshwar Gujarat	Sand	Eocene
Coal	Bokaro Jharkhand	Shale	Permian
Lead-Zinc	Zawar Rajasthan	Dolomite	Proterozoic
Phosphate	Jamarkotra Rajasthan	Limestone, Schist	Proterozoic

85. Materials	**Resistivity (Ωm)**
Sandstone	100-1000
Freshwater	3-100
Oil	1
Seawater	1-0.1
Sea Ice	50-1000

104. Map scale = 1 : 50,000

1 cm = 50000 cm.

1 cm = 500 meter

1 cm = 0.5 km.

The dike length on the map is = 10 cm.

1 cm = 0.5 km (scale)

The length of the dike is = 0.5*10 = 5 km

112. Replacement Time with Respect to River Addition, τ_r, for Some Major and Minor Dissolved Species in Seawater

Component	Concentration (uM)		τ_r[a]
	River Water	Seawater	(1000 yr)
Cl^-	230	558,000	87.000
Na^+	315	479,000	55,000
Mg^{++}	150	54,300	13,000
SO_4^{--}	120	28,900	8700
Ca^{++}	367	10,500	1000
K^+	36	10,400	10,000
HCO_3^-	870	2000	83
H_4SiO_4	170	100	21
NO_3^-	10	20	72
Orthophosphate	0.7	1	50

[a] τ_r = ([SW]/[RW]) τ_u where τ_u = replacement (residence) time of H_2O = 36,000 yr; RW = river water; SW = seawater, and [ ] = concentration in µmoles per liter = µM.

122. Materials	API
Water	10
Heavy oil	< 22.3
Average oil	22.3 – 31.1
Light oil	> 31.1
Extra heavy	< 10

123. Clay	Uses
China clay	Ceramic
Fullers clay	Mold binder
Bentonite clay	Drilling mud
Fire clay	Refractory

126. Different types of Aquifer :

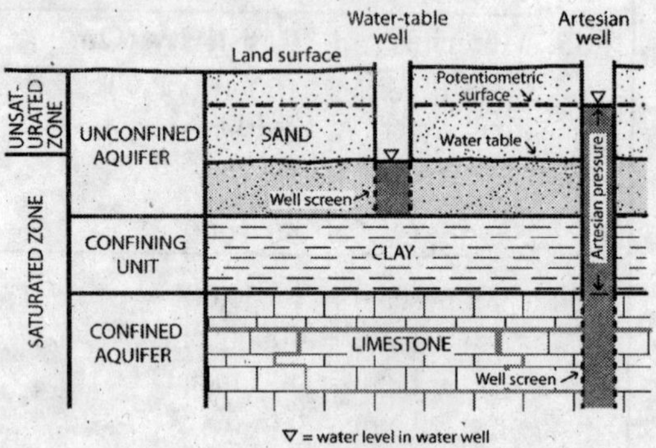

Perched Water table:

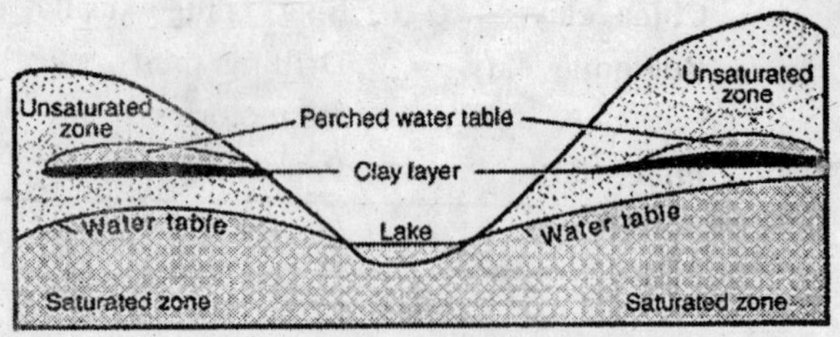

133. Energy scattering in the atmosphere:

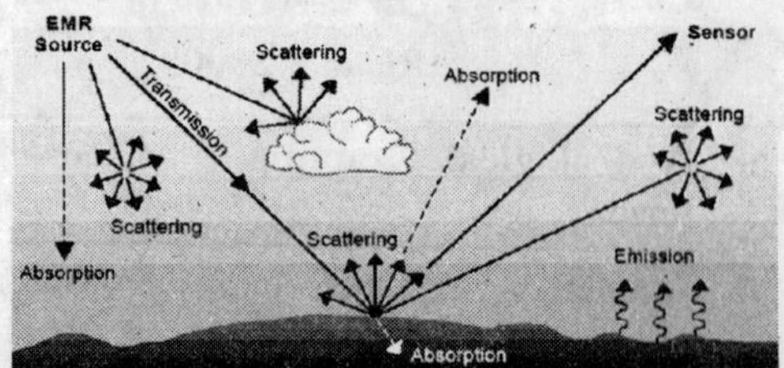

Releigh scattering : In this type of scattering the particle size smaller than the wavelength of the radiation.

Non-selective: The particle size is equal to the wavelength of the radiation.

Mie scattering: In this type of the scattering the particle size larger than the wavelength of the radiation.

136. Hydrological properties	Example	Characteristics
Aquitard	Sandy clay	Porous but less permeable
Aquifuge	Solid granite	Neither porous nor permeable
Aquiclude	Clay bed	Porous but not permeable
Aquifer	Sandy formation	Porous and permeable